METAL OXIDES

Chemistry and Applications

CHEMICAL INDUSTRIES

A Series of Reference Books and Textbooks

Consulting Editor

HEINZ HEINEMANN
Berkeley, California

METAL OXIDES

Chemistry and Applications

EDITED BY

J. L. G. FIERRO

CRC Press
Taylor & Francis Group
Boca Raton London New York

CRC Press is an imprint of the
Taylor & Francis Group, an **informa** business

A TAYLOR & FRANCIS BOOK

CRC Press
Taylor & Francis Group
6000 Broken Sound Parkway NW, Suite 300
Boca Raton, FL 33487-2742

First issued in paperback 2019

© 2006 by Taylor & Francis Group, LLC
CRC Press is an imprint of Taylor & Francis Group, an Informa business

No claim to original U.S. Government works

ISBN-13: 978-0-8247-2371-2 (hbk)
ISBN-13: 978-0-367-39222-2 (pbk)

Library of Congress Cataloging-in-Publication Data

Catalog record is available from the Library of Congress

**Visit the Taylor & Francis Web site at
http://www.taylorandfrancis.com**

**and the CRC Press Web site at
http://www.crcpress.com**

Preface

With advances in the area of metal oxide systems progressing sharply over the past decade, there has been an increasing need for comprehensive surveys and texts that can serve both as introductions for newcomers to the field and as reference materials for the already established investigator. There is no doubt that this timely and interdisciplinary work will emerge as an important milestone and will make a significant impact.

Metal oxides belong to a class of widely used catalysts. They exhibit acidic or basic properties, which make them appropriate systems to be used as supports for highly dispersed metal catalysts or as precursors of a metal phase or sulfide, chloride, etc. Simple metal oxides range from essentially ionic compounds with the electropositive elements to covalent compounds with the nonmetals. However, taking into account the large variety of metal oxides, the principal objective of this book is to examine only metal oxides that are more attractive from the catalytic point of view, and most specifically transition metal oxides (TMO). In particular, TMO usually exhibit nonstoichiometry as a consequence of the presence of defective structures. The interaction of TMO with surfaces of the appropriate carriers develop monolayer structures of these oxides. The crystal and electronic structure, stoichiometry and composition, redox properties, acid–base character and cation valence sates are major ingredients of the chemistry investigated in the first part of the book. New approaches to the preparation of ordered TMO with extended structure of texturally well defined systems are also included.

The second part of the book compiles some practical aspects of metal oxides, with emphasis in catalytic applications. Metal oxides represent an expanding class of compounds with a wide range applications in several areas such as materials science and catalysis, chemical sensing, microelectronics, nanotechnology, environmental decontamination, analytical chemistry, solid-state chemistry, and fuel cells. Our basic knowledge on the metal oxide chemistry is relatively far from that for metals, and as yet, little is known about fundamental relationships between reactivity of oxide compounds and their chemical compositions, crystal structures, and electronic properties at the surface. When examining the importance of metal oxides, and specifically TMOs, in several reactions such as dehydration, selective oxidations, olefin metathesis, VOCs removal, photocatalysis, water splitting, and electrocatalysis, attempts will be made in order to connect properties of the oxides and their reactivity. Since the catalytic phenomenon is confined to the external surface of the solids where molecules or atoms interact, the study of this interaction requires contributions from inorganic and physical chemistry, solid-state chemistry, quantum chemistry, surface science, reaction kinetics, and other branches of science.

In summary, each chapter begins with an excellent introduction to the topic concerned, which is followed by a good overview of the subject and more details for the expert in the area. The book is intended to be used mainly as a research monograph by a vast community of those working in the field of catalysis. However, it may also serve as a supplementary text for postgraduate students working in the fields of industrial chemistry, catalysis, chemical technology, and physical as well as in general chemistry.

Editor

J.L.G. Fierro is a senior research scientist and professor at the Institute of Catalysis and Petrochemistry of the National Council for Scientific Research (CSIC), Madrid, Spain. He is the editor of seven books and the author and coauthor of 600 professional papers, most of which concerned his research on metal oxides, including their synthesis, characterization, and application in catalytic reactions. He is a member of the Spanish Society of Chemistry and the Petroleum Division of the American Chemical Society. Professor Fierro received the B.S. (1973) in chemistry from the University of Oviedo, Spain, and a doctorate (1976) in chemistry from the Complutense University of Madrid.

Contributors

Masakazu Anpo
Department of Applied Chemistry
Graduate School of Engineering
Osaka Prefecture University
Osaka, Japan

Nicolae Bârsan
Institute of Physical and Theoretical
Chemistry
University of Tübingen
Tübingen, Germany

Elisabeth Bordes-Richard
Laboratoire de Catalyse de Lille
ENSCL-USTL, France

Laura E. Briand
Centro de Investigación y Desarrollo en
Ciencias Aplicadas-Dr. Jorge J. Ronco
University Nacional de La Plata
Buenos Aires, Argentina

Guido Busca
Dipartimento di Ingegneria Chimica e
di Processo "G.B. Bonino"
Laboratorio di Chimica delle
Superfici e Catalisi Industriale
Genova, Italy

Gabriele Centi
Department of Industrial Chemistry
and Engineering of Materials and
ELCASS
University of Messina
Messina, Italy

M. Cherian
Department of Chemical Engineering
Indian Institute of Technology
Kanpur, India

P. Courtine
Départment de Génie Chimique
Université de Technologie de
Compiègne
Compiègne, France

Goutam Deo
Department of Chemical Engineering
Indian Institute of Technology
Kanpur, India

Dhammike P. Dissanayake
Department of Chemistry
University of Colombo
Colombo, Sri Lanka

Satoru Dohshi
Department of Applied Chemistry
Graduate School of Engineering
Osaka Prefecture University
Osaka, Japan

Jose Luis G. Fierro
Institute of Catalysis and
Petrochemistry
CSIC, Cantoblanco
Madrid, Spain

Alexander Gurlo
Institute of Physical and Theoretical
Chemistry
University of Tübingen
Tübingen, Germany

Y. Hu
Department of Applied Chemistry
Graduate School of Engineering
Osaka Prefecture University
Osaka, Japan

Yasunobu Inoue
Department of Chemistry
Nagaoka University of Technology
Nagaoka, Japan

John T.S. Irvine
School of Chemistry
University of St. Andrews,
St. Andrews, Scotland, U.K.

Sumio Ishihara
Department of Physics
Tohoku University
Sendai, Japan

Y.-I. Kim
Department of Chemistry
The Ohio State University
Columbus, Ohio, U.S.A.

M. Kitano
Department of Applied Chemistry
Graduate School of Engineering
Osaka Prefecture University
Osaka, Japan

Angela Kruth
School of Chemistry
University of St. Andrews
St. Andrews, Scotland, U.K.

Ian S. Metcalfe
Department of Chemical Engineering
University of Manchester
Institute of Science and Technology
Manchester, England, U.K.

H. Mizoguchi
Department of Chemistry
The Ohio State University
Columbus, Ohio, U.S.A.

J.C. Mol
Institute of Molecular Chemistry
University of Amsterdam
Amsterdam, The Netherlands

R.M. Navarro
Institute of Catalysis and
Petrochemistry
CSIC, Cantoblanco
Madrid, Spain

Ekaterina K. Novakova
CenTACat, Queen's University
Belfast
David Keir Building
Belfast, Northern Ireland, U.K.

Barbara Pawelec
Institute of Catalysis and
Petrochemistry
CSIC, Cantoblanco
Madrid, Spain

M.A. Peña
Institute of Catalysis and
Petrochemistry
CSIC, Cantoblanco
Madrid, Spain

S. Perathoner
Department of Industrial Chemistry
and Engineering of Materials and
ELCASS
University of Messina
Messina, Italy

T.V.M. Rao
Department of Chemical Engineering
Indian Institute of Technology
Kanpur, India

Benjaram M. Reddy
Inorganic and Physical Chemistry
Division
Indian Institute of Chemical
Technology
Hyderabad, India

Stephan A. Schunk
The Aktiengesellschaft
Heidelberg, Germany

Ferdi Schüth
MPI für Kohlenforschung
Mülheim, Germany

M.W. Stoltzfus
Department of Chemistry
The Ohio State University
Columbus, Ohio, U.S.A.

S. Tao
School of Chemistry
University of St Andrews
St. Andrews, Scotland, U.K.

A. Thursfield
Department of Chemical Engineering
University of Manchester Institute of
Science and Technology
Manchester, England, U.K.

Jacques C. Védrine
Laboratoire de Physico-Chimie des
Surfaces
Ecole Nationale Supérieure de Chimie
de Paris
Paris, France

Israel E. Wachs
Operando Molecular Spectroscopy &
Catalysis Laboratory
Department of Chemical Engineering
Lehigh University
Bethlehem, Pennsylvania, U.S.A.

Zhong Lin Wang
School of Materials Science and
Engineering
Georgia Institute of Technology
Atlanta, Georgia, U.S.A.

Udo Weimar
Institute of Physical and Theoretical
Chemistry
University of Tübingen
Tübingen, Germany

Michael S. Wong
Department of Chemical Engineering
Department of Chemistry
Rice University
Houston, Texas, U.S.A.

Patrick M. Woodward
Department of Chemistry
The Ohio State University
Columbus, Ohio, U.S.A.

Contents

Contents

1 Molecular Structures of Surface Metal Oxide Species: Nature of Catalytic Active Sites in Mixed Metal Oxides

Israel E. Wachs
Operando Molecular Spectroscopy & Catalysis Laboratory,
Department of Chemical Engineering, Lehigh University,
Bethlehem, PA, USA

CONTENTS

1.1   Introduction

Metal oxide catalytic materials currently find wide application in the petroleum, chemical, and environmental industries, and their uses have significantly expanded since the mid-20th century (especially in environmental applications) [1,2]. Bulk mixed metal oxides are extensively employed by the chemical industries as selective oxidation catalysts in the synthesis of chemical intermediates. Supported metal oxides are also used as selective oxidation catalysts by the chemical industry, as environmental catalysts, to selectively transform undesirable pollutants to nonnoxious forms, and as components of catalysts employed by the petroleum industry. Zeolite and molecular sieve catalytic materials are employed as solid acid catalysts in the petroleum industry and as aqueous selective oxidation catalysts in the chemical industry, respectively. Zeolites and molecular sieves are also employed as sorbents for separation of gases and to trap toxic impurities that may be present in water supplies. Significant molecular spectroscopic advances in recent years have finally allowed the nature of the active surface sites present in these different metal oxide catalytic materials to be determined in different environments. This chapter examines our current state of knowledge of the molecular structures of the active surface metal oxide species present in metal oxide catalysts and the influence of different environments upon the structures of these catalytic active sites.

1.2   Supported Metal Oxides

Supported vanadium oxide catalysts are employed as catalysts for o-xylene oxidation to phthalic anhydride [3], ammoxidation of pyridine to picoline [4,5], methanol oxidation to formaldehyde [6], methane oxidation to formaldehyde [7], ethane oxidative dehydrogenation (ODH) to ethylene [8], propane ODH to propylene [9,10] n-butane oxidation to maleic anhydride [11], SO_2 oxidation to SO_3 [12], and oxidesulfurization (ODS) of organosulfur compounds [13–15]. Supported vanadium oxide–tungsten oxide and supported vanadium oxide–molybdenum oxide catalysts are extensively employed as catalysts for the selective catalytic reduction (SCR) of NO_x with NH_3 to N_2 and H_2O [16–18]. Supported tungsten oxide and sulfated catalysts are efficient solid acid catalysts for hydrocarbon isomerization reactions [19–22]. Supported rhenium oxide and tungsten oxide find application as olefin metathesis catalysts [23,24]. Supported chromium oxide, vanadium oxide, and molybdenum oxide catalysts are employed to catalyze olefin polymerization reactions [25–27], of there, supported chromium oxide catalysts are commercially employed as alkane dehydrogenation catalysts [28]. Supported molybdenum oxide and tungsten oxide are precursors to their corresponding sulfides that are formed during hydrodesulfurization (HDS) of organosulfur compounds [29,30]. Thus, the applications of supported metal oxide catalysts have significantly expanded since their first applications in the mid-20th century.

It is important to know the molecular structures of the active sites present in supported metal oxide catalysts in order to fully understand their fundamental characteristics. Supported metal oxide catalysts consist of an active metal oxide phase dispersed on a high surface area oxide support [2,31]. The dispersed metal oxide active phase is typically present as a two-dimensional metal oxide overlayer on the high surface area oxide substrate. The molecular structures of the surface metal oxide species have been found to be different than their pure metal oxide phases [31]. For example, supported VO_x possesses VO_4 coordination and bulk V_2O_5 consists of distorted VO_5 coordination [32,33]. Furthermore, the molecular structures of the surface metal oxide species are *dynamic* and strongly depend on the specific environment (e.g., gas phase composition, temperature, and pressure). This portion of the chapter will review what is currently known about the molecular structures of the surface metal oxide species present in supported metal oxide catalysts and the influence of different environments on the structures. Subsequent sections of this chapter will show how these findings can be extended to other mixed metal oxide catalytic materials.

1.2.1 Hydrated Surface Metal Oxide Species

Supported metal oxide species are hydrated when exposed to moist environments and low temperatures ($<230°C$). Thus, all calcined supported metal oxide species are hydrated at ambient conditions (room temperature and air exposed) [34,35]. The hydrated surface layer corresponds to a thin aqueous film that corresponds to multiple layers of moisture [32]. The hydrated surface metal oxide species equilibrate with the pH of the aqueous layer. The pH of the aqueous film is determined by the pH at point zero charge (PZC) of the hydrated surface [36,37]. The net pH at PZC is defined as the equilibrated pH of a hydrated surface when the net charge is zero (protonated positive surface sites are balanced by an equal number of deprotonated negative surface sites). At pH values above the PCZ, the hydrated surface becomes negatively charged, while for pH values below the PCZ, the hydrated surface becomes positively charged. Thus, hydrated surfaces always equilibrate at the pH at PZC in order to preserve charge balance.

The net pH at PZC for a supported metal oxide catalyst possessing monolayer surface coverage is dependent on the pH at PZC value of the oxide support substrate and the pH at PZC value of the pure metal oxide that is in the dispersed metal oxide phase:

$$\text{pH at PZC}_1(\text{monolayer}) \sim (\text{PZC}_{\text{support}} + \text{PZC}_{\text{dispersed oxide}})/2 \qquad (1.1)$$

Both the oxide support and the dispersed oxide exert an influence on the net pH at PZC because the thin aqueous film is in contact with both components (especially when clusters of the surface metal oxides are present). Below monolayer surface coverage, the pH at PZC is a function of the surface coverage of the dispersed oxide and monotonically decreases from the value of the oxide support to the value at monolayer coverage given by Equation (1.1) with increasing surface coverage. The individual values of the pH at PZC for pure oxides at room temperature are well documented in the literature [36–38] and are presented in Table 1.1 for typical oxides encountered as oxide supports and active metal oxide phases.

TABLE 1.1
pH at PZC for Oxide Supports and Active Surface Metal Oxides

Support	pH at PZC	Active surface oxide	pH at PZC
MgO	12.4	V_2O_5	1.4
γ-Al_2O_3	8.8	Nb_2O_5	4.3
CeO_2	6.8	CrO_3 (Cr_2O_3)	ws (7.0)
ZrO_2	6.7	Ta_2O_5	~4
TiO_2	6.3	MoO_3	1.2
Nb_2O_5	4.3	WO_3	0.7
SiO_2	1.8	Re_2O_7	ws

ws — water soluble metal oxide.

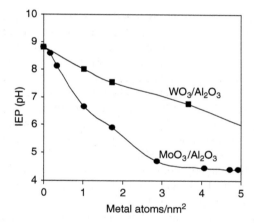

FIGURE 1.1 The influence of the surface coverage of different surface metal oxides on the net pH at PZC for a series of Al_2O_3 supported metal oxides

The influence of the surface coverage of different surface metal oxides on the net pH at PZC for a series of Al_2O_3 supported metal oxides is shown in Figure 1.1. The pH at PZC of the Al_2O_3 support is ~8.8 and continuously decreases as the surface coverage of metal oxides with low values of pH at PZC is increased. Note that at monolayer surface coverage the net pH at PZC of the different supported metal oxide catalysts asymptotically reaches values intermediate between that of the pure alumina support and the pure dispersed metal oxide phase.

The molecular structures of hydrated metal oxide species in aqueous solution are well documented and depend on the solution pH and the metal oxide aqueous concentration with the solution pH having the dominant effect [39]. For high pH values, the hydrated metal oxides tend to be present as isolated MO_4 units in

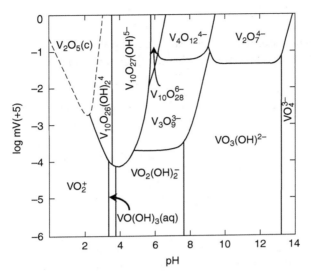

FIGURE 1.2 The V^{+5} aqueous phase diagram

solution (e.g., VO_4, CrO_4, MoO_4, WO_4, ReO_4, etc.). For low pH values, the hydrated metal oxides tend to be present as linear polymeric chains (e.g., $(CrO_3)_n$ with n equal to 2 or greater) and clusters (e.g., $V_{10}O_{28}$, Mo_7O_{26}, Nb_6O_{19}, Ta_6O_{19}, $W_{12}O_{39}$). One exception to this trend is aqueous rhenium oxide that is present as isolated ReO_4 species at all pH values and concentrations. The aqueous phase diagram of vanadium oxide is shown in Figure 1.2. The vanadium oxide molecular structure is very sensitive to the aqueous pH and forms VO_4 (orthovanadate), V_2O_7 (pyrovanadate), V_3O_{10} (trimer), V_4O_{13} (metavanadate or tetramer), $V_{10}O_{28}$ (decavanadate), and $V_2O_5 \cdot nH_2O$ (V_2O_5 gel) complexes.

The molecular structures of the hydrated surface metal oxides on oxide supports have been determined in recent years with various spectroscopic characterization methods (Raman [34,37,40–43], IR [43], UV-Vis [44,45], solid state NMR [32,33], and EXAFS/XANES [46–51]). These studies found that the surface metal oxide species possess the same molecular structures that are present in aqueous solution at the same net pH values. The effects of vanadia surface coverage and the different oxide supports on the hydrated surface vanadia molecular structures are shown in Table 1.2. As the value of the pH at PZC of the oxide support decreases, the hydrated surface vanadia species become more polymerized and clustered. Similarly, as the surface vanadia coverage increases, which decreases the net pH at PZC, the hydrated surface vanadia species also become more polymerized and clustered. Consequently, only the value of the net pH at PZC of a given hydrated supported metal oxide system is needed to predict the hydrated molecular structure(s) of the surface metal oxide species.

The finding that only one parameter, the net pH at PZC, controls the hydrated molecular structures of surface metal oxide species also has very

TABLE 1.2

Hydrated Molecular Structures for Supported Vanadium Oxide Catalysts as a Function of Surface Coverage and Specific Support

		Observed molecular structures	
Oxide support	pH at PZC	Low surface coverage	High surface coverage
MgO	12.4	VO_4, V_2O_7, $(VO_3)_n$	VO_4, V_2O_7, $(VO_3)_n$
Al_2O_3	8.8	$(VO_3)_n$	$(VO_3)_n$, $V_{10}O_{28}$
ZrO_2	6.7	V_2O_7, $(VO_3)_n$, $V_{10}O_{28}$	$V_{10}O_{28}$
TiO_2	6.3	$(VO_3)_n$, $V_{10}O_{28}$	$V_{10}O_{28}$
SiO_2	1.8	$V_2O_5 . nH_2O$	V_2O_5

important implications for the synthesis of supported metal oxide catalysts since all preparation methods, for a given composition and catalyst system, must equilibrate at the same net pH at PZC upon hydration. This means that the preparation method cannot influence the final hydrated, as well as the subsequent dehydrated, surface metal oxide molecular structures. This has been demonstrated for various supported MoO_3/TiO_2 [52], MoO_3/SiO_2 [53], and V_2O_5/TiO_2 [52] catalytic systems synthesized with different precursors and sources of oxide supports. Furthermore, the specific phase of the oxide support (e.g., TiO_2 [anatase], TiO_2 [rutile], TiO_2 [brookite], and TiO_2 [B]) also did not affect the hydrated molecular structures of the surface metal oxide species for the same surface metal oxide coverage [54]. The series of samples examined in these studies originated in many different catalysis laboratories around the world and confirmed that the hydrated molecular structures are independent of the origin of the supported metal oxide catalysts. These conclusions are further confirmed by a careful examination of the catalysis literature containing reproducible structural characterization information of supported metal oxide species.

The only different molecular structures were found for supported metal oxide catalysts where the oxide support contained surface impurities such as Ca [55], Na [55,56], Ca [53,56], and K [34,56]. These basic impurities alter the hydrated molecular structures by increasing the net pH at PZC on the thin aqueous film or directly reacting with the surface metal oxide species to form nanocrystalline compounds (e.g., $CaMoO_4$, Na_2MoO_4, K_2MoO_4, etc.). The presence of nanocrystalline metal oxide phases (e.g., V_2O_5, MoO_3, etc.) in addition to the hydrated surface metal oxide species *below monolayer surface coverage* typically results from preparations employing precursors that have limited or low solubility in the impregnating solvents (e.g., NH_4VO_3 in water, V_2O_5 in aqueous oxalic acid solution). In such instances, the metal oxide precursors are not well dispersed over the oxide support surface and tend to form the crystalline metal oxide phases upon calcination. For some supported metal oxide systems, it was observed that the surface metal oxides were initially able to form hydrated complexes with the oxide support cations (e.g., silicomolybdic acid [57] and $AlMo_6O_x$ clusters [58,59]) for specific preparation sequences, but such hydrated clusters are not stable at 300°C and higher temperatures owing to the loss of waters of hydration, and decompose during calcination to the conventional surface metal oxide species. Thus, the final hydrated supported metal oxide catalysts *after calcination* have no memory effect of the prior presence of such hydrated complexes with the oxide support since they decompose during calcination.

As the supported metal oxide catalyst temperature is increased, the thin aqueous film evaporates and desorbs, ~100 to 200°C, from oxide surfaces to yield dehydrated surfaces. If sufficient moisture is present in the environment at the elevated temperatures, however, it is still possible to maintain an extensively hydrated surface up to ~230°C [35]. At higher temperatures, the desorption rate of the adsorbed moisture from oxide surfaces is very fast and the surfaces are essentially dehydrated (<5% of the surface contains adsorbed moisture at steady-state when moisture is present) [35].

1.2.2 Dehydrated Surface Metal Oxide Species

The dehydrated surface metal oxide species are not coordinated to water and, therefore, their molecular structures are not related to those present in aqueous solutions. Consequently, the pH at PZC model cannot be employed to predict the dehydrated surface metal oxide structures. The molecular structures of the dehydrated surface metal oxide species, however, possess similarity to the structural inorganic chemistry of bulk metal oxides because of the absence of water ligands in both systems [60–62]. Instead of being solvated by coordinated water in the aqueous solution complexes, the bulk metal oxide structures are coordinated to various cations (e.g., K, Na, Ca, Mg, Fe, Al, Ce, Zr, Ti, etc.). Prior to discussing the current understanding of the molecular structures of the dehydrated surface metal oxide species, a brief review of the structural inorganic chemistry of bulk metal oxides and their determination methods are presented to highlight the molecular structural similarities, as well as differences, between these two- and three-dimensional metal oxide systems.

1.2.2.1 Structural determination methods

The bulk metal oxide structures have been determined with extensive and highly accurate x-ray diffraction crystallographic studies [60]. Unfortunately, the structural inorganic chemistry of dehydrated surface metal oxides on oxide supports cannot be determined with x-ray diffraction crystallography because of the absence of long-range order (>4 nm) in the surface metal oxide overalyers. Information about the local structures of the dehydrated surface metal oxides, however, can be obtained with *in situ* molecular spectroscopic techniques of dehydrated supported metal oxides: Raman [31,63], IR [64], UV-Vis [44,50,65,66], XANES/EXAFS [46–51,67,68], chemiluminescence [69], and solid state NMR for certain nuclei (e.g., ^{51}V, ^{95}Mo, ^1H, etc.) [32,33,70,71]. UV-Vis, XANES/EXAFS, chemiluminescence, and solid state NMR provide structural details about the number of O atoms coordinated to a cation (e.g., MO_4, MO_5, or MO_6) and the presence of adjacent neighbors (M–O–M). The bridging M–O–M bonds are also easily detectable with Raman spectroscopy and occasionally also in the IR overtone region. Coupled Raman, IR, and isotopic oxygen exchange studies can establish the number of terminal M=O bonds (e.g., monoxo M=O, dioxo O=M=O or tri-oxo $M(=O)_3$) [64]. For isolated mono-oxo units, the M=O symmetric stretch, v_s, appears at the same frequency in both the Raman and IR spectra. In addition, the IR overtone region exhibits only one band at $\sim 2v_s$. For isolated dioxo structures, the O=M=O functionality possesses both symmetric, v_s, and asymmetric, v_{as}, stretching modes that can be separated by about ~ 10 cm^{-1} and the IR overtone region exhibits three bands at $\sim 2v_s$, $v_s + v_{as}$, and $\sim 2v_{as}$ that span over a ~ 20 cm^{-1} range. For isolated trioxo functionalities, the vibrational spectra are more complex and multiple bands will generally be present in the stretching and overtone regions. For dimeric monoxo species, where the M=O bonds are in the *cis* configuration, the v_s and v_{as} stretching modes are separated by ~ 10 to 50 cm^{-1} and a triplet of bands is also present in the overtone region. For dioxo dimers, the stretching modes are separated by more than 50 cm^{-1}. For polymeric monoxo and polymeric dioxo species, the fundamental stretching vibrations are not coincident and the overtone region reflects the multiplicity of the fundamental stretching vibrations. Raman is generally more sensitive to v_s and IR is generally more sensitive to v_{as}. In the event that the O=M=O bonds are separated by at a $90°$, then the vibrations will degenerate and the splitting of the bands will not be observed. Isotopic $^{16}O/^{18}O$ exchange studies are able to split such degenerate vibrations by scrambling of the oxygen isotopes. For monoxo structures, two symmetric stretching bands will be present due to M=^{16}O and M=^{18}O vibrations. For dioxo structures, three symmetric stretching bands will appear due to ^{16}O=M=^{16}O, ^{18}O=M=^{18}O, and ^{16}O=M=^{18}O vibrations, and four symmetric stretching bands should appear for trioxo functionalities ($M^{16}O_3$, $M^{18}O^{16}O_2$, $M^{18}O_2^{16}O$, and $M^{18}O_3$). In addition, the isotopic shifts due to the substitution of the heavier ^{18}O for ^{16}O can also be calculated for diatomic oscillators and compared with the observed isotopic shifts [62]. Thus, the combination of these molecular spectroscopic measurements coupled with isotopic oxygen exchange studies are required to obtain the complete dehydrated surface metal oxide structures.

1.2.2.2 Vanadium (+5) oxides

The bulk structural inorganic chemistry of vanadium (+5) oxides is the most varied among the bulk metal oxides, and has been determined from extensive x-ray crystallographic studies [60]. Bulk VO_4 vanadate ions consist of isolated (VO_4^{3-} orthovandate), dimeric ($V_2O_7^{4-}$ pyrovanadate) or polymeric chain (($VO_3)_n^{n-}$ metavanadate) structures. These four-coordinated vanadate ions are distinguished by the number of bridging V–O–V bonds that are present in the orthovanadate (0), pyrovanadate (1), and metavandate (2) structures, and are charge balanced by cations (e.g., Na_3VO_4, $Na_4V_2O_7$, and $Na_n(VO_3)_n$). Bulk VO_6

vanadates are also very common structures, and are typically found in extended vanadia structures. For example, the decavandate cluster in $Na_6V_{10}O_{28}$ consists of five distinct distorted VO_6 sites [72]. The highly distorted VO_6 structures usually possess one terminal V=O bond (monoxo) with bond lengths between 0.158 and 0.162 nm. For some highly distorted VO_6 oxides, the sixth O is located very far from the V atom that these compounds are effectively considered to possess VO_5 coordination. This is the case for bulk V_2O_5 that contains its 6 O atoms at 0.158, 0.178, 0.188, 0.188, 0.202, and 0.278 nm, with the most distant oxygen usually not considered to be in the V coordination sphere. Several gas phase monoxo $X_3V=O$ halide species are also known and their vanadyl vibrations vary from 1025 to 1058 cm^{-1} with increasing electronegativity of the halides (Br < Cl < F) [61]. The dixo $F_2VO_2^-$ and $Cl_2VO_2^-$ oxyhalide vibrations are observed at 970/962 and 970/959 cm^{-1}, respectively. In summary, the rich inorganic chemistry of bulk vanadium (+5) oxide is built up from VO_4, VO_5, and VO_6 coordinated structures.

Spectroscopic characterization studies employing solid state ^{51}V NMR [32,33], XANES/EXAFS [46,47,73], UV-Vis [44], and chemiluminescence [69] have revealed that the dehydrated surface VO_x species consist of highly distorted VO_4 units up to monolayer surface coverage. Above the monolayer surface vanadia coverage, V_2O_5 crystallites, possessing VO_5 coordination, are also present on top of the surface VO_x monolayer. *In situ* Raman and IR spectroscopic studies have demonstrated that the dehydrated surface VO_4 species possess only one terminal V=O bond [43,64,74] because of the same fundamental vibrational band position detected by both techniques. The dehydrated terminal V=O bond exhibits its fundamental vibration in the 1015 to 1040 cm^{-1} region and its overtone vibration at ~2036 cm^{-1} [64,74,75]. The fundamental vibration is dependent on the specific oxide support and the surface vanadia coverage. Isotopic oxygen exchange studies, further confirm the presence of only monoxo surface VO_4 species since the V=O band splits into doublets [43,63,76]. The remaining three oxygen atoms are coordinated to the oxide support cation, bridging V–O–S where S represents the support, when the dehydrated VO_4 site is isolated. EXAFS/XANES analysis concluded that the dehydrated surface VO_4 species on SiO_4 is isolated and contains a monoxo vanadyl structure, $O-V(-O-Si)_3$ [46,77].

For supported V_2O_5/SiO_2, Raman and UV-Vis spectroscopy reveal only the presence of isolated surface VO_4 species with a sharp band at ~1035 cm^{-1} from the terminal V=O bond of the dehydrated surface VO_4 species [77,78]. The overtone region exhibits a single band at ~2055 cm^{-1} consistent with a monoxo V=O functionality [79]. The monoxo nature of the V=O bond for the isolated surface VO_4 species on SiO_2 was further confirmed by the splitting of this band into a doublet during isotopic oxygen exchange [80]. The experimentally observed isotopic shift of 43 cm^{-1} agrees well with the theoretically determined isotopic shift of 45 cm^{-1} for a diatomic V=O oscillator [76]. The ~1035 cm^{-1} vibration also falls in the range observed for the vibrations of vanadyls in gas phase monoxo halides (1025 to 1058 cm^{-1}). The low reactivity of the surface Si–OH functionality is responsible for the low surface coverage and isolated nature of surface metal oxides dispersed on the SiO_2 support [81].

For non-SiO_2 supported vanadia catalysts, the simultaneous presence of dehydrated polymeric surface VO_4 species and dehydrated isolated surface VO_4 species is detected by *in situ* UV-Vis as a shift in the $O^{2-} \rightarrow V^{5+}$ charge transfer transition from $\sim$250 to $\sim$300 nm, as well as a decrease in the band gap energy, E_g, from 3.6 to 2.6 eV [44,82]. In addition, Raman detects vibrations originating from bridging V–O–V bonds at $\sim$500–600 (v_s), 700–800 (v_{as}), and 200–300 cm^{-1} (bending mode) when the oxide support does not obscure these vibrational regions [75,83]. For example, the bridging V–O–V vibrational modes are readily observed on the Al_2O_3 support with Raman spectroscopy [74,75]. For non-SiO_2 supports, the isolated monoxo structure of the dehydrated surface VO_4 species responsible for the $\sim$1030 cm^{-1} band is confirmed by (1) the coincidence of this band in Raman and IR, (2) splitting of this band into a doublet during isotopic oxygen exchange (both in Raman and IR), (3) the presence of only one band in the IR overtone region, and (4) its vibration falls in the vibrational region of 1025 to 1058 cm^{-1} for monoxo gas phase oxy halides [43,61,76]. The relative concentration of dehydrated polymeric to isolated surface vanadia species generally increases with surface vanadia coverage as reflected in the decrease in the band gap values [44,82]. From *in situ* UV-Vis spectroscopy it appears that the dehydrated polymeric surface vanadia species on the nonsilica supports probably form extended polymeric $(VO_3)_n$ structures with $n \gg 2$ at monolayer surface coverage because of the low measured E_g values [44,82,84].

Weak IR bands due to V=O bonds are also observed at $\sim$1015 to 1017 cm^{-1} at monolayer surface coverage and have been assigned to surface polymeric VO_4 species [74]. This vibrational position is just slightly lower than that for the monoxo vanadyls of the gas phase halides, 1025 to 1058 cm^{-1}, and is consistent with greater delocalization of monoxo V=O electrons over a more extended surface polyvanadate species. The absence of the additional expected multiple bands for surface polymeric monoxo species may be due to their overlap with the much stronger V=O vibration of the isolated surface monoxo VO_4 species. An exception to this general observation, however, is found for the supported V_2O_5/CeO_2 system where IR detects two small shoulders at 1022 and 1029 cm^{-1} that would correspond to polymeric surface VO_4 species [74]. Furthermore, reactivity studies with supported 1% V_2O_5/CeO_2 have shown that the bands for the isolated, $\sim$1030 cm^{-1}, and polymeric species, $\sim$1015 cm^{-1}, do not behave the same and, therefore, must originate from different surface VO_4 species [63]. The band associated with the surface polymeric VO_4 species is selectively, partially reduced in reactive environments. Detailed EXAFS analysis studies were not able to account for the presence of the polymeric surface vanadate species since the spectra could be fitted with only an isolated surface VO_4 species [73]. However, a minor residual component in the EXAFS spectra may be due to the polymeric surface vanadates [73]. Thus, the fraction of polymeric surface VO_4 species in the surface vanadia monolayer may be small because (1) the surface polymeric V=O bond generally does not give rise to strong vibrations relative to the surface isolated monoxo VO_4 species [74], (2) the overtone region only exhibits the V=O vibration from the isolated surface monoxo VO_4 species [64,74,85], and (3) the EXAFS analysis cannot

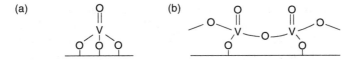

FIGURE 1.3   Structures of (a) dehydrated isolated and (b) polymeric surface monoxo VO_4 species

account for a minor residual component that may be due to the polymeric surface VO_4 species [73].

The molecular structures of the dehydrated, isolated, and polymeric surface vanadia species are depicted in Figure 1.3. The surface vanadia structural chemistry under dehydrated conditions reflects the known bulk vanadium oxide inorganic chemistry, in that isolated and polymeric VO_4 units form with only monoxo terminal V=O bonds.

Although monoxo isolated and dimeric VO_4 bulk structures are known [33], polymeric monoxo VO_4 bulk structures have not been reported in the literature and appear to be unique to the dehydrated surface vanadia structures.

1.2.2.3   Chromium (+6) oxides

Bulk chromates possess CrO_4 coordination in isolated (CrO_4 monochromate), dimer (Cr_2O_7 dichromate), trimer (Cr_3O_{10} trichromate), and tetramer (Cr_4O_{13} tetrachromate) infinite chain (CrO_3 metachromate or polychromate) structures [60]. Unlike the corresponding bulk vanadates, bulk non-CrO_4 containing structures are unknown (e.g., CrO_5 and CrO_6). The crystalline CrO_3 structure is built up of infinite chains by linking CrO_4 units (two short bonds at 0.160 nm and two longer bonds at 0.175 nm) that are only held together by van der Waal forces. The unusually low melting point of CrO_3, 197°C, reflects the weak van der Waal interactions among the polychromate chains. The low thermal stability of bulk CrO_3 is also reflected in its facile reduction and decomposition to bulk Cr_2O_3, which consists of only Cr(+3) cations [27,61]. The Cr(+6) oxidation state is usually stabilized by the presence of nonreducible cations (e.g., K, Na, Rb, P, and As). Gas phase chromium oxy halides are also known and monoxo $F_4Cr=O$ vibrates at 1028 cm^{-1}, dioxo $F_2Cr(=O)_2$ vibrates at 1006 (v_s) and 1016 (v_{as}) cm^{-1}, dioxo $Cl_2Cr(=O)_2$ vibrates at 984 (v_s) and 994 (v_{as}) cm^{-1}, and trioxo $CsBrCr(=O)_3$ vibrates at 908 (v_s), 933 (v_{as}), 947 (v_{as}), and 955 (v_{as}) cm^{-1} [86]. These vibrational frequency shifts as a function of the M=O bonds are significantly beyond that expected for the different halide ligands since the gas phase vanadyl oxy halide complexes shifted downward 23 cm^{-1} in going from F to Cl ligands and downward 10 cm^{-1} in going from Cl to Br ligands. Therefore, increasing the number of chromyl bonds shifts the vibrations to lower wave numbers and progressively increases the number of vibrational bands. In summary, the inorganic structural chemistry of Cr(+6) chromates essentially consists of CrO_4 units with different extents of polymerization.

Spectroscopic characterization of the dehydrated supported chromates with UV-Vis [45,65], chemiluminescence [69], and EXAFS/XANES [87] revealed that the dehydrated surface chromates possess CrO_4 coordination and are stabilized as $Cr(+6)$ at elevated temperatures by the oxide supports below monolayer surface coverage. Above the monolayer surface coverage, the excess chromium oxide that resides on the surface chromia monolyer becomes reduced at elevated temperatures in oxidizing environments and forms $Cr(+3)$ Cr_2O_3 crystallites [41,87]. Thus, the surface $Cr(+6)$ species are only stabilized at elevated temperatures by coordination to the oxide substrates. For non-SiO_2 supports, the Raman and the IR vibrational spectra exhibit two strong bands in the fundamental (1005–1010 and 1020–1030 cm^{-1}) as well as the overtone (1986–1995 and 2010–2015 cm^{-1}) vibrational regions. These bands occur at the exact same fundamental vibrations as well as relative intensities in the Raman and the IR spectra of the Al_2O_3, ZrO_2, and TiO_2 supported chromates, and are separated by 15 to 20 cm^{-1} [88]. The vibrational difference is consistent with dioxo functionality, but is slightly on the high side [64]. However, the Raman and the IR relative intensities of the ν_s and ν_{as} modes should vary inversely because Raman is more sensitive to symmetric stretches and IR is more sensitive to asymmetric stretches, but the observed band intensities are the same in IR and Raman [64]. Furthermore, the very high position of the Cr–O vibrations is consistent with that of the isolated monoxo chromyl structures, 1028 cm^{-1}, and not of the isolated dioxo chromyl structure, 984 to 994 cm^{-1}, present for the isolated gas phase isolated chromium oxyhalides (see earlier). Additional insights were obtained from isotopic oxygen exchange experiments that showed that both bands split into doublets [76]. This reveals that these two bands originate from two independent surface monoxo chromyl species since a single dioxo species would be expected to only give rise to a triplet. This conclusion was further substantiated by reactivity studies that demonstrated that the two surface chromate species reduce at different rates [89–91]. Comparison of the vibrational bands at different extents of reduction revealed that the 1010 cm^{-1} band decreased in the same ratio as bridging Cr–O–Cr vibrations and, consequently, was assigned to dehydrated surface polychromate species [91]. The reduction extent of the 1030 cm^{-1} band did not parallel any of the other bands and was, thus, assigned to the isolated monoxo surface chromate [91].

It is not possible to obtain the chromyl vibrations with IR for low surface coverage of chromia because the SiO_2 support absorbs the fundamental and overtone regions for low surface coverage of surface CrO_x on SiO_2 [64,88]. The corresponding Raman spectrum reveals that the dehydrated surface CrO_x species on SiO_2 is isolated since no bridging Cr–O–Cr vibrations were detected [88,91]. The somewhat low Raman vibration of $\sim$986 cm^{-1} of the supported CrO_3/SiO_2 catalyst suggests (recall that gas phase dioxo Cl_2CrO_2 exhibits Raman vibrations at 984 (ν_s) and 994 (ν_{as}) cm^{-1}) that it may arise from dioxo surface chromate species and that the two expected vibrations are degenerated. Model sesquoxide (sp???) chromia silica model compounds have been proposed to possess dioxo chromate species [Feher reference]. Only isotopic oxygen exchange studies and comparative Raman and IR characterization studies can clearly discriminate

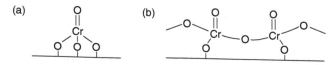

FIGURE 1.4 Structures of (a) dehydrated isolated and (b) polymeric surface monoxo CrO_4 species

between monoxo and dioxo surface chromate species for the dehydrated supported CrO_3/SiO_2 catalyst, however, such successful studies have not been reported till date in the literature.

The presence of bridging Cr–O–Cr bonds is revealed by the $O^{2-} \rightarrow Cr^{6+}$ ligand to metal charge transfer transition [92] and by Raman vibrations at 770 (v_{as}), 600 (v_s) and bending modes in the 300 to 400 cm^{-1} range when not obscured by the support vibrations [88]. The expected vibrational band splitting for the polymeric surface monoxo CrO_4 species was not observed and the additional band may overlap the adjacent band of the isolated monoxo CrO_4 species. The bridging Cr–O–Cr vibrations are easily detected on Al_2O_3, and the ratio of polymeric to isolated surface CrO_4 species appears to be relatively constant with surface coverage. No studies have addressed the issue of the fraction of the surface chromate species that are present as polymeric CrO_4 species.

The molecular structures of the dehydrated surface chromate species are schematically presented in Figure 1.4. The dehydrated surface CrO_4 structures have much in common with their corresponding bulk chromates, CrO_4 coordination and different extents of polymerization, but the surface chromates are monoxo, with the possible exception of the SiO_2 support, and the bulk chromates approach dioxo coordination upon extensive polymerization ($n \gg 4$ as in bulk CrO_3). Thus, monoxo chromates are unique to surface chromate species on oxide supports and some gas phase oxyhalides. Furthermore, the oxide supports stabilize the surface chromate species in the Cr(+6) oxidation state, and chromia in excess of monolayer surface coverage becomes reduced to Cr(+3) Cr_2O_3 crystalline particles upon calcination at elevated temperature.

1.2.2.4 Rhenium (+7) oxides

The bulk inorganic chemistry of rhenium (+7) oxides is rather sparse [60,93]. Several ortho-rhenate compounds containing isolated ReO_4 units are rather common: NH_4ReO_4, $KReO_4$, and $NaReO_4$. Bulk Re_2O_7 possesses a layered structure consisting of alternating ReO_4 and ReO_6 groups, with subunits of rings composed of two ReO_4 and two ReO_6 groups. The weak bonding between the rhenium oxide units in the layered Re_2O_7 structure results in the efficient vaporization of Re_2O_7 dimers that contain two ReO_4 units bridged by one oxygen atom (gaseous O_3Re–O–ReO_3). The ReO_3 groups in the gas phase Re_2O_7 dimer consist of trioxo terminal Re=O bonds that vibrate at 1009 (v_s of terminal Re=O), 972 (v_{as} of terminal Re=O), 456 (v_s of bridging O–Re–O), 341 (bending of

O–Re–O), and 185 cm^{-1} (bending of bridging Re–O–Re). The expected splitting of the asymmetric vibrations due to the C_{3v} symmetry of the –O–Re(=O)$_3$ units was not observed because of the degeneracy of these vibrational modes. Trioxo gas phase rhenium oxyhalides of XReO$_3$ are also known and exhibit the terminal Re=O v_s vibrations at 997 to 1009 cm^{-1}, which increases with the halide electronegativity (Br < Cl < F) [94]. The gas phase dioxo F$_3$ReO$_2$ oxyhalide exhibits its terminal Re=O vibrations at 1026 (v_s) and 990 (v_{as}) cm^{-1} [95] and the monoxo F$_5$Re=O possesses its terminal Re=O v_s vibration at 990 cm^{-1} [62]. The lower Re=O v_s vibration of the monoxo F$_5$Re=O molecule relative to the dioxo F$_3$ReO$_2$ is somewhat surprising since the bond order of the M=O terminal functionality generally increases upon decreasing the number of terminally coordinated oxygen atoms (see gas phase chromate oxyhalides). Thus, the rhenium oxide structural inorganic chemistry is composed of isolated ortho-rhenates (ReO$_4$), ReO$_4$ dimers (O$_3$Re–O–ReO$_3$), and polyrhenates composed of mixtures of alternating ReO$_4$ and ReO$_6$ groups.

The maximum attainable surface ReO$_x$ coverage on oxide supports is always less than monolayer coverage because the surface ReO$_x$ species combine to form volatile Re$_2$O$_7$ dimers at high surface coverage [42]. Furthermore, crystalline Re$_2$O$_7$ is never observed because this metal oxide is not stable to high temperature calcination as well as to exposure to ambient moisture. Consequently, monolayer surface ReO$_x$ coverage is never reached because volatilization and crystalline Re$_2$O$_7$ is never present. Thus, supported ReO$_x$ catalysts are unique among the supported metal oxide catalysts in that only surface ReO$_x$ coverage below monolayer can be achieved without the presence of crystallites.

The dehydrated supported ReO$_x$/Al$_2$O$_3$ system has received most attention among the different supported ReO$_x$ catalysts in the literature. The coordination of the dehydrated surface ReO$_x$ species was determined from XANES [51] and UV-Vis [96] to be distorted ReO$_4$. The complementary Raman and IR spectra revealed that two different dehydrated surface ReO$_x$ species are present in Al$_2$O$_3$ with Re=O v_s at ~1015 and ~1004 cm^{-1} and corresponding Re=O v_{as} at ~980 and 890 cm^{-1}, respectively. The only difference between the two surface ReO$_4$ species is the slightly different Re–O bond lengths that caused the vibrational shifts. The Raman and IR vibration bands are coincident and the IR overtone region reveals two distinct bands at 1972 and 1994 cm^{-1}, which is consistent with the presence of two different surface ReO$_4$ species [41,97]. IR isotopic oxygen exchange studies resulted in the shifting of the terminal Re=O vibrations in the overtone region to ~1896 cm^{-1} and the broadness of the resulting band prevented further resolution of the overtone modes [98]. The IR fundamental vibrations also shifted to lower wave numbers due to the heavier mass of the ^{18}O atom and were masked by the strong IR absorption of the Al$_2$O$_3$ support in this lower vibration region [98]. Raman spectroscopy revealed that the surface ReO$_x$ species on Al$_2$O$_3$ were isolated since no vibrations originating from a bridging Re–O–Re functionality were detected (v_s~456 and bending ~185 cm^{-1}) and only the expected bending vibrations of O–Re–O functionally were present at 340 cm^{-1} with a shoulder at 310 cm^{-1} [42,97]. The Re=O v_s and v_{as} vibrations of the dehydrated surface ReO$_4$/Al$_2$O$_3$

species on Al_2O_3 are consistent with the vibrations exhibited by trioxo mono-meric gas phase oxyhalides $XRe(=O)_3$, possessing C_{3v} symmetry, at 997–1009 and 963–980 cm^{-1}, and the $^-SRe(=O)_3$ monomer, at 953 and 917 cm^{-1} [94]. The gas phase dioxo $F_3Re(=O)_2$ exhibits its ν_s and ν_{as} vibrations at 1026 and 990 cm^{-1}, with both stretching modes higher than that observed for surface ReO_4/Al_2O_3 spe-cies. The gas phase monoxo $F_5Re=O$ exhibits its ν_s vibration at 990 cm^{-1}, which is lower than that measured for the dehydrated surface ReO_4/Al_2O_3 species. Further-more, the monoxo $F_5Re=O$ does not possess the ν_{as} stretching vibration observed for the surface ReO_4/Al_2O_3 species. The vibrations of the dehydrated surface ReO_4 species on Al_2O_3 (ReO_4-I: 1015, 980, 340, and 310 cm^{-1} and ReO_4-II: 1004, 890, 340, and 310 cm^{-1}) are also remarkably similar to those of the gas phase Re_2O_7 ReO_4-containing dimer (1009, 972, 341, and 322 cm^{-1}) without the associated bridging Re–O–Re vibrational modes at 356 and 185 cm^{-1}. The similarity of the vibrations of gaseous Re_2O_7, as well as the $XRe(=O)_3$ oxyhalides, and the dehyd-rated surface ReO_x species on Al_2O_3 strongly suggests that the same coordination is present for both systems: trioxo ReO_4 with one bridging Re–O–bond. The vibrational similarity of monoxo ReO_4 and trioxo ReO_4 structures occur because both structures possess C_{3v} symmetry and the apparent degeneracy of the asym-metric vibrations of the $–O–Re(=O)_3$ functionality in gas phase dimeric Re_2O_7. Isotopic oxygen exchange studies with Raman spectroscopy may provide further insights into the structure of the dehydrated surface $–O–Re(=O)_3$ species on Al_2O_3 because the Raman signal, unlike IR, will not be absorbed by the Al_2O_3 support below 1000 cm^{-1}.

Essentially, the same two dehydrated surface ReO_x species found to be present on Al_2O_3 are also present on ZrO_2 and TiO_2 oxide supports [42]. The coincident Raman and IR vibrations of the dehydrated ReO_x/ZrO_2 sample at low surface coverage occur at $\sim$995 (ν_s) and $\sim$885 (ν_{as}) cm^{-1} for the first surface ReO_x species and $\sim$1008 (ν_s) and $\sim$980 (ν_{as}) cm^{-1} for the second surface ReO_x species at the highest surface coverage (3.3 Re/nm^2). The coincident Raman and IR bands for the dehydrated ReO_x/TiO_2 sample appear at $\sim$1005 (ν_s) cm^{-1} at low surface coverage and $\sim$1011 (ν_s) and 975 (ν_{as}) cm^{-1} at the highest surface coverage (2.4 Re/nm^2). Only very low surface coverage of ReO_x on SiO_2 was achieved (0.54 Re/nm^2) and evidence for only one dehydrated surface ReO_x species with vibrations at $\sim$1015 (ν_s) and $\sim$985 (ν_{as}) cm^{-1} was detected on the SiO_2 support with Raman and IR, respectively. Isotopic oxygen exchange Raman studies with supported ReO_x/ZrO_2 revealed only two vibrations for $Re=^{16}O$ and $Re=^{18}O$ [76]. These results are, at first, surprising since the isotopic exchange was expected to yield four different permutations for the vibrationally coupled trioxo functionality. However, Re=O bonds that are at 90° to each other are not vibrationally coupled and would behave as independent Re=O bonds and, consequently, only give rise to splitting of the Re–O vibrations.

The molecular structure of the dehydrated surface rhenate species is shown in Figure 1.5. The surface rehenate species is always isolated and possesses three ter-minal Re=O bonds and one bridging Re–O support bond. The molecular structure of the dehydrated surface rhenium oxide species is consistent with the known

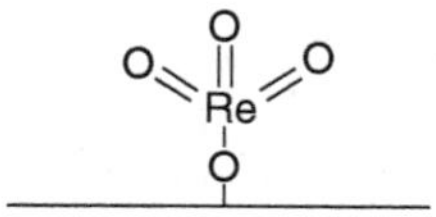

FIGURE 1.5   Structure of dehydrated isolated surface monoxo ReO_4 species

inorganic coordination chemistry of rhenium oxide (preference for isolated ReO_4 units with three Re=O bonds and one Re–O bond).

1.2.2.5 Molybdenum (+6) oxides

Bulk polymolybdate chains usually contain MoO_6 coordinated groups, which is unlike the polyvanadate and polychromate chains that are composed of VO_4 and CrO_4 units, respectively [60]. This reflects the preference of molybdates for higher coordination compared with vanadates and chromates in polymeric structures. However, some exceptions exist to this trend in the bulk molybdate structural chemistry. Low coordinated molybdates are present in $MgMo_2O_7$ (dimer of MoO_4) and in $NaMo_2O_7$ (chain of alternating MoO_4 and MoO_6 units). Isolated MoO_4 coordination is, however, rather common for ortho-molybdates (e.g., K_2MoO_4, Na_2MoO_4, $CaMoO_4$, $MgMoO_4$, $MnMoO_4$, $CuMoO_4$, etc.). Highly distorted, isolated MoO_4 coordination is found in $Al_2(MoO_4)_3$, $Fe_2(MoO_4)_3$, $Cr_2(MoO_4)_3$, and $Gd_2(MoO_4)_3$ [99]. Highly distorted MoO_5 units are present in $Bi_2(MoO_4)_3$ [99]. Polymolybdate clusters composed of 6 to 8 MoO_6 coordinated units are also known (e.g., $[NH_3P_3(NMe_2)_6]_2Mo_6O_{19}$, $(NH_4)_6Mo_7O_{24}$, and $(NH_4)_4Mo_8O_{26}$) [60,61]. Bulk alpha MoO_3 is composed of a 3D structure made up of highly distorted MoO_6 units. The large distortion present in bulk alpha-MoO_3 causes the sixth oxygen atom to be located very far from Mo and, consequently, the bulk alpha-MoO_3 structure is better described as consisting of MoO_5 units [60]. The bulk beta-MoO_3 crystalline phase is another MoO_3 3D structure built up of less distorted MoO_6 units [53]. Several gas phase monoxo molybdenum $X_4Mo{=}O$ oxyhalides are also known and the Mo=O vibrations vary from 1008 to 1039 cm^{-1} with increasing electronegativity of the halide (Cl < F) [61]. The gas phase dioxo $Br_2Mo(=O)_2$ gives rise to bands at 995 (v_s) and 970 (v_{as}) cm^{-1} (electronegativity of Br < Cl < F) [61]. Thus, the structural inorganic chemistry of molybdenum oxides consists of MoO_4, MoO_5, and MoO_6 coordinated groups, with a preference for MoO_6 latter in polymolybdates.

The coordination of the dehydrated surface MoO_x species on different oxide supports was determined with XANES and found to be dependent on the specific oxide support and surface MoO_x coverage [40,100,101]. Above the monolayer surface MoO_x coverage, crystalline MoO_3 was also present. For supported MoO_3/SiO_2, crystalline MoO_3 was also observed to be present significantly below monolayer surface coverage because of the low reactivity of the surface Si–OH groups. The presence of monoxo Mo=O bonds in the surface MoO_x species was revealed by the coincidence of the fundamental Raman and IR vibrations in the

980 to 1006 cm^{-1} region and the appearance of only one band in the IR overtone region [64,85,98]. This was further confirmed by oxygen exchange studies that showed only splitting of the terminal Mo=O vibration during isotopic scrambling [76,102]. The remaining oxygen ligands in the surface MoO$_x$ species are bonded either to the oxide support cations or to the adjacent surface MoO$_x$ species.

For dehydrated supported MoO$_3$/SiO$_2$, both Raman and IR confirmed the absence of bridging Mo–O–Mo bonds and, thus, the presence of only isolated surface MoO$_x$ species at low surface coverage [40,100–102]. EXAFS analysis of the dehydrated surface MoO$_x$ species on SiO$_2$ is consistent with the isolated nature of the supported MoO$_x$/SiO$_2$ because of the almost complete absence of Mo–Mo neighbors in the coordination sphere [100,102]. The EXAFS analysis finds only one short Mo=O bond at 0.169 nm in the coordination sphere around Mo. The monoxo Mo=O nature of the surface MoO$_x$ species on SiO$_2$ is also revealed from the identical Raman/IR vibrations of this functionality at 986 cm^{-1} [103] and the appearance of only two IR bands during isotopic oxygen exchange (Mo=^{16}O and Mo=^{18}O) [102]. The terminal Mo=O vibrations of the dehydrated surface MoO$_x$ species are slightly lower than the reported vibrations of the gas phase monoxo (1008 to 1039 cm^{-1}) and dioxo (995 cm^{-1}) oxyhalides. A dioxo structure would expect to exhibit a doublet in this region and only one band is observed, which is also consistent with the monoxo structure. XANES [40,100,101] and UV-Vis [104] analysis of the coordination of the surface MoO$_x$ species on SiO$_2$ has been found to be neither that of pure MoO$_4$ nor that of pure MoO$_6$, and this dehydrated surface MoO$_x$ species most probably possesses O=Mo(–O–Si)$_4$ coordination [40,102].

For non-SiO$_2$ supported MoO$_x$ catalysts, the dehydrated surface MoO$_x$ coordination depends on the surface molybdena coverage and the specific oxide support. At low surface molybdena coverage (5 to 15% of monolayer), primarily surface MoO$_4$ coordinated units are present on Al$_2$O$_3$ and TiO$_2$ [40,101]. The corresponding Raman spectra of these catalysts also indicate that the surface MoO$_4$ species are isolated on both oxide supports at low surface coverage [40,101]. This is also substantiated by the UV-Vis spectra that exhibit a high band gap value associated with isolated species [104]. At monolayer surface molybdena coverage, however, the surface MoO$_x$ coordination is different and also dependent on the specific oxide support [40]. At monolayer surface molybdena coverage, supported MoO$_3$/TiO$_2$ was found to possess MoO$_6$ coordinated species and supported MoO$_3$/Al$_2$O$_3$ was found to possess a mixture of MoO$_4$ and MoO$_6$ coordinated groups. The additional presence of surface MoO$_6$ for monolayer MoO$_3$/Al$_2$O$_3$ was also reflected in the lower band gap energy of this catalyst [104]. The UV-Vis and Raman spectra for the dehydrated MoO$_3$/ZrO$_2$ and MoO/Al$_2$O$_3$ samples were very similar and suggest that the same surface MoO$_x$ species exist on both supports at a given surface coverage [40,104]. The Raman spectra also exhibits the characteristics of bridging Mo–O–Mo bonds present in polymolybdates [40,101]. The coincidence of the Raman and the IR fundamental vibrations at $\sim$1000 cm^{-1} for supported MoO$_3$/Al$_2$O$_3$ and the presence of only one band in the overtone region is consistent with monoxo Mo=O functionality [64,85,98]. This is also confirmed

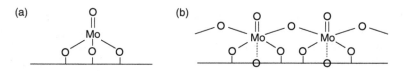

FIGURE 1.6   Structures of dehydrated surface monoxo MoO_x species. (a) Isolated monoxo MoO_4/MoO_5 and (b) polymeric monoxo MoO_6

with isotopic oxygen exchange studies that reveal the presence of only two Mo=O vibrations that arise from terminal Mo=^{16}O and Mo^{18}O bonds [76].

The different molecular structures of the dehydrated surface molybdates are presented in Figure 1.6. At low surface molybdena coverage, the preferred coordination is isolated monoxo MoO_4, with the exception of supported MoO_3/SiO_2 that appears to possess the isolated monoxo MoO_5 structure. At high surface coverage, highly distorted and polymeric monoxo MoO_6 structures are always present. The presence of surface monoxo MoO_4 species at high surface coverage depends on the specific support ($Al_2O_3 \sim ZrO_2 \gg TiO_2 \sim Nb_2O_5$) [40]. For high surface coverage of supported MoO_3/SiO_2, crystalline MoO_3 is also present in addition to the surface MoO_x species. The coordination chemistry of the dehydrated surface MoO_x species parallels that of its known inorganic structural chemistry: (1) monoxo nature of the surface MoO_x species, (2) MoO_4 coordinated isolated species, (3) MoO_6 polymolybdates, (4) mixture of MoO_4/MoO_6 polymolybdates, and (6) presence of isolated MoO_5 on SiO_2.

1.2.2.6   Tungsten (+6) oxides

The structural inorganic chemistry of tungsten oxide closely mirrors that of molybdenum oxide [60,61]. Many ortho-tungstate (Li_2WO_4, Na_2WO_4, Na_2WO_4, Rb_2WO_4, and Cs_2WO_4) compounds possessing isolated WO_4 sites are known. Tungstates rarely form polymeric WO_4 compounds, and one such exception is MgW_2O_7 that consists of a pair of sharing WO_4 units. Alternating polymeric WO_4 and WO_6 sites are present in the polytungstate chains of $Na_2W_2O_7$ and $(NH_4)_2W_2O_7$. An isolated WO_5 coordinated site has been determined to be present in the $Ca_3(WO_5)Cl_2$ compound. Isolated WO_6 coordinated units are present in the Wolframite structure ($FeWO_4$, $MnWO_4$, $CoWO_4$, $NiWO_4$, and $ZnWO_4$). Polytungstate chains composed of WO_6 coordinated units are present in $Li_2W_2O_7$ and $Ag_2W_2O_7$. Tungsten oxide clusters composed of polymeric WO_6 units have been identified with varying number of tungstate units: 4-membered ($Ag_8W_4O_{16}$), 6-membered ($NBu_4)_2W_6O_{19}$, 10-membered ($NH_4BuW_{10}O_{32}$), and 12-membered (paratungstate-$(NH_4)_{10}(H_2W_{12}O_{42}.10H_2O)$ and metatungstate-$(NH_4)_6(H_2W_{12}O_{40})$). Bulk WO_3 is built up a 3D structure of slightly distorted WO_6 units. Several gas phase monoxo tungsten X_4W=O oxyhalides are known (X=F, Cl, and Br) [62]. The F_4W=O gas phase complex exhibits its W=O symmetric stretch at 1055 cm^{-1}. Unfortunately, the vibrations of the gas phase monoxo complexes $Cl_4W=O$ and $Br_4W=O$ have not been experimentally determined. However, it is possible to estimate their vibrational frequency by analogy with the corresponding $X_4Mo=O$ and $X_3V=O$ oxyhalides that are similarly influenced by the electronegativity of the halide ligands. Such a comparison suggests the monoxo $W=O$ vibrations for $Cl_4W=O$ and $Br_4W=O$ oxyhalides should occur at $\sim$1024 and $\sim$1010 cm^{-1}, respectively. The vibrational spectra of dioxo $X_2W(=O)_2$ oxyhalides has not been determined, but the IR spectra for $Br_2Mo(=O)_2$ and the ions $[Se_2Mo(=O)_2]^{2-}$ and $[Se_2W(=O)_2]^{2-}$ has been reported and exhibit their v_s/v_{as} vibrations at 995/970, 864/834, and 888/845 cm^{-1}, respectively [62]. Note that the selenium-containing dioxo ions exhibit similar vibrations, but the W-containing ion vibrates $\sim$10 to 24 cm^{-1} higher than the corresponding Mo-containing ion. This suggests that gas phase $Br_2W(=O)_2$ oxyhalide would vibrate at $\sim$1020/980 cm^{-1} by analogy with $Br_2Mo(=O)_2$. This value can increase further as the electronegativity of the halides increases ($Br < Cl < F$), which may shift these bands to $\sim$1030/990 to 1050/1010 cm^{-1}.

The coordination of dehydrated surface WO_x species on different oxide supports was examined with XANES and found to vary with the surface tungsten oxide coverage and oxide support [48–50]. Above the monolayer coverage for the non-SiO_2 supports, crystalline WO_3 particles are present on top of the surface tungsten oxide monolayer [50,83,105,106]. In the case of supported WO_3/SiO_2, crystalline WO_3, and surface WO_x species are simultaneously present below monolayer surface coverage due to the low reactivity of the Si–OH groups [106,107]. For the supported WO_x systems that were extensively examined with Raman and IR, it was concluded that the surface WO_x species contain only one terminal $W=O$ bond [85,98]. The fundamental vibrations of the terminal $W=O$ bond was found to be coincident in the Raman and IR spectra, and only one band was observed in the IR overtone region. Isotopic oxygen exchange studies further confirmed the monxo $W=O$ nature of the surface WO_x structures since only two bands due to $W={}^{16}O$ and $W={}^{18}O$ vibrations were detected [108]. The vibrations of the terminal $W=O$ bond of the dehydrated surface WO_x species occur at 985 to 1025 cm^{-1} and are in the vibrational range expected for gas phase oxyhalide monoxo tungstanyls (1022 to 1055 cm^{-1}). Thus, the surface WO_x species contain one terminal $W=O$ bond and bridging W–O support and W–O–W bonds for polytungstates.

For dehydrated supported WO_3/SiO_2, the Raman spectra exhibits a single band at $\sim$983 cm^{-1} that has been assigned to the vibration of the terminal $W=O$ bond [108]. The position of this vibration is below that of the gas phase monoxo and dioxo tungsten oxyhalides (1010 to 1055 cm^{-1}). Unfortunately, no corresponding IR and isotopic oxygen exchange studies have yet been reported for this system to allow discrimination between monoxo and dioxo functionalities. In addition, no XANES and UV-Vis characterization studies about the coordination of the surface WO_x species on SiO_2 have been reported till date. Although, much of the critical structural spectroscopic data about the surface WO_x species on SiO_2 are currently not available, comparison with the corresponding surface MoO_x/SiO_2 system tentatively suggests that the surface WO_x on SiO_2 most likely possesses the isolated monoxo $O=W(-O)_4$ structure. The actual details of the surface WO_x structure on SiO_2 will be determined once the additional characterization information becomes available.

For dehydrated non-SiO_2 supported WO_x, there currently exists sufficient Raman and IR data to conclude that these structures possess monoxo $W=O$ surface WO_x species. For supported WO_3/Al_2O_3, the coincidence of the Raman and IR bands at ~ 1010 cm^{-1} for low surface coverage and ~ 1020 cm^{-1} for high surface coverage and the presence of only one band in the IR overtone region are consistent with the presence of a monoxo $W=O$ functionality [85,98]. This was further substantiated with isotopic oxygen exchange studies that only revealed the presence of only $W=^{16}O$ and $W=^{18}O$ vibrations [108]. The coordination of the surface WO_x species on Al_2O_3 was found to be WO_4 at low surface coverage and WO_6 at monolayer coverage [48]. Raman spectroscopy also revealed the presence of bridging W–O–W bonds with $\nu_s = 590$ cm^{-1} and the corresponding 215 cm^{-1} bending mode at high surface coverage, but not at low surface coverage [83]. The above information suggests that at low surface coverage the dehydrated surface WO_x species are isolated and at high surface coverage they are present as polytungstates. The analogous XANES and IR results were found for supported WO_3/TiO_2, with the only difference being the proposed surface WO_5 structure at high surface coverage [49,109]. For supported WO_3/ZrO_2, the surface WO_x species have been proposed to be present as isolated monoxo surface WO_6 species at all coverage from combined Raman, UV-Vis, and XANES/EXAFS studies [50]. Isotopic oxygen exchange Raman studies are consistent with the monoxo surface WO_6 structure on ZrO_2 since isotopic exchange only resulted in splitting of the $W=O$ bond and the isotopic shift was consistent with that predicted for a monoxo structure [76]. The proposed isolated nature of the surface WO_6 species on ZrO_2 needs to be further confirmed since the absence of polymeric WO_6 species at monolayer surface coverage is a rather surprising result for surface metal oxides on ZrO_2.

The dehydrated molecular structures of surface WO_x are presented in Figure 1.7. Analogous to the dehydrated surface MoO_x species, dehydrated isolated surface WO_4/WO_5 and dehydrated polytungstate surface WO_6 species are also found. The isolated monoxo WO_4 and WO_5 species are primarily observed at low surface coverage and the polytungstate monoxo WO_6 species are found to be present at high surface coverage. These dehydrated surface WO_x structures are consistent with the known inorganic chemistry of tungsten oxide compounds.

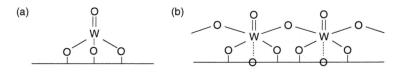

FIGURE 1.7 Structures of dehydrated surface monoxo WO_x species. (a) Isolated surface monoxo (WO_4 and WO_5) and (b) polymeric surface monoxo surface

1.2.2.7 Niobium (+5) oxides

Bulk niobium (+5) oxides primarily possess distorted NbO_6 coordination with different extents of distortion [60,110]. The NbO_7 and NbO_8 coordination are also known (e.g., $Nb_2O_5 \cdot nH_2O$ is composed of NbO_6, NbO_7, and NbO_8 units). Isolated NbO_4 coordination is not common and only found in several rare earth niobates: $LaNbO_4$, $SmNbO_4$, $YbNbO_4$, and $YNbO_4$. Consequently, unlike the isolated MO_4 sites present in bulk $AlVO_4$, $Al_2(MoO_4)_3$, and $Al_2(WO_4)_3$ discussed earlier, the bulk $AlNbO_4$ structure contains highly distorted polymeric NbO_6. The Nb_6O_{19} cluster is present in $K_8Nb_6O_{19}$, as well as in other Nb salts, and is composed of distorted NbO_6 groups. The decaniobate $Nb_{10}O_{28}$ cluster, analogous to the decavanadate cluster, is built up of distorted NbO_6 groups and has been isolated in the $[N(CH_3)_4]_6Nb_{10}O_{28}$ salt. Layered niobia salts, $KCa_2Na_{n-3}Nb_nO_{3n+1}$ with $n = 3$ to 5 layers, are composed of distorted monoxo NbO_6 groups [111–113]. The various bulk Nb_2O_5 crystalline phases are built up of 3D structures made up of distorted NbO_6 units with minor NbO_7 and NbO_8 sites (TT and T phases) and minor NbO_4 sites are also found in the holes of $H-Nb_2O_5$ (1 NbO_4 per 27 NbO_6). The Nb=O vibrations of monoxo $Cl_3Nb=O$ and dioxo $S_2Nb(=O)_2^{3-}$ oxyhalides occur at 997 and 897/872 cm^{-1}, respectively [61]. Thus, the inorganic structural chemistry of niobia compounds is mostly made up of distorted NbO_6 units, with some structures also containing NbO_7 and NbO_8 groups, and the isolated NbO_4 structure is rare.

The XANES/EXFAS data are only available for the dehydrated supported Nb_2O_5/Al_2O_3 and Nb_2O_5/SiO_2 systems [67]. The XANES/EXAFS analyses suggest that surface NbO_4 species are present at low surface coverage on both oxide supports. At high surface coverage on Al_2O_3, dehydrated surface NbO_6 species become the dominant surface NbO_x species [114]. At high surface coverage on SiO_2, crystalline Nb_2O_5 coexists with the surface NbO_4 species because of the low reactivity of the surface Si–OH bonds [114,115]. Above the monolayer surface coverage, crystalline Nb_2O_5 particles form on all oxide supports after calcination. The highest Raman vibration observed for the surface NbO_4 species is detected at ~ 980 to 990 cm^{-1} and occurs at almost the same vibration for the gaseous monoxo $Cl_3Nb=O$ oxyhalide, 997 cm^{-1}, which is significantly greater than the dioxo vibrations observed at 897 and 872 cm^{-1} [62,115]. The corresponding IR bands are coincident with the Raman vibrations and only one Nb=O vibration is present in the overtone region [116]. This collective spectroscopic information points to the presence of dehydrated isolated, monoxo surface O=Nb(–O)$_3$ species at low surface coverage. At high surface NbO_x/Al_2O_3 coverage, the XANES/EXAFS analyses suggest NbO_6 coordination and additional strong bands are present in the Raman spectra at ~ 950, 880, and 645 cm^{-1} associated with bridging Nb–O–Nb bonds. The Raman and IR bands at ~ 985 cm^{-1} are essentially coincident, and only one band is observed in the overtone region. This collective spectroscopic information suggests a dehydrated, distorted monoxo NbO_6 surface species at high surface coverage. Very similar Raman and IR vibrations are also present for the dehydrated supported Nb_2O_5/ZrO_2 and Nb_2O_5/TiO_2 systems. Isotopic oxygen

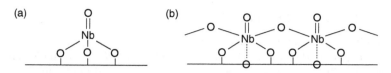

FIGURE 1.8 Structures of dehydrated surface monoxo NbO_x species. (a) Isolated surface monoxo niobate (NbO_4) and (b) polymeric surface monoxo niobate (NbO_6)

exchange Raman studies of monolayer supported Nb_2O_5/ZrO_2 reveal only two bands at $\sim$980 and $\sim$930 cm^{-1} due to Nb=^{16}O and Nb=^{18}O vibrations, and further support the presence of monoxo surface NbO_6 species [76]. Thus, under dehydrated conditions, isolated monoxo surface NbO_4 and polyniobate monoxo NbO_6 species are present on the oxide supports at low and high surface coverage, respectively. The proposed structures of the dehydrated surface NbO_x species are depicted in Figure 1.8.

The dehydrated surface NbO_x coordination chemistry is consistent with the known Nb oxide inorganic structural chemistry. Whereas NbO_4 coordination is rare in Nb inorganic structural chemistry, the surface NbO_4 structure appears to be prevalent at low surface coverage on oxide supports under dehydrated conditions.

1.2.2.8 Tantalum (+5) oxides

The bulk structural inorganic chemistry of tantalum oxide essentially mirrors that of niobium oxide [60,61,68]. Isolated TaO_4 sites are rare (e.g., $YbTaO_4$) and TaO_6 coordinated compounds are most common. Like niobia, tantala can also exhibit TaO_7 and TaO_8 coordination. The salt $K_8(Ta_6O_{19})\cdot16H_2O$ contains the $Ta_6O_{19}^{8-}$ cluster, similar to the $Nb_6O_{19}^{8-}$ cluster described in Section 1.2.2.7, made up of distorted TaO_6 groups. Bulk Ta_2O_5 exhibits a phase transition at 1360°C and the stable phases below and above this temperature are referred to as L–Ta_2O_5 and H–Ta_2O_5, respectively. Poorly crystalline Ta_2O_5 is present below $\sim$800°C and is present as $Ta_2O_5\cdot nH_2O$ (tantalum oxhydrate). Both the L–Ta_2O_5 and H–Ta_2O_5 crystalline phases consist of a three-dimensional network of distorted TaO_6 and TaO_7 units. The vibrational spectra of gaseous Ta oxyhalide complexes have not been reported in the literature [61,62]. It is possible, however, to estimate the vibrational band position of a monoxo $Cl_3Ta=O$ bond by comparing with the corresponding $Cl_3Nb=O$ monoxo complex and correcting for the heavier mass of Ta. Such an estimation gives a value of $\sim$950 cm^{-1} for the Ta=O vibration [68]. An analogous estimation of the vibrations for dioxo O=Ta=O suggests two vibrations in the 800 to 860 cm^{-1} region. Thus, the structural chemistry of tantala compounds rarely consists of TaO_4 coordination and is mostly made up of distorted TaO_6 and TaO_7 (TaO_8) units.

The coordination of the dehydrated surface TaO_x species on several different oxide supports was determined with XANES [68]. Highly distorted TaO_6 species were primarily found to be present on Al_2O_3, ZrO_2, and TiO_2 supports at high surface coverage. UV-Vis characterization of the dehydrated supported Ta_2O_5/Al_2O_3

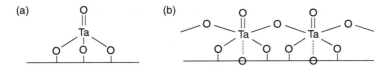

FIGURE 1.9 Structures of dehydrated surface monoxo TaO_x species. (a) Isolated surface monoxo tantalate (TaO_4) and (b) polymerized surface monoxo tantalate (TaO_6)

and Ta_2O_5/ZrO_2 catalysts also resulted in similar assignments [117]. The polymeric nature of these dehydrated surface TaO_6 species is reflected in the presence of bridging Ta–O–Ta vibrations at $\sim$740 (v_{as}) and $\sim$610 (v_s) cm^{-1} in the corresponding Raman spectra. For dehydrated Ta_2O_5/SiO_2, however, XANES and UV-Vis analyses revealed only the presence of TaO_4 units on the SiO_2 surface and Raman does not exhibit bands typically associated with bridging Ta–O–Ta bonds [68,117]. The Raman bands for the Ta=O functionality appeared at $\sim$940 cm^{-1}, which is consistent with the monoxo functionality of the dehydrated surface tantala species. Thus, for high surface coverage the dehydrated surface tantala species on Al_2O_3, ZrO_2, and TiO_2 primarily consist of polymeric TaO_6 units and as isolated TaO_4 units on SiO_2. The dehydrated surface TaO_4 coordination is probably also present at low surface coverage on the oxide supports, as found for the dehydrated surface NbO_x species, but such data has not been reported in the literature till date. The proposed structures for the dehydrated surface TaO_x species are shown in Figure 1.9.

The molecular structures of the dehydrated surface TaO_x species reflect the inorganic chemistry of tantalum oxide and mirror that of the analogous surface niobium oxide species. Isolated surface TaO_4 is present on SiO_2 and probably also at low surface coverage on all the oxide supports. Polytantalate TaO_6 species are the dominant species at high surface tantalum oxide surface coverage.

1.2.3 Surface Metal Oxide Species in Reactive Environments

In situ characterization studies of the surface metal oxide species under reactive environments for supported metal oxide catalysts have only appeared in the literature over the past decade [63,66,75,83,93,114]. For surface metal oxide species that usually do not undergo redox processes (acidic WO_3, Nb_2O_5, and Ta_2O_5), the surface metal oxide molecular structures under reactive conditions are identical to those present for the dehydrated conditions discussed earlier. For surface metal oxide species that can undergo redox processes (V_2O_5, CrO_3, MoO_3, and Re_2O_7), both, fully oxidized and partially reduced surface metal oxide species can be present. The fraction of reduced surface species is dependent on the reduction potential of the specific surface metal oxide (Re > V > Cr > Mo), and the specific reactive environment (partial pressures of reducing reactant/O_2 and the specific reducing reactant (e.g., propylene > *n*-butane > propane > ethane > methane) and the specific reactive environment (temperature and ratio of reducing agent/O_2)).

At present, the molecular structures of the dehydrated reduced surface metal oxide species present for supported metal oxide catalysts under reactive environments are not well-known and, hopefully, will receive more attention in the coming years. Fortunately, the fully oxidized surface metal oxide species are the predominant species found to be present under typical reaction conditions employed for redox supported metal oxide catalysts.

1.3 Molecular Sieves and Zeolites

Molecular sieves and zeolites are highly porous, crystalline metal oxides made up of three-dimensional channel structures where every atom is on the surface and exposed to the reactive environment (100% dispersed) [118]. Zeolites consist of AlO_4 and SiO_4 units with or without other cations and the term molecular sieves is reserved for SiO_4-based systems that are free of Al and typically doped with other cations. Zeolites find wide application in the petroleum industry as catalytic acidic materials for fluid catalytic cracking (FCC) of crude oil. Molecular sieves have been shown to exhibit surface redox properties and find application as catalysts for liquid phase epoxidation of olefins with H_2O_2. Recent characterization studies comparing cation-containing molecular sieves (e.g., Ti, V, and Nb) have revealed that essentially the same dehydrated surface metal oxide species are present in the doped molecular sieves and the amorphous SiO_2 supported metal oxide analogs. For example, isolated VO_4 sites were found to be present for both dehydrated V-silicalite and supported V_2O_5/SiO_2 catalytic systems [119]. For dehydrated Ti-silicalite and supported TiO_2/SiO_2 at low surface coverage, isolated TiO_4 units have been shown to be present for both systems [120]. Similarly, both dehydrated Nb-MCM-41 and supported Nb_2O_5/SiO_2 catalytic materials contain isolated NbO_4 sites [114]. The same situation also occurs for surface acidic sites generated by zeolites such as Al-ZSM5 and supported Al_2O_3/SiO_2 [121]. Thus, the above molecular structural descriptions for the surface metal oxide species present in SiO_2 supported metal oxide catalysts also apply to isolated metal oxide active sites present in molecular sieve catalytic materials.

1.4 Bulk Mixed Metal Oxides

Bulk mixed metal oxide catalytic materials consist of multiple metal oxide components. Such mixed metal oxide catalysts find wide application as selective oxidation catalysts for the synthesis of chemical intermediates. For example, bulk iron–molybdate catalysts are employed in the selective oxidation of CH_3OH to H_2CO [122], bulk bismuth-molybdates are the catalysts of choice for selective oxidation of $CH_2=CHCH_3$ to acrolein ($CH_2=CHCHO$) and its further oxidation to acrylic acid ($CH_2=CHCOOH$) [123], selective ammoxidation of $CH_2=CHCH_3$ to acrylonitrile ($CH_2=CHCN$) [123], and selective oxidation of linear $CH_3CH_2CH_2CH_3$ to cyclic maleic anhydride consisting of a five-membered ring (four carbons and one O atom) [124]. The characterization of the surface metal oxide sites present for the bulk mixed metal oxide catalytic materials has been one of the most challenging undertakings because the same elements are present both on the surface and in the bulk of these materials. Consequently, very few characterization methods exist that are able to selectively structurally probe the outermost surface metal oxide sites present in bulk mixed metal oxides. The typical surface science techniques based on electron spectroscopic methods probe the surface region 0.5 to 3 nm and are not limited to the ousstermost surface layer (<0.5 nm). The usual catalyst characterization methods of Raman, IR, UV-Vis, NMR, and XANES/EXAFS are all bulk characterization methods and their spectra are usually dominated by signals from the bulk rather than the outermost surface layer. In the past few years, however, characterization studies employing low energy ion scattering spectroscopy (LEISS) [125,126] chemical probe molecules [127,128], synchrotron-based surface XPS [129] are revealing that the surfaces of bulk mixed metal oxides possess compositions that are different than those found in the bulk of these catalytic materials. Surprisingly, the surfaces of many molybdate and vanadate mixed metal oxides have revealed the exclusive presence of surface molybdenum oxide and vanadium oxide monolayers. At present, the molecular structures of these outermost surface metal oxide monolayers have not been determined. It is possible, however, that these outermost surface metal oxide monolayers of bulk mixed metal oxides also possess the same molecular structures found for the surface metal oxide species present in supported metal oxide catalysts. Hopefully, these issues can be resolved in the coming years.

1.5 Conclusions

Significant advances in determining the molecular structures of the surface metal oxide species of metal oxide materials, the catalytic active sites, have been achieved in recent years. These advances have been made possible by the application of modern molecular spectroscopic characterization techniques (Raman, IR, UV–Vis, XANES, EXAFS, and NMR). A very important aspect of these molecular spectroscopic techniques is their ability to collect spectroscopic data under different environmental conditions since the molecular structures of the surface metal oxide species are dynamic and environmentally dependent. The surface metal oxide species are presented as hydrated metal oxide species under ambient conditions that are essentially the same as the well-known aqueous metal oxide species. Upon dehydration, these metal oxide species coordinate with the oxide support substrates to form surface metal oxide species, for the group 5 to 7 metal oxides, that terminate with monoxo M=O terminal bonds. The dehydrated surface metal oxides can be present as isolated species with only bridging M–O Support bonds as well as polymeric species with additional bridging M–O–M bonds. Above the monolayer surface coverage, crystalline metal oxides form on top of the surface metal oxide monolayer. The surface molecular structures found to be present under reactive environments are the dehydrated surface metal oxide species. Depending on the reducing/oxidizing environmental conditions, reduced surface metal oxide species may also be present. At present, very little structural information is available about these reduced surface metal oxide species. Essentially the same surface metal oxide structures are found to be present for corresponding supported metal oxide, molecular sieve, and zeolite materials. The surface molecular structures present for bulk mixed metal oxides are not known at present. In spite of this impressive recent progress, the molecular structures of many oxidized and reduced surface metal oxide species have not been completely determined or are not even known at present. Consequently, many more fundamental structural studies are still required to obtain complete determination of the molecular structures of surface metal oxide species under different environmentally conditions.

ACKNOWLEDGMENT

The author would like to acknowledge the assistance of Professor Goutam Deo, Indian Institute of Technology, Kanpur, for his assistance in preparing this detailed chapter and his positive discussion of its contents.

REFERENCES

1. C.L. Thomas, *Catalytic Processes and Proven Catalysts*, Academic Press, New York (1970).
2. I.E. Wachs, Ed. Special Issue on "Applications of Supported Metal Oxide Catalysts," *Catal. Today* 51 (1999) 201–348.
3. M.S. Wainwright and N.R. Foster, *Catal. Rev. Sci. Eng.* 33 (1991) 1.
4. B.N. Reddy and M. Subrahmanyam, *Langmuir* 8 (1992) 2072.
5. D. Heinz, W.F. Hoelderich, S. Krill, W. Boeck, and K. Huthmacher, *J. Catal.* 192 (2000) 1.
6. G. Deo and I.E. Wachs, *J. Catal.* 146 (1994) 323.
7. Q. Sun, J.-M. Jehng, H. Hu, R.G. Herman, I.E. Wachs, and K. Klier, *J. Catal.* 165 (1997) 101.
8. M.A. Banares, *Catal. Today* 51 (1999) 319.
9. K.D. Chen, A. Khodakov, J. Yang, A. Bell, and E. Iglesia, *J. Catal.* 186 (1999) 325.
10. X. Gao, J.-M. Jehng, and I.E. Wachs, *J. Catal.* 209 (2004) 43.
11. V.V. Guliants, *Catal. Today* 51 (1999) 255.
12. J. Dunn, H. Stenger, and I.E. Wachs, *Catal. Today* 51 (1999) 301.
13. S. Choi and I.E. Wachs, *Fuel Chem. Div. Prepr.* 47 (2002) 138.
14. N. Moretti, *Pollut. Eng.*, January 2002, 24.
15. M. Jacoby, *Chem. Eng. News*, March 21, 2002, 39.
16. L. Lietti, J. Scachula, P. Forzatti, G. Busca, G. Ramis, and F. Bregnani, *Catal. Today* 17 (1993) 131.
17. N.-Y. Topsoe, H. Topsoe, and J.A. Dumesic, *J. Catal.* 151 (1995) 226.
18. I.E. Wachs, G. Deo, B.M. Weckhuysen, A. Adreini, M.A. Vuurman, M. de Boer, and M.D. Amiridis, *J. Catal.* 161 (1997) 211.
19. M. Hino and K. Arata, *J. Chem. Soc. Chem. Commun.* 1259 (1987).

20. D.G. Barton, S.L. Soled, G.D. Meitzner, G.A. Fuentes, and E. Iglesia, *J. Catal.* 181 (1999) 57.
21. J.G. Santiesteban, J.G. Vartuli, S. Han, R.D. Bastion, and C.D. Chang, *J. Catal.* 168 (1997) 431.
22. G. Larsen and L.M. Petrovick, *J. Mol. Catal. A: Chem.* 113 (1996) 517.
23. J.C. Mol, *Catal. Today* 51 (1999) 289.
24. B.E. Leach, *Appl. Ind. Catal.* 3 (1984) 215.
25. B.E. Leach, *Appl. Ind. Catal.* 1 (1983) 149.
26. M.P. McDaniel, *Adv. Catal.* 33 (1985) 47.
27. B.M. Weckhuysen, I.E. Wachs, and R.A. Schoonehydt, *Chem. Rev.* 96 (1996) 3327.
28. B.M. Weckhuysen and R.A. Schoonheydt, *Catal. Today* 51 (1999) 223.
29. H. Topsoe, *Hydrotreating Catalysts: Science and Technology*, Springer-Verlag, Berlin (1996).
30. R. Prins, in *Characterization of Catalytic Materials*, I.E. Wachs (Ed.), Butterworth-Heinemann, Stoneham, MA (1992) p. 89.
31. I.E. Wachs, *Catal. Today* 27 (1996) 437.
32. H. Eckert and I.E. Wachs, *J. Phys. Chem.* 93 (1989) 6796.
33. N. Das, H. Eckert, H. Hu, J.K. Waltzer, F. Feher, and I.E. Wachs, *J. Phys. Chem.* 97 (1993) 8240.
34. G. Deo and I.E. Wachs, *J. Phys. Chem.* 95 (1991) 5889.
35. J.-M. Jehng, G. Deo, B.M. Weckhuysen, and I.E. Wachs, *J. Mol. Catal. A: Chem.* 110 (1996) 41.
36. F.J. Gil-Llambias, A.M. Escudy, J.L.G. Fierro, and A.L. Agudo, *J. Catal.* 95 (1985) 520.
37. S.D. Kohler, J.G. Ekerdt, D.S. Kim, and I.E. Wachs, *Catal. Lett.* 16 (1992) 231.
38. G.A. Park, *Chem. Rev.* 65 (1965) 177.
39. C.F. Baes, Jr. and R.E. Mesmer, *The Hydrolysis of Cations*, Wiley, New York (1970).
40. H. Hu, I.E. Wachs, and S.R. Bare, *J. Phys. Chem.* 99 (1995) 10897.
41. F.D. Hardcastle and I.E. Wachs, *J. Mol. Catal.* 46 (1988) 173.
42. M.A. Vuurman, D.J. Stufkens, A. Oskam, and I.E. Wachs, *J. Mol. Catal.* 76 (1992) 263.
43. G. Ramis, C. Cristiani, P. Forzatti, and G. Busca, *J. Catal.* 124 (1990) 574.
44. X. Gao and I.E. Wachs, *J. Phys. Chem. B* 104 (2000) 1261.
45. J.-M. Jehng, I.E. Wachs, B.M. Weckhuysen, and R.A. Schoonheydt, *J. Chem. Soc., Faraday Trans.* 91 (1994) 953.
46. T. Tanaka, H. Yamashita, R. Tsuchitani, T. Funabiki, and S. Yoshida, *J. Chem. Soc., Faraday Trans. 1* 84 (1988) 2987; S. Yoshida, T. Tanaka, T. Hiraiwa, and H. Kanai, *Catal. Lett.* 12 (1992) 277.
47. M.V. Martinez-Huerta, J.M. Coronado, M. Fernandez-Garcia, A. Iglesias-Juez, G. Deo, J.L.G. Fierro, and M.A. Banares, *J. Catal.* 225 (2004) 240.
48. J.A. Horsley, I.E. Wachs, J.M. Brown, G.H. Via, and F.D. Hardcastle, *J. Phys. Chem.* 91 (1987) 4014.
49. F. Hilbrig, H.E. Gobel, H. Knozinger, H. Schmelz, and B. Lengeler, *J. Phys. Chem.* 95 (1991) 6973.
50. D.G. Barton, S.L. Soled, G.D. Meitzner, G.A. Fuentes, and E. Iglesia, *J. Catal.* 181 (1999) 57.

51. F.D. Hardcatle, I.E. Wachs, J.A. Horsley, and G.H. Via, *J. Mol. Catal.* 46 (1988) 15.
52. T. Machej, J. Haber, A.M. Turek, and I.E. Wachs, *Appl. Catal.* 70 (1991) 115.
53. M.A. Banares, H. Hu, and I.E. Wachs, *J. Catal.* 150 (1994) 407.
54. G. Deo, A.M. Turek, I.E. Wachs, T. Machej, J. Haber, N. Das, and H. Eckert, *Appl. Catal. A: Gen.* 91 (1992) 27.
55. G. Deo, F.D. Hardcastle, M. Richards, I.E. Wachs, and A.M. Hirt, in *Novel Materials in Heterogeneous Catalysis*, R.T.K. Baker and L.L. Murrell (Eds.), ACS Symposium Series 437, American Chemical Society, Washington, D.C. (1990).
56. M.A. Banares, N.D. Spencer, M.D. Jones, and I.E. Wachs, *J. Catal.* 146 (1994) 204.
57. M.A. Banares, H. Hu, and I.E. Wachs, *J. Catal.* 155 (1995) 249 .
58. X. Carrier, J.F. Lambert, and M. Che, *J. Am. Chem. Soc.* 119 (1997) 10137.
59. L. Le Bihan, P. Blanchard, M. Fournier, J. Grimblot, and E. Payen, *J. Chem. Soc., Faraday Trans.* 94 (1998) 937.
60. A.F. Wells, *Structural Inorganic Chemistry*, Oxford University, London (1984).
61. N.N. Greenwood and A. Earnshaw, *Chemistry of the Elements*, Pergamon Press, Elmsford, NY (1989).
62. K. Nakamoto, *Infrared and Raman Spectra of Inorganic and Coordination Compounds* (5th ed), Wiley, New York (1997).
63. M.A. Banares and I.E. Wachs, *J. Raman Spectrosc.* 33 (2002) 359.
64. G. Busca, *J. Raman Spectrosc.* 33 (2002) 348.
65. B.M. Weckhuysen, A.A. Verberckmoes, A.L. Buttiens, and R.A. Schoonheydt, *J. Phys. Chem.* 98 (1994) 579.
66. A. Bruckner, *Catal. Rev.* 45 (2003) 97.
67. T. Tanaka, T. Yoshida, H. Yoshida, H. Aritani, T. Funabiki, S. Yoshida, J.-M. Jehng, and I.E. Wachs, *Catal. Today* 28 (1996) 71.
68. T. Tanaka, H. Nojima, T. Yamamoto, S. Takenaka, T. Funabiki, and S. Yoshida, *Phys. Chem. Chem. Phys.* 1 (1999) 5235; Y. Chen, J.L.G. Fierro, T. Tanaka, and I.E. Wachs, *J. Phys. Chem. B* 107 (2003) 5243.
69. M. Anpo, M. Sunamoto, and M. Che, *J. Phys. Chem.* 93 (1989) 1187; M.F. Hazenkamp and G. Blasse, *J. Phys. Chem.* 96 (1992) 3442.
70. J.C. Edwards, R.D. Adams, and P.D. Ellis, *J. Am. Chem. Soc.* 112 (1990) 8349.
71. V.M. Mastikhin, A.V. Nosov, V.V. Terskikh, K.I. Zamaraev, and I.E. Wachs, *J. Phys. Chem.* 98 (1994) 13621.
72. F.D. Harcastle and I.E. Wachs, *J. Phys. Chem.* 95 (1991) 5031.
73. M. Ruitenbeek, A.J. Van Dillen, F.M.F. de Groot, I.E. Wachs, J. Geus, and D.C. Koningsberger, *Top. Catal.* 10 (2000) 241.
74. L.J. Burcham, G. Deo, X. Gao, and I.E. Wachs, *Top. Catal.* 11/12 (2000) 85.
75. G.T. Went, S.T. Oyama, and A.T. Bell, *J. Phys. Chem.* 94 (1990) 4240.
76. B.M. Weckhusyen, J.-M. Jehng, and I.E. Wachs, *J. Phys. Chem. B* 104 (2000) 7382.
77. X. Gao, S.R. Bare, B.M. Weckhuysen, and I.E. Wachs, *J. Phys. Chem. B* 102 (1998) 10842.
78. S.T. Oyama, G.T. Went, K.B. Lewis, A.T. Bell, and G.A. Somorjai, *J. Phys. Chem.* 93 (1989) 6786.
79. C. Resini, T. Montanari, G. Busca, J.-M. Jehng, and I.E. Wachs, *Catal. Today* (in press).

80. J.H. Cardoso, Ph.D. thesis, Universidade Federal de Sao Carlos, Sao Paulo, Brazil (1998).
81. R.D. Roark, S.D. Kohler, and J.G. Ekerdt, *Catal. Lett.* 16 (1992) 71.
82. M.D. Argyle, K. Chen, C. Resini, C. Krebs, A.T. Bell, and E. Iglesia, *J. Phys. Chem. B* 108 (2004) 2345.
83. M.A. Vuurman and I.E. Wachs, *J. Phys. Chem.* 96 (1992) 5200.
84. X. Gao, J.-M. Jehng, and I.E. Wachs, *J. Catal.* 209 (2002) 43.
85. M.A. Vuurman, D.J. Stufkens, A. Oskam, G. Deo, and I.E. Wachs, *J. Chem. Soc., Faraday Trans.* 92 (1996) 3259.
86. M. Cieslak-Golonka, *Coord. Chem. Rev.* 109 (1991) 223.
87. M.A. Vuurman, F.D. Hardcastle, and I.E. Wachs, *J. Mol. Catal.* 84 (1993).
88. M.A. Vuurman, I.E. Wachs, D.J. Stufkens, and A. Oskam, *J. Mol. Catal.* 80 (1993) 209.
89. A. Cimino, D. Cordischi, S. Febraro, D. Gazzoli, D. Indovina, M. Occhiuzzi, and M. Valigli, *J. Mol. Catal.* 55 (1989) 23.
90. V. Indovina, D. Cordiscji, S. de Rossie, G. Ferraris, G. Ghiotti, and A. Chiorino, *J. Mol. Catal.* 68 (1991) 53.
91. B.M. Weckhuysen and I.E. Wachs, *J. Phys. Chem. B* 100 (1996) 14437.
92. B.M. Weckhuysen, R.A. Schoonheydt, J.-M. Jehng, I.E. Wachs, F.J. Cho, R. Ryoo, S. Kijlstra, and E. Poels, *J. Chem. Soc., Faraday Trans.* 91 (1995) 3245.
93. I.R. Beatti and G. Ozin, *J. Chem. Soc. A* (1969) 2615.
94. R.A. Johnson, M.T. Rogers, and G.E. Leroi, *J. Chem. Phys.* 56 (1972) 789.
95. A. Muller, B. Krebs, and W. Holtje, *Spectrochim. Acta* 23 (1967) 2753.
96. G. Deo and I.E. Wachs, to be published.
97. L. Wang and W.K. Hall, *J. Catal.* 82 (1983) 177.
98. L. Wang, Ph.D. dissertation, Universiteit Wisconsin-Milwaukee, USA (1982).
99. F.D. Hardcastle and I.E. Wachs, *J. Raman Spectrosc.* 21 (1990) 683.
100. M. de Boer, A.J. van Dillen, D.C. Koningsberger, J.W. Geus, M.A. Vuurman, and I.E. Wachs, *Catal. Lett.* 11 (1991) 227.
101. R. Radhakrishnan, C. Reed, S.T. Oyama, M. Seman, J.N. Kondo, K. Domen, Y. Ohminami, and K. Asakura, *J. Phys. Chem. B* 105 (2001) 8519.
102. M. Cornac, A. Janin, and J.C. Lavalley, *Polyhedron* 5 (1986) 183.
103. M. de Boer, Ph.D. dissertation, Universiteit Utrecht, The Netherlands (1992).
104. H. Tian and I.E. Wachs, to be published.
105. S.S. Chan, I.E. Wachs, L.L. Murrell, and N.C. Dispenziere, *J. Catal.* 92 (1985) 1.
106. D.S. Kim, M. Ostromecji, and I.E. Wachs, *J. Mol. Catal. A: Chem.* 106 (1996) 93.
107. D.S. Kim, M. Ostromecki, I.E. Wachs, s.D. Kohler, and J.G. Ekerdt, *Catal. Lett.* 33 (1995) 209.
108. J.M. Stencel, L.E. Makovsky, J.R. Diehl, and T.A. Sarkus, *J. Raman Spectrosc.* 15 (1984) 282.
109. G. Ramis, G. Busca, C. Christiani, L. Lietti, P. Forzatti, and F. Bregnani, *Langmuir* 8 (1992) 1744.
110. J.-M. Jehng and I.E. Wachs, *Chem. Mater.* 3 (1992) 100.
111. M. Dion, M. Ganne, and M. Tournoux, *Mater. Res. Bull.* 16 (1981) 1429.
112. A.J. Jacobson, J.W. Johnson, and J.T. Lowendowski, *Inorg. Chem.* 24 (1985) 3727.
113. A.J. Jacobson, J.T. Lowendowski, and J.W. Johnson, *J. Less-Common Met.* 116 (1986) 137.

114. X. Gao, I.E. Wachs, M.S. Wong, and J.Y. Ying, *J. Catal.* 203 (2002) 18.
115. J.-M. Jehng and I. E. Wachs, *J. Phys. Chem.* 95 (1991) 7373.
116. L.J. Burcham, J. Datka, and I.E. Wachs, *J. Phys. Chem. B* 103 (1999) 6015.
117. M. Baltes, A. Kytokivi, B.M. Weckhuysen, R. Schoonheydt P.V.D. Voort, and E.F. Vansant, *J. Phys. Chem. B* 105 (2001) 6211.
118. M.E. Davis, in *Characterization of Catalytic Materials*, I.E. Wachs (Ed.), Butterworth-Heinemann, Stoneham, MA (1992) 129.
119. C.-B. Wang, G. Deo, and I.E. Wachs, *J. Catal.* 178 (1998) 640.
120. G. Deo, A.M. Turek, I.E. Wachs, D.R.C. Huybrechts, and P.A. Jacobs, *Zeolites* 13 (1993) 365.
121. X. Gao and I.E. Wachs, *J. Catal.* 192 (2000) 18.
122. H. Adkins and W.R. Pederson, *J. Am. Chem. Soc.* 53 (1931) 1512.
123. R.K. Grasselli and J.D. Burrington, *Adv. Catal.* 30 (1981) 133.
124. G. Centi, Ed. *Catal. Today* 16 (1993) 1, Special issue devoted to "Vanadyl Pyrophosphate Catalysts."
125. L.E. Briand, O.P. Tkachenko, M. Guraya, I.E. Wachs, and W. Gruenert, *Surf. Interface Anal.* 36 (2004) 238.
126. L.E. Briand, O.P. Tkachenko, M. Guraya, X. Gao, I.E. Wachs, and W. Gruenert, *J. Phys. Chem. B* 108 (2004) 4823.
127. L.E. Briand, A.M. Hirt, and I.E. Wachs, *J. Catal.* 202 (2001) 268.
128. L.E. Briand, J.-M. Jehng, L. Cornaglia, A.M. Hirt, and I.E. Wachs, *Catal. Today* 78 (2003) 257.
129. M. Havecker, A. Knop-Gericke, H. Bluhm, E. Kleimenov, R.W. Mayer, M. Fait, and R. Schlogl, *Appl. Surf. Sci.* 230 (2004) 272.
130. O.B. Lapina, V.M. Mastikhin, L.G. Simonova, and Y.O. Bulgakova, *J. Mol. Catal.* 69 (1991) 61; Z. Sobalik, M. Markvart, P. Stopka, O.B. Lapina, and V.M. Mastikhin, *J. Mol. Catal.* 71 (1992) 69.

2 Nanostructured Supported Metal Oxides

M.S. Wong
Department of Chemical and Biomolecular Engineering,
Department of Chemistry, Rice University, Houston, TX, USA

CONTENTS

2.1 INTRODUCTION

Supported catalysts comprise a very general solid catalyst design in which the active phase (the supported layer) is located on the surface of an underlying solid (the support). The motivation for supported catalysts is that the dispersion and stabilization of the supported layer on a high surface area support leads to a highly active and robust composite catalyst. Metals, metal oxides, metal

31

sulfides, organometallic complexes, and enzymes can be supported on inorganic supports such as metal oxides, zeolites, and clays. This subject matter has been well documented by a number of books [1–4] and more recently by *Preparation of Solid Catalysts* [5], a monograph based on the comprehensive *Handbook of Heterogeneous Catalysis* volumes edited by Ertl et al. [6].

Supported catalysts in which a metal oxide is supported on a different metal oxide find tremendous use in commercial chemical production, petroleum refining, and environmental remediation [7]. Table 2.1 lists a number of important industrial reactions that are catalyzed by supported metal oxides. The active phase consists of single metal oxides, metal oxide mixtures, or complex metal oxides. Some supported metal catalysts can, technically, be considered supported metal oxide catalysts if the metal is partially oxidized under reaction conditions, such as alumina-supported silver epoxidation catalysts and palladium-on-stabilized alumina combustion catalysts. Clearly, supported metal oxide compositions are diverse, especially considering that the range of relative amounts of the active phase to the support is large, and that additional metal oxides and modifiers can be introduced.

It can be argued that some mixed metal oxides can also be technically considered as supported metal oxide catalysts because the surface is discernibly different from the underlying mixed metal oxide in terms of composition and molecular structure. For example, the vanadium phosphorus oxide (VPO) catalyst is used in the commercial production of maleic anhydride from butane [12]. The most active crystal phase is the vanadium pyrophosphate $(VO)_2P_2O_7$, and the surface structure proposed to be the active phase is a nanometer-thick amorphous VPO layer enriched in phosphorus [12,15]. As another example, Wachs and coworkers [16]

TABLE 2.1
Commercial Processes [3,4]

Supported metal oxide compositions	Catalytic reaction
V_2O_5/(Kieselguhr or SiO_2)	Oxidation of SO_2 for sulfuric acid production
$(V_2O_5+WO_3)$/TiO_2	Selective catalytic reduction of NO_x using ammonia [8]
$(V_2O_5+MoO_3)$/Al_2O_3	Oxidation of benzene to maleic anhydride
V_2O_5/TiO_2	Oxidation of o-xylene to phthalic anhydride [9]
Cr_2O_3/Al_2O_3	Dehydrogenation of light alkanes to alkenes [10]
CrO_x/SiO_2	Ethylene polymerization [11]
$(K_2O+Cr_2O_3)$/Fe_2O_3	Dehydrogenation of ethylbenzene to styrene
$Bi_9PMo_{12}O_{52}$/SiO_2	Ammoxidation of propene to acrylonitrile [12]
Re_2O_7/Al_2O_3	Olefin metathesis [13]
TiO_2/SiO_2	Dehydration of phenylethanol to styrene
TiO_2/SiO_2	Epoxidation of propylene with ethylbenzene hydroperoxide [14]
AgO_x/γ-Al_2O_3	Oxidation of ethylene to ethylene oxide
(PtO_x+PdO_x)/$(La_2O_3+Al_2O_3)$	Catalytic combustion of volatile organic compounds [12]

provided methanol chemisorption data which indicate that mixed metal molybdates and supported molybdenum oxides have similar surface structures.

In this chapter, the common methods of supported metal oxide catalyst preparation are presented. The molecular and nanoscale structure (i.e., textural properties) of supported metal oxides, specifically tungstated zirconia, are also discussed. Finally, several new synthesis techniques are presented, through which simultaneous control of the molecular and nanoscale structures may be possible.

2.2 METHODS OF PREPARATION

The basic principle behind supported metal oxide catalyst preparation involves deposition followed by activation. According to The International Union of Pure and Applied Chemistry (IUPAC) recommendations, deposition is defined as "the application of the catalytic component on to [sic] a separately produced support," and activation as "the transformation of the precursor to the active phase," usually entailing the calcination of the composite material [17]. There are several well-studied deposition steps, with impregnation being the most common method for preparing industrial supported metal oxides.

2.2.1 Impregnation

Contacting the support material with a solution that contains the supported metal oxide precursor is known as impregnation. The incipient wetness method is impregnation in which the support is contacted with a volume of precursor solution slightly in excess of the total pore volume of the support, and then the slightly wet powder is allowed to dry. This particular method allows a wide range of loadings to be possible, regardless of the surface interactions between the precursor and the support that would limit the overall loading (e.g., adsorption and ion exchange).

If the support material is initially in the dry state, then the deposition process is called capillary, or dry, impregnation. The speed of pore filling is fast and the process is simple, but there are potential problems of heat release and internal pore collapse due to capillary pressure by entrapped air [18]. If the support is already in contact with the solvent (without the precursor), then the subsequent contact with the precursor solution is known as diffusional, or wet, impregnation.

2.2.2 Ion exchange

Impregnation through ion exchange (or equilibrium adsorption) involves immersing of the support material in an aqueous solution of the supported metal oxide precursor, and recovering the solid from the solution after a period of time. The charged salt precursor electrostatically binds to the oppositely charged surface of the support (typically a metal oxide). The surface charge arises from the difference in the pH of the aqueous solution and the point-of-zero charge (PZC) of the metal oxide support. If the pH > PZC, then the terminating hydroxyl groups at the surface deprotonate. The net surface charge is negative, and the surface can adsorb cationic species. Conversely, if the pH > PZC, the hydroxyl groups protonate and the surface can adsorb anionic species. It is difficult to differentiate between electrostatic attraction and strict ion exchange of the precursor and the surface-bound counterion. Other chemical interactions can also take place, such as metal coordination.

2.2.3 Grafting

Grafting involves covalent bond formation between the metal precursor and the hydroxyl group(s) on the support. This is more readily controlled as a preparation method in which the support is contacted with the precursor solubilized in a nonaqueous solvent. Common precursors are metal alkoxides, metal chlorides, and organometallic complexes. Both ion exchange and grafting methods lead to coverage by the precursor up to a monolayer, but the latter allows a greater control of dispersion and molecular structure of the supported metal species. Grafting can also be carried out using precursors vaporized in the gas-phase, and this method of catalyst preparation is called chemical vapor deposition (CVD).

2.2.4 Deposition–Precipitation

Like incipient wetness impregnation, this method provides supported metal oxides with high loadings. The support material is suspended in a precursor solution, and then precipitation of the supported metal oxide is induced such that the metal oxide nanoparticles (NPs) are nucleated and grown on internal and external surfaces of the support.

2.2.5 Co-Precipitation

Supported metal oxides can be prepared through co-precipitation, in which metal precursors to both the support and the supported metal oxide are induced to form the support material and the supported layer simultaneously. The support and the supported metal oxide are more spatially distributed than materials derived from the various deposition methods, but a fraction of the supported metal oxide may be located below the surface, leading to overall lower metal oxide dispersion.

2.2.6 Other Methods

Other methods for preparing supported metal oxides have been reported. An interesting approach is the use of bulk MoO_3 in the preparation of MoO_x/Al_2O_3. In the solid–solid wetting or thermal spreading approach, MoO_3 and Al_2O_3 powders are mechanically ground together and heated at elevated temperatures ($\sim$500°C) for 24 h to yield MoO_x/Al_2O_3 [19]. The ability of MoO_3 to spread across the Al_2O_3 surface as MoO_x surface species is attributed to the relatively low melting point and Tammann temperatures of MoO_3 (795 and 261°C, respectively), the affinity of the acidic MoO_3 surface for the basic Al_2O_3 surface, and the reduction in surface energy of the bulk solids to form a supported metal oxide structure. Instead of calcination, refluxing MoO_3 and Al_2O_3 powders in water for several hours can also yield the supported material [20]. In this slurry impregnation method, MoO_3 partially decomposes in water and the solubilized Mo species deposit on the support. The resulting material is structurally similar to conventionally prepared MoO_x/Al_2O_3.

2.3 CATALYST MOLECULAR STRUCTURE

After the deposition (or co-precipitation) step, the resultant material is calcined to decompose the metal precursor to form the supported metal oxide layer. Several different surface metal oxide structures can emerge (Figure 2.1), which depends on the preparation method, the metal oxide loading, the specific surface area of the underlying supporting substrate, and the solid–solid chemical interactions between the support and the supported layer. At low loadings, the metal oxide precursor can be dispersed as isolated metal (monomeric) species. Increased loadings leads to the adjoining mono oxo species forming oligomeric or polymeric species via metal–oxygen–metal linkages, and eventually, to surface-bound (nano)crystalline metal oxide domains. In addition to these structures, the metal oxide precursor can form a solid solution with the support (e.g., $Al_2(WO_4)_3$ resulting from the preparation of WO_x/Al_2O_3 [21] and $MgWO_4$ resulting from the preparation of WO_x/MgO [22]) or a surface species of a mixed metal oxide composition (e.g., supported aluminotungstate clusters on WO_x/Al_2O_3 [23]). Identification of surface structures and elucidation of structure–property relationships for many supported metal oxide compositions have been enabled by *in situ/operando* Raman and other surface-sensitive spectroscopic techniques in recent years [24–29].

An important parameter in determining the structure of the supported layer is the surface density of the metal center (ρ_{surf}). This is experimentally determined from the measured weight loading of the supported metal oxide and the overall BET surface area of the catalyst. Quantitatively,

$$\rho_{surf} = \frac{\text{Weight fraction of supported metal oxide} \div \text{molecular weight} \times \text{Avogadro's number}}{\text{Surface area} \times 10^{18}}$$

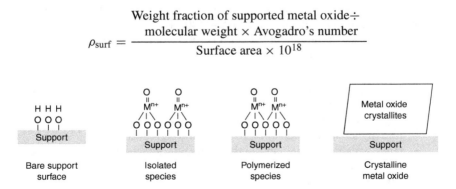

| Bare support surface | Isolated species | Polymerized species | Crystalline metal oxide |

FIGURE 2.1 Representations of possible molecular structures of supported metal oxide

where the units of ρ_{surf} and surface area are (metal atom)/nm^2 and m^2/(g of material), respectively, and the molecular weight is that of the supported metal oxide. The surface density represents an average value and not the actual metal distribution. There is the implicit assumption that the supported layer is localized at the support surface, which may not hold for all supported metal oxides. Thus, the actual surface density values of co-precipitated supported metal oxides may be overestimated by ρ_{surf} calculations if a portion of the supported phase is entrapped within the support structure. Nevertheless, it is a useful metric for describing supported metal oxides.

2.4 CASE STUDY: TUNGSTATED ZIRCONIUM OXIDE

Tungstated zirconia is discussed as an illustration of some of the important concepts of supported metal oxides. This supported catalyst composition has exceptional low-temperature activity for acid-catalyzed reactions, as first reported by Hino and Arata [30] for light alkane isomerization over 15 years ago. There has been a continued interest in these WO_x/ZrO_2, materials as a robust alternative to sulfated zirconia and other acid catalysts in industrial acid catalytic processes [31–33], and progress has been made in understanding the nature of acidity and the relationship between the surface structure and catalytic behavior. WO_x/ZrO_2 can be doped with iron, platinum, sulfate, and other species for improved catalytic properties [34–39]. The short-hand notations "WO_3/ZrO_2" and "WO_x/ZrO_2" are commonly used, with the latter term more appropriate as it indicates the variable surface structures of the supported tungstate species (i.e., monomeric, polymeric, and crystalline).

Different routes have been studied in the synthesis of WO_x/ZrO_2, since Hino and Arata's original report, with many variations of the incipient wetness impregnation theme (Table 2.2). The sol–gel approach is not well studied, and will be discussed in Section 2.5.3. For impregnation, crystalline ZrO_2 (tetragonal or monoclinic phases, or a mixture of both) and amorphous ZrO_2 (variously written as $Zr(OH)_4$, $ZrO(OH)_2$, or $ZrO_x(OH)_{4-2x}$) have been used. Prepared through a proprietary process, uncalcined WO_x/ZrO_2 is commercially available from MEL Chemicals (Magnesium Elektron Inc., USA) [40,41].

Common observations about WO_x/ZrO_2 can be drawn, which is independent of the preparation method:

- WO_x species suppress the sintering of the ZrO_2 support and the crystallization of amorphous ZrO_2.
- Higher calcination temperatures lead to lower surface areas and therefore higher surface densities, for a given weight loading.
- High weight loadings and high calcination temperatures lead to the formation of WO_3 crystals.
- WO_x/ZrO_2 with the same surface density can have different WO_3 weight loadings.

TABLE 2.2
Conventional Preparation Methods for WO_x/ZrO_2

Authors	Method	Precursor source and activation procedure
Hino and Arata, 1988 [30]	Incipient wetness impregnation	ZrO_2 source: $Zr(OH)_4$ precipitated from zirconyl chloride using NH_4OH WO_x source: ammonium metatungstate solution Calcined in air, 600 to 1000°C
Santiesteban et al., 1997 [42]	Incipient wetness impregnation	ZrO_2 source: $Zr(OH)_4$ precipitated from zirconyl chloride using NH_4OH; dried at 95°C; refluxed in NH_4OH solution overnight WO_x source: ammonium tungstate solution Calcination at 825°C for 3 h in air
	Co-precipitation	ZrO_2 source: Zirconyl chloride solution added to NH_4OH/ammonium tungstate solution WO_x source: ammonium tungstate solution Precipitate was placed in a steambox at 100°C for 16 h before calcination at 825°C for 3 h in air
Vaudagna et al., 1997 [43]	Incipient wetness impregnation	ZrO_2 source: $Zr(OH)_4$ precipitated from zirconyl chloride using NH_4OH; dried at 110°C WO_x source: ammonium tungstate solution Calcination at 800°C for 3 h in air
Yori et al., 1997 [44]	Incipient wetness impregnation; ion exchange	ZrO_2 source: $Zr(OH)_4$ precipitated from zirconyl chloride using NH_4OH; dried at 110°C WO_x source: ammonium tungstate solution Calcination at 800°C for 3 h in air
Boyse and Ko, 1997 [45]	Incipient wetness impregnation	ZrO_2 source: ZrO_2 aerogel derived from zirconium n-propoxide through CO_2 supercritical drying at 70°C WO_x source: ammonium tungstate solution Calcination at 700 to 1000°C for 2 h in air
	Co-gelation	ZrO_2 source: Zirconium n-propoxide gelled with ammonium tungstate solution WO_x source: ammonium tungstate solution Gel was supercritical dried in CO_2 at 70°C, dried at 110 and 250°C, and calcined at 700 to 1000°C in air
Scheithauer et al., 1998 [46]	Modified incipient wetness impregnation	ZrO_2 source: $Zr(OH)_4$ supplied by MEL Chemicals WO_x source: ammonium tungstate solution Suspension was refluxed at 110°C for 16 h, dried at 110°C for 12 h, calcined between 500 and 1000°C for 3 h
Barton et al., 1999 [47]	Incipient wetness impregnation	ZrO_2 source: $Zr(OH)_4$ precipitated from zirconyl chloride using NH_4OH; dried at 150°C WO_x source: ammonium tungstate solution Samples calcined between 500 and 1010°C for 3 h in air

TABLE 2.2
Continued.

Authors	Method	Precursor source and activation procedure
Naito et al., 1999 [48]	Incipient wetness impregnation	ZrO_2 source: $Zr(OH)_4$ precipitated from zirconium oxynitrate using NH_4OH WO_x source: ammonium tungstate solution Calcination at 650°C for 4 h in air
Valigi et al., 2002 [49]	Incipient wetness impregnation; ion exchange	ZrO_2 source: $Zr(OH)_4$ precipitated from zirconyl chloride using NH_3 gas; dried at 110°C WO_x source: ammonium tungstate solution Calcination at 800°C for 5 h in air

2.4.1 Surface Coverage

Monolayer or saturation coverage can be defined as the surface density above which nanocrystals of the supported phase can be detected through Raman spectroscopy. Wachs and coworkers reported that the monolayer coverage was ~ 4 W/nm^2 for WO_x/ZrO_2 (prepared through incipient wetness impregnation) [22]. Other supports such as TiO_2, Al_2O_3, Nb_2O_5, and Ta_2O_5 were reported to have similar values for W monolayer coverages [22,25,50]. The term "monolayer catalyst" is occasionally used to refer to a supported metal oxide catalyst with monolayer coverage [9,51].

Different surface structure models lead to a range of theoretical monolayer coverages. Zhao et al. calculated that a close-packing of WO_3 units in which the oxygens form a complete monolayer on a ZrO_2 (or any) surface would have a theoretical surface density of 0.19 g WO_3 per 100 m^2, or 4.93 W/nm^2 [52,53]. Scheithauer et al. [46] considered 6.5 W/nm^2 to be the theoretical monolayer, based on the surface area of a WO_3 unit derived from the bulk density of crystalline WO_3. Iglesia and coworkers [47] indicated that a range of 7 W/nm^2 [54,55] to 7.8 W/nm^2 can be calculated for the theoretical monolayer coverage for polytungstates from "the density of WO_x species in a two-dimensional plane of corner-shared WO_6 octahedra with W–O bond distances corresponding to those in low-index planes of monoclinic WO_3 crystallites" [47].

From a molecular standpoint, WO_x surface species can be thought to titrate the ZrO_2 surface hydroxyl groups. Scheithauer et al. [56] performed *in situ* fourier transform infrared (FTIR) spectroscopy on a series of WO_x/ZrO_2 prepared through a modified incipient wetness impregnation method. They found that surface Zr–OH groups (corresponding to an infrared (IR) peak between 3765 and 3782 cm^{-1}) of the ZrO_2 support decreased in population with increasing WO_x loading and disappeared at 4.0 W/nm^2. They also detected IR peaks in the O–H stretching region at lower wavelengths (between 3625 and 3682 cm^{-1}), which remained at all WO_x loadings. Based on WO_x/ZrO_2 samples prepared using incipient wetness impregnated Degussa ZrO_2, Vaidyanathan et al. attributed the lower-wavelength

IR peaks to bridging OH groups (Zr–OH–Zr). They concluded that the WO_x species preferentially bonded to the ZrO_2 surface through the isolated OH groups rather than the bridging OH groups, implying incomplete coverage of the underlying support [57]. However, it cannot be ruled out that the bridging OH groups are located below the surface and not accessible to the WO_x surface species, as suggested by proton nuclear magnetic resonance (NMR) studies of alumina-supported metal oxides by Mastikhin et al. [58,59].

Still, other data indicate that the ZrO_2 surface is exposed, even at surface coverages >7.8 W/nm^2. Through combined low-temperature CO chemisorption (CO adsorbs on coordinately unsaturated Zr cations) and ion scattering spectroscopy (with a surface sensitivity of 0.3 nm), Vaidyanathan et al. reported that, for samples prepared through impregnation of crystalline ZrO_2, surface coverage increased approximately linearly, up to a WO_x loading of 1.7 wt% (or 1.2 W/nm^2) and reached a plateau of $\sim54\%$ above 6 wt% (or 4.6 W/nm^2) [57,60]. Ferraris et al. [61] found a very similar trend for materials prepared through ion-exchange and incipient wetness impregnation (Figure 2.2[a]). Scheithauer et al. [56] presented CO adsorption/FTIR data suggesting that coordinatively unsaturated Zr (Lewis acid) sites were present at surface densities as high as 14.1 W/nm^2.

Using benzaldehyde as a selective probe molecule for the basic sites of ZrO_2, Naito et al. [48] reported that there was near-100% coverage of ZrO_2 by WO_x species at 4 W/nm^2 for materials calcined at 650°C. However, for materials calcined at a lower temperature of 300°C, they found that $\sim55\%$ coverage by WO_x was measured at 4 W/nm^2; full surface coverage was not reached until 10 W/nm^2

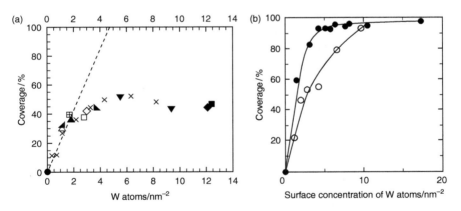

FIGURE 2.2 WO_x surface coverage of WO_x/ZrO_2 determined from (a) CO chemisorption at 77 K (The [×] marks are data from Vaidyanathan et al. [60]. Taken from Ferraris, G., De Rossi, S., Gazzoli, D., Petliti, I., Valigi, M., Magnacca, G., and Morterra, C., *Appl. Catal. A: Gen.* 2003, *240*, 119–128. With permission.) and (b) benzaldehyde chemisorption at 250°C (The open and closed symbols represent samples calcined at 300 and 650°C, respectively. Taken from Naito, N., Katada, N., and Niwa, M., *J. Phys. Chem. B* 1999, *103*, 7206–7213. With permission.)

(Figure 2.2[b]). Materials with the same surface density ρ_{surf} values can thus have different coverages of the ZrO_2 surface, highlighting the notion that monolayer coverage by the supported layer does not necessarily correspond to a completely inaccessible support.

2.4.2 Molecular Structure of WO_x Species

Wachs and coworkers prepared WO_x/ZrO_2 with different WO_x loadings through incipient wetness impregnation, using Degussa ZrO_2 (mixture of monoclinic and tetragonal crystal phases, 39 m^2/g) and ammonium metatungstate. Samples exhibited a sharp Raman band in the $\sim$1000 to 1010 cm^{-1} region, which is due to a stretching mode of a terminal tungsten oxo (W=O) bond (Figure 2.3). The broader Raman bands at 804 and 875 cm^{-1} were attributed to W–O–W bonds, indicating that polytungstates coexist with monotungstates below monolayer coverage. Direct quantification of the distribution of these species and WO_3 crystals through Raman spectroscopy is not possible, since the Raman signals for the polytungstates (and polymerized metal oxide species, in general) are very weak [25,62].

Monotungstates have only one W=O oxo bond under dehydrated conditions, according to oxygen isotope exchange studies [63]. The tungsten center is

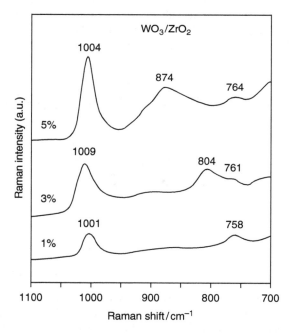

FIGURE 2.3 *In situ* Raman spectra taken after dehydrating the materials (previously calcined) at 500°C. The weight loadings of 1, 3, and 5% correspond to 0.7, 2.1, and 3.5 W/nm^2, respectively. For reference, crystalline WO_3 has sharp and intense bands at $\sim$715 and $\sim$ 800 cm^{-1} (Taken from Kim, D.S., Ostromecki, M., Wachs, I.E. *J. Mol. Catal. A: Chem.* 1996, *106*, 93–102. With permission.)

coordinated totally to six oxygens in distorted octahedral symmetry, as indicated by EXAFS and CO adsorption/FTIR studies [46,47,56,61,64]. Thus, isolated WO_6 octahedra presumably bind to the zirconia support through five W–O–Zr bonds. Polytungstates are oligomers of WO_6, which connect via W–O–W linkages (Figure 2.4). The number of WO_6 units, extent of connectivity, and physical dimensions of a polytungstate domain are not known and cannot be quantified yet [47].

2.4.3 Relationship Between WO_x Surface Structure and Catalytic Properties

An understanding of how the molecular structure of WO_x relates to observed catalytic properties has increased since Hino and Arata's work. In their report, they calcined their impregnation-derived WO_x/ZrO_2 samples at several temperatures (Table 2.3), and found that those calcined at 800 and 850°C were the most active for pentane isomerization [30]. The tungsten content was ~13 wt%; no surface areas were reported, and the surface densities could not be calculated.

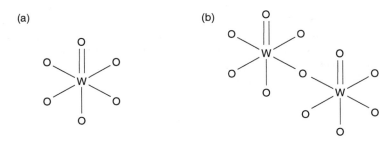

FIGURE 2.4 Schematic of (a) monotungstate and (b) polytungstate surface species

TABLE 2.3
Structural Properties of Two 825°C-Calcined $WO_x ZrO_2$ Materials Prepared by Santiesteban *et al.*

Preparation method	Surface area (m²/g)	WO₃ content (wt%)	ρ_{surf} (W/nm²)	Strong Brønsted site amount	
				(meq/g)	Site/nm²
Impregnation of refluxed Zr(OH)₄	62	21.3	8.9	0.002	0.019
Co-precipitation	62	19.5	8.2	0.0039	0.037

Source: Calculations are based on values taken from Santiesteban, J.G., Vartuli, J.C., Han, S., Bastian, R.D., and Chang, C.D. *J. Catal.* 1997, *168*, 431–441.

They concluded that the active (super)acid sites were generated from the crystallization of the amorphous $Zr(OH)_4$ support into the tetragonal phase, though they did not comment on the molecular nature of the active sites [65].

Mobil (now ExxonMobil) researchers compared WO_x/ZrO_2 materials prepared through impregnation and co-precipitation for pentane isomerization [42]. By titrating the Brønsted acid sites with 2,6-dimethylpyridine and examining the catalytic activity of the poisoned samples, they determined that the impregnated and co-precipitated WO_x/ZrO_2 samples had a similar amount of total Brønsted acid sites (∼0.0094 meq/g, or ∼0.09 site/nm^2). However, the latter contained twice as many strong Brønsted sites active for pentane isomerization (Table 2.3). Co-precipitation led to a more active material due to a higher surface density of strong Brønsted sites, though the strong Brønsted site amount did not appear to be related to tungsten surface density. They reported a maximum in isomerization activity as a function of tungsten oxide content in co-precipitated materials. This maximum activity behavior of WO_x/ZrO_2 was observed by others for n-butane isomerization [41,48].

Iglesia and coworkers [47,55,66] observed a similar maximum in reaction rates for o-xylene isomerization over WO_x/ZrO_2. Like Hino and Arata, they observed that reaction rates (in terms of turnover frequencies or TOF's, normalized to total W content) were greatest in value at an intermediate calcination temperature (Figure 2.5[a]). WO_x/ZrO_2 samples with different tungsten loadings exhibited a different maximum in o-xylene TOFs as a function of calcination temperature. By accounting for surface areas at different calcination temperatures, they were able to collapse the reaction rates roughly onto a single volcano-shaped curve when plotted against W surface density (Figure 2.5[b]). They found that WO_x/ZrO_2

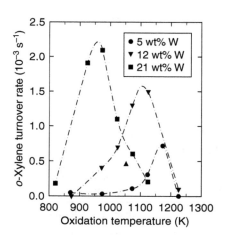
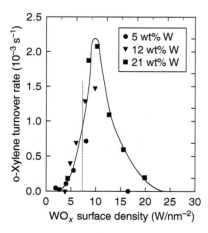

FIGURE 2.5 o-Xylene reaction rates over WO_x/ZrO_2 with different tungsten loadings plotted as a function of (a) calcination temperature and (b) tungsten surface density (Taken from Barton, D.G., Soled, S.L., Meitzner, G.D., Fuentes, G.A., and Iglesia, E., *J. Catal.* 1999, *181*, 57–72. With permission.)

exhibited similar catalytic behavior for 2-butanol dehydration in terms of initial TOF's [67].

Barton et al. [47,55] attributed the high reaction rates to the presence of polytungstates, as inferred from UV-Vis, x-ray diffraction (XRD), and Raman characterization results. Monotungstates and polytungstates are found at $\rho_{surf} < 4$ W/nm^2, and crystalline WO$_3$ species are detected through XRD at $\rho_{surf} > 8$ W/nm^2. In the intermediate range of 4 to 8 W/nm^2, polytungstates and crystalline WO$_3$ nanodomains (which cannot be detected through XRD but can be detected through Raman spectroscopy) are found. They suggested that the TOF increases with surface density and polytungstate population until $\sim$10 W/nm^2, above which additional tungsten content contributes to the formation of WO$_3$ crystals containing inactive and inaccessible tungsten atoms. They proposed that polytungstates reduce to form bronze-like WO$_x^{\delta-}$ domains, charge balanced by spatially delocalized H$^{\delta+}$, or Brønsted acid sites (Figure 2.6). The polytungstates are reducible by hydrogen gas or by the reactant hydrocarbon (e.g., 2-butanol) during the reaction, and the ease in reduction increases with domain size [54]. Two important findings are noted: calcination temperatures above 800°C are not needed to yield active WO$_x$/ZrO$_2$ materials, and the reaction rates over WO$_x$/ZrO$_2$ depend solely on W surface density.

In their modification of incipient wetness impregnation, Knözinger et al. [46] prepared WO$_x$/ZrO$_2$ by refluxing a suspension of Zr(OH)$_4$ in an aqueous solution of ammonium metatungstate for 16 h, and then evaporating the suspension to dryness. After calcining at 650°C, these refluxed WO$_x$/ZrO$_2$ materials have surface densities that exceed the monolayer coverage of 4 W/nm^2 without the formation of WO$_3$ crystals (Table 2.4). One would expect that calcination temperatures above the Tammann temperature of bulk WO$_3$ (600°C) lead to sufficient mobility of the surface WO$_x$ species for coalescence and crystallization to occur [68]. The calcination temperature of 825°C is sufficiently high to induce WO$_3$ crystal formation at most surface densities above 4 W/nm^2. The refluxing step apparently yields WO$_x$/ZrO$_2$ with polytungstates which are more thermally stable than impregnated materials; Scheithauer et al. [46,56] speculated the polytungstate domains

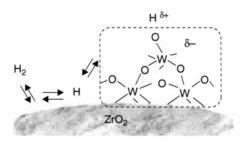

FIGURE 2.6 Proposed mechanisms for activation of polytungstate to form acid site (Taken from Baertsch, C.D., Soled, S.L., and Iglesia, E. *J. Phys. Chem. B* 2001, *105*, 1320–1330. With permission.)

TABLE 2.4

WO$_3$ content, surface area, and surface densities for WO$_x$/ZrO$_2$ calcined at 650 and 825°C

WO$_3$ weight loading (wt%)	WO$_x$/ZrO$_2$ calcined at 650°C		WO$_x$/ZrO$_2$ calcined at 825°C	
	Surface area (m^2/g)	ρ_{surf} (W/nm^2)	Surface area (m^2/g)	ρ_{surf} (W/nm^2)
0	~30	0	~ 18	0
3.6	54	1.7	30	3.1
5.9	64	2.4	40	3.8
8.6	69	3.2	46	4.9
10.5	82	3.3	46	5.9[a]
13.6	88	4.0	42	8.4[a]
19.0	96	5.1	35	14.1[a]
23.9	70	8.9	30	20.7[a]
32.0	43	19.3	28	29.7[a]

[a] Raman bands for WO$_3$ crystals observed.

Source: Values taken from Scheithauer, M., Grasselli, R.K., and Knözinger, H. *Langmuir* 1999, *14*, 3019–3029; Scheithauer, M., Cheung, T.K., Jentoft, R.E., Grasse R.K., Gates, B.C., and Knözinger, H., *J. Catal.* 1998, *180*, 1–13.

to contain Zr cations in a heteropolyacid-like form. Kuba et al. [69,70] reported that pentane molecules formed radicals over polytungstate-containing WO$_x$/ZrO$_2$ under reaction conditions, leading to the reduction of W^{6+} to W^{5+} cations and the formation of hydroxyl groups. This redox initiation step may be responsible for the induction period observed in pentane conversion with time [56].

2.5 NEW SYNTHESIS APPROACHES

The concept of polymerized WO$_x$ surface species as the active sites for acid cata-lyzed molecular conversions upon partial reduction provides the motivation to learn how to control the molecular structure of these surface species more ration-ally at the synthesis level. The following synthesis routes have been reported for WO$_x$/ZrO$_2$ materials with desirable structural properties (such as high surface areas, high thermal stability, and controlled pore sizes) and may provide greater control over polytungstate formation with further study.

2.5.1 Surfactant Templating

Surfactant templating chemistry involves the hydrolysis and condensation of solu-bilized precursors of metal oxides in the presence of surfactant molecules that form the eventual pores in the inorganic matrix. Mobil (now ExxonMobil) researchers reported in 1992 that surfactants could be used to template highly ordered pore channels of uniform sizes in silicate structures (Figure 2.7) [71–84]. A solution of hexadecyltrimethylammonium bromide contains aggregates of surfactant molecules called micelles, and addition of a silicate precursor in basic solution causes immediate precipitation of an organic/inorganic mesostructure. Formation involves the electrostatic binding of negatively charged silicate oligomers to the positively charged ammonium head groups on the spherical micelle surface, the elongation of micelles into rods, the packing of the rods into a hexagonal array, and the condensation of the silicate between the rod-shaped micelles. The mesostructure is calcined to remove the surfactant, thereby creating a honeycomb-like structure. The resulting mesoporous silica material has high surface areas (>1000 m^2/g); uniform pore sizes (2 to 10 nm); liquid crystal-like, long-range ordering of the pores, and high thermal stability ($>800°$C). Materials with larger pore sizes can be prepared using longer surfactant molecules, organic liquids to swell the micelles, and amphiphilic polymers.

Surfactant templating chemistry can be extended to many nonsilicate compositions after modifications to the synthesis route. These materials are less structurally stable than the mesoporous silicates, which is attributed to the thinness of the amorphous pore walls ($\sim$1 to 2 nm). Stucky and coworkers [85,86] showed that this problem could be mitigated by preparing the materials with thicker walls. To prepare mesoporous WO$_3$, they dissolved a poly(ethylene oxide)–poly(propylene oxide)–poly(ethylene oxide) triblock copolymer and WCl$_6$ salt in ethanol, and dried the resulting solution in open air. The tungsten salt reacted with moisture to undergo hydrolysis and condensation reactions. These chemical reactions caused the eventual formation of amorphous WO$_3$ around triblock copolymer micelle-like domains, and after calcination at 400°C, a mesoporous WO$_3$ with thick, nanocrystalline walls ($\sim$5 nm) and surface area of 125 m^2/g was formed.

A mesoporous WO$_x$/ZrO$_2$ was prepared in an analogous fashion, in which ZrCl$_4$ salt was added with WCl$_6$ prior to drying of the ethanolic solution [85]. After calcination at 400°C, the resulting mesoporous material was amorphous, with a pore wall thickness of 4.5 nm and a surface area of 170 m^2/g. The distribution of W and Zr atoms was not reported, although the W can be presumed to be distributed on the surface and throughout the pore walls.

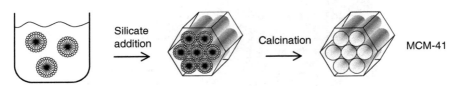

FIGURE 2.7   Illustration of two-step preparation of hexagonally ordered mesoporous silica MCM-41

2.5.2 NP Surfactant Templating

Wong et al. [87] reported the use of metal oxide NPs in surfactant templating, which allowed the synthesis of tungstated zirconia, tungstated titania, and tungstated alumina. In the preparation of WO_x/ZrO_2, a clear suspension of ZrO_2 NPs (diameter ~3 to 10 nm) is added to a solution of ammonium metatungstate and (ethylene oxide)$_{20}$–(propylene oxide)$_{70}$–poly(ethylene oxide)$_{20}$ triblock copolymer, which leads to immediate precipitation. It is thought that the metatungstate anion binds to the surface of the positively charged NPs, and the coated NPs aggregate around the polymer micelles (Figure 2.8).

Calcination of the dried precipitate at 600°C removes the organics from the inorganic framework and transforms the metatungstate into WO_x surface species. The resulting material has a surface of 130 m^2/g, a pore size in the 4 to 6 nm range, and a pore wall thickness of ~4 to 6 nm (Figure 2.9). There are no crystalline phases of ZrO_2 or WO_3, as characterized through XRD. The material does not crystallize into distinct ZrO_2 and WO_3 crystal phases until heated above 700°C.

The use of ZrO_2 NPs ensures that the WO_x species are on the surface. With a WO_3 content of ~30 wt%, the surface density is calculated to be 6 W/nm^2.

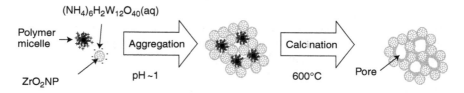

FIGURE 2.8 Illustration of two-step preparation of NP/surfactant-templated tungstated zirconia

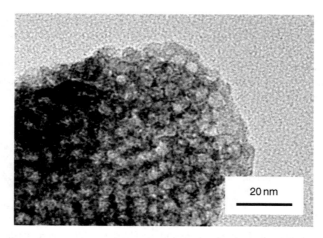

FIGURE 2.9 TEM image of NP/surfactant-templated WO_x/ZrO_2 (calcined at 600°C)

NP/surfactant-templated WO_x/ZrO_2 contains no crystalline WO_3 according to *in situ* laser Raman spectroscopy [88,89]. It is hypothesized that the ZrO_2 NPs stabilize polytungstates from crystallizing by trapping them in the gaps between adjacent NPs, allowing for the high content of amorphous WO_x. Some batch-to-batch variation is observed due to sensitivity of the final structure to synthesis conditions, which affects the WO_3 content, surface area, and surface density (as high as 12 W/nm^2).

2.5.3 Modified Sol–Gel Chemistry

To reduce the variability in the final structure, a new synthesis strategy involving sol–gel chemistry is being studied. Knowles et al. [89] posited that, if ZrO_2 NPs act to stabilize polytungstates by entrapment in the NP gaps, then a network of ZrO_2 NPs would also stabilize WO_x in a similar fashion. A zirconia gel prepared from the gelation of a zirconium alkoxide-derived sol is one such example of an NP network. This gel was prepared through controlled hydrolysis of zirconium *n*-propoxide, and subsequently mixed with a solution of ammonium metatungstate. To avoid catastrophic pore collapse upon drying to form a xerogel, ascorbic acid was added to the mixture. This organic is thought to bind to the gel surface and to prop up the internal gel framework during drying. Calcination at 600°C led to mesoporous WO_x/ZrO_2 with an open structure similar to that of NP/surfactant-templated WO_x/ZrO_2 (Figure 2.10). XRD and Raman spectroscopic data indicate the presence of tetragonal ZrO_2 with grain sizes ~16 nm and the absence of WO_3 crystals for a WO_x/ZrO_2 with 10.6 W/nm^2. This two-step gelation method allows the WO_3 content of the final material to be controlled independent of the support formation, unlike NP/surfactant templating, which require the simultaneous presence of metatungstate and ZrO_2 NPs.

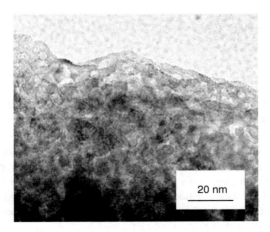

FIGURE 2.10 TEM image of sol–gel-derived WO_x/ZrO_2 (calcined at 600°C)

The two forms of NP-supported WO_x/ZrO_2 materials (prepared from NP/surfactant templating and sol–gel chemistry) are active for the dehydration of methanol to dimethylether [89], according to methanol oxidation studies [90,91]. It is hypothesized that these NP-supported materials contain a larger amount of polytungstates than conventionally prepared WO_x/ZrO_2 and therefore are more active [89].

Boyse and Ko [45] prepared WO_x/ZrO_2 through a one-step co-gelation (Table 2.2). An acidic alcoholic solution of zirconium n-propoxide and ammonium metatungstate was prepared, which gelled in $\sim$1 min. The resulting gel was supercritically dried in CO_2 to remove the solvent without collapsing the pore structure. Calcination led to a material with mesopores in the range of 10 to 20 nm. The co-gelled WO_x/ZrO_2 contained more thermally stable WO_x species compared with the impregnated material. As in the case of co-precipitated supported metal oxides, it could not be ruled out that tungsten atoms were trapped within the ZrO_2 support.

Signoretto et al. [92] reported a one-step co-gelation method, in which zirconium n-propoxide and tungsten n-propoxide was combined with nitric acid, water, and isopropanol to create a homogeneous solution that gelled overnight. Supercritically dried in CO_2 and calcined, the final material was found to contain more entrapped tungsten than those prepared with zirconium n-propoxide and ammonium metatungstate.

In a very different approach, Melezhyk et al. [93] prepared gel-like particles using zirconyl chloride and ammonium metatungstate. A solution of the two salt precursors and a solution of poly(vinyl alcohol) were prepared at 100°C, and after combining, the resulting mixture was diluted with a buffer solution to yield transparent gel particles. These particles were carefully dried and calcined. By calcining the particles under inert atmosphere before calcining in air, higher surface areas could be achieved. This pyrolysis step converted the poly(vinyl alcohol) into carbonaceous residue that kept the framework from collapsing. The poly(vinyl alcohol) also acted as a porogen during gel formation and drying processes, which resulted in 10 nm pore sizes.

2.6 CONCLUSIONS

The challenge of controlling the nanostructure of WO_x/ZrO_2, namely the molecular surface structure and the nanometer-scale framework properties, can be generalized to supported metal oxides. It is currently addressed through the development of new material synthesis chemistries, such as surfactant templating and sol–gel chemistry. Surfactant templating leads to relatively unstable porous materials, but thicker pore walls through a different chemistry or using NPs overcomes this problem. This method and sol–gel chemistry can lead to mixed metal oxides but the active metal oxide could be trapped below the support surface. Here, too, the use of NPs (either preformed or formed *in situ*) is a solution, by ensuring the localization of the active phase on the support surface. It appears that NPs in the form of a porous framework can stabilize the active phase as thermally stable polymeric surface species, providing a new and interesting type of nanostructured supported metal oxide. Ultimately, the ability to engineer the catalyst structure at the nanoscale and the fundamental understanding of the synthesis–structure–property relationships of supported metal oxides will further enable the rational design and preparation of supported metal oxides.

ACKNOWLEDGMENTS

M.S.Wong acknowledges Rice University and the Oak Ridge Associated Universities Ralph E. Powe Junior Faculty Enhancement Award for financial support.

REFERENCES

1. Richardson, J.T. *Principles of Catalyst Preparation*. New York: Plenum Press, 1989.
2. Satterfield, C.N. *Heterogeneous Catalysis in Industrial Practice*, 2nd ed. New York: McGraw-Hill, 1991.
3. Twigg, M.V. Ed. *Catalyst Handbook*, 2nd ed. London: Manson Publishing, 1996.
4. Rase, H.F. *Handbook of Commercial Catalysts: Heterogeneous Catalysts*. Boca Raton, FL: CRC Press, 2000.
5. Ertl, G., Knözinger, H., and Weitkamp, J. Eds. *Preparation of Solid Catalysts*. Weinheim: Wiley-VCH, 1999.
6. Ertl, G., Knözinger, H., and Weitkamp, J. Eds. *Handbook of Heterogeneous Catalysis*. Weinheim: Wiley-VCH, 1997.
7. Wachs, I.E. Preface. *Catal. Today* 1999, *51*, 201.
8. Busca, G., Lietti, L., Ramis, G., and Berti, F. Chemical and mechanistic aspects of the selective catalytic reduction of NO_x by ammonia over oxide catalysts: a review. *Appl. Catal. B: Environ.* 1998, *18*, 1–36.
9. Centi, G. Nature of active layer in vanadium oxide supported on titanium oxide and control of its reactivity in the selective oxidation and ammoxidation of alkylaromatics. *Appl. Catal. A: Gen.* 1996, *147*, 267–298.
10. Weckhuysen, B.M. and Schoonheydt, R.A. Alkane dehydrogenation over supported chromium oxide catalysts. *Catal. Today* 1999, *51*, 223–232.
11. Weckhuysen, B.M. and Schoonheydt, R.A. Olefin polymerization over supported chromium oxide catalysts. *Catal. Today* 1999, *51*, 215–221.
12. Hodnett, B.K. *Heterogeneous Catalytic Oxidation: Fundamental and Technological Aspects of the Selective and Total Oxidation of Organic Compounds*. Chichester: John Wiley & Sons, 2000.
13. Mol, J.C. Olefin metathesis over supported rhenium oxide catalysts. *Catal. Today* 1999, *51*, 289–299.
14. Sheldon, R.A. and van Vliet, M.C.A. Oxidation. In *Fine Chemicals through Heterogeneous Catalysis*, Sheldon, R.A. and van Bekkum, H., Eds. Weinheim: Wiley-VCH, 2001.
15. Hävecker, M., Mayer, R.W., Knop-Gericke, A., Bluhm, H., Kleimenov, E., Liskowski, A., Su, D., Follath, R., Requejo, F.G., Ogletree, D.F., Salmeron, M.,

Lopez-Sanchez, J.A., Bartley, J.K., Hutchings, G.J., and Schlögl, R. *In situ* investigation of the nature of the active surface of a vanadyl pyrophosphate catalyst during *n*-butane oxidation to maleic anhydride. *J. Phys. Chem.* B 2003, *107*, 4587–4596.

16. Briand, L.E., Hirt, A.M., and Wachs, I.E. Quantitative determination of the number of surface active sites and the turnover frequencies for methanol oxidation over metal oxide catalysts: Application to bulk metal molybdates and pure metal oxide catalysts. *J. Catal.* 2001, *202*, 268–278.

17. Haber, J. Manual on catalyst characterization. *Pure Appl. Chem.* 1991, *63*, 1227–1246.

18. Che, M., Clause, O., and Marcilly, Ch. Impregnation and ion exchange. In *Preparation of Solid Catalysts*, Ertl, G., Knözinger, H., and Weitkamp, J. Eds. Weinheim: Wiley-VCH, 1999.

19. Del Arco, M., Carrazan, S.R.G., Rives, V., Gilllambías, F.J., and Malet, P. Surface species formed upon supporting molybdena on alumina by mechanically mixing both oxides. *J. Catal.* 1993, *141*, 48–57.

20. Hillerová, E., Morishige, H., Inamura, K., and Zdrazil, M. Formation of monolayer of molybdena over alumina by unconventional slurry impregnation or solvent assisted spreading method. *Appl. Catal. A: Gen.* 1997, *156*, 1–17.

21. Chan, S.S., Wachs, I.E., Murrell, L.L., and Dispenziere, N.C. Laser Raman characterization of tungsten oxide supported on alumina — Influence of calcination temperatures. *J. Catal.* 1985, *92*, 1–10.

22. Kim, D.S., Ostromecki, M., and Wachs, I.E. Surface structures of supported tungsten oxide catalysts under dehydrated conditions. *J. Mol. Catal. A: Chem.* 1996, *106*, 93–102.

23. Carrier, X., de la Caillerie, J.B.D., Lambert, J.F., and Che, M. The support as a chemical reagent in the preparation of WO_x/γ-Al_2O_3 catalysts: Formation and deposition of aluminotungstic heteropolyanions. *J. Am. Chem. Soc.* 1999, *121*, 3377–3381.

24. Wachs, I.E. and Hardcastle, F.D. Applications of Raman spectroscopy to heterogeneous catalysis. In *Catalysis*. Cambridge: The Royal Society of Chemistry, Vol. 10, 1993.

25. Wachs, I.E. Raman and IR studies of surface metal oxide species on oxide supports: Supported metal oxide catalysts. *Catal. Today* 1996, *27*, 437–455.

26. Wachs, I.E. *In situ* Raman spectroscopy studies of catalysts. *Top. Catal.* 1999, *8*, 57–63.

27. Banares, M.A., Guerrero-Perez, M.O., Fierro, J.L.G., and Cortez, G.G. Raman spectroscopy during catalytic operations with on-line activity measurement (operando spectroscopy): A method for understanding the active centres of cations supported on porous materials. *J. Mater. Chem.* 2002, *12*, 3337–3342.

28. Weckhuysen, B.M. Snapshots of a working catalyst: Possibilities and limitations of *in situ* spectroscopy in the field of heterogeneous catalysis. *Chem. Commun.* 2002, 97–110.

29. Wachs, I.E. *In situ* Raman spectroscopy of catalysts: past, present and future. In *Proceedings 2nd International Congress on Operando Spectroscopy*, Lunteren: The Netherlands, March 2–6, 2003, p. 30.

30. Hino, M. and Arata, K. Synthesis of solid superacid of tungsten oxide supported on zirconia and its catalytic action for reactions of butane and pentane. *J. Chem. Soc. Chem. Commun.* 1988, 1259–1260.

31. Song, X. and Sayari, A. Sulfated zirconia-based strong solid-acid catalysts: Recent progress. *Catal. Rev. Sci. Eng.* 1996, *38*, 329–412.
32. Yadav, G.D. and Nair, J.J. Sulfated zirconia and its modified versions as promising catalysts for industrial processes. *Micropor. Mesopor. Mater.* 1999, *33*, 1–48.
33. Arata, K., Matsuhashi, H., Hino, M., and Nakamura, H. Synthesis of solid superacids and their activities for reactions of alkanes. *Catal. Today* 2003, *81*, 17–30.
34. Larsen, G., Lotero, E., Raghavan, S., Parra, R.D., and Querini, C.A. A study of platinum supported on tungstated zirconia catalysts. *Appl. Catal. A: Gen.* 1996, *139*, 201–211.
35. Larsen, G. and Petkovic, L.M. Effect of preparation method and selective poisoning on the performance of platinum supported on tungstated zirconia catalysts for alkane isomerization. *Appl. Catal. A: Gen.* 1996, *148*, 155–166.
36. Santiesteban, J.G., Calabro, D.C., Borghard, W.S., Chang, C.D., Vartuli, J.C., and Tsao, Y.P., Natal-Santiago, M.A., and Bastian, R.D. H-spillover and SMSI effects in paraffin hydroisomerization over $Pt/WO_x/ZrO_2$ bifunctional catalysts. *J. Catal.* 1999, *183*, 314–322.
37. Arribas, M.A., Márquez, F., and Martínez, A. Activity, selectivity, and sulfur resistance of $Pt/WO_x–ZrO_2$ and Pt/Beta catalysts for the simultaneous hydroisomerization of *n*-heptane and hydrogenation of benzene. *J. Catal.* 2000, *190*, 309–319.
38. Santiesteban, J.G., Calabro, D.C., Chang, C.D., Vartuli, J.C., Fiebig, T.J., and Bastian, R.D. The role of platinum in hexane isomerization over $Pt/FeO_y/WO_x/ZrO_2$. *J. Catal.* 2001, *202*, 25–33.
39. Kuba, S., Gates, B.C., Grasselli, R.K., and Knözinger, H. An active and selective alkane isomerization catalyst: Iron- and platinum-promoted tungstated zirconia. *Chem. Commun.* 2001, 321–322.
40. MEL Chemicals *Sulphated and Tungstated Zirconia — Solid, Strong Acid Catalysts*. MELCat Doc. 2000. Machester: England, 2000.
41. Boyse, R.A. and Ko, E.I. Commercially available zirconia–tungstate as a benchmark catalytic material. *Appl. Catal. A: Gen.* 1999, *177*, L131–L137.
42. Santiesteban, J.G., Vartuli, J.C., Han, S., Bastian, R.D., and Chang, C.D. Influence of the preparative method on the activity of highly acidic WO_x/ZrO_2 and the relative acid activity compared with zeolites. *J. Catal.* 1997, *168*, 431–441.
43. Vaudagna, S.R., Comelli, R.A., and Figoli, N.S. Influence of the tungsten oxide precursor on $WO_x–ZrO_2$ and $Pt/WO_x–ZrO_2$ properties. *Appl. Catal. A: Gen.* 1997, *164*, 265–280.
44. Yori, J.C., Vera, C.R., and Parera, J.M. *n*-Butane isomerization on tungsten oxide supported on zirconia. *Appl. Catal. A: Gen.* 1997, *163*, 165–175.
45. Boyse, R.A. and Ko, E.I. Crystallization behavior of tungstate on zirconia and its relationship to acidic properties. 1. Effect of preparation parameters. *J. Catal.* 1997, *171*, 191–207.
46. Scheithauer, M., Grasselli, R.K., and Knözinger, H. Genesis and structure of WO_x/ZrO_2 solid acid catalysts. *Langmuir* 1998, *14*, 3019–3029.
47. Barton, D.G., Shtein, M., Wilson, R.D., Soled, S.L. and Iglesia, E. Structure and electronic properties of solid acids based on tungsten oxide nanostructures. *J. Phys. Chem. B* 1999, *103*, 630–640.

48. Naito, N., Katada, N., and Niwa, M. Tungsten oxide monolayer loaded on zirconia: Determination of acidity generated on the monolayer. *J. Phys. Chem. B* 1999, *103*, 7206–7213.

49. Valigi, M., Gazzoli, D., Pettiti, I., Mattei, G., Colonna, S., De Rossi, S., and Ferraris, G. WO_x/ZrO_2 catalysts Part 1. Preparation, bulk and surface characterization. *Appl. Catal. A: Gen.* 2002, *231*, 159–172.

50. Chen, Y.S. and Wachs, I.E. Tantalum oxide-supported metal oxide (Re_2O_7, CrO_3, MoO_3, WO_3, V_2O_5, and Nb_2O_5) catalysts: Synthesis, Raman characterization and chemically probed by methanol oxidation. *J. Catal.* 2003, *217*, 468–477.

51. Bond, G.C. and Tahir, S.F. Vanadium-oxide monolayer catalysts-preparation, characterization and catalytic activity. *Appl. Catal.* 1991, *71*, 1–31.

52. Zhao, B.Y., Xu, X.P., Gao, J.M., Fu, Q., and Tang, Y.Q. Structure characterization of WO_3/ZrO_2 catalysts by Raman spectroscopy. *J. Raman Spectrosc.* 1996, *27*, 549–554.

53. Xie, Y.C. and Tang, Y.Q. Spontaneous monolayer dispersion of oxides and salts onto surfaces of supports: Applications to heterogeneous catalysis. *Adv. Catal.* 1990, *37*, 1–43.

54. Baertsch, C.D., Soled, S.L., and Iglesia, E. Isotopic and chemical titration of acid sites in tungsten oxide domains supported on zirconia. *J. Phys. Chem. B* 2001, *105*, 1320–1330.

55. Barton, D.G., Soled, S.L., Meitzner, G.D., Fuentes, G.A., and Iglesia, E. Structural and catalytic characterization of solid acids based on zirconia modified by tungsten oxide. *J. Catal.* 1999, *181*, 57–72.

56. Scheithauer, M., Cheung, T.K., Jentoft, R.E., Grasselli, R.K., Gates, B.C., and Knözinger, H. Characterization of WO_x/ZrO_2 by vibrational spectroscopy and *n*-pentane isomerization catalysis. *J. Catal.* 1998, *180*, 1–13.

57. Vaidyanathan, N., Hercules, D.M., and Houalla, M. Surface characterization of WO_3/ZrO_2 catalysts. *Anal. Bioanal. Chem.* 2002, *373*, 547–554.

58. Mastikhin, V.M., Nosov, A.V., Terskikh, V.V., Zamaraev, K.I. and Wachs, I.E. ^{1}H MAS NMR studies of alumina-supported metal oxide catalysts. *J. Phys. Chem.* 1994, 98, 13621–13624.

59. Mastikhin, V.M., Terskikh, V.V., Lapina, O.B., Filimonova, S.V., Seidl, M., and Knözinger, H. Characterization of (V_2O_5–WO_3) on TiO_x/Al_2O_3 catalysts by [1]H-, ^{15}N-, and [51]V-solid state NMR spectroscopy. *J. Catal.* 1995, *156*, 1–10.

60. Vaidyanathan, N., Houalla, M., and Hercules, D.M. Surface coverage of WO_3/ZrO_2 catalysts measured by ion scattering spectroscopy and low temperature CO adsorption. *Surf. Interface Anal.* 1998, *26*, 415–419.

61. Ferraris, G., De Rossi, S., Gazzoli, D., Pettiti, I., Valigi, M., Magnacca, G., and Morterra, C. WO_x/ZrO_2 catalysts Part 3. Surface coverage as investigated by low temperature CO adsorption: FT-IR and volumetric studies. *Appl. Catal. A: Gen.* 2003, *240*, 119–128.

62. Busca, G. Differentiation of mono-oxo and polyoxo and of monomeric and polymeric vanadate, molybdate and tungstate species in metal oxide catalysts by IR and Raman spectroscopy. *J. Raman Spectrosc.* 2002, *33*, 348–358.

63. Weckhuysen, B.M., Jehng, J.M., and Wachs, I.E. *In situ* Raman spectroscopy of supported transition metal oxide catalysts: $^{18}O_2$–$^{16}O_2$ isotopic labeling studies. *J. Phys. Chem. B* 2000, *104*, 7382–7387.

64. Iglesia, E., Barton, D.G., Soled, S.L., Miseo, S., Baumgartner, J.E., Gates, W.E., Fuentes, G.A., and Meitzner, G.D. Selective isomerization of alkanes on supported tungsten oxide acids. *Stud. Surf. Sci. Catal.* 1996, *101*, 533–542.
65. Arata, K. Solid superacids. *Top. Catal.* 1990, *37*, 165–211.
66. Barton, D.G., Soled, S.L., and Iglesia, E. Solid acid catalysts based on supported tungsten oxides. *Top. Catal.* 1998, *6*, 87–99.
67. Baertsch, C.D., Komala, K.T., Chua, Y.H., and Iglesia, E. Genesis of Bronsted acid sites during dehydration of 2-butanol on tungsten oxide catalysts *J. Catal.* 2002, *205*, 44–57.
68. Knözinger, H. and Taglauer, E. Spreading and Wetting. In *Preparation of Solid Catalysts*, Ertl, G., Knözinger, H., and Weitkamp, J. Eds. Weinheim: Wiley-VCH, 1999.
69. Kuba, S., Heydorn, P.C., Grasselli, R.K., Gates, B.C., Che, M., and Knözinger, H. Redox properties of tungstated zirconia catalysts: Relevance to the activation of *n*-alkanes. *Phys. Chem. Chem. Phys.* 2001, *3*, 146–154.
70. Kuba, S., Lukinskas, P., Grasselli, R.K., Gates, B.C., and Knözinger, H. Structure and properties of tungstated zirconia catalysts for alkane conversion. *J. Catal.* 2003, *216*, 353–361.
71. Kresge, C.T., Leonowicz, M.E., Roth, W.J., Vartuli, J.C., and Beck, J.S. Ordered mesoporous molecular sieves synthesized by a liquid-crystal template mechanism. *Nature* 1992, *359*, 710–712.
72. Beck, J.S., Vartuli, J.C., Roth, W.J., Leonowicz, M.E., Kresge, C.T., Schmitt, K.D., Chu, C.T.W., Olson, D.H., Sheppard, E.W., McCullen, S.B., Higgins, J.B., and Schlenker, J.L. A new family of mesoporous molecular sieves prepared with liquid crystal templates. *J. Am. Chem. Soc.* 1992, *114*, 10834–10843.
73. Zhao, X.S., Lu, G.Q.M., and Millar, G.J. Advances in mesoporous molecular sieve MCM-41. *Ind. Eng. Chem. Res.* 1996, *35*, 2075–2090.
74. Ying, J.Y., Mehnert, C.P., and Wong, M.S. Synthesis and applications of supramolecular-templated mesoporous materials. *Angew. Chem.:Int. Edit.* 1999, *38* 56–77.
75. He, H. and Antonelli, D. Recent advances in synthesis and applications of transition metal containing mesoporous molecular sieves. *Angew. Chem.:Int. Edit.* 2001, *41*, 214–229.
76. Schüth, F. Non-siliceous mesostructured and mesoporous materials. *Chem. Mat.* 2001, *13*, 3184–3195.
77. Sayari, A. and Liu, P. Non-silica periodic mesostructured materials: Recent progress. *Micropor. Mater.* 1997, *12*, 149–177.
78. Corma, A. From microporous to mesoporous molecular sieve materials and their use in catalysis. *Chem. Rev.* 1997, *97*, 2373–2419.
79. Sayari, A. Catalysis by crystalline mesoporous molecular sieves. *Chem. Mater.* 1996, *8*, 1840–1852.
80. Pang, J.B., Qiu, K.Y., and Wei, Y. Recent progress in research on mesoporous materials I: Synthesis. *J. Inorg. Mater.* 2002, *17*, 407–414.
81. Vartuli, J.C., Shih, S.S., Kresge, C.T., and Beck, J.S. Potential applications for M41S type mesoporous molecular sieves. In Proceedings of the 1st International Symposium on *Mesoporous Molecular Sieves*, 10–12 July, 1998, Baltimore, MD; Bonneviot, L., Beland, F., Danumah, C., Giasson, S., and Kaliaguire, S. Eds. New York: Elsevier, 1998 pp. 13–21.

82. On, D.T., Desplantier-Giscard, D., Danumah, C., and Kaliaguine, S. Perspectives in catalytic applications of mesostructured materials. *Appl. Catal. A: Gen.* 2001, *222*, 299–357.

83. Ciesla, U. and Schüth, F. Ordered mesoporous materials. *Micropor. Mesopor. Mater.* 1999, *27*, 131–149.

84. Wong, M.S. and Knowles, W.V. Surfactant-templated mesostructured materials: Synthesis and compositional control. In *Nanoporous Materials — Science and Engineering*, Lu, G.Q. and Zhao, X.S. Eds. London: Imperial College Press, 2004, *2*, 30–60.

85. Yang, P., Zhao, D., Margolese, D.I., Chmelka, B.F. and Stucky, G.D. Generalized syntheses of large-pore mesoporous metal oxides with semicrystalline frameworks. *Nature* 1998, *396*, 152–155.

86. Yang, P., Zhao, D., Margolese, D.I., Chmelka, B.F., and Stucky, G.D. Block copolymer templating syntheses of mesoporous metal oxides with large ordering lengths and semicrystalline framework. *Chem. Mater.* 1999, *11*, 2813–2826.

87. Wong, M.S., Jeng, E.S., and Ying, J.Y. Supramolecular templating of thermally stable crystalline mesoporous metal oxides using nanoparticulate precursors. *Nano Lett.* 2001, *1*, 637–642.

88. Wong, M.S., Jehng, J.M., Wachs, I.E., and Ying, J.Y., unpublished results.

89. Knowles, W.V., Ross, E.I., Wachs, I.E., and Wong, M.S. Comparative study of variously prepared WO_x/ZrO_2 materials for methanol oxidation (in preparation).

90. Tatibouët, J.M. Methanol oxidation as a catalytic surface probe. *Appl. Catal. A: Gen.* 1997, *148*, 213–252.

91. Badlani, M. and Wachs, I.E. Methanol: A 'smart' chemical probe molecule. *Catal. Lett.* 2001, *75*, 137–149.

92. Signoretto, M., Scarpa, M., Pinna, F., Strukul, G., Canton, P., and Benedetti, A. WO_3/ZrO_2 catalysts by sol–gel processing. *J. Non-Cryst. Solids* 1998, *225*, 178–183.

93. Melezhyk, O.V., Prudius, S.V., and Brei, V.V. Sol–gel polymer-template synthesis of mesoporous WO_3/ZrO_2. *Micropor. Mesopor. Mater.* 2001, *49*, 39–44.

3 Defect Chemistry and Transport in Metal Oxides

A. Thursfield and I.S. Metcalfe

Department of Chemical Engineering, University of
Manchester Institute of Science and Technology,
Manchester, England, UK

A. Kruth and J.T.S. Irvine

School of Chemistry, University of St. Andrews,
St. Andrews, Scotland, UK

CONTENTS

3.1 INTRODUCTION

Heterogeneous catalysis occurs at the surface of solids. To describe the overall catalytic process it is obvious that one must consider fluid-phase transport of reactants to, and products away from, the solid surface. However, it is also true that, in the case of oxide materials, solid-phase transport of defects must be considered if we are to have a full description of the processes occurring at the surface. Furthermore, such defects, when present at the surface of the oxide, play a very important role in determining the catalytic activity of the solid. The nature of these surface defects is strongly related to the bulk defect chemistry of the solids. In this chapter we describe the bulk defect chemistry of solids and discuss the role of this defect chemistry in determining defect transport and electrical conductivity. We conclude the chapter by looking at the use of oxide membranes in chemical reactors.

Such membranes rely upon the ability of the oxide material to selectively transport defects which themselves undergo catalytic reactions at the free membrane surfaces. We focus on the versatile perovskite family of metal oxides which exhibits both ionic and electronic conductivity (pure ion conductors or solid electrolytes are dealt with elsewhere in this book).

3.2 TYPES OF DEFECTS

In an ideal ionic crystal, each atom is located at its regular lattice site and all sites are occupied, however, due to entropy considerations all real crystals deviate from the perfect atomic arrangements. Important physical properties of a metal oxide such as electrical transport, diffusion and diffusion-controlled processes such as sintering and phase separation, catalytic activity, melting point and various optical properties depend on the presence of defects. Numerous types of lattice defects are recognized: electrons and positive holes, excitons (excited electrons that are accompanied by their holes), vacant lattice sites and interstitial atoms, impurity atoms in interstitial or substitutional positions, dislocations and stacking faults. In addition to these primary imperfections, there are transient imperfections such as electromagnetic quanta and charged and neutral radiation. For convenience, we will concentrate on atomic defects and some electronic defects; however, it is important to understand that all different types of imperfections interact among themselves.

3.2.1 Point Defects

Point defects are missing, substituted or interstitial ions and occur in all crystalline materials. They are electronically charged and can be intrinsic or extrinsic. Intrinsic point defects are thermally generated in a crystal, whereas extrinsic defects are formed by the addition of an impurity or dopant.

The most common types of intrinsic point defects are Schottky and Frenkel defects (Figure 3.1). A Schottky defect consists of a vacant cation lattice site and a vacant anion lattice site. To form a Schottky defect, ions leave their normal lattice positions and relocate at the crystal surface, preserving overall charge neutrality. Hence, for a metal monoxide, MO, vacant sites must occur equally in the cation and anion sublattice and form a Schottky pair, whereas in binary metal oxides, MO_2, a Schottky defect consists of three defects: a vacant cation site and two vacant anion sites. A Frenkel defect forms when a cation or anion is displaced from its regular site onto an interstitial site, where, the resulting vacancy and interstitial atom form a Frenkel defect pair.

The presence of Schottky defects lowers the overall density of the material because the volume is increased at constant mass. Frenkel defects do not change the volume and the density of the material therefore remains the same. Hence, density measurements can be a useful tool for assessing the type of defect disorder in oxides if the defect content is high enough.

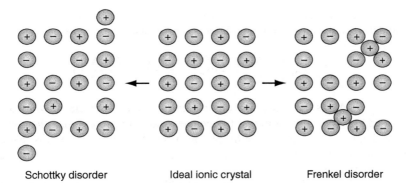

| Schottky disorder | Ideal ionic crystal | Frenkel disorder |

FIGURE 3.1   Formation of Schottky and Frenkel disorders in an ionic crystal

During thermal activation of an intrinsic defect in a lattice, the amplitude of lattice vibrations is increased and, subsequently, atoms become more likely to be displaced off their regular sites. The concentration of point defects is in thermal equilibrium with the crystal, given by:

$$G = G_0 + n\Delta g - T\Delta S_c \qquad (3.1)$$

where G is the free energy of the crystal upon forming n defect pairs, each defect pair consuming an energy Δg during formation, and ΔS_c is the configurational entropy of the crystal. The change in free energy is therefore:

$$\Delta G = n\Delta g - T\Delta S_c = n\Delta g - Tk \ln P \qquad (3.2)$$

with P being the probability or the number of possible ways in which the defect can be arranged. For instance, N Frenkel pairs consist of n_i interstitials and n_v vacancies. Since the defects are formed in pairs, $n = n_i = n_v$, the configurational entropy is given by:

$$\Delta S_c = k \left(\frac{N!}{(N - n_v)!n_v!} \right) \left(\frac{N!}{(N - n_i)!n_i!} \right) = 2k \ln \left(\frac{N!}{(N - n)!n!} \right) \qquad (3.3)$$

The Stirling approximation, $\ln N! = N \ln N - N$, can be applied for large numbers of N and n, giving:

$$\Delta S_c = 2k[N \ln N - (N - n) \ln (N - n) - n \ln n] \qquad (3.4)$$

When Equation (3.4) is substituted for ΔS_c in Equation (3.2), the total free energy is given by:

$$\Delta G = n\Delta g - 2kT \left[N \ln \left(\frac{N}{N - n} \right) + n \ln \left(\frac{N - n}{n} \right) \right] \qquad (3.5)$$

and, when the derivative $(\partial\Delta G/\partial n)_{T,P} = 0$, is evaluated with $(N - n) \sim N$ for dilute defects, the concentration of defects is given as:

$$\frac{n}{N} = \exp\left(-\frac{\Delta g}{2kT}\right) = \exp\left(-\frac{\Delta s}{2kT}\right)\exp\left(-\frac{\Delta h}{2kT}\right) \qquad (3.6)$$

The entropy, Δs, is the nonconfigurational entropy that is associated with lattice strains and changes in vibrational frequencies and is much less than the configurational entropy. The most significant conclusion from relation (3.6) is that the defect concentration depends exponentially on the formation of free energy, Δg, and on temperature, T.

Point defect equilibria can be described with the "concept of defect pairing," however this concept only applies to materials with low concentrations of defects, <1%, where interactions between defects may be ignored and the defects may be considered as randomly distributed. In the concept of defect pairing, ω_{vv} is the energy of interaction between a pair of cation vacancies and ω_{ii} the energy of interaction between a pair of interstitial cations. If ω_{vv} and ω_{ii} are negative, then the defects exert an attractive force on each other and there is a tendency to form groups or clusters, which, if sufficiently large, will nucleate a new phase. At a critical composition, the crystal structure will break up into two phases (=saturation of a solution with a solute).

At higher defect concentrations, however, when substantial defect interactions are involved, the concept of point defects is mostly inadequate to explain nonstoichiometry. Large deviations from stoichiometry are accommodated by the formation of clusters of point defects and alternative models involve ordering of point defects to give rise to planar faults or new structural features, which eliminate point defects.

Defects and defect chemical equations are best described using a standard notation, the Kröger–Vink notation [1,2], which contains three parts: the main body identifies whether the defect is a vacancy "V" or an ion such as "Al." The subscript denotes the site that the defect occupies which is either a regular lattice site or an interstitial site, i. The superscript indicates the effective charge of the defect relative to the perfect crystal lattice; a dot represents a positive charge and a dash a negative charge whereas an × is used to indicate neutrality. For instance, $V_{Al}^{///}$ is a vacant aluminium site that has a triple negative effective charge and $Al_{Mg}^{\bullet}$ is an aluminium atom at a magnesium site with a positive effective charge relative to the perfect lattice. The Frenkel reaction for the system AgCl can be written using the Kröger–Vink notation as: $Ag_{Ag}^{X} \rightarrow Ag_i^{\bullet} + V_{Ag}^{/}$, indicating that a silver atom located at its regular site moves to an interstitial site giving a positive effective charge and subsequently generating a vacancy at its regular site that is negatively charged.

Extrinsic defects are formed not only by the presence of impurities or introduction of dopant or solutes into the metal oxide but also by oxidation or reduction processes. Whether a defect structure is considered as intrinsic or extrinsic depends on the relative concentration of intrinsic and extrinsic defects, however, at higher temperature, intrinsic defects are likely to dominate the largely temperature-independent extrinsic defects.

3.2.2 Electronic Defects

The probability that an electron occupies an energy state is given by the Fermi–Dirac distribution function:

$$F(E) = \frac{n_e}{N} = \frac{1}{1 + \exp[(E - E_F)/kT]} \tag{3.7}$$

where n_e is the number of electrons, N the number of available energy states and E_F is the Fermi energy. At 0 K, the probability that energy states which are smaller than E_F are occupied by electrons approaches unity, whereas for energies larger than E_F, the probability is close to zero and the probability function has a step-like shape. At $T > 0$ K, the probability function becomes more diffuse as some electrons are excited to energy levels above E_F (Figure 3.2).

For an intrinsic semiconductor, such as pure silicon, the number of electrons in the conduction band is equal to the number of holes in the valence band and is given by:

$$n = n_e = n_h = n_0 \exp\left(\frac{-E_G}{2kT}\right) \tag{3.8}$$

where n_0 is constant and E_G is the band gap. Given the fact that most oxides have a large band gap and contain high levels of impurities, most electronic defect properties are extrinsic. Extrinsic defects can introduce carriers into localized

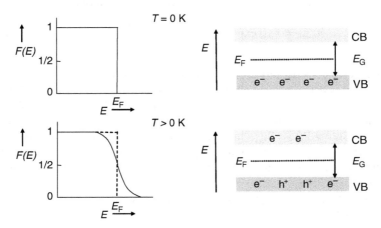

FIGURE 3.2 Fermi–Dirac distribution function and corresponding distribution of electrons and holes in valence and conduction band at $T = 0$ K and $T > 0$ K

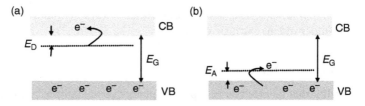

Figure 3.3 (a) Creation of a donor band below the conduction band and (b) acceptor band above the valence band on doping

energy levels within the band gap and, in such cases, are electrically active. An electronic defect in an energy level just below the conduction band edge is considered as a donor because it will donate electrons to the conduction band and therefore increase n-type conductivity. For example, when a phosphorus atom with a valence state of $+5$ is added to silicon that has a $+4$ valence state, the additional electron introduces a donor state. The crucial energy required to excite electrons is now E_D rather than E_G and therefore electrons are more easily excited into the conduction band (Figure 3.3[a]). An acceptor is a dopant that has an energy level near the valence band and hence can accept an electron, giving rise to increased p-type conductivity. An example of acceptor doping is the addition of a gallium atom with valence state $+3$ to silicon, which creates a hole in the valence band and an acceptor level just above the valence band; above the valence band. Electrons are much more easily excited than pure Silicon because E_A is the controlling energy rather than E_G [3], Figure 3.3(b).

For a donor-doped material, the total number of charge carriers is given by:

$$n_{total} = n_e(\text{dopant}) + n_e(\text{intrinsic}) + n_h(\text{intrinsic}) \tag{3.9}$$

$$n_{total} = n_{0,D} \exp\left(\frac{-E_D}{kT}\right) + 2n_0 \exp\left(\frac{-E_G}{2kT}\right) \tag{3.10}$$

In band theory considerations, electrons are strictly delocalized, however, in many materials, it is better to view electronic defects as located at particular sites. The additional electron from the phosphorus donor atom in silicon, for example, could be viewed as not completely delocalized but located on the Si atom according to: $P_{Si}^{/} + Si_{Si}^{\times} \rightarrow P_{Si}^{\times} + Si_{Si}^{/}$. In such cases, the electronic model moves from band-type to polaronic-type model (a polaron is an electronic point defect and is formed when an electron has a bound state in the potential created by a distorted lattice).

From a polaronic viewpoint, the incorporation of an acceptor dopant such as Al^{3+} in $SrTiO_3$ can be written as:

$$Al_2O_3 + 2Ti_{Ti}^{\times} \rightarrow 2Al_{Ti}^{/} + V_O^{\bullet\bullet} + 2TiO_2 \tag{3.11}$$

The negative charge of the acceptor center, Al'_{Ti}, is compensated by the positive charge of the oxygen vacancies [4]. Alternatively, it is possible that the acceptor dopant is electronically compensated by the generation of holes, contributing to p-type conductivity:

$$Al_2O_3 + 2Ti^\times_{Ti} + \tfrac{1}{2}O_2 \rightarrow 2Al'_{Ti} + 2h^\bullet + 2TiO_2 \tag{3.12}$$

A trivalent ion, such as La^{3+}, can be incorporated in $SrTiO_3$ as a donor atom and the positive charge of the donor center can be compensated by negatively charged strontium vacancies (3.13) or by electrons (3.14), giving rise to n-type conductivity:

$$La_2O_3 + 3Sr^\times_{Sr} \rightarrow 2La^\bullet_{Sr} + V''_{Sr} + 3SrO \tag{3.13}$$

$$La_2O_3 + 3Sr^\times_{Sr} \rightarrow 2La^\bullet_{Sr} + 3SrO + 2e' \tag{3.14}$$

Another example is the superconductor, $(Nd_{1-x}Ce_x)_2CuO_4$ with $0 \le x \le 0.20$, of the T'-type structure (a K_2NiO_4-related structure), where the aliovalent substitution of Nd^{3+} by Ce^{4+} is compensated by reduction of some of the Cu^{2+} to Cu^+ and formation of Cu'_{Cu} donor defects:

$$2Cu^\times_{Cu} + 2Nd^\times_{Nd} + 2CeO_2 \rightarrow 2Ce^\bullet_{Nd} + 2Cu'_{Cu} + Nd_2O_3 + \tfrac{1}{2}O_2 \tag{3.15}$$

As a result, superconductivity is induced by donor doping and $(Nd_{1-x}Ce_x)_2CuO_4$ is one of the rare examples of an n-type conducting superconductor, which distinguishes it from other high-T_C superconductors [5].

Electronic defects can also occur on reduction and oxidation of metal ions at different oxygen partial pressures (pO_2). At low pO_2, the material loses oxygen and generates electrons that contribute to n-type conductivity:

$$O_O \leftrightarrow \tfrac{1}{2}O_2 + V^{\bullet\bullet}_O + 2e'' \tag{3.16}$$

At high pO_2, oxygen is incorporated at an oxygen vacancy and takes two electrons from the valence band, leaving two holes to contribute to p-type conduction:

$$\tfrac{1}{2}O_2 + V^{\bullet\bullet}_O \leftrightarrow O_O + 2h^\bullet \tag{3.17}$$

However, redox processes can also lead to the formation of acceptor or donor dopants, for instance, in $SrMnO_{3-\delta}$ where some of the Mn^{4+} are reduced to Mn^{3+} at low pO_2 or high temperatures:

$$2Mn^\times_{Mn} + O^\times_O \rightarrow 2Mn'_{Mn} + V^{\bullet\bullet}_O \tag{3.18}$$

Alternatively, in $LaMnO_3$, Mn^{3+} can be oxidized to Mn^{4+}, hence leading to the formation of a donor dopant, $Mn_{Mn}^{\bullet}$, and the formation of a Mn vacancy:

$$3Mn_{Mn}^{\times} \rightarrow 3Mn_{Mn}^{\bullet} + V_{Mn}^{///} \tag{3.19}$$

In some metal oxides such as barium cerate-based perovskites, water can be incorporated at oxygen vacancies as well as oxygen atoms, leading to the formation of singly charged protonic defects:

$$H_2O + V_O^{\bullet\bullet} + O_O^{\times} \rightarrow 2OH_O^{\bullet} \tag{3.20}$$

At low temperatures, the holes are effectively trapped by the acceptor centers and electrons are trapped by donor centers:

$$A^{/} + h^{\bullet} \leftrightarrow A^{\times} \tag{3.21}$$

$$D^{\bullet} + e^{/} \leftrightarrow D^{\times} \tag{3.22}$$

where $A^{\times}$ is an acceptor with a trapped hole and $D^{\times}$ is a donor with a trapped electron. In the band model, a similar localization occurs that can be described in, for example, Equation (3.9). Holes can also be trapped by vacancies that occur in NiO where a hole may be trapped by a doubly charged nickel vacancy, resulting in the formation of a singly charged nickel vacancy:

$$V_{Ni}^{//} + h^{\bullet} \leftrightarrow V_{Ni}^{/} \tag{3.23}$$

A consequence of the trapping of electronic defects is that materials become insulating at low temperatures. Defect association can also occur between oxygen vacancies and acceptor dopants, as it was observed in Y-doped zirconia, where $V_O^{\bullet\bullet}$ are trapped by $Y_{Zr}^{/}$ defects at low temperature, which led to a drastic decrease in oxide ion conductivity [6–8]. Similarly, protonic defects are likely to form associations with acceptor dopants, which hence would limit protonic conduction in metal oxides [9].

As all metal oxides contain impurities, the defect mechanisms of "pure" oxides may subsequently be similar to that of doped oxides. For example, "pure" strontium titanate, $SrTiO_3$, has a similar defect chemistry to acceptor-doped $SrTiO_3$, because the impurities are usually metals that occur in lower oxidation states such as Fe^{3+}, Al^{3+}, Mg^{2+}, Na^{1+}, etc., and their concentrations, typically around 10^2 to 10^3 ppm, are sufficient to dominate the defect chemistry, especially at higher pO_2.

Defect concentrations vary with temperature, solute concentration, and gas activity (usually pO_2). In order to establish defect equilibria, it is always necessary to establish bulk electrical neutrality. For example, intrinsic MgO has four dominant defects, $V_{Mg}^{//}$, $V_O^{\bullet\bullet}$, $e^{/}$, and $h^{\bullet}$ with the following defect-forming reaction:

$$\text{null} \rightarrow V_{Mg}^{//} + V_O^{\bullet\bullet} \tag{3.24}$$

$$\text{null} \rightarrow e^{/} + h^{\bullet} \tag{3.25}$$

$$O_O^{\times} \rightarrow \tfrac{1}{2}O_2 + V_O^{\bullet\bullet} + 2e^{/} \tag{3.26}$$

The equilibrium constants for reactions (3.24), (3.25), and (3.26) are given by:

$$K_S = [V_{Mg}^{//}][V_O^{\bullet\bullet}] \tag{3.27}$$

$$K_i = np \tag{3.28}$$

$$K_R = n^2[V_O^{\bullet\bullet}]\sqrt{pO_2} \tag{3.29}$$

where K_S is the equilibrium constant for the formation of the Schottky pair, K_i the equilibrium constant for the formation of intrinsic electronic defects, and K_R the equilibrium constant for the formation of oxygen vacancies on reduction. The electroneutrality condition is given by:

$$2[V_{Mg}^{//}] + n = 2[V_O^{\bullet\bullet}] + p \tag{3.30}$$

This system of equations, (3.24) to (3.30), can now be solved for each defect, provided that the equilibrium constants are known. If a single defect of each sign has a concentration that is much higher than that of other defects of the same sign, the electroneutrality condition can be simplified using the Brouwer approximation in which there is just one positive and one negative defect. For MgO, four Brouwer approximations are principally possible: (i) $n = 2V_O^{\bullet\bullet}$, (ii) $2V_{Mg}^{//} = p$, (iii) $[V_{Mg}^{//}] = [V_O^{\bullet\bullet}]$, and (iv) $n = p$. The intrinsic Schottky concentration (iii) is usually greater than the electronic charge carrier concentration (iv), hence only three Brouwer regimes — (i), (ii), and (iii) — have to be considered. Brouwer regime (i) applies at low pO_2 because reduction increases the concentration of oxygen vacancies and electrons. At high pO_2, the electron concentration is decreased and the hole concentration is increased, hence Brouwer approximation (ii) becomes important. Variation in defect concentration with ambient gas activity is usually displayed in a Brouwer diagram, which plots log [concentration] against log pO_2 at a constant temperature (Figure 3.4).

For Brouwer region (iii), which is located in the intermediate pO_2 region, the concentration of both vacancies is constant, $[V_{Mg}^{//}] = [V_O^{\bullet\bullet}] = \sqrt{K_S}$. When we apply Equation (3.29), the pO_2 dependence of the electron concentration in region (iii) can be obtained:

$$n = K_R^{1/2}K_S^{-1/2}pO_2^{-1/4} \tag{3.31}$$

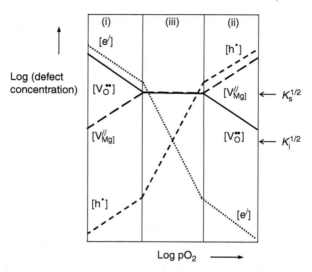

FIGURE 3.4 Brouwer diagram for MgO with three distinguished Brouwer regions: defect concentration as a function of pO_2 [4]

In the diagram, the line representing n therefore has a slope of $-\frac{1}{4}$. From Equation (3.28), the hole concentration in region (iii) must hence be given by:

$$p = \frac{K_i}{n} = K_i K_R^{-1/2} K_S^{1/2} pO_2^{1/4} \tag{3.32}$$

and therefore, the plot for p has a slope of $+\frac{1}{4}$. As discussed before, under reducing conditions, Brouwer approximation (i) can be applied and by substituting (i) in Equation (3.29), the concentration of electrons at low pO_2 can be obtained:

$$n = 2[V_O^{\bullet\bullet}] = 2K_R^{1/3} pO_2^{-1/6} \tag{3.33}$$

and from this relation, the lines that represents n and $[V_O^{\bullet\bullet}]$ have a slope of $-\frac{1}{6}$. Similarly, at high pO_2, Brouwer approximation (ii) becomes most relevant and substitution of (ii) in (3.29) gives a $+\frac{1}{6}$ slope of the lines presenting p and $[V_{Mg}^{//}]$. Since Brouwer diagrams are isothermal, the dependence of defect concentrations on the temperature can be obtained from the equilibrium constant:

$$K = \exp\left(-\frac{\Delta G}{kT}\right) = \exp\left(\frac{\Delta S}{k}\right) \exp\left(-\frac{\Delta H}{kT}\right) \tag{3.34}$$

From this, the defect concentration varies exponentially with temperature and the activation energy for different defect regimes can be calculated. In MgO, for instance, the concentration of electrons has an activation energy of

$E_A = (\Delta h_R/2k - \Delta h_S/2k)$ in region (iii) and in region (i), the activation energy is given by $E_A = (\Delta h_R/3k)$, with Δh_j representing the enthalpy of the corresponding defect formation reaction, (3.24) or (3.26) [4].

3.2.3 Nonstoichiometry

Simple and complex metal oxides based on elements of the first transition series display a wide variety of nonstoichiometric phenomena, which have their origin in the unfilled 3d electron shell.

Simple metallic oxides, that is, oxides formed from a single element, MO, MO_2, MO_3, etc., are rarely stoichiometric and gross deviations from nominal cation or anion stoichiometry can occur. For instance, up to 18% of the Fe sites can be vacant in FeO with the rocksalt structural framework being retained over the entire range $Fe_{1-x}O, 0.05 \leq x \leq 0.18$ [4].

Nonstoichiometry in perovskite oxides can arise from cation deficiency in A or B sites or oxygen deficiency or oxygen excess. A classical example for A-cation deficient perovskites is tungsten bronzes, A_xWO_3. In A_xWO_3, A-cations (typically alkali ions) can be missing either partially or wholly and the resulting structures are therefore intermediate between the ABO_3 perovskite structure and the ReO_3 structure.

The A-cation content has a great influence on the properties of the materials. WO_3 is insulating; at low x, A_xWO_3 is semiconducting and at high x, it becomes metallic [10]. The B–O interactions are usually stronger than the A–O interactions, and B-site vacancies in an ABO_3 lattice are therefore energetically unfavorable. Examples for B-site deficient perovskites are $Ba_2Sm_{1-x}UO_6$ and $Ba_2Ce_{1-x}SbO_6$ [11]. In general, B-site vacancies are much more favored in structures with AO_3 close-packed layers in hexagonal stacking sequence rather than in a cubic arrangement, giving rise to sequences of corner-sharing and face-sharing octahedra and hexagonal polytype structures (Figure 3.5). Examples are the homologous series $(Ba,La)_nTi_{n-1}O_{3n}$ [12] and $BaCo_{1-x}O_3$ [13].

Oxygen-deficient perovskites have attracted much attention because some desirable physical properties of such a material can depend directly on the oxygen stoichiometry. Oxygen stoichiometry depends on temperature and pO_2 and physical properties can therefore easily be controlled by reduction or oxidation treatments. If the cation framework remains intact, such redox processes are highly reversible and render the family of oxygen-deficient perovskite very attractive as oxidation catalysts and oxygen electrode materials. Depending on the nature of the B-cation and the overall concentration of oxygen defects, various arrangements of oxygen vacancies are possible. BO_6 octahedra may be reduced to lower-coordinated polyhedra, such as BO_5 squared pyramids, BO_4 tetrahedra, and BO_4 planar squares as well as linear BO_2 arrangements. The reduced $BO_{6-\delta}$ polyhedra may order giving rise to large supercells. Oxygen defects may also cluster and form locally segregated regions or microdomains. For example, in $WO_{3-\delta}$-type structures, oxygen vacancies are eliminated by the formation of crystallographic shear (CS) planes where octahedra are linked through edges rather than

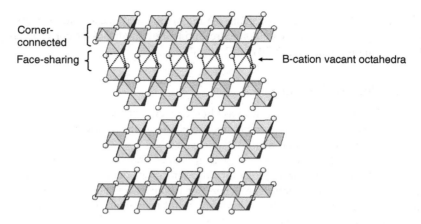

Corner-connected {

Face-sharing {

← B-cation vacant octahedra

Figure 3.5 Alternating sequences of face-sharing and corner-connected octahedra in hexagonal perovskite, $La_4Ti_3O_{12}$

corners [14]. Two well-characterized oxygen-deficient systems are $CaMnO_{3-\delta}$ and $CaFeO_{3-\delta}$ [15–18]. In the fully oxidized compositions, $CaMnO_{3.0}$ and $CaFeO_{3.0}$, the transition metal is in valence state +4.0. In both systems, the transition metal can be reduced to valence state +3.0 and octahedral frameworks can accommodate up to ~17% oxygen vacancies without decomposition of the perovskite structure. However, the two systems differ greatly in the arrangement of oxygen vacancies and resulting superstructures: in oxygen-deficient $CaMnO_{3-\delta}$ phases, Mn^{4+} cations are reduced to Mn^{3+} and MnO_6 octahedra transformed into MnO_5 square pyramids (Figure 3.6). Five different structures with ordered vacancies have been reported for different values of $(3 - \delta)$ and intermediate compositions are complicated intergrowths of oxidized and reduced structures. In $CaFeO_{3-\delta}$, FeO_6 octahedra are reduced to FeO_4 tetrahedra rather than FeO_5 square pyramids and the crystal structure of $Ca_2Fe_2O_5$ ($=2CaFeO_{2.5}$) is built up from chains of alternating octahedra and tetrahedra, running parallel to the b-axis with Ca atoms in the holes between the polyhedra (Figure 3.6). This structure is known as the brownmillerite structure, named after the mineral, Ca_2AlFeO_5. Partially reduced compositions of ferrite perovskite oxides contain $Ca_2Fe_2O_5$ brownmillerite microdomains, which are intergrown three-dimensionally with the parent structure. Individual domains can be as large as ~10^6 Å^3 and decrease with increasing oxygen content [19].

$ACoO_{3-\delta}$ is an oxygen-deficient system that, depending on the A-cation, may exhibit different ordering schemes of oxygen vacancies; when A = Ca, CoO_6 octahedra are reduced to CoO_5 squared pyramids and the reduced compositions have structures similar to that obtained for $CaMnO_{3-\delta}$. However, in $SrCoO_{3-\delta}$, oxygen vacancies order in brownmillerite fashion. The $Sr_{1-x}La_xCo_{1-y}Fe_yO_{3-\delta}$ series with brownmillerite-type oxygen defects exhibits high mixed electronic/oxygen ion conductivities and members of this series are promising candidates for application in solid oxide fuel cells [20]. The number and the arrangement of oxygen vacancies

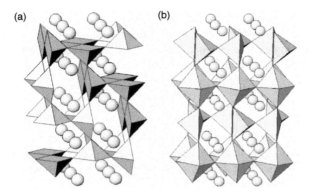

FIGURE 3.6 Ordering of oxygen vacancies in nonstoichiometric perovskite oxides. (a) $CaMnO_{2.5}$ and (b) Brownmillerite, $CaFeO_{2.5}$

have direct consequences for the magnitude of oxygen transport. In general, large unit cells and open channels such as those present in the brownmillerite structure increase carrier mobilities, while a high concentration of randomly distributed (disordered) oxygen vacancies lead to high carrier densities.

In perovskite oxides, a small oxygen excess is accommodated by the creation of cation vacancies at A- and B-sites leaving an intact oxygen sublattice. $LaMnO_{3+\lambda}$ is a system where such "oxygen excess" is described by the formation of A- and B-cation vacancies. The Mn^{4+} content can be as high as 33%, then the formula will approximately correspond to $La_{0.945}Mn_{0.945}O_3$; however, $La_{1-\delta}MnO_3$ and $LaMn_{1-\delta'}O_3$ have also been observed [21,22]. Where there is a large oxygen excess, layered perovskite-related structures are formed. The cubic perovskite is sliced parallel to (110) to give slabs containing n {110} perovskite layers joined by CS along the $(100)_p$ direction forming a homologous series, $A_nB_nO_{3n+2}$ with $n = 4 - \alpha$ [23].

3.3 DIFFUSION AND ELECTRICAL CONDUCTIVITY IN METAL OXIDES

We now proceed by describing the diffusion of defects, and hence their contribution to conductivity in metal oxides. Fick's first law can be used to relate a diffusive flux to a concentration gradient,

$$j_x = -D\frac{\partial C}{\partial x} \tag{3.35}$$

where j_x is the flux in the x-direction and D is the diffusion coefficient that depends upon jump frequency, distance of the jump and whether such jumps are correlated or not (i.e., whether the probability of a jump taking place depends upon the previous path of the particle). To determine the diffusion coefficient, a concentration gradient may be imposed and the flux measured. This can be achieved, for example, using a membrane configuration. Alternatively, the response of the material to a change in concentration can be monitored as a function of time. Such an experiment is described by Fick's second law of diffusion,

$$\frac{\partial C}{\partial t} = D\frac{\partial^2 C}{\partial x^2} \tag{3.36}$$

First, let us consider the case of random diffusion. Consider two parallel planes in a solid with a separation distance between them, ℓ. The concentration of the diffusing species in Plane 1 is C_1 and in Plane 2 is C_2. If there are n jumps for each particle in unit time, t,

$$\frac{n}{t} = f \tag{3.37}$$

where f is the jump frequency. If diffusion is a random process then there is a 50% probability that a jump from Plane 1 will end in Plane 2 (in an infinite medium) and vice versa. Therefore, the net flux between the two planes can be given by,

$$j = \tfrac{1}{2}C_1\ell f - \tfrac{1}{2}C_2\ell f \tag{3.38}$$

However,

$$C_2 = C_1 + \ell\frac{dC}{dx} \tag{3.39}$$

and therefore,

$$j = -\frac{1}{2}f\ell^2\frac{dC}{dx} \tag{3.40}$$

Hence, for random diffusion in one dimension the diffusion coefficient is,

$$D_{r1} = \tfrac{1}{2}f\ell^2 \tag{3.41}$$

where the subscript 'r' refers to the random nature of the diffusion and the subscript '1' refers to one-dimensional diffusion. For three-dimensional diffusion it can easily be shown that,

$$D_{r3} = \tfrac{1}{6}f\ell^2 \tag{3.42}$$

Diffusive fluxes can also be expressed in terms of a chemical potential driving force. Consider a species moving through the lattice with a drift velocity, v, as a result of a driving force, $d\mu/dx$. Then,

$$v = -m\frac{d\mu}{dx} \tag{3.43}$$

where m is a mobility coefficient. The flux then depends upon the concentration of the species multiplied by the drift velocity,

$$j = -mC\frac{d\mu}{dx} \tag{3.44}$$

The diffusion coefficient (for random diffusion) is given by,

$$D = mRT \tag{3.45}$$

and therefore,

$$j = -\frac{D}{RT}C\frac{d\mu}{dx} \tag{3.46}$$

The chemical potential of the diffusing species can be related to its activity in the usual way,

$$\mu = \mu^0 + RT\ln a \tag{3.47}$$

and therefore,

$$j = -DC\frac{d\ln a}{dx} = -DC\frac{d\ln a}{d\ln C}\frac{dC}{dx} \tag{3.48}$$

For a dilute solution the activity approaches the value of the concentration and therefore,

$$\frac{d\ln a}{d\ln C} = 1 \tag{3.49}$$

and Fick's first law is recovered.

As these considerations describe diffusion in terms of chemical potential driving forces, the approach can be used to describe diffusion in nonideal systems provided that the chemical potential of the diffusing species can be adequately described (such nonidealities are beyond the scope of this chapter).

If Equation (3.46) is rewritten in terms of the electrochemical potential driving force, $d\bar{\mu}/dx$, we can investigate the effect of an electrical field on the migration of ionic defects in the solid,

$$j = -\frac{D}{RT}C\frac{d\bar{\mu}}{dx} \tag{3.50}$$

The electrochemical potential for an ion of charge, z, is related to the potential, ϕ, by,

$$\bar{\mu} = \mu + zF\phi \tag{3.51}$$

and therefore,

$$j = -\frac{D}{RT}C\frac{\mathrm{d}\mu}{\mathrm{d}x} - \frac{D}{RT}CzF\frac{\mathrm{d}\phi}{\mathrm{d}x} \tag{3.52}$$

If we assume a uniform composition for the solid, that is, uniform chemical composition, then there is no diffusive flux and the current density is given by,

$$i = zFj = -\frac{D}{RT}Cz^2F^2\frac{\mathrm{d}\phi}{\mathrm{d}x} \tag{3.53}$$

Conductivity, σ, is equal to current density divided by the electrical field gradient,

$$\sigma = \frac{i}{-(\mathrm{d}\phi/\mathrm{d}x)} = \frac{D}{RT}Cz^2F^2 \tag{3.54}$$

If we employ an electrical mobility, u,

$$u = \frac{zF}{RT}D \tag{3.55}$$

then,

$$i = zFj = -zFuC\frac{\mathrm{d}\phi}{\mathrm{d}x} \tag{3.56}$$

and

$$\sigma = zFuC \tag{3.57}$$

or for an individual specie,

$$\sigma_i = z_i Fu_i C_i \tag{3.58}$$

The total conductivity, σ, of the material depends upon the sum of all individual conductivities,

$$\sigma = F\sum_i z_i u_i C_i \tag{3.59}$$

Hence, individual conductivities can be expressed as a fraction of the total conductivity by using a transport number, t_i,

$$\sigma_i = t_i \sigma \tag{3.60}$$

Let us now consider the diffusion mechanism. A vacancy diffusion mechanism involves an atom or an ion at a normal site hopping into an unoccupied site or vacancy. An interstitial mechanism involves the ion hopping from one interstitial site to another identical interstitial site (there are other diffusional mechanisms but they are beyond the scope of this chapter). Here, for the purpose of illustration we will focus on the vacancy mechanism as this is believed to be responsible for diffusion in many oxides. Diffusion of oxygen in the lattice can only occur if the oxygen species are adjacent to an oxygen vacancy. In, for example, a simple binary oxide material the fraction of vacancies in the material is given by:

$$\text{fraction of vacancies} = \frac{C_V}{C'} \tag{3.61}$$

where C_V is the oxygen vacancy concentration (here we use C_V as the vacancy concentration and not $[V_O^{\bullet\bullet}]$ as is commonly used when discussing defect equilibria — see Section 3.1) and C' is the concentration of oxygen lattice sites. The fraction of occupied sites is then given by:

$$\text{fraction of occupied sites} = 1 - \frac{C_V}{C'} \tag{3.62}$$

The diffusion of the oxygen species in the lattice depends upon the concentration of vacancies and the number of nearest neighbors, N,

$$D_O = \frac{1}{6} f \ell^2 N \frac{C_V}{C'} \tag{3.63}$$

For an fcc structure this reduces to,

$$D_O = fa^2 \frac{C_V}{C'} \tag{3.64}$$

as there are six nearest neighbors and the lattice parameter, a, is equal to the jump distance (the same expression is obtained for a bcc structure). Likewise, the diffusion coefficient for oxygen vacancies is given by:

$$D_V = fa^2 \left(1 - \frac{C_V}{C'} \right) \tag{3.65}$$

Note that,

$$D_V C_V = D_O(C' - C_V) = D_O C_O \tag{3.66}$$

In Section 3.1, we have already discussed how the vacancy concentration depends upon temperature, solute concentration and gas activity. However, the jump frequency will also depend upon temperature. A simple application of transition state theory to the hopping process gives,

$$f = f' \exp\left(\frac{-\Delta G_h}{RT}\right)$$ (3.67)

where ΔG_h is the change in free energy required to form the transition state during the hopping process.

Until recently, we have assumed that diffusion is a random process. However, in real experiments diffusion coefficients are often determined by the use of isotopic tracers. If the tracer (in this case an oxygen isotope) has jumped into a vacancy it is clearly in the right environment to jump immediately back to the unoccupied vacancy that it leaves behind. Such correlated jumps serve to reduce the diffusion coefficient to a value lower than that which would be expected for a random walk. The resulting diffusion coefficient in the presence of such correlated jumps is known as the tracer diffusion coefficient (as its value can be determined from tracer experiments) or the self-diffusion coefficient.

Such tracer diffusion coefficients tend to be determined through dynamic experiments rather than steady state experiments. Accordingly, one must solve Fick's second law with appropriate boundary conditions, to obtain the tracer diffusion coefficient, D_t,

$$\frac{\partial C}{\partial t} = D_t \frac{\partial^2 C}{\partial x^2}$$ (3.68)

These boundary conditions will usually include an initial condition, a symmetry condition and a surface flux condition that will depend upon the surface exchange coefficient and a surface, a surface rate constant controlling the rate of exchange between the tracer in the solid and the tracer in the surrounding atmosphere. The tracer diffusion coefficient is related to the diffusion coefficient through a correlation factor or the Haven ratio (the tracer diffusion coefficient is equal to the diffusion coefficient divided by the Haven ratio).

3.4 Mixed Ionic and Electronic Conducting Materials

Although oxides have a wide range of catalytic applications their transport properties are most obviously critical when they are used in the form of a membrane within a chemical or electrochemical reactor. As such their ionic conductivity must be high if they are going to support a reasonable ion flux. Such materials fall broadly into two classes: those materials that exhibit a very low electronic conductivity and, if the electronic transport number is <0.01, are generally termed solid electrolytes (solid electrolytes are covered in a separate chapter); and those materials that exhibit an appreciable or high electronic conductivity as well as ionic conductivity and are hence termed mixed conductors. In the rest of this chapter we will focus on such mixed ionic and electronic conducting (MIEC) materials. First, we will address transport in MIEC membranes from a theoretical perspective before reviewing work that has been performed on preparing and characterizing such membranes.

3.4.1 Transport in MIEC Membranes

In order for permeation of oxygen to take place through a ceramic membrane, adsorption and dissociation of oxygen molecules must occur on the surface at active adsorption sites. The adsorbed oxygen must be reduced and incorporated into the crystal lattice as oxide anions. Such surface processes can be lumped together and described by an oxygen exchange coefficient, k. Similar processes will also occur on the oxygen-evolving side of the membrane. Under steady operation, the flux of oxygen, j_O, through a membrane surface can be expressed in terms of a partial pressure, chemical potential or concentration driving force, for example,

$$j_O = k(C_{O,g} - C_O) \tag{3.69}$$

where C_O is the solid-state oxygen concentration at the surface and the subscript g is used to denote a virtual solid-state oxygen concentration that would be in equilibrium with the oxygen in the gas phase.

The MIEC membrane itself exhibits both ionic and electronic conductivity with the total conductivity given by the sum of these two contributions,

$$\sigma = \sigma_{ion} + \sigma_e \tag{3.70}$$

Substituting (3.54) into (3.50) and expressing for individual specie,

$$j_i = -\frac{\sigma_i}{z_i^2 F^2} \frac{d\bar{\mu}_i}{dx} \tag{3.71}$$

therefore, the individual current density is given by:

$$i_i = -\frac{\sigma_i}{z_i F} \frac{d\bar{\mu}_i}{dx} \tag{3.72}$$

The ionic current can be described by:

$$i_{ion} = -\frac{\sigma_{ion}}{z_{ion} F} \frac{d\bar{\mu}_{ion}}{dx} \tag{3.73}$$

and the electronic current, if carried by electrons (n-type conductor),

$$i_e = \frac{\sigma_e}{F} \frac{d\bar{\mu}_e}{dx} \tag{3.74}$$

In the absence of an external electrical circuit,

$$i_{ion} + i_e = 0 \tag{3.75}$$

and therefore,

$$\frac{\sigma_{ion}}{z_{ion}F}\frac{d\bar{\mu}_{ion}}{dx} = \frac{\sigma_e}{F}\frac{d\bar{\mu}_e}{dx} \tag{3.76}$$

In oxygen-deficient perovskites (general formula of $ABO_{3-\delta}$) the ion flux is carried by oxygen-ion vacancies and we may write the above equation in terms of oxygen vacancy conductivity, σ_V, and vacancy electrochemical potential, $\bar{\mu}_V$,

$$\frac{\sigma_V}{2F}\frac{d\bar{\mu}_V}{dx} = \frac{\sigma_e}{F}\frac{d\bar{\mu}_e}{dx} \tag{3.77}$$

Defect concentrations are determined by reaction between gas-phase oxygen, oxygen vacancies, and electrons on both sides of the membrane through the reaction,

$$O_2 + 2V_O^{\bullet\bullet} + 4e' = 2O_O \tag{3.78}$$

The oxygen chemical potential difference across the membrane results in an electrochemical potential difference in oxygen-ion vacancies and electrons such that,

$$\Delta\mu_{O_2} = 2\Delta\mu_O = -2\Delta\bar{\mu}_V - 4\Delta\bar{\mu}_e \quad \text{or} \quad d\mu_{O_2} = 2\,d\mu_O = -2\,d\bar{\mu}_V - 4\,d\bar{\mu}_e \tag{3.79}$$

Combining with Equation (3.77), we obtain,

$$d\bar{\mu}_V = -\frac{1}{2}\frac{\sigma_e}{\sigma}d\mu_{O_2} \tag{3.80}$$

and therefore using Equation (3.77),

$$\frac{\sigma}{\sigma_{ion}}d\bar{\mu}_e = -\frac{1}{4}d\bar{\mu}_{O_2} \tag{3.81}$$

It should be noted that the emf developed across the membrane depends upon $\Delta\bar{\mu}_e$. Equation (3.81) can therefore, on integration, be used to calculate the emf. This is simply performed if the ionic and electronic transport numbers are constant and results in,

$$E = t_{ion}E_{elec} \tag{3.82}$$

where E_{elec} is the emf developed in the case of a solid electrolyte membrane with an ion transport number of unity.

The flux across the membrane can now be evaluated,

$$j_i = -\frac{\sigma_i}{z_i^2 F^2}\frac{d\bar{\mu}_i}{dx} \tag{3.83}$$

Therefore, the ionic flux is given by,

$$j_V = -\frac{\sigma_V}{4F^2}\frac{d\bar{\mu}_V}{dx} \tag{3.84}$$

or, using Equation (3.80),

$$j_V = \frac{1}{8F^2}\frac{\sigma_V\sigma_e}{\sigma}\frac{d\bar{\mu}_{O_2}}{dx} \tag{3.85}$$

We now assume that the electron transport number approaches unity. Substituting for electrochemical potential, in the absence of a gradient in electrical potential, in terms of pO_2 with the electron transport number at unity gives,

$$j_V = \frac{RT}{8F^2}\sigma_V\frac{d\ln P_{O_2}}{dx} \tag{3.86}$$

Expressing the above equation in terms of an oxygen-vacancy concentration driving force,

$$j_V = \frac{RT}{8F^2}\frac{\sigma_V}{C_V}\frac{d\ln P_{O_2}}{d\ln C_V}\frac{dC_V}{dx} \tag{3.87}$$

$(d\ln P_{O_2})/(d\ln C_V)$ can be evaluated from the defect equilibria or from experimental determination of the degree of nonstoichiometry and is known as the thermodynamic factor. For an oxygen deficient material with local equilibrium of reaction (3.78) this factor is given by (see Equation [3.33]),

$$\frac{d\ln P_{O_2}}{d\ln C_V} = -6 \tag{3.88}$$

Expressing conductivity in terms of the vacancy diffusion coefficient (see Equation [3.54]),

$$\sigma_V = \frac{D_V}{RT}4F^2 C_V \tag{3.89}$$

and substituting Equations (3.88) and (3.89) in Equation (3.87) gives,

$$j_V = -3D_V\frac{dC_V}{dx} \tag{3.90}$$

The chemical diffusion coefficient, $\tilde{D}$, determined by oxygen permeation experiments is defined by,

$$j_O = -\tilde{D}\frac{dC_O}{dx} = -j_V = \tilde{D}\frac{dC_V}{dx} \tag{3.91}$$

and the chemical diffusion coefficient is three times the vacancy diffusion coefficient,

$$\tilde{D} = 3D_V \tag{3.92}$$

The oxygen diffusion coefficient can then be calculated from the vacancy diffusion coefficient using Equation (3.66).

3.4.2 Fabrication and Characterization of MIEC Membranes

The acceptor-doped perovskites (general formula of $ABO_{3-\delta}$) have shown much promise as MIECs. Much attention has been focused on these versatile metal oxides as they exhibit catalytic activity [24] and therefore their use in some applications can avoid the need for catalytic modification of the membrane surface. This activity has meant that MIECs can be used as electrode materials in solid-oxide fuel cell systems for the reduction of oxygen; simultaneously, this inherent catalytic activity means that problems with chemical stability can arise.

When used as a dense membrane in a reactor system the MIEC acts as a barrier between two isolated chambers allowing only the ionically conducted species to pass through the membrane under a chemical potential gradient. Air can be used as a freely available oxidant that is supplied to one chamber with no mixing of nitrogen with the product stream from an oxidation reaction as depicted in Figure 3.7,

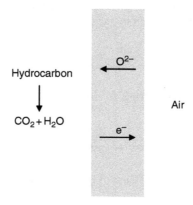

FIGURE 3.7   Schematic of a MIEC membrane employed in an oxidation reaction

thereby eliminating the need to have a separate and expensive cryogenic oxygen production unit leading to reduction in the overall system cost.

The most important consideration for a MIEC membrane is the delivery of a stable continuous transmembrane flux. This flux is central to membrane performance and therefore oxygen permeation studies in MIEC membrane characterization are fundamental. From an economic standpoint, an oxygen flux of 1 to 10 ml cm^{-2} min^{-1} (STP) has been cited as the requirement for future needs [25,26]. In order to provide a broad overview of this research area we will briefly outline synthesis and characterization methods applied to the MIEC perovskites before moving on to look at selected MIEC membranes currently used in laboratory-scale tests.

The perovskite-type metal oxides can be synthesized by a number of different routes. The solid-state reaction uses high purity oxides or carbonates of the component metal elements, which are thoroughly mixed to obtain a homogeneous powder. The powder is then fired to promote the solid-state reaction of the oxides. Another common method is the sol–gel route that involves the production of a gel from a liquid followed by calcination to form the product. Metal nitrates or alkoxides of the constituent elements are dissolved in water and a chelating agent is added. The mixture is heated and stirred causing polymerization followed by condensation. Evaporation of the water causes the gel to form, which is then heated to remove the organics leaving the ceramic powder product. Coprecipitation is a well-established method of preparing mixed oxides. An aqueous solution of the metal cations is mixed with another solution that contains a precipitating agent. The precipitated product is filtered, dried, and thermally decomposed to isolate the ceramic powder. The novel solid-state combustion route can be subcategorized into the self-propagating high temperature and volume combustion routes. A pellet of the requisite metal oxides is ignited by an external source, the combustion wave passes through the pellet producing the ceramic powder. In spray pyrolysis, a solution of the metal nitrates of the required stoichiometry is fed as a fine spray into a reaction chamber. Spray drying involves the rapid vaporization of a solution of the metal cations while in freeze drying fine droplets are sprayed into liquid nitrogen followed by slow sublimation of the solvent. These methods allow control of impurities and produce fine homogeneous particles. Other more specialized and less common methods include hydrothermal methods, plasma spraying, electron beam evaporation, and arc vaporization.

X-ray diffraction is invariably employed for determination of crystallinity and phase purity of the ceramic powder. Scanning electron microscopy (SEM) and transmission electron microscopy (TEM) are used to determine particle size and morphology of the ceramic powder. Thermogravimetric analysis (TGA) allows one to obtain the degree of oxygen nonstoichiometry, δ, of the perovskite. Less common techniques include solid-state nuclear magnetic resonance (NMR), which has been used to calculate the chemical diffusion coefficient, $\tilde{D}$ [27], to observe oxygen vacancy motion [28] and for structural investigations [29–31].

The three basic shaping processes for ceramics powders are pressing, casting, and plastic forming. Die pressing consists of the uniaxial compaction of a ceramic powder while confined in a compression die and is the most widely used shaping technique for laboratory scale studies such as oxygen permeation and conductivity tests. Development of this simple method is isostatic pressing that produces a membrane that is more uniformly compacted, hot isostatic pressing gives additional densification and increased membrane strength. In slip casting, a low viscosity suspension of micrometer size ceramic particles is poured into a porous mould. For solid ceramic objects the slip is supplied continuously into the mould and the suspension is left as a soft solid. Tape casting facilitates the production of ceramic sheets, again a slip is used, which is poured onto a rolling film. Plastic forming with extrusion processes involves forcing a plastic mix of a ceramic powder through a constricting die to produce elongated shapes of constant cross-section. Following shape forming the membrane undergoes densification via sintering, this process also gives rise to grain boundaries [32,33] between the individual crystals (or "grains"). SEM can give the grain size distribution, show the grain boundaries, reveal imperfections such as cavities and porosity of the membrane surface, and estimate surface areas.

Diffusion coefficients and surface exchange coefficients that govern oxygen flux can be calculated by a number of techniques. A commonly used method is the ^{18}O isotope tracer method coupled with secondary ion mass spectroscopy (SIMS). ^{18}O is incorporated into the membrane sample that is sectioned and the ^{18}O diffusion profile is measured by SIMS line scanning [34]. This gives the tracer diffusion coefficient which can then be converted to the diffusion coefficient. The surface exchange coefficient is more difficult to obtain as it involves many processes and is very sensitive to surface conditions. AC impedance spectroscopy (IS) can be used to obtain conductivities of membrane samples but application of this method to MIECs requires the use of selective blocking electrodes. This technique has been used to obtain chemical diffusion and surface exchange coefficients [35]. Conductivity relaxation involves changing the oxygen partial pressure surrounding a sample and monitoring the change in electrical conductivity versus time and can yield chemical diffusion coefficients and surface exchange coefficients. The solution of Fick's second law is then used to obtain the chemical diffusion coefficient.

However, the direct observation of membrane performance is provided by oxygen flux measurements. At elevated temperatures, oxygen permeates through the membrane under an oxygen chemical potential gradient. The oxygen permeate leaves the cell and is then analyzed, usually by GC, to obtain the oxygen flux. The results of recent oxygen permeation studies using perovskite-type MIEC membranes, including salient experimental conditions, are summarized in Table 3.1 for ease of comparison. We have quoted the molar flux where fluxes at standard temperature and pressure (STP) have been reported in the literature; in a number of articles fluxes are not quoted at STP and so we omit using the molar flux unless quoted by the authors.

Various members of the La–Sr–Co–Fe oxide system have received great attention with particular interest in the robust $La_{0.6}Sr_{0.4}Co_{0.2}Fe_{0.8}O_{3-\delta}$ (LSCF6428) (note that the chemical formula of complex perovskites can be abbreviated by using the first letter of the chemical symbol of each of the metal constituent

TABLE 3.1
Data Summary of Cited Perovskite MIECs in Section 3.3.2

Perovskite-type	Membrane form	Thickness (mm)	Temperature (°C)	Flux JO_2^a ml cm^{-2} min^{-1}	mol cm^{-2} sec^{-1}	$\bar{D}$ cm^2 sec^{-1}	k_{chem} cm sec^{-1}	Atmospheres used low pO$_2$ | high pO$_2$	References
La$_{0.6}$Sr$_{0.4}$Co$_{0.2}$Fe$_{0.8}$O$_{3-\delta}$	Disk	0.96	800 and 1000	0.04	NA	10^{-6}–10^{-5}	10^{-5}–10^{-4}	He | Air	36
La$_{0.6}$Sr$_{0.4}$Co$_{0.2}$Fe$_{0.8}$O$_{3-\delta}$	Tabular	1.5	900	0.13	0.1	NA	NA	He | Air	37
LSCF6428	Disk	1–2	600–800	0.1–0.2	0.07–0.14	NA	NA	He | Air	38
asymmetric membrane									
La$_{0.6}$Sr$_{0.4}$Co$_{0.2}$Fe$_{0.8}$O$_{3-\delta}$	Disk	1.53	900	NA	0.08	1.7 × 10^{-5}	3.6 × 10^{-4}	Ar | Air	39
La$_{0.1}$Sr$_{0.9}$Co$_{0.9}$Fe$_{0.1}$O$_{3-\delta}$	Disk	1	900	2.0	1.48	NA	NA	He | Air	40
La$_{0.2}$Sr$_{0.8}$Co$_{0.8}$Fe$_{0.2}$O$_{3-\delta}$	Disk	1	900	1.1	0.81	NA	NA	He | Air	40
La$_{0.6}$Sr$_{0.4}$Co$_{0.8}$Fe$_{0.2}$O$_{3-\delta}$	Disk	1	900	0.2	0.14	NA	NA	He | Air	40
La$_{0.6}$Sr$_{0.4}$Co$_{0.6}$Fe$_{0.4}$O$_{3-\delta}$	Disk	2.5	650–980	NA	NA	10^{-6}–5 × 10^{-5}	10^{-5}–10^{-2}	N$_2$| 0.03–1 atm O$_2$	41
La$_{0.2}$Sr$_{0.8}$Co$_{0.2}$Fe$_{0.8}$O$_{3-\delta}$	Disk	2	950	0.81	0.63	1.7 × 10^{-5}	NA	He | Air	42
La$_{0.2}$Ba$_{0.8}$Co$_{0.2}$Fe$_{0.8}$O$_{3-\delta}$	Disk	2	950	0.4	0.29	8.5 × 10^{-6}	NA	He | Air	42
La$_{0.2}$Ca$_{0.8}$Co$_{0.2}$Fe$_{0.8}$O$_{3-\delta}$	Disk	1	950	0.12	0.09	2.9 × 10^{-6}	NA	He | Air	42
La$_{1-x}$Sr$_x$FeO$_{3-\delta}$ (x = 0.1–0.4)	Disk	0.5–2	850–1050	NA	2.5	5.3–9.3 × 10^{-6}	NA	He | 0.01–1 atm O$_2$	43
La$_{0.4}$Ca$_{0.6}$Co$_x$Fe$_{1-x}$O$_{3-\delta}$ (x = 0, 0.25, 0.5)	Disk	0.72–1.73	900	NA	0.04–0.09	NA	NA	Ar | Air	44
La$_{0.4}$Ca$_{0.6}$Co$_{0.25}$Fe$_{0.75}$O$_{3-\delta}$	Disk	0.72	850–900	NA	0.6	NA	NA	0.1–0.2 atm H$_2$/Ar | Air	44
Ba$_{0.5}$Sr$_{0.5}$Co$_{0.8}$Fe$_{0.2}$O$_{3-\delta}$	Disk	1.8	950	1.4	NA	NA	NA	He | 0.01–1 atm O$_2$	45
Ba$_{0.5}$Sr$_{0.5}$Co$_{0.8}$Fe$_{0.2}$O$_{3-\delta}$	Tubular	3	900	3.0	NA	2.8 × 10^{-5}	NA	He | 0.09–1 atm O$_2$	46
La$_{0.8}$Sr$_{0.2}$Ga$_{0.8}$Mg$_{0.2-x}$Co$_x$O$_{3-\delta}$ (x = 0–0.2)	Disk	1.5	600–1000	NA	0.16–0.6	4.20 × 10^{-7}	7.93 × 10^{-7}	N$_2$| Air	47
SrCo$_{0.8}$Fe$_{0.2}$O$_{3-\delta}$	Disk	1	900	2.0	1.4	NA	NA	He | Air	49
SrCo$_{0.8}$Fe$_{0.2}$O$_{3-\delta}$	Tubular	2 and 2.5	797–890	NA	0.6–6.0^b	2.11–9.31 × 10^{-6}	0.36–2.1 × 10^{-4}	He | 0.01–1 atm O$_2$	49
SmFe$_{0.5}$Co$_{0.5}$O$_{3-\delta}$	Tubular	2.8	800–937	NA	0.2–1.0^b	0.21–1.25 × 10^{-6}	2.5–7.5 × 10^{-5}	He | 0.01–1 atm O$_2$	49
SrCo$_{0.4}$Fe$_{0.6-x}$Zr$_x$O$_{3-\delta}$ (0 ≤ x ≤ 0.2)	Disk	1.78	727–927	NA	0.05–0.45	NA	NA	He | Air	51
SrCo$_{0.95-x}$Fe$_x$Zr$_{0.05}$O$_{3-\delta}$ (0 ≤ x ≤ 0.8)	Disk	1.78	727–927	NA	0.01–0.6	NA	NA	He | Air	51

NA: not given/investigated.
^{a} Where STP conditions have been stated the molar flow is also given. ^{b} Not given per cm^2.

elements followed by digits corresponding to the stoichiometry of the relevant element). This material has been studied in disk [36] and tubular form [37] using ^{18}O tracer with SIMS, conductivity relaxation and oxygen permeation techniques. A successful attempt at improving oxygen flux involved applying a thick porous coating of the same perovskite to make an asymmetric disk membrane [38]. Fluxes three times that obtained for the disk alone were reported. AC IS has been used to determine the chemical diffusion coefficient and surface exchange on disk membranes [39]. It was found that permeation was under control of both bulk diffusion and surface exchange. Studies using LSCF powder coatings on LSCF membrane disks have been undertaken using LSCF1991, LSCF2882, LSCF6482, and $SrCo_{0.8}Fe_{0.2}O_{3-\delta}$ [40]. The disks were treated with porous coatings of $LaCoO_3$, $La_{0.8}Sr_{0.2}CoO_{3-\delta}$, and $SrCo_{0.8}Fe_{0.2}O_{3-\delta}$ perovskites. The transport parameters of LSCF6464 have been investigated using electrical conductivity relaxation and high temperature coulometric titration techniques [41]. The strontium substituted $La_{0.2}A_{0.8}Co_{0.2}Fe_{0.8}O_{3-\delta}$ (A = Sr, Ba, Ca) [42] perovskite has been shown to give the highest oxygen flux compared to the calcium and barium analogues. However, under low oxygen partial pressures the material decomposed. The barium-substituted material was found to be more stable. Studies on other La-containing perovskites have included disks of $La_{1-x}Sr_xFeO_{3-\delta}$ ($x = 0.1$ to 0.4) [43]. Treatment of the lower oxygen partial pressure side with a CO-containing atmosphere at high temperature led to improved oxygen fluxes. Experiments using membranes of various thicknesses revealed the presence of bulk diffusion limitations. The $La_{0.4}Ca_{0.6}Co_xFe_{1-x}O_{3-\delta}$ ($x = 0$, 0.25, and 0.5) series was tested for oxygen permeation and stability [44]. The perovskite with highest cobalt content was mechanically the weakest during long-term testing. Experiments in which hydrogen was added to the permeate side gave an improvement in flux.

It has been reported that barium instead of lanthanum at the A site improves phase stability of the perovskite by preventing oxidation of the B site cation without adversely affecting oxygen permeability. The $Ba_{0.5}Sr_{0.5}Co_{0.8}Fe_{0.2}O_{3-\delta}$ (BSCF5582) system has been prepared in disk [45] and tubular membrane [46] form for permeation studies. At low oxygen partial pressures and lower temperatures there was evidence of membrane decomposition. However, this was found to be reversible with an increase in temperature. In tubular form, a stable oxygen flux was maintained over a working period of 150 h.

Strontium and magnesium doped $LaGaO_3$ produces a family that possesses good oxide anion conductivity, further doping with cobalt at the gallium-site in the $La_{0.8}Sr_{0.2}Ga_{0.8}Mg_{0.2-x}Co_xO_{3-\delta}$ [47] system increases oxide ion and electronic conductivity. There is also interest in the lanthanum nickelate based oxides. A small increase in phase stability has been obtained when nickel was used as the dopant in the $La_{0.90}Sr_{0.10}Ga_{0.65}Mg_{0.15}Ni_{0.2}O_{3-\delta}$ and $La_{0.5}Pr_{0.5}Ga_{0.65}Mg_{0.15}Ni_{0.2}O_{3-\delta}$ systems at low oxygen partial pressures. The praseodymium analog was found to be more stable [48].

The Sr–Co–Fe perovskite systems have the highest oxygen permeabilities of all the perovskite MIECs but can suffer from degradation as a result of the formation of a brownmillerite phase. Experimental and modeling work concerning the transport mechanism of oxygen in $SrCo_{0.8}Fe_{0.2}O_{3-\delta}$ and for comparison $SmCo_{0.5}Fe_{0.5}O_{3-\delta}$ has been undertaken [49]. Attempts to increase stability by doping in divalent and tetravalent cations have been made [50]. Success was obtained with the larger divalent cations, Ba^{2+} provided the highest stability, but the reduction in free volume also hindered oxygen permeation. Further attempts to solve the instability problems in the Sr–Co–Fe system have resulted in the study of two new MIEC families, $SrCo_{0.4}Fe_{0.6-x}Zr_xO_{3-\delta}$ ($0 \leq x \leq 0.2$) and $SrCo_{0.95-x}Fe_xZr_{0.05}O_{3-\delta}$ ($0.1 \leq x \leq 0.8$) [51]. For the first family, 5 mol% of zirconia improved stability while high cobalt content in the second family improved both phase stability and oxygen flux.

The surface condition of the membrane and behavior of oxygen on the metal oxide surface play a crucial role in performance and greatly influence interfacial processes. This influence can be evidenced by the wide-ranging surface exchange coefficient values often obtained by different workers for a given MIEC material. An additional complication is the segregation of constituent metals leading to differences in the composition of the surface and bulk of the metal oxides which can influence the catalytic properties. Adsorption of species from the gas phase can also be a contributory factor in segregation.

Changes in the oxidation state, coordination, and symmetry of the A- and B-cations of the MIEC metal oxide at the surface can facilitate the presence of a number of surface oxide species. The different oxide species include lattice oxygen, various chemisorbed oxygen species [52–55] (e.g., O, O^-, O_2^-, O_3^-, O_4^-, and O^{2-}) and in addition, in the presence of water or hydrogen, surface hydroxyls. Techniques such as electron paramagnetic resonance (EPR), temperature programmed desorption (TPD), infrared spectroscopy (IR) and x-ray photoelectron spectroscopy (XPS) have indicated the presence of these oxide species [52,56–58]. These normally unstable oxide species can be stabilized when adsorbed on metal oxide surfaces and have the ability to promote either complete oxidation or partial oxidation depending on the type of metal oxide surface employed. They can influence oxidation reactions such as syngas (hydrogen and carbon monoxide) production from methane and oxygen and oxidative coupling of methane to higher hydrocarbon products. The oxide species can participate in a reaction in a number of ways such as proton abstraction, direct attack, or interaction with the metal oxide by rejuvenation of spent lattice oxygen anions. The electronic conductivity of the MIEC metal oxide is central to oxygen exchange and can play a promoting role, for example, CeO_2 is superior in this role than ZrO_2 due to its higher electronic conductivity [59].

The major application of MIECs in dense membrane form is envisaged to be chemical production and air separation. Currently the La–Sr–Co–Fe and more recently Ba–Sr–Co–Fe perovskite metal oxide systems have drawn most attention for chemical production, notably for the production of syngas. However, the large oxygen chemical potential gradients established during reaction conditions often lead to membrane failure. A more detailed discussion of recent MIECs investigated for their oxygen permeation performance and use in laboratory-scale chemical production studies is provided elswhere [60].

REFERENCES

1. Kröger, F.A. *The Chemistry of the Imperfect Crystals*. Amsterdam: North Holland, 1974, p. 207.
2. Kröger, F.A. and Vink, H.J. *Solid State Physics*. Seitz, F. and Turnbull, D., Eds. New York: Academic Press, 1956, Vol. 3, p. 307.
3. Askeland, D.R. *The Science and Engineering of Materials*. London: Chapman & Hall, 1996, p. 39.
4. Chiang, Y.-M., Birnie III, D., and Kingery, W.D. *Physical Ceramics, Principles for Ceramic Science and Engineering*. Canada: John Wiley & Sons, 1997, p. 222.
5. James, A.C.W.P. and Murphy, D.W. *Chemistry of Superconductor Materials*, Vanderah, T.A., Ed. New Jersey: Noyes Publications, 1992, p. 427.
6. Kondoh, J., Kawashima, T., Kikuchi, S., Tomii, Y., and Ito,Y. Effect of aging on yttria-stabilized zirconia I. A study of its electrochemical properties. *J. Electrochem. Soc.* 1998, *145*, 1527–1536.
7. Kondoh, J., Kawashima, T., Kikuchi, S., Tomii, Y., and Ito,Y. Effect of aging on yttria-stabilized zirconia III. A study of the effect of local structures on conductivity. *J. Electrochem. Soc.* 1998, *145*, 1550–1560.
8. Goff, J.P., Hayes, W., Hull, S., Hutchings, M.T., and Clausen, K.C. Defect structure of yttria-stabilized zirconia and its influence on the ionic conductivity at elevated temperatures. *Phys. Rev. B* 1999, *59*, 14202–14219.
9. Islam, M.S. Ionic transport in ABO_3 perovskite oxides: a computer modelling tour. *J. Mater. Chem.* 2000, *10*, 1027–1038.
10. West, A.R. *Basic Solid State Chemistry*. Sussex: John Wiley & Sons, 1999, p. 56.
11. Woodward, P.M. Octahedral tilting in perovskites. 2. Structure stabilizing forces. *Acta Cryst.* 1997, *B53*, 44–66.
12. Teneze, N., Mercurio, D., Trolliard, G., and Frit, B. Cation-deficient perovskite-related compounds $(Ba,La)_n Ti_{n-1} O_{3n}$ ($n = 4$, 5, and 6): a Rietveld refinement from neutron powder diffraction data. *Mater. Res. Bull.* 2000, *35*, 1603–1614.
13. Verela, A., Parras, M., Boulahya, K., and Gonzalez-Calbert, J.M. Ordering of Anionic Vacancies in the $BaCoO_2$. 94 hexagonal related Perovskite. *J. Solid State Chem.*, 1997, *128(1)*, 130–136.
14. Rao, C.N.R., Gopalakrishnan, J., and Vidyasagar, K. Superstructures, ordered defects and nonstoichiometry in metal-oxides of perovskite and related structure. *Indian J. Chem.* 1984, *23A*, 265–284.
15. Bertaut, E.F., Blum, P., and Sagnières A. Structure du Ferrite Bicalcique et de la Brownmillerite. *Acta Cryst.*, 1959, *12*, 149–159.
16. Reller, A., Jefferson, D.A., Thomas, J.M., Beyerlein, B.A., and Poeppelmeier, K.R. Three new ordering schemes for oxygen vacancies in $CaMnO_{3-x}$ superlattices based on square-pyramidal coordination of Mn^{3+}. *J. Chem. Soc., Chem. Commun.* 1982, *24*, 1378–1380.
17. Poeppelmeier, K.R., Leonowitz, M.E., and Longo, J.M. $CaMnO_{2.5}$ and $Ca_2 MnO_{3.5}$ — New oxygen-defect perovskite-type oxides. *J. Solid State Chem.* 1982, *44*, 89–98.
18. Poeppelmeier, K.R., Leonowitz, M.E., Scalon, J.C., Longo, J.M., and Yelon, W.B. Structure determination of $CaMnO_3$ and $CaMnO_{2.5}$ by X-ray and neutron methods. *J. Solid State Chem.* 1982, *45*, 71–79.

19. Alario-Franco, M.A., González-Calbet, J.M., and Vallet-Regi, M.J. Brownmillerite-type microdomains in the calcium lanthanum ferrites: $Ca_xLa_{1-y}FeO_{3-y}$, $1.2/3 < x < 1$. *Solid State Chem.* 1983, *49*, 219–231.

20. Liu, L.M., Lee, T.H., Qui, L., Yang, Y.L., and Jacobson, A.J. A thermogravimetric study of the phase diagram of strontium cobalt iron oxide, $SrCo_{0.8}Fe_{0.2}O_{3-\delta}$. *Mater. Res. Bull.* 1996, *31*, 29–35.

21. Mahesh, R., Kannan, K.R., and Rao, C.N.R. Electrochemical synthesis of ferromagnetic $LaMnO_3$ and metallic $NdNiO_3$. *J. Solid State Chem.* 1995, *114*, 294–296.

22. Rao, C.N.R., Cheetham, A.K., and Mahesh, R. Giant magnetoresistance and related properties of rare-earth manganates and other oxide systems. *Chem. Mater.* 1996, *8*, 2421–2432.

23. Lichtenberg, F., Hernberger, A., Wiedemann, K., and Mannhart, J. Synthesis of perovskite-related layered $A_nB_nO_{3n+2} = ABO_x$ type niobates and titanates and study of their structural, electric and magnetic properties. *Prog. Solid State Chem.* 2001, *29*, 1–70.

24. Gellings, P.J. and Bouwmeester, H.J.M. Ion and mixed conducting oxides as catalysts. *Catal. Today* 1992, *12*, 1–105.

25. Steele, B.C.H. Oxygen ion conductors and their technological applications. *Mat. Sci. Eng. B* 1992, *13*, 79–87.

26. Bouwmeester, H.J.M. Dense ceramic membranes for methane conversion. *Catal. Today* 2003, *82*, 141–150.

27. Brinkmann, D. Magnetic resonance in superionic conductors. *Magn. Reson. Rev.* 1989, *14*, 101–156.

28. Fuda, K., Kishio, K., Yamuchi, S., and Fueki, K. Study on vacancy motion in Y_2O_3-doped CeO_2 by O-17 NMR technique. *J. Phys. Chem. Sol.* 1985, *46*, 1141–1146.

29. Adler, S.B., Reimer, J.A., Baltisberger, J., and Werner, U. Chemical structure and oxygen dynamics in $Ba_2In_2O_5$. *J. Am. Chem. Soc.* 1994, *116*, 675–681

30. Fukamachi, T., Kobayashi, Y., Miyashita, T., and Sato, M. La-139 NMR studies of layered perovskite systems $La_3Ni_2O_{7-\delta}$ and $La_4Ni_3O_{10}$. *J. Phys. Chem. Sol.* 2001, *62*, 195–198.

31. Kim, N. and Grey, C.P. O-17 MAS NMR study of the oxygen local environments in the anionic conductors $Y_2(B_{1-x}B'_x)_2O_7$(B, B' = Sn, Ti, Zr). *J. Solid State Chem.* 2003, *175*, 110–115.

32. Flewitt, P.J.E. and Wild, R.K. *Grain Boundaries: Their Microstructure and Chemistry*. Chichester: John Wiley & Sons, 2001.

33. Bowen, P. and Carry, C. From powders to sintered pieces: forming, transformations and sintering of nanostructured ceramic oxides. *Powder Tech.* 2002, *128*, 248–255.

34. Chater, R.J., Carter, S., Kilner, J.A., and Steele, B.C.H. Development of a novel SIMS technique for oxygen self-diffusion and surface exchange coefficient measurements in oxides of high diffusivity. *Solid State Ion.* 1992, *53*, 859–867.

35. Diethelm, S., Closset, A., van Herle, J., and Nisancioglu, K.J. Determination of chemical diffusion and surface exchange coefficients of oxygen by electrochemical impedance spectroscopy. *Electrochem. Soc.* 2002, *149*, E424–E432.

36. Lane, J.A., Benson, S.J., Waller, D., and Kilner, J.A. Oxygen transport in $La_{0.6}Sr_{0.4}Co_{0.2}Fe_{0.8}O_{3-\delta}$. *Solid State Ion.* 1999, *121*, 201–208.

37. Li, S., Qi, H., Xu, N., and Shi, J. Tubular dense perovskite type membranes. Preparation, sealing, and oxygen permeation properties. *Ind. Eng. Chem. Res.* 1999, *38*, 5028–5033.

38. Jin, W., Li, S., Huang, P., Xu, N., and Shi, J. Preparation of an asymmetric perovskite-type membrane and its oxygen permeability. *J. Membr. Sci.* 2001, *185*, 237–243.

39. Diethelm, S. and van Herle, J. Oxygen transport through dense $La_{0.6}Sr_{0.4}Fe_{0.8}Co_{0.2}O_{3-\delta}$ perovskite-type permeation membranes. *J. Eur. Ceram. Soc.* 2004, *24*, 1319–1323.

40. Teraoka, Y., Hombe, Y., Ishii, J., Furukawa, H., and Moriguchi, I. Catalytic effects in oxygen permeation through mixed-conductive LSCF perovskite membranes. *Solid State Ion.* 2002, *152*, 681–687.

41. Ten Elshof, J.E., Lankhorst, M.H.R., and Bouwmeester, H.J.M. Chemical diffusion and oxygen exchange of $La_{0.6}Sr_{0.4}Co_{0.6}Fe_{0.4}O_{3-\delta}$. *Solid State Ion.* 1997, *99*, 15–27.

42. Li, S., Jin, W., Huang, P., Xu, N., Shi, J., Lin, Y.S., Hu, M.X.-C., and Payzant, E.A. Comparison of oxygen permeability and stability of perovskite type $La_{0.2}A_{0.8}Co_{0.2}Fe_{0.8}O_{3-\delta}$ (A = Sr, Ba, Ca) membranes. *Ind. Eng. Chem. Res.* 1999, *38*, 2963–2972.

43. Ten Elshof, J.E., Bouwmeester, J.M., and Verweij, H. Oxygen-transport through $La_{1-x}Sr_xFeO_{3-\delta}$ membrane. 1. Permeation in air/He gradients. *Solid State Ion.* 1995, *81*, 97–109; Oxygen transport through $La_{1-x}Sr_xFeO_{3-\delta}$ membranes. 2. Permeation in air/CO, CO_2 gradients. *Solid State Ion.* 1996, *89*, 81–92.

44. Diethelm, S., van Herle, J., Middleton, P.H., and Favrat, D. Oxygen permeation and stability of $La_{0.4}Ca_{0.6}Fe_{1-x}Co_xO_{3-\delta}$ ($x = 0, 0.25, 0.5$) membranes. *J. Power Sources* 2003, *118*, 270–275.

45. Shao, Z., Xiong, G., Cong, Y., and Yang, W. Synthesis and oxygen permeation study of novel perovskite-type $BaBi_xCo_{0.2}Fe_{0.8-x}O_{3-\delta}$ ceramic membranes. *J. Membr. Sci.* 2000, *164*, 167–176.

46. Shao, Z., Yang, W., Cong, Y., Dong, H., Tong, J., and Xiong, G. Investigation of the permeation behavior and stability of a $Ba_{0.5}Sr_{0.5}Co_{0.8}Fe_{0.2}O_{3-\delta}$ oxygen membrane. *J. Membr. Sci.* 2000, *172*, 177–188.

47. Ishihara, T., Furutani, H., Honda, M., Yamada, T., Shibayama, T., Akbay, T., Sakai, N., Yokokawa, H., and Takita, Y. Improved oxide ion conductivity in $La_{0.8}Sr_{0.2}Ga_{0.8}Mg_{0.2}O_3$ by doping Co. *Chem. Mater.* 1999, *11*, 2081–2088.

48. Kharton, V.V., Yaremchenko, A.A., Shaula, A.L., Patrkeev, M.V., Naumovich, N.D., Logvinovich, E.N., Frade, J.R., and Marques, F.M.B. Transport properties and stability of Ni-containing mixed conductors with perovskite- and K_2NiF_4-type structure. *J. Solid State Chem.* 2004, *177*, 26–37.

49. Kim, S., Yang, Y.L., Jacobson, A.J., and Abeles, B. Diffusion and surface exchange coefficients in mixed ionic electronic conducting oxides from the pressure dependence of oxygen permeation. *Solid State Ion.* 1998, *106*, 189–195.

50. Tan, L., Yang, L., Gu, X., Jin, W., Zhang, L., and Xu, N. Influence of the size of doping ion on phase stability and oxygen permeability of $SrCo_{0.8}Fe_{0.2}O_{3-\delta}$ oxide. *J. Membr. Sci.* 2004, *230*, 21–27.

51. Yang, L., Tan, L., Gu, X., Jin, W., Zhang, L., and Xu, N. A new series of $Sr(Co,Fe,Zr)O_{3-\delta}$ perovskite-type membrane materials for oxygen permeation. *Ind. Eng. Chem. Res.* 2003, *42*, 2299–2305.

52. Che, M. and Tench, A.J. Characterisation and reactivity of mononuclear oxygen species on oxide surfaces. *Adv. Catal.* 1982, *31*, 77–133.
53. Che, M. and Tench, A.J. Characterisation and reactivity of mononuclear oxygen species on oxide surfaces. *Adv. Catal.* 1983, *32*, 1–148.
54. Jacox, M. and Milligan, D.E. Spectrum and structure of the O_3^- and O_4^- anions isolated in an argon matrix. *Chem. Phys. Lett.* 1972, *14*, 518–521.
55. Lunsford, J.H. ESR of adsorbed oxygen species. *Catal. Rev. Sci. Eng.* 1973, *8*, 135–157.
56. Takasu, Y., Yoko-o, T., Matsui, M., Matsuda, Y., and Toyoshima, I. Catalytic reactivity of the lattice oxygen-atoms of terbium oxide. *J. Catal.* 1982, *77*, 485–490.
57. Li, C., Domen, K., Marayu, K., and Onishi, T. Oxygen-exchange reactions over cerium oxide — An FT-IR study. *J. Catal.* 1990, *123*, 436–442.
58. Xu, Q., Huang, D-P., Chen, W., Wang, H., Wang, B-t., and Yuan, R.-Z. X-ray photoelectron spectroscopy investigation on chemical states of oxygen on surfaces of $La_{0.6}Sr_{0.4}Co_{1-y}Fe_yO_3$ mixed electronic-ionic conducting ceramics. *App. Surf. Sci.* 2004, *228*, 110–114.
59. Kurumchin, E. Kh. and Perfil'ev, M.V. An isotope exchange study of the behaviour of electrochemical systems. *Solid State Ion.* 1990, *42*, 129–133.
60. Thursfield, A. and Metcalfe, I.S. The use of mixed ionic and electronic conducting membranes for chemical production. *J. Mater. Chem.*, 2004, *14*, 2475–2485.

4 Cation Valence States of Transitional Metal Oxides Analyzed by Electron Energy-Loss Spectroscopy

Zhong Lin Wang
School of Materials Science and Engineering, Georgia
Institute of Technology, Atlanta, GA

CONTENTS

4.1 INTRODUCTION

Transition and rare earth metal oxides are the fundamental ingredients for the advanced smart and functional materials. Many functional properties of inorganic materials are determined by the elements with *mixed valences* in the structure unit [1], by which we mean that an element has two or more different valences while forming a compound. The discovery of high-temperature superconductors is a successful example of the mixed valence chemistry, and the colossal magnetoresistivity (CMR) [2,3] observed in the perovskite structured $La_{1-x}A_xMnO_3$ (A = Ca, Sr, or Ba) is another example. Transition and rare earth metal elements with mixed valences are mandatory for these materials to stimulate electronic, structural, and chemical evolution leading to specific functionality.

The valence states of metal cations in such materials can certainly be determined chemically using the redox titration, but it is inapplicable to nanophase or nanostructured materials, such as thin films. The wet chemistry approaches usually do not provide any spatial resolution. X-ray photoelectron spectroscopy (XPS) can provide information on the average distribution of cation valences for nanostructured materials with certain spatial resolution, but the spatial resolution is nowhere near the desired nanometer scale, and the information provided is limited to a surface layer of 2 to 5 nm in thickness.

Electron energy-loss spectroscopy (EELS), a powerful technique for material characterization at nanometer spatial resolution, has been widely used in chemical microanalysis and the studies of solid-state effects [4]. In EELS, the L ionization edges of transition metal and rare earth elements usually display sharp peaks at the near-edge region, known as *white lines*. For transition metals with unoccupied 3d states, the transition of an electron from 2p state to 3d levels lead to the formation of white lines. The L_3 and L_2 lines are the transitions from $2p^{3/2}$ to $3d^{3/2}3d^{5/2}$ and from $2p^{1/2}$ to $3d^{3/2}$, respectively, and their intensities are related to the unoccupied states in the 3d bands [5,6].

Numerous EELS experiments have shown that a change in valence state of cations introduces a dramatic change in the ratio of the white lines, leading to the possibility of identifying the occupation number of 3d orbital using EELS. Morrison et al. [7] have applied this technique to study the valence modulation in Fe_xGe_{1-x} alloy as a function of Ge doping. The 3d and 4d occupations of transition and rare earth elements have been studied systematically [8–10]. The crystal structure of a new compound $Mn_{7.5}Br_3O_{10}$ has been refined in reference to the measured Mn valences [11]. The oxidation states of Ce and Pr have been determined in an orthophosphate material, in which the constituents of Ce and Pr are in the order of 100 ppm [12]. Llord et al. [13] and Yuan et al. [14] have demonstrated the sensitivity of the Fe white lines to the magnetic momentum of the Fe layers.

In this chapter, we review our current progress made in applying EELS for quantitative determination of valence states of Mn and Co oxides. The fundamental experimental approach is given first. The applications of EELS will be demonstrated for quantifying the valence transition in Mn and Co oxides, determining the concentration of oxygen vacancies, refining the crystal structure of an anion deficient perovskite, and identifying the crystal structure of nanoparticles (CoO and Co_3O_4). With the use of energy-filtered transmission electron microscopy (EFTEM), a new experimental approach is introduced for mapping the spatial distribution of a cation by its valence. The valence state map is almost independent of the specimen thickness and can be used to directly read out the local valence state from the image. A spatial resolution of $\sim$2 nm has been demonstrated.

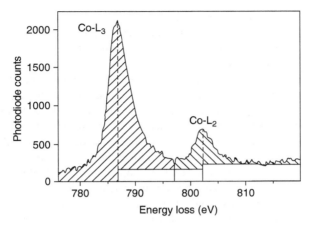

FIGURE 4.1   An EELS spectrum acquired from a Co oxide, showing the technique used to extract the intensities of white lines

4.2 PRINCIPLE OF EELS MEASUREMENTS

Figure 4.1 shows an EELS spectrum of Co oxide acquired at 200 kV using a Hitachi HF-2000 transmission electron microscope equipped with a Gatan 666 parallel-detection electron energy-loss spectrometer. The EELS spectra were acquired in the image mode at a magnification of 40 to 100 K depending on the required spatial resolution and signal intensity. Also, the EELS data must be processed first to remove the gain variation introduced by the detector channels. A low-loss valence spectrum and the corresponding core-shell ionization edge EELS spectrum were acquired consecutively from the same specimen region. The low energy-loss spectrum was used to remove the multiple inelastic scattering effect in the core-loss region using the Fourier ratio technique. Consequently, the data presented here are the results of single inelastic scattering.

Several techniques have been proposed to correlate the observed EELS signals with the valence states; the ratio of white lines, the normalized white-line intensity in reference to the continuous state intensity located ~50 to 100 eV beyond the energy of the L_2 line, and the absolute energy shift of the white lines. In this study, we use the white-line intensity ratio that is calculated using the method demonstrated in Figure 4.1 [9]. The background intensity was modeled by step functions in the threshold regions. A straight line over a range of approximately 50 eV was fitted to the background intensity immediately following the L_2 white line. This line was then modified into a double step of the same slope with onsets occurring at the white-line maxim. The ratio of the step heights is chosen to be 2:1 in accordance with the multiplicity of the initial states (four $2p^{3/2}$ electrons and two $2p^{1/2}$ electrons) [9,10,13,15]. Although there exist some disagreements in the literature about the calculation of the normalized white-line intensity because the theory behind the white line and their continuous background is rather complex [16], it appears, based on our experience, that the ratio of the white-line intensities is likely to be a reliable and a sensitive approach. This background subtraction procedure is followed consistently for all of the acquired spectra. The calculated result of L_3/L_2 is rather stable and is not sensitive to either the specimen thickness or the noise level in the spectrum.

The EELS analysis of valence state is carried out in reference to the spectra acquired from standard specimens with known cation valence states. Since the intensity ratio of L_3/L_2 is sensitive to the valence state of the corresponding element, if a series of EELS spectra are acquired from several standard specimens with known valence states, an empirical plot of these data serves as the reference for determining the valence state of the element present in a new compound.

The L_3/L_2 ratio for a few standard Co compounds are plotted in Figure 4.2(a). EELS spectra of Co-$L_{2,3}$ ionization edges were acquired from $CoSi_2$ (with Co^{4+}), Co_3O_4 (with $Co^{2.67+}$), $CoCO_3$ (with Co^{2+}), and $CoSO_4$ (with Co^{2+}). Figure 4.2(b) shows a plot of the experimentally measured intensity ratios of white line L_3/L_2 for Mn. The curves clearly show that the ratio of L_3/L_2 is very sensitive to the valence state of Co and Mn. This is the basis of our experimental approach for measuring the valence states of Co or Mn in a new material.

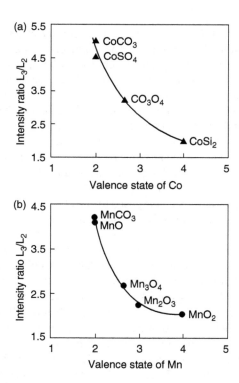

FIGURE 4.2   Plots of the intensity ratios of L_3/L_2 calculated from the spectra acquired from (a) Co compounds and (b) Mn compounds as a function of the cation valence. A nominal fit of the experimental data is shown by a solid curve

4.3 *In Situ* Observation of Valence State Transition

For demonstrating the sensitivity and the reliability of using white-line intensity to determine the valence states in mixed valence compounds [17], the *in situ* reduction behavior of Co_3O_4 is first examined. A Gatan TEM (transmission electron microscopy) specimen heating stage was employed to carry out the *in situ* EELS experiments, and the specimen temperature could be increased continuously from room temperature to 1000°C. The column pressure was kept at 3×10^{-8} torr or lower during the *in situ* analysis.

Figure 4.3 shows the Co L_3/L_2 ratio and the relative composition of n_O/n_{Co} for the same piece of crystal as the specimen temperature was increased. The specimen composition was determined from the integrated intensities of the O–K and Co-$L_{2,3}$ ionization edges with the use of ionization cross-sections calculated using the SIGMAK and SIGMAL programs [4]. The L_3/L_2 ratios corresponding to Co^{2+} determined from the EELS spectra of $CoSO_4$ and $CoCO_3$ at room temperature, and $Co^{2.67+}$ obtained from Co_3O_4 are marked by a shadowed band, the width of which represents the experimental error and the variation among different compounds.

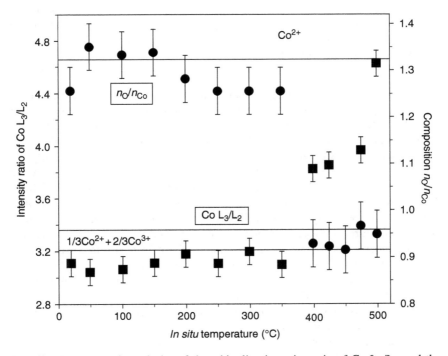

Figure 4.3 An overlapped plot of the white-line intensity ratio of Co L_3/L_2 and the corresponding chemical composition of n_O/n_{Co} as a function of the *in situ* temperature of the Co_3O_4 specimen, showing the abrupt change in valence state and oxygen composition at 400°C. The error bars are determined from the errors introduced in background subtraction and data fluctuation among spectra

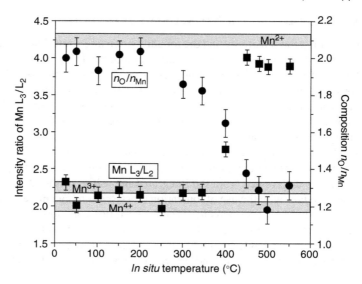

FIGURE 4.4 An overlapped plot of the white-line intensity ratio of Mn L_3/L_2 and the corresponding chemical composition of n_O/n_{Mn} as a function of the *in situ* temperature of the MnO_2 specimen based on EELS spectra, showing that the change in Mn valence state is accompanied with the variation in oxygen content

The Co L_3/L_2 ratio and the composition, n_O/n_{Co}, simultaneously experience a sharp change at $T = 400°C$. The chemical composition changes from O:Co = 1.33 ± 0.5 to O:Co = 0.95 ± 0.5 corresponding to the change of the average valence state of Co from 2.67+ to 2+ when the temperature is above 400°C.

The second experiment is performed on the reduction of MnO_2. Similarly, the plot of composition, n_O/n_{Mn}, and white-line intensity, Mn L_3/L_2, are shown in Figure 4.4, where the shadowed bands indicate the white-line ratios for Mn^{2+}, Mn^{3+}, and Mn^{4+} as determined from the standard specimens of MnO, Mn_2O_3, and MnO_2, respectively. The reduction of MnO_2 occurs at 300°C. As the specimen temperature increases, the O/Mn ratio drops and the L_3/L_2 ratio increases, which indicates the valence state conversion of Mn from 4+ to lower valence states. At $T = 400°C$, the specimen contains the mixed valences of Mn^{4+}, Mn^{3+}, and Mn^{2+}. As the temperature reaches 450°C, the specimen is dominated by Mn^{2+} and Mn^{3+} and the composition is O/Mn = 1.3 ± 0.5, in correspondence of Mn_3O_4, which is consistent with the mixed valence of Mn cations and implies the incomplete reduction of MnO_2.

To trace the relationship between the valence transition and the evolution of crystal structure, electron diffraction patterns were recorded *in situ* at different temperatures, as shown in Figure 4.5. The crystal structure is MnO_2 (with rutile structure), and no visible change in crystallography is observed up to 400°C. From 400° to 450°C, the crystal structure experiences a rapid change from rutile to spinel, and the final phase at 500°C is identified to be dominated by Mn_3O_4, with

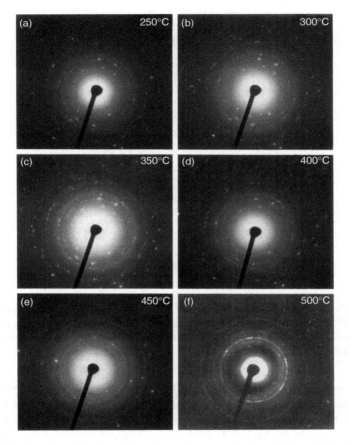

FIGURE 4.5 A series of electron diffraction patterns recorded *in situ* from MnO_x during the thermal induced reduction of MnO_2 to Mn_3O_4. Note that a small fraction of other phases do exist at the final stage. The final phase is identified in reference to the x-ray powder diffraction data

the presence of a small fraction of other phases, consistent with the composition measured by EELS in Figure 4.4.

Similar analysis has been performed for $MnFe_2O_4$ spinel structured nanocrystals [17,18]. The AB_2O_4 spinel structure has two types of cation lattice sites: a tetrahedral site A^{2+} formed by four nearest-neighbor oxygen anions, and an octahedral B^{3+} site formed by six oxygen anions. In $MnFe_2O_4$, the percentage of the A sites occupied by Fe specifies the degree of valence inversion. For a general case, the ionic structure of $MnFe_2O_4$ is written as $(Mn_{1-x}^{2+}Fe_x^{2+})$ $(Fe_{1-y}^{3+}Mn_y^{3+})O_4$, in which the A and B sites can be occupied by either Mn or Fe. The magnetic property of this material strongly depends on the degree of inversion because the $Fe^{2+}A-Fe^{3+}B$ super-exchange interaction is much stronger than the $Mn^{2+}A-Fe^{3+}B$ interaction [19]. An experimental measurement of the

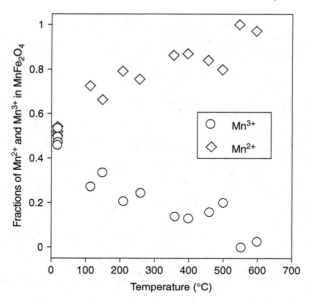

FIGURE 4.6   Fractions of the Mn^{2+} and Mn^{3+} ions in $MnFe_2O_4$ measured by quantitative fitting of the experimental EELS spectra with the standard spectra of the oxides containing Mn^{2+} and Mn^{3+} ions. Five repeated measurements at 25°C are shown that give a consistent result

valence conversion of Mn in this material can provide concrete information on the distribution of Fe in A and B sites, possibly leading to a better understanding of its magnetic property. Figure 4.6 shows the EELS measured fractions of Mn^{2+} and Mn^{3+} ions in the $MnFe_2O_4$ specimen as a function of the *in situ* specimen temperature in TEM. The fraction was calculated by fitting the experimentally observed L_3 and L_2 EELS spectra by a linear summation of the spectra acquired from MnO and Mn_2O_3, and the coefficients for the linear combination give the percentages of the Mn ions of different valence states in the material. It is clear that the fractions of Mn^{2+} and Mn^{3+} ions at room temperature are 0.5:0.5, while a complete conversion into divalent Mn occurs at 600°C. These data explicitly illustrate the evolution in the valence state of the Mn ions, leading to temperature dependent magnetic properties of $MnFe_2O_4$.

4.4 QUANTIFICATION OF OXYGEN VACANCIES IN CMR OXIDES

Many properties of advanced materials are determined by the elements with mixed valences in the structure [20], by which we mean an element that has two or more different valences while forming a compound. In the perovskite structured $La_{1-x}A_xMnO_3$ (A = Ca, Sr, or Ba) CMR materials, for example, the residual charges induced by a partial substitution of trivalent La^{3+} by divalent element A^{2+} are balanced by the conversion of Mn valence states between Mn^{3+} and

Mn^{4+} (or Co^{3+} and Co^{4+} for Co) and the creation of oxygen vacancies as well. The ionic structure of $La_{1-x}A_xMnO_{3-y}$ is

$$La_{1-x}^{3+} A_x^{2+} Mn_{1-x+2y}^{3+} Mn_{x-2y}^{4+} O_{3-y}^{2-} V_y^O \qquad (4.1)$$

provided there is no residual charge trapped in the vacancy sites, where V_y^O stands for the fraction of oxygen vacancies. Valence transition between Mn^{3+} and Mn^{4+} is responsible for the transition from insulator to conductor and possibly the magnetoresistance. This ionization formula is proposed with an assumption that there is no residual charge trapped in the vacancy sites.

In practice, quantifying oxygen vacancies is a challenge to existing microscopy techniques although x-ray and neutron diffuse scattering can be used to determine vacancies in large bulk single crystalline specimens. Moreover, for thin films grown on a crystalline substrate the diffraction analysis may be strongly affected by the defects of the substrate–film interface and the surface disordering. In this section, we show the application of EELS for quantifying oxygen vacancies.

From Equation (4.1), the mean valence state of Mn is

$$\langle Mn \rangle_{vs} = 3 + x - 2y \qquad (4.2)$$

The amount of doping x is usually known from energy dispersive x-ray microanalysis. The $\langle Mn \rangle_{vs}$ can be determined using EELS based on the white-line intensity as illustrated in Section 4.2. Therefore, the content of oxygen vacancies y can be obtained [21].

For a $La_{0.67}Ca_{0.33}MnO_{3-y}$ thin film grown by metal–organic chemical vapor deposition, the L_3/L_2 ratio was measured to be 2.05–2.17, thus, the average valence state of Mn is 3.2–3.5 according to the empirical plot shown in Figure 4.2(b). Substituting this value into Equation (4.2) for $x = 0.33$, yields $y \leq 0.065$, which is equivalent to <2.2% of the oxygen content. At the maximum oxygen vacancy $y_{max} = 0.065$, the atom ratio of Mn^{4+} to Mn^{3+} in the specimen is 0.25, thus, the charge introduced by Mn valence conversion is $(x - 2y) = 0.2^+$, the charge due to oxygen vacancy is $2y = 0.13^-$, which means that 60% of the residual charge introduced by Ca doping is balanced by the conversion of Mn^{3+} to Mn^{4+} and 40% by oxygen vacancies. Therefore, a small percentage of oxygen vacancy can introduce a large effect in balancing the charge. Quantification of oxygen vacancies by this technique may have higher sensitivity than the conventional EELS microanalysis for such a small percentage of vacancies.

4.5 REFINING THE CRYSTAL STRUCTURES OF NONSTOICHIOMETRIC OXIDES

$La_{0.5}Sr_{0.5}CoO_{3-y}$ is a magnetic oxide that has potential applications in fuel cells and ionic conductivity. The cation structure of this material can be determined by

FIGURE 4.7 A high-magnification (100) TEM image of $La_{0.5}Sr_{0.5}CoO_{2.25}$, where the white spots correspond to the projected atom columns with La the strongest contrast, Sr strong, Co weak and oxygen invisible. The rectangular box indicates the (100) projection of the unit cell. The image was recorded at 300 kV

high-resolution TEM. Figure 4.7 shows a high-magnification TEM image of the $La_{0.5}Sr_{0.5}CoO_{3-y}$ crystal oriented along (100), exhibiting c-axis directional aniso-tropy structure. This type of image can directly give the projected position of the cations in the unit cell [22,23], while no information can be provided about the dis-tribution of oxygen anions. The image is also insensitive to the valence state of Co.

For perovskite structured oxides, the oxygen deficiency, if any, is rather small, thus, the quantification of oxygen content is difficult using either EELS or energy dispersive spectroscopy (EDS) microanalysis technique. Alternatively, one can use EELS to measure the mean valence state of Co, then apply the result to determine the oxygen deficiency. For the specimen $La_{0.5}Sr_{0.5}CoO_{3-y}$ used to record the TEM image given in Figure 4.7, the mean valence of Co is determined to be 2+, hence the ionic structure of this crystal is

$$La_{0.5}^{3+}\ Sr_{0.5}^{2+}\ Co^{2+}O_{2.25}^{2-}\ V_{0.75}^{O}$$

To confirm that the valence state of Co in $La_{0.5}Sr_{0.5}CoO_{3-y}$ is 2+, the *in situ* EELS measurement is carried out. As the specimen temperature increases, a reduction of oxide would lead to a reduction in the valence state of Co if the Co has a valence state other than 2+. Figure 4.8 shows the experimentally measured L_3/L_2 ratio as a function of the specimen temperature. The partial pressure of oxygen

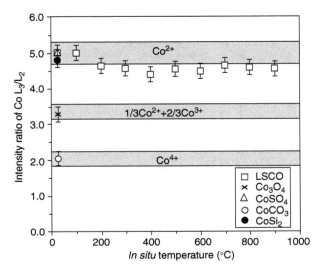

FIGURE 4.8   A relationship between the Co L_3/L_2 intensity ratio and the *in situ* temperature of $La_{0.5}Sr_{0.5}CoO_{2.25}$, proving the divalent state of Co

is rather low in TEM, thus, oxidation is unlikely to occur based on our studies of MnO_x and CoO_y (see Figure 4.3 and Figure 4.4). It is anticipated that a significant change in L_3/L_2 ratio would be observed if the valence state of Co changes. In contrast, the experimentally observed L_3/L_2 ratio has little dependence on the temperature and the ratio remains in the Co^{2+} range even when the oxide is a totally changed crystallographically at 900°C. Therefore, the valence state of Co is undoubtedly 2+. This information is important to confirm the reliability of the structural mode proposed earlier. The surprisingly high stability of $La_{0.5}Sr_{0.5}CoO_{2.25}$ is likely to be very useful for ionic conductor because of the maximum density of oxygen vacancies.

Quantitative determination of the structure of this crystal needs the support of data from x-ray diffraction, electron diffraction, and high-resolution transmission electron microscopy (HRTEM) imaging. More importantly, the valence state of Co measured by EELS is indispensable for refining the crystal structure because the compound is chemically nonstoichiometric. From electron diffraction, we also know that the oxygen vacancies are ordered in the crystal. Figure 4.9 gives the structural model proposed, based on all of the known information about the structure [17]. The unit cell is made of two fundamental structural modules M_1 and M_2 and its crystal structure is $La_8Sr_8Co_{16}O_{36}$, while the entire structure still preserves the characteristics of perovskite framework and is a superstructure induced by an ordered structure of oxygen vacancies. The polyhedra formed by the oxygen anions that coordinate a Co atom can be a planar square (coordination number, CN = 4), a square-based pyramid (CN = 5), or a octahedron (CN = 6). These modules are required to balance the chemical structure of the crystal.

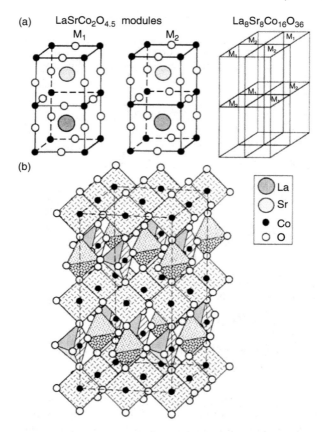

FIGURE 4.9 (a) Two anion deficient modules of $LaSrCo_2O_{4.5}$, and the corresponding stacking to form a complete unit cell of $La_8Sr_8Co_{16}O_{36}$ (or $La_{0.5}Sr_{0.5}CoO_{2.25}$). (b) The three-dimensional model of the structure proposed for $La_8Sr_8Co_{16}O_{36}$, where the Co with coordination numbers of 4, 5, and 6 are shown and La and Sr cations are omitted for clarity

4.6 IDENTIFYING THE STRUCTURE OF NANOPARTICLES

Determining the crystal structure of nanoparticles is a challenge, particularly, when the particles are <5 nm. The intensity-maxim observed in the x-ray or electron diffraction patterns of such small particles are broadened due to the crystal shape factor, which greatly reduced the accuracy of structure refinement. The quality of the HRTEM images of the particles is degraded because of the strong effect from the substrate. This difficulty arises in our recent study of CoO nanocrystals whose shape is dominated by tetrahedral of sizes <5 nm. Electron diffraction indicates that the crystal has a NaCl type of structure. To confirm that the synthesized nanocrystals are CoO, EELS is used to measure the valence state of Co. Figure 4.10

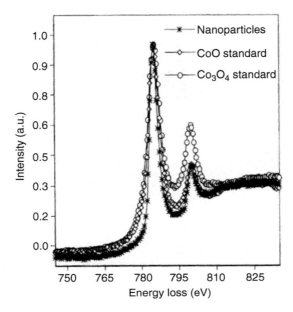

FIGURE 4.10   A comparison of EELS spectra of Co-$L_{2,3}$ ionization edges acquired from Co_3O_4 and CoO standard specimens and the as-synthesized nanocrystals, proving that the valence state of Co is 2+ in the nanocrystals. The full width at half maximum of the white lines for the Co_3O_4 and CoO standards is wider than that for the nanocrystals, possibly due to size effect

shows a comparison of the spectra acquired from Co_3O_4 and CoO standard specimens and the synthesized nanocrystals. The relative intensity of Co-L_2 to Co-L_3 for the nanocrystals is almost identical to that of standard CoO, while the Co-L_2 line of Co_3O_4 is significantly higher, indicating that the Co valence in the nanocrystals is 2+, confirming the CoO structure of the nanocrystals [24].

Ex situ annealing of the CoO nanoparticles in an oxygen atmosphere is likely to convert the particles into Co_3O_4. This structural evolution is verified by EELS, as shown in Figure 4.11, where the EELS spectra are acquired from the as-synthesized CoO nanocrystals, a standard bulk Co_3O_4 specimen, and the post *ex situ* annealed nanoparticles. The spectrum of the postannealed nanoparticles fits very well with that of Co_3O_4, strongly support the conversion from CoO to Co_3O_4.

4.7 EXPERIMENTAL APPROACH FOR MAPPING THE VALENCE STATES OF Co AND Mn

We now present the experimental approach for mapping the valence state of transition metal oxides. Figure 4.12(a) shows an EELS spectrum of Co_3O_4. Several

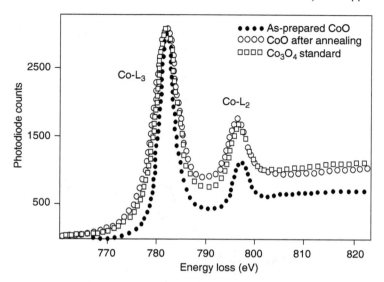

FIGURE 4.11 A comparison of EELS spectra of Co-L$_{2,3}$ ionization edges acquired from the as-prepared CoO nanoparticles, postannealed CoO in oxygen atmosphere, and the Co$_3$O$_4$ standard, proving that the CoO nanoparticles of 5 nm in size have been transformed into Co$_3$O$_4$ after annealing at 250°C for 18 h in oxygen

techniques have been proposed to correlate the observed EELS signals with the valence states, the ratio of white lines (the normalized white-line intensity in reference to the continuous state intensity located ~50 to 100 eV beyond the energy of the L$_2$ line), and the absolute energy shift of the white lines. In this study, we use the white-line intensity ratio that is calculated using the method demonstrated in Figure 4.12(b) [9], as explained in fig. 4.1.

The valence state was discerned by referring to spectra acquired from standard specimens with known cation valence states. Since the intensity ratio of L$_3$/L$_2$ is directly related to the valence state of the corresponding element, a series of EELS spectra were acquired from several standard specimens with known valence states, and an empirical plot of these data serves as a calibration for determining the valence state of the element present in a new compound (see fig. 4.2).

To map the distribution of ionization states, an energy window of ~10 eV in width is required to isolate L$_3$ from L$_2$ white lines of Mn or Co. A five-window technique is introduced (see Figure 4.12[a]): two images are acquired at the energy-loss prior to the L ionization edges, and they are to be used to subtract the background for the characteristic L edge signals; two images are acquired from L$_3$ and L$_2$ white lines, respectively, and the fifth image is recorded as loss of region right after the L$_2$ line (used to subtract the continuous background underneath the L$_3$ and L$_2$ lines). To extract the L$_3$/L$_2$ image that is most sensitive to the valence state of Mn or Co, a background, as illustrated in Figure 4.12(b), needs to

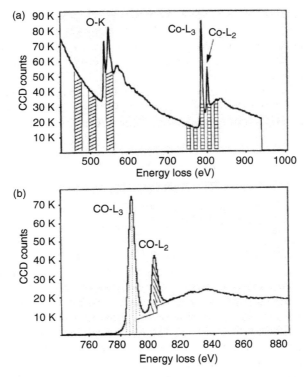

FIGURE 4.12 (a) An EELS spectrum acquired from Co_3O_4, showing the five-window technique used to extract the intensities of the white lines and the three-window technique for O–K edge. (b) The Co-L edge after subtraction of the background, illustrating the presumed background underneath L_3 and L_2 lines

be subtracted. This procedure can be done easily in the EELS spectrum, but for the energy-filtered image acquired in TEM under parallel illumination a different approach has to be taken, as given by

$$L_3/L_2 = \frac{I(L_3) - c_1 I(\text{postlines})}{I(L_2) - c_2 I(\text{postlines})} \quad (4.3)$$

where $I(L_3)$, $I(L_2)$, and $I(\text{postlines})$ are the images recorded by positioning the energy selection window at the L_3, L_2, and the post L_2 line energy-losses, respectively, after subtracting the conventional preedge background as ascribed by A $\exp(-r\Delta E)$; c_1 and c_2 are the adjustable parameters that represent the fractions of the continuous background below the L_3 and L_2 lines, respectively, as contributed by the single atomic scattering (as illustrated in Figure 4.12[b]). The choice of the c_1 and c_2 factors depend on the specimen thickness since the $I(\text{postlines})$ image is strongly affected by the multiple scattering effect, while $I(L_3)$ and $I(L_2)$ are less affected. Equation (4.1) represents the optimum choice for L_3/L_2 mapping in EFTEM under the data collection conditions allowed by the Gatan Imaging Filter (GIF). A more accurate data treatment can be adopted in scanning transmission electron microscopy (STEM).

To confirm the information provided by the L_3/L_2 images, the specimen composition is mapped from the integrated intensities of O–K and Mn-$L_{2,3}$ (or Co-$L_{2,3}$) ionization edges by following the routine procedure of EELS microanalysis [4]

$$\frac{n_O}{n_{Mn}} = \frac{I_O(\Delta)}{I_{Mn}(\Delta)} \frac{\sigma_{Mn}(\Delta)}{\sigma_O(\Delta)} \tag{4.4}$$

where $I_O(\Delta)$ and $I_{Mn}(\Delta)$ are the integrated intensities of O–K and Mn-L (or Co-L) edges for an energy window Δ, respectively, above the ionization thresholds; $\sigma_{Mn}(\Delta)$ and $\sigma_O(\Delta)$ are the integrated ionization cross-sections for the corresponding energy window Δ, and they can be calculated by the SIGMAK2 and SIGMAL2 programs in the hydrogen-like atomic model. From the energy-filtered images, the distribution map of the atomic ratio O/Mn or O/Co can be calculated.

The EFTEM experiments were performed using a Philips CM30 (300 kV) TEM, equipped with a GIF system. This TEM provides a high-beam current needed for chemical imaging. The energy window width was selected to be 10 eV for Mn or 12 eV for Co. Energy-filtered images were acquired with a 1024×1024 CCD, $2\times$ pinning, and gain normalized. Exposure times were 10 to 30 sec, depending on specimen thickness, to achieve images having satisfactory signal-to-noise ratios. The selection of the energy window width depends on the energy separation between L_3 and L_2 lines. It took 2 to 4.5 min to acquire a complete set of images. Specimen drift between different images was corrected after the acquisition, but it was important to minimize the drift of the specimen during data acquisition.

4.8 MAPPING THE VALENCE STATES OF Co USING THE WHITE-LINE RATIO

The first specimen selected for illustrating the experimental approach is a directionally solidified eutectic ZrO_2/CoO [25], which is composed of trilayer structures of ZrO_2, Co_3O_4, and CoO after heat treatment in a high-oxygen partial pressure. This is an ideal geometry for studying CoO–Co_3O_4 interfaces; as the differences in crystal structure, the coordination configuration of cations and the valence states result in dramatic differences in EELS spectra of CoO and Co_3O_4 [26]. Shown in Figure 4.13 is an one-dimensional spatially dispersed EELS spectra across a CoO–Co_3O_4 interface. The valence-loss spectra of the two phases are distinctly different. The O–K edge exhibits a double split peak for Co_3O_4 while no splitting

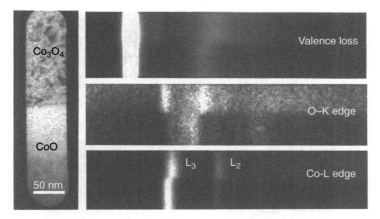

FIGURE 4.13   Zero-loss bright-field TEM image of $CoO–Co_3O_4$ interface, showing the selection area aperture for forming the EELS spectra; spectrum lines for the low-loss region, the oxygen K edge, and the cobalt L edge, exhibiting distinct differences in the intensities and energy positions of the characteristic peaks between CoO and Co_3O_4

is associated with CoO. The white lines of the two phases differ not only in their relative intensity, but also in having a slight chemical shift (1.5 eV). The width and position of the energy-selection window were carefully justified. The peak-to-peak energy between L_3 and L_2 is 15 eV and the full width of the line at 10% intensity cut-off is 7 to 8 eV, thus, the energy window width is chosen to be $\Delta = 12$ eV to separate the two lines and do ensure the signal-to-noise ratio.

Figure 4.14 shows a group of energy-filtered images from a triple point in the $CoO–Co_3O_4$ specimen. The bright-field TEM image (Figure 4.14[a]) shows that one of the grains was strongly diffracting. The energy-filtered images using L_3 and L_2 lines (Figure 4.14[b] and [c]) show the distinct difference in contrast due to a difference in the relative white-line intensities. The L_3/L_2 image given by Equation (4.1), clearly shows the distribution of cobalt oxides having different valence states (Figure 4.14[d]), where the diffraction contrast disappears. The relative fractions of the ions were determined by comparing the local average L_3/L_2 intensity with the intensity obtained using EELS from the standard specimens. The region with lower oxidation state (Co^{2+}) shows brighter contrast and the one with high oxidation states show darker contrast. Although the energy-filtered O–K edge image exhibits some diffraction contrast, the O/Co compositional ratio image greatly reduces the effect. The O/Co image was calculated from the images recorded from the O–K edge and the $L_3 + L_2$ white lines for an energy window width of $\Delta = 24$ eV. The high-intensity region in the O/Co image indicates the relative high-local concentration in oxygen (e.g., higher Co oxidation states), the low-intensity region contains relatively less oxygen (e.g., lower Co valence state), entirely consistent with the information provided by the L_3/L_2 image.

Under the single scattering approximation, the intensities of L_3 and L_2 lines scale in proportion to the specimen thickness, thus, their ratio L_3/L_2 has little

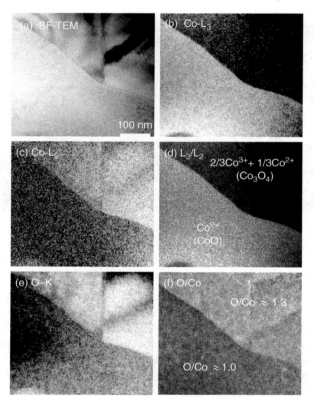

FIGURE 4.14 A group of images recorded from the same specimen region using signals of (a) the zero-loss bright-field, (b) the Co-L$_3$ edge, (c) the Co-L$_2$ edge, (d) the L$_3$/L$_2$ ratio, (e) the O–K edge, and (f) the atomic concentration ratio of O/Co. The continuous background contributed from the single atom scattering has been removed from the displayed Co-L$_3$ and Co-L$_2$ images by choosing the factors in Equation (4.1) to be $c_1 = 0.3$ and $c_2 = 0.7$. Each gain normalized image was acquired with an energy window width of $\Delta = 12$ eV except for O–K at $\Delta = 24$ eV

dependence on the specimen thickness. This result holds even for slightly thicker specimens because the near-edge structure is less affected by the multiple plasmon scattering effect and the energies of the characteristic plasmon peaks are larger than the energy split between the L$_3$ and L$_2$ lines. Therefore, in the conventional thickness range for performing EELS microanalysis of $t/\Lambda < 0.8$, where t is the specimen thickness and Λ is the mean-free-path length of electron inelastic scattering, the L$_3$/L$_2$ image truly reflects the distribution of valence state across the specimen.

Figure 4.15 shows another group of images recorded from the CoO–Co$_3$O$_4$ specimen. The L$_3$ and L$_2$ images display some contrast across the phases, while the image recorded from the postline energy-loss region shows a small contrast

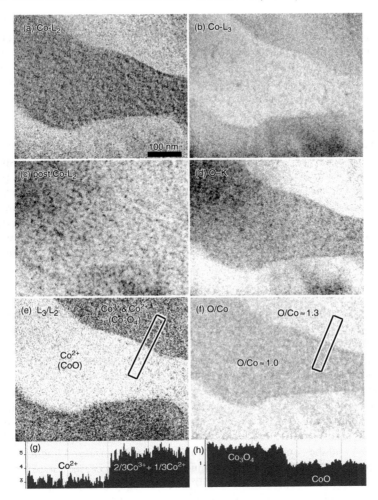

FIGURE 4.15   A group of images recorded from the same specimen region using signals of (a) the Co-L$_2$ edge, (b) the Co-L$_3$ edge, (c) the post Co-L$_2$ continuous energy-loss region, (d) the O–K edge, (e) the L$_3$/L$_2$ ratio, and (f) the atomic concentration ratio of O/Co. The continuous background contributed from the single atom scattering has been removed from the displayed Co-L$_3$ and Co-L$_2$ images. The L$_3$/L$_2$ ratio image was calculated by taking $c_1 = 0.3$ and $c_2 = 0.8$. (g) A line scan across the CoO–Co$_3$O$_4$ interface in the L$_3$/L$_2$ image averaged over a 30 pixel width parallel to the interface, showing the local average white-line ratio. Based on comparison of the displayed numbers with the values measured from standard specimens (see fig. 4.2), the regions corresponding to CoO and Co$_3$O$_4$ are apparent. (h) A line scan across the CoO–Co$_3$O$_4$ interface in the O/Co ratio image averaged over a 30 pixel width parallel to the interface, from which the local compositions match very well to CoO and Co$_3$O$_4$. Each raw image was acquired with an energy window width of $\Delta = 12$ eV except for O–K at $\Delta = 24$ eV

variation (Figure 4.15[c]) possibly because the postline region is dominated by the single atomic scattering properties and it is less affected by the solid-state effects, provided the thickness-projected density of Co atoms is fairly uniform across the grains. The image from the O–K edge show some variation due to diffraction contrast as well as specimen thickness (Figure 4.15[d]). In contrast, the L_3/L_2 image (Figure 4.15[e]) and the O/Co image (Figure 4.15[f]) show little dependence on the specimen thickness and the diffracting condition. This is uniquely suited for mapping the valence state distribution across the specimen.

To determine the optimum spatial resolution achieved in the L_3/L_2 image, a line scan across the CoO–Co_3O_4 interface averaged over a 30 pixel width is displayed in Figure 4.15(g). The half width of the profile at the interface is about 2 pixels, which corresponds to a resolution of $\sim$1.8 nm. The image across the interface in the O/Co image shows a half width of 3 pixels, which corresponds to a resolution of $\sim$2.8 nm. It must be pointed out that a better resolution achieved in the L_3/L_2 image is likely due to the smaller width of the energy window ($\Delta = 12$ eV) than the $\Delta = 24$ eV used for O/Co image as well as the sharp shape of the white lines.

4.9 In Situ Observation of Valence State Transition of Mn

For demonstrating the application of the technique for a more complex case, a reduced MnO_x powder was prepared by in situ annealing [27]. A Gatan single-tilt TEM specimen heating holder was employed to carry out the in situ EELS experiments to provide continuous temperature control from room temperature to 600°C. The column pressure was maintained at 2×10^{-7} torr or lower during the in situ analysis. Upon reduction of the oxide, cations of multi-valences would be anticipated in the system. A detailed analysis of this reduction process by EELS has been reported previously [21]. The results indicated that the reduction of MnO_2 occurs at 300°C. As the temperature reaches 450°C, the specimen is dominated by Mn^{2+} and Mn^{3+} and the composition is O/Mn = 1.3 ± 0.5, corresponding to Mn_3O_4.

To obtain a specimen containing multi-valences, the in situ annealing of one MnO_2 specimen was performed at 350°C. The resulting reduced phases were a mixture of oxides of Mn with valences of 2+, 3+, and 4+, and this model system was used for mapping the valence state distribution of Mn. Figure 4.16 shows a group of images recorded from an agglomeration of MnO_x with different valences. The bright-field image does not indicate any information about the valence states of Mn. The EFTEM Mn-L_3 image reflects the distribution of Mn phases, but its contrast is approximately proportional to the local projected thickness of the specimen. The L_3/L_2 ratio image (Figure 4.16[d]) directly gives the distribution of Mn^{4+}, Mn^{2+}, and Mn^{3+}. The low intensity regions are Mn^{4+}, and the high intensity regions are the mixed valences of Mn^{2+} and Mn^{3+}, in correspondence to the formation of Mn_3O_4. To confirm this observation, the atomic ratio O/Mn image is calculated from images acquired from O–K and Mn-L edges, and the result is given in Figure 4.16(c). The image clearly indicates that the regions with Mn^{4+} have higher O atomic concentration because of the balance of the cation charge. This is an excellent proof of the information provided by the L_3/L_2 image.

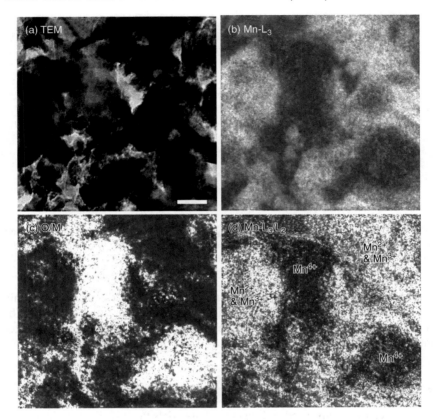

FIGURE 4.16 A group of energy-filtered images acquired from the same specimen region of mixed phases of MnO_2 and Mn_3O_4. (a) The conventional bright-field TEM image, (b) the energy-filtered TEM images of Mn-L_3 white-line, (c) the distribution of O/Mn in the region calculated according to Equation (4.4) using the energy-filtered images from O–K and Mn-L edges, and (d) the calculated Mn L_3/L_2 ratio image. The complimentary contrast of (c) to (d) proves the experimental feasibility of valence state mapping using the white-line ratio. Each raw image was acquired with an energy window width of $\Delta = 10$ eV except for O–K at $\Delta = 20$ eV

4.10 PHASE SEPARATION USING THE NEAR-EDGE FINE STRUCTURE

The near-edge fine structure observed in EELS is closely related to the solid-state effect and it is most sensitive to bonding and near-neighbor coordination configurations. Phase and bonding mapping using the near-edge structure have been carried out for diamond [28], in which the images were formed using the π^* and σ^* peaks in the C–K edge and the distribution of diamond bonding was retrieved. Similarly, the O–K edge displayed in Figure 4.17 clearly shows a difference in the near-edge structure of CoO from Co_3O_4. The first peak observed in the O–K edge of Co_3O_4 is separated by 12 eV from the main peak. Using an energy selection

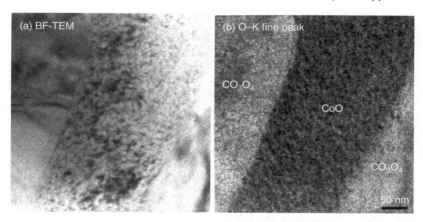

FIGURE 4.17 (a) The bright-field TEM image and (b) the energy-filtered image recorded by selecting the sharp peak located at 532 eV in the O–K, displaying the distribution of the Co_3O_4 phase. Energy window width $\Delta = 6$ eV

window of 6 eV in width it is possible to map the Co_3O_4 regions that generate this peak. A bright-field image of Co_3O_4 is given in Figure 4.16(a), and Figure 4.16(b) shows an energy-filtered image and it is clear that the Co_3O_4 regions show greater intensity, just as expected.

4.11 SUMMARY

For characterizing advanced and functional materials that usually contain cations with mixed valences, EELS is a very powerful approach with a spatial resolution higher than any other spectroscopy techniques available. Based on the intensity ratio of white lines, it has been demonstrated that the valence states of Co and Mn in oxides can be determined quantitatively. This information is important in studying valence transition in oxides.

The EELS is most sensitive to the divalent and trivalent Mn and Co ions, while the difference between Mn^{3+} and Mn^{4+} or Co^{3+} and Co^{4+} is small, leading to a larger error in the identification of the valence state because of experimental error. The only possible solution is to acquire high-quality EELS spectra. From the experimental point of view, we found that Mn and Co are the only transition metal elements whose white-line ratios are most sensitive to valence state variation, while the white lines of Fe are almost independent of its valence states. Therefore, more theoretical research is required to explore the origin of white lines.

Based on the intensity ratio of white lines, a new experimental approach has been demonstrated for mapping the valence states of Co and Mn in oxides using EFTEM. Resulting L_3/L_2 images are almost independent of either the specimen thickness (provided $t/\Lambda < 0.8$) or the diffraction effects, and are reliable for mapping the distribution of cation valences. An optimum spatial resolution of

~2.0 nm has been attained experimentally. This is a new application of the EFTEM for characterizing the electronic structure of magnetic oxides.

ACKNOWLEDGMENTS

We are grateful to Dr. J. Zhang and Dr. Z.J. Zhang for providing the $La_{0.5}Sr_{0.5}CoO_{3-y}$ and the $MnFe_2O_4$ specimens, respectively. Thanks also to Dr. A. Revcolevschi, Dr. S. McKernan, and Dr. C.B. Carter for kindly providing the oxidized CoO specimen.

REFERENCES

1. Wang, Z.L. and Kang, Z.C. *Functional and Smart Materials — Structural Evolution and Structure Analysis.* New York: Plenum Press, 1998.
2. Jin, S., Tiefel, T.H., McCormack, M., Fastnacht, R.A., Ramech, R., and Chen, L.H. Thousandfold change in resistivity in magnetoresistive La–Ca–Mn–O films. *Science* 1994, *264*, 413–415.
3. von Helmolt, R., Wecker, J., Holzapfel, B., Schultz L., and Samwer, K. Giant negative magnetoresistance in perovskite like $La_{2/3}Ba_{1/3}MnO_x$ ferromagnetic films. *Phys. Rev. Lett.* 1994, *71*, 2331–2334.
4. Egerton, R.F. *Electron Energy-Loss Spectroscopy in the Electron Microscope* 2nd ed. New York: Plenum Press, 1996.
5. Krivanek, O.L. and Paterson, J.H. ELNES of 3d transition-metal oxides: I. Variations across the periodic table. *Ultramicroscopy* 1990, *32*, 313–318.
6. Pease, D.M., Bader, S.D., Brodsky, M.B., Budnick, J.I., Morrison, T.I., and Zaluzec, N.J. Anomalous L_3/L_2 white line ratios and spin pairing in 3d transition metals and alloys: Cr metals and $Cr_{20}Au_{80}$. *Phys. Lett.* 1986, *114A*, 491–494.
7. Morrison, T.I., Brodsky, M.B., Zaluzec, N.J., and Sill, L.R. Iron d-band occupancy in amorphous Fe_xGe_{1-x}. *Phys. Rev. B* 1985, *32*, 3107–3111.
8. Pearson, D.H., Fultz, B., and Ahn, C.C. Measurement of 3d state occupancy in transition metals using electron energy-loss spectroscopy. *Appl. Phys. Lett.* 1988, *53*, 1405–1407.
9. Pearson, D.H., Ahn, C.C., and Fultz, B. White lines and d-electron occupancies for the 3d and 4d transition metals. *Phys. Rev. B* 1993, *47*, 8471–8478.
10. Kurata, H. and Colliex, C. Electron-energy loss core-edge structures in manganese oxides. *Phys. Rev. B* 1993, *48*, 2102–2108.
11. Mansot, J.L., Leone, P., Euzen, P., and Palvadeau, P. Valence of manganese, in a new oxybromide compound determined by means of electron energy loss spectroscopy. *Microsc. Microanal. Microstruct.* 1994, *5*, 79–90.
12. Fortner, J.A. and Buck, E.C. The chemistry of the light rare-earth elements as determined by electron energy-loss spectroscopy. *Appl. Phys. Lett.* 1996, *68*, 3817–3819.
13. Lloyd, S.J., Botton, G.A., and Stobbs, W.M. Changes in the iron white-line ratio in the electron energy-loss spectrum of iron-copper. *J. Microsc.* 1995, *180*, 288–293.
14. Yuan, J., Gu, E., Gester, M., Bland, J.A.C., and Brown, L.M. Electron energy-loss spectroscopy of Fe thin-films on GaAs(001). *J. Appl. Phys.* 1994, *75*, 6501–6503.

15. Botton, G.A., Appel, C.C., Horsewell, A., and Stobbs, W.M. Quantification of the EELS near-edge structures to study Mn doping in oxides. *J. Microsc.* 1995, *180*, 211–216.

16. Thole, B.T. and van der Laan, G. Branching ratio in X-ray absorption spectroscopy. *Phys. Rev. B* 1988, *38*, 3158–3169.

17. Wang, Z.L. and Yin, J.S. Cobalt valence and crystal structure of $La_{0.5}Sr_{0.5}CoO_{2.25}$. *Phil. Mag. B* 1998, *77*, 49–65.

18. Zhang, Z.J., Wang, Z.L., Chakoumakos, B.C., and Yin, J.S. Temperature dependence of cation distribution and oxidation state in magnetic Mn–Fe ferrite nanocrystals. *J. Am. Chem. Soc.* 1998, *120*, 1800–1804.

19. Goodenough, J.B. Metallic oxides. *Prog. Solid State Chem.* 1971, *5*, 145–399. In *Proceedings of the Microscopy Society of America*, Cleveland.

20. Wang, Z.L. and Kang, Z.C. *Functional and Smart Materials — Structural Evolution and Structure Analysis.* New York: Plenum Press, 1998.

21. Wang, Z.L., Yin, J.S., Jiang, Y.D., and Zhang, J. Studies of Mn valence conversion and oxygen vacancies in $La_{1-x}Ca_xMnO_{3-y}$ using electron energy-loss spectroscopy. *Appl. Phys. Lett.* 1997, *70*, 3362–3364.

22. Wang, Z.L. and Zhang, J. Tetragonal domain structure magnetoresistance of $La_{1-x}Sr_xCoO_3$. *Phys. Rev. B* 1996, *54*, 1153–1158.

23. Wang, Z.L. and Zhang, J. Microstructure of conductive $La_{1-x}Sr_xCoO_3$ grown on MgO(001). *Phil. Mag. A* 1995, *72*, 1513–1529.

24. Yin, J.S. and Wang, Z.L. Ordered self-assembling of tetrahedral oxide nanocrystals. *Phys. Rev. Lett.* 1997, *79*, 2570–2573.

25. Bentley, J., McKernan, S., Carter, C.B., and Revcolevschi, A. Microanalysis of directionally solidified cobalt oxide-zirconia eutectic. In *Microbeam Analysis*, Armstrong, J.T. and Porter, J.R., Eds., 1993, Vol. 2 (Suppl.), pp. S286–S287.

26. Bentley, J. and Anderson, I.M. Spectrum lines across interfaces by energy-filtered TEM. In *Proceedings of the Microscopy and Microanalysis*, Bailey G.W., Corbett J.M., Dimlich R.V.W., Michael, J.R., and Zaluzec, N.J., Eds. San Francisco: San Francisco Press, 1996, pp. 532–533.

27. Wang, Z.L., Yin, J.S., Zhang, Z.J., and Mo, W.D. *In-situ* analysis of valence conversion in transition metal oxides using electron energy-loss spectroscopy. *J. Phys. Chem. B* 1997, *101*, 6793–6798.

28. Mayer, J. and Plitzko, J.M. Mapping of ELNES on a nanometre scale by electron spectroscopic imaging. *J. Microsc.* 1996, *183*, 2–8.

5 Surface Processes and Composition of Metal Oxide Surfaces

B. Pawelec
Institute of Catalysis and Petrochemistry, CSIC, Madrid, Spain

CONTENTS

5.1 INTRODUCTION

Metal oxides constitute a class of inorganic materials that have an extraordinary range of properties and are therefore used in many technological applications involving the properties of bulk and surface layers. An account of the properties of metal oxides in catalysis, gas sensing, and fuel cells can be found in Section 5.2.

As with most materials, many of the properties of metal oxides depend strongly on the preparation procedure employed in their synthesis. During the life of these materials, the properties of surface and interface usually play an important role in the overall properties and behavior; this holds for both preparation and use. This is why the chemical and structural features of surfaces and interfaces must be analyzed carefully in order to maintain a permanent control over reactivity. With a few exceptions, the properties of many metal oxides are still poorly known and under-exploited in comparison with metals. Nevertheless, there is currently great

111

interest in exploring this huge field of materials science. This growing activity is related to the impressive development of instrumental tools over the last two decades, making it possible to investigate — at local scale — chemical structures, bonding, chemical composition, and the bonding of adsorbates to the surface layer. Additionally, the more demanding functionalities for high performances and tailor-made materials require accurate characterization of the internal and the external surfaces of solids, particularly in advanced materials such as oxide-based composites, oxide catalysts, oxide coatings in protecting layers, chemical sensors, oxide electrodes for high temperature fuel cells, among others [1–3].

The surface of many metal oxides is altered when exposed to the external environment. Since the symmetry and coordination of the metal ions is lost at the surface, they show a strong tendency to become saturated by reacting with gas molecules. One major process that takes place is hydroxylation, which is the result of a true chemical reaction between surface M–O bonds and water molecules. The kinetics of this process depends on many variables, among which temperature plays an important role, and the resulting hydroxyl group densities are basically dictated by the nature of the metal oxide. Another process that occurs simultaneously with hydroxylation — specifically on metal oxides of basic character — is carbonation.

Many studies have addressed the chemical reactivity of surface hydroxyl groups in relation to applications such as adsorbents, catalysts, as well as ion exchangers. The presence of hydroxyl groups with different energies has been established using infrared (IR) spectroscopy [4,5], and it has been suggested that such groups are heterogeneous. Evidence for the formation of surface hydroxyl groups on the surface of metal oxides also comes from other techniques. For instance, analysis of the O1s profile in the photoelectron spectra of a broad variety of oxides exposed to moisture or reduced in a hydrogen environment indicates the appearance of more than one O-species [6–8]; that is, the lattice O^{2-} ions are, in general, accompanied by other less electron-rich oxygen species such as hydroxyl groups. Accordingly, in this chapter selected examples are considered in order to illustrate the high reactivity of a metal oxide surface when exposed to the surrounding atmosphere for prolonged periods. The hydroxylation and carbonation of surfaces are discussed, particularly in the case of basic metal oxides.

Another important factor to be considered is the surface composition of metal oxides. A common characteristic of solid solutions or complex metal oxide systems is the appearance of a segregated phase on the surface as a consequence of the inability of the lattice to accommodate homogeneously all the ions in the oxide lattice [9–13]. This situation is even more critical in metal oxides supported on a porous substrate. When the precursors of the metal oxides are being deposited within the intricate porous networks of currently used supports (SiO_2, Al_2O_3, ZrO_2, etc.), mass transfer limitations are usually involved. This makes it difficult, if not impossible, for equilibrium to be reached, which leads to a concentration gradient of the supported oxide between the external and the internal surface. Most supported metal oxides prepared by methodologies involving the chemistry of solutions show inhomogeneities in their distribution across the pore network.

Thus, the differences in the distribution of a metal oxide in a solid solution and in supported metal oxides are also discussed in this chapter, referring to highly surface-sensitive tools to reveal the atomic composition of the surface layer of selected systems.

5.2 HYDROXYLATION/DEHYDROXYLATION OF METAL OXIDE SURFACES

Oxide surfaces usually terminate in oxide ions due to the larger size of the O^{2-} ion as compared with the M^{n+} cation and low polarizing power. In the crystal of a MO_x oxide, the symmetry and coordination of the M^{n+} cations are lost at the surface. This surface unsaturation tends to be compensated by reaction with gases, and particularly with moisture. Upon exposure to water, the surface of a metal oxide undergoes a series of chemical reactions that are largely dictated by the chemistry of the M^{n+} ion. The first step in surface hydration involves the formation of surface hydroxyl groups (dissociative adsorption) followed by molecular adsorption. Using high resolution electron energy loss spectroscopy (HREELS), it has been demonstrated that H_2O adsorbed on the (110) layer of TiO_2 originates a first layer, composed of dissociated molecules characterized by the stretching frequency of OH groups at 3690 cm^{-1}, and a second layer due to molecularly adsorbed water, as evidenced by its stretching and bending modes at 3420–3505 and 1625 cm^{-1}, respectively [14]. H-bonding between molecular H_2O and OH groups also occurs.

In other oxides, the hydroxylation reaction is strongly limited by a high activation barrier. For instance, in the case of silica, the totally dehydroxylated surface is highly hydrophobic and hydroxylation can only be achieved under severe pretreatments such as high temperature conditions and pretreatment with aggressive chemicals [15].

5.2.1 Surface Hydroxyl Site Densities

Most metal oxides react with adsorbed water molecules to form hydroxyl groups, which can be regarded as the ion-exchange sites [16–20]. The two types of surface hydroxyl groups (acid and base) formed by the donation of protons from the adsorbed water to oxide ions are the anion- and cation-exchange sites, and their capacities and properties depend on the surface hydroxyl groups [21]. The population of hydroxyl site densities in metal oxides can be measured by different methods, including surface acid–base, ion-exchange reactions [22,23], reaction with Grignard reagents [21], IR spectroscopy [4,5,24], photoelectron spectroscopy [6,25–28], dehydration upon heating [29,30], and crystallographic calculations [31]. Among these, acid–base exchange sometimes affords far smaller values than those obtained with other methods.

Owing to its simplicity, the reaction of hydroxyl groups with Grignard reagents has frequently been used to measure surface densities. Hydroxyl groups (–OH) on metal oxides react with methyl magnesium bromide (CH_3MgBr), yielding CH_4

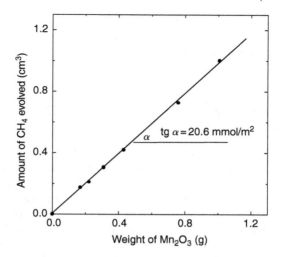

FIGURE 5.1 Amount of the methane evolved from a bulk Mn_2O_3 sample upon reaction of its hydroxyl groups with methyl magnesium bromide (CH_3MgBr) reagent

according to the equation [21]:

$$\vdash OH + CH_3MgBr \rightarrow \dashv OMgBr + CH_4 \qquad (5.1)$$

The amount of CH_4 evolved allows determination of the surface density of the hydroxyl groups. Figure 5.1 shows the volume of CH_4 (STP) evolved by the Grignard reaction (Equation [5.1]) as a function of the weight of a bulk Mn_2O_3 (Aldrich) sample. From the slope of the straight line, the amount of hydroxyl sites per unit weight can be derived, and the surface OH density ($mmol/m^2$) can be readily obtained by dividing the slope by the specific area (m^2/g) of the oxide. The surface densities of the hydroxyl groups of several metal oxides thus calculated are listed in Table 5.1 together with their specific areas.

The surface densities of the hydroxyl groups included in Table 5.1 agree reasonably well with the values reported by other authors. For TiO_2, the value of $17.3\ mmol/m^2$ is only slightly lower than the $20.7\ mmol/m^2$ obtained by tritium exchange with OH groups [31]. Also, there is good agreement in the OH surface densities obtained by the Grignard method for the Fe_2O_3 (Aldrich) sample ($22.9\ mmol/m^2$) listed in Table 5.1 and a Fe_2O_3 (Kanto A) sample using the exchange reaction between OH and NaOH ($23.8\ mmol/m^2$) [21]. The surface OH site densities compiled in Table 5.1 and many other data from the literature are quite similar for several oxides with di-, tri-, and tetravalent ions. However, assuming that the chemisorption of water molecules occurs only on the Lewis acid cation sites, differences in hydroxyl site densities would be expected, since the number of lattice metal ions in oxides changes with valence. To explain these differences, a different mechanism of surface hydroxylation must be considered.

TABLE 5.1

Specific BET Areas and Hydroxyl Site Densities of Pure Metal Oxide Samples

Metal oxide	BET area (m^2/g)	[OH] $(mmol/m^2)$
ZrO_2 (MEL)	104	22.9
TiO_2 (Degussa)	25	17.3
Cr_2O_3 (Aldrich)	5.9	24.5
MnO_2 (CI1)[a]	43	22.5
Mn_2O_3 (Aldrich)	2.0	20.6
Mn_3O_4 (Aldrich)	2.7	29.4
Fe_2O_3 (Aldrich)	7.6	22.9
Fe_3O_4 (Aldrich)	1.4	25.7
CoO (Aldrich)	3.4	31.0

[a] Taken from Tamura, H., Tanaka, A., Mita, K., and Furuichi, R. *J. Colloid Interface Sci.* 1999, *209*, 225–231.

Metal oxide surfaces usually expose oxide (O^{2-}) ions, which are strong bases and cannot exist unchanged in aqueous solution. If an O^{2-} ion is brought into an aqueous solution as a free ion, it will be instantaneously neutralized by a water molecule, yielding two hydroxide ions according to the reaction:

$$O^{2-} + H_2O \rightarrow 2OH^- \tag{5.2}$$

The entire surface of O^{2-} ions of metal oxides exposed to water molecules may undergo this neutralizing reaction to form OH groups, and the water molecules turn into the other type of OH groups by losing protons. These two kinds of OH groups are conjugated acids (OH_a) of lattice oxide ions and conjugated bases (OH_b) of water molecules, and hence their chemical nature is different. The two kinds of hydroxyl groups are considered to form two layers, as depicted in Figure 5.2. From the top view sketched in this figure, quantification of surface OH site densities can be achieved. Assuming a radius r for hydroxyl groups closely packed in two layers, the area of the hexagon is $6\sqrt{3}r^2$. Since, in the hexagon, there are three OH groups in the first layer and another three in the second, the hydroxyl group site density is $N_{OH} = 1/(\sqrt{3}r^2 N_A)$, where N_A is Avogadro's number. By using a value of 0.145 nm for the radius of the hydroxyl group [21], a value of 45.6 $\mu mol/m^2$, almost twice the value of the measured ones, is obtained. This difference can be explained by taking into account that the ionic radius of OH groups was determined for bulk ionic crystals, where hydroxide ions alternate with lattice cations. As stated earlier, terminal OH groups lie adjacent to each other on the oxide surface, and their electric charges are not symmetrically neutralized by the cations due to a

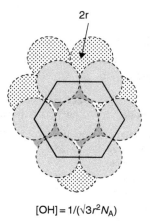

$$[OH] = 1/(\sqrt{3}r^2 N_A)$$

FIGURE 5.2 Closest packing of hydroxide anions with radius r. Each shade corresponds to one layer

local imbalance of anion and cation arrangements at the surface. As a result, there would be electrostatic repulsion between OH, leading to a looser packing. This simple argument could explain why the experimental N_{OH} values are smaller than the calculated ones.

5.2.2 Dehydroxylation of Metal Oxide Surfaces

As shown earlier, hydroxyl groups are developed on metal oxide surfaces as a consequence of the reaction between water molecules and exposed oxide (O^{2-}) ions. Removal of these OH groups can be achieved by thermal treatment of the oxides under controlled atmospheres. The process can be described schematically by the equation:

$$\left| \begin{array}{l} -OH \\ -OH \end{array} \right. \longrightarrow \triangleright O + H_2O \tag{5.3}$$

The extent of the dehydroxylation process is a critical issue, particularly when metal oxides are being used for practical applications. Although quantification of remaining (and removed) hydroxyl groups can be achieved by different techniques [21–29], differences in the binding energies of the O1s core levels of O^{2-} and OH^- surface species are frequently observed, which makes them a simple and accurate procedure for monitoring surface dehydroxylation. To illustrate this, and taking into account the differences in binding energies of the O1s level of surface O^{2-} and OH^- species, a bulk Mn_2O_3 sample exposed to ambient atmosphere has been examined by photoelectron spectroscopy. The O1s core levels of the Mn_2O_3 sample recorded upon degassing at temperatures between 303 and 673 K are shown

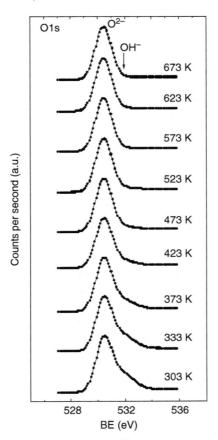

FIGURE 5.3 O1s core-level spectra of a bulk Mn_2O_3 sample degassed at increasing temperatures

in Figure 5.3. At lower degassing temperatures, two components at 530.4 and 531.7 eV associated with lattice oxide ions and hydroxyl groups, respectively, can be identified. The possibility that some carbonate structures might be involved in the O1s component at 531.7 eV can be ruled out since no C-species lying in the energy region at about 289 eV are detected. Upon degassing at increasing temperatures, the higher binding energy component decreases in intensity and disappears at 673 K.

Although the O1s spectra in Figure 5.3 clearly show the progress of the dehydroxylation processes upon increasing the degassing temperature, a more precise understanding of the process can be gained by plotting the hydroxyl groups-to-oxide ion intensity ratios as a function of the degassing temperature (Figure 5.4). The [OH]/[O] intensity ratio decreases almost exponentially with increasing degassing temperature. This ratio drops strongly in the range of 303 to 450 K ; at about 400 K, one-half of the surface hydroxyl groups are removed, while

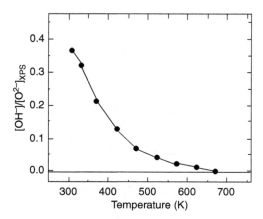

FIGURE 5.4   Evolution of the relative surface concentration of hydroxyl groups as a function of degassing temperature

temperatures of 673 K are required for a completely dehydroxylated surface to be achieved. It can be said that the qualitative and quantitative information collected in Figure 5.3 and Figure 5.4, respectively, is a power tool for developing a protocol to obtain reproducible metal oxide surfaces, which is an essential requisite if samples prepared by different routes are to be compared.

5.3 CARBONATION

The surface of metal oxides are usually hydroxylated and carbonated when exposed to air, a phenomenon that may further progress in subsurface layers depending on the reactivity of the oxide and the ageing time. This kind of process takes place, to a much greater extent, in the oxides of basic cations, such as alkaline, alkaline earth, and lanthanide oxides. The nature and the extent of the carbonated/hydroxylated surfaces can be readily determined by photoelectron spectroscopy since this technique is highly surface sensitive. Many investigators have addressed the analysis of the surface of basic metal oxides [6,25–28,32–36]. The information gained is derived from the analysis of the O1s and C1s line profiles.

The O1s and C1s photoelectron spectra of a powdered La_2O_3 sample heated under high vacuum at 303, 623, and 923 K are shown in Figure 5.5 and Figure 5.6, respectively. The fit of the O1s profile was done assuming two components: one at low binding energy (529.0 eV) (hereafter O_L), which is typical of lattice oxide ions (O^{2-}), and another at high binding energy (531.8 eV) (hereafter O_H), which can be attributed to different species such as carbonates, hydroxyls, O–, etc., all of them with very similar binding energy [28,36]. The C1s profile also shows two components at 284.9 and 289.8 eV (C_H), unambiguously assigned to adventitious carbon from sample contamination and carbonate structures, respectively (Figure 5.6). Heating of the sample at increasing temperatures (623 and 923 K) leads to a progressive decrease in the intensity of the O_H component. This

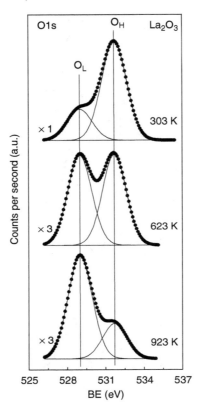

FIGURE 5.5 O1s core-level spectra of a La_2O_3 sample degassed at different temperatures under high vacuum conditions

observation indicates that these O-species are partially eliminated upon thermal treatment under vacuum. Similar behavior can be observed in C_H at 289.8 eV, although in this case the decrease in peak intensity with temperature is less pronounced than in O_H. These observations were also confirmed with a ZnO sample, although complete dehydration and decarbonation took place at temperatures of about 773 K (see Table 5.2).

Quantitative evaluation of O1s and C1s spectra allows one to obtain a better understanding of the evolution of oxygen- and carbon-containing species on the surface of La_2O_3 sample. The surface atomic ratios and peak percentages calculated from the area of O1s and C1s components are summarized in Table 5.2. This table also includes the surface ratios and peak percentages for a ZnO sample, subjected to thermal treatments under high vacuum between 303 and 773 K, and a $CaCO_3$ reference sample. These results indicate that the carbonate peaks of La_2O_3 and ZnO, and at least some of the OH contribution, are due to the presence of carbonate species in the surface region sampled by the photoelectron techniques. The fact that the O_H/C_H atomic ratio in the $CaCO_3$ reference sample approaches

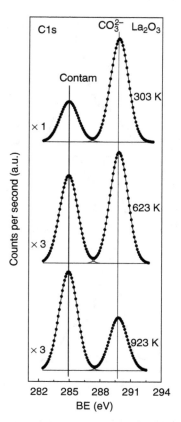

FIGURE 5.6   Photoelectron spectra of C1s level of a La_2O_3 sample degassed at different temperatures under high vacuum conditions

the stoichiometric composition ($O/C = 3$) means that similar ratios should be observed if the surface of the two oxides analyzed were covered by a carbonate layer. However, the O_H/C_H ratio only approaches 3 upon thermal treatment of L_2O_3 at 923 K and ZnO at 673 K. Thus, the large deviation in these ratios in the samples outgassed at 303 K must be considered as indicative of the presence of O-containing species other than carbonates.

Many other mixed La–M–O (M, first transition row element) oxides analyzed by x-ray diffraction and thermogravimetry methods [25,27] can be hydroxylated and carbonated by moisture, giving $La(OH)_3$ and $La(OH)_x(CO_3)_y$ compounds. These structures, which are well characterized, undergo dehydroxylation at temperatures above 673 K, while their decarbonation takes place at somewhat higher temperatures: typically above 873 K [37]. This finding agrees with the O_H/C_H ratios and the percentage of O_H reported in Table 5.2, indicating that dehydroxylation progresses substantially at 623 K while decarbonation requires higher temperatures (923 K). The same applies for the ZnO sample, although

TABLE 5.2
Surface Ratios and Peak Percentages Derived from
the Area of O1s and C1s Spectra

Sample/treatment (K)	$O_H{}^a$	O_H/C_H	$O(C_H)/O_H \times 100^b$
La$_2$O$_3$			
303	70	5.4	51
623	49	5.2	62
923	34	3.3	84
ZnO			
303	37	3.7	81
673	16	3.2	92
773	0	0	—
CaCO$_3$			
303	100	3.1	100

^{a} Percentage of the OH species referred to total oxygen.
^{b} Percentage of carbonate species with respect to total oxygen.

both dehydroxylation and decarbonation processes take place at temperatures substantially lower than in La$_2$O$_3$.

The examples discussed here elegantly illustrate the fact that the surface of La$_2$O$_3$ and ZnO is not only hydroxylated but also carbonated when these oxides are exposed to the ambient environment. The results also show that the carbonate structures are very stable, particularly in the case of La$_2$O$_3$ and other La-containing mixed oxides [25,27]. It should also be stressed that basic oxides, such as the oxides of alkaline, alkaline earth, and lanthanide elements, are readily carbonated/hydroxylated when exposed in air. This, as a consequence of the difficulties in controlling their level of carbonation hydroxylation, represents an important source of error in the evaluation of the final stoichiometry of binary or more complex combinations of these oxides.

5.4 DEVIATIONS OF SURFACE COMPOSITION

5.4.1 Solid Solutions

The ability to tailor metal oxide systems for physical and chemical applications represents an obvious advantage for improved performance. Interesting examples are cobalt and nickel oxides, which find applications in many oxidation reactions and also as promoters of Mo and W oxide catalysts for the hydrodesulfurization reactions of middle distillates. NiO and CoO can be mixed in all proportions to form homogeneous solid solutions of the type $Co_xNi_{1-x}O$, $0 \leq x \leq 1$, in a well-ordered rock salt crystal structure in the bulk material. When used in redox reactions in alkaline media [38], these solid solutions exhibit electrochemical properties. The reason why CoO and NiO form solid solutions lies in the close match of their individual unit cell structures. Both oxides are fcc rock salt structures and have comparable lattice parameters of 0.4168 and 0.4267 nm for NiO and CoO, respectively, and at intermediate compositions the lattice parameter varies linearly as a function of x [39]. In addition, a thermodynamically stable mixed $NiCo_2O_4$ spinel exists, in which all cobalt ions are in the +3 oxidation state. The $NiCo_2O_4$ phase, which adopts an inverse spinel structure has been shown to undergo phase separation at the surface into NiO and a cobalt-enriched spinel of the type $Ni_xCo_{2+x}O_4$. What is interesting in the mixed oxide is the difference in oxidation properties of NiO and CoO. Thus, while CoO forms a layer of the Co_3O_4 even under relatively low partial pressures of oxygen, the Ni-containing oxides are thermodynamically less accessible, and the spinel Ni_3O_4 is much harder to obtain [40].

Quantitative analyses of mixed $Co_xNi_{1-x}O$ oxides by surface-sensitive Auger and x-ray photoelectron spectroscopic techniques provide evidence of differences in surface and bulk composition for $x < 0.6$ [39]. The line profile and binding energies of the most intense Co $2p_{3/2}$ and Ni $2p_{3/2}$ core-level spectra of samples with $x < 0.5$ are characteristics of rock salt CoO and NiO. The constant values of the binding energy of the Ni $2p_{3/2}$ level at 854.6 eV and of the Co $2p_{3/2}$ level at 780.2 eV for solid solutions with $x < 0.5$ indicates that the chemical environment of the cations remains indistinguishable by photoelectron spectroscopy, and they agree with those of the $2p_{3/2}$ levels observed for the bulk CoO and NiO single crystals [39]. In addition, the satellite structures of Co^{2+} and Ni^{2+} ions are characteristic of photoemission processes from CoO and NiO species. Deviation from rock salt stoichiometry in either direction to form O-vacancies or through oxidation to generate Ni^{3+} and Co^{3+} ions has been shown to decrease strongly the intensity of the satellites.

An evaluation of the Ni and Co concentration can be made by measuring the peak intensities. Sensitivity factors for the various transitions can be estimated by assuming an exponential escape-depth energy dependence for the photoelectron:

$$I_M = I_o \cdot \Sigma[M] \cdot \sigma_M \cdot \exp[-n \cdot x/\lambda_M \cdot \cos\theta] \qquad (5.4)$$

where I_M is the integral peak intensity of Co or Ni, I_o is the x-ray photon intensity, [M] is the concentration of Ni or Co in the nth layer in the solid, σ_M is the photoemission cross-section, x is the distance between layers in the crystal lattice, λ_M is the mean free path for an electron with kinetic energy appropriate to the transition of interest, and $\cos\theta$ corrects the analyzer detection angle. Assuming the concentration of Ni and Co to be homogeneous in the near-surface region, a sensitivity factor may be defined, and the concentration proportional to peak intensity

$$[M] = I_M/(I_o \cdot S_M) \qquad (5.5)$$

can be used to obtain the surface concentrations of Ni and Co ions. It should be noted that I_o remains constant within a given series of measurements, and is cancelled by taking ratios of intensities.

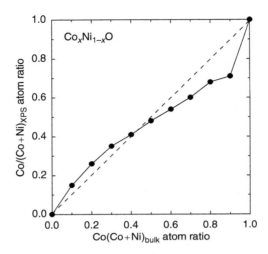

FIGURE 5.7 Surface cobalt concentration in $Co_xNi_{1-x}O$ solid solutions as a function of the bulk value (Reproduced with permission from Elsevier)

Surface atomic ratios of cobalt to the sum of cobalt and nickel [Co/(Co+Ni)], are plotted as a function of bulk composition of the $Co_xNi_{1-x}O$ samples in Figure 5.7. From this plot, it is evident that the surface of the $Co_xNi_{1-x}O$ mixed oxides becomes Ni-enriched for $x > 0.6$. The Ni $2p_{3/2}$ binding energies remain essentially constant (854.6 $\pm$ 0.2 eV) and — simultaneously — the solid solution concentration varies across the series. Ni^{3+} ions in solid oxides usually display binding energies almost 1.4 eV higher than this value [41]. Therefore, the data are inconsistent with the presence of Ni^{3+} as a dominant species in the near-surface region for the $Co_xNi_{1-x}O$ solid solutions, regardless of composition. Tetrahedrally coordinated Ni^{2+} in $NiAl_2O_4$ and $NiCr_2O_4$ has also been reported, with binding energies 855.4–856.2 eV. Thus, although the spinel phase appears to be developed in the $x = 0.6$–0.7 region, the nickel species is predominantly octahedrally coordinated Ni^{2+}. This suggests that the Ni-enriched oxide forming at approximately $x \approx 0.6$ bulk cobalt concentration is an inverse spinel structure with the classic cationic $Co^{3+}[Ni^{2+},Co^{2+}]O_4$ distribution, where the cations in brackets are octahedrally coordinated and those preceding the bracket are tetrahedrally coordinated. While the bulk $NiCo_2O_4$ is classified as an inverse spinel, variations in site occupancy have been proposed and bulk magnetic properties have been interpreted as more in keeping with a distribution of site occupancies given by $Co_\delta^{2+}Co_{1-\delta}^{3+}[Ni_{1-\delta}^{2+}Ni_\delta^{3+}Co^{3+}]O_4$, with the majority of nickel as octahedrally coordinated Ni^{2+}. This may represent a difference in the spinel-like surface formed on the $Co_xNi_{1-x}O$ solid solution as compared with native $NiCo_2O_4$ surfaces.

More complex mixed oxides such as $La_{1-x}Ce_xCoO_3$ usually display important deviations in surface composition with respect to bulk composition. In the family of $La_{1-x}Ce_xCoO_3$ ($x = 0.0$ to 0.5) oxides calcined at 1173 K, it was found that

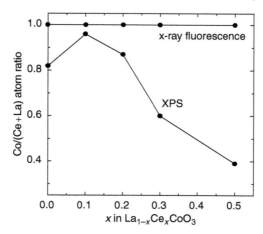

FIGURE 5.8 Surface and bulk concentration of cobalt in $La_{1-x}Ce_xCoO_3$ mixed oxides as a function of the substitution degree (x) (Reproduced with permission from Springer)

surface composition, as determined by photoelectron spectroscopy, varies to a large extent with substitution (x) [42]. The cobalt surface ratio increased gradually with cerium substitution, showing a maximum at $x = 0.1$, and then decreased again (Figure 5.8). However, the composition, as determined by x-ray fluorescence, agreed well with stoichiometric values for all the samples.

The activity profile of $La_{1-x}Ce_xCoO_3$ samples in the oxidation of CO at 523 K and of CH_4 at 733 K follow a close parallelism with the cobalt surface ratio. Since the activity of the individual CeO_2 and La_2O_3 oxides is negligible in both reactions, it can be inferred that activity must be associated with surface Co^{3+} ions. In addition, the constant and stoichiometric Co/(La+Ce) ratio determined by x-ray fluorescence, which probes more layers than photoelectron spectroscopy, leads to the conclusion that the surface region of the mixed oxide crystal is contaminated by the La_2O_3, CeO_2, and Co_3O_4 oxides. Therefore, the maximum in the Co/(La+Ce) ratio at $x = 0.1$ may be reasonably assigned to the presence of a Co_3O_4 phase on the surface of the mixed $La_{1-x}Ce_xCoO_3$ oxide, which in turn catalyzes the oxidation of CO and CH_4. This, and many other examples, illustrates the relationship between catalytic behavior and the presence of an individual oxide segregated on the surface of a mixed oxide catalyst.

5.4.2 Inhomogeneous Distribution of Metal Oxides in Porous Media

The incorporation of metal oxides on mesoporous and microporous substrates often results in inhomogeneities in the distribution of the oxide phase across the pore network. This is particularly critical in the case of zeolites, in which the dimensions of the channels are too small to facilitate the diffusion of the oxide phases during the preparation steps. As a result of this, molybdenum oxide-loaded zeolites are excellent examples for showing how the external surface usually becomes enriched in the oxide phase.

In the case of Y-type zeolite, the application of conventional methods to incorporate molybdenum precursors is hindered because of its micropore structure and low ion-exchange capacity. Other less common preparation methods, such as solid–solid ion exchange with $MoCl_5$, decomposition of the neutral $Mo(CO)_6$ complex and impregnation with ammonium heptamolybdate precursor followed by thermal decomposition at high temperature under vacuum, are required [43, and references therein]. Since these methodologies may vary in the chemical environment, symmetry, and location of the molybdenum oxide in the zeolite, and because these factors exert a strong effect on reactivity, precise knowledge of the chemical structures and electronic characteristics of Mo ions generated by the preparation method and pretreatments are of prime importance.

The XANES spectra of Mo L_{III} x-ray absorption edges indicate that there is no edge shifting of the Mo edge in the three Mo-loaded zeolites (Figure 5.9). At the L_{III} edge, the atomic-like electric dipole transitions from the 2p initial states to the 4d final states produce the largest white lines observed. In all the samples, the white line at the Mo L_{III} edges is split into a doublet as a consequence of the ligand field splitting of the final-state d orbitals. For tetrahedral symmetry, the magnitude of the splitting is less than the octahedral field. The range of values

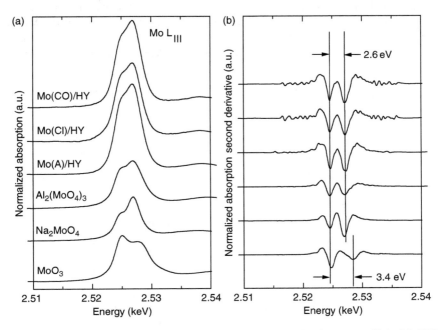

FIGURE 5.9 Mo L_{III} edge XANES spectra (a) and second-derivative curves (b) for Mo/HY catalysts and reference compounds (Reproduced with permission from American Chemical Society)

for the splitting at the Mo L_{III} edge is 1.8 to 2.6 eV for tetrahedral coordination and 3.1 to 4.6 eV for Mo octahedrally coordinated with oxide anions (O^{2-}). The splitting values, as determined in the second derivative spectra, of the MoO$_3$-zeolites and reference compounds (Figure 5.9) clearly indicate that molybdenum atoms are tetrahedrally surrounded by O^{2-} ions. The number of available orbitals is reflected in the relative intensity of the L_{III} edge absorption and can be predicted through simple ligand-field theory. At the Mo L_{III} edge in compounds with octahedrally coordinated molybdenum, the first maximum is higher than the second, and vice versa for tetrahedral symmetry. With respect to the differences in white-line intensity between MoO$_3$ zeolites and reference compounds, the fact that the white-line intensity in MoO$_3$ zeolites is higher than in the reference samples could be related to the stabilization of small MoO$_3$ clusters interacting with exchange sites in the HY zeolite cages.

Comparison of Mo/Si surface ratios derived from photoelectron spectroscopy with Mo/Si ratios derived from chemical analysis (Figure 5.10) indicates that molybdenum oxide is quite well dispersed within the pore network of the HY zeolite. However, all Mo/Si surface ratios are slightly higher than Mo/Si bulk ratios. This finding indicates a certain Mo-enrichment in the near-surface layers of the zeolite crystals. If Mo-loaded zeolites were subjected to a consecutive impregnation, the distribution of MoO$_3$ species and their location along the pore channels would change to a significant extent. The data plotted in Figure 5.10 for three calcined samples subjected to a consecutive impregnation with a Pd^{2+} salt clearly indicate that the Mo/Si bulk ratio decreases and also that the Mo/Si surface ratio approaches the Mo/Si ratio determined by chemical analysis. These data allow one to conclude that a fraction of the MoO$_3$ loading is lost during

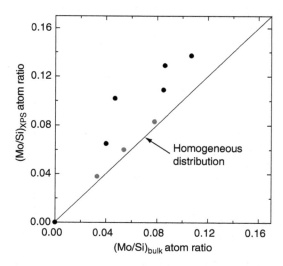

FIGURE 5.10   Surface versus bulk composition of Mo-loaded zeolites. White symbols refer to samples subjected to a consecutive impregnation with an aqueous solution of Pd^{2+} ions

the second impregnation step. However, the remaining MoO_3 species maintain an almost uniform distribution within the zeolite crystal.

5.4.3 Aggregation Phenomena in Supported Oxides

The structures of supported metal oxides change strikingly when exposed to moisture and pretreated at high temperatures in air. Although the chemistry of the processes involved in such transformations is described in detail in other chapters of this book, here emphasis is placed on the changes in the oxidation state of the metal and also on the aggregation state.

Among the many supported oxides that exhibit redox properties (V_2O_5, CrO_3, MnO_2, Fe_2O_3, WO_3, MoO_3, etc.), chromium oxide is an interesting example since it appears in different oxidation states. In the most widely investigated CrO_x/SiO_2 system (Phillips polymerization catalyst), it is accepted that CrO_3 could be strongly dispersed and stabilized as surface chromate species through reaction with surface hydroxyl groups of silica during the calcination step in oxygen or dry air [44]. A highly dispersed state of surface-stabilized chromate species can be achieved via many redispersion cycles of sublimation, volatilization, spreading, deposition, and stabilization of CrO_3 structures on the silica surface during calcination. Unfavorable phenomena can also take place during calcination. The induced reduction of surface-stabilized hexavalent surface chromates into lower valence states and the formation of aggregated Cr_2O_3 structure oxides usually occur simultaneously.

Careful analysis by photoelectron spectroscopy and electron microprobe analysis [44] affords a plausible mechanism regarding the formation of Cr_2O_3 microcrystals during the calcinations of a SiO_2-supported Cr-oxide catalyst containing 0.4 Cr/nm^2 according to the following scheme:

$$CrO_3 \xrightarrow[\text{$-H_2O$}]{\text{Dispersion}} Cr^{VI}O_{x,\,surf} \xrightarrow[\text{high T}]{\text{Reduction}} Cr^{III}O_{x,\,surf} \xrightarrow[\text{H_2O}]{\text{Cleavage}} Cr_2O_{3\,crystals}$$

In the first stage, the bulk CrO_3 oxide is deposited on the substrate and stabilized as chromate species, including monochromate, dichromate, and sometimes even polychromate, through reaction with surface hydroxyl groups of silica during calcination in dry air at temperatures of about 1000 K. Under these conditions, a highly dispersed state of chromate species is gradually achieved through many redispersion cycles of sublimation, volatilization, spreading, and stabilization. These redispersion cycles may be facilitated by the presence of traces of moisture generated from the deep dehydroxylation of silica in the early stage of dispersing bulk CrO_3 during the calcinations. The dehydroxylation also results in increasing strain in surface siloxane groups, thus increasing the reduction potential of the surface chromate species. At a certain critical point, the calcination-induced reduction of chromate species to $Cr^{III}O_{x,surf}$ species would be expected. Subsequently, further traces of moisture evolved from dehydroxylation might detach some $Cr^{III}O_{x,surf}$ species from the silica surface, leading to the formation of three-dimensional Cr_2O_3 structures. Higher temperature, longer duration, and higher moisture content in the late stage of calcination can induce a stronger aggregation of surface chromium oxide species.

The example discussed and many others highlight the need to carefully monitor all the changes brought about by sublimation, volatilization, spreading, and stabilization processes that often take place along the pretreatment steps employed to develop specific metal oxide structures.

5.5 Summary

Precise knowledge of atomic scale of the chemical structures, bonding, chemical composition, and bonding of adsorbates onto the surface layer of metal oxides is of prime importance. Many properties of metal oxides strongly depend on the preparation procedure employed in their synthesis and play an important role in their overall properties and behavior. This is why the chemical and structural features of surfaces and interfaces must be analyzed carefully in order to maintain a permanent control of reactivity. With a few exceptions, for many metal oxides these properties are still poorly known and under-exploited in comparison with metals.

Owing to high reactivity, the surface of many metal oxides becomes altered when exposed to ambient environment. Since the symmetry and coordination of the metal ions are lost at the surface, the ions show a strong tendency to be saturated by reaction with gas molecules. One major process taking place is hydroxylation, which is the result of a true chemical reaction between surface M–O bonds and water molecules. Another process that occurs simultaneously with hydroxylation, and specifically on metal oxides of basic character, is carbonation. Quantification of these processes is a critical issue since the overall properties of oxides often depend on the surface properties of the latter. Accordingly, these phenomena must be considered carefully, particularly, in the case of oxides used for technological applications.

Another important factor to be considered is the surface composition of metal oxides. Although in many cases there is evidence that the surface composition closely follows that of the bulk, the different examples discussed in this chapter illustrate the fact that segregated phases on the surface unavoidably occur as a consequence of the inability of the lattice to accommodate homogeneously all the ions in the oxide lattice. Additionally, changes in the oxidation state of the surface elements often occur with respect to that of the bulk, particularly when exposed to moisture, thermal treatments, and cycling operations in reducing and oxidative environments.

References

1. Henrich, V.E. and Cox, P.A. *The Surface Science of Metal Oxides*. Cambridge: Cambridge University Press, 1994.

2. Freund, H.J., Kuhlenbeck, H., and Staemmler, V. Oxide surfaces. *Rep. Prog. Phys.* 1996, *59*, 283–347.

3. Dufour, L.C., Bertrand, G.L., Caboche, G., Decorse, P., El Anssari, A., Poirson, A., and Vareille, M. Fundamental and technological aspects of the surface properties and reactivity of some metal oxides. *Solid State Ion.* 1997, *101–103*, 661–666.

4. Peri, J.B. Infrared and gravimetric study of surface hydration of gamma-alumina. *J. Phys. Chem.* 1965, *69*, 211–218.

5. Morrow, B.A. Surface groups on oxides. *Stud. Surf. Sci. Catal.* 1990, *57A*, A161–A224.

6. Fierro, J.L.G. and Gonzalez Tejuca, L. Nonstoichiometric surface behavior of $LaMO_3$ oxides as evidenced by XPS. *Appl. Surf. Sci.* 1987, *27*, 453–457.

7. Kohiki, S., Ozaki, S., and Hamada, T. Photoemission from small palladium clusters supported on various substrates. *Appl. Surf. Sci.* 1987, *28*, 85–91.

8. Fierro, J.L.G. Structure and composition of perovskite surface in relation to adsorption and catalytic properties. *Catal. Today* 1990, *8*, 153–174.

9. Rogynskaya, Y.E., Morozova, O.V., Lubnin, E.N., Ulitina, Y.E., Lopukhova, G.V., and Trasatti, S. Characterization of bulk and surface composition of $Co_xNi_{1-x}O_y$ mixed oxides for electrocatalysis. *Langmuir* 1997, *13*, 4621–4627.

10. Pyke, D., Mallich, K.K., Reynolds, R., and Bhattacharya, A.K. Surface and bulk phases in substituted cobalt oxide spinels. *J. Mater. Chem.* 1998, *8*, 1095–1098.

11. Nydegger, M.W., Couderc, G., and Langell, M.A. Surface composition of $Co_xNi_{1-x}O$ solid solutions by X-ray photoelectron and Auger spectroscopies. *Appl. Surf. Sci.* 1999, *147*, 58–66.

12. Langell, M.A., Gevrey, F., and Nydegger, M.W. Surface composition of $Mn_xCo_{1-x}O$ solid solution by X-ray photoelectron and Auger spectroscopies. *Appl. Surf. Sci.* 2000, *153*, 114–127.

13. Larachi, F., Pierre, J., Adnot, A., and Bernis, A. Ce 3d XPS study of composite $Ce_xMn_{1-x}O_{2-y}$ wet oxidation catalysts. *Appl. Surf. Sci.* 2002, *195*, 236–250.

14. Kurtz, R.L., Stockbauer, R., Madey, T.E., Roman, E., and de Segocia, J.L. Synchrotron radiation studies of H_2O adsorption on $TiO_2(110)$. *Surf. Sci.* 1989, *218*, 178–189.

15. Lasaga, A.C. Mineral water interface geochemistry. In *Mineralogical Society of America*, Hochella Jr., M.F. and White, A.F., Eds. Washington, DC, 1990, 17 pp.

16. Stumm, W. *Chemistry of the Solid-Water Interface*. New York: Wiley, 1992, 13 pp.

17. Blesa, M.A., Morando, P.J., and Regazzoni, A.E. *Chemical Dissolution of Metal Oxides*. Boca Raton, FL: CRC Press, 1994, 127 pp.

18. Sparks, D.L. *Environmental Soil Chemistry*. San Diego, CA: Academic Press, 1995, 100 pp.

19. Cornell, R.M. and Schwertmann, U. *The Iron Oxides*. Weinheim: VCH, 1996, 207 pp.

20. Drever, J.I. *The Geochemistry of Natural Waters*. Upper Saddle River, NJ: Prentice Hall, 1997, 91 pp.

21. Tamura, H., Tanaka, A., Mita, K., and Furuichi, R. Surface hydroxyl site densities on metal oxides as a measure for the ion-exchange capacity. *J. Colloid Interface Sci.* 1999, *209*, 225–231.

22. Müller, B. and Sigg, L. Adsorption of lead(II) on the goethite surface-voltammetric evaluation of surface complexation parameters. *J. Colloid Interface Sci.* 1992, *148*, 517–532.

23. Laiti, E., Öhman, L.O., Nordin, J., and Sjöberg, S. Acid-base properties and phenylphosphonic acid complexation at the aged γ-Al_2O_3 water interface. *J. Colloid Interface Sci.* 1995, *175*, 230–238.

24. Gorski, D., Klemm, E., Fink, P., and Hörhold, H.H. Investigation of quantitative SiOH determination by the silane treatment of disperse silica. *J. Colloid Interface Sci.* 1988, *126*, 445–449.

25. Gonzalez Tejuca, L., Fierro, J.L.G., and Tascon, J.M.D. Structure and reactivity of perovskite-type oxides. *Adv. Catal.* 1989, *36*, 237–328.

26. Gonzalez-Elipe, A.R., Espinos, J.P., Fernández, A., and Munuera, G. XPS study of the surface carbonation hydroxylation state of metal-oxides. *Appl. Surf. Sci.* 1990, *45*, 103–108.

27. Peña, M.A. and Fierro, J.L.G. Chemical structures and performance of perovskite oxides. *Chem. Rev.* 2001, *101*, 1981–2017.

28. Garcia de la Cruz, R., Falcon, H., Peña, M.A., and Fierro, J.L.G. Role of bulk and surface structures of $La_{1-x}Sr_xNiO_3$ perovskite-type oxides in methane combustion. *Appl. Catal. B: Environ.* 2001, *33*, 45–55.

29. Pajares, J.A., Fierro, J.L.G., and Weller, S.W. Kinetics of chemisorption of CO_2 on scandia — new rate equation. *J. Catal.* 1978, *52*, 521–530.

30. Morishige, K., Kittaka, S., and Morimoto, T. The thermal-desorption of surface hydroxyls on tin(IV) oxide. *Bull. Chem. Soc. Jpn.* 1980, *53*, 2128–2132.

31. Westall, J. and Hohl, H. Comparison of electrostatic models for the oxide-solution interface. *Adv. Colloid Interface Sci.* 1980, *12*, 265–294.

32. Marcos, J.A., Buitrago, R.H., and Lombardo, E.A. Surface-chemistry and catalytic activity of $La_{1-Y}Sr_YCoO_3$, $La_{1-Y}Th_YCoO_3$ perovskite. 1. Bulk and surface reduction studies. *J. Catal.* 1987, *105*, 95–106.

33. Tejuca, G.L., Bell, A.T., Fierro, J.L.G., and Peña, M.A. Surface behavior of reduced $LaCoO_3$ as studied by TPD of CO, CO_2 and H_2 probes and by XPS. *Appl. Surf. Sci.* 1988, *31*, 301–316.

34. Gourieux, T., Krill, G., Maurer, M., Ravet, M.F., Menuy, A., Tolentino, H., and Fontaine, A. Oxygen-stoichiometry dependence of electronic structure of $Yba_2Cu_3O_{7-\delta}$ ($0 < \delta < 0.7$) — possibility of a highly correlated mixed-valent state. *Phys. Rev. B* 1988, *37*, 7516–7524.

35. Salvador, P. and Fierro, J.L.G. XPS study of the dependence on stoichiometry and interaction with water of copper and oxygen valence states in the $Yba_2Cu_3O_{7-X}$ compound. *J. Solid State Chem.* 1988, *81*, 240–249.

36. Ponce, S., Peña, M.A., and Fierro, J.L.G. Surface properties and catalytic performance in methane combustion of Sr-substituted lanthanum manganites. *Appl. Catal. B: Environ.* 2000, *24*, 193–205.

37. Bernal, S., Botana, F.J., Garcia, R., and Rodríguez-Izquierdo, J.M. Behavior of rare-earth sesquioxides exposed to atmospheric carbon-dioxide and water. *Reactivity Solids* 1987, *4*, 23–40.

38. Tarasevich, M.A. and Efernov, B.N. *Electrodes of Conductive Metal Oxides.* Amsterdam: Elsevier, 1980, 221 pp.

39. Nydegger, N.W., Couderc, G., and Langell, M.A. Surface composition of $Co_xNi_{1-x}O$ solid solutions by X-ray photoelectron and Auger spectroscopies. *Appl. Surf. Sci.* 1999, *147*, 58–66.

40. Pyke, D., Mallick, K.K., Reynolds, R., and Bhattacharya, A.K. Surface and bulk phases in substituted cobalt oxide spinels. *J. Mater. Chem.* 1998, *8*, 1095–1098.

41. Briggs, D. and Seah, M.P. *Practical Surface Analysis*. New York: Wiley, 1990, 641 pp.
42. Tabata, K., Matsumoto, I., Kohiki, S., and Misono, M. Catalytic properties and surface-states of $La_{1-x}Ce_xCoO_3$. *J. Mater. Sci.* 1987, *22*, 4031–4035.
43. Lede, E.J., Requejo, F.J., Pawelec, B., and Fierro, J.L.G. XANES Mo L-edges and XPS study of Mo loaded in HY zeolite. *J. Phys. Chem. B* 2002, *106*, 7824–7831.
44. Liu, B. and Terano, M. Investigation of the physico-chemical state and aggregation mechanism of surface Cr species on a Phillips CrO_x/SiO_2 catalyst by XPS and EPMA. *J. Mol. Catal. A: Chem.* 2001, *172*, 227–240.

6 The Electronic Structure of Metal Oxides

P.M. Woodward, H. Mizoguchi, Y.-I. Kim, and M.W. Stoltzfus
Department of Chemistry, The Ohio State University,
Columbus, Ohio

CONTENTS

6.1 INTRODUCTION

Metal oxides play an important role in many fields, including inorganic and materials chemistry, condensed matter physics, geology, materials science, and mechanical and electrical engineering. The importance of metal oxides lies in their variety, their chemical stability, and their fascinating chemical and physical properties. Metal oxides display properties ranging from piezoelectricity to super-conductivity, from negative thermal expansion to ionic conductivity. Metal oxides are used as gemstones, transparent conductors, gas sensors, and catalysts to name but a few applications. They are used in computers, Li-ion batteries, fluorescent lights, cellular phones, and fuel cells. An important focus of scientists who synthesize and utilize metal oxides is to understand the origin of their properties and to learn how to manipulate these properties through modification of the composition, the crystal structure, and/or the defects. This requires an understanding of the electronic band structure, which, on the one hand, is the link between composition and crystal structure and on the other hand with chemical and physical properties. The optical and electrical transport properties can be understood directly from the electronic structure. The chemical reactivity and catalytic properties depend upon the energy levels and symmetry of electronic states near the Fermi level. Even the dielectric and mechanical properties can be traced to the chemical bonding that is the origin of the electronic structure.

The purpose of this chapter is to show how the electronic band structure arises from the fundamental properties of atomic orbitals and the arrangement of atoms within a crystal. The treatment is described in the language of chemical bonding and the corresponding electronic states are derived from linear combinations of atomic orbitals (LCAOs). This approach, often referred to as the tight binding approach, differs from the free electron approach, which is more widely used in physics and engineering. The free electron approach works well in compounds where the valence electrons exhibit a high degree of delocalization, such as metals and semiconductors. However, the inherent ionicity in most metal–oxygen bonds tends toward localized or nearly localized states for the valence electrons. In fact, many of the interesting electronic properties of metal oxides arise from the close competition between delocalized and localized electronic states. The LCAO model is well suited to compounds that exist near this boundary. Furthermore, the LCAO model allows to trace the key features of the electronic structure back to the composition and crystal structure, thereby facilitating the chemical design and manipulation of materials.

An examination of the chapter will quickly reveal that the compounds considered here do not require a detailed consideration of electron–electron repulsion. This has been done intentionally to keep the focus on the relationships between the composition, the crystal structure, and the electronic structure. A thoughtful treatment of concepts, such as intrasite and intersite exchange, magnetic ordering, the Hubbard model, superconductivity, etc., would probably double the length of this already long contribution. Many readers will already be familiar with these concepts; however, those readers who are less familiar with these concepts are encouraged to seek descriptions of these effects, which can be found readily in the solid-state chemistry literature. In either case, the concepts presented have equal relevance to oxides where electron exchange effects play an important role.

6.2 METAL–OXYGEN BONDS: ELECTRONIC ENERGY LEVELS AND CHEMICAL BONDING

As a starting point let us consider the energetics of the metal–oxygen bond. To understand the electronic structure of metal oxides it is necessary to develop a semiquantitative notion of the energy levels, sizes, and symmetry of atomic orbitals.

6.2.1 The Ionic Model

In simplified treatments, metal oxides are often taken to be completely ionic compounds. While this is a rather inaccurate view of most metal oxides, it is worthwhile to briefly consider the implications of the ionic model and consider how this picture changes when covalency is taken into consideration. Let us take MgO as a prototypical ionic metal oxide. Figure 6.1(a) shows a Born–Haber cycle for MgO. It is interesting to note that the formation of a gas-phase O^{2-} atom is an endothermic process. Its existence is only stabilized by the ionic energy that comes from the positively charged cations surrounding it, the lattice energy. This is the driving force for the formation of ionic compounds. A simple calculation of the lattice energy for MgO based on electrostatic interactions of point charges (the Madelung energy) is 47.9 kJ/mol. This result is in reasonable agreement with the reported experimental value of 40.0 kJ/mol [1].

Figure 6.1(b) shows how the band gap, the amount of energy required to transfer an electron from a filled O 2p orbital to an empty Mg 3s orbital, varying as different factors are introduced. As already discussed, it would be energetically downhill to transfer an electron from a gas-phase O^{2-} ion to a gas-phase Mg^{2+} ion. Upon taking the Madelung potential into account the picture changes dramatically.

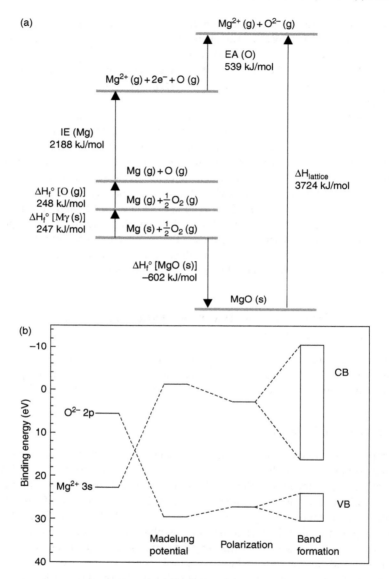

FIGURE 6.1 (a) Born–Haber cycle for MgO. (b) Band gap energy in MgO as the Madelung site potential, polarization, and bandwidth are considered in turn

The Madelung site potential is given by the equation [2]:

$$V_M = (A_M Z e)/(4\pi\varepsilon_0 r) \tag{6.1}$$

where A_M is the Madelung constant, Z is the charge ($+2$ for Mg, -2 for O), e is the charge of an electron (1.602×10^{-19} C), ε_0 is the permitivity of free space

$(8.854 \times 10^{-12} \, \mathrm{Fm}^{-1})$, and r is the distance to the nearest-neighbor ions of opposite charge. The Mg^{2+} cation is surrounded by six negatively charged O^{2-} ions thereby destabilizing the electronic energy levels on Mg by $V_M = 23.96 \, \mathrm{eV}$ (the Madelung site potential), whereas the oxide ion is surrounded by the positively charged Mg^{2+} cations lowering its electronic energy levels by an equivalent amount.

The discussion would be complete at this point if the ions were idealized point charges, but they are not. To complete the picture, two additional effects must be taken into account: polarization and bandwidth. Polarization, which describes the reorganization of electron density in response to a missing electron ($O^{2-} \rightarrow O^-$) or an extra electron ($Mg^{+2} \rightarrow Mg^+$), lowers the binding energy of an electron in an O 2p orbital (the highest occupied molecular orbital, HOMO) and raises the binding energy associated with placing an electron in a Mg 3s orbital (the lowest unoccupied molecular orbital, LUMO), as shown in Figure 6.1. These values are estimated to be 2.43 eV for O^{2-} and 3.98 eV for Mg^{+2} [3]. Finally, we need to consider the overlap of atomic orbitals to form bands. This is an important effect that lowers the band gap, E_g, significantly, even in "ionic" oxides such as MgO. The width of the O 2p valence band has been estimated, from experimental measurements, to be 6.5 eV [4]. The width of the Mg 3s conduction band can be hypothetically estimated from the band gap and width of the valence band to be 26.7 eV, lowering the band gap by 16 to 17 eV. The importance of bandwidth in understanding the electronic structure of solids cannot be underestimated. Bandwidth and the factors that determine the bandwidth will be discussed in considerable detail in Section 6.7.

6.2.2 Orbital Energies and Covalent Interactions

Our investigation of the electronic structures of metal oxides begin by considering the covalency in metal–oxygen bonds. The degree of covalency in a metal–oxygen bond depends upon the position of the metal in the periodic table, as well as its oxidation state and coordination environment. In this section we consider periodic trends in the orbital energies of main group ions. The effects of changing the oxidation state and coordination environment in transition metal ions are examined in Section 6.3.

Traditionally, the gradient from ionic to covalent bonding is estimated from the electronegativity difference between atoms. Electronegativity as originally defined by Linus Pauling is "the power of an atom in a molecule to attract electrons to itself," or stated in another way "it is the electron withdrawing power of an atom" [5,6]. Electronegativity is a valuable concept but unfortunately it is not a quantity that can be directly measured. Over the years, it has been defined in a variety of ways, ranging from the original electronegativity scale developed by Pauling, to alternate scales put forward by Allred and Rochow [7], Mulliken [8], Sanderson [9], Phillips and van Vechten [10], among others. The details of these scales vary, but the periodic trends tend to be similar from one scale to another. The electronegativity of a metal atom increases, thereby increasing the covalency of the metal–oxygen bond, upon moving toward the upper right-hand corner of the periodic table.

The electronegativity of an atom is intrinsically linked to the energy levels of its atomic orbitals. If the energies of the frontier orbitals are low lying the atom will attract electrons to itself and the electronegativity will be high, and vice versa. Table 6.1 shows calculated orbital energies for the valence orbitals of main group ions. These values were taken from the NIST website of atomic reference data for electronic structure calculations [11]. They are calculated for neutral gas-phase atoms within the framework of Kohn–Sham theory using the scalar relativistic local-density approximation. The transition metals are not included because their electronic configurations and orbital energies are very sensitive to changes in oxidation state and coordination environment. They will be discussed in detail in Section 6.3.

The basic periodic trends observed in electronegativity scales are also seen in the orbital energies, namely the orbital energy becomes increasingly negative as one moves up and to the right in the periodic table. However, there are a number of bumps in the smooth periodicity that merit further comment. This is particularly true of the vertical trends in the s orbital energies of the p block elements. First of all, note that upon moving from the 3rd period (Al–Cl) to the 4th period (Zn–Br) the s orbital energies become more negative. This is in contradiction to the general expectation that elements become less electronegative when moving down a group. This anomaly can be attributed to the incomplete shielding of the nuclear charge associated by filling the d subshell for the first time. The same effect leads to an irregular progression in metallic radii and is often referred to as the transition metal contraction [12]. A similar discontinuity is seen on moving from the 5th period (Cd–I) to the 6th period (Hg–At). While the unexpectedly low orbital energies of the 6th period elements can be partially traced to incomplete shielding of the nuclear charge by the 4f electrons (lanthanide contraction), calculations show that this deviation from regular periodicity originates primarily in relativistic effects.* This is illustrated by considering the calculated orbital energies for tin and lead, both with and without scalar relativistic effects included. If relativistic effects are not included the 5s and 5p orbital energies for Sn are calculated to be -10.1 and -3.9 eV, respectively, while the 6s and 6p orbital energies for Pb are -9.7 and -3.8 eV. When compared with values given in Table 6.1, where scalar relativistic effects have been taken into account, we see that relativistic effects lower the energy of the Pb 6s orbitals significantly with respect to the position of the Sn 5s orbitals. Similar trends are observed for other members of the 5th and 6th periods.

An important feature of the orbital energies shown in Table 6.1 is the distinctly different orbital energies of the valence shell s and p orbitals. There is much to be learnt from matching the energy level between metals and oxygen. The covalency of a bond will be optimized when orbitals on neighboring atoms have favorable

---

* As the charge on the nucleus becomes large the velocity of an electron in the vicinity of the nucleus approaches the speed of light and its mass increases due to relativistic effects. These effects are largest for core electrons, but they have a nonnegligible influence on the valence electrons of the heavier elements. The effect is most pronounced for s orbitals of elements in the 6th and 7th periods.

TABLE 6.1

Calculated Orbital Energies for the Valence Orbitals of Main Group Ions in eV

	Li	Be	Sc	Zn	B	C	N	O	F
2s	−2.9	−5.6			−9.4	−13.6	−18.4	−23.7	−29.6
2p	—	—			−3.7	−5.4	−7.2	−9.2	−11.3
	Na	Mg			Al	Si	P	S	Cl
3s	−2.8	−4.8			−7.8	−10.9	−14.0	−17.3	−20.7
3p	—	—			−2.8	−4.2	−5.6	−7.1	−8.7
	K	Ca	Sc	Zn	Ga	Ge	As	Se	Br
4s	−2.4	−3.9	−4.3	−6.2	−9.2	−11.9	−14.7	−17.4	−20.3
4p	—	—	—	—	−2.7	−4.0	−5.3	−6.7	−8.0
	Rb	Sr	Y	Cd	In	Sn	Sb	Te	I
5s	−2.4	−3.6	−4.2	−5.9	−8.5	−10.8	−13.0	−15.3	−17.6
5p	—	—	—	—	−2.7	−3.9	−5.0	−6.1	−7.2
	Cs	Ba	La	Hg	Tl	Pb	Bi	Po	At
6s	−2.2	−3.3	−3.8	−7.1	−9.8	−12.3	−14.7	−17.1	−19.6
6p	—	—	—	—	−2.6	−3.7	−4.8	−5.8	−6.8

Source: As taken from the Kotochigova, S., Levine, Z.H., Shirley, E.L., Stiles, M.D., and Clark, C.W. *Atomic Reference Data for Electronic Structure Calculations.* http://physics.nist.gov/PhysRefData/DFTdata/contents.html

spatial and energetic overlap. It is important to note that the oxygen 2s orbitals are considerably deeper (more negative) in energy than practically all the metal valence orbitals. While this does not preclude metal orbitals interacting with the oxygen 2s orbitals, and we will see the effects of this interaction later, it does imply that the predominant interaction will be between the metal orbitals and the oxygen 2p orbitals. The s orbital energies for the alkali, alkaline-earth, and rare-earth (not shown) metals are quite high and their energetic overlap with the oxygen 2p orbitals is relatively poor. As a consequence, these cations largely form ionic bonds with oxygen. We will see later that in complex oxides where such ions are combined with transition metals (i.e., $KTaO_3$) or p block metals (i.e., $BaSnO_3$) to a first approximation, the covalent interactions with the "ionic" cations (alkali, alkaline-earth, or rare-earth) can be neglected. In a similar vein, note that for the posttransition metals the valence shell p orbital energies are similar to the s orbital energies of the group 1 to 3 elements. Consequently, for these ions the electronic structure near the Fermi level will be determined primarily by the interaction between metal s and oxygen 2p orbitals.[†]

[†] This is not true for cations with $ns^2 np^0$ configuration (i.e., Sn^{2+}, Sb^{3+}, Te^{4+}), where the conduction band/LUMO states will originate from antibonding interactions between metal p and oxygen 2p orbitals.

TABLE 6.2

Values of R_{max}, the Radius at which the Magnitude of the Atomic Wavefunction Reaches a Maximum, given in Å

	Sc	Ti	V	Cr	Mn	Fe	Co	Ni	Cu	Zn
4s	1.716	1.622	1.544	1.626	1.419	1.366	1.319	1.276	1.374	1.200
3d	0.592	0.528	0.479	0.455	0.409	0.382	0.358	0.338	0.325	0.304
	Y	Zr	Nb	Mo	Tc	Ru	Rh	Pd	Ag	Cd
5s	1.891	1.785	1.819	1.753	1.574	1.647	1.604	—	1.532	1.368
4d	0.939	0.840	0.789	0.729	0.668	0.639	0.604	0.58	0.729	0.517
	La	Hf	Ta	W	Re	Os	Ir	Pt	Au	Hg
6s	1.867	1.779	1.709	1.650	1.600	1.555	1.515	1.591	1.560	1.417
5d	0.950	0.876	0.820	0.776	0.739	0.706	0.678	0.659	0.635	0.608

R_{max} values are based on Hartree–Fock calculations.

Source: Taken from Mann, J.B. *Atomic Structure Calculations II*. California: Los Alamos Scientific Laboratory, 1968.

The situation becomes more complicated for the transition metals. The valence-shell s and p orbitals of the transition metal elements have a greater spatial extent than the d orbitals. The s and p orbitals form significant bonding and antibonding interactions with oxygen, leading to significant destabilization of the antibonding states that have predominant metal character. The properties of the transition metal s and p orbitals are intermediate between the s block and p block elements as are the electronegativities. However, the relatively subtle periodic trends observed across the transition metals, either in orbital energies or electronegativity, do not begin to adequately describe the varied and complex behavior of transition metal oxides. The small size of the d orbitals leads to a smaller, but nonnegligible, spatial overlap with the oxygen orbitals. As a consequence, the antibonding states that originate from this interaction are not highly destabilized and can usually be found in close proximity to the Fermi level. These states are responsible for many of the fascinating properties observed in transition metal oxides. To get a feel of the relative sizes of atomic orbitals, values of R_{max} for the ns and $(n - 1)d$ atomic orbitals are given in Table 6.2 [13]. The R_{max} value is the radius at which the magnitude of the atomic wavefunction reaches a maximum value (as calculated using Hartree–Fock wavefunctions). The contracted nature of the d orbitals is particularly pronounced for the 3d transition metals (Sc–Zn). This is responsible for the many distinctive differences between the behavior of 1st row transition metal oxides (the occurrence of high spin states, the stability of the +2 oxidation state, etc.) and their 2nd and 3rd row transition metal brethren. For all three periods, the d orbitals contract more rapidly (in a relative sense) than the s orbitals upon moving from the early transition metals to the late transition metals (left to right). The implications of these trends will be explored in Section 6.3.

The nature of the metal–oxygen interaction varies dramatically with changes in the oxidation state and d electron count. To illustrate this concept consider the comparison between two compounds containing Mn–O bonds: manganese (II) oxide, MnO, containing Mn^{+2} and potassium permanganate; $KMnO_4$, containing Mn^{+7}. MnO adopts the rock salt structure with octahedral coordination at both manganese and oxygen sites. MnO is an extended solid that is both antiferromagnetic and insulating below 120 K. It is not unreasonable to consider the bonding in MnO to be more ionic than covalent, similar to the case of MgO discussed in Section 6.2.1. Whereas, in $KMnO_4$ the manganese atoms are tetrahedrally coordinated by oxygen with a Mn–O bond distance, 1.61 Å, that is dramatically shorter than the Mn–O distance found in MnO, 2.22 Å. $KMnO_4$ is a diamagnetic salt that dissolves in water to yield K^+ ions and the highly oxidizing permanganate anions, MnO_4^-. Clearly, the Mn–O bonding in permanganate is highly covalent, more reminiscent of perchlorate than MgO. The actual charge on the manganese ion in MnO_4^- is estimated from density functional theory calculations to be approximately +1.5 [14]. While not everyone would agree on what this number means, there is universal agreement that the true charge on manganese in the permanganate ion is not even remotely close to +7. The formal oxidation state is a powerful tool from an electron counting perspective, but it is not meant to represent the actual positive charge on the metal. The changes in metal–oxygen bonding that accompany changes in oxidation state and coordination environment of the metal make it somewhat futile to discuss absolute properties of a given transition metal. Section 6.4 explores the molecular orbital energies of some transition metal oxide building blocks in detail. This treatment while not comprehensive is intended to bring out the important periodic trends of the transition metal ions.

6.3 CALCULATIONS

In the rest of this chapter, calculations of various types are used to highlight important features in the electronic structure. The details of how these calculations were carried out are given here. The lowest level computations were semiempirical calculations based on the extended Hückel tight binding (EHTB) method [15,16]. EHTB calculations were carried out on molecules and extended solids using the program CAESAR [17]. The quantitative accuracy of EHTB computations is limited; however, they can be useful tools in understanding the symmetry and orbital overlap at critical points in the electronic structure. The orbital energies and ζ-coefficients used in the EHTB calculations were modified in some cases to improve agreement with more sophisticated computational results. The remaining calculations were done using density functional methods and can be divided into two classes: molecular orbital (MO) calculations for molecular complexes and band structure calculations for extended solids.

The MO calculations were performed using the Amsterdam density functional (ADF) program, version 2002.03, developed by Te Velde and Baerends [18]. The atomic electronic configurations were described using triple-ζ STO basis functions, with frozen core levels. Nonlocal exchange and correlation effects were treated using the corrections of Perdew and Wang (PW91) [19]. The electronic band structure calculations were carried out using the linear muffin-tin orbital (LMTO) method with the atomic sphere approximation (ASA), including the combined correction (CC). The LMTO–ASA code was developed in Stuttgart by Andersen et al. [20]. k space integrations employed the tetrahedron method using irreducible k points within the Brillouin zone (BZ). The basis sets consisted of the valence s, p, and d orbitals of cations and the 2s and 2p orbitals of oxygen. As with ADF calculations the Perdew–Wang (PW91) exchange correlation was utilized [19].

6.4 THE ELECTRONIC STRUCTURES OF METAL CENTERED POLYHEDRA

The crystal structures of metal oxides are extremely varied, ranging from the open frameworks of zeolites to the dense packing of perovskites. One approach to organizing this structural variety and complexity is to classify structures according to the types of cation centered coordination polyhedra they contain (octahedra, tetrahedra, square planes, trigonal bipyramids, cubes, etc.) and how those polyhedra are connected (corner-sharing sheets, face-sharing chains, etc.). For transition metal and p block cations in oxides the tetrahedron and octahedron are the dominant building units. In this section, we will examine the MO diagrams and energy levels of these building units. Calculations and experimental data are given for MO_4^{m-} and $M(OH)_6^{n-}$ polyatomic ions, both in isolation and in oxides where the octahedral or tetrahedral units are isolated from each other and surrounded by electropositive "ionic" (alkali, alkaline-earth, and rare-earth) cations.

6.4.1 Electronic Structures of Metal Centered Tetrahedra

There are thousands of metal oxides whose structures contain tetrahedral building units. Examples include (i) the alkali metal permanganates and alkaline-earth chromates, that were among the first inorganic molecules whose electronic structures were studied in detail [21], (ii) the scheelites AWO_4, that have long been utilized for their intrinsic photoluminescence [22], (iii) the versatile ZnO, that finds applications as a transparent conductor [23] and a nanowire laser [24], (iv) the broad class of naturally occurring and synthetic silicate, aluminosilicate and zinc phosphate minerals (including zeolites).

The calculated MO diagram for the permanganate (MnO_4^-) ion is shown in Figure 6.2, along with images of select molecular orbitals. The HOMO is a nonbonding linear combination of oxygen 2p orbitals with t_1 symmetry. The LUMO is the doubly degenerate set of orbitals with e symmetry, while the next highest energy orbital is the triply degenerate t_2 set. The unoccupied e and t_2 orbitals owe their existence to antibonding interactions between the Mn 3d orbitals (e $\rightarrow$ dz^2, d$x^2 - y^2$; $t_2 \rightarrow$ dxy, dxz, dyz) and oxygen (primarily O 2p). At even higher energies, the strongly antibonding Mn 4s–O 2p and Mn 4p–O 2p interactions give rise to the high lying a_1 and t_1 orbitals. If the central element is a p block element,

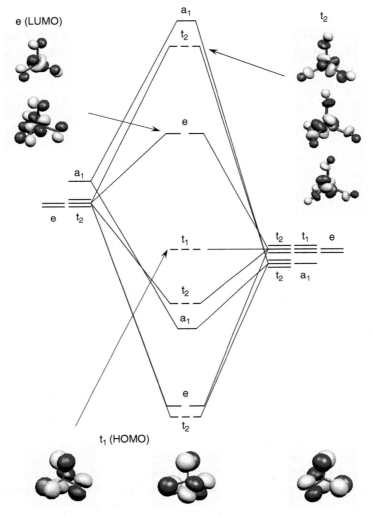

e (LUMO)

a_1

t_2

t_2

e

a_1

e t_2

t_1

t_2 t_1 e

t_2 a_1

t_2

a_1

e

t_2

t_1 (HOMO)

FIGURE 6.2   MO diagram for a tetrahedral MnO_4^- complex ion

such as is the case in SO_4^{2-} or PO_4^{3-}, the d orbitals are filled core states and the empty a_1 and t_1 orbitals become the LUMO and second LUMO, respectively. The presence of the weakly antibonding empty orbitals with metal 3d character gives rise to low lying unoccupied states that are responsible for the vibrant color of the permanganate (purple) and chromate (yellow) ions, whereas the absence of these states in ions such as phosphate, sulfate, and perchlorate explains their lack of color.

In order to understand the periodic trends associated with the transition metal ion it is instructive to consider the relative energies, the unoccupied e and t_2 orbitals

TABLE 6.3

Calculated energy gap between the nonbonding O 2p HOMO with t_1 symmetry and bond distances were chosen to be in line with experimental values [14]

Complex	M–O Distance (Å)	$t_1 \rightarrow e$ Calculated (eV)	$t_1 \rightarrow e$ Observed (eV)[a]	$t_1 \rightarrow e$ Calculated (eV)[a]
VO_4^{3-}	1.72	4.3	4.5	4.8
CrO_4^{2-}	1.64	3.4	3.3	3.5
MnO_4^-	1.63	2.2	2.2	2.6
MoO_4^{2-}	1.76	5.1	5.3	5.3
TcO_4^-	1.72	4.1	4.3	4.1
RuO_4	1.71	2.9	3.1	3.1
WO_4^{2-}	1.77	6.0	6.2	6.1
ReO_4^-	1.72	5.1	5.3	5.3
OsO_4	1.70	4.1	4.0	4.1

[a] The last two columns show the antibonding M d–O 2p LUMO with e symmetry for tetrahedral MO_4^{n-} complexes. Ideal M–O distances were calculated from bond valences [26], which tended to be in good agreement with experimental bond distances. For MnO_4^-, TcO_4^-, RuO_4, and OsO_4 the discrepancy was largest considering the observed and calculated values given in Reference 14.

with respect to the nonbonding t_1 HOMO. The calculated and experimentally measured energy levels of the e and t_2 levels for a variety of MO_4^{n-} complexes, where M is a d^0 ion, are given in Table 6.3. The values calculated in Reference 14 for the same set of complexes are listed for comparison. Both sets of calculations were carried out using the ADF code. The lack of exact agreement stems from two factors. First, the results reported in Reference 14 correspond to complexes where the M–O bond length was geometry optimized, whereas our calculations are based on idealized bond distances. Second, the authors in Reference 14 applied the local-density approximation (LDA) [25] to treat exchange correlations, whereas as mentioned in Section 6.3 our calculations employ the generalized gradient approximation of Perdue and Wang. Regardless of these differences, the periodic trends are the same for both sets of calculated values as well as the observed values.

Two important periodic trends can be extracted from this data. First, the HOMO–LUMO splitting decreases by roughly 1 eV per step upon moving from left-to-right across the periodic table while maintaining the d^0 electron configuration. The increase in the effective electronegativity of the central metal cation is more a function of the increase in the formal oxidation state than it is a function of moving from left-to-right across the periodic table. The increased electronegativity of the metal also translates into an increase in the covalency of the metal–oxygen bonds as demonstrated in the metal–oxygen bond distances. One might expect this increased covalency to destabilize the energy of the antibonding e and t_2 levels, thereby raising the HOMO–LUMO gap. Such a trend is seen in the HOMO–LUMO splitting for phosphate, sulfate, and perchlorate groups, where the calculated separations between the t_1 HOMO and the a_1 LUMO are 6.4 eV for PO_4^{3-}, 8.1 eV for SO_4^{2-}, and 8.2 eV for ClO_4^-. However, the 3d orbitals are only partially responsible for the increased covalency of the M–O bond. The empty valence-shell s orbital of the cation is also involved in bonding to oxygen, as are the cation p orbitals but to a lesser extent. The decrease in HOMO–LUMO splitting that accompanies the increase in covalency among the transition metal complexes can be traced to the fact that the symmetry of the d orbitals is not optimal for overlap with the tetrahedral ligand geometry. Consequently, the e and t_2 MOs are relatively weakly antibonding, and energy stabilization of these levels due to increased electronegativity of the central atom is larger than the energy destabilization that comes from increased covalency. The net result is a decrease in the energy of the oxygen-to-metal charge transfer excitation.[‡]

The second important periodic trend is a significant increase in the energy of the HOMO–LUMO splitting upon moving down a group (i.e., $CrO_4^{2-} \rightarrow MoO_4^{2-} \rightarrow WO_4^{2-}$). This jump is particularly pronounced upon going from the 3d series (i.e., CrO_4^{2-}) to the 4d series (i.e., MoO_4^{2-}). This effect originates largely in the relative sizes of the 3d, 4d, and 5d orbitals (see Table 6.2). As previously discussed, the 4d orbitals are larger than the 3d orbitals, leading to a stronger interaction with oxygen and a subsequent increase in the antibonding character of the e and t_2 MOs. A similar but smaller effect is seen upon replacing a 4d ion with the 5d ion immediately below it in the periodic table. If both the horizontal and vertical periodic trends are taken into account, we see that in an approximate sense the HOMO–LUMO splitting is similar for cations located diagonally from each other, such as V(V)–Mo(VI)–Re(VII) or Cr(VI)–Tc(VII)–Os(VIII).

The calculations presented in the preceding paragraphs were carried out on isolated tetrahedral ions and molecules. How do the relative energies change when the polyatomic ions are surrounded by "ionic" cations in a salt crystal, such as is the case in YVO_4, $SrCrO_4$, or $KMnO_4$? While surrounding the polyatomic anions with positively charged cations will undoubtedly stabilize the absolute energies of the electronic states, it is not unreasonable to assume that the shift in orbital energies will be roughly constant, thereby leaving the energy separations in the MO diagram essentially unchanged? However, the validity of this assumption degrades as the covalent interactions between the electropositive cations and the polyatomic anions increase, or at least when they differ strongly from interactions between water and the anions. Experimental band gap values are available for some salts of the anions in Table 6.3. The alkaline-earth tungstates and molybdates with the Scheelite structure have been studied for their self-luminescence. The band gaps of the tungstates, AWO_4 (A = Ba, Sr, Ca), range from 5.0 to 5.3 eV, while those

---

[‡] The calculations of Stückl et al. indicate that the e and t_2 orbitals have significant oxygen character and that the formal charge transfer is quite small (0.06 for MnO_4^-), nonetheless the ligand-to-metal charge transfer description is commonly used to describe these transitions [14].

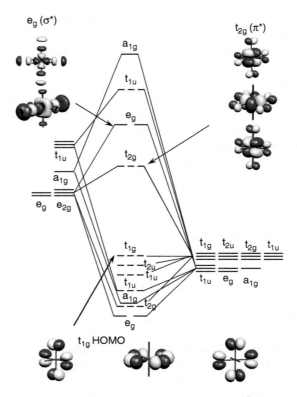

FIGURE 6.3 MO diagram for an octahedral TiO_6^{8-} ion

of the corresponding molybdates range from 4.2 to 4.4 eV [22]. Measurements in our lab estimate the band gap of YVO_4 to be 3.5 eV, in good agreement with reports in the literature [27]. These values are consistently smaller than the free ion values given in Table 6.3 by roughly 1 eV, but the periodic trends are very much the same.

6.4.2 Electronic Structures of Metal Centered Octahedra

The majority of oxides with interesting electronic properties contain octahedral building units. Examples can be found among many structure types, including rock salt, rutile, corundum, bixbyite, perovskite, pyrochlore, and spinel. The MO diagram for a TiO_6^{8-} cluster with O_h symmetry is shown in Figure 6.3, along with images of selected MOs. The HOMO is a nonbonding linear combination of oxygen 2p π orbitals[§] with t_{1g} symmetry. A second set of nonbonding oxygen

---

[§] Throughout the course of this chapter the term oxygen 2p σ orbitals will be used to refer to oxygen p orbitals oriented parallel to the M–O bonds in an octahedron, and the term oxygen 2p π orbitals will be used to refer to oxygen p orbitals oriented perpendicular to the M–O bonds in an octahedron.

2p orbitals with t_{2u} symmetry lies slightly lower in energy. The LUMO is a triply degenerate set of orbitals with t_{2g} symmetry that arises from the π^* interaction between the Ti 3d orbitals (dxy, dxz, dyz) and oxygen 2p orbitals. The next highest energy orbital is the doubly degenerate e_g set that arises from the σ^* interaction between Ti 3d orbitals (dz^2, $dx^2 - y^2$) and oxygen. At higher energies the strongly σ-antibonding Ti 4s–O 2p and Ti 4p–O 2p interactions give rise to high lying unoccupied states with a_{1g} and t_{1u} symmetry. When the central element is a p block element, such as is the case in $Te(OH)_6$, the d orbitals are filled core states and the empty a_{1g} and t_{1u} orbitals become the LUMO and second LUMO, respectively.

In contrast to the tetrahedral case, stable octahedral MO_6^{n-} complexes are not common. This is presumably due in part to the large negative charge that would be associated with such a cluster. The unrealistically large negative charge of the MO_6^{n-} cluster also poses some potential problems for density functional calculations. Therefore, MO calculations were carried out on $M(OH)_6^{n-}$ clusters, where the O_h point group symmetry constrains the M–O–H bonds to be linear. Is it realistic to assume that, the electronic structures of hydroxide clusters will be a good approximation to the electronic structures of oxide clusters. While it is true that hydrogen participates in bonding, the spherical symmetry of the hydrogen 1s orbital does not allow mixing with oxygen 2p orbitals oriented perpendicular to the linear M–O–H bond (the O 2p π orbitals). Fortunately, the oxygen orbitals that participate in both the nonbonding t_{1g} HOMO and the π^* t_{2g} LUMO are O 2p π orbitals. Thus, calculations based on $M(OH)_6^{n-}$ complexes are in fact good model systems for probing the HOMO–LUMO splitting in d^0 metal centered octahedra.

Calculated values of the HOMO–LUMO splitting in select $M(OH)_6^{n-}$ complexes are given in Table 6.4. As stated earlier, isolated molecular hydroxide complexes are not available for comparison. However, the results of Section 6.4.1 suggest that it might be reasonable to compare isolated complexes with salts containing MO_6^{m-} ions surrounded by electropositive "ionic" cations. Such compounds exist among the ordered perovskites, $A_2MM'O_6$, where A and M are electropositive cations and M' is the d^0 transition metal ion of interest. The optical band gaps of ordered perovskites were recently investigated in detail by Eng et al. [28]. Their results are shown in Table 6.4 for comparison. The observed band gaps of the ordered perovskites and the values calculated for both perovskites and $M(OH)_6^{n-}$ complexes are in good agreement, particularly for the 4d and 5d metal cations. Once again, the assertion that s block and lanthanide cations do not play an active role in the electronic structure near the Fermi level seems to be reasonably valid. It is interesting to compare the periodic trends seen for octahedral complexes with those found earlier for tetrahedral complexes. The HOMO–LUMO splitting decreases significantly with increasing oxidation state, and increases as the principle quantum number of the valence shell d orbitals increases: $E_g(3d) < E_g(4d) < E_g(5d)$, mirroring the trends seen for tetrahedral complexes. It is important to note that in octahedral complexes the t_{2g} LUMO with π^* character lies much closer in energy to the nonbonding oxygen orbitals than does the e LUMO in the tetrahedral complexes. This can be clearly seen by inspection of the group six ions (Cr^{+6}, Mo^{+6}, W^{+6}) where the HOMO–LUMO

TABLE 6.4

Calculated Energy Gap between the nonbonding O 2p HOMO with t_{1g} Symmetry and the M d–O 2p π^* LUMO with t_{2g} Symmetry for Octahedral $M(OH)_6^{n-}$ Complexes

Complex	M–O distance (Å)	$t_{1g} \to t_{2g}$ calculated (eV)	Observed E_g $A_2MM'O_6$ salt (eV)[a]	Calculated E_g $A_2MM'O_6$ salt (eV)[a]
$Ti(OH)_6^{2-}$	1.965	3.0	~3.9	3.3
$V(OH)_6^{-}$	1.871	1.8	—	—
$Cr(OH)_6$	1.794	0.9	—	—
$Nb(OH)_6^{-}$	1.979	3.9	3.8–3.9	4.0
$Mo(OH)_6$	1.907	2.8	2.7	2.6
$Ta(OH)_6^{-}$	1.988	4.2	4.5–4.7	4.5
$W(OH)_6$	1.917	3.7	3.4–3.7	3.5

Ideal M–O distances were calculated from bond valences.

[a] The last two columns give the observed and calculated values for ordered perovskites, $A_2MM'O_6$, where the d^0 metal centered octahedron is surrounded by electropositive "ionic" cations, as taken from Eng, H.W., Barnes, P.W., Auer, B.M., and Woodward, P.M. *J. Solid State Chem.* 2003, *175*, 94–109. With permission.

gaps of the octahedral complexes are smaller by 1.5 to 2.5 eV than those calculated for the tetrahedral complexes. The presence of a low-lying LUMO in the octahedral geometry has an important effect not only on the optical and electrical properties, but also on the crystal chemistry of metal oxides containing d^0 cations. The implications of this are discussed in Section 6.4.4.

6.4.3   Octahedra and Tetrahedra Containing d^n ($n \neq 0$) Transition Metal Cations

The treatment of Section 6.4.1 and Section 6.4.2 focused exclusively on polyhedra containing d^0 transition metal cations. How does the situation change for complexes that contain a transition metal cation with partially filled d orbitals? Similar considerations apply, but the presence of unpaired electrons necessitates accurate treatment of exchange effects and increases the complexity of the analysis. The properties of d^0 metal oxides depend primarily upon the low-lying set of metal–oxygen antibonding orbitals, e for tetrahedral coordination and t_{2g} for octahedral coordination. In transition metal oxides where the d orbitals are partially occupied it is often necessary to consider the more highly antibonding states: t_2 for tetrahedral coordination and e_g for octahedral coordination. In molecular complexes, coordination chemists have extensively studied the energy separations between these states. The separation in energy between these two levels, often referred to as the ligand or crystal field splitting, depends upon the bonding characteristics of the ligand as well as the oxidation state and identity of the transition metal cation. Extensive discussions of ligand field splitting can be found in the literature [12,29]. A few salient points related to the nature of the transition metal ion are summarized below:

- The separation between the e and t_2 levels in a tetrahedron is smaller than the separation between t_{2g} and e_g levels in an octahedron. Within the assumptions of crystal field theory the e–t_2 splitting in a tetrahedron is 4/9 (44%) of the value of the t_{2g}–e_g splitting in an octahedron. While many assumptions used in simple crystal field theory are not completely valid, a value of 40 to 50% is a reasonable rough approximation. As an example the t_{2g}–e_g splitting in $Ti(H_2O)_6^{3+}$ is 2.5 eV, while the e–t_2 splitting in $Ti(H_2O)_4^{3+}$ is 1.1 eV.
- Increasing the cation oxidation state reduces the metal-ligand distance, thereby increasing the ligand field splitting. For example in octahedral $M(H_2O)_6^{n+}$ complexes the $t_{2g} - e_g$ separation is 2.2 eV for M = V^{3+} and 1.7 eV for M = Fe^{3+}, while separation is notably smaller in the divalent complexes, 1.5 eV for M = e^{2+} and 1.2 eV for M = Fe^{2+}.
- The ligand field splitting increases upon moving down a group. For example, the t_{2g}–e_g separations in the low spin d^6 complexes $Co(NH_3)_6^{3+}$, $Rh(NH_3)_6^{3+}$, and $Ir(NH_3)_6^{3+}$ are 2.8, 4.2, and 5.1 eV, respectively. Replacing ammonia with water leads to a decrease in the ligand field splitting, as expected from the spectrochemical series, but the trends are the same: $Co(H_2O)_6^{3+}$ and $Rh(H_2O)_6^{3+}$ have ligand field splitting of 2.3 and 3.3 eV, respectively. Huheey estimates that among octahedral complexes the ligand field splitting increases by ∼50% upon replacing a 1st row transition metal ion with the analogous 2nd row ion. The splitting increases by roughly another 25% upon moving the 3rd row of the transition metal series [29].

The underlying reality of these trends is that the higher lying set of orbitals (tetrahedron → t_2, octahedron → e_g) has a better spatial overlap with the surrounding ligands. Therefore, any change that increases the bonding interaction (increasing the cation oxidation state, increasing the spatial extent of the d orbitals, etc.) will destabilize both sets of orbitals, but the effect will be larger for the higher lying set of orbitals, thereby increasing the ligand field splitting. The difference in overlap is more pronounced for octahedral geometry, where the e_g orbitals are perfectly aligned for σ-bonding with the surrounding ligands. This difference in the overlap of the t_{2g} and e_g orbitals with oxygen has profound and interesting effects on the electrical and magnetic properties of 1st row transition metal oxides. Finally, bear in mind that the increased ligand field splitting seen in octahedral complexes does not imply that the metal–ligand covalency is greater in an octahedron than it is in a tetrahedron. In fact, the calculations presented in Section 6.4.1 and

Section 6.4.2, as well as bond valence considerations, strongly suggest the opposite assertion is closer to the truth.

6.4.4 Electronically Driven Geometric Distortions: 1st and 2nd Order Jahn–Teller Distortions

There are certain combinations of cation electron configurations and ligand geometries that are electronically unstable with respect to symmetry lowering geometric distortions. The best-known examples of electronically driven distortions are those classified as 1st order Jahn–Teller distortions. The Jahn–Teller theorem states that any nonlinear molecule with an incompletely filled, degenerate HOMO should undergo a structural distortion that removes the degeneracy and lowers the energy [30]. While there are numerous electron configurations predicted to undergo distortions by the Jahn–Teller theorem, the most important examples occur for octahedral coordination of either a d^9 (i.e., Cu^{2+}) or a high spin d^4 (i.e., Mn^{3+}, Cr^{2+}) ion. These two cases are strongly prone to undergo large Jahn–Teller distortions because the partially filled HOMO is the e_g set of orbitals that have a strong σ-antibonding overlap with oxygen, and the energy stabilization that results from the distortion increases as the antibonding character of the HOMO increases. In the octahedral case, the degeneracy can be removed by elongating the octahedron (expansion of the M–O bonds in the z-direction and contraction of the M–O bonds in the xy-plane), which stabilizes the dz^2 orbital and destabilizes the $dx^2 - y^2$ orbital. The opposite distortion, compression of the octahedron, shifts the energies of the e_g orbitals in the reverse direction but still satisfies the Jahn–Teller theorem. Another possibility is a more complicated distortion leading to three sets of bond lengths. The Jahn–Teller theorem does not provide any clues regarding the magnitude or the preferred mode of distortion. However, Burdett has examined the energy consequences of these competing distortions [31]. Burdett's insightful analysis shows that the key to identifying the most favorable distortion mechanism is to recognize that in D_{4h} symmetry the empty 4s orbital is of the appropriate symmetry to mix with the $3dz^2$ orbital, but not with the $3dx^2 - y^2$ orbital. For Cu^{2+} an octahedral elongation is favored because it leads to double occupation of the $3dz^2$ orbital, whose energy is further stabilized by mixing with the empty Cu 4s orbital. Similar arguments, albeit with different electron counts, can be used to rationalize the preference of Mn^{3+} for octahedral elongation. Burdett uses the same logic to explain the tendency for d^{10} ions to adopt linear coordination. He argues that linear coordination in many extended solids is simply an extreme version of an octahedral compression (two short axial bonds, and four long equatorial bonds), and this geometry is favored because it allows maximum mixing between the doubly occupied $(n-1)dz^2$ orbital and the empty ns cation orbitals.

There are many electronically driven structural distortions that meet the spirit but not the letter of the Jahn–Teller theorem, because they do not possess a partially occupied HOMO. These distortions are often referred to as 2nd order Jahn–Teller (SOJT) distortions. Formally, a SOJT is defined as a structural distortion arising from a 2nd order energy change in the HOMO [32]. While, in principle, almost any molecule could undergo a SOJT, in practice, molecules with small energy gaps between the HOMO and LUMO are the most susceptible to this type of distortion. In metal oxides the most common SOJT distortions can be divided into two categories: those involving a d^0 cation in octahedral coordination (i.e., Ti^{4+}, Nb^{5+}, Mo^{6+}), and those involving a main group ion with an s^2p^0 configuration (i.e., Sn^{2+}, Sb^{3+}, Te^{4+}). These two classes are considered in the following paragraph.

The molecular orbital diagram for an octahedrally coordinated d^0 cation was considered in detail in Section 6.4.2. The LUMO is the t_{2g} orbital set with M nd–O $2p \pi^*$ character and the HOMO has nonbonding O $2p$ character. SOJT distortions from the ideal geometry are characterized by a shift of the cation out of the center of the octahedron. The shift can be along the C_4 axis toward a corner of the octahedron, along the C_2 axis toward an edge of the octahedron, or along the C_3 axis toward a face of the octahedron (or something in between). The details of the energetics behind this distortion are somewhat complex. They involve not only consideration of the local M–O covalent bonding [31], but also consideration of parameters such as bond strains, and cation repulsion [33]. Well-known examples of this type of distortion include the perovskites $BaTiO_3$ and $KNbO_3$. The technologically important properties of these materials, ferroelectricity and piezoelectricity in the case of $BaTiO_3$ and its solid solutions, and nonlinear optical behavior in the case of $KNbO_3$, are a direct consequence of the SOJT distortion. The distortion lowers the symmetry of the cation and anion sites thereby enabling the unoccupied t_{2g} orbitals to hybridize with filled states in the oxygen $2p$ band [34,35]. The driving force for this type of SOJT distortion is expected to increase as the energy separation between the t_{2g} orbitals and the nonbonding oxygen $2p$ states decreases. Kunze and Brown [33] have examined the tendency for various d^0 cations to undergo SOJT distortions. They find that the magnitude of the SOJT distortion increases in the following sequence: $Zr^{+4} < Ta^{+5} < Nb^{+5} < W^{+6} < V^{+5} < Mo^{+6}$, while the behavior of Ti^{+4} has a broad distribution. Note that the parallel in this series with the periodic trends in HOMO–LUMO gap for d^0 metal octahedral complexes discussed in Section 4.2, particularly for the 4d and 5d transition metal ions. The small HOMO–LUMO gap seen for Cr^{+6} destabilizes octahedral geometry to the extent that six fold coordination for these ions is rarely seen in oxides. While the crystal chemistry of V^{+5} is quite variable, the regular octahedral coordination is quite rare. Finally, note that SOJT distortions of this type are not generally seen when the d^0 cation is tetrahedrally coordinated. This can be at least partially attributed to the larger HOMO–LUMO gap that results from tetrahedral coordination.

6.5 BASIC CONCEPTS OF ELECTRONIC BAND THEORY

Section 6.4 covered the electronic energy levels in metal centered octahedra and tetrahedra, setting the stage for a discussion of electronic band structures of metal oxides. How does the electronic structure change upon connecting these units to form extended structures? One approach to this question would be to consider a finite crystal as a huge molecule and go about calculating energy levels in the standard manner to generate a molecular orbital diagram. In principle, this can be done for model systems such as chains or rings of hydrogen atoms, and the results extrapolated as the size of the chain or ring goes to infinity. In general though, it is not practical to use the giant molecule approach to calculate the electronic structure of a crystal. This would be the equivalent of a molecular chemist using C_1 symmetry for all calculations regardless of the molecular geometry. In fact, it would be much worse because the computational cost of treating even a small crystal as a macromolecule is prohibitive. A more elegant and practical approach is to take advantage of translational symmetry to limit the number of atoms and orbitals that must be considered. Excellent descriptions of how to go from the molecular orbital approach to the electronic band structure formalism can be found in books such as "Solids and Surfaces: A Chemists View of Bonding in Extended Structures" by Hoffmann [36], "Orbital Interactions in Chemistry" by Albright et al. [32], "Chemical Bonding in Solids," by Burdett [31], and "The Electronic Structure of Solids" by Cox [2]. In the section that follows, this treatment is illustrated through extended Hückel calculations on a one-dimensional metal oxygen chain, a two-dimensional metal oxygen sheet, and a three-dimensional metal oxygen network.

6.5.1 The Electronic Band Structure of a One-Dimensional MO Chain

Consider a hypothetical linear chain with composition MO. It was previously shown that as far as properties and structures of the molecule are concerned the most important orbital interaction is between the d orbitals on the metal and the 2p orbitals on oxygen. For the sake of simplicity, the metal 4s and 4p orbitals as well as the O 2s orbital are neglected in the treatment that follows. If the chain axis is defined as the z-direction, the M dz^2–O 2pz interaction is of the σ-type, the M dxz–O 2px and M dyz–O 2py interactions are π interactions, and the M dx^2–y^2 and M dxy orbitals are nonbonding.** The band structure can be built up considering each of these interactions in turn. The one-dimensional unit cell contains one atom of each type. Therefore, eight bands are expected from the eight atomic orbitals present in the unit cell (remember we are only considering the d orbitals on the metal and the 2p orbitals on oxygen). An important concept to remember is that *the number of bands is always equal to the number of atomic orbitals contained in the unit cell*, regardless of the complexity of the structure.

Let us begin by considering the wavefunctions that describe the interaction *within the unit cell* between M dz^2 and O 2pz:

$$\chi_{ab} = [\chi(M\ dz^2) - \chi(O\ 2pz)]/\sqrt{2} \qquad (6.2)$$

$$\chi_b = [\chi(M\ dz^2) + \chi(O\ 2pz)]/\sqrt{2} \qquad (6.3)$$

** Strictly speaking the Ti $3dx^2 - y^2$ interaction with the oxygen 2p π orbitals is a δ-interaction, but their overlap will be minimal and this interaction will be ignored.

These functions are shown graphically in Figure 6.4. Applying translational symmetry constraints gives rise to what a chemist would refer to as symmetry adapted linear combinations of the basis functions, χ_b and χ_{ab}. In the language of electronic band structure, these wavefunctions are called Bloch functions. The Bloch functions, Ψ_k, are periodic waves delocalize throughout the crystal and can be mathematically expressed as follows:

$$\Psi_k = \Sigma_n \exp(ikna)\chi_n \qquad (6.4)$$

Where a is the length of the unit cell, χ_n is the basis set (χ_b or χ_{ab}) for the nth unit cell, and k is the wavevector because it gives the repeat period of the modulation of the basis set. In fact the wavelength of the electronic wavepacket is given by the expression, $\lambda = 2\pi/k$. Alternatively, k can be thought of as a quantum number that acts as a label for each symmetry adapted linear combination of the basis functions. More formally, k labels the different irreducible representations of the translation group that Ψ transforms as. The number of discreet values of k is equal to the number of unit cells in the crystal. In a macroscopic crystal the number of unit cells is very large, so that k is essentially continuous. However, unique states are generated only for $|k| \leq \pi/a$, a range referred to as the first BZ. The delocalized periodic waves that are the Bloch functions, undergo constructive and destructive interference that leads to electron density distributions that are the equivalent of atomic orbitals and chemical bonds. That is, the connection between these two seemingly disparate theories.

There is a less formal, but more descriptive way to understand the meaning of k. The value of k tells us how rapidly the orbitals that make up the basis set change phase. When $k = 0$ the phase of the basis set is invariant from one unit cell to another, whereas when $k = \pi/a$ the phase of the basis set is inverted every time a translation of one unit cell is applied. This can be seen most clearly in the graphical representation shown in Figure 6.4. At $k = 0$ the orbital phases are invariant from one unit cell to the next. At this value of k the symmetries of the O 2pz and M dz^2 orbitals are not appropriate for mixing, and the Bloch functions representing this overlap are nonbonding. Because the O 2pz orbital is more electronegative, χ_b becomes O 2pz nonbonding, while χ_{ab} becomes M dz^2 nonbonding. This picture reflects the general tendency at $k = 0$ for interactions between oxygen p orbitals with *ungerade* symmetry and either metal s or d orbitals, which have *gerade* symmetry. At the edge of the BZ, where $k = \pi/a$ the basis sets reverse phase every unit cell so that all M–O interactions within the Bloch function of χ_b are fully bonding, stabilizing its energy, and the M–O interactions within the Bloch function of χ_{ab} are fully antibonding, destabilizing its energy.

Using similar logic, the M dxz–O 2px overlap can be analyzed and the shape and approximate energy levels of the two bands that arise from this overlap can be deduced. The lower band goes from nonbonding O 2px at $k = 0$ to M–O π bonding at $k = \pi/a$, while the upper band goes from nonbonding M dxz to antibonding π^* over the same interval, as shown in Figure 6.5. The M dyz–O 2py interaction is analogous and gives rise to a degenerate set of bands. The orientation of the

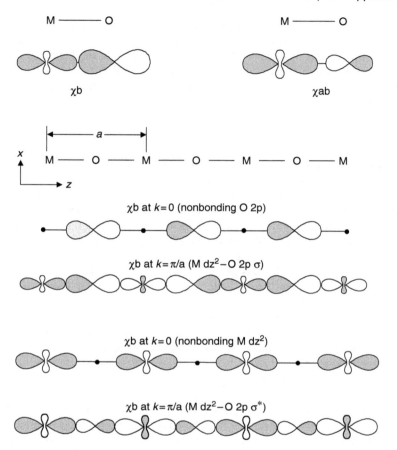

FIGURE 6.4 M dz^2–O 2pz σ/σ^* overlap in a one-dimensional metal oxygen, MO, chain

M d$x^2 - y^2$ and M dxy orbitals is such that there is minimal interaction with the O 2p orbitals. These orbitals are nonbonding and insensitive to the phase of the orbitals in the neighboring unit cells. Consequently, the two bands associated with these orbitals are expected to have very similar energies at $k = 0$ and $k = \pi/a$.

The band structure as calculated using the Extended Hückel method incorporated in the program CAESAR is shown in Figure 6.6(a). Given the bonding analysis contained in the previous paragraphs the electronic band structure of the MO chain can be readily understood. The three lowest energy bands run downhill because the orbital interactions go from nonbonding O 2p at $k = 0$ to bonding M d–O 2p at $k = \pi/a$. By the same token the three highest energy bands run uphill because the orbital interactions go from nonbonding M d at $k = 0$ to antibonding M d–O 2p at $k = \pi/a$. The σ band of M 3dz^2–O 2p origin (shown by – – – line) is wider than the π bands (shown by – $\cdot$ – $\cdot$ line), because the strength of σ-bonding is larger than that of π-bonding. The nonbonding M dxy and M d$x^2 - y^2$ bands

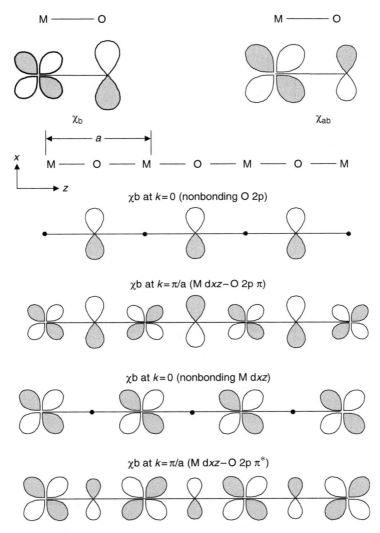

FIGURE 6.5   M dxz–O 2p$x\pi/\pi^*$ overlap in a one-dimensional metal oxygen, MO, chain

(shown by ———) are almost perfectly flat as expected. As a general principle the width of a band depends upon the change in bonding strength as the value of k changes.

Still some features of the band structure are not explained by the arguments presented thus far. If the upper five bands ($-12 < E < -8$) have nonbonding d character at $k = 0$ then why does the σ^* band reside at a distinctly higher energy ($E \sim -9.7$ eV) than the other four bands with predominantly d character ($E \sim -12$ eV)? Furthermore, why is the σ^* band not wider than the π^* bands? While neither feature is expected from the orbital drawings in

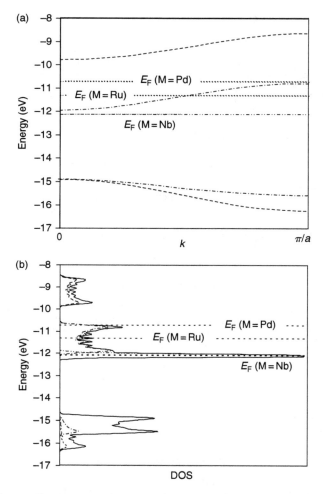

FIGURE 6.6 (a) The electronic band structure of a one-dimensional metal oxygen, MO, chain as calculated using the EHTB method. The bands that arise due to M dz^2–O $2pz\sigma$ interactions are shown as – – – lines, the doubly degenerate bands that arise due to M dxz–O $2px$ and M dyz–O $2py\pi$ interactions are shown as – · – · lines, and the doubly degenerate nonbonding bands attributed to M dxy and dx^2-y^2 orbitals are shown as – · – · line. The Fermi level is marked as · · · · · · . for different 2nd row transition metal ions. (b) The electronic density of states (DOS) of the same MO chain. The total DOS is shown as —— line, while the M dz^2 partial density of states (PDOS) is shown as – – – line, the M dxz/dyz PDOS is shown as – · – · line and the M $dxy/dx^2 - y^2$ PDOS is shown as – · – · line

Figure 6.4 and Figure 6.5, they can readily be understood once the influence of oxygen 2s orbitals is taken into account. The M dz^2–O 2s interaction will be antibonding at $k = 0$, thereby raising the energy of the σ^* band at $k = 0$. This effect is not present at $k = \pi/a$ due to the symmetry reversal that prevents mixing between the O 2s and the M dz^2 orbitals. The net effect is a reduction in the width of this band. The remaining d orbitals are oriented to overlap in either a π or a δ fashion with oxygen and cannot mix with the spherically symmetric O 2s orbitals. Therefore, the O 2s orbitals have no impact on the width of these bands.

An important parameter in any band structure is the Fermi level, which is denoted for different metals in Figure 6.6 by dotted lines. The Fermi level, E_F, denotes the separation between filled and empty states at absolute zero. In molecular terms, the Fermi level marks the position of the HOMO. At finite temperatures, the electron distribution in a metal or a semiconductor is smeared out so that some electrons are thermally excited from states below E_F to states above E_F. The position of the E_F can be determined by simple electron counting arguments. For oxides, the electron counting is most easily done using formal oxidation states. In this case, the oxide ion has a completely filled 2p shell accounting for six electrons (recall that we have neglected the O 2s orbitals). Each band can hold two electrons so that the lower energy set of three bonding bands are completely filled by these electrons. The filling of the remaining bands depends upon the electron count of the M^{2+} transition metal ion. In Figure 6.6 the Fermi level is shown for $M = Nb^{+2}$ (d^3), Ru^{+2} (d^6), and Pd^{+2} (d^8). When $M = Nb^{+2}$ the two nonbonding bands are three-fourth filled. When $M = Ru^{+2}$ the two nonbonding bands are completely filled and the two π^* bands are half-filled. Finally, when $M = Pd^{+2}$ the nonbonding and π^* bands are filled and the σ^* band is empty.

6.5.2 Density of States Plots

Even for a simple one-dimensional chain with a mere two atoms in the unit cell there is considerable information contained in the band structure diagram. In essence, it is like an MO diagram at each point in k-space. In a three-dimensional crystal with numerous atoms in the unit cell the complexity can increase to the point where it looks like a confusing jumble of curvy lines. It is not surprising then that nonspecialists often refer to the band structure (or E versus k curves) as a spaghetti diagram. Sometimes it will be useful to depict the electronic structure in a less complicated manner using a density of states (DOS) plot. The DOS plot for the one-dimensional MO chain is shown in Figure 6.6(b). The y-axes of the DOS and E versus k plots are identical; they denote the energy levels of the electronic states. The DOS, N(E), is plotted on the x-axis of the DOS plot. The quantity N(E)dE is the number of allowed energy levels per unit volume of the solid in the energy range E to $E+dE$. Figure 6.6(b) shows not only the total electronic DOS (as ———), but also the partial DOS representing the various metal d orbital contributions to the electronic structure.

There are several useful pieces of information that can be extracted from the DOS plot. The PDOS contributions allow us to divide the electronic structure into four energy regions: (i) the bonding states covering an energy range of roughly -14.5 to -16.5 eV, (ii) the nonbonding states peaked in a narrow energy range just below -12 eV, (iii) the π^* antibonding states covering an approximate energy range of -12 to -10.5 eV, and (iv) the σ^* antibonding states covering an approximate energy range of -10 to -8.5 eV. The area contained under each region of the DOS plot is proportional to the number of bands that are contained within that energy range. In this case, the areas of the regions (i)–(iv) defined earlier will have a 3:2:2:1 intensity ratio. The large peak in the DOS plot corresponds to the energy where the nonbonding M dxy and M$x^2 - y^2$ states fall. The large DOS at this energy follows directly from the fact that four electrons per unit cell can occupy states covering a very narrow energy interval. As a general concept flat bands give rise to peaks in the DOS plot, while wide (disperse) bands give rise to broad shallow features in the DOS plot. The forbidden energy regions or band gaps can also be readily seen from the DOS plot because N(E) goes to zero in those regions. For example, if the composition of the chain were PdO the DOS plot would tell us that this compound is a semiconductor with a band gap, E_g, a little larger than 0.5 eV. Finally, note that the metal PDOS contributions go to zero at the top of the σ/π bonding region. That is, the electronic states at the top of this band are oxygen 2p nonbonding, for reasons that were explained in the preceding section. This feature is common to the electronic structures of most metal oxides.

In summary, the DOS plot is a very useful way of depicting the electronic structure. DOS plots provide information about the allowed energy levels, filling, and widths of bands in a way that can be assimilated without knowledge of k-space notation. The use of PDOS plots allows one to break down the electronic structure in order to get a sense of how the individual atomic orbitals contribute. In fact, for depicting the electronic structure of an extended solid, the DOS plot is essentially the solid-state equivalent of the MO diagram used by molecular chemists. Given this correspondence, it is possible to sketch DOS plots for metal oxides from knowledge of the cation(s) coordination environment, electron count, and electronegativity, in the same way that approximate MO diagrams can be constructed from this information [36]. We will explore this correspondence in detail in Section 6.5.4 and Section 6.5.5. Nonetheless, there is some information that is lost upon transforming the E versus k curves into the DOS plot, particularly in three dimensions. For example, the presence of many narrow bands closely separated in energy may give the appearance of a set of wide bands in the DOS plot. Yet the transport properties of these two cases will be quite different. In an anisotropic crystal structure, the bands may be wide in some directions and narrow in others, indicating that electronic transport only occurs in certain directions. Such information will be lost in the DOS plot. The carrier mobility and nature of the band gap (direct or indirect) cannot be determined directly from the DOS plot.

6.5.3 Conductivity, Carrier Mobility, and Their Relationship to the Band Structure

At the simplest level, band theory predicts that a compound will be metallic if the Fermi level cuts through a band. Metallic conductivity results from the fact that within a partially filled band there is an infinitesimal energy difference between the uppermost filled state and the lowest energy unoccupied state. This enables electrons to move through the crystal by accessing nearly degenerate filled and empty states near E_F via thermal excitations. Within this simple picture, the NbO and RuO chains discussed earlier should be metallic conductors, while the PdO chain would be a semiconductor. A slightly more detailed analysis paints a somewhat different picture. The conductivity of a material, σ, is given by the expression:

$$\sigma = ne^2\tau/m^* = ne\mu \tag{6.5}$$

where n is the concentration of charge carriers (electrons or holes) per unit volume, e is the charge of an electron, τ is the average time between electron scattering events, m^* is the effective mass of the charge carriers, and $\mu = e\tau/m^*$ is the carrier mobility. Let us consider in more detail the three quantities that determine conductivity: carrier concentration, scattering time, and effective mass. The carrier concentration, n, is neither the total number of electrons in a solid, nor the total number of valence electrons. Even in the best conductor, the vast majority of electrons reside in localized states far from the Fermi level. The concentration of electrons capable of carrying charge by moving through the lattice is determined by the concentration of states located within a few kT of E_F. For a metal, where E_F cuts through a band(s), the carrier concentration is independent of temperature to a first approximation. For a semiconductor, where E_F falls between bands, the carrier concentration increases exponentially as the temperature increases. Therefore, the conductivity of a semiconductor increases as the temperature increases. Electrons are scattered primarily by lattice vibrations (phonons) and crystalline imperfections, including impurities, defects, dislocations, stacking faults, etc. Scattering by lattice vibrations increases strongly with increasing temperature, while scattering by impurities and defects is largely independent of temperature. The dependence of phonon scattering on temperature causes the conductivity of a metal to decrease with increasing temperature.

Of the three parameters that determine the conductivity, the effective carrier mass, m^*, is most closely related to the details of the electronic band structure. For a carrier at a given point in a band structure, the effective mass is related to the curvature of the $E(k)$ function by the equation:

$$m^* = (h/2\pi)^2(d^2E/dk^2)^{-1} \tag{6.6}$$

while this formula is only strictly valid for parabolic bands that come out of the free electron derivation of the band structure, it is qualitatively accurate for more complex band structures. The implications of this relationship are important, and can be stated in simple language. Wide bands have large curvature, which causes the second derivative of the E verus k curve to be large, which in turn leads to a small effective mass. Carriers with small effective mass are highly delocalized and possess high mobilities. The converse is that narrow bands have little curvature, which leads to a large effective mass, localized carriers, and low carrier mobility. Coming back to the electronic structure of the NbO chain, recall that the partially occupied bands are flat. Therefore, the effective mass will be very small and the conductivity is expected to be small as well. Intuitively, this makes sense because the electrons reside in nonbonding orbitals and one would expect localized, rather than highly delocalized carriers. Having said that, there is a very small gap between the $dxy/dx^2 - y^2$ bands and optical and thermal excitation into the π^* bands can lead to conductivity. The properties of MoO would be similar because the extra electron per unit cell is soaked up by the nonbonding d bands. The situation is different in RuO because the last two electrons occupy antibonding π^* states, leading to an upward movement of Fermi level as indicated in Figure 6.6. The carrier mobility and electron delocalization in the RuO chain would be considerably higher because the Fermi level cuts a band that has significant contributions from both M d and O 2p orbitals. Of course, a secondary consequence of changing the electron count is that occupation of antibonding states will lead to an increase in the metal–oxygen distance that will in turn decrease orbital overlap and decrease bandwidth to a certain extent.

6.5.4 The Electronic Band Structure of a Two-Dimensional CuO_2^{2-} Layer

To discuss the electronic structures of real materials, it is necessary to extend the analysis of the preceding sections into three-dimensional space. As an intermediate step, we will consider the electronic structure of a two-dimensional CuO_2^{2-} layer. The two-dimensional case is not just a conceptual intermediate on the way to three dimensions, but is an interesting case in its own right. Surfaces are two-dimensional entities and the two-dimensional electronic structures of surfaces are useful constructs for understanding heterogeneous catalytic processes. Furthermore, many compounds are made up of covalently bonded layers that are held together by ionic or even dispersion forces. The electronic structures of such materials are essentially two-dimensional in character. In fact, the two-dimensional CuO_2^{2-} layer that is analyzed later is widely accepted to be the structural motif that is responsible for superconductivity in the large family of high T_C cuprate superconductors. That is not to say that interactions between layers are of no consequence; however, it is true that many of the dominant features of the electronic structures of these materials can be captured from a two-dimensional analysis of the CuO_2^{2-} layer fragment. As is usually the case, the underlying concepts do not change upon increasing the dimensionality, but the mathematics of the analysis and the representation of the results become more complicated. In this treatment, we are leaving the mathematical analysis to computers, so the details will not concern us. However, it is necessary to understand how to represent electronic structures in two and three dimensions. The most important difference is that k is now a vector rather than a scalar quantity. In two dimension, the value of k will be defined as $\mathbf{k} = k_x\mathbf{x} + k_y\mathbf{y}$, where $\mathbf{x}$ and $\mathbf{y}$ are the translation vectors in reciprocal space.

The crystal structure of a planar CuO_2^{2-} layer is shown in Figure 6.7(a). Each Cu^{+2} ion is surrounded by four O^{-2} ions in a square planar geometry, while the oxygen ions are linearly coordinated by copper. The MO diagram for a Cu^{+2} ion in square planar coordination is illustrated in Figure 6.7(b). Focusing on the states derived from the Cu 3d–O 2p antibonding interactions we see that four of the five

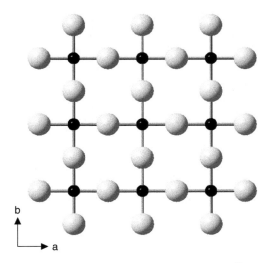

FIGURE 6.7 Structure of a two-dimensional CuO_2^{2-} layer

states are grouped together in a narrow range of energy. The Cu $3dz^2$–O 2p inter-action gives rise to a weakly antibonding state with a_{1g} symmetry that lies at -13.8 eV. As mentioned in Section 6.4.4 in D_{4h} symmetry both Cu $3dz^2$ and 4s orbitals have a_{1g} symmetry and the energy of the Cu $3dz^2$ level is lowered by the admixture of some Cu 4s–O 2p bonding character. The dxz and dyz orbitals have a π^* interaction with oxygen giving rise to a doubly degenerate set of orbitals with e_g symmetry at -13.6 eV, while the Cu $3dxy$–O 2p π^* interaction has a different symmetry, b_{2g}, but a similar energy, -13.4 eV. Finally, the Cu $3dx^2 - y^2$–O 2p interaction has strong σ^* character and thus is split off to higher energy, -11.5 eV.

The calculated electronic band structure for the CuO_2^{2-} layer is shown in Figure 6.8(a). The DOS plot is shown in Figure 6.8(b), with the energy levels from the MO diagram of the CuO_4^{6-} unit superimposed on the right-hand side. Note how clearly the basic features of the MO diagram are reproduced in the DOS plot. This illustrates the relative ease with which the DOS plot can be estimated from an accurate knowledge of the MO diagram of the cation centered building unit. Given the d^9 electron configuration of the Cu^{+2} ion the Fermi level would be expected to cut the Cu $dx^2 - y^2$ band in half, in good agreement with the calculated results. Within the assumptions of one electron band theory, the CuO_2^{2-} layer should be a metallic conductor. This view is too simplistic, and in reality electron–electron interactions stabilize an antiferromagnetically insulating ground state.

The shape of the bands in Figure 6.8(a) can be derived from orbital overlap considerations for various values of k. The x-axis no longer has numerical values, instead it has letters labeling various points. These labels refer to different values of the wavevector: the Γ point corresponds to $k_x = 0$, $k_y = 0$; the X point corresponds to $k_x = \pi/a$, $k_y = 0$, and the M point corresponds to $k_x = \pi/a$, $k_y = \pi/a$. To show the entire electronic structure, a three-dimensional plot with axes of k_x, k_y and

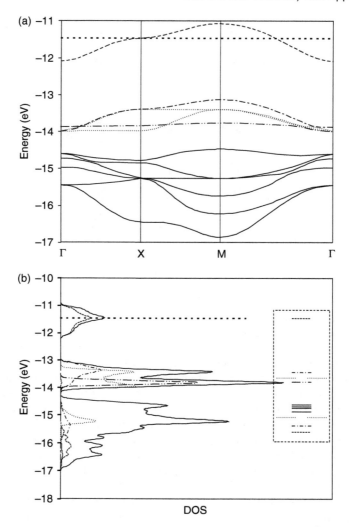

FIGURE 6.8 (a) The electronic band structure of a two-dimensional CuO_2^{2-} layer as calculated using the EHTB method. The bands with predominant Cu 3d character are coded as follows: the Cu $dx^2 - y^2$–O 2p σ^* band is shown – – –, the Cu dxy–O 2p π^* band is shown as – · – ·, the Cu dxz–O 2p and Cu dyz–O 2p π^* bands are shown in · · · · · ·, and the Cu dz^2 band is shown as – – –. The Fermi level is marked with a dashed line. (b) The electronic DOS. The total DOS is shown as ——, while the Cu $dx^2 - y^2$ PDOS is shown as – – –, the Cu dxy PDOS is shown as – · – ·, the Cu dxz/dyz PDOS are shown as · · · · · ·, and the Cu dz^2 PDOS is shown as – – –. The energy levels from the MO diagram of a CuO_4^{6-} square planar complex are shown in the dashed box for comparison. The coding is the same as used in (a)

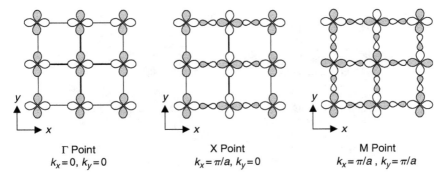

Γ Point
$k_x = 0, k_y = 0$

X Point
$k_x = \pi/a, k_y = 0$

M Point
$k_x = \pi/a, k_y = \pi/a$

FIGURE 6.9 Orbital overlap in the Cu $3dx^2 - y^2\sigma^*$ band at the Γ, X, and M points

energy would be needed. Although this could be done for a two-dimensional band structure, it becomes impossible to show everything in a single plot for a three-dimensional crystal, for which a four-dimensional plot (k_x, k_y, k_z, E) would be needed. Instead, slices from one high symmetry point in k-space to another are shown. The orbital overlap associated with the Cu $3dx^2 - y^2$–O 2p σ^* band at Γ, X, and M are shown in Figure 6.9. At Γ mixing between the Cu $3dx^2 - y^2$ and O 2p σ orbitals is forbidden by symmetry and the orbital character is largely nonbonding Cu $3dx^2 - y^2$. As with the one-dimensional MO chain we see that the energy of this band does not degenerate with the other Cu 3d bands at Γ, due to the antibonding Cu $3dx^2 - y^2$–O 2s interaction. This leads to a significant reduction in the width of this band.[††] Upon moving from Γ to X, the interactions with one-half of the O 2p orbitals become antibonding. Upon moving to the M point, the Cu $3dx^2 - y^2$–O 2p interactions become completely antibonding. Naturally, the band energy increases as the percentage of antibonding interactions increases.

6.5.5 The Electronic Band Structure of a Three-Dimensional WO₃ Cubic Lattice

Completing the progression to three dimensions let us consider the band structure of WO_3 with the cubic ReO_3 structure type, shown in Figure 6.10(a). While the actual crystal structure of WO_3 is distorted from the aristotype structure by tilting of the octahedra and 2nd order Jahn–Teller displacements of the W^{+6} cations [37,38], here we will consider the idealized cubic structure, which maintains the same topology and has a similar but less complicated band structure. The ReO_3 structure, which can be described as a cubic network of corner-sharing octahedra, contains octahedrally coordinated cations and linearly coordinated anions. If a large cation is inserted in the center of the unit cell the structure becomes cubic perovskite,

---

[††] It would seem from comparison with more sophisticated calculations, where this band is found to be wider, that the default orbital parameters used in CAESAR overestimate the interaction between the O 2s orbitals and the cations.

(a)   (b)

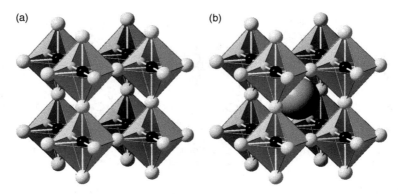

Figure 6.10 (a) Cubic ReO_3 crystal structure with the cation shaded black and the anion grey. (b) Cubic perovskite crystal structure, with large A-site cation in the cuboctahedral cavity

as shown in Figure 6.10(b). The perovskite structure class is arguably the largest and most important family of oxides. A large number of oxides with interesting electronic properties are perovskites. They will be discussed in some detail in Section 6.6.

The electronic band structure and DOS plot for cubic WO_3 are shown in Figure 6.11. To understand the basic features of the electronic structure, consider the MO diagram for a d^0 cation in octahedral coordination (see Section 6.4.2). The first nine bands (shown as ——) have primarily oxygen 2p orbital character. Among the oxygen 2p valence bands those that have the lowest energies are W 5d–O 2p bonding in character, while the flat bands that make up the top of the valence band are primarily O 2p nonbonding. The five unoccupied bands that lie above E_F are W 5d–O 2p antibonding in character. The lower three bands (shown as $- \cdot - \cdot$) arise from the antibonding π^* interactions between the triply degenerate t_{2g} set of W 5d orbitals and O 2p π orbitals. The upper two bands (shown as ——) arise from the antibonding σ^* interactions between the doubly degenerate e_g set of W 5d orbitals and O 2p σ orbitals. For a primitive cubic structure the special points in k-space are as follows: the Γ point corresponds to $k_x = 0$, $k_y = 0$, $k_z = 0$; the X point corresponds to $k_x = \pi/a$, $k_y = 0$, $k_z = 0$; the M point corresponds to $k_x = \pi/a$, $k_y = \pi/a$, $k_z = 0$; and the R point corresponds to $k_x = \pi/a, k_y = \pi/a, k_z = \pi/a$. The orbital overlap of the W $5dxy$–O 2p π^* band and the W $5dx^2 - y^2$–O 2p σ^* band at Γ and M are shown in Figure 6.12. From the previous analysis of one-dimensional and two-dimensional structures the overlap and band structure of cubic WO_3 can be readily understood. At Γ the t_{2g} set of bands is strictly W 5d nonbonding, and the e_g set of bands is W 5d–O 2p nonbonding, with a weak W 5d–O 2s σ^* component.[‡‡] Moving from Γ to X the dxy,

---

[‡‡] In order to obtain a proper shape for the e_g bands the zeta coefficients of the oxygen 2s orbitals had to be increased to 4.000, significantly contracting their radial extension. If this was not done the strength of the W e_g–O 2s interaction at the Γ point was too strong and the band shape was inverted.

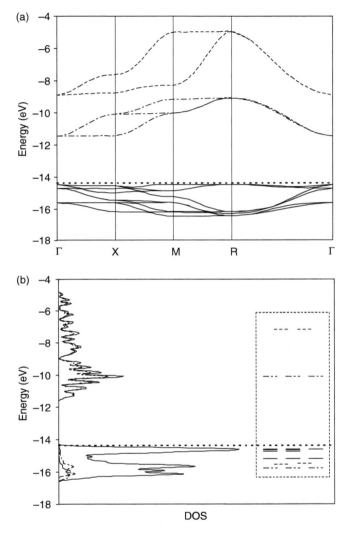

FIGURE 6.11 (a) The electronic band structure of WO_3 lattice with the ideal cubic ReO_3 structure as calculated using the EHTB method. The bands with predominant W 5d character that cover the energy range -5 to -12 eV are coded as follows: the W 5d e_g–O 2p σ^* bands are shown as $---$ and the W 5d t_{2g}–O 2p π^* bands are shown as $-\cdot-\cdot$. The Fermi level is marked as $---$. (b) The electronic DOS. The total DOS is shown as ———, while the W 5d e_g PDOS is shown as $-\cdot-\cdot$ and the W 5d t_{2g} PDOS is shown as $---$. The energy levels from the MO diagram of a WO_6^{6-} octahedron are shown in the dashed box for comparison. The coding is the same as used in (a)

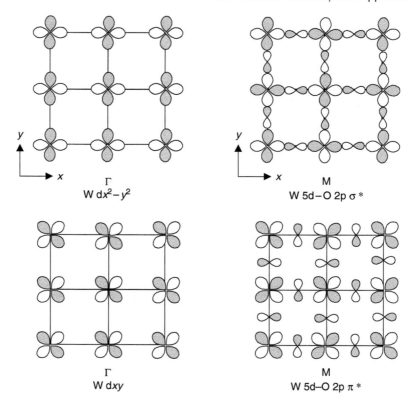

FIGURE 6.12 The W $5dx^2 - y^2$–O 2p σ^* and the W $5dxy$–O 2p π^* orbital overlap at the Γ and M points in cubic WO_3

dxz, and $dx^2 - y^2$ bands become antibonding in one direction, while the dz^2 band remains mostly nonbonding and dyz band remains completely nonbonding, giving rise to a flat band along the Γ to X line. At the M point, dxy and $dx^2 - y^2$ bands have become completely antibonding as shown in Figure 6.12. Finally, at the R point all five W 5d bands are completely antibonding, with each band reaching its maximum value.

The symmetry of the cubic ReO_3 (and perovskite) structure gives rise to a number of special features in the electronic band structure. At the Γ point, the states at the top of the valence band are oxygen 2p nonbonding, while those at the bottom of the conduction band are completely metal t_{2g} nonbonding. Therefore, the direct band gap energy represents a charge transfer excitation in the truest sense. This is not the case with isolated tetrahedra and octahedra, because while the HOMO is oxygen 2p nonbonding, the LUMO is metal–oxygen antibonding, as discussed in Section 6.4.1. Another point of interest is the separation between the π^* and σ^* bands at the R point. This energy separation is the solid-state equivalent of the ligand field splitting of an isolated octahedron.

6.6 ELECTRONIC BAND STRUCTURES OF PEROVSKITES

It is far beyond the scope of this chapter to review the electronic structures and properties of all metal oxides, or even all of the important metal oxide structure types. Instead, this section covers some features of one structural family, perovskite, in some detail. In doing so, it is hoped that the important concepts will be illustrated in such a way that they can be widely applied. Of course, the choice of the perovskite structure as an illustrative example is not a random choice. The perovskite family of compounds is very extensive, encompassing most of the periodic table. Furthermore, perovskites exhibit nearly every type of interesting electronic or magnetic behavior seen in oxides (ferromagnetism, ferroelectricity, piezoelectricity, nonlinear optical behavior, metallic conductivity, superconductivity, colossal magnetoresistance, ionic conductivity, photoluminescence, etc.). One important property that is not readily found among perovskites, transparent conductivity, is the focus of Section 6.7.

The perovskite structure is illustrated in Figure 6.10(b). The ideal stoichiometry of a perovskite is ABX_3, where X is an anion, B is a cation with octahedral coordination, and A is a "large" cation with cuboctahedral coordination. The two cation sites are commonly referred to as the B-site (coordination number = 6) and the A-site (coordination number = 12). In order for the ideal cubic structure to be realized the size of the A- and B-site cations must be well matched. The size match or lack thereof between the two cations is given by the Goldschmidt tolerance factor, t:

$$t = (R_A + R_X)/\{\sqrt{2(R_B + R_X)}\} \tag{6.7}$$

where R_A, R_B, and R_X are the ionic radii of the A, B, and X ions, respectively. Alternatively, the tolerance factor can be calculated from the A–X and B–X distances predicted by bond valence calculations [39]. For a perfect match, $t = 1$, and the cubic perovskite structure is typically realized. When $t > 1$ the A-site cation is too large and the B–X bonds are stretched. This can lead to stabilization of other structure types, in particular, the so-called hexagonal perovskites. However, when B is a d^0 cation this can lead to 2nd order Jahn–Teller driven displacements of the B-site cation. The classic example is $BaTiO_3$. By comparison, $SrTiO_3$ has a tolerance factor very close to unity and adopts the undistorted cubic structure at room temperature. A tolerance factor less than unity is an indication that the A-site cation is too small for the cuboctahedral site. This situation typically leads to a structural distortion via cooperative tilting of the octahedra. This type of distortion, commonly referred to as an octahedral tilting distortion, is very common. Prototypical cases include the mineral perovskite, $CaTiO_3$, and the compound $GdFeO_3$. The details and driving forces behind octahedral tilting distortions can be found in the literature [40–43], for our purposes the important ramifications are as follows: (i) the immediate anion environment about the B-site cation usually remains close to a perfect octahedron, (ii) the M–X–M bonds bend away from the linear coordination of the cubic structure, (iii) the A-site polyhedron becomes very distorted leading to a decrease in the effective coordination number of the A-site cation.

In analyzing the electronic structures of perovskites to a first approximation the influence of the A-site cation on the states near the Fermi level can often be neglected. This can be done because: (i) the A-site cation is normally highly electropositive, and (ii) the A–O bonds are very long. When these conditions are not met, for example if the A-site cation is Cd or Pb and/or there is a significant octahedral tilting distortion, this assumption is no longer valid. In the treatment that follows the role of the A-site cation will be largely ignored, apart from its role in achieving the proper electron count. In Section 6.6.3 the electronic structure of $CdSnO_3$ will be examined as an example of a compound where the A-site cation plays an important role in the electronic structure. In the absence of A-site cation effects, the electronic structure of a cubic perovskite containing a d^0 cation on the B-site is perfectly analogous to the electronic structure of cubic WO_3. Having deconstructed the orbital overlap of cubic WO_3 using EHTB calculations in Section 6.5.5, we now turn to more accurate LMTO calculations to analyze the electronic structures of perovskites. While the LMTO approach does not allow one to explicitly look at the atomic orbital coefficients within the LCAO formalism, it is possible to see the individual atomic orbital contributions to the total electronic structure through "fatband" plots [44]. A fatband plot is an E versus k plot of the band structure where hatching is used to denote the bands where an individual atomic orbital makes a contribution. The width of the hatching increases as the orbital contribution to a band increases. Fatbands are a useful tool to quantify the change in orbital contribution from band to band, and within a given band upon moving through reciprocal space.

6.6.1 Cubic Perovskites with Transition Metal Ions: Semiconductors and Metals

Figure 6.13 shows the calculated band structure of the cubic perovskite $KTaO_3$. The features seen in the EHTB calculated electronic structure of cubic WO_3 are perfectly mirrored. The fatband plots confirm the earlier assertion that the valence band has predominantly O 2p character, ranging from Ta–O bonding at the bottom of the valence band to O 2p nonbonding at the top. The Ta 5d t_{2g}–O 2p π^* bands are the next highest energy group, followed by the Ta 5d e_g–O 2p σ^* bands. Note that the O 2p contribution to both these bands goes to zero at Γ. At the Γ point Ta 5d–O 2p mixing is symmetry forbidden and therefore both sets of bands realize a minimum value at Γ. The calculations show that the K 4s band at an energy level similar to the Ta e_g bands. Despite the apparent energetic overlap between these two sets of bands, the spatial overlap between the A-site cation and the Ta–O σ bonding interactions are minimal. The orbital overlap at critical points in the electronic structure has been described in detail in Section 6.5.5 and will not be repeated here. Instead, our focus will be on semiquantitative aspects of the band structure. The valence band is roughly 5 eV wide, with maxima at the M and R points. The π^* and σ^* bands are 3.9 and 6.4 eV wide, respectively, and both bands

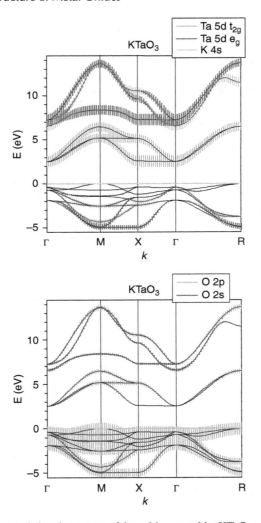

FIGURE 6.13 Electronic band structure of the cubic perovskite $KTaO_3$ as calculated using LMTO–ASA calculations. The Fermi level is set to zero energy and marked with light grey. Fatband hatching indicates the contributions of specific orbitals to the electronic structure. The upper plot shows the Ta 5d t_{2g} (dark grey), Ta 5d e_g (thick line) and K 4s (dark grey) fatbands, while the lower plot shows the O 2p (dark grey) and O 2s (thick line) fatbands

reach a minimum at the Γ point. The band gap is calculated to be an indirect gap of 2.6 eV. There is also a direct gap at Γ (and X) of 3.0 eV. Experimentally the band gap of $KTaO_3$ is reported to be 3.5 eV [28]. This level of agreement is quite reasonable, although it is not as good as observed in Section 6.4.2 for LMTO calculations on ordered perovskites containing electronically isolated octahedra. In general, the LMTO calculations tend to underestimating the band gaps of oxides with extended structures by anywhere from approximately 0.5 to 2.5 eV [28,45,46]. The reason for this is not fully understood, although the observation that the calculations are accurate for isolated polyhedra suggests that the discrepancy stems from an overestimation of the bandwidth. Nevertheless, the calculated band gaps reproduce observed trends in band gap energy and are sufficiently accurate to allow useful semiquantitative comparisons between compounds to be made.

In Section 6.5.5, the band gap of a cubic perovskite was shown to be equal to the energy difference between oxygen 2p orbitals and nonbonding metal d orbitals. This offers the unusual opportunity to explore orbital energies in an absolute sense. Unfortunately, there are a limited number of compounds with the cubic ReO_3/perovskite structure type that contain a d^0 ion on the octahedral site. $KTaO_3$ is one such example. Other examples include: $SrTiO_3$ with a band gap of 3.1 eV [28] and $BaZrO_3$ with a band gap of 5.3 eV [47]. $KNbO_3$ ($E_g = 3.1$ eV) and WO_3 ($E_g = 2.4$ eV) have the same topology, but their structures are distorted from the ideal cubic structure, by 2nd order Jahn–Teller displacements of the B-site cations in both compounds and octahedral tilting distortions in WO_3 [37]. The band gaps of these two compounds would be smaller if they adopted the undistorted cubic structure. A reasonable lower limit would be the calculated band gaps of the hypothetical cubic structures. Those values are 2.6 eV for $KNbO_3$ and 1.1 eV for WO_3 [28]. The trends in cation orbital energies parallel the trends observed in Section 6.4 for the HOMO–LUMO splitting in MO_4^{n-} and MO_6^{m-} species.

Perovskites with a d^0 ion on the octahedral site are electrical insulators due to the band gap between the O 2p valence band and the π^* conduction band. Replacing the d^0 ion with a d^1 or a d^2 ion leads to a partially occupied π^* band and in many cases metallic conductivity. Perhaps the best example is ReO_3, whose resistivity of $\sim 4 \times 10^{-5}$ Ω cm, is among the lowest recorded for an oxide. By comparison, copper metal is only about 20 times more conducting. Other examples include $BaMoO_3$ [48], $SrCrO_3$ [49], and $SrVO_3$ [49]. The electronic structure of $SrVO_3$ is shown in Figure 6.14. The band structure of $SrVO_3$ is qualitatively similar to $KTaO_3$, with two differences of note. First, the Fermi level position has changed and it now cuts across the three V 3d t_{2g}–O 2p π^* bands. This gives rise to a large concentration of conduction electrons and accounts for the metallic conductivity of this compound. Second, the width of the π^* and σ^* bands, 2.5 and 4.5 eV respectively, are not as wide as they are in $KTaO_3$, where they were calculated to be 3.9 and 6.4 eV, respectively. This feature stems from the smaller size of the 3d orbitals (see Section 6.2). The reduced bandwidth of the 3d transition metal oxide perovskites makes the metallic state less robust than it is in 4d and 5d perovskites. There are a number of compounds in the former group that have partially filled d orbitals, but are not metallic. One example is the isoelectronic compound, $LaTiO_3$ [50], another is the closely related compound $LaVO_3$ [51]. A full explanation of the reasons why these compounds are not metals is beyond the scope of this chapter, but conduction bandwidth is unquestionably a contributing factor. In reducing the oxidation state of the B-site cation from +4 to +3 the B–O bond distance increases and the bond covalency decreases. Furthermore, the tolerance factor in these latter two compounds is less than one and octahedral tilting distortions are present. Both

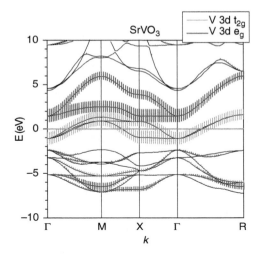

FIGURE 6.14 Electronic band structure of the cubic perovskite $SrVO_3$ as calculated using LMTO–ASA calculations. The Fermi level is set to zero energy and marked with light grey line. Fatband hatching indicates the contributions of specific orbitals to the electronic structure. The V 3d t_{2g} (dark grey) and V 3d e_g (thick line) fatbands are shown

factors will reduce the width of the partially filled π^* band, thereby reducing carrier mobility and favoring localized electronic states. Many other interesting phenomena can be found among the 3d transition metal oxide perovskites that are not covered here. John Goodenough has written extensively on this topic over the past fifty years and the interested reader is encouraged to see his relatively recent review with Zhao and coworkers [49] as a starting point into this fascinating field.

6.6.2 Cubic Perovskites with Main Group Ions

The electronic band structure for the cubic perovskite $BaSnO_3$ is shown in Figure 6.15. The upper valence band features are similar to those seen for $KTaO_3$ with maxima located at both the M and R points. The conduction band minimum is located at the Γ point. The calculations predict that $BaSnO_3$ should be a semiconductor with an indirect band gap of 0.7 eV. This value is considerably smaller than the measured value of 3.1 eV [45]. As observed for $KTaO_3$, the orbital character at the top of the valence band is predominantly nonbonding O 2p. However, the orbital overlap that gives rise to the conduction band in $BaSnO_3$ is quite different from that described for $KTaO_3$. In $BaSnO_3$ there is a single conduction band, which has Sn 5s–O 2p σ^* character as seen clearly in the fatband representation in Figure 6.15. The oxygen fatbands show that at the Γ point where the conduction band minimum is found the oxygen 2p contribution falls to zero. The absence of Sn 5s–O 2p σ^* interactions at Γ is responsible for the significant lowering of the energy of the conduction band near Γ. The minimum would be even deeper

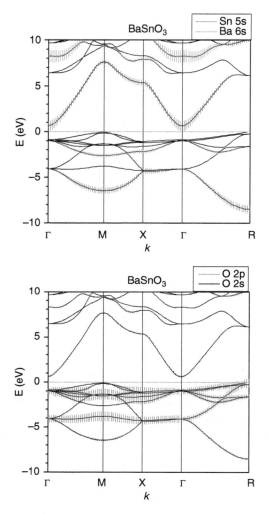

FIGURE 6.15   Electronic band structure of the cubic perovskite BaSnO$_3$ as calculated using LMTO–ASA calculations. The Fermi level is set to zero energy and marked with light grey. Fatband hatching indicates the contributions of specific orbitals to the electronic structure. The upper plot shows the Sn 5s (dark grey) and Ba 6s (light grey) fatbands, while the lower plot shows the O 2p (dark grey) and O 2s (thick line) fatbands

(leading to a wider conduction band and a smaller band gap) if not for the weak Sn 5s–O 2s σ^* interaction that is fully symmetry allowed at Γ. The orbital overlap that occurs at the Γ and M points of the conduction band is shown in Figure 6.16. The conduction bandwidth is $\sim$7 eV, and would be even wider if not for an avoided crossing near the R point. This value is almost double the calculated conduction bandwidth of KTaO$_3$. Further analysis of the BaSnO$_3$ band structure can be found in Reference 45.

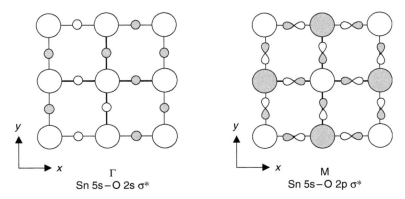

FIGURE 6.16 The Sn 5s–oxygen interactions in cubic $BaSnO_3$ at the Γ and M points. At Γ Sn 5s–O 2p interactions are symmetry forbidden; however, Sn 5s–O 2s σ^* interactions are allowed, while at M Sn 5s–O 2p σ^* interactions are allowed, but Sn 5s–O 2s interactions (in the xy plane) are symmetry forbidden. The atoms shown are those found in the (001) plane

The perovskite structure is not as common among main group metal oxides as it is among transition metal oxides. One of the principle reasons behind this fact, is the absence of π-bonding in main group metal oxides, which decreases the stability of the perovskite structure with respect to competing structure types. A good example of this is seen in the crystal chemistry of $NaSbO_3$. $NaSbO_3$ adopts the ilmenite structure, which has layers of edge-sharing octahedra in violation of Pauling's rules [52], despite the fact that it has a tolerance factor that is close to unity. Recently, the perovskite polymorph of $NaSbO_3$ has been prepared using high pressure high temperature synthesis methods [53]. In the same work, the authors provide a detailed analysis of the competition among cubic perovskite, distorted perovskite, and ilmenite structures. The conclusion for main group oxides is that ionic interactions favor perovskite, whereas strong σ-bonding favors structures where the M–O–M angles are strongly distorted from linear, such as ilmenite. Therefore, perovskite becomes increasingly unstable as the oxidation state of the main group ion on the octahedral site increases.

6.6.3 Distorted Perovskites: $NaTaO_3$, $CaSnO_3$, and $CdSnO_3$

As stated earlier octahedral tilting distortions are common in perovskites. In fact, it is estimated that only $\sim 10\%$ of perovskites adopt the ideal cubic structure [39]. How do these distortions affect the electronic structure? To investigate this question the calculated electronic structures of three distorted perovskites are shown in Figure 6.17 to Figure 6.19. Figure 6.17 shows the electronic structure of $NaTaO_3$, which undergoes an octahedral tilting distortion lowering the symmetry from Pm3m cubic to Pnma orthorhombic. The distortion does not affect the Ta–O distances and O–Ta–O angles significantly, but the Ta–O–Ta angles distort from $180°$

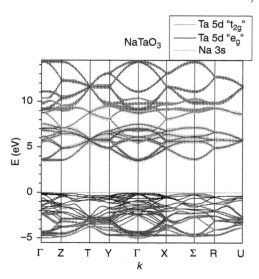

FIGURE 6.17 Electronic band structure of the orthorhombic perovskite $NaTaO_3$ as calculated using LMTO–ASA calculations. The Fermi level is set to zero energy and marked with a light grey line. Fatband hatching indicates the contributions of specific orbitals to the electronic structure. The Ta $5dxz$, $5dxz$, and $5dx^2 - y^2$ fatband contributions are shown as dark grey. These orbitals closely approximate the t_{2g} orbitals in the cubic perovskite structure. The Ta $5dxy$ and $5dz^2$ fatband contributions are shown as thick line. These orbitals closely approximate the e_g orbitals in the cubic perovskite structure. The Na $3s$ fatbands are shown as light grey.

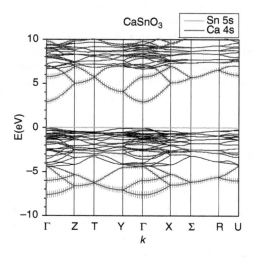

FIGURE 6.18 Electronic band structure of the orthorhombic perovskite $CaSnO_3$ as calculated using LMTO–ASA calculations. The Fermi level is set to zero energy and marked as $\times\times\times\times$ line. Fatband hatching indicates the contributions of specific orbitals to the electronic structure. The Sn $5s$ $(- \cdot - \cdot)$ and Ca $3s$ $(- - -)$ fatbands are shown

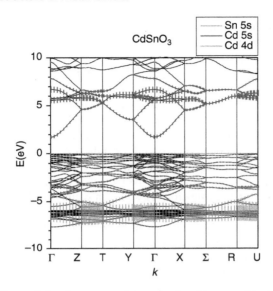

FIGURE 6.19 Electronic band structure of the orthorhombic perovskite CdSnO$_3$ as calculated using LMTO–ASA calculations. The Fermi level is set to zero energy and marked with a light grey. Fatband hatching indicates the contributions of specific orbitals to the electronic structure. The Sn 5s (dark grey), Cd 5s (thick line), and Cd 4d (light grey) fatbands are shown

in KTaO$_3$ to 158–160° in NaTaO$_3$. The electronic structure looks considerably more complicated, and in many ways it is. For starters, there are now four formula units in the unit cell, which leads to a fourfold increase in the number of bands. There is more twisting and tangling of bands and many more special points on the k-axis. The increased complexity is primarily a consequence of increasing the size of unit cell, and the additional bands that appear can be explained through the concept of band folding [36]. The reciprocal space mapping from Pm3m to Pnma is described in Reference 45 and will not be repeated here. Though it should be noted that band folding moves the valence band maximum to the Γ point and changes the optical gap from indirect in KTaO$_3$ to direct in NaTaO$_3$.

The fatbands are a particularly useful tool for comparing the electronic structures of NaTaO$_3$ and KTaO$_3$. In both compounds, the bands can be broken down into the same groupings upon increasing energy: (i) a set of valence bands with predominantly O 2p character, (ii) a set of conduction bands with Ta t$_{2g}$–O 2p π^* character, (iii) a higher lying set of conduction bands with Ta e$_g$–O 2p σ^* character. The K 4s and Na 3s bands lie high in energy, well above the bottom of the conduction band, and do not play a significant role near E_F. The most significant difference between the two compounds lies in the curvature of the conduction bands. While the total energy range covered by the Ta t$_{2g}$ fatbands is not terribly different in the two compounds the curvature of the individual bands is smaller in NaTaO$_3$, which leads to a decrease in carrier mobility. Furthermore, the transition from linear to bent Ta–O–Ta bonds introduces a small but not insignificant amount of antibonding character at the conduction band minimum. This raises the energy of the conduction band minimum, thereby increasing the band gap. The calculated and observed band gaps for $NaTaO_3$ are 3.5 and 4.0 eV, respectively. These values are 0.9 (calculated) and 0.5 eV (observed) larger than those found for $KTaO_3$ [28].

$CaSnO_3$ and $CdSnO_3$ are distorted perovskites, isostructural with $NaTaO_3$. These compounds possess even larger octahedral tilting distortions, as seen from the Sn–O–Sn angles: 146–148° in $CaSnO_3$ and 142–145° in $CdSnO_3$. Their electronic band structures are shown in Figure 6.18 and Figure 6.19. The comparison between these two compounds and $BaSnO_3$ parallels the $KTaO_3/NaTaO_3$ analysis given in the preceding paragraph. The increased complexity of the $CaSnO_3$ and $CdSnO_3$ band structures follows from the increased number of atoms in the unit cell and the reduction in symmetry. As was seen with $NaTaO_3$, the orthorhombic band structures can be derived from the cubic perovskite band structure through band folding. Details of this process can be found in Reference 45; however, a qualitative summary will be given here. The calculations predict that $CaSnO_3$ will be a direct-gap semiconductor. The change from an indirect to a direct gap semiconductor originates from the band-movements that accompany the band folding. In all three compounds, the Sn 5s fatbands split into two regions: the Sn 5s–O 2p σ-bonding states are located at the bottom of the valence band and the antibonding states make up the conduction band. In $CaSnO_3$, the conduction band curvature is noticeably smaller than observed for $BaSnO_3$. The effect is most pronounced at the bottom of the CB, where it has the most significant impact on the optical and electrical properties. The calculated and observed band gaps for $CaSnO_3$ are 2.9 eV and 4.4 eV, respectively [45]. These values are 2.2 (calculated) and 1.3 eV (observed) larger than those found for $BaSnO_3$. Analysis of the orbital overlap at the conduction band minimum shows that the octahedral tilting distortion introduces a significant degree of antibonding Sn 5s–O 2p σ^* interaction, giving rise to an increase of the band gap. In contrast, the influence of distortion on the Sn 5s–O 2s σ^* overlap at the CB minimum is small, because of the isotropic character of the s orbital. We will return to these points in more detail in Section 6.7. It is worth noting that the $BaSnO_3$ to $CaSnO_3$ octahedral tilting distortion leads to a significantly larger increase in the band gap (1.3 eV) than seen for $KTaO_3$ to $NaTaO_3$ (0.5 eV). While it is true that, the octahedral tilting distortion is larger in $CaSnO_3$ most of this difference stems from the fact that the conduction band in $BaSnO_3$ is much wider than it is in $KTaO_3$. Therefore, decreasing its width has a more pronounced effect on the band gap.

The electronic structure of $CdSnO_3$ (Figure 6.19) is qualitatively similar to the electronic structure of $CaSnO_3$. $CdSnO_3$ is a direct gap semiconductor with the conduction band minimum at the Γ point. Although the lack of dispersion at the top of the valence band makes the differences in energy between the direct and indirect gap transitions very small. Quantitatively though the calculated band gap of $CdSnO_3$ (1.7 eV) is much smaller than that of $CaSnO_3$ (2.9 eV). The calculations accurately reproduce the trends seen in the observed band gaps of $CdSnO_3$ (3.0 eV) and $CaSnO_3$ (4.4 eV) [45]. This effect cannot be explained using the Sn 5s–O 2p orbital overlap arguments of the preceding paragraph, because the octahedral tilting distortion is somewhat larger in $CdSnO_3$ than it is $CaSnO_3$. Yet the conduction band dispersion in $CdSnO_3$ is clearly greater than seen in $CaSnO_3$. The origin of this effect can be seen in the Ca 4s and Cd 5s fatband plots. The Ca 4s orbitals make a small contribution at the conduction band minimum, and while this contribution is more pronounced than the Ba 6s contribution in $BaSnO_3$ it is still relatively minor. In contrast, the Cd 5s and Sn 5s orbitals contribute almost equally to the lower region of the conduction band in $CdSnO_3$. Thus, we see an inductive effect (active participation of the A-site cation in bonding near E_F) not observed in the other $ASnO_3$ perovskites. This interaction introduces Cd 5s–O 2p σ-bonding interactions at the conduction band minimum, stabilizing the Sn 5s–O 2p interaction and lowering its energy. As a result, the conduction band dispersion increases and the band gap decreases. There are several reasons for the strong inductive effect in $CdSnO_3$. First, the Cd 5s orbital has a good energetic overlap with the Sn 5s–O 2p σ*-based conduction band. Second, the large octahedral tilting distortion leads to several short Cd–O contacts. Examination of the coordination environment about oxygen reveals that the short Cd–O distances [2.21 Å for O(4c) and 2.26 Å for O(8d)] are only about 10% longer than the Sn–O distances (2.04 to 2.08 Å). Thus, in $CdSnO_3$ we get both a good spatial and energetic overlap of the Cd 5s orbitals and the antibonding Sn 5s–O 2p σ* CB states.

6.7 ELECTRONIC BAND STRUCTURES OF BINARY OXIDES USED AS TRANSPARENT CONDUCTORS

It was pointed out in Section 6.5 that an approximate DOS plot can be constructed from the MO diagram of the cation centered building unit, provided the assumption that the contribution of the electropositive (alkali, alkaline-earth, and rare-earth) cations can be neglected is valid. Following this logic SnO_2 (rutile), $BaSnO_3$ (cubic perovskite), and $CaSnO_3$ (orthorhombic perovskite) should all have similar electronic structures. To a certain extent there is some truth to this statement. All three compounds have a valence band with significant O 2p character and a conduction band with Sn 5s–O 2p antibonding character. However, important details such as the band gap (SnO_2 $E_g = 3.8$ eV; $BaSnO_3$ $E_g = 3.1$ eV; $CaSnO_3$ $E_g = 4.4$ eV) and the carrier mobility vary considerably across this group. These differences can largely be attributed to differences in the conduction bandwidth. The importance of bandwidth, particularly conduction bandwidth, was illustrated several times in Section 6.6. In order to be able to estimate electronic band structures with semi-quantitative accuracy we need a simple criterion for estimating bandwidth. Does such a criterion exist? That question is the focus of the final section of this chapter.

The issue of bandwidth is particularly important for oxides that are used as transparent conductors. Transparent conducting oxides (TCOs) are used in a wide variety of applications including flat panel displays, energy-efficient window coatings, photovoltaics, organic and inorganic light emitting diodes, and antistatic coatings. In order to maintain a high degree of transparency across the visible region of the spectrum, the carrier concentration cannot be too high or the plasma

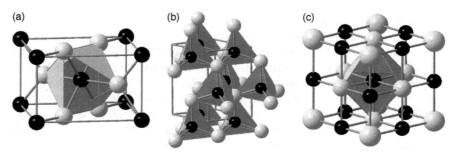

FIGURE 6.20   The crystal structures of (a) SnO_2 rutile, (b) ZnO wurtzite, and (c) CdO rock salt

excitations (collective excitations of the free carriers) will extend into the visible region of the spectrum. Yet in order to meet performance standards, while minimizing the thickness of the TCO film, high electrical conductivity is needed. The simultaneous demands for low carrier concentration and high conductivity can only be met if the carrier mobility is large. Consequently, only those compounds with wide conduction bands can act as effective transparent conductors. As seen in Section 6.6, main group metal oxides generally have much wider conduction bands than isostructural transition metal analogues. Compare, for example, the conduction bandwidths of the cubic perovskites $BaSnO_3$ with $KTaO_3$. Therefore, it is not surprising to find that the best-known and most extensively studied transparent conductors are n-doped main group metal oxides, in particular, binary main group metal oxides such as In_2O_3, ZnO, CdO, and SnO_2. In the following sections, the electronic band structures of SnO_2, ZnO, and CdO are analyzed and compared. The crystal structures of these three compounds, rutile, wurtzite, and rock salt respectively, are shown in Figure 6.20. Surprisingly a simple criterion for estimating the conduction bandwidth emerges from this analysis.

6.7.1 The Electronic Structure of SnO_2

SnO_2 adopts the rutile structure as shown in Figure 6.20. The tin cations are surrounded by six oxide ions that form a slightly distorted octahedron, while the oxide ions are coordinated by three tin cations in a planar geometry that is slightly distorted from trigonal planar. The rutile structure contains infinite linear chains of edge-sharing octahedra running parallel to the c-axis. Each chain is connected to four neighboring chains, each shifted by one-half unit cell along the c-axis. This produces a structure where each oxygen ion acts as a bridging ligand across the shared edge of one chain, while at the same time axially coordinating to a cation in a neighboring chain. The electronic band structure of SnO_2 is shown in Figure 6.21. The VB maximum is located at the R point, while the conduction band minimum is located at the Γ point, indicating that SnO_2 should be a semiconductor with an indirect band gap of 2.3 eV. The experimentally observed band gap, 3.8 eV, is considerably larger as expected [46]. The fatband plots show that the valence

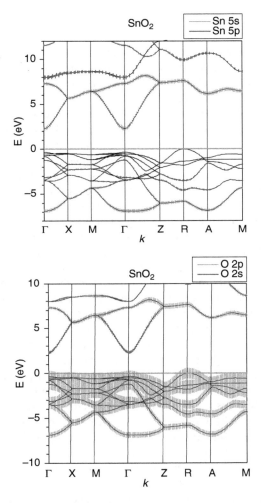

FIGURE 6.21 Electronic band structure of rutile SnO_2 as calculated using LMTO–ASA calculations. The Fermi level is set to zero energy and marked with light grey. Fatband hatching indicates the contributions of specific orbitals to the electronic structure. The upper plot shows the Sn 5s (dark grey) and Sn 5p (thick line) fatbands, while the lower plot shows the O 2p (dark grey) and O 2s (thick line) fatbands

band character is primarily O 2p. The nonbonding character of these bands is particularly pronounced over the region 0.0 to −2.0 eV. The small amount of dispersion that is present comes from weak antibonding O 2p–O 2p interactions in the closed packed oxygen lattice. There are four bands that have significant Sn 5s character. The lower set of bands that make up the bottom of the valence band arise from bonding interactions with oxygen, while the upper set of bands that form the conduction band arise from antibonding interactions with oxygen. There are two bonding and two antibonding Sn 5s–O 2p bands because there are two Sn atoms in the unit cell.

Let us focus our attention on the two antibonding Sn 5s–O 2p σ^* bands. It is the energy placement and width of these bands that determine the band gap and the carrier mobility, respectively. Figure 6.21 shows that the upper conduction band has a significant O 2p contribution throughout the 1st BZ, indicative of strong antibonding Sn 5s–O 2p σ^* character. Because the strength of the antibonding interaction is relatively insensitive to the wavevector, **k**, the width of this band is not large (<1.5 eV). While it is tempting to associate strong covalent bonding with large bandwidth, the behavior of this band shows that this assumption is not always valid. A high degree of covalency in the metal–oxygen interaction is necessary but not sufficient criterion for the attainment of wide bands. The energy of the lower conduction band is degenerate with the upper band throughout much of the 1st BZ, but their energies are strongly divergent near the Γ point. As the wavevector approaches Γ the O 2p contribution to the lower band decreases. At the Γ point, its contribution is minimal, but a small contribution from the O 2s orbitals can be seen. This is similar to the behavior seen earlier in the $BaSnO_3$ conduction band. The energy of this band drops precipitously toward the valence band because the strong Sn 5s–O 2p σ^* interaction is disappearing. The minimum would be even lower if it were not for the fact that it is replaced by a much weaker Sn 5s–O 2s σ^* interaction.[§§] This peculiar feature of the electronic structure gives the lower conduction band a large dispersion, particularly near its minimum at the Γ point. This feature leads to high carrier mobility in n-doped SnO_2, a characteristic that is of fundamental importance for its performance as a transparent conductor.

Let us examine the orbital overlap of these two bands at Γ in more detail. The unit cell of SnO_2 is made of two chains of edge-sharing octahedra that run parallel to the c-axis. Each oxygen ion is coordinated by two tin atoms within a chain (intrachain bonding) and one tin atom from a neighboring chain (interchain bonding). The coordination environment can be described as a Y-shaped planar $Sn'OSn_2$ unit, where the Sn' symbolizes a tin atom from the neighboring chain. The bonding overlap between the 2s and 2p orbitals on the central oxygen with the Sn 5s orbitals is illustrated in Figure 6.22. The O 2s orbital can form bonding and antibonding interactions with tin as long as the Sn 5s orbitals on neighboring chains (Sn and Sn') are in phase with each other, whereas the orbital labeled O 2p $\perp$ needs the Sn 5s orbitals on neighboring chains to be out of phase with each other in order to form effective bonding and antibonding interactions. The symmetry of the orbital labeled O 2p $\parallel$ allows bonding interactions with Sn but not with Sn'. Consequently, this orbital will contribute to the intrachain bonding, but not to the interchain bonding. Finally, the symmetry of the orbital labeled O 2p nb is such that it cannot mix with the Sn 5s orbitals.

[§§] It is possible that the tendency of the LMTO calculations to underestimate the band gap originates in part from an underestimation of the strength of the M–O 2s σ^* interaction. We note that the underestimation is much smaller for cubic transition metal oxide perovskites such as $KTaO_3$, where the O 2s orbital does not contribute to the conduction band states.

The notation used in the previous paragraph will be helpful in describing the orbital overlap of the two conduction bands at Γ, as shown in Figure 6.23. At Γ translational symmetry dictates that the phase of all Sn 5s orbitals within a given chain must be the same. Consequently, intrachain Sn 5s–O $2p_\parallel$ orbital interactions are symmetry forbidden, and as explained above the O $2p_\parallel$ orbitals do not contribute to the interchain bonding. As a result, the O $2p_\parallel$ orbitals do not contribute to either band at Γ. Next, consider the interaction between the Sn 5s and O $2p_\perp$ orbitals. When the phases of the Sn 5s orbitals on neighboring chains are opposite, the O $2p_\perp$ orbital can form strong σ^* interactions with all three of the surrounding Sn atoms, as shown in Figure 6.23. This is the orbital overlap corresponding

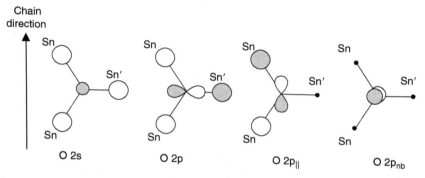

FIGURE 6.22 Labeling of the oxygen 2s and 2p orbitals showing the orbital phases that give rise to the strongest antibonding interactions with the surrounding Sn 5s orbitals. The Sn atoms reside in the same rutile chain, while the Sn′ atom belongs to a neighboring chain

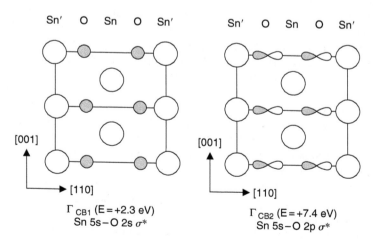

FIGURE 6.23 The orbital interactions for the two lowest energy conduction bands of SnO_2 showing the Sn 5s–O 2p σ^* and Sn 5s–O 2s σ^* orbital interactions at the Γ point. The atoms shown are those found in the (110) plane

to the upper conduction band. The strong antibonding interaction is responsible for the relatively high energy level of this point ($+7.4$ eV). On the other hand, when the tin atoms on neighboring chains have the same phase, the O 2p $\perp$ orbital makes a σ^* intrachain interaction and a σ interchain interaction. The two interactions tend to cancel out and the net interaction is close to a nonbonding one. As a result, the contribution of the O 2p $\perp$ orbital to the lower conduction band at Γ is minimal and the energy level of this point is much closer to the valence band ($+2.3$ eV). Thus, we see that interchain interactions are responsible for essentially eliminating the Sn 5s–O 2p antibonding interactions at the conduction band minimum. In contrast, the Sn 5s–O 2s σ^* interaction at Γ is fully allowed by symmetry for the lower conduction band. However, the strength of this interaction is diminished by the relatively poor energetic overlap of the Sn 5s and O 2s orbitals. A more thorough analysis of the electronic structure of SnO_2 can be found in Reference 46.

6.7.2 The Electronic Structure of ZnO

ZnO adopts the wurtzite structure as shown in Figure 6.20. Both zinc and oxygen are tetrahedrally coordinated, although the hexagonal layer stacking prevents the atoms from attaining perfect tetrahedral site symmetry. Among transparent conducting oxides, tetrahedral cation coordination is very unusual, ZnO being one of the rare examples. The electronic band structure of ZnO is shown in Figure 6.24. The valence band once again is predominantly O 2p in character, particularly toward the top, with a maximum at the Γ point. There is somewhat more dispersion in the valence band than observed in SnO_2 and $BaSnO_3$. This can be attributed to three factors: (i) the shorter O–O contacts that come with tetrahedral cation coordination, (ii) the tetrahedral anion coordination environment that prevents the formation of oxygen 2p nonbonding lone pairs, and (iii) mixing of filled Zn 3d and oxygen 2p orbitals. A recent soft x-ray emission study of ZnO identified a feature attributed to Zn 3d–O 2p mixing located approximately 8 eV below the Fermi level [54]. The LMTO calculations place the Zn 3d levels somewhat closer to E_F. They are the flat bands concentrated over the range -4 to -6 eV. The conduction band minimum also falls at the Γ point (as it does in every main group metal oxide we have examined). The calculated band gap is a direct gap of 1.0 eV, which is considerably smaller than the observed value of 3.4 eV [55]. Like SnO_2 there are two cations in the unit cell, and similar to SnO_2 the Zn 4s–O 2p σ^* interaction is primarily responsible for the two lowest conduction bands that span a bandwidth of ~ 10 eV.

As explained in Section 6.7.1 the electronic structure near the conduction band minimum will have the largest influence on the TCO properties. As was observed for SnO_2 and $BaSnO_3$ near the Γ point, the O 2p contribution to the lowest energy conduction band becomes minimal, and is replaced by a small contribution from the O 2s orbitals. Analysis of the orbital coefficients obtained from an extended Hückel calculation of the electronic structure show that at the conduction band minimum all of the Zn 4s orbitals have the same phase. Given the tetrahedral

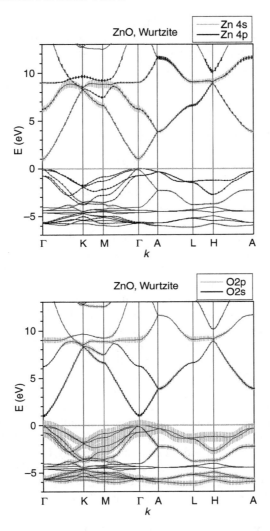

FIGURE 6.24 Electronic band structure of wurtzite ZnO as calculated using LMTO–ASA calculations. The Fermi level is set to zero energy and marked with light grey. Fatband hatching indicates the contributions of specific orbitals to the electronic structure. The upper plot shows the Zn 4s (dark grey) and Zn 4p (thick line) fatbands, while the lower plot shows the O 2p (dark grey) and O 2s (thick line) fatbands

coordination of oxygen it is easy to see that a 2p orbital on oxygen cannot mix to any significant extent with the s orbitals on the four neighboring Zn atoms if they all have the same phase. If the oxygen site symmetry were perfectly tetrahedral (T_d) O 2p–Zn 4s mixing at the conduction band minimum would be explicitly forbidden by symmetry. Conduction bands also cross the Γ point at energies of

+6.2 and +9.1 eV. The orbital overlap at these points in the electronic structure involves Zn 4s/Zn 4pz and O 2s/O 2pz mixing to form sp-hybrid orbitals that interact with each other in an antibonding fashion. The s/pz mixing is possible because of the hexagonal symmetry of the crystal system. The orbital mixing at the A point ($k_x = k_y = 0$, $k_z = \pi/a$) involves strong antibonding interactions between Zn 4s and O 2pz orbitals.

6.7.3   The Electronic Structure of CdO

CdO adopts the rock salt structure, with perfect octahedral coordination of both cadmium and oxygen, as shown in Figure 6.20. The electronic band structure of CdO is shown in Figure 6.25. The calculation was performed on the primitive unit cell that contains a single atom of each type, which results in a simple band diagram. The five lowest energy bands, which fall between −6 to −8 eV, are Cd 4d states. These bands are quite flat because the 4d orbitals are quite contracted and the overlap with oxygen is small, but not completely negligible. McGuinness et al. [54] observed the Cd 4d–O 2p states in an approximate energy window of −8 to −11 eV below E_F, which is in reasonably good agreement with our calculations. Within the energy window from −5 to +15 eV there are seven bands. The lower three bands that make up the valence band have predominantly O 2p character, while the orbital character of the lowest energy conduction band is mostly Cd 5s–O 2p σ^*. The upper three conduction bands have predominantly Cd 5p–O 2p σ^* orbital character. The calculated band structure is that of a gapless semimetal rather than a semiconductor, although there is a direct gap of ∼1.3 eV at Γ. Experimentally there is some debate regarding the size of the band gap in CdO. There is general agreement that the direct band gap is 2.3 eV and that there is a smaller indirect gap, but the size of the indirect gap is not clear due to the influence of nonstoichiometry on the optical properties. In a recent soft x-ray emission and absorption study McGuinness et al. [54] concluded that is approximately 1.2 eV [54]. In any event, the LMTO calculations once again underestimate the magnitude of the band gap. It should be noted that our calculated band structure is in qualitative agreement with the earlier calculations, such as those of Asahi et al. [56], based on the full-potential linearized augmented plane wave (FLAPW) method with the screen exchanged LDA treatment [56], although the absolute agreement with experimental values is better for the more sophisticated calculations.

The octahedral site symmetry and the fact that the primitive unit cell contains only two atoms make the orbital overlap analysis of the conduction band straightforward. At Γ point Cd 5s–O 2p mixing is symmetry forbidden, but the Cd 5s–O 2s σ^* interaction is fully allowed. Although the details differ, this parallels the situation observed at the conduction band minima in BaSnO$_3$, SnO$_2$, and ZnO. While the Cd 5s orbital with *gerade* symmetry and the O 2p orbitals with *ungerade* symmetry cannot mix at Γ, the translational symmetry constraints are such that the Cd 5p–O 2p interaction is fully antibonding. Thus, the energies of the upper three conduction bands at Γ are so high they are off-scale in Figure 6.25. The translational symmetry considerations at the L point, which is the face centered cubic

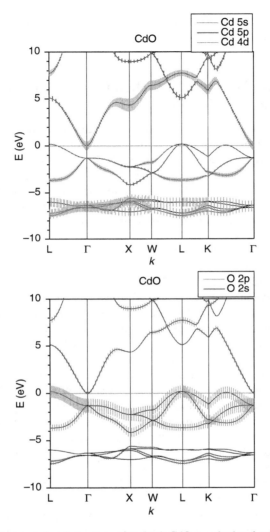

FIGURE 6.25   Electronic band structure of rocksalt CdO as calculated using LMTO–ASA calculations. The Fermi level is set to zero energy and marked as ××× × line. Fatband hatching indicates the contributions of specific orbitals to the electronic structure. The upper plot shows the Cd 5s (– · – ·), Cd 5p (– – –), and Cd 4d (××× ×) fatbands, while the lower plot shows the O 2p (– · – ·) and O 2s (– – –) fatbands

equivalent of the R point in primitive cubic symmetry, are reversed. That is to say, the orbital phases are inverted each time a unit translation is applied (analogous to the 1D MO chain at $k = \pi/a$). Consequently, the Cd 5s–O 2p interaction becomes antibonding, destabilizing its energy by ∼7.8 eV with respect to the nonbonding Γ point energy, and one of the Cd 5p–O 2p interactions becomes nonbonding. The

LMTO results suggest this effect is sufficiently large that one of the Cd 5p bands dips below the Cd 5s band at L.

6.7.4 The Link between Conduction Band Width and Oxygen Coordination

It is striking how many similarities exist in the band structures of $BaSnO_3$, SnO_2, ZnO, and CdO considering the structural variety of this group. The polyhedral connectivity ranges from corner-sharing octahedra ($BaSnO_3$) to corner-sharing tetrahedra (ZnO) to edge-sharing octahedra (CdO). The oxygen coordination number is different in each case, and the space group symmetries encompass hexagonal, tetragonal, and cubic crystal systems. Is there an unseen commonality that ties these compounds together or are these band structures simply representative of all main group metal oxides?

Recent published and unpublished works by our group have explored the electronic structures of a wide range of (mostly) ternary main group oxides [45,46,53,57]. The results of those studies in combination with the results presented in Section 6.7.1 to Section 6.7.3 reveal an unmistakable link between the electronic and crystal structures of main group metal oxides. Main group metal oxides where the anion coordination environment is symmetric tend to have a wide conduction band. In such compounds, the conduction band dispersion is most pronounced near the conduction band minimum, which falls at the Γ point. This statement assumes a symmetric cation environment, but for main group cations in their highest oxidation state that condition is usually met. While this relationship seems simple its implications are quite important. A large dispersion near the conduction band minimum is the key to obtaining high carrier mobility in n-doped metal oxides. Therefore, this postulate can be used to rapidly evaluate new compounds or libraries of existing compounds for their potential application as transparent conductors.

It should be kept in mind that a highly symmetric anion environment is a necessary but not sufficient condition for the occurrence of transparent conductivity. Some compounds that meet this criterion will not be effective transparent conductors because the metal–oxygen bonds are too ionic. For example, the oxygen coordination in MgO (rock salt) is perfectly octahedral, but Mg–O bonds are too ionic to support delocalized carrier transport. At the other extreme, the energy of the cation s orbital can become sufficiently negative that the band gap collapses. This is particularly true for the main group elements of the 6th period where relativistic effects significantly lower the energy of the 6s orbital (see Section 6.2.2). Compounds exhibiting this behavior include $BaPbO_3$ (perovskite) and β-PbO_2 (rutile), both of which are metallic/semimetallic. Finally, it is possible to have a wide conduction band but it would not be possible to attain a reasonable carrier concentration through doping, due to accompanying defect compensation mechanisms. Our studies to date suggest that $BaSn_{1-x}Sb_xO_3$ falls into this category.

What are the underlying principles behind this postulate, and how can it be applied? The underlying principle is a based on the following symmetry and energy considerations. Consider a generic metal oxide with stoichiometry $A_x M_y O_z$, where A is an electropositive s or f block cation and M is a p block cation in its maximum oxidation state. In compounds of this type the lowest energy conduction band will originate from antibonding M ns–O 2p and M ns–O 2s interactions. However, the M ns–O 2p interactions will be much stronger than the M ns–O 2s interactions based on energetic overlap considerations. For many values of the wavevector, $\mathbf{k}$, and the M ns–O 2p interactions will be symmetry allowed and this interaction will be strongly antibonding. Invariably the orbital picture at the conduction band minimum will involve a state where all of the cation s orbitals have the same phase. This state will have the lowest energy because the antibonding M ns–O 2p interactions will be at their weakest when the symmetry adapted linear combination (SALC) of metal s orbitals adopts its most symmetric arrangement. This is a direct consequence of the fact that cation ns and oxygen 2p orbitals have opposite symmetries, *gerade*, and *ungerade*, respectively. Furthermore, the conduction band minimum will always fall at the Γ point, because this is the only location in the BZ where translational symmetry constraints dictate that the M ns orbitals in all neighboring unit cells are of the same phase.

Given this picture of the conduction band minimum, consider the importance of the anion environment. This can be most easily visualized by picturing an oxygen atom surrounded by an increasing number of ligands that are generic sigma acceptors. In this construct, the ligands are the valence shell s orbitals of the metal cations. Consider the mixing between the 2p orbitals on oxygen with the most symmetric ligand SALC that we have already said represents the M ns orbitals at the conduction band minimum. Table 6.5 gives the irreducible representation for the most symmetric ligand SALC and the 2p orbitals on the central oxygen atom for the various coordination environments and site symmetries. By definition, the O 2s orbital will have the same symmetry as the symmetric ligand SALC in all cases, but for most of the symmetric coordination environments the O 2p orbitals cannot mix with the surrounding M ns orbitals. Mixing is allowed in less symmetric environments such as bent, trigonal pyramidal, and square pyramidal. When the oxygen environment falls into one of these categories, M ns–O 2p σ^* character will be introduced at the conduction band minimum. This is illustrated in Figure 6.26 for two-coordinate oxygen upon distorting the oxygen coordination environment from linear to bent. The net effect will be an energetic destabilization at the conduction band minimum, which will lead to an increase in the band gap, a decrease in the conduction band dispersion, and a decrease in the carrier mobility. The effects in the two-coordinate case are effectively illustrated by a comparison of $BaSnO_3$ (cubic perovskite, Sn–O–Sn $\angle = 180°$), $CaSnO_3$ (orthorhombic perovskite, Sn–O–Sn $\angle \sim 147°$), and $La_2 Sn_2 O_7$ (pyrochlore, Sn–O–Sn $\angle \sim 132°$). $BaSnO_3$ has a band gap of 3.1 eV, while $CaSnO_3$ has a band gap of 4.4 eV, and $La_2 Sn_2 O_7$ has a band gap of 4.5 eV [45]. When using Table 6.5 keep in mind that even though the crystallographic site symmetry may be low,

TABLE 6.5

Point Group and Site Symmetries for Various Oxygen Coordination Environments and the Irreducible Representations of the most Symmetric Sigma Donor Ligand Symmetry Adapted Linear Combination, and the Oxygen 2p Orbitals

Geometry	Coordination number	Point group[a]	Site symmetric[b]	SALC irreproducible	O 2p Irreducible representations
Linear	2	$D_{\infty h}$	—	Σ_g^+	$\Sigma_g^+(p_z)$, $\Pi_u^+(p_x, p_y)$
Bent	2	C_{2v}	2 mm	**A_1**	**$A_1(p_z)$**, $B_1(p_x)$, $B_2(p_y)$
Trigonal planar	3	D_{3h}	−6 m2	A_1'	$A_2''(p_z)$, $E'(p_x, p_y)$
Trigonal pyramidal	3	C_{3v}	3 m	**A_1**	**$A_1(p_z)$**, $E(p_x, p_y)$
Tetrahedral	4	T_d	-432	A_1	$T_2(p_x, p_y, p_z)$
Square planar	4	D_{4h}	4/mmm	A_{1g}	$B_{1u}(p_z)$, $B_{3u}(p_x)$, $B_{2u}(p_y)$
Square pyramidal	4–5	C_{4v}	4 mm	**A_1**	**$A_1(p_z)$**, $E(p_x, p_y)$
Trigonal prismatic	5	D_{3h}	−6 m2	A_1'	$A_2''(p_z)$, $E'(p_x, p_y)$
Trigonal bipyramidal	6				
Octahedral	6	O_h	m3m	A_{1g}	$T_{1u}(p_x, p_y, p_z)$
Cubic	8				

The irreproducible representations shown in bold indicate cases where an oxygen 2p orbital can participate in bonding to the ligand SALC.

[a] Schoenflies notation.

[b] Crystallographic site symmetry.

the effective site symmetry may be close to one of the more symmetric cases. SnO_2 and ZnO are two examples of this. The strict oxygen site symmetry in SnO_2 is C_{2v} (3m), but the environment is close to trigonal planar, while in ZnO the oxygen site symmetry is C_{3v} (3m), but the environment is close to tetrahedral. In both cases there will be a small M ns–O 2p antibonding interaction at the conduction band minimum, but it becomes very small as the environment approaches the higher symmetry. Another example of this is found in the host structure of the most widely used TCO, In_2O_3. In_2O_3 adopts the cubic bixbyite structure with distorted octahedral coordination at the indium sites and distorted tetrahedral coordination at the oxygen sites. Despite the fact that the oxygen resides on the general position with C_1 symmetry the approximate tetrahedral environment is sufficient to eliminate most of the In 5s–O 2p σ^* interactions at the conduction band minimum, enabling the excellent TCO behavior of this material when doped with tin.

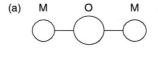

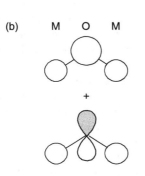

FIGURE 6.26   Bonding interactions between a central oxygen atom and the most symmetric SALC of cation s orbital acceptor orbitals, for (a) linear ($D_{\infty h}$) geometry where only the O 2s orbital has appropriate symmetry for mixing, and (b) bent (C_{2v}) geometry where both the O 2s and O $2p_z$ can mix

6.7.5 Generality of the Anion Coordination — Bandwidth Connection

The arguments put forward in the preceding section to explain the link between conduction bandwidth and oxygen coordination were based on bonding in main group metal oxides. How general is this concept? Can it be applied to compounds other than oxides? Can it be applied to transition metal oxides? Although we have not investigated these questions in detail it is possible to make some general statements based on knowledge of orbital symmetry and energy.

The symmetry concepts will be identical for compounds containing p block cations and anions other than oxygen, but the energy matching will be different. The arguments put forward in Section 6.7.4 are based on the premise that the M ns–O 2s interaction is much weaker than the M ns–O 2p interaction, due to poor energetic overlap. Consider the atomic energy levels given in Table 6.1. If we replace oxygen with fluorine the energy of the anion 2s orbital becomes even more negative and the energetic overlap of the M ns–F 2s interaction with respect to the M ns–F 2p interaction will be even smaller than it is in oxides. Therefore, this concept should work at least as well if not better in main group metal fluorides. On the other hand, the cation s–anion s energetic overlap will be enhanced for all other choices of anion. In such compounds, while the cation s–anion p mixing will still be symmetry forbidden at Γ, the relative strength of the cation s–anion s antibonding interaction will be increased, thereby destabilizing the energy of the conduction band at Γ. This will result in a reduction of the conduction band dispersion near Γ.

For transition metal oxides the situation is more complicated, because the critical interaction is M $(n-1)$d–O 2p rather than M ns–O 2p. Like the s orbitals, the d orbitals have *ungerade* symmetry, so similar symmetry arguments should apply. In fact for the transition metal oxide band structures discussed in Section 6.5 and Section 6.6 the conduction band minimum was always found at the Γ point, where the M $(n-1)$d–O 2p mixing was at a minimum. Although in some cases there were conduction band minima or near minima at other k-points as well. The same observation can be made of more extensive studies of the electronic structures of d^0 transition metal oxides from the perovskite, pyrochlore, and Ruddlesden-Popper families [28]. These studies and many others in the literature leave little doubt that in perovskites and related structures the width of the π^* and σ^* bands will be largest when the M–O–M bonds are linear. However, these features are not universal to all transition metal oxides. For example, Felser, Thieme, and Seshadri have investigated the electronic structures of transition metal compounds based on hexagonal nets, and have found that the minima of the d-based bands do not fall at Γ [58]. In these structures as well as others where edge- and/or face-sharing polyhedra are present the bonding analysis is complicated by the presence of metal–metal interactions. Thus, the simple link between anion coordination and conduction bandwidth can be extended to transition metal oxides where the metal oxygen framework is based on corner-sharing octahedra, but other factors must be taken into account when edge-sharing and/or face-sharing octahedra are present.

6.8 SUMMARY

The relationships between the chemical bonding and the electronic structure of metal oxides have been explored in this chapter, starting from the properties of atomic orbitals and extending to the electronic structures of a number of important oxide structure types. While this chapter is by no means an exhaustive treatment of the electronic structures of metal oxides the concepts discussed are general and should find widespread applicability. Of particular note is the relatively little explored link between anion coordination and conduction bandwidth.

REFERENCES

1. Brown, T.L., LeMay, H.E., Jr., and Bursten, B.E. *Chemistry, The Central Science*, 8th ed. Prentice Hall: Upper Saddle River, NJ, 2000, p. 265.
2. Cox, P.A. *Transition Metal Oxides: An Introduction to Their Electronic Structure and Properties*. Clarendon Press: Oxford, 1992.
3. du Pre, F.K., Hunter, R.A., and Rittner, E.S. Concerning the work of polarization in ionic crystals of the NaCl type. 3. Numerical results for a single charge in the rigid lattice. *J. Chem. Phys.* 1950, *18*, 379–380.
4. Kowalczyk, S.P., McFeely, F.R., Ley, L., Gritsyna, V.T., and Shirley, D.A. Electronic-structure of SrTiO$_3$ and some simple related oxides (MgO, Al$_2$O$_3$, SrO, TiO$_2$). *Solid State Commun.* 1977, *23*, 161–169.
5. Pauling, L. *The Nature of the Chemical Bond*, 3rd ed. Cornell University Press: Ithaca, NY, 1960.

6. Housecroft, C.E. and Sharpe, A.G. *Inorganic Chemistry*. Prentice Hall: London, 2001, p. 35.
7. Allred, A.L. and Rochow, E.G. A scale of electronegativity based on electrostatic force. *J. Inorg. Nucl. Chem.* 1958, *5*, 264–268.
8. Mulliken, R.S. *J. Chem. Phys.* 1934, *2*, 782.
9. Sanderson, R.T. *J. Chem. Educ.* 1952, *29*, 539.
10. Phillips, J.C. Ionicity of chemical bond in crystals. *Rev. Mod. Phys.* 1970, *42*, 317–318.
11. Kotochigova, S., Levine, Z.H., Shirley, E.L., Stiles, M.D., and Clark, C. W. *Atomic Reference Data for Electronic Structure Calculations*. Located at http://physics.nist.gov/PhysRefData/DFTdata/contents.html.
12. Mingos, D.M.P. *Essential Trends in Inorganic Chemistry*. Oxford University Press: Oxford, 1998, p. 36.
13. Mann, J.B. *Atomic Structure Calculations II. Hartree–Fock Wave Functions and Radial Expectation Values: Hydrogen to Lawrencium*. LA-3691, Los Alamos Scientific Laboratory, 1968.
14. Stückl, A.C., Daul, C.A., and Güdel, H.U. Excited-state energies and distortions of d(0) transition metal tetraoxo complexes: A density functional study. *J. Chem. Phys.* 1997, *107*, 4606–4617.
15. Whangbo, M.H., and Hoffmann, R. Band-structure of tetracyanoplatinate chain. *J. Am. Chem. Soc.* 1978, *100*, 6093–6098.
16. Whangbo, M.-H., Hoffmann, R., and Woodward, R. B. Conjugated one and 2-dimensional polymers. *Proc. Roy. Soc. London Ser. A* 1979, *366*, 23–46.
17. Ren, J., Liang, W., and Whangbo, M.H. *PrimeColor Software Raleigh*, 1998, http://www.PrimeC.com/.
18. Te Velde, G. and Baerends, E.J. Numerical-integration for polyatomic systems. *J. Comput. Phys.* 1992, *99*, 84–98.
19. Perdew, J.P., Chevary, J.A., Vosko, S.H., Jackson, K.A., Pederson, M.R., Singh, D. J., and Fiolhais, C. Atoms, molecules, solids, and surfaces — Applications of the generalized gradient approximation for exchange and correlation. *Phys. Rev. B* 1992, *46*, 6671–6687.
20. Andersen, O.K., Pawlowska, Z., and Jepsen, O. Illustration of the linear-muffin-tin orbital tight-binding representation — Compact orbitals and charge-density in Si. *Phys. Rev. B* 1986, *34*, 5253–5269.
21. Wolfsberg, M. and Helmholz, L. The spectra and electronic structure of the tetrahedral ions MnO_4^{2-}, CrO_4^{2-} and ClO_4^-. *J. Chem. Phys.* 1952, *20*, 837–843.
22. Kröger, F.A. *Some Aspects of the Luminescence of Solids*. Elsevier: New York, 1948.
23. Wang, R.P., Sleight, A.W., Platzer, R., and Gardner, J.A. Nonstoichiometric zinc oxide and indium-doped zinc oxide: Electrical conductivity and In-111-TDPAC studies. *J. Solid State Chem.* 1996, *122*, 166–175.
24. Yan, P.D., Yan, H.Q., Mao, S., Russo, R., Johnson, J., Saykally, R., Morris, N., Pham, J., He, R.R., and Choi, J. ZnO nanoribbon microcavity lasers. *Adv. Funct. Mater.* 2002, *15*, 1907–1917.
25. Vosko, H.S., Wilk, L., and Nusair, N. Accurate spin-dependent electron liquid correlation energies for local spin-density calculations — A critical analysis. *J. Phys.* 1980, *58*, 1200–1211.
26. Brown, I.D. *The Chemical Bond in Inorganic Chemistry*. IUCr Monographs on Crystallography, 2002, Vol. 12. Oxford University Press: Oxford.

27. Polity, A., Schwabe, D., Ackermann, L., and Dupré, K. Transmission spectra of crystals at elevated temperatures for the calculation of internal radiant heat transport during crystal growth — Part 2: Spectra of YAG: Cr, YVO$_4$: Nd and the bandgap variation of various materials. *Cryst. Res. Technol.* 2003, *38*, 874–880.

28. Eng, H.W., Barnes, P.W., Auer, B.M., and Woodward, P.M. Investigations of the electronic structure of d(0) transition metal oxides belonging to the perovskite family. *J. Solid State Chem.* 2003, *175*, 94–109.

29. Huheey, J.E. *Inorganic Chemistry*, 3rd ed. Harper Collins: New York, 1983.

30. Jahn, H.A. and Teller, E. *Proc. Roy. Soc.* 1937, *A161*, 220.

31. Burdett, J.K. *Chemical Bonding in Solids*. Oxford University Press: Oxford, UK, 1995.

32. Albright, T.A., Burdett, J.K., and Whangbo, M.H. *Orbital Interactions in Chemistry*. Wiley: New York, 1985.

33. Kunz, M. and Brown, I.D. Out-of-center distortions around octahedrally coordinated d^0 transition metals. *J. Solid State Chem.* 1995, *115*, 395–406.

34. Hughbanks, T. Superdegenerate electronic-energy levels in extended structures. *J. Am. Chem. Soc.* 1985, *107*, 6851–6859.

35. Wheeler, R.A., Whangbo, M.H., Hughbanks, T., Hoffmann, R., Burdett, J. K., and Albright, T.A. Symmetrical vs asymmetric linear M–X–M linkages in molecules, polymers, and extended networks. *J. Am. Chem. Soc.* 1986, *108*, 2222–2236.

36. Hoffmann, R. *Solids and Surfaces: A Chemists View of Bonding in Extended Structures.*

37. Vogt, T., Woodward, P.M., and Hunter, B.A. The high-temperature phases of WO$_3$. *J. Solid State Chem.* 1999, *144*, 209–215.

38. Howard, C.J., Luca, V., and Knight, K.S. High-temperature phase transitions in tungsten trioxide — the last word? *J. Phys. Condens. Matter* 2002, *14*, 377–387.

39. Lufaso, M.W. and Woodward, P.M. Prediction of the crystal structures of perovskites using the software program SPuDS. *Acta Crystallogr. B* 2001, *57*, 725–738.

40. Woodward, P.M. Octahedral tilting in perovskites. 1. Geometrical considerations. *Acta Crystallogr. B* 1997, *53*, 32–43.

41. Woodward, P.M. Octahedral tilting in perovskites. 2. Structure stabilizing forces. *Acta Crystallogr. B* 1997, *53*, 44–66.

42. Howard, C.J. and Stokes, H.T. Group-theoretical analysis of octahedral tilting in perovskites. *Acta Crystallogr. B* 1998, *54*, 782–789.

43. Mitchell, R.H. *Perovskites Modern and Ancient*. Almaz Press Inc.: Thunder Bay, Ontario, 2002.

44. Jepsen, O. and Andersen, O.K. Calculated electronic-structure of the sandwich d^1 metals LAI2 and CEI2 — Application of new LMTO techniques. *Z. Phys. B* 1995, *97*, 35–47.

45. Mizoguchi, H., Eng, H.W., and Woodward, P.M. Probing the electronic structures of ternary perovskite and pyrochlore oxides containing Sn^{4+} or Sb^{5+}. *Inorg. Chem.* 2004, *43*, 1667–1680.

46. Mizoguchi, H. and Woodward, P.M. *Chem. Mater.* 2004, (in press).

47. Robertson, J. Band offsets of wide-band-gap oxides and implications for future electronic devices. *J. Vac. Sci. Technol. B* 2000, *18*, 1785–1791.

48. Kurosaki, K., Oyama, T., Muta, H., Uno, M., and Yamanaka, S. Thermoelectric properties of perovskite type barium molybdate. *J. Alloys Compd.* 2004, *372*, 65–69.

49. Goodenough, J.B. and Zhou, J.S. Localized to itinerant electronic transitions in transition-metal oxides with the perovskite structure. *Chem. Mater.* 1998, *10*, 2980–2993.

50. Arima, T. and Tokura, Y. Optical study of electronic-structure in perovskite-type RMO_3 (R = La, Y; M = Sc, Ti, V, Cr, Mn, Fe, Co, Ni, Cu). *J. Phys. Soc. Jpn.* 1995, *64*, 2488–2501.

51. Inoue, I.H., Goto, O., Makino, H., Hussey, N.E., and Ishikawa, M. Bandwidth control in a perovskite-type $3d^1$-correlated metal $Ca_{1-x}Sr_xVO_3$. I. Evolution of the electronic properties and effective mass. *Phys. Rev. B* 1998, *58*, 4372–4383.

52. Pauling, L. *J. Am. Chem. Soc.* 1929, *51*, 1010.

53. Mizoguchi, H., Woodward, P.M., Byeon, S.H., and Parise, J.B. Polymorphism in $NaSbO_3$: structure and bonding in metal oxides. *J. Am. Chem. Soc.* 2004, *126*, 3175–3184.

54. McGuinness, C., Stagarescu, C.B., Ryan, P.J., Downes, J.E., Fu, D., Smith, K.E., and Egdell, R.G. Influence of shallow core-level hybridization on the electronic structure of post-transition-metal oxides studied using soft X-ray emission and absorption. *Phys. Rev. B* 2003, *68*, 165104.

55. Matz, R. and Lueth, H. Ellipsometric spectroscopy of the polar ZnO(1100) surface. *Appl. Phys. (Berlin)* 1979, *18*, 123–130.

56. Asahi, R., Wang, A., Babcock, J.R., Edleman, N.L., Metz, A.W., Lane, M.A., Dravid, V.P., Kannewurf, C.R., Freeman, A.J., and Marks, T.J. First-principles calculations for understanding high conductivity and optical transparency in In_xCd_{1-x} films. *Thin Solid Films* 2002, *411*, 101–105.

57. Mizoguchi, H., Woodward, P.M., Park, C.H., and Keszler, D.A. Strong near-infrared luminescence in $BaSnO_3$. *J. Am. Chem. Soc.* 2004, *126*, 9796–9800.

58. Felser, C., Thieme, K., and Seshadri, R. Electronic instabilities in compounds with hexagonal nets. *J. Mater. Chem.* 1999, *9*, 451–457.

59. Waghmare, U.V., Spaldin, N.A., Kandpal, H.C., and Seshadri, R. First-principles indicators of metallicity and cation off-centricity in the IV–VI rock-salt chalcogenides of divalent Ge, Sn, and Pb. *Phys. Rev. B* 2003, *67*, 125111–125117.

60. Seshadri, R. Visualizing lone pairs in compounds containing heavier congeners of the carbon and nitrogen group elements. *Proc. Ind. Acad. Sci. (Chem. Sci.)* 2001, *113*, 487–496.

61. Watson, G.W. The origin of the electron distribution in SnO. *J. Chem. Phys.* 2001, *114*, 758–763.

62. Balamurugan, B., Aruna, I., Mehta, B.R., and Shivaprasad, S.M. Size-dependent conductivity-type inversion in Cu_2O nanoparticles. *Phys. Rev. B* 2004, *69*, 165419.

7 Optical and Magnetic Properties of Metal Oxides

Sumio Ishihara
Department of Physics, Tohoku University, Sendai, Japan

CONTENTS

7.1 INTRODUCTION

It is widely known that metal oxides show a variety of phenomena, such as magnetism, dielectrics, superconductivity, etc., which are remarkably sensitive to their chemical compositions, crystal structures, carrier concentrations, and also the applied external fields. The studies of magnetic and optical properties in metal oxides long time ago, from the view point of fundamental aspects as well as applications [1]. The discovery of the high transition-temperature superconducting (HTSC) cuprates in 1986 [2] was a milestone in the study of metal-oxides. The method of synthesing single large crystals has been greatly developed, and systematic transport, optical and diffraction measurements have also become a matter of concern. Also, a lot of exotic physical concepts have been proposed by the theoretical examinations. Following the discovery of HTSC cuprates, metal oxides has been studied from the modern point of view. One of the most attractive compounds after HTSC cuprates is the manganites with the perovskite structure. Study of the perovskite manganite started as early as the 1950s [3], and recently the massive reduction of the electric resistivity applying the magnetic field — the so-called colossal magnetoresistance (CMR) — was discovered [4–7]. A common physics concept in these two oxides is the electron correlation, that is, the many-body nature of $3d$ electrons in the metal ions. In this chapter, among a number of metal oxides, we consider two representative oxides, that is, the HTSC cuprates and the CMR manganites, and introduce the magnetic and the optical properties from the view point of the electron correlation.

Before the detailed introduction of the magnetic and the optical properties, we briefly review the fundamental physics and chemistry of cuprates and manganites. The crystal structure of a series of HTSC cuprates is termed as the layered-perovskite structure, where the characteristic two-dimensional CuO_2 sheets and other sheets are stacked along the c axis. The chemical composition of the initially discovered HTSC cuprates is $R_{2-x}A_xCuO_4$, where R and A are the trivalent rare-earth ion and the divalent alkaline-earth ion, respectively. At $x = 0$, the nominal valence of the Cu ion is 2+, and the electron configuration of this ion is $3d^9$. As explained later, the $3d_{x^2-y^2}$ orbital in the Cu ion is occupied by one hole at each site. Therefore, in the conventional band picture, the highest band is half-filled and the system is expected to become a metal. However, in reality, the electrical resistivity monotonically increases with decreasing temperature, that is, the system is an insulator. This implies that the conventional band theory is broken down and the electron correlation effects are important. By replacing the R ion by A, hole carriers are doped in the system and the superconductivity appears around $x \sim 0.05$. The superconducting transition-temperature T_c reaches about 30 K in $La_{2-x}Sr_xCuO_4$. Through a large amount of theoretical and experimental investigations, several electric, magnetic, and optical properties of the normal and superconducting phases have been revealed. Although, one of the key ingredients for the origin of HTSC is the electron correlation and the spin degree of freedom, this issue is still under debate.

The manganite with the perovskite structure $R_{1-x}A_xMnO_3$ was first studied by Jonker and van Santen in 1950s [3]. Here, the strong relation between the metallic conductivity and the ferromagnetic ordering was discovered. In the mother compound of CMR, $LaMnO_3$, the nominal valence of Mn ion is Mn^{3+} with the $3d^4$ electron configuration. It is also expected, in the simple band theory, that the system should be metallic, although the actual $LaMnO_3$ shows the insulating behavior in electric and magnetic measurements. The ferromagnetic metallic state appears with the doping of the hole by changing the A ion concentration. The massive drop of the electric resistivity by applying the external magnetic field is observed near the ferromagnetic transition temperature. The recent extensive and intensive researches concluded that (i) the magnetoresistance ratio reaches 10^{11} in $Pr_{1-x}Ca_xMnO_3$ [8] (ii) the CMR effect is not explained within the conventional theory based on the double-exchange interaction introduced later [9], and (iii) the various novel exotic phenomena have stimulated the development of new physical concepts.

7.2 MAGNETIC PROPERTIES OF OXIDES

7.2.1 Electronic Structure of Metal Oxides

We start from the electronic structure of the metal-oxide solids. Consider the d electrons of a single metal ion, such as Cu and Mn ions, termed M ion from now on, located in solids. In addition to the Coulomb potential from the nucleus and other electrons in the inner shells, the Coulomb interaction between the d electrons plays a crucial role to determine the electronic structure. In the second quantization scheme, the generic form of this interaction is given by:

$$\mathcal{H}_{\text{Coulomb}} = \frac{1}{2} \sum_{i,\gamma_1,\gamma_2,\gamma_3,\gamma_4,\sigma_1,\sigma_2} V_{\gamma_1\gamma_2\gamma_3\gamma_4} d^{\dagger}_{i\gamma_1\sigma_1} d^{\dagger}_{i\gamma_2\sigma_2} d_{i\gamma_3\sigma_2} d_{i\gamma_4\sigma_1} \quad (7.1)$$

where $d_{i\gamma\sigma}$ is an annihilation operator of the $3d$ electron with orbital γ ($= 3z^2 - r^2$, $x^2 - y^2$, xy, yz, zx) and spin σ ($= \uparrow, \downarrow$) at site i, and $V_{\gamma_1\gamma_2\gamma_3\gamma_4}$ is the matrix element of the Coulomb interaction $e^2 / |\vec{r}_1 - \vec{r}_2|$ between the electrons at $\vec{r}_1$ and $\vec{r}_2$. Among the several types of Coulomb interactions, the four dominant interactions are

$$\mathcal{H}_{\text{Coulomb}} = U \sum_{i,\gamma} n_{i,\gamma\uparrow} n_{i,\gamma\downarrow} + U' \frac{1}{2} \sum_{i,\gamma_1 \neq \gamma_2} n_{\gamma_1} n_{\gamma_2}$$

$$+ J \frac{1}{2} \sum_{i,\gamma_1 \neq \gamma_2,\sigma_1,\sigma_2} d^{\dagger}_{i\gamma\sigma_1} d_{i\gamma\sigma_2} d^{\dagger}_{i\gamma_2\sigma_2} d_{i\gamma_2\sigma_1}$$

$$+ I \sum_{i,\gamma_1 \neq \gamma_2} d^{\dagger}_{i\gamma_1\uparrow} d^{\dagger}_{i\gamma_1\downarrow} d_{i\gamma_2\downarrow} d_{i\gamma_2\uparrow} \quad (7.2)$$

$n_{\gamma\sigma}(= d^{\dagger}_{\gamma\sigma} d_{\gamma\sigma})$ and $n_{\gamma}(= \sum_{\sigma} d^{\dagger}_{\gamma\sigma} d_{\gamma\sigma})$ are the number operators. Here, U ($= V_{\gamma\gamma\gamma\gamma}$) and $U'(= V_{\gamma\gamma'\gamma'\gamma})$ are the direct Coulomb interactions between electrons, which occupy the same orbital and the different orbitals, respectively, J ($= -\frac{1}{2} V_{\gamma\gamma'\gamma\gamma'}$) is the exchange interaction and I ($= V_{\gamma\gamma\gamma'\gamma'}$) is termed the pair-hopping interaction. In the atomic limit, there are the relations $U = U' + 2J$ and $I = J$. Because of these Coulomb interactions, the 10 ($=_{10}C_9$) and 210 ($=_{10}C_4$) fold degeneracies in the electronic states at Cu^{2+} and Mn^{3+}, respectively, are split into the LS multiplets. The lowest LS multiplets are given by following the Hund rule as $^2D_{5/2}$ for Cu^{2+} and $^5D_0$ for Mn^{3+}, respectively. The next relevant interaction for the $3d$ electrons in solids is the crystalline field effects, originating from the long-range Coulomb potential from the O^{2-} ions surrounding the metal ions as well as the covalency between the O $2p$ and the Cu (Mn) $3d$ orbitals. In an O_6 octahedral cage with the cubic symmetry for manganites, the fivefold degeneracy in the $3d$ orbitals is lifted into the double degenerated e_g orbitals with higher energy and the triple degenerated t_{2g} ones with lower energy. In the tetragonal crystalline field in an elongated octahedron along the c axis, as seen in the cuprate, the e_g and t_{2g} orbitals are further split as $e_g \rightarrow a_{1g} + b_{1g}$ and $t_{2g} \rightarrow e_g + b_{2g}$. As a result, the lowest multiplets have a double degeneracy for cuprates and are tenfold degeneracy for manganites. One remarkable difference between the two ions is the orbital degree of freedom, that is, the twofold degeneracy in the orbital sector remains in Mn^{3+} and the orbital degree of freedom in Cu^{2+} is quenched due to the tetragonal symmetry of the crystalline field.

The relativistic spin–orbit interaction, that is, the LS coupling, is smaller than the crystalline field potential in the $3d$ transition-metal ions. This is in contrast to the f electrons systems, where the LS coupling should be considered before the crystalline field. In the Mn^{3+} ion where the twofold orbital degeneracy remains under the crystalline field effects, one may think that the spin–orbit interaction further lifts the orbital degeneracy. However, this interaction is irrelevant in this case; the matrix elements of the angular momentum operator $\vec{L}$ are zero for the e_g orbital wave functions, because of the spatial symmetry of the e_g wave function. After all, the lowest electronic configurations in the single Cu^{2+} and Mn^{3+} ions are the state where one hole occupies the $d_{x^2-y^2}$ orbital with $S = \frac{1}{2}$ and the $(t_{2g})^3(e_g)^1$ state with the $S = 2$ high spin state, respectively.

Now, we consider the regular alignments of the $3d$ orbitals in the metal ions and the $2p$ orbitals in the O^{2-} ions in a crystalline solid. The inter-ionic interactions are considered based on the atomic-like wave functions (the Wannier functions) located around the ions. This is the tight-binding model for the ionic compounds. The general electronic Hamiltonian model is represented as:

$$
\mathcal{H}_{p-d} = \sum_{i,\gamma,\sigma} \varepsilon_{d\gamma} d^{\dagger}_{i\gamma\sigma} d_{i\gamma\sigma} + \sum_{i,\alpha,\sigma} \varepsilon_{p\alpha} P^{\dagger}_{j\alpha\sigma} P_{j\alpha\sigma}
$$

$$
+ \sum_{\langle ij\rangle,\gamma,\alpha,\sigma} \left(t^{\gamma\alpha}_{ij} d^{\dagger}_{i\gamma\sigma} P_{j\alpha\sigma} + \text{H.c.} \right) + \mathcal{H}_{\text{Coulomb}} \tag{7.3}
$$

where $p_{j\alpha\sigma}$ is the electron annihilation operator for the $O\,2p_\alpha$ electron at site j with the Cartesian coordinate α. This Hamiltonian is termed as the generalized $p-d$ model. The first and the second terms are the electronic energy levels at M and O sites, respectively, and the third term describes the electron hopping between the nearest-neighboring M and O sites with the hopping amplitude $t^{\gamma\alpha}_{ij}$.

Consider the state where the nominal $O\,2p$ electrons at each site are six, that is, the $2p$ orbitals are fully occupied and the $3d$ orbitals are partially occupied by an integer number of electrons per site. The electron energy gap is considered based on the above Hamiltonian model, without the electron hopping term. The energy gap is defined by the energy required to remove an electron from a site and add this electron at another site, which is far away from the initial site in the system:

$$
E_{\text{gap}} = \Delta E(N \to N + 1) + \Delta E(N \to N - 1) \tag{7.4}
$$

where the interaction between the extra electron and the remaining hole is not included. $\Delta E(N \to N + 1)$ corresponds to $-$ (ionization energy) + (Madelung energy) in the M site and (electron affinity energy) + (Madelung energy) in the

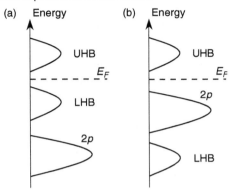

FIGURE 7.1 Schematic density of states for (a) the Mott–Hubbard insulator and (b) the charge-transfer insulator. LHB and UHB indicate the lower Hubbard band and upper Hubbard band, respectively. E_F represents the fermi level

O site. There are two possibilities of the energy gap [10]: (i) One electron is removed from a M site and added to another M site, that is, $d^n d^n \rightarrow d^{n-1} d^{n+1}$. This is termed the Mott–Hubbard gap. This energy denoted by E_{gap}^{MH} is of the order of the Coulomb interaction energy U. (ii) One electron that is removed from a O site is added to the M site, that is, $d^n p^6 \rightarrow d^{n+1} p^5$. This is termed the charge-transfer gap and its energy gap, denoted by E_{gap}^{CT}, is given by the energy difference Δ between the O $2p$ and M $3d$ orbitals. When E_{gap}^{MH} is smaller (larger) than E_{gap}^{CT}, the minimum excitation energy is given by $E_{gap}^{MH}(E_{gap}^{CT})$, and the system is termed the Mott–Hubbard (charge-transfer) type insulator. The schematic picture of these two insulating states are shown in Figure 7.1. Due to the strong electron correlation, the d band is split to the two components; the occupied band termed the lower Hubbard band and the unoccupied band termed the upper Hubbard band. The insulating nature in both the systems is attributed to the on-site Coulomb interaction, which is in contrast to the conventional band insulating state. As explained later, the insulating cuprates and manganites belong to the class of the charge-transfer type insulator. In the case of the Mott–Hubbard system, the electronic states in the low energy regime is studied by the following generalized Hubbard Hamiltonian:

$$\mathcal{H}_{Hubbard} = \sum_{i,\gamma,\sigma} \varepsilon_{d\gamma} d_{i\gamma\sigma}^\dagger d_{i\gamma\sigma} + \sum_{\langle ij \rangle,\gamma,\gamma',\sigma} \left(\tilde{t}_{ij}^{\gamma\gamma'} d_{j\alpha\sigma}^\dagger d_{i\gamma'\sigma} + \text{H.c.} \right) + \mathcal{H}_{Coulomb} \quad (7.5)$$

Here, the $2p$ electrons are not explicitly taken into account and $\tilde{t}_{ij}^{\gamma\gamma'}$ is the effective electron hopping between the M ions through the O $2p$ orbitals. Even in the charge-transfer type insulators, it is believed that this mode is relevant in the low energy region around the fermi level where the electronic structure is mapped onto the effective single band model with strong electron correlation.

7.2.2 Magnetic Interactions in Metal Oxides

It is well known that there are several types of magnetic interactions among the magnetic metal ions in solids. Here, we introduce two representative magnetic interactions in the HTSC cuprates and CMR manganites: the superexchange interaction and the double-exchange interaction, which are the relevant interactions in the insulating and metallic states, respectively.

The magnitude of the magnetic interactions is usually, at least, one-order smaller than the energy parameters in the Hamiltonian models in Equations (7.3) and (7.5). The direct-exchange interaction between the nearest-neighboring d electron spins is very weak to explain the experimental observation because of the small overlap of the wave functions. The exchange interaction through the O $2p$ orbitals termed the superexchange interaction is one of the dominant magnetic interaction in the insulating metal oxides [11–13]. The explicit form of the superexchange interaction is obtained by the perturbational process with respect to the inter-site electron transfer. Let us consider the state where the double degenerated e_g orbitals are located at each M site in the perovskite crystal and the total electron number is the same with that of the M ions. This is relevant for $LaMnO_3$, although the t_{2g} electrons and their interactions with the e_g electrons are neglected, and also for La_2CuO_4 where an electron is replaced by a hole. Start from the generalized Hubbard Hamiltonian Equation (7.5), where the O $2p$ degree of freedom are interpreted to be included in the hopping matrix $t_{ij}^{\gamma\gamma'}$ and the electronic operator $d_{i\gamma\sigma}$ is considered to annihilate an electron in a Wannier function including the atomic M $3d$ and O $2p$ components. The lowest energy states described by the first and third terms in Equation (7.5) is that one of the two e_g orbitals at each site are occupied by an electron. Through the second-order perturbational process with respect to the second term of Equation (7.5), we obtain the following effective Hamiltonian [14,15]:

$$\mathcal{H}_J = -2\sum_{\langle ij \rangle} J_1 \left(\tfrac{3}{4}n_i n_j + \vec{S}_i \cdot \vec{S}_j\right) \left(\tfrac{1}{4}n_i n_j - \tau_i^l \tau_j^l\right)$$

$$-2\sum_{\langle ij \rangle} J_2 \left(\tfrac{1}{4}n_i n_j + \vec{S}_i \cdot \vec{S}_j\right) \left(\tfrac{3}{4}n_i n_j + \tau_i^l \tau_j^l + \tau_i^l n_j + n_i \tau_j^l\right) \qquad (7.6)$$

Here $n_i \ (= \sum_{\gamma\sigma} d_{i\gamma\sigma}^\dagger d_{i\gamma\sigma})$ is the number operator, $\vec{S}_i$ is the spin operator with $S = \tfrac{1}{2}$, and the τ_i^l operator describes the orbital state, defined by:

$$\tau_i^l = \cos\left(\frac{2\pi n_l}{3}\right) T_{iz} - \sin\left(\frac{2\pi n_l}{3}\right) T_{ix} \qquad (7.7)$$

with the direction of the bond l connecting site i and site j. We introduce the pseudo-spin operator $\vec{T}_i$ for the orbital degree of freedom with a magnitude $\tfrac{1}{2}$. The eigen state for the z component of the pseudo-spin operator T_z with the eigen energy

$+\frac{1}{2}$ $(-\frac{1}{2})$ corresponds to the state where one electron occupies the $d_{3z^2-r^2}$ $(d_{x^2-y^2})$ orbital, and T_x operator represents the transition between the two. J_1 and J_2 are the magnitudes of the superexchange interactions represented by $J_1 = t_0^2/(U'-I)$ and $J_2 = t_0/U$ with $t_0 = t_{ii+a\hat{z}}^{3z^2-r^2 3z^2-r^2}$. Because $U'-I$ is less than U, J_1 is the dominant interaction. When we assume that the electron transfer integrals between the different orbitals are zero, that is, $t_{ij}^{\gamma\gamma'\neq\gamma} = 0$, $\mathcal{H}_J$ becomes a more compact form

$$\mathcal{H}_J = -2J_1 \sum_{\langle ij \rangle} \left(\tfrac{3}{4} n_i n_j + \vec{S}_i \cdot \vec{S}_j \right) \left(\tfrac{1}{4} n_i n_j - \vec{T}_i \cdot \vec{T}_j \right)$$

$$- 2J_2 \sum_{\langle ij \rangle} \left(\tfrac{1}{4} n_i n_j - \vec{S}_i \cdot \vec{S}_j \right) \left(\tfrac{3}{4} n_i n_j + \vec{T}_i \cdot \vec{T}_j \right) \qquad (7.8)$$

Within the mean field approximation, the ferromagnetic ordered state $\langle \vec{S}_i \cdot \vec{S}_j \rangle = \frac{1}{4}$ associated with the alternating orbital ordered state $\langle \vec{T}_i \cdot \vec{T}_j \rangle = -\frac{1}{4}$ becomes the ground state. That is, these superexchange processes cause the ferromagnetic interaction between the nearest-neighboring spins. This is attributed to the Hund coupling at the same M site, which reduces the energy of the intermediate state where the two electrons with parallel spins occupy different orbitals. In the case of cuprates, there is no orbital degree of freedom due to the tetragonal crystalline field and one hole occupies the $d_{x^2-y^2}$ orbital at each site. When we take $\langle \vec{T}_i \cdot \vec{T}_j \rangle = \frac{1}{4}$, the Hamiltonian is then reduced to the conventional Heisenberg type

$$\mathcal{H}_J = -2J_2 \sum_{\langle ij \rangle} \left(\tfrac{1}{4} n_i n_j - \vec{S}_i \cdot \vec{S}_j \right) \qquad (7.9)$$

This magnetic interaction is antiferromagnetic and is attributed to the Pauli principle in the intermediate perturbational state.

When we introduce the hole carriers in the insulating manganites, another type of the magnetic interaction appears. This is termed the double-exchange interaction caused by the combination effects of the electron hopping and the Hund coupling between e_g and t_{2g} spins [16,17]. It is considered that the itinerancy of t_{2g} orbital is much weaker than that of e_g electron. This is because the hybridization bond between the t_{2g} orbitals and the nearest-neighboring O $2p$ orbitals is the π bond in contrast to the σ bond between the e_g and $2p$ orbitals. In this scheme, the t_{2g} electrons are treated as the localized spin with a magnitude $S = \frac{3}{2}$ and the Hund coupling is described as:

$$\mathcal{H}_{\text{Hund}} = J_H \sum_i \vec{S}_i \cdot \vec{S}_i^{(t)} \qquad (7.10)$$

where $\vec{S}_i^{(t)}$ is the spin operator for the t_{2g} spin and J_H is negative. In order to derive the double-exchange interaction, consider a pair of the nearest-neighboring Mn^{3+} and Mn^{4+} ions where the orbital degeneracy is neglected for simplicity. At a Mn^{3+}

site, where one e_g electron and a localized t_{2g} spin with $S = \frac{3}{2}$ exist, possible spin configurations are classified by the total spin quantum number: the $S_{tot} = S + \frac{1}{2}$ state with energy $-J_H S$ and the $S_{tot} = S - \frac{1}{2}$ state with $J_H(S + 1)$. When the Hund coupling is much larger than the inter-site electron transfer, it is enough to consider the state with the maximum spin quantum number $S_{tot} = S + \frac{1}{2}$. Consider the matrix element of the electron transfer between a pair of Mn ions where the t_{2g} spin at the Mn^{4+} ion (site i) is canted from the Mn^{3+} ion (site j) by θ. We consider the local coordinates for the e_g electron where the z axis in the spin space is defined to be parallel to the t_{2g} spin. The up spin wave function defined at site i is expressed by a linear combination of the up- and down-states defined at site j:

$$|\uparrow_i\rangle = \cos\left(\frac{\theta}{2}\right)|\uparrow_j\rangle + \sin\left(\frac{\theta}{2}\right)|\downarrow_j\rangle \tag{7.11}$$

Then, the matrix elements with respect to $|d_i\rangle|\uparrow_i\rangle$ and $|d_j\rangle|\uparrow_j\rangle$ states are given by

$$E = \begin{pmatrix} -J_H S & b\cos\left(\frac{\theta}{2}\right) \\ b\cos\left(\frac{\theta}{2}\right) & -J_H S \end{pmatrix} \tag{7.12}$$

with $b = \langle d_i|\mathcal{H}_t|d_j\rangle$. Here, $|d_i\rangle$ represents the state where the e_g orbital at site i is occupied by an electron and $\mathcal{H}_t$ is the transfer term of the Hamiltonian. By diagonalizing this matrix, the lowest energy is given by

$$E_0 = -J_H S - b\cos\left(\frac{\theta}{2}\right) \tag{7.13}$$

which becomes minimum at $\theta = 0$. Therefore, this combination effect of the Hund coupling and the electron transfer induces a ferromagnetic interaction between the Mn spins. This interaction is termed the double-exchange interaction. It is noted that the energy gain due to the double-exchange interaction is proportional to $\cos(\theta/2)$, which is in contrast to that due to the superexchange interaction as $\cos\theta$.

7.2.3 Magnetic Properties in Cuprates and Manganites

The electronic phase diagrams of HTSC cuprates and CMR manganites are presented in Figure 7.2(a) and (b), respectively. It is considered that the magnetism in La_2CuO_4 is well described by the two-dimensional $S = \frac{1}{2}$ antiferromagnetic Heisenberg model introduced in Equation (7.10). In the theoretical aspect, this model shows the antiferromagnetic order at $T_N = 0$ K, that is, there are no long-range orders at finite temperature due to the two dimensionality and the small magnitude of spin. In the actual compounds, due to the weak magnetic interaction between the CuO_2 plane, the long-range magnetic order appears at $T_N = 320$ K, as shown in the figure. The magnitude of the exchange interaction J is obtained by the analyses of the spin wave dispersion relation. The energy of the spin wave

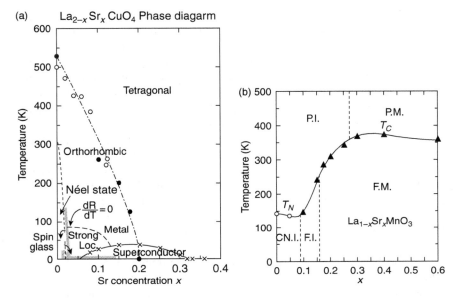

FIGURE 7.2 Phase diagram of (a) $La_{2-x}Sr_xCuO_4$ [18] and (b) $La_{1-x}Sr_xMnO_3$ [19]. The narrow dashed line ($dR/dT = 0$) in (a) separate the region of metallic linear resistance from that of logarithmically increasing resistance. The abbreviations in (b) mean paramagentic insulator (PI), paramagnetic metal (pm), spin-canted insulator (CNI), ferromagnetic insulator (FI), and ferromagnetic metal (FM)

around the Brillouin zone center is obtained by $\hbar\omega = 1.17\sqrt{2}Ja|\vec{q}|$ where a is the nearest-neighbor Cu–Cu distance and $|\vec{q}|$ is the spin wave momenta. Through the inelastic neutron scattering experiments, J is estimated to be 135 meV which is a larger value than that in other metal oxides [20]. Characteristic magnetic features in La_2CuO_4 are seen in the paramagnetic phase; the large spin correlation length remains due to the two dimensionality of the CuO_2 sheet and the large value of J. By the theoretical calculation in the two dimensional Heisenberg model based on the scaling hypothesis, the temperature, T, dependence of the spin correlation length ξ is obtained as [21]:

$$\frac{\xi}{a} = \frac{e}{8} \frac{\hbar c/a}{2\pi\rho_s} \exp\left(\frac{2\pi\rho_s}{k_b T}\right) \left[1 - \frac{1}{2}\frac{T}{2\pi\rho_s}\right] \tag{7.14}$$

where c and ρ_s are the spin wave velocity and the spin stiffness constant, respectively. It is shown that the observed ξ is well reproduced by Equation (7.14) in the temperature region between 800 K and the Neel temperature [22]. With doping of holes, T_N rapidly goes down. The three-dimensional magnetic order disappears at $x \sim 0.02$ and the spin glass phase appears. Then, the superconductivity comes up around $x \sim 0.05$. It is experimentally confirmed that the introduced hole is mainly doped into the O $2p$ band and this hole spin couples with the d hole spin antiferromagnetically. This bound state, termed the Zhang–Rice singlet, is a fundamental object in the doped HTSC cuprates. In the under-doped metallic region ($x < 0.15$), several anomalous features are experimentally observed. One of the attractive phenomena is the so-called pseudo-gap suggested in several experimental observation, in particular, in the magnetic excitation. This was first observed in nuclear magnetic resonance (NMR) experiments where the knight shift reflecting the spin susceptibility, as well as the inverse of the spin relaxation time $1/T_1$, decreases with decreasing temperature, far above the superconducting transition temperature [23]. The inelastic neutron scattering intensity being proportional to the dynamical spin correlation function also shows the gap like feature below 150 K in $YBa_2Cu_3O_{6.69}$ ($T_c = 59$ K) [24]. The related behaviors are also observed in the photonemission spectra, the tunneling spectra, the electric resistivity, etc. One of the interpretation of this pseudo-gap phenomena is a formation of the spin singlet state due to the large superexchange interaction. However, the origin of this phenomena and its implication to the superconductivity are still open to the question.

As for the manganites, $LaMnO_3$ is the charge-transfer type insulator. The long-range magnetic order occurs at $T_N = 141$ K below which the spins ferromagnetically align in the ab plane and antiferromagnetically along the c axis. This type of the anisotropic magnetic structure is termed the A type antiferromagnetic structure. This magnetic order originates from the long-range orbital order: the two different orbitals are alternately occupied by an electron in the ab plane and the same orbitals are aligned along the c axis. These two orbitals are approximately identified as $d_{3x^2-r^2}$ and $d_{3y^2-r^2}$ orbitals which are represented by the linear combinations of $d_{3z^2-r^2}$ and $d_{x^2-y^2}$ orbitals. This kind orbital ordering occurs at $T_{OO} = 780$ K. The lattice distortion in a MnO_6 octahedron, the so-called cooperative Jahn–Teller distortion, is associated with the orbital ordering. The ferromagnetic interaction together with the alternate orbital order in the ab plane is well understood by the superexchange electronic process under the orbital degeneracy introduced in Equation (7.6). The antiferromagnetic spin alignment along the c-axis is caused by the second term of the Hamiltonian Equation (7.6) and the superexchange interaction between the t_{2g} spins. This interpretation is directly supported by the measurements of the exchange interaction by the inelastic neutron scattering [25]. Through the observation of the spin wave dispersion relation in the A-type antiferromagnetic phase, it is found that the antiferromagnetic exchange interaction along the c axis is much smaller than the ferromagnetic in the ab plane. This is reproduced by the theoretical calculation based on Equation (7.6) and is understood based on the long-range orbital ordered state.

By introducing the hole carriers in $LaMnO_3$, the A type antiferromagnetic state is changed into the ferromagnetic insulating phase. Between the two phases, several magnetic states are proposed experimentally and theoretically, such as the spin canting state, the spin spiral state, the flux state, the spin-lattice polaronic state, the spin and orbital phase separations, etc. However, the situation in this region is still controversy. With further doping of holes by increasing the Sr concentration, the ferromagnetic metallic phase appears around $x \sim 0.175$. The fundamental features in this ferromagnetic metallic phase, for example, the reduction of the resistivity at the Curie temperature, are well understood by the double-exchange interaction introduced in Equation (7.13). Above T_c around $x \sim 0.2$, the electric resistivity increases with decreasing temperature. Therefore, at T_c, the paramagnetic insulating phase is transformed into the ferromagnetic metallic phase. Thus, the large reduction of the electric resistivity by applying the magnetic field, that is, the CMR observed near T_c. The insulating feature in the paramagnetic phase is not explained by the conventional double-exchange interaction scenario [9], and a variety of spin, charge, orbital, and lattice states are proposed in the phase. In order to uncover the origin of the paramagnetic insulating state and its relation to the CMR, further intensive and extensive researches are required.

7.3 OPTICAL PROPERTIES OF OXIDES

The optical studies in metal oxides have directly revealed the ground state electronic structures as well as several excitations of charge, spin, orbital, and lattice degrees of freedom. The recent progress of the experimental technique develops the optical experiments as direct probes for the electronic structure in metal oxides, for example, the angular resolved photoemission spectroscopy, the x-ray diffraction, and/or absorption spectroscopy by utilizing the high brilliant synchrotron x-ray and so on. The optical properties are also important for a wide variety of the technological applications, such as the optical and optoelectric devices by utilizing the large nonlinear optical responses, magneto-optical effect, photo-refractive effect, elasto-optic effect and so on. It is widely known that the large applications of several metal oxides have been done for a long time, for example, the famous ruby ($Cr : Al_2O_3$) and $Nd : YAG$ ($Y_3Al_5O_{12}$) lasers, the well-developed nonlinear optical crystals $LiNbO_3$ and $Ba_2NaNb_5O_{15}$, and the photo-catalytic materials TiO_2. Since the optical and dielectric properties in conventional dielectric and ferroelectric metal oxides have been reviewed in several books and articles [26], in this section, we introduce the optical properties in the correlated metal oxides, in particular, in HTSC cuprates and CMR manganites which have large potentials for the new optical devises in the generations to come.

7.3.1 Optical Conductivity

The optical conductivity $\sigma(\omega)$ is one of the fundamental optical properties in solids. As it is well known, there is a relation between $\sigma(\omega)$ and the dielectric function $\sigma(\omega)$ in the optical region

$$\varepsilon(\omega) = 1 + \frac{4\pi i}{\omega}\sigma(\omega) \qquad (7.15)$$

By utilizing this relation, $\sigma(\omega)$ is deduced from the reflectivity and absorption experiments. In the microscopic point of view, $\sigma(\omega)$ is represented as:

$$\text{Re}\,\sigma(\omega) = D\delta(\omega) + \frac{\pi e^2}{N}\sum_{m \neq 0}\frac{|\langle m|j|0\rangle|^2}{E_m - E_0}\delta(\omega - E_m + E_0) \qquad (7.16)$$

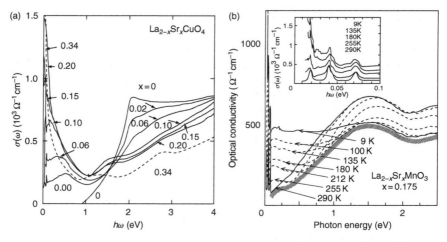

FIGURE 7.3 Optical conductivity spectra in (a) $La_{2-x}Sr_xCuO_4$ [27] and (b) $La_{1-x}Sr_xMnO_3$ with $x = 0.175$ [28]. The inset in (b) shows a magnification of the far-infrared part

where $|m\rangle$ and $|0\rangle$ are the m-th excited state and the ground state with energies E_m and E_0, respectively, j indicates the current, and N is the number of the sites. D is termed the Drude weight, which is given in the nearly free electron model by $D = Ne^2\pi/m^*$ with the effective mass of electron m^*. Thus, $\sigma(\omega)$ represents the dipole transition amplitudes in solids.

In Figure 7.3(a) and (b), the optical conductivity spectra for cuprates and manganites are shown. The hole concentration dependence of the optical spectra is presented for $La_{2-x}Sr_xCuO_4$ [27]. The spectra at $x = 0$, that is, La_2CuO_4, have an absorption edge around 2 eV. This is the charge-transfer excitation corresponding to the excitation from the O $2p$ occupied level to the Cu $3d$ unoccupied level, that is, $2p^63d^9 \rightarrow 2p^53d^{10}$. By introducing the hole carrier, this charge-transfer excitation intensity is rapidly collapsed and the spectral weights are transferred into the lower energy region. Around the zero excitation energy, a finite spectral intensity corresponding to the Drude weights is confirmed with doping of holes. It is worth mentioning that these changes of the spectra cannot be explained by the conventional rigid-band shift, that is, a simple chemical potential shift to the valence band by doping of holes. This may indicate a creation of new electronic state between the band gap. A doped hole cannot move easily in the antiferromagnetic spin background and has a large damping due to a strong coupling with the spin fluctuation. The broad spectra appearing in the mid-infrared region is considered to be related to the dynamics of doped holes associated with the spin excitations.

As for the CMR manganites, the temperature dependence of the optical spectra is shown in the figure. In $La_{1-x}Sr_xMnO_3$ with $x = 0.175$, the ferromagnetic transition occurs at $T_c = 300$ K [28]. Above T_c the system is an insulator and the optical spectra does not show the Drude weight. The broad peak around 1.5 eV corresponds to the charge-transfer excitation described as $2p^63d^4 \rightarrow 2p^53d^5$. Below T_c, in addition to the remnant of this charge-transfer excitation, the broad spectra appear around 0.5 eV associated with the small component of the Drude weight at $\omega = 0$. Through the detailed data analyses, the integrated intensity of the broad spectra and the Drude component continue to change to very low temperature despite the fact that the magnetic moment is already saturated. This fact suggests that other nonspin degrees of freedom still remain active in this temperature region. One of the plausible candidate contributing such broad spectra is the charge carrier strongly coupled with the orbital fluctuations [29]. This kind of orbital fluctuation is proposed by some theoretical calculations and is also consistent with the absence of the long-range orbital order in the ferromagnetic metallic phase suggested by magnetic and transport experiments.

7.3.2 Raman Scattering

Raman scattering is well known to be the powerful optical method to detect the several excitations in molecules and solids. This scattering is described by the second-order processes with respect to the interaction between the electric dipole and photons. The scattering cross-section is given as:

$$\frac{d^2\sigma}{d\Omega\, d\omega_f} = \sigma_T \frac{\omega_f}{\omega_i} \sum_f \sum_{\alpha\beta} \left| (\vec{e}_{k_i\lambda_i})_\alpha S_{\alpha\beta} (\vec{e}_{k_f\lambda_f})_\beta \right|^2 \delta(\varepsilon_f + \omega_f - \varepsilon_i - \omega_i) \quad (7.17)$$

with

$$S_{\alpha\beta} = \frac{m}{e_2} \sum_m \left[\frac{\langle f|\vec{j}_\alpha|m\rangle \langle m|\vec{j}_\beta|i\rangle}{\varepsilon_i - \varepsilon_m - \omega_f + i\eta} + \frac{\langle f|\vec{j}_\beta|m\rangle \langle m|\vec{j}_\alpha|i\rangle}{\varepsilon_i - \varepsilon_m - \omega_i + i\eta} \right] \quad (7.18)$$

We consider the scattering of photon with momentum $\vec{k}_i$, energy $\omega_i = c|\vec{k}_i|$, and polarization λ_i, to $\vec{k}_f$, ω_f and λ_f. $\vec{e}_{k_l\lambda_l}$ ($l = i,f$) is the polarization vector, and $|i\rangle$, $|m\rangle$, and $|f\rangle$ are the electronic states in the initial, intermediate, and final scattering states with energies ε_i, ε_m, and ε_f, respectively. A factor $\sigma_T = (e^2/mc^2)^2$ is the total scattering cross-section of the Thomson scattering and η in the denominators is an infinitesimal positive constant. $\vec{j}$ is defined by the current density with the zero momentum. The Raman scattering is often utilized to detect the vibrational excitation, that is, the phonon since its selection rule is complement to that in the infrared spectra. Here we introduce the electronic Raman scattering, in particular, focus on the spectra from the collective electronic excitation in cuprates and manganites.

The Raman spectra in La_2CuO_4 is presented in Figure 7.4(a) below the Neel temperature [30]. The polarization configuration is chosen to be the B_{1g} symmetry. A peak structure is observed around 3000 cm^{-1} with a broad width. This is identified as the two magnon Raman scattering. Consider the antiferromagnetic order of the $S = \frac{1}{2}$ spins in CuO_2 plane and focus on a Cu(i)-O-Cu(j) bond where i and j indicate Cu sites. The incident photon excites an electron with up spin from

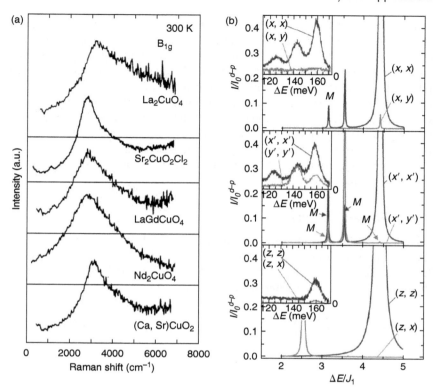

FIGURE 7.4   (a) Raman spectra in insulating cuprates [31] and (b) theoretical Raman spectra in LaMnO₃ [33]. The insets show the experimental spectra in LaMnO₃ [32]

the O site to the Cu(*j*) site. In the intermediate state, Cu(*j*) is doubly occupied by electrons and two holes at O and Cu(*i*) sites from a spin singlet state term the Zhang-Rice singlet state. Through the interaction with the second photon, the down spin electron at site *j* comes back to the O site. As the result, the spins at two Cu sites are exchanged, and the two magnons are created by photons. Because photon cannot flip the electron spin directly and the photon momentum is almost zero in comparison with the magnon momentum, the created magnon have the momenta of $\vec{k}$ and $-\vec{k}$. The magnitude of the exchange energy J is estimated from the analyses of the peak position of the two magnon Raman spectra. The estimated value is about $1030\,cm^{-1}$, which is consistent with the estimation by the spin wave dispersion relation introduced previously.

As the magnon Raman scattering in HTSC cuprates, the orbital excitations are expected to be observed in orbital ordered LaMnO₃. The orbital degree of freedom is a quantum variable of an electron, as spin and charge degrees of freedom are, and this degree of freedom has its own dynamics. The collective excitation for the orbital degree of freedom is termed as orbital wave, which is in analogy to the spin wave in the magnetically ordered state. A quantized object of the orbital wave is termed orbiton. The orbiton is excited by the Raman scattering through the electronic-exchange process between a Mn site and a nearest-neighbor O site. Recently, new peak structures in the Raman scattering spectra are found around 120 to 170 cm^{-1} in LaMnO$_3$ [32]. The experimental results are shown in the insets of Figure 7.4(b). The relative intensity of the peak structures strongly depend on the polarization configurations. The Raman spectra observed in such a high energy region are usually interpreted to the multi-photon excitations. However, it is difficult to explain the energy and temperature dependences of the new spectra by the multi-photon excitations. The magnon Raman scattering is not expected in the *ab* plane where spins align ferromagnetically. As shown in Figure 7.4 [33], the polarization and energy dependences and relative intensity of the new Raman scattering peaks are well explained by the theoretical calculations based on the orbital wave scattering. It is reasonable to interpret that the new observed peak structure originates from the orbital wave.

7.3.3 Resonant X-Ray Scattering

Resonant x-ray scattering (RXS) is the x-ray diffraction method where the incident x-ray energy is tuned around an absorption edge of a certain ion in a solid. This method was developed in 1970s by Templeton and Templeton [34], and was first applied to observation of the orbital orderings in La$_{0.5}$Sr$_{1.5}$MnO$_4$ and LaMnO$_3$ by Murakami et al. [35,36]. As we know well, it is a hard task to detect the orbital degree of freedom and its ordering by the conventional x-ray scattering, because the different charge distribution for the different occupied orbitals is less than the unit charge. In Murakami's RXS experiment, the incident x-ray energy was tuned near the Mn K-edge of 6.552 KeV. In the orbital ordered state in LaMnO$_3$ (Figure 7.5[a]), the $d_{3x^2-r^2}$ and $d_{3y^2-r^2}$ type orbitals are alternately aligned in the *ab* plane. A new unit cell appears below the orbital ordering temperature, that is, the orbital unit cell with the volume of $\sqrt{2}a \times \sqrt{2}a$ where a is the nearest-neighbor Mn–Mn bond length. Murakami et al. focus on the the reflection corresponding to the orbital superlattice reflection. The experimental results of the incident x-ray energy, which depends on the scattering intensity, are presented in Figure 7.5(b), where the scattering intensity is resonantly enhanced near the K-edge. The remarkable difference of RXS from the conventional x-ray scattering is seen in the observation of the polarization dependence of the scattering intensity. After the first observation, this method has been widely utilized to study of the orbital degree of freedom in several manganites, other transition-metal oxides and f-electron systems [37].

The differential cross-section of x-ray from electrons is obtained by perturbational calculation in terms of the electron–photon interaction:

$$\frac{d^2\sigma}{d\Omega\, d\omega_f} = \sigma_T \frac{\omega_f}{\omega_i} \sum_f \left| S_1 \vec{e}_{k_i\lambda_f} \cdot \vec{e}_{k_f\lambda_f} + \sum_{\alpha\beta} (\vec{e}_{k_i\lambda_f})_\alpha S_{2\alpha\beta} (\vec{e}_{k_f\lambda_f})\beta \right|^2$$

$$\times\ \delta(\varepsilon_f + \omega_f - \varepsilon_i - \omega_i) \tag{7.19}$$

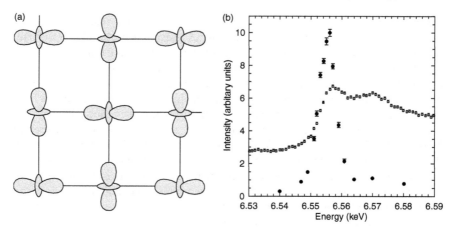

FIGURE 7.5 (a) Schematic orbital ordered state in *ab* plane in LaMnO$_3$. (b) Energy dependence of the integrated RXS intensity (closed circles) and fluorescence (open circles) in LaMnO$_3$ [36]

where

$$S_1 = \langle f | \rho_{\vec{k}_i - \vec{k}_f} i \rangle | \tag{7.20}$$

and

$$S_{2\alpha\beta} = \frac{m}{e^2} \sum_m \left[\frac{\langle f | \vec{U}_{-\vec{k}_i\alpha} | m \rangle \langle m | \vec{U}_{\vec{k}_f\beta} | i \rangle}{\varepsilon_i - \varepsilon_m - \omega_f} + \frac{\langle f | \vec{U}_{\vec{k}_f\beta} | m \rangle \langle m | \vec{U}_{-\vec{k}_i\alpha} | i \rangle}{\varepsilon_i - \varepsilon_m + \omega_i + i\Gamma} \right] \tag{7.21}$$

In Equation (7.20), $\rho_{\vec{k}}$ is the Fourier transform of the charge density, and other notations are the same with those in Equation (7.18). Although S_2 is the similar form with that in the Raman scattering in Equation (7.18), the wave length of x-ray is much smaller than that of the conventional light. Therefore, the momentum dependent scattering and the diffraction experiments become possible in the present case. When the incident x-ray energy ω_i is far from the energy difference of the initial and intermediate electronic states $\Delta\varepsilon_{mi} = \varepsilon_m - \varepsilon_i$, S_1 dominates the scattering. On the other hand, when ω_i is close to $\Delta\varepsilon_{mi}$, the real part of the denominator in the second term of S_2 becomes zero and the scattering increases divergently; that is, the resonant scattering. In comparison with other methods, RXS has the following advantage to detect the orbital ordering: (i) The wave length of x-ray, which is tuned near the K absorption edge of a transition-metal ion, is shorter than a lattice constant of the perovskite unit cell. Thus, the diffraction experiments can be carried out for the orbital superlattice, unlike Raman scattering and optical reflection/absorption experiments. (ii) As shown in Equation (21), S_2 is a tensor with respect to the incident and scattered x-ray polarizations. On the contrary, S_1 does not show any polarization dependence. (iii) By tuning the x-ray energy at the absorption edge, the scattering from a specific element, such as an Mn ion, can be identified.

Recently, beyond the x-ray diffraction to observe the static order, the technique of the resonant inelastic x-ray scattering (RIXS) has been rapidly developed. This is an inelastic version of RXS. This experimental method has been utilized as a momentum resolved probe to detect the bulk electronic structures in solids [38–40], and has rapidly progressed accompanied with the recent advances of the third-generation synchrotron light source. One of the remarkable results is obtained in the recent experiment in insulating cuprates $Ca_2CuO_2Cl_2$ [39]. An observed peak around 2.5 to 3 eV is interpreted as an electronic excitation from the so-called Zhang–Rice singlet band to the upper Hubbard band across the Mott gap. Dispersion relation for this peak is clearly observed by the momentum dependence of RIXS experiment and is well reproduced by the theoretical calculation. Recently, RIXS is also applied to $LaMnO_3$ [40]. When comparing the excitations in cuprates and manganites, both the upper and lower Hubbard bands in insulating cuprates have the $3d_{x^2-y^2}$ orbital character. The electron and hole are brought about by x-ray in the upper and lower Hubbard bands, respectively, and move around in the antiferromagnetic background. On the other hand, in manganites, the upper and lower Hubbard bands consist of different orbital characters, that is, $[3d_{x^2-z^2}/3d_{y^2-z^2}]$ and $[3d_{3x^2-r^2}/3d_{3y^2-r^2}]$ orbitals, respectively. The electronic excitations across the Mott gap are recognized to be excitations associated with the changing orbital symmetry. Three peak structures are found in the recent RIXS experiments in $LaMnO_3$ [40]. The lowest peak around the 2.5 eV energy transfer is interpreted to be the excitation across the Mott gap and its nature is well described by the theoretical calculations.

7.4 SUMMARY

In this chapter, we reviewed the magnetic and the optical properties in metal oxides, in particular, we focus on the HTSC cuprates and the CMR manganites. One of the common key words in the two representative metal oxides is the electron correlation effects due to the local nature of the $3d$ electrons. This effect provides several highly exotic and nontrivial phenomena, which attract the attention of a lot of theoretical and experimental researchers. The recent intensive and extensive studies opened new magnetic and optical properties in metal oxides as well as its large potential for the technological and chemical devises in the next generation to replace conventional semiconductors and metals.

REFERENCES

1. Goodenough, J.B. *Magnetism and the Chemical Bond*. New York: John Wiley & Sons, 1963; Adler, D. Insulating and metallic states in transition metal oxides. In *Solid State Physics*, Seitz, F., Turnbull, D., and Ehrenreich, H., Eds., New York: Academic, 1986, Vol. 21, pp. 1–113; Tsuda, N., Nasu, K., Fujimori, A., and Siratori, K. *Electronic Conduction in Oxides*, Springer Series in Solid-State Sciences. Berlin: Springer-Verlag, 2000.

2. Bednorz, J.G. and Muller, K.A. Possible high T_c superconductivity in the Ba–La–Cu–O system. *Z. Phys. B* 1986, *64*, 189–193.

3. Jonker, G.H. and van Santen, H. Ferromagnetic compounds of manganese with perovskite structure. *Physica* 1950, *16*, 337–349.

4. Chahara, K., Ohono, T., Kasai, M., Kanke, M., and Kozono, Y. Magnetoresistance effect of $La_{0.72}Ca_{0.25}MnO_z$/$YBa_2Cu_3O_y$/$La_{0.72}Ca_{0.25}MnO_z$ trilayered films. *Appl. Phys. Lett.* 1990, *62*, 780–782.

5. von Helmolt, R., Wecker, J., Holzapfel, R., Schultz, L., and Samwer, K. Giant negative magnetoresistance in perovskitelike $La_{2/3}Ba_{1/3}MnO_x$ ferromagnetic films. *Phys. Rev. Lett.* 1993, *71*, 2331–2333.

6. Tokura, Y., Urushibara, Y., Moritomo, Y., Arima, T., Asamitsu, A., Kido, G., and Furukawa, N. Giant Magnetotransport phenomena in filling-controlled Kondo lattice system: $La_{1-x}Sr_xMnO_3$. *J. Phys. Soc. Jpn.* 1994, *63*, 3931–3935.

7. Jin, S., Tiefel, T.H., McCormack, M., Fastnacht, R.A., Ramesh, R., and Chen, L.H. Thousandfold change in resistivity in magnetoresistive La–Ca–Mn–O films. *Science* 1994, *264*, 413–415.

8. Tomioka, Y., Asamitsu, Y., Moritomo, Y., and Tokura, Y. Anomalous magnetotransport properties of $Pr_{1-x}Ca_xMnO_3$. *J. Phys. Soc. Jpn* 1995, *64*, 3626–3630.

9. Millis, A.J., Littlewood, P.B., and Shrainman, B.I. Double exchange alone does not explain the resistivity of $La_{1-x}Sr_xMnO_3$. *Phys. Rev. Lett.* 1995, *74*, 5144–5147.

10. Zannen, J., Sawatzky, G.A., and Allen, J.W. Band gaps and electronic structure of transition-metal compounds. *Phys. Rev. Lett.* 1985, *55*, 418–421.

11. Anderson, P.W. New approach to the theory of superexchange interactions. *Phys. Rev.* 1959, *115*, 2–13, Theory of magnetic exchange interactions: exchange in insulators and semiconductors. In *Solid State Physics*, Seitz, F. and Turnbull, D., Eds. 1963, Vol. 14, pp. 99–214.

12. Goodenough, J.B. Theory of the role of covalence in the perovskite-type manganites [La,M(II)]MnO$_3$. *Phys. Rev.* 1955, *100*, 564–573.

13. Kanamori, J. Superexchange interaction and symmetry properties of electron orbitals. *J. Phys. Chem. Solid* 1959, *10*, 87–98.

14. Kugel, K.I. and Khomskii, D.I. Superexchange ordering of degenerate orbitals and magnetic structure of dielectrics with Jahn-Teller ions. *Sov. Phys.: JETP Lett.* 1972, *15*, 446–448.

15. Ishihara, S., Inoue, J., and Maekawa, S. Effective Hamiltonian in manganites: study of the orbital and spin structures. *Phys. Rev. B* 1997, *55*, 8280–8286.

16. Zener, C. Interaction between the d-shells in the transition metals. II. Ferromagnetic compounds of Manganese with perovskite structure. *Phys. Rev.* 1951, *82*, 403–405.

17. Anderson, P.W. and Hasegawa, H. Considerations on double exchange. *Phys. Rev.* 1955, *100*, 675–681.

18. Keimer, B., Belk, N., Birgeneau, R.J., Cassanho, A., Chen, C.Y., Greven, M,, Kastner, M.A., Aharony, A., Endoh, Y., Erwin, R.W., and Shirane, G. Magnetic excitations in pure, lightly doped, and weakly metallic La_2CuO_4. *Phys. Rev. B* 1992, *46*, 14034–14053.

19. Urushibara, A., Moritomo, Y., Arima, T., Asamitsu, A., Kido, G., and Tokura, Y. Insulator-metal transition and giant magnetoresistance in $La_{1-x}Sr_xMnO_3$. *Phys. Rev. B* 1995, *51*, 14103–14109.

20. Aeppli, G., Hayden, S.M., Mook, H.A., Fisk, Z., Cheong, S.W., Rytz, D., Remeika, J.P., Espinosa, G.P., and Cooper, A.S. Magnetic dynamics of La_2CuO_4 and $La_{2-x}Ba_xCuO_4$. *Phys. Rev. Lett.* 1989, *62*, 2052–2055.

21. Chakravarty, S., Halperin, B.I., and Nelson, D.R. Low-temperature behavior of two-dimensional quantum antiferromagnets. *Phys. Rev. Lett.* 1988, *60*, 1057–1060.

22. Greven, M., Birgeneau, R.J., Endoh, Y., Kaster, M.A., Matsuda, M., and Shirane, G. Neutron scattering study of the two-dimensional spin $S = 1/2$ square-lattice Heisenberg antiferromagnet $Sr_2CuO_2Cl_2$. *Z. Phys. B* 1995, *96*, 465–477.

23. Yasuoka, H., Imai, T., and Shimizu, T. In *Strong Correlation and Superconductivity*; Fukuyama, H., Maekawa, S., and Malozemoff, A.P., Eds., Springer Series in Solid-State Sciences. Berlin: Springer-Verlag, 1984.

24. Rossat-Mignod, J., Regnault, L.P., Vettier, C., Burlet, P., Henry, Y.J., and Lapertot, G. Investigation of the spin dynamics in $YBa_2Cu_3O_{6+x}$ by inelastic neutron scattering. *Physica B* 1991, *169*, 58–65.

25. Hirota, K., Kaneko, N., Nishizawa, A., and Endoh, Y. Two-dimensional planar ferromagnetic coupling in $LaMnO_3$. *J. Phys. Soc. Jpn* 1996, *65*, 3736–3739.

26. See e.g., Lines, M.E. and Glass, A.M. *Principles and Applications of Ferroelectrics and Related Materials*. Oxford: Clarendon Press, 1977.

27. Uchida, S., Ido, T., Takagi, H., Arima, T., Tokura, Y., and Tajima, S. Optical spectra of $La_{2-x}Sr_xCuO_4$: effect of carrier doping on the electronic structure of the CuO_2 plane. *Phy. Rev. B* 1991, *43*, 7942–7954.

28. Okimoto, Y., Katsufuji, T., Ishikawa, T., Urushibara, A., Arima, T., and Tokura, Y. Anomalous variation of optical spectra with spin polarization in double-exchange ferromagnet: $La_{1-x}Sr_xMnO_3$. *Phys. Rev. Lett.* 1995, *75*, 109–112.

29. Ishihara, S., Yamanaka, M., and Nagaosa, N. Orbital liquid in perovskite transition-metal oxides. *Phys. Rev. B* 1997, *56*, 686–692.

30. Lyons, K.B., and Fleury, P.A., Schneemeyer, L.F., and Waszczak, J.V. Spin fluctuations and superconductivity in $Ba_2YCu_3O_{6+\delta}$. *Phys. Rev. Lett.* 1988, *60*, 732–735.

31. Tokura, Y., Koshihara, S., Arima, T., Takagi, H., Ishibashi, S., Ido, T., and Uchida, S. Cu–O network dependence of optical charge-transfer gaps and spin-pair excitations in single-Cu–O_2-layer compunds. *Phys. Rev. B* 1990, *41*, 11657–11660.

32. Saitoh, E., Okamoto, S., Takahashi, K.T., Tobe, K., Yamamoto, K., Kimura, T., Ishihara, S., Maekawa, S., and Tokura, Y. Observation of orbital waves as elementary excitations in a solid. *Nature* 2001, *410*, 180–183.

33. Okamoto, S., Ishihara, S., and Maekawa, S. Theory of Raman scattering from orbital excitations in manganese oxides. *Phys. Rev. B* 2002, *66*, 014435-1-9.

34. Templeton, D.H. and Templeton, L.K. X-ray dichroism and polarized anomalous scattering of the uranyl ion. *Acta Crystallogr. A* 1982, *38*, 62–67.

35. Murakami, Y., Kawada, H., Kawata, H., Tanaka, M., Arima, T., Moritomo, H., and Tokura, Y. Direct observation of charge and orbital ordering in $La_{0.5}Sr_{1.5}MnO_4$. *Phys. Rev. Lett.* 1998, *80*, 1932–1935.

36. Murakami, Y., Hill, J.P., Gibbs, D., Blume, M., Koyama, I., Tanaka, M., Kawata, H., Arima, T., Tokura, Y., Hirota, K., and Endoh, Y. Resonant x-ray scattering from orbital ordering in $LaMnO_3$. *Phys. Rev. Lett.* 1998, *81*, 582–585.

37. As a review see, Ishihara, S. and Maekawa, S. Resonant x-ray scattering in manganites: study of the orbital degree of freedom. *Rep. Prog. Phys.* 2002, *65*, 561–598.

38. Ederer, D.L. and McGuire, J.H. *Raman Emission by X-Ray Scattering*. Singapore: World Scientific, 1996.

39. Hasan, M.Z., Isaacs, E.D., Shen, Z.-X., Miller, L.L., Tsutsui, K., Tohyama, T., and Maekawa, S. Electronic structure of Mott insulators studied by inelastic x-ray scattering. *Science* 2000, *288*, 1811–1814.

40. Inami, T., Ishihara, S., Kondo, H., Mizuki, J., Fukuda, T., Maekawa, S., Nakao, H., Matsumura, T., Hirota, K., Murakami, Y., and Endoh, Y. Orbital excitations in $LaMnO_3$ studied by resonant inelastic x-ray scattering. *Phys. Rev. B* 2003, *67*, 045108-1-6.

8 Redox Properties of Metal Oxides

Benjaram M. Reddy
Inorganic and Physical Chemistry Division, Indian Institute of Chemical Technology, Hyderabad, India

CONTENTS

8.1 INTRODUCTION

Metal oxides represent one of the most important and widely employed classes of solid catalysts, either as active phases or as supports. Metal oxides are used for both their acid–base and redox properties and constitute the largest family of catalysts in heterogeneous catalysis [1–6]. The three key features of metal oxides, which are essential for their application in catalysis, are (i) coordination environment of the surface atoms, (ii) redox properties of the oxide, and (iii) oxidation state of the surface. Surface coordination environment can be controlled by the choice of crystal plane exposed and by the preparation procedures employed; however, specification of redox properties is largely a matter of choice of the oxide. The majority of oxide catalysts correspond to more or less complex transition metal oxides containing cations of variable oxidation state. These cations introduce redox properties and, in addition, acid–base properties. The acid–base properties of the oxides are usually interrelated to their redox behavior. Many attempts were made in the literature to find correlations between acid–base and redox characteristics, and the catalytic properties. It is obvious that both characteristics are not independent since cations are Lewis acids while lattice oxygen anions are basic and hydroxyl groups could be either acidic (Brønsted site) or basic. The acidic character of the cations depends on their positive charge and size while the basic character of the lattice oxygen anions depends on the ionic character of the metal–oxygen bonds [1,5,6].

Metal oxides are usually formed from their corresponding hydroxides, through calcination. Electrostatically, neutral oxide surfaces expose both cations as well as oxygen atom anion sites, whereas in the bulk the formal charge on oxygen can often be considered to be equal to -2, this is not the case on metal or metal oxide surfaces. The decreased Madelung energy on the surface makes charge transfer between cation and anion less favorable. This can give rise to reactive O^- sites, or in the case of the adsorption of O_2, O_2^- anions which are known to be active in nonselective combustion pathways for reacting hydrocarbons. Dehydrated surfaces possess both Lewis acid cationic sites as well as Lewis base anionic sites. Water normally physisorbs at low temperature and readily dissociates on nonpolar surfaces. This occurs on reducible as well as on nonreducible oxides. Protons attach to bridging oxygen sites and behave as Brønsted acids, where as the OH^- fragments adsorb to the cation sites and behave as Brønsted bases.

Oxides commonly studied as catalytic materials belong to the structural classes of corundum, rocksalt, wurtzite, spinel, perovskite, rutile, and layer structure. These structures are commonly reported for oxides prepared by normal methods under mild conditions [1,5]. Many transition metal ions possess multiple stable oxidation states. The easy oxidation and reduction (redox property), and the existence of cations of different oxidation states in the intermediate oxides have been thought to be important factors for these oxides to possess desirable properties in selective oxidation and related reactions. In general terms, metal oxides are made up of metallic cations and oxygen anions. The ionicity of the lattice, which is often less than that predicted by formal oxidation states, results in the presence of charged adsorbate species and the common heterolytic dissociative adsorption of molecules (i.e., a molecule AB is adsorbed as A^+ and B^-). Surface exposed cations and anions form acidic and basic sites as well as acid–base pair sites [1]. The fact that the cations often have a number of commonly obtainable oxidation states has resulted in the ability of the oxides to undergo oxidation and reduction, and the possibility of the presence of rather high densities of cationic and anionic vacancies. Some of these aspects are discussed in this chapter. In particular, the participation of redox sites in oxidation and ammoxidation reactions and the role of redox sites in various oxides that are currently pursued in the literature are presented with relevant references.

8.2 REDOX CATALYSTS

The catalysts in the given reaction conditions that undergo reduction and reoxidation simultaneously by giving out the surface lattice oxide ions and taking in oxygen from the gas phase are called redox catalysts. The phenomenon is known as redox catalysis. Catalytic redox processes on metal oxides are often described in terms of a general redox mechanism:

$$Cat-O + Red \rightarrow Cat + Red-O$$
$$Cat + Ox-O \rightarrow Cat-O + Ox$$

Here, the oxide catalyst surface (Cat–O) is reduced by a reductant (Red) and reoxidized by an oxidant (Ox–O) to its initial state. The net result of this two-step reaction is the transfer of oxygen from one species to the other. Here, reduction and reoxidation are a relative phenomenon and the oxidation of the substrate is done at the expense of the reduction of the surface by losing surface oxide ions and vice versa. Redox reactions are part and parcel of oxidation (both selective and nonselective), ammoxidation, and oxidative dehydrogenation catalysis. It can occur on the surfaces of pure metals, metal oxides, and metal sulfides. In the given limitations of the present text, we would confine our discussion to oxide surfaces only. Redox processes necessarily demand microscopic reversibility. Therefore, it is widely recognized that an oxide catalyst in the course of the reaction and during the activation is a dynamic system ("living" or "breathing" moiety) in which various centers may be formed or disappear involving diffusion phenomenon within the catalyst and on its surface. In majority of the redox reactions occurring on the surface of oxide catalysts, the source of oxygen is the surface of the metal oxide lattice, although oxygen can also be supplied from the gas phase. The principal role of surface oxygen in these reactions is that of a nucleophile, attacking electron deficient centers in the adsorbate.

The commonly accepted mechanism of catalysis in oxidation processes by metal oxide catalysts is the redox mechanism developed many years ago by Mars and van Krevelen [7]. According to this mechanism the substrate is oxidized by the catalyst and not directly by molecular oxygen of the gaseous phase. The role of dioxygen is to regenerate or maintain the oxidized state of the catalyst. The oxygen species introduced in the substrate (or giving H_2O for oxidative dehydrogenation reactions) stems from the lattice. The mechanism involves the presence of two types of distinct active sites: an active cationic site, which oxidizes the substrate and another site active for dioxygen reduction. Such a mechanism necessitates a catalyst that contains a redox couple as, for instance, transition metal ions, which exhibit high-electrical conductivity to favor electron transfer and which has a high-lattice oxygen anion mobility within the material to ensure the reoxidation of the reduced catalyst. It was later shown that the process of selective oxidation of a hydrocarbon molecule starts with the abstraction of a proton, accompanied by transfer of two electrons, which formally reduces the transition metal cations of the catalyst [8–14]. This step is followed by a nucleophilic addition of an oxide ion from the catalyst surface to the oxidized hydrocarbon molecule with the formation of an oxygen vacancy in the crystal lattice of the catalyst. The vacancy is then filled by oxygen from the gas phase with a simultaneous reoxidation of the cation. Thus, the process is composed of two separate redox steps: oxidation of the hydrocarbon by transition metal cations of the catalyst and reoxidation of the catalyst by gas-phase oxygen. The withdrawal of oxygen from the surface and reinsertion of an oxygen atom may be separated with time. It was even considered to conduct the catalytic oxidative dehydrogenation of hydrocarbons in a cyclic manner, alternatively applying the substrate oxidation and the catalyst reoxidation cycles to prevent the immediate contact of the reaction substrate with oxygen, thus avoiding total oxidation [15]. As the final products of a catalytic reaction are neutral species and oxygen in the crystal lattice of an oxide appears as an ion in both processes, the exchange of charge between the surface and the adsorbed molecules must occur. For example, the reinsertion of oxygen to the crystal lattice may be described by the equation

$$O_{2(\text{adsorbed})} + 4e^- \rightarrow 2O^{2-}_{(\text{lattice})}$$

with the participation of valence band electrons. The possible detailed scheme for the above reaction can be written as

$$O_2 + e^- \rightarrow O_2^- + e^- \rightarrow 2O^- + 2e^- \rightarrow 2O^{2-}$$

This process has its own kinetics related to the reactivity of the sites with oxygen, their concentration, the efficiency of electron transfer, the partial pressure of oxygen, etc. Usually, it is much faster than the oxidation of the substrate. It is generally admitted that the rate-determining step is the substrate activation. It is established well beyond doubt that a single and isolated metallic ion site cannot take into account all the necessary transformations involved in the reaction since several steps, such as replenishing of oxygen anion vacancies, H-atom extraction, and electron transfer, are concerned. For instance n-butane oxidation reaction to maleic anhydride necessitates 7-lattice oxide ions, 8-hydrogen atom abstraction from the substrate, 3-oxygen atoms insertion, and 14-electron transfer [12]. The reinsertion of oxygen does not necessarily occur at the created vacant site — at elevated temperatures the diffusion of vacancies and free electrons within the crystal lattice may be sufficiently fast, even in the case of poorly conducting oxides. Thus, it may be anticipated that the conditions of the charge transfer process at the surface strongly influence the functioning of catalyst [13]. Among the class of oxide catalysts, the ones that strongly interact with O_2, are excellent redox catalysts. Others such as Al_2O_3, SiO_2, and MgO, which do not interact much with oxygen, are poor oxidation catalysts (they catalyze dehydration).

A reaction mechanism and the description of the active site as a molecular ensemble of atoms is shown in Scheme 8.1, as proposed by Haber and coworkers [8–10] and established by Grasselli et al. [11,12] for the (amm)oxidation of propene to (acrylonitrile) acrolein. A selective oxidation reaction involves one (or several) H-atom abstraction(s) from the hydrocarbon molecule, one (or several)

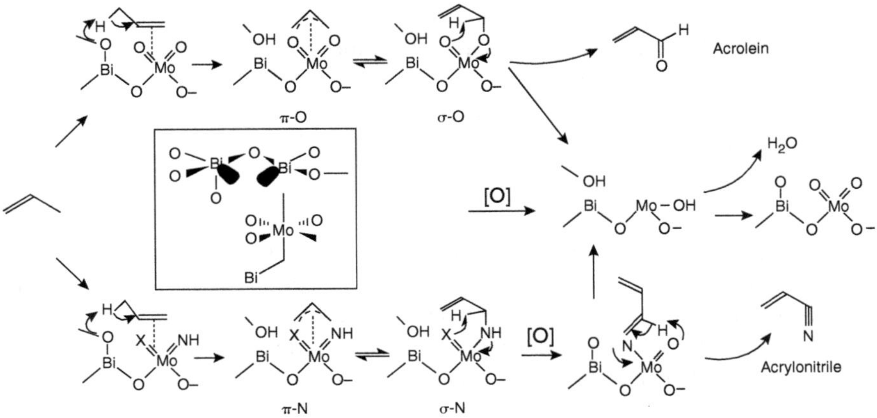

Scheme 8.1 Propane oxidation and ammoxidation mechanism and active site (From Védrine, J.C., *Top. Catal.* 2002, *21*, 97–106. With permission.)

O atom(s) insertion(s) from the lattice, and several electron transfers. In the collective properties of solid surfaces, in particular, lattice oxygen mobility and electron conductivity are involved. In general, the oxygen requirement of the selective reactions is less demanding than that of the nonselective reactions. It follows that restricting the availability of lattice oxygen anions around the active site will favor the selective oxidation reactions over the total oxidation reactions. This was clearly demonstrated, for example, for USb_3O_{10} catalyst for propene oxidation to acrolein by Grasselli et al. [16,17] several years ago.

8.3 ACID–BASE AND REDOX PROPERTIES

More than 130 acid–base catalysts are known, many of which are currently used in industrial processes [18]. The estimation of strength, structure, and concentration of acidic and basic sites at the surface of such catalysts is consequently a topic of primary interest [19–21]. This aspect has been dealt more lucidly in a separate chapter of this book. Oxides are used in a great variety of catalytic applications. Those having only s or p electrons in their valence orbitals tend to be effective only for acid–base catalyzed reactions, including those having carbocationic intermediates, while those having d or f outer electrons find a wider range of uses, including selective and nonselective oxidations, ammoxidation, alkene metathesis, and polymerization.

As mentioned earlier, oxides possess redox properties in addition to the acidic and basic properties. This is particularly true for solids containing transition metal ions, because the interaction with reactant molecules such as CO, H_2, and O_2 can lead to electron transfer from the surface to the adsorbed species and to the modification of the valence state of the metal centers. For example, an important role in surface redox processes involving CO is played by the most reactive oxygen on the surface (e.g., those located at the most exposed positions such as corners), which can react with CO as follows:

$$O^{2-} + CO \rightarrow CO^{2-} + e^-$$

$$2O_2 \rightarrow CO_3^{2-} + 2e^-$$

These reactions lead to the formation (transformation) of surface carboxylate and carbonate-like species and to the two electron reduction of the (electrons that can reduce) transition metal ions located in nearest-neighbor positions. On oxidic surfaces that do not contain transition metal ions, redox reactions accompanied by electron transfer from the surface to the adsorbed molecule (or vice versa) are much less probable.

The reduction of an oxide can be accomplished by removal of lattice oxygen or by dissolution of the reductant into the lattice. The former is common to all oxides, while the latter occurs only in selected systems. An example of the latter is the dissolution of hydrogen in WO_3 to form H_xWO_3, tungsten bronze. Removal of lattice oxygen can be achieved by many different reducing agents. The common ones include hydrogen, carbon monoxide, ammonia gas, and carbon dioxide. Usually the thermodynamic driving force is the formation of water and carbon dioxide. However, while the use of different reducing agents may involve nearly equivalent thermodynamic driving forces, the kinetics and the mechanism of reduction can be very different. The rates of reduction may depend strongly on the presence of surface defects and bulk grain boundaries, the orientation of the exposed surface planes, the nature of support if any, the presence of hydroxyl groups, and the presence of other metals [1].

Hydrogen is also a species that can produce surface reduction of metal oxides by attacking the most reactive surface oxygen species, with the formation of water and reduced metal centers. Important oxidation processes involving the metal centers can occur at the surface of transition metal oxides (e.g., α-Cr_2O_3) upon dissociative oxygen adsorption. In some cases the (nondissociative) adsorption of oxygen can lead to the formation of superacidic O_2^- or peroxide O_2^{2-} species with simultaneous oxidation of the surface metal cation centers [1].

8.4 Nature of Oxygen Species

There are a number of oxygen species: O_2 (adsorbed molecule), O (adsorbed neutral atom), O_2^- (superoxide), O_2^{2-} (peroxide), O_3^- (ozonide), O^- (ion radical), etc., which may be present on the metal oxide surfaces. Bielanski and Haber [22] presented a scheme, as shown in Figure 8.1, of the different types of oxygen of significance in catalysis. Depending on the nature, oxidation state of the metal ion and its environment (coordination structure), the metal–oxygen bonds may be more or less polarized and therefore the oxygen ion may exhibit electrophilic or nucleophilic properties. Based on isotope exchange experiments, the redox reactions

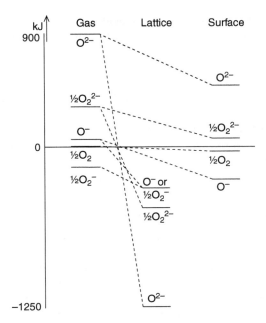

FIGURE 8.1 Survey of oxygen species (From Gellings, P.J. and Bouwmeester, H.J.M., *Catal. Today* 2000, *58*, 1–53. With permission.)

are more specifically divided into (i) extrafacial reactions, in which only adsorbed surface oxygen reacts and lattice oxygen does not participate (electrophilic reactions), and (ii) interfacial reactions, where lattice oxygen is extracted and oxygen vacancies are created (nucleophilic reactions) [11,22,23]. Stability of the oxide and the type of active oxygen species might also determine selectivity of the oxidation reactions. The redox reactions, which involve lattice oxygen, result in the partial oxidation of hydrocarbons (nucleophilic reactions), whereas active surface species lead to complete combustion products (electrophilic reactions). Electrophilic oxygen comprise electron deficient adsorbed species such as superoxide O_2^-, peroxide O_2^{2-}, and ion radical species O^-, where nucleophilic oxygen includes saturated species such as terminal oxygen groups $M=O$, or μ-oxo bridging groups $M-O-M$, both with the oxygen atom in a nominal O^{2-} state. There is perhaps one generalization, which is the fact that the ionic radii of transition metals are smaller than that of O^{2-}. Thus, the oxygen ions are usually close-packed with the smaller metal ions situated in the octahedral and tetrahedral holes among the oxygen ions.

There is enormous evidence in the literature that nucleophilic O^{2-} oxygen species is capable of carrying out selective oxidation, for example, from the observation that catalytic activity and selectivity persist at the same level even after gas-phase oxygen is cut off. However, one cannot rule out the existence of an equilibrium between this nucleophilic species and another type of oxygen. In the case of electrophilic oxygen species, the evidence for their involvement in deep oxidation is even more tenuous. The species have been observed by electron spin resonance spectroscopy, but generally at subambient temperatures. At the catalytic conditions the charge on the oxygen species has been determined by electrical capacitance methods, but this has been a deduction from total surface space charge measurements. There is a need for conclusive spectroscopic and kinetic characterization of these species at reaction conditions. Surface potential measurements that can be carried out at catalytic conditions seem to offer a means of determining the charged nature of oxygen species on surfaces [24]. However, they have not been widely applied.

8.5 STUDIES OF REDOX CHARACTERISTICS

The thermodynamics of redox reactions in solids can be studied by various techniques. Most often, the composition of the oxide is determined as a function of the partial pressure of oxygen and temperature and the energies of the redox reactions deduced from the variation of the equilibrium P_{O_2} with composition [25,26]. Alternatively, calorimetric determination of enthalpies of oxidation can be made [27,28]. A thermodynamic model that describes the redox energies of nonstoichiometric perovskite-related oxides has been reported recently [29], in which the redox properties are rationalized in terms of the relative stability of the oxidation states involved. A number of systems considered were adequately described with an enthalpy of oxidation that is independent of oxygen stoichiometry. The stability of a given oxidation state is related to the structure of the oxide, and the large difference in redox behavior between hexagonal and cubic $SrMnO_{3-\delta}$, for example, has been described [29]. Enthalpies of oxidation obtained directly by calorimetry and indirectly from equilibrium data were found to be consistent, and the combined data revealed that the enthalpy of oxidation of selected $La_{1-x}Ae_xMO_{3-\delta}$ phases varies linearly with x. While the entropy of oxidation is a less important contributor to the Gibbs energy of the reactions considered, it varies considerably from one system to another. The variations are largely related to the vibrational characteristics of the phases.

The redox properties of oxides can be investigated by $^{16}O_2/^{18}O_2$ exchange. This method has been extensively used in the literature to understand the redox behavior of various oxides and the activation of oxygen on different oxide surfaces [30–40]. A ranking order as shown in Figure 8.2 was also envisaged, for example, with an effort to design a better catalyst system for methane selective oxidation to methanol, more recently [41]. The virtual mechanism approach envisaged by Hutchings and Scurrell [41] was based on the consideration of three important functions of the catalyst namely, activation of C–H bond of methane, activation of dioxygen, and stability of product under the reaction conditions. The activation of C–H bond was probed by using CH_4/D_2 exchange, the activation of oxygen by $^{16}O_2/^{18}O_2$ exchange, and the stability of methanol was investigated under oxygen-rich conditions. Based on these concepts the designed catalyst containing Ga_2O_3

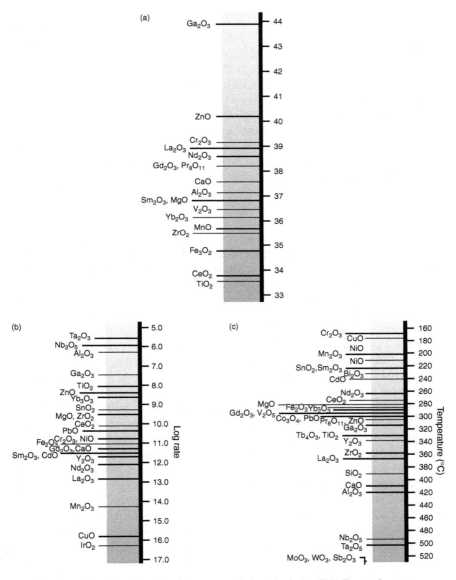

FIGURE 8.2 The virtual mechanism approach based on (a) CH_4/D_2 surface area normalized exchange rates at 773 K ($CH_4/D_2 = 1:1$, 700 h^{-1}). (b) Relative ranking order for the exchange of $^{18}O_2/^{16}O_2$ with oxide surfaces normalized at 773 K. (c) Ranking order for methanol stability based on the temperature at which 30% of the methanol feed was converted to carbon oxides. The same rank order found if different methanol conversion levels were chosen (From Hutchings, G.J. and Scurrell, M.S. *CATTECH* 2003, *7*, 90–103.)

and MoO_3 ($^{16}O_2/^{18}O_2$ exchange is very rapid on MoO_3 by a different mechanism, hence not included in this figure) exhibited significantly, enhanced yield of methanol. This method was extended to design total oxidation catalysts [42] and the selective oxidation of ethane to acetic acid [43]. In complex oxide catalyst systems, synergistic combination exists in which one component is principally responsible for the hydrocarbon activation and the other for oxygen activation/insertion. The currently accepted view of the mode of operation of the bismuth molybdate (SOHIO) catalyst for propene oxidation and ammoxidation is that the bismuth component is responsible for hydrocarbon activation while the molybdenum component effects oxygen activation and insertion [12,44]. This clearly signifies the role of different metal oxide components in a complex catalyst system.

Another extensively used method is temperature programmed reduction/oxidation (TPR/TPO) of oxides with different reductants such as H_2 and CO and oxidants such as O_2. For example, the redox property of oxide materials of third period transition metals was studied systematically by TPR/TPO method, recently [45]. The activation energies of TPR/TPO exhibited in the range of 33 to 149 kJ/mol, while for the oxides of V, Mo, and W, showed relatively higher values. The change of activation energies of TPR/TPO with various metal oxides exhibited a similar trend to the change of their metal–oxygen bond strengths. The change of activation energies of o-xylene oxidation for various metal oxides was proportional to the difference (ΔE_a) between the activation energy of TPR and TPO. From these results, it was concluded that the oxidation of o-xylene over various metal oxide catalysts follows the Mars–van Krevelen mechanism including the surface reduction and oxidation of the metal oxide itself. Several interesting publications can be found in the literature on the use of TPR/TPO techniques to understand the redox behavior of various oxides employed for different catalytic purposes [45–55].

One of the best methods for determining oxygen transport in electronically conducting materials like perovskites is the solid-state potentiostatic technique, which was recently used and critically analyzed by Sunde et al. [56]. These authors investigated the conduction of $SrFe_{1-x}Fe_xO_{3-\delta}$ and found that in this material there are high-diffusivity paths in the sample. The overall transport was found to be controlled by an oxygen exchange reaction at the grain boundaries. This complication prevented the determination of the oxygen diffusion coefficient for these materials. Yokokawa et al. [57] discussed the thermodynamic stabilities of perovskites, including the energies of formation of point defects in perovskites and the effect of dopants thereon. They observed that oxygen vacancies are more easily formed upon doping with strontium and barium than calcium. Depending on the desired direction of the formation of defects, this gives a guideline for the choice of dopant.

8.6 REDOX CATALYSTS — CASE STUDIES

Most of the redox processes occurring during oxidation (both selective and nonselective), ammoxidation, and oxidative dehydrogenation reactions can be explained by the well-known Mars and van Krevelen mechanism [7]. In the following sections a few redox processes occurring on some of the extensively studied metal oxides will be discussed in detail. Basically two types of surface reactions are possible over the metal oxide surfaces; acid–base reactions and oxidation–reduction reactions; both are interdependent and one has to invariably involve the latter to describe the former and vice versa. Nevertheless, the two broad reaction classes noted earlier provide a framework for comparing the behavior of widely different oxide materials and sometimes of apparently unrelated reactants.

8.6.1 Vanadium Oxide-Based Catalysts

Vanadium oxide catalysts are complex inorganic materials that play an important role in heterogeneous redox catalysis in both the gas and the liquid phases. Pure vanadium pentoxide is active in most of the redox reactions though usually the selectivity to the desired product is low and mostly favors complete oxidation products. Addition of other elements is known to improve or alter the redox abilities. Generally, addition of more acidic oxide (with respect to V_2O_5) components such as Mo, P, etc., to vanadia are known to yield acidic products, whereas the ODH (oxidative dehydrogenation) reactions proceed on systems in which other elements, beside vanadium (Mg, alkaline earth) form basic oxides [58]. The formal valence state of vanadium in different vanadia-based systems varies between V^{5+} (V_2O_5, $VOPO_4$, magnesium vanadate, VO_4 tetrahedra, and VO_6 octahedra on MgO, hydrotalcite, sepiolite, and Al_2O_3) and V^{4+} (vanadyl pyrophosphate, solid solution of MoO_3 on V_2O_5) and the mixture of V^{3+} and V^{4+} in vanadium antimonate. In the stationary state of oxidation reactions, all catalysts contain both oxidized and reduced vanadium cations, the ratio depending on the reaction conditions and the nature of hydrocarbon, which is oxidized. The V^{5+}–V^{4+} couple plays a key role in all the redox processes catalyzed by vanadia-based catalysts [58].

The V–Mo–O oxides are well-known industrial catalysts for the synthesis of acrylic acid from acrolein and maleic anhydride from benzene; more recently, V–P–O systems are being utilized for maleic anhydride production from n-butane. The V_2O_5/TiO_2 combination was employed for phthalic acid production from o-xylene. V–Fe–O catalyzes oxidation of polycyclic aromatic hydrocarbons to dicarboxylic acids and quinones. Methyl formate is produced by the oxidation of methanol over V–Ti–O catalysts [58]. For many of these processes, it has been experimentally proved that the catalytic reaction follows a Mars–van Krevelen mechanism. The surface coverage with active oxygen θ in the steady state of the redox reaction following Mars–van Krevelen mechanism is given by

$$\theta = \frac{1}{(k_r P_H / k_0 P_{O_2}) + 1}$$

where P_H and P_{O_2} are the partial pressures of hydrocarbons and oxygen, respectively, and k_r and k_0 are the rate constants of the reduction and reoxidation steps, respectively [58]. Thus, a redox catalytic cycle invariably involves changes of the vanadium valence state and the coordination geometry as well as the structure of the reactant molecule, which can only be visualized by using different *in situ* methods.

Among various vanadia-based catalysts, the vanadium phosphorus oxides (VPOs) have been proved to be excellent catalysts for selective O- and N-insertion reactions of aliphatic and methylaromatics, in particular for the oxidation of *n*-butane to maleic anhydride and the ammoxidation of methylaromatics and heteroaromatics to their corresponding aldehydes and nitriles [12,44,59–65]. Various VPO precursors of different structure and vanadium valence state were studied in recent years with an aim to elucidate their reaction behavior and to improve their catalytic performance in the earlier mentioned processes. Regarding the nature of active phase in these catalysts and the way its structure influences the catalytic activity and selectivity of these catalysts, comprehensive investigations were made by direct observation of the catalysts under reaction conditions with the help of various spectroscopic *in situ* methods. More recently, a comprehensive picture of structure–reactivity relationship for the industrially important ammoxidation of toluene to benzonitrile was demonstrated experimentally for VPO catalysts [5]:

$$C_6H_5\text{--}CH_3 + NH_3 + 1.5H_2O \rightarrow C_6H_5\text{--}CN + 3H_2O$$

The integrated evaluation of the EPR (electron paramagnetic resonance) results [59,66,67] combined with information from other *in situ* experiments using FT-IR spectroscopy [68,69], laser Raman spectroscopy [70], XRD [59,71], TAP [72], UV-VIS DRS [59], and XPS [59] allowed the authors to derive a possible scheme for the ammoxidation reaction as shown in Scheme 8.2. In the first step, oxidative hydrogen abstraction converts the toluene molecule adsorbed at a Lewis site of the surface into a methylene intermediate and reduces the neighboring vanadyl ion. In due course, the lattice is reoxidized by oxygen from the gas phase and water is eliminated, leaving behind adsorbed benzaldehyde. Accompanied by further changes of the valance-state of the interacting vanadium, this intermediate reacts with a NH_4^+ ion from the catalyst structure, probably via benzylimine to benzonitrile, which desorbs from the catalyst surface. From this scheme, the reaction is highly evident in the presence of NH_3, air, and water vapor. The electrical conductivity measurements indicated that such redox processes do not only involve V sites in superficial surface positions but also penetrate within some depth into the bulk of the catalyst [73]. This is understandable by taking into account the fact that lattice oxide ions at elevated temperatures become highly mobile and can diffuse between bulk and surface. This is a general phenomenon observed for many transition-metal oxides that catalyze selective oxidation and related processes via a Mars–van Krevelen redox cycle. In fact, it is necessary for guaranteeing selectivity since nonselective electrophilic O species adsorbed on the surface are converted by this diffusion into nucleophilic lattice oxygen species. It is clear that oxygen transport proceeding via anionic diffusion must be related to changes of electron

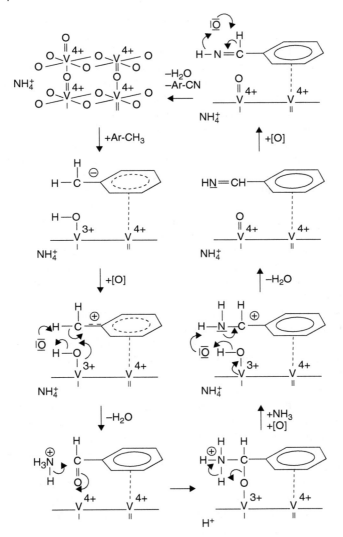

SCHEME 8.2 Reaction mechanism for the ammoxidation of toluene over VPO catalysts (From Brückner, A., *Catal. Rev.* 2003, *45*, 97–150. With permission.)

density at the corresponding vanadium sites. This, in turn, influences the spin–spin exchange behavior of the latter and was clearly observed by EPR technique [5].

The direct oxidation of propane to acrylic acid using molecular oxygen as an oxidant has recently attracted a lot of attention in both academia and industry, both for fundamental reasons, which concern the understanding of alkane activation and oxygen insertion, and for economical reasons, such as the high abundance of propane in natural gas and its lower price than propene. Till date, the industrial production of acrylic acid involves a two-step process, which consists of propene oxidation to acrolein over multicomponent Mo–Bi–Co–Fe-based oxide catalysts followed by the oxidation of acrolein to acrylic acid over Mo–V-based oxide catalysts [74]. No catalyst system has been reported for the direct oxidation of propane that is active and selective enough to substitute the existing industrial process; this is mainly due to the higher reaction temperatures required for the activation of the paraffin, which results in the enhancement of total oxidation reactions. The oxidation of propane takes place via several different reaction pathways, leading to the formation of many partial oxidation products such as propene, acrolein, acetone, acrylic, propionic, and acetic acids as well as carbon oxides and water [75,76]. The reaction is generally believed to proceed via the Mars–van Krevelen mechanism with the incorporation of lattice oxygen to form the aforementioned products followed by catalyst reoxidation by molecular oxygen. The oxidation of propane to acrylic acid requires four lattice oxygen atoms, abstraction of four hydrogen atoms from the substrate, insertion of two oxygen atoms from the lattice, and transfer of eight electrons, that is, a coordinated action of the active sites as well as balanced redox properties of the catalyst to complete the catalytic cycle. It is an established fact in the literature that the reaction network and product distribution are very sensitive toward the catalyst used [77]. Significant differences in the reaction pathways were observed when the reaction was carried out over V–P–O [78], Te–P/NiMoO [79], and Mo–V–Te–Nb–O catalysts [80]. More recently, a detailed kinetics of propane oxidation over Mo–V–Sb–Nb mixed-oxide catalyst was also envisaged [81].

In the case of supported vanadia-based systems, the nature of substrate support is known to have a huge influence on the redox properties of the resultant catalytic system. According to Grzybowska-Swierkosz, the nature of V–O–Me^{n+} (Me^{n+} = support) bonds effects the electronic cloud densities around vanadium and oxygen atoms, which is reflected in the degree of ionicity of the V–O bond, that is, nucleophilicity of oxygen and electrophilicity of vanadium ions [58]. These latter parameters are related to the acido–basic properties and the redox ability (reducibility) of the system. As in the case of V–P–O systems, the higher electronegativity of the P renders oxygen atom in the V–O–Me^{n+} unit less nucleophilic (less basic) and vanadium atom more electrophilic (acidic), thereby exhibiting high acidity and low basicity. The inverse is true for the V–Mg–O systems [58].

The oxidation of CH_3OH to $HCHO$ is considered as a probe reaction for other selective oxidation reactions such as butane to maleic anhydride, o-xylene to phthalic anhydride, and ODH of alkanes to alkenes. Consequently, the concepts developed for the selective oxidation of methanol over vanadium oxide catalysts can be easily transferred to other catalytic reactions. Weckhuysen and Keller [82] carried out methanol oxidation as a probe reaction over various V_2O_5/S oxides (S = TiO_2, ZrO_2, Nb_2O_5, CeO_2, and Al_2O_3). The relative independence of turnover frequency (TOF) to vanadia loading on amorphous oxide supports indicated that the reaction was first order with respect to surface vanadium oxide site, that is, catalytically active site. Such scheme is a typical example of the Mars–van Krevelen redox mechanism. *In situ* Raman studies using isotopic oxygen-18 have established beyond doubt that the V–O–S bonds form the active sites and V–O–V and V=O bonds do not play any substantial role in the oxidation of methanol to formaldehyde. Thus, establishing the metal oxide support effect in heterogeneous catalysis and by a careful choice of the support oxide one can tune the catalyst redox abilities [82].

A multitechnique approach comprising *in situ* investigation by UV/VIS-DRS, FT-IR, and laser-Raman spectroscopy was used by Wachs and coworkers to study supported VO_x/metal oxide catalysts (metal oxides: TiO_2, ZrO_2, CeO_2, Nb_2O_5, Al_2O_3, and SiO_2) during the selective oxidation of methanol and to derive a conclusive description of the redox behavior of active V sites in these systems [83]. Catalytic tests revealed very different TOFs in methanol oxidation depending on the support material and this has been attributed to the different nature of V–O-support bonds. By *in situ* investigations these authors wanted to differentiate whether the intrinsic activity of individual vanadium sites or the number of V–O-support bonds (i.e., the percentage of V^{5+} reduction under study-state conditions) are responsible for these differences. Isopropanol chemisorption on the surface of various pure metal oxides at 383 K resulted in the formation of surface isopropoxide species. The surface species either underwent dehydrogenation or loss of a hydroxyl group to yield acetone or propylene, respectively. The selectivity of various oxide catalysts to different products, under differential conversion conditions (<10% conversion), reflects the nature of the active surface sites, with redox sites yielding acetone and acidic sites yielding propylene. Surface isopropoxide species is the common reaction intermediate for isopropanol oxidation and isopropanol chemisorption. Thus, the number of chemisorbed surface isopropoxide species at 383 K used to determine the TOF for acidic catalysts varied by eight-orders of magnitude (10^1–10^{-7} sec^{-1}) and TOFs for redox catalysts varied by six-orders of magnitude (10^2–10^4 sec^{-1}). There is a weak inverse correlation between the variation of TOF (redox) and the bulk heats of formation of metal oxides per oxygen atom at low $-\Delta H_f$. There is no correlation between TOFs and TPR-H_2 reduction onset temperatures. Variations of the TOFs (redox) with the surface isopropoxide intermediate decomposition temperatures revealed an inverse trend for the least stable surface intermediates and there is very little variation in TOFs (redox) with the decomposition temperature for the moderately and less reactive surface intermediates. The selectivity of the metal oxide catalysts is independent of the TOFs and is an inherent property of the nature of the active surface site.

The application of vanadium oxide catalysts in partial oxidation reactions is widely practiced. Several studies were carried out to investigate the promotional effect of alkali compounds or basic oxides to V_2O_5 on the catalytic performance in different reactions [58,84–88]. Generally, the promotion of V_2O_5 bulk and supported catalysts with alkali compounds improves the selectivity to partial oxidation products. The doping effect of mainly potassium compounds has been explained: (i) by lowering of the acid sites, and/or (ii) by changing the redox properties of the catalysts. Furthermore, the addition of potassium influences the surface concentration of electrophilic oxygen, which is responsible for deep oxidation, and the concentration of nucleophilic oxygen, responsible for the partial oxidation [87]. Recent studies of the partial oxidation of substituted toluenes to their corresponding aldehydes on alkali metal-containing vanadia catalysts have shown that toluene conversion and aldehyde selectivity change with alkali cation size [83]. Increasing cation size from Li to Cs causes declining toluene conversion but increasing aldehyde selectivity. Further investigations of these catalysts revealed [89] that the onset temperatures of the reduction decrease in the order $V_2O_5 > Li-V_2O_5 > K-V_2O_5 \approx Cs-V_2O_5$. The ability to form crystalline mixed valence bronze-like phases $M_xV_2O_5$ ($x = 0.2–0.5$) beside V_2O_5 increases in the same order. It was assumed that the bronze phases stabilize V^{4+} oxidation-state and improve the redox properties of the catalysts.

Alkali-promoted V_2O_5 catalysts essentially exhibit higher conversion and higher paramethoxybenzaldehyde selectivity than vanadyl pyrophosphate catalysts which is caused by their lower surface acidity and better redox properties indicated by the formation of mixed-valence alkali vanadia phases during reaction [88]. Many groups reported on the role of acid–base properties of vanadium oxide catalysts containing base metal oxides (M = K, Rb, Cs, Tl, Ag) for the partial oxidation of methyl aromatics to their corresponding aldehydes [90–95]. These studies revealed that the selectivity to substituted benzaldehydes is closely related to the basic properties of the catalyst whereas activity strongly depends on the amount and strength of acid sites. Additionally, pyridinium cations generated during reaction are able to incorporate into the structure of the alkali vanadia bronze phases. These mixed-valence bronze phases stabilize V^{4+} oxidation-state and improve the redox properties. The potassium-promoted vanadia catalyst revealed the best catalytic results, obviously caused by the efficient combination of low surface acidity and high reducibility realized by the structural variability of the potassium vanadia bronzes. These findings agree with the results of oxidation of methyl aromatics to their corresponding benzaldehydes on the rather acidic $(VO)_2P_2O_7$ catalysts [69,88,96]. It was found that Brønsted acid OH-groups formed during the reaction by interaction of V–O–P bonds with water are responsible for a restricted aldehyde desorption, which leads to poor aldehyde selectivities by consecutive over-oxidation. The addition of the nonreactive base pyridine to the feed significantly improves the aldehyde selectivity by an effective blockade of these acidic sites [96].

In contrast to the above discussion, vanadium oxide is also known to play a crucial role in the deactivation of ultra stable Y (USY) zeolite in fluid catalytic cracking (FCC) processes. Vanadium is known to enhance the removal of sodium ions from zeolite Y, thereby acting as a promoter/catalyst. These Na^+ ions released are known to attack and destroy the Si–O–Si frame in the zeolite there by deactivating it. The source of vanadium in FCC processes comes from crude oil (feed) which contain V-porphyrin and organo-vanadium compounds. The released vanadium migrates to the surface of zeolite. In this process, vanadium is known to fluctuate between V^{4+}/V^{5+} oxidation states there by making it a redox process [82].

In a recent article Grzybowska-Swierkosz [58] compared the relative efficiencies of MoO_3 and V_2O_5 based catalysts and attributed the variance in their catalytic activities toward a given organic substrate with their inherent redox properties. The lower reducibility of the MoO_3 as compared with V_2O_5; the lower Taman temperatures, that is, the mobility of the lattice of V_2O_5; the difference in the redox potentials of V^{5+}/V^{4+} and Mo^{6+}/Mo^{5+} couples are proposed to be responsible for the variance in their catalytic activities, which in turn are related to their inherent redox properties. During the course of a redox process, the fate of an organic substrate depends on two factors namely (i) the number of oxygen atoms in their vicinity and the ease of the extraction of these atoms from the catalyst, and (ii) rate of diffusion of oxygen in the catalyst lattice relative to the residence time of organic species on the catalyst surface. If the number of easily removable oxygen ions around an adsorbed organic species is too high, the over-oxidation with formation of carbon oxides is possible. If mobility of the oxygen ions is high, it will help to replenish the oxygen vacancies formed on insertion of oxygen into an organic species and a long residence time of the latter species would lead to products of higher number of oxygen, and eventually to the formation of carbon oxides [58].

8.6.2 Cerium Oxide-Based Catalysts

The active form of cerium (rare earth) is CeO_2, it is easily formed when cerium salts are calcined in air or oxygen containing environments [97–107]. Ce_2O_3 with trivalent Ce(III) can be prepared in strongly reducing conditions but is unstable in air and readily converts to cerium dioxide. Cerium has one of the highest free energies of formation for an oxide [108]. Gandhi et al. [109] reported the role of ceria in the automotive exhaust TWC processes. The addition of ceria promotes dynamic performance in purifying CO, NO_x, and HC under conditions of rich and lean air-to-fuel ratios in automotive exhaust, which is referred in terms of oxygen storage/release capacity (OSC/ORC). The OSC/ORC is the ability of CeO_2 containing oxides to adsorb and release oxygen under respective fuel-lean and fuel-rich conditions [97–101]. The catalyst formulation for automotive exhaust treatment consists of finely divided ceria dispersed either on the surface of alumina pellets or on an alumina wash-coat anchored to a monolithic ceramic substrate [99]. The feature of ceria to form a continuum of O-deficient nonstoichiometric compositions of the type CeO_{2-x} ($0 < x < 0.178$) under oxygen-lean high-temperature environments with the retention of fluorite crystal structure is highly unique and crucial to its utilization in redox catalysis [110]. If the atmospheric environment in which ceria operates changes continuously from a net oxidizing to a net reducing composition, the cerium cation shifts alternately from a 4+ to a 3+ oxidation state [99].

$$4CeO_2 \leftrightarrow 2Ce_2O_3 + O_2$$

$$(Ce^{4+}) \leftrightarrow (Ce^{3+})$$

The textural stability of the ceria alone systems is not high enough to meet the requirements of high-temperature gas-phase catalytic reactions. Therefore, recently the mono dispersed ceria is replaced by $CeZrO_4$ solid solution over the surface of Al_2O_3 wash-coat in the monolithic catalysts [111]. Ozawa et al. [112,113] investigated the OSC characteristics of various catalytic ceria-based mixed oxide systems by thermogravimetry (TG) technique under cyclic heat treatment in flowing air. The recovery of weight during cooling cycle and loss of weight during second heating cycle corresponded to the potential oxygen release capacity of the corresponding materials. Using these experiments they showed that OSC of $CeZrO_2/Al_2O_3$ system is seven times more than that of CeO_2/Al_2O_3 at a given temperature. Thus, proving evidence that the addition of ZrO_2 is able to increase the amount of active/mobile oxygen content in the ceria lattice and/or lowering the redox potential (to practical reducing state versus temperature) of Ce ions [112,113].

CeO_2 is well known for its large deviation from stoichiometry at low oxygen partial pressures and temperatures above 773 K. As already emphasized, ceria doped with other valent cations is reported to be an excellent catalyst for direct oxidation of SO_2 by CO [101]. The Cu–Ce–O catalyst system exhibits complete conversion of SO_2 and >95% selectivity to elemental sulfur [114]. The same system also exhibits high activity for methane oxidation and low temperature CO oxidation [115]. Manganese–ceria composite oxides are being widely used in sub- and super critical catalytic wet oxidation for the treatment of wastewater containing toxic organic pollutants [116]. Very recently, Mn–Ce mixed oxide based redox catalysts have also been employed for low temperature NO_x abatement processes, which exhibited a 100% NO conversion at 273 to 323 K [117]. The performance of such catalysts depends on the redox processes involving CeO_{2-x} and MnO_x oxides. Recently, catalytic oxidation of n-hexane, benzene, and 2-propanol was investigated over $Au/CeO_2/Al_2O_3$ catalysts. Catalytic results show that ceria improves the activity of gold particles in the oxidation of the tested volatile organic compounds (VOCs), probably by increasing the mobility of the lattice oxygen and controlling and maintaining the adequate oxidation state of the active gold particles [118].

Tschope et al. [101] have synthesized ultrafine nonstoichiometric cerium oxide-based catalysts by inert gas condensation method [101]. Nonstoichiometric metal oxides can be produced by first synthesizing metallic nanoclusters, followed by controlled oxidation. These nonstoichiometric CeO_2-based catalysts, along with precipitated stoichiometric samples, were investigated for SO_2 reduction by CO to elemental sulfur, CO oxidation, and total oxidation of methane. In general, the nonstoichiometric CeO_2-based materials exhibited better catalytic activity, their light off temperatures for various reactions were 273 to 453 K lower than the respective precipitated materials. The presence of oxygen vacancies in abundance and the availability of quasi-free electrons in the nonstoichiometric materials were anticipated for the enhanced catalytic activity [101]. These examples reveal an intricate relationship between the redox and defect chemistry of a given metal oxide system. The defect chemistry of bulk ceria is well-established in the literature [119]. A partial substitution of Ce in CeO_2 with isovalent Hf lead to an increase in the overall activity of methane combustion and the presence of defective fluorite structured oxide is recognized as the key factor in the activity enhancement [120]. The defective lattice is known to yield high-oxygen vacancies and high-ionic mobility in turn leading to enhanced redox ability and thus enhanced activity. According to Cho [121], two types of oxygen vacancies can be created in ceria: intrinsic and extrinsic. Intrinsic vacancies are created on reduction of ceria according to the following process:

$$CeO_2 + \delta R \leftrightarrow CeO_{2-\delta} + \delta RO + \delta V_o$$

Extrinsic vacancies are defects that are formed by a charge-compensating mechanism when a bivalent or trivalent cation M is introduced into the CeO_2 lattice according to the following two mechanisms:

$$(1 - \alpha)CeO_2 + 0.5\alpha M_2O_3 \leftrightarrow Ce_{1-\alpha}M_\alpha O_{2-0.5\alpha} + \delta' V_o$$

$$(1 - \alpha)CeO_2 + \alpha MO \leftrightarrow Ce_{1-\alpha}M_\alpha O_{2-\alpha} + \delta'' V_o$$

The oxygen storage capacity of ceria can benefit from either type of defect. In some cases, such as with the introduction of Ca^{2+}, Mg^{2+}, Gd^{3+}, La^{3+}, and $Pr^{3,4+}$, the increased OSC is obtained chiefly through the creation of extrinsic type defects. In other cases, such as when Zr^{4+} is used as a dopant, promoting the kinetics of Ce^{4+} reduction, increases the intrinsic oxygen vacancies and enhanced oxygen storage [99].

In Ce–Zr–Ca ternary mixed oxide systems, both intrinsic and extrinsic oxygen vacancies are possible. On the one hand, distortion of oxygen sublattice, by the substitution of smaller Zr^{4+} (0.084 nm) ions into the ceria lattice, leads to an enhancement of the anionic mobility (intrinsic oxygen vacancies). On the other hand, doping with cations whose oxidation state is lower than 4+ leads to the formation of extrinsic oxygen vacancies to maintain electroneutrality. Thus, Fernandez-Garcia et al. [122]. observed that Ca introduction into Ce–Zr oxidic network strongly modifies surface and bulk oxygen handling properties. In this context, the presence of Ca^{2+} was shown to maximize the OSC of ceria with a 40% increment.

An adequate synthesis methodology is a fundamental starting point for developing any viable/durable oxide catalytic material. The textural stability and structural properties of the resultant material plays an important role on the redox properties [106]. There are several reports in the literature that highlight the intricate relationship between the method of preparation and resultant redox properties [101,123–138]. Influence of the particle size on the redox properties of resultant supported ceria-based mixed oxide has also been attempted in the literature [139–142]. The formation of nanoparticles or nanostructures over the surface of supports is known to enhance the catalytic redox ability. Ceria has been used as a solid electrolyte in fuel cells. The larger lattice parameter in nanocrystalline ceria implies the migration enthalpy of an oxygen vacancy is smaller, resulting in a higher ionic conductivity and fuel cells that are more efficient. The systematic increase in lattice parameter with decreasing particle size provides specific control of the surface oxygen spacing and will affect ceria catalysis of the water gas shift and three-way auto exhaust clean up reactions [141,142].

Our work on various mixed oxides of ceria (CeO_2–MO_2, M = Si^{4+}, Ti^{4+}, Zr^{4+}) as investigated by XRD, Raman, and XPS provided information about the stabilization of ceria cubic lattice, particle size, and phase purity as a function of temperature [143–146]. The XPS measurements, in particular, revealed that the presence of zirconium affects the average oxidation state of the cerium. The addition of V_2O_5 to CeO_2 was also reported to show a strong influence on the redox properties of CeO_2 as investigated by TPR and ESR techniques [147]. At low vanadium contents polymeric V–O–V chains are stabilized on the ceria surface and at higher contents formation of $CeVO_4$ and V_2O_5 phases were observed. The V–O–V chains and V_2O_5 species are more easily reducible than $CeVO_4$ [148]. The CeO_2–TiO_2 system under suitable treatments gave way to a wealth of mixed oxides with improved redox properties for various catalytic purposes [135]. Rocchini et al. [149,150] have investigated the ceria–silica combination (coprecipitation) and attributed the enhanced textural and thermal stability of these mixed oxides due to the formation of an intermediate $Ce_{9.33}(SiO_4)_6O_2$ (characterized by TEM and XRD) phase between ceria and silica which on suitable treatment decomposes into amorphous silica and smaller crystallites of ceria.

8.6.3 Redox Behavior of Mixed Oxides

The substitutions at A or B site is known to influence the performance of perovskite-type oxides (ABO_3) for catalytic reactions, such as those involved in the pollution abatement [151–154]. Similarly, the catalytic properties of oxygen-rich orthovanadates, which have a sheelite-type structure like that of the molybdates have received some attention recently [155]. The rare earth orthovanadate, $LaVO_4$ exhibited a very limited activity for CO oxidation reaction, however, the substitution of V^{5+} by Mn^{4+} led to a remarkable improvement not only in the catalytic activity but also in its structural stability over repetitive redox cycles [155]. Oxygen nonstoichiometry generated in this system on introduction of Mn^{4+} was found to be responsible for this improvement. The substitution of vanadia in $LaVO_4$ by Fe^{3+} gave rise to mixed phases of $LaVO_4$, $LaV(Fe)O_{4-\delta}$, and $LaFeO_3$ in varying concentrations. The high-catalytic activity and stability of substituted samples over repeated redox cycles were attributed to a synergy between these phases along with the nonstoichiometry generated because of iron substitution [156].

The selective catalytic reduction (SCR) of NO by CO over supported noble metal (Pt, Pd, or Rh) catalysts is an active area of interest because of the high prices and scarcity of noble metals. Therefore, it is an important task to find alternative catalytic components to reduce even replace the noble metals. Recent research reveals that the transition-metal oxides have high catalytic activity for reduction of NO by CO [154–162]. Copper oxide has been demonstrated to be an active species among the various transition-metal oxides for this reaction [158,163–166].

In particular the $CuO/Ce_{0.8}Zr_{0.2}O_2$ combination catalyst exhibited very good activity for the reduction of NO by CO. The structural modification and imperfection of the CeO_2 lattice by insertion of copper cations causes the enhancement of redox properties and the resulting activity [166].

Among the mixed oxides of interest for oxidation catalysis, MnO_x-based materials exhibited great potential. These oxides have been identified as active phases in several processes, such as oxidative coupling of methane [167], CO oxidation [168], ethyl-benzene oxidative dehydrogenation [168,169], as well as other classes of catalytic oxidation processes [170–176]. In general, MnO_x are compounds with a typical berthollide structure that contain labile lattice oxygen. Their catalytic properties are attributed to the capacity for manganese to form oxides with variable oxidation states (MnO_2, Mn_2O_3, Mn_3O_4, or MnO), and to their OSC in the crystalline lattice. Due to its labile oxidation state, Mn is capable of playing the role of either a reducing agent ($Mn^{2+} - e^- = Mn^{3+} - e^- = Mn^{4+}$) or an oxidizing agent ($Mn^{4+} + e^- = Mn^{3+} + e^- = Mn^{2+}$), in both cases acting as an active component of the redox system. Structural characterization of unsupported MnO_x catalysts identified the presence of MnO_x or mixed MnO_2/Mn_2O_3 as the active catalyst component [168,169,173,174]. A combination of MnO_x with other oxides deposited on a high-surface area support and used as catalysts in oxidation processes exhibits different catalytic activity as compared with the support as well as other components present in the catalyst, which also significantly influences its bulk and surface structure [168,174–178]. Recent studies reveal that when MnO_2 is placed in the proximity of La_2O_3 or CeO_2, the oxygen mobility from the MnO_x structure is strongly affected [168,175,176]. In a Mn–Ce mixed oxide, Ce provides oxygen to Mn at low temperature and, in contrast, withdraws oxygen at elevated temperatures (>773 K). Thus, cerium provides the activity of MnO_x in oxidation processes at lower temperature and decreases its activity at high temperatures [176,178]. The activity results reveal that Mn^{3+} (as Mn_2O_3) can provide effective adsorption sites for CO and supply oxygen from its oxide structure for oxidation, leading to Mn^{2+} (as MnO) formation. The catalytic results also indicate that labile lattice oxygen from the yttrium-stabilized zirconia plays an important role in the oxidation process and the CO is being adsorbed on the active sites (formation of a Zr^{4+}–CO bond).

High oxidation state transition-metal oxide ions isolated and sparsely distributed within the Al^{3+} sublattice of open-structure metal microporous aluminophosphate (MAlPOs) solids ($M = Co^{3+}$, Mn^{3+}, Fe^{3+}) function as powerful redox, catalytically active centers in the selective oxyfunctionalization of alkanes. Important chemical commodities are also conveniently prepared by using such microporous catalysts in solvent free conditions, and using oxygen or air as oxidants [179,180].

8.7 CONCLUSIONS

The reducibility and reoxidizability (redox property) of the metal oxide catalyst, electron and lattice oxygen mobilities within the bulk or its upper surface layers, surface acid–base properties, mode of activation of hydrocarbon and molecular oxygen, desorption rates of the products, structural state of the surface layers, etc., are all important parameters which control the reaction mechanism and catalytic activity and selectivity. In general, all these parameters are important and interconnected and should be considered while designing a new generation metal oxide catalyst or modifying the existing ones. In particular, the redox behavior of a metal oxide under consideration and the influence of various additives on its redox properties are essential factors for the formulation of a catalyst for oxidation and related reactions. Finally, the term "redox" plays and continues to play a significant role in the area of catalysis.

ACKNOWLEDGMENTS

I would like to thank Mr. A. Khan for his extensive help in organizing this chapter. A major portion of this chapter was written during my sabbatical at Institut de Recherches sur la Catalyse — CNRS, 69626 Villeurbanne, France.

REFERENCES

1. Kung, H.H. *Transition Metal Oxides: Surface Chemistry and Catalysis.* Studies in Surface Science Catalysis; Elsevier: Amsterdam, 1989; Vol. 45, pp. 1–277.

2. Henrich, V.E. and Cox, P.A. *The Surface Science of Metal Oxides*, Cambridge University Press: Cambridge, UK, 1994.

3. Noguera, C. *Physics and Chemistry at Oxide Surface*; Cambridge University Press: Cambridge, U.K., 1996.

4. Zecchina, A., Scarano, D., Bordiga, S., Spoto, D., and Lamberti, C. Surface structures of oxides and halides and their relationships to catalytic properties. *Adv. Catal.* 2001, *46*, 265–397.

5. Brückner, A. Looking on heterogeneous catalytic systems from different perspectives: multitechnique approach as a new challenge for in situ studies. *Catal. Rev.* 2003, *45*, 97–150; and references therein.

6. Védrine, J.C. The role of redox, acid–base and collective properties and of crystalline state of heterogeneous catalysts in the selective oxidation of hydrocarbons. *Top. Catal.* 2002, *21*, 97–106; and references therein.

7. Mars, P. and van Krevelen, D.W. Oxidations carried out by means of vanadium oxide catalysts. *Chem. Eng. Sci.* 1954, *3*, 41–59.

8. Bielanski, A. and Haber, J. *Oxygen in Catalysis*; Chemical Industries Series; Marcel Dekker: New York, 1991; Vol. 43, pp. 1–448.

9. Haber, J. and Grzybowska, B. Mechanism of the oxidation of olefins on mixed oxide catalysts. *J. Catal.* 1973, *28*, 489–505.

10. Grzybowska, B., Haber, J., and Janas, J. Interaction of allyl iodide with molybdate catalysts for the selective oxidation of hydrocarbons. *J. Catal.* 1977, *49*, 150–163.

11. Grasselli, R.K. and Burrington, J.D. Selective oxidation and ammoxidation of propylene by heterogeneous catalysis. *Adv. Catal.* 1981, *30*, 133–163.

12. Grasselli, R.K. Genesis of site isolation and phase cooperation in selective oxidation catalysis. *Top. Catal.* 2001, *15*, 93–101.

13. Védrine, J.C., Millet, J.M.M., and Volta, J.C. Molecular description of active sites in oxidation reactions, acid–base and redox properties, and role of water. *Catal. Today* 1996, *32*, 115–123.

14. Védrine, J.C., Coudurier, G., and Millet, J.M.M. Molecular design of active sites in partial oxidation reactions on metallic oxides. *Catal. Today* 1997, *33*, 3–13.

15. Mamedov, E.A. and Cortes Corberan, V. Oxidative dehydrogenation of lower alkanes on vanadium oxide-based cattalysts: the present state of art and outlooks. *Appl. Catal. A: Gen.* 1995, *127*, 1–40.

16. Grasselli, R.K., Suresh, D.D., and Knox, K. Crystalline structures of USb_3O_{10} and $USbO_5$ in acrylonitrile catalysts. *J. Catal.* 1970, *18*, 356–358.

17. Grasselli, R.K. and Suresh, D.D. Aspects of structure and activity in U-Sb-oxide acrylonitrile catalysts. *J. Catal.* 1972, *25*, 273–291.

18. Tanabe, K. and Hölderich, W.F. Industrial application of solid acid-base catalysts. *Appl. Catal. A: Gen.* 1999, *181*, 399–434.

19. Knözinger, H. Infrared spectroscopy for the characterization of surface acidity and basicity. In *Handbook of Heterogeneous Catalysis*; Ertl, G., Knözinger, H., and Weitkamp, J., Eds., Wiley-VCH: Weinheim, 1997; Vol. 2, pp. 707–732.

20. Hall, W.K. Acidity and basicity. In *Handbook of Heterogeneous Catalysis*; Ertl, G., Knözinger, H., and Weitkamp, J. Eds., Wiley-VCH: Weinheim, 1997; Vol. 2, pp. 689–692.

21. Zecchina, A., Lamberti, C., and Bordiga, S. Surface acidity and basicity: general concepts. *Catal. Today* 1998, *41*, 169–177.

22. Bielanski, A. and Haber, J. *Catal. Rev.-Sci. Eng.* 1979, *19*, 1–41.

23. Gellings, P.J. and Bouwmeester, H.J.M. Solid state aspects of oxidation catalysts. *Catal. Today* 2000, *58*, 1–53.

24. Barbaux, Y., Elamrani, A., and Bonnelle, J.-P. Catalytic oxidation of methane on MoO_3–SiO_2: mechanism of oxidation with O_2 and N_2O studied by surface potential measurements. *Catal. Today* 1987, *1*, 147–156.

25. Mizusaki, J., Yoshihiro, M., Yamauchi, S., and Fueki, K. Thermodynamic quantities and defect equilibrium in the perovskite-type oxide solid solution $La_{1-x}Sr_xFeO_{3-\delta}$. *J. Solid State Chem.* 1987, *67*, 1–8.

26. Lankhorst, M.H.R., Bouwmeester, H.J.M., and Verweij, H. Thermodynamics and transport of ionic and electronic defects in crystalline oxides. *J. Am. Ceram. Soc.* 1997, *80*, 2175–2198.

27. Navrotsky, A. *Physics and Chemistry of Earth Materials*; Cambridge University Press: Cambridge, U.K., 1994; pp. 1–417.

28. Rørmark, L., Mørch, A.B., Wiik, K., Stølen, S., and Grande, T. Enthalpies of oxidation of $CaMnO_{3-\delta}$, $CaMnO_{4-\delta}$ and $SrMnO_{3-\delta}$ deduced redox properties. *Chem. Mater.* 2001, *13*, 4005–4013.

29. Bakken, E., Norby, T., and Støle, S. Redox energies of perovskite related oxides. *J. Mater. Chem.* 2002, *12*, 317–324.

30. Novakova, J. Isotopic exchange of oxygen-18 between the gaseous phase and oxide catalysts. *Catal. Rev.* 1970, *4*, 77–113.

31. Boreskov, G.K. Forms of oxygen bonds on the surface of oxidation catalysts. *Disc. Faraday Soc.* 1966, *41*, 263–276.

32. Grasselli, R.K. Selectivity and activity factors in bismuth–molybdate oxidation catalysts. *Appl. Catal.* 1985, *15*, 127–139.

33. Cunningham, J., Cullinane, D., Farell, F., and Gibson, C. Oxygen isotope equilibrium and exchange as probes for differing dioxygen interactions with pure and rhodia promoted CeO_2 and Al_2O_3. *Top. Catal.* 1999, *8*, 179–187.

34. Ueda, W., Morooka, Y., and Ikawa, T. ^{18}O tracer study of the active species of oxygen on Bi_2MoO_6 catalyst. *J. Chem. Soc., Faraday Trans. I.* 1982, *78*, 495–500.

35. Ono, T. and Ogata, N. Raman band shifts of γ-Bi_2MoO_6 and α-$Bi_2Mo_3O_{12}$ exchanged with ^{18}O tracer at active sites for reoxidation. *J. Chem. Soc., Faraday Trans. I.* 1994, *90*, 2113–2118.

36. Jehng, J.M., Deo, G., Weckhuysen, B.M., and Wachs, I.E. Effect of water vapour on the molecular structures of supported vanadium oxide catalysts at elevated temperatures. *J. Mol. Catal. A: Chem.* 1996, *110*, 41–54.

37. Gao, X., Deo, G., Wachs, I.E., and Ponec, V. The oxygen isotopic exchange reaction on vanadium oxide catalysts. *J. Catal.* 1999, *185*, 415–422.

38. Yeung, J.A., Chen, K., Bell, A.T., and Iglesia, E. Isotopic studies of methane oxidation pathways on PdO catalysts. *J. Catal.* 1999, *188*, 132–139.

39. Argyle, M.D., Chen, K., Bell, A.T., and Iglesia, E. Ethane oxidative dehydrogenation pathways on vanadium oxide catalysts. *J. Phys. Chem. B* 2002, *106*, 5421–5427.

40. Krauss, K., Drochner, A., Fehlings, M., Kunert, J., and Vogel, H. Oxygen exchange at Mo/V mixed oxides: a transient and ^{18}O isotope study under technical conditions. *J. Mol. Catal. A: Chem.* 2002, *177*, 237–245.

41. Hutchings, G.J. and Scurrell, M.S. Designing of oxidation catalysts: are we getting better. *CATTECH* 2003, *7*, 90–103.

42. Hutchings, G.J., Heneghan, C.S., Hudson, I.D., and Taylor, S.H. Uranium-oxide-based catalysts for the destruction of volatile chloro-organic compounds. *Nature* 1996, *384*, 341–343.

43. Lopez-Sanchez, J.A., Tanner, R., Collier, P., Wells, R.P.K., Rhodes, C., and Hutchings, G.J. Acetic acid stability in the presence of oxygen over vanadium phosphate catalysts: comments on the design of catalysts for the selective oxidation of ethane. *Appl. Catal. A: Gen.* 2002, *226*, 323–327.

44. Grasselli, R.K. Ammoxidation. In *Handbook of Heterogeneous Catalysis*; Ertl, G.; Knözinger, H.; and Weitkamp, J., Eds.; Wiley-VCH: Weinheim, 1997; Vol. 5, pp. 2302–2326.

45. Kim, Y-H. and Lee, H-I. Redox property of transition metal oxides in catalytic oxidation. *Kongop Hwahak* 1999, *10*, 1161–1168.

46. Ferrandon, M., Carno, J., Jaras, S., and Bjornbom, E. Total oxidation catalysts based on manganese or copper oxides and platinum or palladium: characterization. *Appl. Catal. A: Gen.* 1999, *180*, 141–151.

47. Yongnian, Y., Ruili, H., Lin, C., and Jiayu, Z. Redox behavior of trimanganese tetroxide catalysts. *Appl. Catal. A: Gen.* 1993, *101*, 233–252.

48. Shin, M.Y., Park, D.W., and Chung, J.S. Development of vanadia-based mixed oxide catalysts for selective oxidation of H_2S to sulfur. *Appl. Catal. B: Environ.* 2001, *30*, 409–419.

49. Banáres, M.A., Cardoso, J.H., Agulló-Rueda, F., Correa-Bueno, J.M., and Fierro, J.L.G. Dynamic states of V-oxide species: reducibility and performance for methane oxidation on V_2O_5/SiO_2 catalysts as a function of coverage. *Catal. Lett.* 2000, *64*, 191–196.

50. Gao, X., Bare, S.R., Fierro, J.L.G., and Wachs, I.E. Structural characteristics and reactivity/reducibility properties of dispersed and bilayered $V_2O_5/TiO_2/SiO_2$ catalysts. *J. Phys. Chem. B* 1999, *103*, 618–629.

51. Vidal, H., Kaspar, J., Pijolat, M., Colon, G., Bernal, S., Cordon, A., Perrichon, V., and Fally, F. Redox behaviour of CeO_2–ZrO_2 mixed oxides: influence of redox treatment on high surface area catalysts. *Appl. Catal. B: Environ.* 2000, *27*, 49–63.

52. Simonot, L., Garin, F., and Maire, G. A comparative study of $LaCoO_3$, Co_3O_4 and $LaCoO_3$–Co_3O_4: preparation, characterization and catalytic properties for the oxidation of CO. *Appl. Catal. B: Environ.* 1997, *11*, 167–179.

53. Brunet, S., Requieme, B., Matouba, E., Barrault, J., and Blanchard, M. Characterization by temperature-programmed reduction and by temperature-programmed oxidation (TPR-TPO) of chromium (III) oxide-based catalysts: correlation with the catalytic activity for hydrofluoro alkane synthesis. *J. Catal.* 1995, *152*, 70–74.

54. Agrell, J., Birgersson, H., Boutonnet, M., Melian-Cabrera, I., Navarro, R.M., and Fierro, J.L.G. Production of hydrogen from methanol over Cu/ZnO catalysts promoted by ZrO_2 and Al_2O_3. *J. Catal.* 2003, *219*, 389–403.

55. Cherian, M., Rao, M.S., and Deo, G. Niobium oxide as support material for the oxidative dehydrogenation of propane. *Catal. Today* 2003, *78*, 397–409.

56. Sunde, S., Nisancioglu, K., and Guer, T.M. Critical analysis of potential step data for oxygen transport in electronically conducting perovskites. *J. Electrochem. Soc.* 1996, *143*, 3497–3503.

57. Yokokawa, H., Sakai, N., Kawada, T., and Dokiya, M. Thermodynamic stabilities of perovskite oxides for electrodes and other electrochemical materials. *Solid State Ionics* 1992, *52*, 43–56.

58. Grzybowska-Swierkosz, B. Vanadia-titania catalysts for oxidation of *o*-xylene and other hydrocarbons. *Appl. Catal. A: Gen.* 1997, *157*, 263–310.

59. Brückner, A. A new approach to study the gas-phase oxidation of toluene: probing active sites in vanadia-based catalysts under working conditions. *Appl. Catal. A: Gen.* 2000, *200*, 287–297.

60. Centi, G. Preface and Vanadyl pyrophosphate — a critical review. *Catal. Today* 1993, *16*, 1–26.

61. Martin, A. and Lücke, B. Ammoxidation of methylaromatics over vanadium phosphate catalysts: effect of the site, position and electronic properties of substituents on the catalytic conversion of methylaromatics to the corresponding nitriles. *Catal. Today* 1996, *32*, 279–283.

62. Reddy, B.M. and Manohar, B. Ammoxidation of 3-picoline to nicotinonitrile on silica-supported VPO catalysts. *Chem. Ind.* 1992, 182–183.

63. Reddy, B.M., Kumar, M.V., and Manohar, B. Vanadium phosphorus oxide catalysts for ammoxidation of 3-picoline to nicotinonitrile and 2-methylpyrazine to 2-cyanopyrazine. In *Catalysis of Organic Reactions*; Scaros, M.G.; Prunier, M.L., Eds.; Marcel Dekker: New York, 1995; pp. 487–491.

64. Reddy, B.N., Reddy, B.M., and Subrahmanyam, M. Dispersion and 3-picoline ammoxidation investigation of V_2O_5/α-Al_2O_3 catalysts. *J. Chem. Soc., Faraday Trans.* 1991, *87*, 1649–1655.

65. Manohar, B. and Reddy, B.M. Ammoxidation of 3-picoline to nicotinonitrile over vanadium phosphorus oxide-based catalysts. *J. Chem. Technol. Biotechnol.* 1998, *71*, 141–146.

66. Brückner, A., Martin, A., Steinfeldt, N., Wolf, G.-U., and Lücke, B. Investigation of vanadium phosphorus oxide catalysts (VPO) during toluene ammoxidation: new mechanistic insights by in-situ EPR. *J. Chem. Soc., Faraday Trans.* 1996, *92*, 4257–4263.

67. Brückner, A., Martin, A., Kubias, B., and Lücke, B. Structure of vanadium sites in VPO catalysts and their influence on the catalytic performance in selective O- and N-insertion reactions. *J. Chem. Soc., Faraday Trans.* 1998, *94*, 2221–2225.

68. Martin, A., Bentrup, U., Brückner, A., and Lücke, B. Catalytic performance of vanadyl pyrophosphate in the partial oxidation of toluene to benzaldehyde. *Catal. Lett.* 1999, *59*, 61–65.

69. Bentrup, U., Brückner, A., Martin, A., and Lücke, B. Selective oxidation of *p*-substituted toluenes to the corresponding benzaldehydes over $(VO)_2P_2O_7$: an *in situ* FTIR and EPR study. *J. Mol. Catal. A: Chem.* 2000, *162*, 391–399.

70. Zhang, Y., Meisel, M., Martin, A., Lücke, B., Witke, K., and Brzezinka, K.W. *In situ* Raman investigation on the structural transformation of oxovanadium hydrogenphosphate hemihydrate in the presence of ammonia. *Chem. Mater.* 1997, *9*, 1086–1091.

71. Steinike, U., Müller, B., and Martin, A. Disordered structure of $VOHPO_4 \times 0.5H_2O$ by mechanical treatment. *Mater. Sci. Forum* 2000, *321–324*, 1078–1085.

72. Martin, A., Brückner, A., Zhang, Y., and Lücke, B. In *Heterogeneous Catalysis and Fine Chemicals IV*; Blaser, H.U.; Baiker, A.; and Prins, A., Eds.; Elsevier Science B.V.: Amsterdam, 1997; Vol. 108, p. 377.

73. Aït-Lachgar, K., Tuel, A., Brun, M., Hermann, J.M., Krafft, J.M., Martin, J.M., Volta, J.-C., and Lücke, B. Selective oxidation of *n*-butane to maleic anhydride on vanadyl pyrophosphate: characterization of the oxygen-treated catalyst by electrical conductivity, Raman, XPS, and NMR spectroscopic techniques. *J. Catal.* 1998, *177*, 224–230.

74. Voge, H.H. and Adams, C.R. Catalytic oxidation of olefins. *Adv. Catal.* 1967, *17*, 151–221.

75. Bettahar, M.M., Costentin, G., Savary, L., and Lavalley, J.C. On the partial oxidation of propane and propylene on mixed metal oxide catalysts. *Appl. Catal. A: Gen.* 1996, *145*, 1–48.

76. Centi, G., Cavani, F., and Trifiro, F. In *Selective Oxidation of Hydrocarbons by Heterogeneous Catalysis*; Kluwer Academic/Plenum: New York, 2001; p. 434.

77. Lin, M.M. Selective oxidation of propane to acrylic acid with molecular oxygen. *Appl. Catal. A: Gen.* 2001, *207*, 1–16.

78. Ai, M. Oxidation of propane to acrylic acid on V_2O_5–P_2O_5-based catalysts. *J. Catal.* 1986, *101*, 389–395.

79. Kaddouri, A.C., Mazzocchia, C., and Tempesti, E. The synthesis of acrolein and acrylic acid by direct propane oxidation with Ni–Mo–Te–P–O catalysts. *Appl. Catal. A: Gen.* 1999, *180*, 271–275.

80. Lin, M., Desai, T.B., Kaiser, F.W., and Klugherz, P.D. Reaction pathways in the selective oxidation of propane over a mixed metal oxide catalyst. *Catal. Today* 2000, *61*, 223–229.

81. Novakova, E.K., Védrine, J.C., and Derouane, E.G. Propane oxidation on Mo–V–Sb–Nb mixed oxide catalysts. *J. Catal.* 2002, *211*, 226–234.

82. Weckhuysen, B.M. and Keller, D.E. Chemistry, spectroscopy and the role of supported vanadium oxides in heterogeneous catalysts. *Catal. Today* 2003, *78*, 25–46.

83. Burcham, L.J., Deo, G., Gao, X., and Wachs, I.E. In situ IR, Raman, and UV–Vis DRS spectroscopy of supported vanadium oxide catalysts during methanol oxidation. *Top. Catal.* 2000, *11/12*, 85–100.

84. Ponzi, M., Duschatzky, C., Carrascull, A., and Ponzi, E. Obtaining benzaldehyde via promoted V_2O_5 catalysts. *Appl. Catal. A: Gen.* 1998, *169*, 373–379.

85. Ono, T., Tanaka, Y., Takeuchi, T., and Yamamoto, K. Characterization of K-mixed V_2O_5 catalyst and oxidative dehydrogenation of propane on it. *J. Mol. Catal. A: Chem.* 2000, *159*, 293–300.

86. Martin, A., Bentrup, U., and Wolf, G.-U. The effect of alkali metal promotion on vanadium-containing catalysts in the vapour phase oxidation of methyl aromatics to the corresponding aldehydes. *Appl. Catal. A: Gen.* 2002, *227*, 131–142.

87. Bulushev, D.A., Kiwi-Minsker, L., Zaikovski, V.I., Lapina, O.B., Ivanov, A.A., Reshetnikov, S.I., and Renken, A. Effect of potassium doping on the structural and catalytic properties of V/Ti-oxide in selective toluene oxidation. *Appl. Catal. A: Gen.* 2000, *202*, 243–250.

88. Bentrup, U., Martin, A., and Wolf, G.-U. Thermal behaviour, redox properties, and catalytic performance of alkali-promoted V_2O_5 catalysts in the selective oxidation of *p*-methoxytoluene: a comparative study. *Catal. Today* 2003, *78*, 229–236

89. Bentrup, U., Martin, A., and Wolf, G.-U. Comparative study of the thermal and redox behaviour of alkali promoted V_2O_5 catalysts. *Thermo. Chim. Acta* 2003, *398*, 131–143.

90. Kung, H.H. Desirable catalytic properties in selective oxidation reactions. *Ind. Eng. Chem. Prod. Res. Dev.* 1986, *25*, 171–178.

91. Reddy, B.M., Ganesh, I., and Chowdhury, B. Vapour-phase selective oxidation of 4-methylanisole to anisaldehyde over V_2O_5/Ga_2O_3–TiO_2 catalyst. *Chem. Lett.* 1997, 1145–1146.

92. Reddy, B.M., Kumar, M.V., and Ratnam, K.J. Selective oxidation of *p*-methoxytoluene to *p*-methoxybenzaldehyde over V_2O_5/CaO–MgO catalysts. *Appl. Catal. A: Gen.* 1999, *181*, 77–85.

93. Reddy, B.M., Ganesh, I, and Chowdhury, B. Design of stable and reactive vanadium oxide catalysts supported on binary oxides. *Catal. Today* 1999, *49*, 115–121.

94. Reddy, B.M., Giridhar, G., and Kumar, M.V. Vapour phase oxidation of 4-methylanisole to anisaldehyde over V_2O_5/MgO–Al_2O_3 catalysts. *Res. Chem. Intermed.* 2001, *27*, 225–236.

95. Reddy, B.M., Kumar, M.V., and Ratnam, K.J. Preparation and characterization of V_2O_5/MgO catalysts for selective oxidation of 4-methylanisole to anisaldehyde. *Res. Chem. Intermed.* 1998, *24*, 919–931.

96. Bentrup, U., Brückner, A., Martin, A., and Lücke, B. Permanent blockade of in situ generated acid Brønsted sites of vanadyl pyrophosphate catalysts by pyridine during the partial oxidation of toluene. *Chem. Commun.* 1999, (13), 1169–1170.

97. Bernal, S., Kasper, J., and Trovarelli, A., Eds. *Recent Progress in Catalysis by Ceria and Related Compounds. Catal. Today* 1999, *50*, 173–443.

98. Trovarelli, A. *Catalysis by Ceria and Related Materials*; Catalytic Science Series; World Scientific Publishing Company: U.K., 2002; Vol. 2, 1–309.

99. Trovarelli, A., de Leitenburg, C., and Dolcetti, G. Design better cerium-based oxidation catalysts. *Chemtech* 1997, 27, 32–37.

100. Taylor, K.C. *Catalysis Science and Technology*; Springer-Verlag: Berlin, 1984; chapter 2.

101. Tschope, A., Liu, W., Stephanopoulos, M. F-., and Ying, J.Y. Redox activity of non-stoichiometric cerium oxide-based nanocrystalline catalysts. *J. Catal.* 1995, 157, 42–50.

102. Sato, T., Dosaka, K., Ishitsuka, M., Haga, E.M., and Okuwaki, A. Sintering behaviour of ceria-doped tetragonal zirconia powders crystallized and dried using supercritical alcohols. *J. Alloys Compd.* 1993, 193, 274–276.

103. Sahibzada, M., Steele, B.C.H., Zheng, K., Rudkin, R.A., and Metcalfe, I.S. Development of solid oxide fuel cells based on a $Ce(Gd)O_{2-x}$ electrolyte film for intermediate temperature operation. *Catal. Today* 1997, 38, 459–466.

104. Rossignol, S., Madier, Y., and Duprez, D. Preparation of zirconia-ceria material by soft chemistry. *Catal. Today* 1999, 50, 261–270.

105. Fornasiero, P., Balducci, G., Dimonte, R., Kaspar, J., Sergo, V., Gubitosa, G., Ferrero, A., and Graziani, M. Modification of the redox behaviour of CeO_2 induced by structural doping with ZrO_2. *J. Catal.* 1995, 164, 173–183.

106. Fornasiero, P., Dimonte, R., Rao, G.R., Kaspar, J., Meriani, S., Trovarelli, A., and Graziani, M. Rh-loaded CeO_2–ZrO_2 solid solutions as highly efficient oxygen exchangers: dependence of the reduction behaviour and the oxygen storage capacity on the structural properties. *J. Catal.* 1995, 151, 168–177.

107. Terribile, D., Trovarelli, A., Llorca, J., de Leitenburg, C., and Dolcetti, G. The preparation of high surface area CeO_2–ZrO_2 mixed oxides by a surfactant assisted approach. *Catal. Today* 1998, 43, 79–88.

108. Kilbourn, B.T. *Cerium, a Guide to its role in Chemical Technology*; Molycorp Inc.: White Plains, NY, 1992.

109. Gandhi, H.S., Graham, G.W., and McCabe, R.W. Automotive exhaust catalysts. *J. Catal.* 2003, 216, 433–442; and references therein.

110. Ricken, M., Nölting, J., and Riess, I. Specific heat and phase diagram of nonstoichiometric ceria (CeO_{2-x}). *J. Solid. State. Chem.* 1984, 54, 89–99.

111. Trovarelli, A., Boaro, M., Rocchini, E., de Leitenburg, C., and Dolcetti, G. Some recent developments in the characterization of ceria-based catalysts. *J. Alloys Compd.* 2001, 323/324, 584–591.

112. Ozawa, M. Role of cerium–zirconium mixed oxides as catalysts for car pollution: a short review. *J. Alloys Compd.* 1998, 275–277, 886–890.

113. Ozawa, M., Matuda, K., and Suzuki, S. Microstructure and oxygen release properties of catalytic alumina-supported CeO_2–ZrO_2 powders. *J. Alloys Compd.* 2000, 303/304, 56–59.

114. Liu, W. and Flytzani-Stephanopoulos, M. Total oxidation of carbon monoxide and methane over transition metal–fluorite oxide composite catalysts; part-I: catalyst composition and activity. *J. Catal.* 1995, 153, 304–316.

115. Liu, W. and Flytzani-Stephanopoulos, M. Total oxidation of carbon monoxide and methane over transition metal–fluorite oxide composite catalysts; part-II: catalyst characterization and reaction kinetics. *J. Catal.* 1995, 153, 317–332.

116. Larachi, F., Pierre, J., Adnot, A., and Bernis, A. Ce 3d XPS study of composite $Ce_xMn_{1-x}O_{2-y}$ wet oxidation catalysts. *Appl. Surf. Sci.* 2002, *195*, 236–250.

117. Qi, G. and Yang, R.T. A superior catalyst for low-temperature NO reduction with NH_3. *J. Chem. Soc., Chem. Commun.* 2003, (7), 848–849.

118. Centeno, M.A., Paulis, M., Montes, M., and Odriozola, J.A. Catalytic combustion of volatile organic compounds on $Au/CeO_2/Al_2O_3$ and Au/Al_2O_3 catalysts. *Appl. Catal. A: Gen.* 2002, *234*, 65–78.

119. Tuller, H.L. In *Non-Stoichiometric Oxides* Sorensen, O.T., Eds., Academic Press; New York, 1981; 271 pp.; and references therein.

120. Zamar, F., Trovarelli, A., de Leitenburg, C., and Dolcetti, G. CeO_2-based solid solutions with the fluorite structure as novel and effective catalysts for methane combustion. *J. Chem. Soc. Chem. Commun.* 1995, (9), 965–966.

121. Cho, B.K. Chemical modification of catalyst support for enhancement of transient catalytic activity: nitric oxide reduction by carbon monoxide over rhodium. *J. Catal.* 1991, *131*, 74–87.

122. Fernandez-Garcia, A., Martinez-Arias, A., Guerrero-Ruiz, A., Conesa, J.C., and Soria, J. Ce–Zr–Ca ternary mixed oxides: structural characteristics and oxygen handling properties. *J. Catal.* 2002, *211*, 326–334.

123. Kaspar, J. and Fornasiero, P. Nanostructured materials for advanced automotive depollution catalysts. *J. Solid State Chem.* 2003, *171*, 19–29.

124. Martinez-Arias, A., Fernandez-Garcia, A., Hungria, A.B., Conesa, J.C., and Soria, J. Surface properties of $CeZrO_4$-based materials employed as catalysts supports. *J. Alloys Compd.* 2001, *323/324*, 605–609.

125. Bozo, C., Gaillard, F., and Guilhaume, N. Characterization of ceria–zirconia solid solutions after hydrothermal ageing. *Appl. Catal. A: Gen.* 2001, *220*, 69–77.

126. Shuk, P., Greenblatt, M., and Croft, M. Hydrothermal synthesis and properties of mixed conducting $Ce_{1-x}Tb_xO_{2-\delta}$ solid solutions. *Chem. Mater.* 1999, *11*, 473–479.

127. Yashima, M., Arashi, H., Kakihana, M., and Yoshimura, M. Raman scattering study of cubic-tetragonal phase transition in $Zr_{1-x}Ce_xO_2$ solid solution. *J. Am. Ceram. Soc.* 1994, *77*, 1067–1071.

128. Zhang, Y., Andersson, S., and Muhammed, M. Nanophase catalytic oxides: synthesis of doped cerium oxides as oxygen storage promoters. *Appl. Catal. B: Environ.* 1995, *6*, 325–337.

129. Lammonier, C., Bennani, A., D'Huysser, A., Aboukais, A., and Wrobel, G. Evidence for different copper species in precursors of copper-cerium oxide catalysts for hydrogenation reactions: an X-ray diffraction, EPR and X-ray photoelectron spectroscopy study. *J. Chem. Soc., Faraday Trans.* 1996, *92*, 131–136.

130. Imamura, S., Shono, M., Okamoto, N., Hamada, A., and Ishida, S. Effect of cerium on the mobility of oxygen on manganese. *Appl. Catal. A: Gen.* 1996, *142*, 279–288.

131. Rynkowski, J., Farbotko, J., Touroude, R., and Hilaire, L. Redox behaviour of ceria-titania mixed oxides. *Appl. Catal. A: Gen.* 2000, *203*, 335–348.

132. Shigapov, A.N., Graham, G.W., McCabe, R.W., and Plummer Jr., H.K. The preparation of high-surface area, thermally stable, metal oxide catalysts and supports by a cellulose templating approach. *Appl. Catal. A: Gen.* 2001, *210*, 287–300.

133. Bozon-Verduraz, F., Bensalem, A., Delamar, M., and Bugli, G. Preparation and characterization of highly dispersed silica-supported ceria. *Appl. Catal. A: Gen.* 1995, *121*, 81–93.

134. Cunningham, J., Cullinane, D., Farrell, F., and Gibson, C. Oxygen isotope equilibration and exchange as probes for differing dioxygen interactions with pure and rhodia-promoted CeO_2 and Al_2O_3. *Top. Catal.* 1999, *8*, 179–187.

135. Rynkowski, J.M., Szynkowska, M.I., Paryjczak, T., and Lewiscki, A. Redox properties of Al_2O_3-supported cerium oxides. *Chem. Environ. Res.* 1999, *8*, 261–270.

136. Spanier, J.E., Robinson, R.D., Zhang, F., Chan, S.-W., and Herman, I.P. Size dependent properties of CeO_{2-y} nanoparticles as studied by Raman-scattering. *Phys. Rev. B* 2001, *64*, 245407–245414.

137. Nelson, A.E. and Schulz, K.H. Surface chemistry and microstructural analysis of $Ce_xZr_{1-x}O_{2-y}$ model catalyst surfaces. *Appl. Surf. Sci.* 2003, *210*, 206–221.

138. Kozlov, A.I., Kim, D.H., Aleksey, Yezerets., Anderson, P., Kung, H.H., and Kung, M.C. Effect of preparation method and redox treatment on the reducibility and structure of supported ceria–zirconia mixed oxide. *J. Catal.* 2002, *209*, 417–426.

139. Yao, H.C. and Yu Yao, Y.F. Ceria in automotive exhaust catalysts: oxygen storage. *J. Catal.* 1984, *86*, 254–265.

140. Yao, M.H., Baird, R.J., Kunz, F.W., and Hoost, T.E. An XRD and TEM investigation of the structure of alumina-supported ceria–zirconia. *J. Catal.* 1997, *166*, 67–74.

141. Tsunekawa, S., Fukuda, T., and Kasuya, A. X-ray photoelectron spectroscopy of monodisperse CeO_{2-x} nanoparticles. *Surf. Sci.* 2000, *457*, L437–L440.

142. Zhang, F., Chan, S.-W., Spanier, J.E., Apak, E., Jin, Q., Robinson, R.D., and Herman, I.P. Cerium oxide nanoparticles: size-selective formation and structure analysis. *Appl. Phys. Lett.* 2002, *80*, 127–129.

143. Reddy, B.M., Khan, A., Yamada, Y., Kobayashi, T., Loridant, S., and Volta, J.C. Raman and X-ray photoelectron spectroscopy study of CeO_2–ZrO_2 and V_2O_5/CeO_2–ZrO_2 catalysts. *Langmuir* 2003, *19*, 3025–3030.

144. Reddy, B.M., Khan, A., Yamada, Y., Kobayashi, T., Loridant, S., and Volta, J.C. Structural characterization of CeO_2–TiO_2 and V_2O_5/CeO_2–TiO_2 catalysts by Raman and XPS techniques. *J. Phys. Chem. B* 2003, *107*, 5162–5167.

145. Reddy, B.M., Khan, A., Yamada, Y., Kobayashi, T., Loridant, S., and Volta, J.C. Structural characterization of CeO_2/SiO_2 and V_2O_5/CeO_2/SiO_2 catalysts by Raman, XPS, and other techniques. *J. Phys. Chem. B* 2002, *106*, 10964–10972.

146. Reddy, B.M., Khan, A., Yamada, Y., Kobayashi, T., Loridant, S., and Volta, J.C. Structural characterization of CeO_2–MO_2 (M = Si^{4+}, Ti^{4+}, and Zr^{4+}) mixed oxides by Raman spectroscopy, X-ray photoelectron spectroscopy, and other techniques. *J. Phys. Chem. B* 2003, *107*, 11475–11484.

147. Matta, J., Abi-Aad, E., Courcot, D., and Aboukais, A. Simultaneous EPR and TPR study of the V–Ce–O catalysts redox properties. In *Magnetic Resonance in Colloid and Interface Science*; Nato Science Series, II: Mathematics, Physics and Chemistry, 2002; Vol. 76, pp. 565–374.

148. Daniell, W., Ponchel, A., Kuba, S., Anderle, F., Weingand, T., Gregory, D.H., and Knözinger, H. Characterization and catalytic behaviour of VO_x–CeO_2 catalysts for the oxidative dehydrogenation of propane. *Top. Catal.* 2002, *20*, 65–74.

149. Rocchini, E., Trovarelli, A., Llorca, J., Graham, G.W., Weber, W.H., Maciejewski, M., and Baiker, A. Relationships between structural/morphological modifications and oxygen storage-redox behaviour of silica-doped ceria. *J. Catal.* 2000, *194*, 461–478.

150. Rocchini, E., Vicario, M., Llorca, J., de Leitenburg, C., Dolcetti, G., and Trovarelli, A. Reduction and oxygen storage behaviour of noble metals supported on silica-doped ceria. *J. Catal.* 2002, *211*, 407–421.

151. Rojas, M.L., Fierro, J.L.G., Tejuca, L.G., and Bell, A.T. Preparation and characterization of $LaMn_{1-x}Cu_xO_{3+\lambda}$ perovskite oxides. *J. Catal.* 1990, *124*, 41–51.

152. Belessi, V.C., Trikalitis, P.N., Ladavos, A.K., Bakas, T.V., and Pomonis, P.J. Structure and catalytic activity of $La_{1-x}FeO_3$ system for the NO + CO reaction. *Appl. Catal. A: Gen.* 1999, *177*, 53–58.

153. Gao, L.Z. and Au, C.T. Perovskite-type halo-oxide $La_{1-x}Sr_xFeO_{3-\delta}$ X_σ (X = F, Cl) catalysts selective for the oxidation of ethane to ethene. *J. Catal.* 2000, *189*, 52–62.

154. Peter, S.D., Garbowski, E., Perrichon, V., Pommier, B., and Primet, M. Activity enhancement of mixed lanthanum–copper–iron–perovskites in the CO + NO reaction. *Appl. Catal. A: Gen.* 2001, *205*, 147–158.

155. Varma, S., Wani, B.N., and Gupta, N.M. Synthesis, characterization, TPR/TPO and activity studies on $LaMn_xVi_{1-x}O_{4-\delta}$ catalysts. *Appl. Catal. A: Gen.* 2001, *205*, 295–304.

156. Varma, S., Wani, B.N., and Gupta, N.M. Redox behaviour and catalytic activity of La–Fe–V–O mixed oxides. *Appl. Catal. A: Gen.* 2003, *241*, 341–348.

157. Shelef, M., Otto, K., and Gandhi, H. The oxidation of CO by O_2 and by NO on supported chromium oxide and other metal oxide catalysts. *J. Catal.* 1968, *12*, 361–375.

158. Kapteijn, F., Stegenga, S., and Dekker, N.J.J. Alternatives to noble metal catalysts for automotive exhaust purification. *Catal. Today* 1993, *16*, 273–287.

159. Wu, Y., Zhao, Z., Liu, Y., and Yang, X.-G. The role of redox property of $La_{2-x}(Sr, Th)_xCuO_{4\pm\lambda})$, playing in the reaction of NO decomposition and NO reduction by CO. *J. Mol. Catal. A: Chem.* 2000, *155*, 89–100.

160. Koebel, M., Madia, G., and Elsener, M. Selective catalytic reduction of NO and NO_2 at low temperatures. *Catal. Today* 2002, *73*, 239–247.

161. Ladavos, A.K. and Pomonis, P.J. Mechanistic aspects of NO + CO reaction on $La_{2-x}Sr_xNiO_{4-\delta}$ (x = 0.00–1.50) perovskite-type oxides. *Appl. Catal. A: Gen.* 1997, *165*, 73–85.

162. Teraoka, Y., Nii, H., Kagawa, S., Jansson, K., and Nygren, M. Influence of the simultaneous substitution of Cu and Ru in the perovskite-type (La, Sr)MoO_3 (M = Al, Mn, Fe, Co) on the catalytic activity for CO oxidation and CO–NO reactions. *Appl. Catal. A: Gen.* 2000, *194/195*, 35–41.

163. Hu, Y.-H., Dong, L., Wang, J., Ding, W.-P., and Chen, Y. Activities of supported copper oxide catalysts in the NO + CO reaction at low temperatures. *J. Mol. Catal. A: Chem.* 2000, *162*, 307–316.

164. Centi, G. and Perathoner, S. Nature of active sites in copper-based catalysts and their chemistry of transformation of nitrogen oxides. *Appl. Catal. A* 1995, *132*, 179–259.

165. Wen, B. and He, M.-Y. Study of Cu–Ce synergism for NO reduction with CO in the presence of O_2, H_2O and SO_2 in FCC operation. *Appl. Catal. B: Environ.* 2002, *37*, 75–82.

166. Ma, L., Luo, M.-F., and Chen, S.-Y. Redox behaviour and catalytic properties of $CuO/Ce_{0.8}Zr_{0.2}O_2$ catalysts. *Appl. Catal. A: Gen.* 2003, *242*, 151–159.

167. Nohman, A.K.H., Duprez, D., Kappenstein, C., Mansour, S.A.A., and Zaki, M.I. In *Preparation of Catalysts V*; Delmon, B.; Jacobs, P.A.; and Poncelet, G., Eds.; Elsevier: Amsterdam, 1991; p. 617.

168. Cracium, R. and Dulamita, N. Ethylbenzene oxidative dehydrogenation on MnO_x/SiO_2 catalysts. *Ind. Eng. Chem. Res.* 1999, *38*, 1357–1363.

169. Yamashita, T. and Vannice, A. NO decomposition over Mn_2O_3 and Mn_3O_4. *J. Catal.* 1996, *163*, 158–168.

170. Nishino, A. Household applications using catalysis. *Catal. Today* 1991, *10*, 107–118.

171. Moggridge, G.D., Rayment, T., and Lambert, R.M. An *in situ* XRD investigation of singly and doubly promoted manganese oxide methane coupling catalysts. *J. Catal.* 1992, *134*, 242–252.

172. Kapteijn, F., Singoredjo, L., van Driel, M., Andreini, A., Moulijn, J.A., Ramis, G., and Busca, G. Alumina-supported manganese oxide catalysts: surface characterization and adsorption of ammonia and nitric oxide. *J. Catal.* 1994, *150*, 105–116.

173. Chen, A., Xu, H., Yue, Y., Hua, W., Shen, W., and Gao, Z. Support effect in hydrogenation of methyl benzoate over supported manganese oxide catalysts. *J. Mol. Catal. A: Chem.* 2003, *203*, 299–306.

174. Strohmeier, B.R. and Hercules, J. Surface spectroscopic characterization of Mn/Al_2O_3 catalysts. *J. Phys. Chem.* 1984, *88*, 4922–4929.

175. Kapteijn, F., van Langeveld, D., Moulijn, J.A., Andreini, A., Vuurman, M.A., Turek, A.M., Jehng, J.M., and Wachs, I.E., Alumina-supported manganese oxide catalysts: characterization: effect of precursor and loading. *J. Catal.* 1994, *150*, 94–104.

176. Craciun, R. Structure/activity correlation for unpromoted and CeO_2 promoted MnO_2/SiO_2 catalysts. *Catal. Lett.* 1998, *55*, 25–31.

177. Wöllner, A., Lange, F., Schmetz, H., and Knozinger, H. Characterization of mixed copper–manganese oxides supported on titania catalysts for selective oxidation of ammonia. *Appl. Catal. A: Gen.* 1993, *94*, 181–203.

178. Craciun, R., Nentwick, B., Hadjiivanov, K., and Knozinger, H. Structure and redox properties of MnOx/yttrium-stabilized zirconia (YSZ) catalyst and its uses in CO and CH_4 oxidation. *Appl. Catal. A: Gen.* 2003, *243*, 67–79.

179. Thomas, J.M., Raja, R., Sankar, G., and Bell, R.G. Molecular-sieve catalysts for the selective oxidation of linear alkanes by molecular oxygen. *Nature* 1999, *398*, 227–230.

180. Thomas, J.M. On the nature of isolated active sites in open-structure catalysts for aerial oxidation of alkanes. *Top. Catal.* 2001, *15*, 85–91.

9 The Surface Acidity and Basicity of Solid Oxides and Zeolites

G. Busca

Dipartimento di Ingegneria Chimica e di Processo
"G.B. Bonino," Laboratorio di Chimica delle Superfici
e Catalisi Industriale, Genova, Italy

Contents

9.1   INTRODUCTION

Powders constituted by metal oxides are relevant products of the inorganic chemical industry and find a number of applications, as such or as precursors of dense sintered ceramics. Surface acid–base interactions are involved in many aspects. They are not only fundamental phenomena involved in heterogeneous catalysis [1] and adsorption [2], but are also of interest in pigment technologies [3] as well as in particle sintering [4], to produce bulky ceramics for electronic, magnetic, and optical applications. For these reasons, the acid–base properties of oxide powders have been the object of many investigations. Some comprehensive papers have been published in the last 10 years [5–9]. "Metal oxides" are a wide family of materials that can differ significantly from the point of view of their structure. In Table 9.1, some industrially applied metal oxide catalysts are summarized to provide evidence for the significant difference of their structure.

In this chapter, we will attempt a systematization of the factors governing acid–base properties of metal oxide catalysts and will focus on some systems to which belong particularly relevant materials from the point of view of their surface acid–base adsorption and catalytic properties.

TABLE 9.1
Families of Solid Oxide Catalysts and Typical Processes

Catalyst family	Example	Example process
Bulk binary mixed oxides	γ-Al_2O_3	Claus process
Modified or "doped" binary oxides	Chlorided γ-Al_2O_3	n-butane isomerization
Solid solutions of binary metal oxides	$(Fe,Cr)_2O_3$	High temperature water–gas shift
Mixed oxide ternary phase	$MgAl_2O_4$	Support of Ni methane steam reforming catalysts
Salts	$(VO)_2P_2O_7$	n-butane oxidation to maleic anhydride
Protonic zeolites	H-ZSM5 zeolite	Xylenes isomerization
Cationic zeolites	Cs-ZSM5	Methylthiazole synthesis
Oxides supported on oxide carriers	V_2O_5/TiO_2	o-xylene oxidation to phthalic anhydride
Oxide-supported inorganic acids	H_3PO_4/kieselguhr	Olefin oligomerization
Multicomponent oxides	Mixed molybdate	Propene ammoxidation

9.2 FUNDAMENTALS AND DEFINITIONS

9.2.1 Definitions of Acidity and Basicity

To discuss the surface acido-basicity of solids we should remember the Brønsted and Lewis definitions of acidity and basicity:

1. According to the concepts proposed by Brønsted [10] and, simultaneously and independently but less precisely, by Lowry [11] in 1923, an acid is any hydrogen-containing species able to release protons and a base is any species capable of combining with protons. This definition does not exclusively imply water as the reaction medium. In this view, acid–base interactions consist of the equilibrium exchange of a proton from an acid HA to a base B giving rise to the conjugated base of HA, A^-, plus the conjugated acid of B, HB^+: $HA + B = A^- + HB^+$.
2. In the same year, 1923, Lewis [12] first proposed a different approach. In his view, an acid is any species that, because of the presence of an incomplete electronic grouping, can accept an electron pair to give rise to a dative or coordination bond. Conversely, a base is any species that possess a nonbonding electron pair that can be donated to form a dative or coordination bond. The Lewis-type acid–base interaction can be depicted as follows: $B : +A = {}^{\delta+}B \rightarrow A^{\delta-}$. This definition is completely independent from water as the reaction medium, and is more general than the former one.

In terms of Lewis acidity, the Brønsted-type acid HA is the result of the interaction of the Lewis-type acid species H^+ with the base A^-. According to the definitions given, Lewis-type acids (typically, but not only, coordinatively unsaturated cations) do not correspond to Brønsted-type acids (typically species with acidic hydroxy groups). On the contrary, Lewis basic species are also Brønsted bases.

Later on, Pearson [13,14] introduced the concept of hard and soft acid and bases (HSABs): hard acids (defined as small-sized, highly positively charged, and not easily polarizable electron acceptors) prefer to associate with hard bases (i.e., substances that hold their electrons tightly as a consequence of large electronegativities, low polarizabilities, and difficulty of oxidation of their donor atoms) and soft acids prefer to associate with soft bases, giving thermodynamically more stable complexes; these hard–hard and soft–soft associations also occur faster.

Acid–base properties of solids are usually discussed, mostly, in terms of Lewis and Brønsted definitions of acidity and basicity, but interesting applications of HSAB theory to solid–gas interfaces can be found in Reference 15 and Reference 16.

9.2.2 Ionicity/Covalency and the Structures of Solid Oxides

According to basic inorganic chemistry, the oxides are the compounds of any element with oxygen. In fact, the oxides that are applied in the field of heterogeneous catalysis and adsorption are those that are solid in the applicable conditions, that is, at temperatures below, for example, 573 to 673 K and atmospheric or higher pressures. From the point of view of the bonding and structure, the binary oxides can be classified as in Table 9.2, the corresponding example structure is being reported in Figure 9.1. A systematization of the structures of more complex mixed oxides is more difficult. A tentative summary is given in Table 9.3.

As discussed previously [9], although the distinction between covalent and ionic bonding is not straightforward, the structure of oxides allows having an indication on it. In typical ionic structures, the coordination around ions is higher than their valency. For example, in magnesium oxide (Figure 9.1[a]), the coordination around both ions is six, while the valency is two. This occurs for oxides of true base metals, and of transition metals (such as, e.g., for ceria, Figure 9.1[b]; and

TABLE 9.2

Classification of Structure-Bonding in Solid Binary Oxides

Structure-bonding	Example (Figure 9.1)
Ionic network	MgO, CeO_2, TiO_2
Covalent network	SiO_2, WO_3
Molecular	
Layered	MoO_3
"Linear polymeric"	CrO_3
Singly molecular	P_4O_{10}

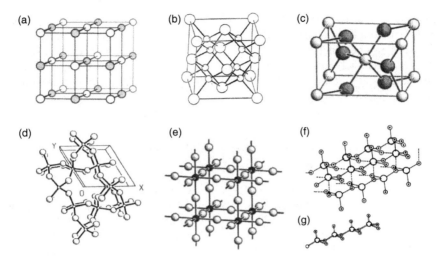

FIGURE 9.1 Crystal structures of metal oxides. (a) MgO: rock salt type, Mg coordination number (CN) 6; O CN 6. (b) CeO_2: fluorite-type, Ce CN 8, O CN 4. (c) TiO_2: rutile, Ti CN 6, O CN 3. (d) SiO_2: quartz, Si CN 4, O CN 2. (e) WO_3, ReO_3 type, actually distorted, W CN 6, O CN 2. (f) MoO_3: layered structure, the coordinations are questionable. (g) CrO_3: linear polymer, Cr CN 4, O CN 1 and 2

TABLE 9.3
Classification of Structure-Bonding in Solid Mixed Oxides

Structure-bonding	Example
Ionic network	
Solid solution	$(Fe,Cr)_2O_3$ (disordered corundum structure)
Compound	$ZnFe_2O_4$ (spinel Structure)
Covalent network	
Solid solutions	WO_3–MoO_3
Compound	GeP_2O_7
Salts	
Layered anion	Phyllosilicates
Linear chain anion	Metavanadates, metaphosphates
Molecular anion	Ortho- and pyro-vanadates, molybdates, tungstates, chromates, phosphates, etc.

---

titania, Figure 9.1[c]) except when they are in very high oxidation states. On the contrary, in covalent network structures the valency and coordination are mostly equal. In silicas, for example, valency and coordination at silicon are four, and at oxygen two (see Figure 9.1[d] for quartz). On tungsta (Figure 9.1[e]), both valency and coordination at tungsten are six. When "double element-oxygen bonds" exist, coordination is even lower than valency, and layered (e.g., MoO_3 [17]), linear polymeric (e.g., CrO_3) or molecular (e.g., P_4O_{10}) structures appear. For example, phosphorus coordination is four in P_4O_{10} and phosphates, while valency is five. This occurs mostly for semimetal or nonmetal oxides and for some oxides of transition metals at very high oxidation state, although in the latter case the definition of coordination number is not so straightforward, frequently being copresent by true short coordination bonds with weak long interactions (e.g., for MoO_3 [17]).

9.2.3 Acidity and Basicity on the Ideal Surface of a Solid Oxide

The ideal surface of the particles of a solid oxide can be constituted by exposed planes with corners and edges that join them. The exposed planes, in principle supposed to be similar to "ideal" lattice planes, can become exposed by cutting a bigger crystal. Doing this with network (nonmolecular) oxides, metal–oxygen bonds (either covalent or ionic) have been broken thus oxide species and metal or nonmetal centers remain exposed and coordinatively unsaturated at the surface. These sites should be associated to very high free energy and, consequently, should be very unstable. To stabilize the surface, reconstruction phenomena as well as reaction with molecules from the environment (e.g., water and CO_2) occur. This would limit the number of coordinatively unsaturated centers and cause the formation of new surface species such as hydroxy-groups and surface carbonates. However, unsaturated centers at the surface can remain or be generated by desorption of adsorbed water (and CO_2) under outgassing or heating. As a consequence of these phenomena, the surface of solid oxides can be constituted by:

1. Exposed coordinatively unsaturated cationic centers, potentially acting as Lewis acid sites.
2. Exposed oxide species, potentially acting as basic sites.
3. Exposed hydroxy-groups, arising from water dissociative adsorption, potentially acting as Brønsted acid sites, or, alternatively, as basic sites.
4. Other surface species arising from the reactivity of the surface with the environment, such as carbonate species, when they have not been decomposed by appropriate pretreatments.

As already mentioned, some "covalent" solid oxides are molecular in nature (i.e., nonframework), formed either by relatively small molecules (e.g., P_4O_{10}) or by macromolecular chains (e.g., CrO_3) or by layers (e.g., V_2O_5 and MoO_3 [17]). In such cases, most of the external surface is formally obtained by breaking van der Waals interactions more than true bonds. This occurs for the 100 and 010 faces of α-MoO_3 [17]. However, also in these cases coordinatively unsaturated atoms can be present, although their presence can be limited to some of the exposed planes (e.g, the nonbasal planes for layered structures such as the V_2O_5 [18] and the 001 face of α-MoO_3 [17]) and to defects. As for the relation to the HSAB theory, we may note that the proton and most metal cations (i.e., the oxide's surface acid sites) are hard acids, excluded the ions of noble metals and the reduced metal centers. Conversely, the hydroxide ion and the oxide ions are quite hard bases, also like covalently bonded oxygen species.

9.2.4 Amount, Strength, and Distribution of Surface Acid and Basic Sites on the Ideal Surface of a Solid Oxide

In acid and base water solutions, the "amount" of acidity and basicity depends not only on the concentration of the acid (base) in solution but also on their strength; therefore, strength and amount of acidity and basicity are somehow linked. The strength of acidity in water solution is determined by the equilibrium constant Ka of the acid, and from it the calculation of the amount of acidity (protons) is possible. For bases, one refers to Ka of the conjugated acid. For gas-phase molecules, Brønsted acid and basic strengths are associated to the enthalpy of deprotonation of the (conjugated) base [19], which can be measured experimentally or estimated by theory. This gives rise to an acidity/basicity scale based on "proton affinity" (PA): $PA = -\Delta H_{protonation}$.

Due to the more complex interactions occurring in water solutions (where solvation effects may be determinant) with respect to the gas phase, the proton affinity scale does not completely correspond to the Ka scales. In the case of surface hydroxy-groups of solids, which are the potential Brønsted acid sites as well as the conjugated acids of the surface basic sites, experimental deprotonation techniques cannot be applied. Deprotonation enthalpies can be estimated according to the Bellamy–Hallam–Williams relation [20] from the wavenumber shift $\Delta \bar{v}$ of the infrared (IR) OH stretching band when a H-bonding interaction occurs. In heterogeneous systems (in particular zeolites [21,22]), the proton affinity of strong Brønsted sites can be calculated from that of the silanols groups of silica, taken as a standard, from the following equation:

$$PA_{Brønsted} = PA_{SiOH} - A \log(\Delta \bar{v}_{Brønsted} / \Delta \bar{v}_{SiOH})$$

the ratio $\Delta \bar{v}_{Brønsted} / \Delta \bar{v}_{SiOH}$ being independent from the base chosen and where

$$PA_{SiOH} = 1390 \, kJ/mol \quad \text{and} \quad A = 442 \, kJ/mol$$

Alternatively, deprotonation enthalpies can be evaluated from probe adsorption calorimetric data or from temperature programmed desorption (TPD) measurements. The strengths of surface Lewis acid sites and of surface basic sites can also be evaluated, in principle, by the heat of probes adsorption or desorption. In all cases, however, probes adsorption on solids can result in multiple interactions: for example, van der Waals interactions can be superimposed to true acid–base interactions, which can also be multiple and finally give rise to some kind of solvation effects, in particular, in the zeolite cavities [23–25]. Thus, the pure acidity/basicity data are difficult to be extracted from experimental results. As an effect of this, although solution phase data of acid–base probes are inappropriate, in principle, to describe gas–solid interactions [26], in practice, sometimes, the scales based on Kas could be even more effective than those based on PAs [27,28].

In heterogeneous catalysis, the catalytic activity (reaction rate) depends on the amount of active sites (e.g., of acidic or base sites having the appropriate strength) that are present on the catalyst as a whole. This means that the "density" of active sites (amount of sites per gram of the solid or per unit surface area) is an important parameter. On solids, amount and strength of acidic or basic sites are quite independent parameters, so both must be analyzed independently for a complete characterization. Additionally, several different families of acidic or basic sites may occur in the same solid surface, so their "distribution" (density of sites of any site family) must be characterized.

Also, both acidic and basic sites can be present in different position (but frequently near each other) on the same solid surface. This provides evidence for the significant complexity of acid–base characterization of solids.

9.3 TECHNIQUES FOR THE CHARACTERIZATION OF SURFACE ACIDITY AND BASICITY

Different techniques can be used to study the surface acidity and basicity of solids, either at the gas–solid interface [29,30] or at the solid–water solution interface [31]. Here, we will comment only on few most widely used and available methods for surface acid–base characterization.

9.3.1 Molecular Probes for Surface Acidity and Basicity Characterization

Most surface characterization techniques are "indirect": they investigate the strength of adsorption or desorption of bases on acid sites and of acids on basic sites. In Table 9.4, some data on the basic probe molecules typically applied in acidity surface characterization are summarized. Similarly, in Table 9.5 some data on the "acidic" (or electrophilic) probe molecules typically applied in basicity surface characterization are summarized. Reasons for choosing one or another probe have been discussed in the above-cited reviews and will be briefly summarized later, in relation to the techniques that are applied and the type of characterization that has to be performed. In the opinion of the present author, a careful analysis of the available data and of the conditions in which they have been obtained shows that the data arising from the different techniques and from the use of different probes generally agree quite satisfactorily. In general, the use of several techniques and of different probes is useful to have a clear picture. Most of the basic probes allow to obtain a good picture for surface acidity characterization, because they behave (at least for classes of catalysts and in the most typical conditions) quite specifically interacting with acid sites, although most of them interact with both Brønsted and Lewis acid sites. 2,6-Diterbutyl pyridine is considered to act

TABLE 9.4
Molecular Probes Applied for Surface Acidity Characterization

Family	Base Example	Conjugated acid	Basic strength pKa	Basic strength PA	Mostly applied techniques
Cyclic amines	Piperidine	$C_5H_{10}NH$	$C_5H_{10}NH_2^+$	11.1	933	IR, ^{15}N NMR, calorimetry, TPD
Alkyl amines	n-butylamine	$n\text{-}C_4H_9\text{-}NH_2$	$n\text{-}C_4H_9\text{-}NH_3^+$	10.9	916	Calorimetry, TPD, ^{15}N NMR, IR
Ammonia		NH_3	NH_4^+	9.2	857	Calorimetry, TPD, ^{15}N NMR, IR
Phosphines	Trimethyl-phosphine	$(CH_3)_3P$	$(CH_3)_3PH^+$	8.65	957	^{31}P NMR, IR
Phosphine oxides	Trimethyl-phosphine oxide	$(CH_3)_3P{=}O$	$(CH_3)_3P{=}OH^+$		907	^{31}P NMR
Heterocyclic amines	Pyridine	C_5H_5N	$C_5H_5NH^+$	5.2	928	IR, ^{15}N NMR, calorimetry, TPD
Ketones	Acetone	$(CH_3)_2C{=}O$	$(CH_3)_2C{=}OH^+$	−7.2	824	^{13}C NMR, IR
Nitriles	Acetonitrile	$CH_3\text{-}C{\equiv}N$	$CH_3\text{-}C{\equiv}NH^+$	−10.4	783	IR, ^{15}N NMR
Hydrocarbons	Ethylene	$H_2C{=}CH_2$	$CH_3\text{-}CH_2^+$		680	IR
Carbon monoxide		CO	$[HCO]^+$		598	IR, calorimetry
Nitrogen		N_2			477	IR
Argon		Ar				TPD

a Range cm^{-1}.
b Fermi resonance doublet.

TABLE 9.5

Molecular Probes Applied for Surface Basicity Characterization

Compound	Formula	Reactivity	Technique
Carbon dioxide	CO_2	Carbonates, bicarbonates, linear coordination on Lewis sites	TPD, IR, calorimetry
Carbon monoxide	CO	Formates, carbonite, polymeric anionic species, dioxycarbene	IR (ESR)
Carbon disulphide	CS_2	S_2CO^-, CO_3^{2-}, other species	IR
Sulfur dioxide	SO_2	Sulphites, disulphites, hydrogensulphite	Calorimetry IR
Pyrrole	H_4C_4NH	H-bonding, pyrrolate anions	IR, ^1H NMR
Chloroform	$H\text{-}CCl_3$	H-bonding, coordination by chlorine on Lewis acid sites, hydrolysis	IR, ^1H NMR
Acetonitrile	$CH_3\text{--}CN$	$^-CH_2\text{--}CN$ anions, previous coordination by nitrogen on Lewis acid sites	IR
Alcohols	R–OH	$R\text{--}O^-$ anions, undissociative coordination, different H-bondings	IR, calorimetry
Thiols	R–SH	$R\text{--}S^-$ anions, undissociative coordination, different H-bondings	IR
Trimethylborate	$(CH_3O)_3B$	Tetrahedral borate, dissociation	^{11}B NMR
Borontrifluoride	BF_3	$F_3B\leftarrow O^{2-}$, dissociation	TPD
Ammonia	NH_3	H-bonding, $^-NH_2$ anions, coordination by nitrogen on Lewis acid sites	IR
Pyridine	H_5C_5N	$H_5C_4N^-$ anions, coordination by nitrogen on Lewis acid sites, dimerization	IR
Nitromethane	CH_3NO_2	$H_2CNO_2^-$, CNO^- formed by decomposition	IR, ^{13}C NMR

as a specific probe for Brønsted sites, due to the steric hindrance of the alkyl groups, which should not allow coordination on Lewis sites. In contrast, the use of "acidic" molecules to probe surface basicity is far less satisfactory. In fact, all "acidic" (or electrophilic) molecules (Table 9.5) also contain well available nucleophilic (basic) atoms. It seems impossible to find a molecule that actually only interacts specifically with basic sites. On the other hand, metal oxides that display significant surface basicity are always very ionic and carry Lewis acidity, although weak. Therefore, in this case the acido-basicity is relevant more than pure basicity properties. This will be discussed later.

9.3.2 Quantitative Adsorption of Probe Molecules from Gas and Liquid Phases

Titration methods [29–32], that is, the study of the interaction of indicator dyes with the solids from solutions, have been proposed as a technique for both qualitative and quantitative characterization of solid surfaces. If a basic indicator B is used, the proton acidity of the surface is expressed by the Hammett acidity function, originally proposed for solution chemistry:

$$H_o = pK_{BH+} + \log[B]/[BH^+]$$

Similarly, the basicity can be defined when an acid is converted by its conjugated base. This allows to define acidity and basicity in the same scale. A similar expression can be proposed for interaction of bases with Lewis sites. The amine titration method, described by Tanabe et al. [33], consists of titrating a solid acid suspended in benzene with n-butylamine using an indicator. Although this technique is widely applied in the fields of colloids and soil sciences [31], it has many limitations to deduce gas–solid phenomena, the surface Hammett acidity function also having doubtful physical meaning [6].

The quantitative measure of the adsorption of probes on the surfaces from the vapor by adsorption volumetry and gravimetry measured at different temperature and in contact with different vapor pressures [30,34,35], is a quite easy technique that gives useful although rough information.

9.3.3 Calorimetric Methods

Adsorption microcalorimetry is the measure of the heat of adsorption evolved when dosing measured small amounts of a vapor probe on a surface. Cardona-Martinez and Dumesic [36] summarized the results obtained for oxides, zeolite, and metal catalysts before 1992. Summaries of the application of these techniques to gas–solid interactions and heterogeneous catalysis have been published recently [37–39]. As done by Auroux and Gervasini [40] for a number of binary metal oxides, calorimetric studies of the acidity and basicity are mostly performed using ammonia as an acidity probe and carbon dioxide as a basicity probe [41].

Calorimetric techniques allow to obtain an interesting quantification and evaluation of the gas–solid interactions. However, they give an integral response of the phenomena occurring upon interactions, without giving information on their quality. Consequently, interactions of amphoteric compounds with acidic or basic sites or interaction of bases with Lewis and Brønsted acid sites cannot be distinguished. In particular, the use of ammonia as a probe for acid sites is dangerous, because ammonia not only adsorbs on both Brønsted and Lewis acid sites but can also react as an acid on basic sites and can be oxidized too [42]. Similarly, the use of CO_2 as a probe for basic sites is not easy, because CO_2 also interacts with Lewis acid sites as such or upon forming carbonate and bicarbonates [43]. On the other hand, by coupling calorimetric data with spectroscopic and theoretical data, complete information on the actually occurring phenomena can be obtained.

Other amines (such as pyridine, butylamine [44], etc.) have been used as probes for acidity. In Figure 9.2, the differential adsorption heats relative to the adsorption of n-butylamine at room temperature on γ-Al_2O_3 and titania-anatase samples after activation at 723 K for 3 h are reported. They show that on alumina a very small number of stronger acid sites are present than on titania, but that on titania a larger number of medium strong sites exist in these conditions. In this

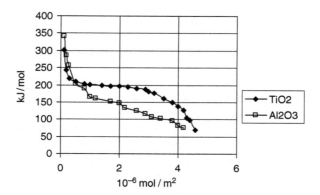

FIGURE 9.2 Differential adsorption heats relative to *n*-butylamine adsorption at room temperature on γ-Al$_2$O$_3$ and TiO$_2$ (anatase) preactivated by outgassing at 673 K

case, IR spectroscopy showed that interaction occurs essentially on Lewis sites in both cases, although some protonation also occurs [45]. In fact, also amines are not always specific probes, in particular concerning Brønsted/Lewis sites [45].

Bolis et al. [46] recently summarized their data concerning calorimetry of CO adsorption on oxides. They concluded that, CO adsorption calorimetry is not a good technique for the characterization of the surface Lewis acidity at least for aluminas because the adsorption exothermic signal is in part combined by an endothermic signal due to surface reconstruction phenomena. Due to the nonspecific interaction of CO$_2$ with basic sites, Rossi et al. [47,48] proposed the use of hexafluoro isopropanol as an alternative for the characterization of basic sites. The use of sulfur dioxide as a probe is also successful [49].

Recently, Zajac et al. [50] suggested that two-cycle ammonia adsorption studies coupled with microcalorimetric data can give additional information on surface acidity. In addition, liquid phase adsorption calorimetric studies can give, with some limitations, interesting data [51].

9.3.4 Temperature Programmed Desorption Methods

The gas-chromatographic or mass spectrometric analysis of the gases evolved from a surface upon a temperature programmed heating ramp after adsorption of probes gives also indication on energetics of adsorption/desorption and, consequently, on the strength of acid–base adsorption sites. In Figure 9.3, the ammonia TPD spectrum of a H-MFI zeolite is reported, showing two main desorption peaks. The higher temperature peak is mainly associated to ammonia desorbing from the Brønsted sites of the zeolite (see later). However, a contribution from Lewis-bonded ammonia cannot be excluded. In fact, TPD techniques suffer of limits similar to those of adsorption microcalorimetry: both techniques do not give qualitative information on the nature of the phenomena detected. Additionally, TPD curves can show features that are artifacts depending on experimental

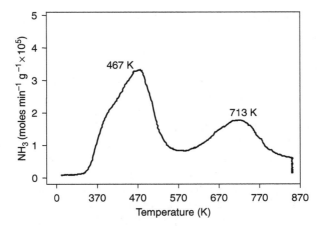

FIGURE 9.3 Ammonia TPD spectrum of H-MFI zeolite

conditions [26]. TPD of ammonia is largely applied for the study of the zeolites acidity [52,53] but is quite dangerous because of the lack of specificity of its adsorption (ex. Brønsted versus Lewis sites). Gorte [26] applied successfully the TPD of reactive amines, which give rise to the Hoffmann elimination to olefins + ammonia when "desorbing" from protonic zeolites. However, such "reactive" amines in principle are as nonspecific as ammonia in adsorption [45]. Ammonia TPD is also dangerous for investigating the surfaces of solids which can give redox reactions, because of the formation of species such as molecular nitrogen [54] and nitrogen oxides. A similar problem can occur with pyridine that can decompose more than desorb like observed on sulfated zirconia [55]. Recently the TPD of Ar has been proposed for characterization of solid acids [56] although according to the theoretical calculations of Bolis et al. it is only sensitive to confinements effects, but almost insensitive to specific interaction with acid sites [24]. TPD of adsorbed carbon dioxide, compared with other techniques, is widely applied for basicity characterization [57–59]. BF_3 has been proposed recently as a probe for basicity by TPD [60].

9.3.5 IR Spectroscopic Methods

Infrared spectroscopy is largely used for surface acidity [61–64] and basicity [62,65] characterization. By using appropriated bases as probes (see Table 9.4 and Table 9.5), this technique allows the separated characterization of Lewis and Brønsted acid sites and (with more difficulty) of basic sites.

Brønsted acidity characterization can be performed by looking at what bases (among strong bases such as ammonia, pyridine, and other amines, including diazines [66]) are protonated by the site (the protonation method), or by investigating the interaction of weak bases via H-bondings with the sites. In this case, the extent of shift of the νOH can be used to measure the acid strength of the sites (the H-bonding method, according to the Bellamy–Hallam–Williams relation [20]). A third method (the olefin oligomerization method) detects the proton initiated oligomerization of olefins, following the rate (or extent) trend: 1,3-butadiene > isobutene > propene > ethylene [9,64].

Lewis acidity can be tested by using bases whose spectrum presents bands whose position (or sometimes intensity) is highly sensitive to the strength of the Lewis acid–base interaction. The low temperature adsorption of CO is today perhaps not only the most popular technique for Lewis acidity characterization but is also used for characterizing Brønsted sites (i.e., of zeolites) by applying the H-bonding method. It actually allows a very detailed analysis of the surface sites as they appear at low temperature without strong perturbations of the surface, having also free access to most cavities and avoiding steric hindrances [67–69]. This is a good opportunity to evaluate "pure acidity" [24] but, together, is a drawback, because additional interactions do occur with most other molecules including the reactants of catalytic reactions. The position of νCO of adsorbed CO is shifted upwards on Lewis acidic d_0 and d_{10} cations as the result of a σ-type donation of the lone pair at the carbon atom, or of a simple polarization. However, it has been shown that, at least on cationic zeolites, also the O lone pair can be involved in an O-bond interaction [70]. Similar interaction also occurs with Brønsted acidic protons. Consequently, the measure of the shift $\Delta\nu$CO can be taken as a measure of the strength of such interactions, while the shift $\Delta\nu$OH can be taken as a measure of the Brønsted acidity of the hydroxy groups. An electron back-donation from d-orbitals of transition metal cations to π-type antibonding CO orbitals can also occur, increasing adsorption strength but decreasing CO bond order and stretching frequency. Consequently, CO allows a good characterization of coordination and oxidation states of adsorbing transition metal ions, but not always a straightforward information on their Lewis acidity.

Two examples of the application of low temperature adsorption of CO are reported in Figure 9.4 and Figure 9.5, to titanium silicalite-1 (TS1) and to a H-ZSM5 zeolite. Adsorption of CO at 120 K (Figure 9.4) causes the shifts of the OH stretching band of the silanol groups of TS1 from 3746 to 3654 cm^{-1} ($\Delta\nu$OH $\sim$ 90 cm^{-1}), typical of silanols of silicas, but the evidence of a shoulder near 3580 cm^{-1} suggests that a population of more acidic silanols likely exists on this solid. The broad feature at 3525 cm^{-1}, assigned to OH's H-bonded in structural defects, is not shifted showing that CO cannot break these interactions. The two νCO components at 2156 and 2138 cm^{-1}, respectively, are assigned to CO interacting with the silanol groups, and to "physisorbed" or liquid-like CO, respectively, but the latter can contain the band of O-bonded CO on silanols [71]. All these features disappear by outgassing at 120 K. This shows that no Lewis bound CO is detectable with this experiments. As we will discuss later, the use of stronger bases such as nitriles [72] and ammonia [73] allows to detect Lewis sites on the same powders, showing that CO fails in the detection of Lewis acidity of titanium silicalite.

Similar experiments performed on a H-ZSM5 zeolite sample (Figure 9.5) show how more acidic are the bridging OH's of zeolites absorbing at 3622 cm^{-1} and shifted to 3290 cm^{-1} ($\Delta\nu$OH $\sim$ 330 cm^{-1}) as well as those due to extraframework

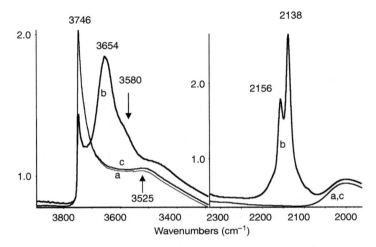

FIGURE 9.4   FT-IR spectrum of titanium silicalite after activation by outgassing at 673 K, and cooling at 120 K (a), in contact with CO at 120 K (b), and after outgassing at 120 K (c)

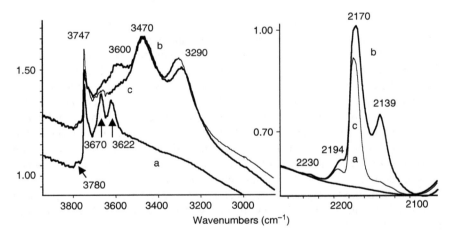

FIGURE 9.5   FTIR spectra of a H-ZSM5 sample containing extraframework material, after activation by outgassing at 673 K, and cooling at 120 K (a), in contact with CO at 120 K (b), and after outgassing at 120 K (c)

material (see later, shifted from 3670 to 3470 cm^{-1}, $\Delta\nu$OH ~ 200 cm^{-1}). Additionally, the νCO components at 2230 and 2194 cm^{-1} are certainly due to CO adsorbed on Al^{3+} ions acting as very strong Lewis sites. As we will discuss later, the use of hindered bases allows to distinguish where these sites are located, whether in the cavities or at the external surface.

Data similar to those arise from CO adsorption, can be obtained with small and very weak bases such as nitrogen or ethylene, and with more hindered and heavy (but still weak) bases such as nitriles. The adsorption of strong bases like pyridine is widely used for quantitative analyses and for the determination of the Lewis/Brønsted sites ratios [74].

The adsorption of strong and heavy bases (such as ammonia, amines, pyridines) certainly perturbs the surface to a relevant extent, rather than do weak and light bases. For example, it has been shown that pyridine and ammonia detect strong Lewis sites associated to pentavalent vanadium [18] and hexavalent tungsten when CO does not show them [75]. A discrepancy from the results obtained by using CO and pyridine for characterizing Lewis sites of sulfated zirconia has been pointed out recently by Morterra et al. [76,77]. According to these authors, this effect is associated to the harder nature of pyridine with respect to CO, which allows it, unlike CO, to displace sulfates as ligands of Zr cations. They conclude that the results obtained with pyridine, as a probe is an artifact. In our opinion, this however, is only the result of the much stronger (more than harder) basic character of pyridine with respect to CO, which allows it to compete (unlike CO) as a base with sulfate ions, which are also bases.

In fact, the use of strong bases and of room or higher temperature adsorption experiments is more "perturbing," likely producing a slight modification of the coordination state of some cations, but frequently gives a more exact view of the surface chemistry of oxides than low temperature CO adsorption. This is particularly true if related to adsorption and catalysis phenomena that involve high temperatures and more polar reactants. As for another example, the characterization of the catalysts for the selective catalytic reduction of NO_x by ammonia (NH_3-SCR) cannot be performed with CO as a probe. In fact, CO is not able to reveal the Lewis acidity of pentavalent vanadium [18], which is well-evidenced by the use of pyridine and ammonia itself as probes. Vanadium centers act as the active sites for this process just by strongly adsorbing ammonia in a coordinative way, as shown by experiments [78] and strongly supported by theoretical calculations [79]. According to Thibault-Starzyk et al. [80], the results of low temperature adsorption of CO on protonic zeolites gives a picture of the surface acidity, which does not correlates well with catalytic activity, in contrast to the high-temperature protonation method using deuterated acetonitrile. Carbon monoxide is reported to be not able to interact with the low frequency OHs of zeolite USY in contrast to nitriles and pyridine [81].

In recent years the tentative to go deeper in the analysis of surfaces (in particular, but not exclusively, of zeolites) suggested the use of sets of molecular probes with similar functionality but different steric hindrance. As for example, different nitriles like those reported in Scheme 9.1 allow to explore the position of the sites in porous materials such as zeolites [82]. Similarly, hindered pyridines, such as 2,6-dimethylpyridine (lutidine, [83]) and 2,6-diterbutyl pyridine can be used [84]. The use of these molecules allow to show that some of the adsorption sites detected by CO and previously supposed to be in the interior of zeolite channels are actually out of the channels [85].

Even more complex is the choice of a suitable base for basicity characterization. This problem has been summarized some years ago by Kustov [63] and Lavalley [65]. IR studies show that all probes have relevant reactivity and poor specificity

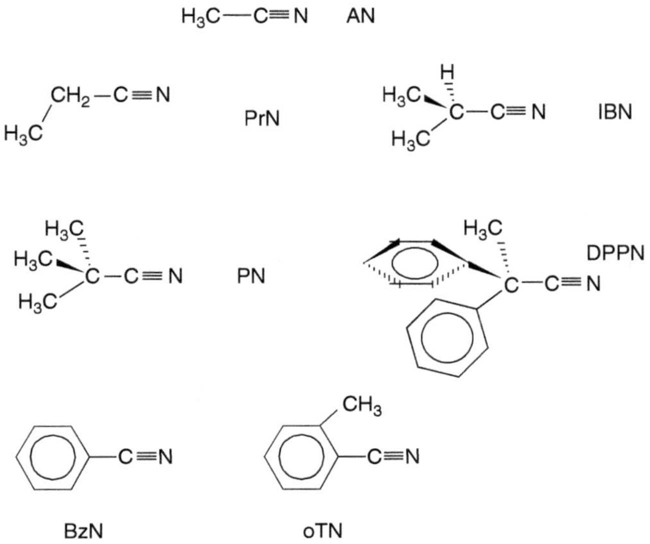

$$H_3C-C{\equiv}N \quad AN$$

$$\underset{H_3C}{\overset{CH_2-C{\equiv}N}{\diagup}} \quad PrN \qquad \underset{H_3C}{\overset{H_3C}{\diagdown}}\overset{\overset{\displaystyle H}{|}}{C}-C{\equiv}N \quad IBN$$

$$\underset{H_3C}{\overset{H_3C}{\diagdown}}\overset{H_3C}{\underset{}{C}}-C{\equiv}N \quad PN \qquad DPPN$$

BzN   oTN

SCHEME 9.1 Structure of nitriles utilized to distinguish the opposition of adsorption sites in low and medium pore zeolites. AN = Acetonitrile, PrN = propionitrile, IBN = isobutironitrile (2-methyl-propionitrile), PN = pivalonitrile (2,2-dimethylpropionitrile), DPPN = 2,2-diphenylpropionitrile, BzN = benzonitrile, oTN = ortho-toluonitrile

for basic sites. In spite of this, interesting results have been obtained [86]. Recently, new probes for IR metal oxide basicity characterization have been proposed like cyanate ions produced by decomposition of diazomethane [87], CS_2 [88], and 2-methyl-3-butyn-2-ol [89].

9.3.6 Nuclear Magnetic Resonance (NMR) Spectroscopic Methods

Magic angle spinning nuclear magnetic resonance spectroscopy came to be considered, in the last 10 years, as one of the most powerful techniques for the investigation of solid catalysts [90–92]. Room temperature ^1H MAS-NMR spectroscopy and liquid Helium temperature broad line ^1H-NMR allow the direct detection of the protons of surface hydroxy groups of many systems as such and upon interaction with probe molecules [93,94]. The ^{27}Al MAS-NMR spectra are very helpful for probing the quantity, coordination, and location of aluminum atoms in Al-containing materials such as aluminas and aluminosilicate catalysts (see Later) and minerals. In Figure 9.6 the ^{27}Al MAS-NMR spectra of (a) NH_4^+-MOR zeolite, (b) γ-Al_2O_3, and (c) silica–alumina are reported. The sharp peak near 54 ppm in the case of NH_4^+-MOR zeolite is due to framework tetrahedral Al^{3+}. The two main components in the spectrum of γ-Al_2O_3 near 60 and 10 ppm are due to octahedral and tetrahedral Al ions, in a ratio, which is near 25:75. In the case of silica-alumina, again both tetrahedral and octahedral Al ions

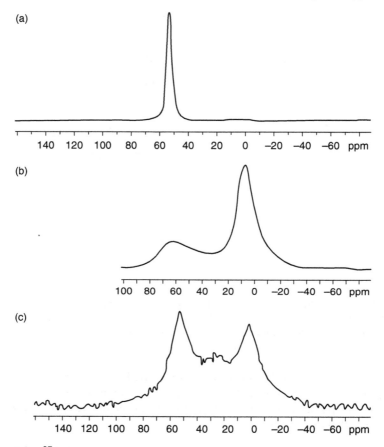

FIGURE 9.6 ^{27}Al MAS NMR spectrum of NH$_4$-MOR zeolite (a), γ-Al$_2$O$_3$ (b), and silica–alumina (c) (Courtesy: Prof. S. Caldarelli [Marseille].)

are detected, but a third species is also observed. The structures of these materials will be discussed later on.

Actually, the quadrupolar nature of the nucleus does not allow the observation of the structurally significant fine structure of the bands, limiting the applicability of ^{27}Al MAS-NMR to little more than the determination of the coordination number at the aluminum atom. However, new techniques such as ^{27}Al MQ MAS NMR [95] and [^{27}Al]-^{1}H REAPOR NMR (REAPOR = rotational echo adiabatic passage double resonance) [96] allowed a deeper investigation of Al containing materials and to study the Brønsted OH's in zeolites. ^{31}P, ^{11}B, ^{29}Si, and ^{51}V MAS-NMR are applied to study phosphate and phosphides [97], borate [98], silicate [99], and vanadate [100] catalysts.

^{13}C, ^{11}B, ^{15}N, ^{19}F, and ^{31}P MAS-NMR have been applied to study the interaction of solid surfaces with molecular probes. ^{13}C and ^{1}H MAS-NMR has been applied to the investigation of adsorption and conversion of a number of organic molecules on catalytic surfaces [26,90–92,101,102]. Nitromethane has been proposed as a probe for ^{13}C MAS-NMR studies of the acid–base of oxides [103]. The adsorption of basic probes such as amines [104], phosphines [105], and phosphine oxides [106] can be followed by ^{15}N and ^{31}P NMR. Recently, ^{11}B NMR studies concerning the adsorption of triethylborate as a probe for basic sites have been reported [107].

9.3.7 Theoretical Methods

Theoretical methods applied to heterogeneous catalysis undergone an impressive development in the last few years [108], and have been largely applied to zeolite and oxide catalysts. Until the end of the 1990s, most studies of zeolite acidity and spectroscopic properties used finite cluster models [109,110]. They provided valuable data on the average position of zeolitic Brønsted sites on a general acidity scale and helped to assign vibrational spectra and NMR spectra. However, cluster models have two limitations [111]: (i) They allow for a relaxation of the atoms on deprotonation, which is not as constrained as would be the case if the modeled site were part of the periodic solid. (ii) They neglect the long-range potential of the periodic solid. In practice, they do not allow detailed studies of sites in different crystallographic positions or in different frameworks; for example, they did not allow to distinguish the sites of protonic zeolites from those of silica–alumina. The periodic approaches can include all these effects and are increasingly applied in spite of the enormous computational effort needed. Several applications of density functional theory (DFT) and of alternative techniques to oxide catalysts appeared recently. DFT calculations are increasingly applied to metal oxide systems and allowed to investigate the surface and bulk structure of γ-Al$_2$O$_3$ [112]. Embedded cluster models have been used to determine the adsorption structures and the energetics of the interaction of CO$_2$, SO$_x$, and NO$_x$ on (001) terrace sites of alkaliearth oxide, concluding that the stability of the surface species is determined by the surface basicity [113]. Cluster models have also been applied to the study of interactions of probes with the Lewis sites of aluminas as compared with interactions in zeolite nanocavities [25]. On the other hand the application of computational methods to zeolites is today one of the main subjects of research. Calculations allowed also to support complete reaction mechanisms proposed, based on the experiments, such as, for example, in the case of the ammonia SCR process over vanadia-based catalysts [78,79].

9.3.8 Catalytic Probe Reactions

Catalytic test reactions [6,33,114] represent a very important tool for acid–base characterization. Conversion of secondary alcohols such as isopropanol [115,116], 2-butanol [59,86], and cyclohexanol [117], either to olefins or to ketones, is considered to be evidence of acidic and basic behavior, respectively. Skeletal and double bond isomerization of n-butene [44,118], and of branched olefins [119] such as 3,3-dimethylbut-1-ene and methylenecyclohexane [117] are applied to evaluate both Brønsted acidity and basicity. For basicity characterization, retroaldolisation of diacetonealcohol [115] and the decomposition of 2-methyl-3-butyn-2-ol [86] have been used successfully.

Skeletal isomerization and cracking of *n*-alkanes [80], have been widely used to characterize and evaluate strong Brønsted acids. However, they are complex reactions, occurring with different mechanisms and affected by deactivation by coking. As for example, *n*-pentane isomerization can occur via monomolecular and bimolecular mechanisms and catalytic activity strongly depends on the presence of hydrogen [120]. Ethylbenzene disproportionation has been recently proposed as a standard reaction for acidity evaluation of zeolites by the International Zeolite Association [121]. Liquid-phase reactions such as benzylation of toluene, rearrangement of α-pinene, and dihydropiran methoxylation can be applied to characterize the acidity, for example, of acid-treated clays [122].

9.4 COMPOSITION RELATED SURFACE PROPERTIES OF OXIDE MATERIALS

9.4.1 Lewis Acid Strength

The actual existence of Lewis acid sites is evident on the surface of all ionic oxides in the most usual conditions. In Table 9.6, we have correlated the shift of the IR sensitive bands of the probe molecules (assumed to be a measure of the Lewis acid strength) with the polarizing power (PP) of the cation involved in the oxide, calculated as the radio of the cationic charge/ionic radius, taking also into account its coordination into the bulk. A correction has been introduced for metal cations in very high oxidation states; usually forming oxo-species with very short metal–oxygen bonds (the -yl cations). In this case, to calculate the PP we took the charge of the -yl oxo-ion more than that of the metal ion itself. This allows to improve the correlation. This correlation, quite obvious indeed, seems to be quite satisfactory.

Lewis acidity is not observed usually for the covalent oxides of nonmetal elements, such as silica [123], germania [124], germanium phosphate [125], and silicon phosphates. This can be seen in Figure 9.7 where the IR spectrum of "solid phosphoric acid," a catalyst produced by impregnating and reacting silica kieselguhr with phosphoric acid (used since decades in the refinery industry for olefin oligomerization processes) is shown. Upon adsorption of a nitrile base, the CN stretching is observed at 2260 cm^{-1}, which indicates that the probe only interacts via H-bonding on acidic OHs.

In the case of silica [123], Lewis acidity appears only after very hard pretreatments under outgassing. The calculated PP of the "cations" in covalent oxide is higher than 8. This shows that the reason for the absence of Lewis acidity on the surface of these oxides is just due to the difficulty of breaking the M–(OH) bonds, which are so covalent due to the too strong PP of the cation that would results by their dissociation. Additionally, when Si(OH) groups actually break, couples of them tend to give rise to "strained" Si–O–Si bridges, which can later reopen by reaction with suitable molecules [126].

TABLE 9.6

Position (cm^{-1}) of the Sensitive Bands of Adsorbed Basic Probe Molecules on Different Catalyst Surfaces

Adsorbate IR Mode → Surface ↓	CO ν CO	Pyridine 8a	Ammonia $\delta_{sym}NH_3$	Adsorbing site Type	PP
AlF_3		1627		$_{IV}Al^{3+}$	7.7
γ-Al_2O_3	2235	1625	1295	$_{IV}Al^{3+}$	7.7
	2210–2190	1615	1265	$_{IV}Al^{3+}$	
	2170	1595	1220	$_{IV}Al^{3+}$	5.7
β-Ga_2O_3	2220	1623	1275	$_{IV}Ga^{3+}$	6.4
	2190	1608			
	2160	1595		$_{VI}Ga^{3+}$	4.8
WO_3	—	1613	1275, 1222	$_{VI}WO^{4+}$	6.9*
MoO_3	—	1613	1270, 1250	$_{VI}MoO^{4+}$	6.78*
γ-Fe_2O_3	2180	1612	1230	$_{IV}Fe^{3+}$	6.1
SnO_2	2210–2180	1610	1230	$_{IV}Sn^{4+}$	5.8
V_2O_5	—	1608	1249	VO^{3+}	5.5*
TiO_2 anatase	2226, 2208	1610	1225	$_V Ti^{4+}$	
	2182		1185	$_{IV}Ti^{4+}$	6.6
BeO		1610	1220	$_{IV}Be^{2+}$	7.4
α-Fe_2O_3	2165	1608	1220, 1180	$_{IV}Fe^{3+}$	5.5
α-Cr_2O_3	2180–2160	1608	1220	$_{IV}Cr^{3+}$	4.9
ZrO_2	2192, 2170	1606	1210, 1160	Zr^{4+}	~5.0
ZnO	2192, 2180	1605	1210, 1183	$_{IV}Zn^{2+}$	3.3
			1160		
α-Al_2O_3	2165	1597		$_{IV}Al^{3+}$	5.7
Mn_3O_4	2190	1608	1220	$_{IV}Mn^{2+}$	
	2180	1600	1170	$_{VI}Mn^{3+}$	3.45
Co_3O_4	2195		1220	$_{IV}Co^{2+}$	3.45
			1175	$_{VI}Co^{3+}$	
MnO	2154	1600		$_{VI}Mn^{2+}$	2.98
ThO_2	2172	1597		$_{VIII}Th^{4+}$	3.8
NiO	2155	1595	1180	$_{VI}Ni^{2+}$	2.90
CoO	2136		1175	$_{VI}Co^{2+}$	2.7
MgO	2158		1110	$_{IV}Mg^{2+}$	3.7
	2149	1593	1070	$_{VI}Mg^{2+}$	2.8
CeO_{2-x}	2170, 2155	1593		Ce^{4+}, Ce^{3+}	~3
La_2O_3	2155			La^{3+}	2.8
CuO	2120	1593	1070	$_{VI}Cu^+$	~2.0
CaO	2160, 2140	1593		$_{VI}Ca^{2+}$	2.0
SrO		1593		$_{VI}Sr^{2+}$	1.69
BaO		1590		$_{VI}Ba^{2+}$	1.48
K_2O/supp		1588		K^+	0.72
Liquids	2143	1583	1054		

PP: polarizing power (charge to radius ratio, see text).

Lewis acid strength decreases from top to bottom.

* Data obtained using an approximate radius of the wolframyl, molybdenyl and vanadyl ions.

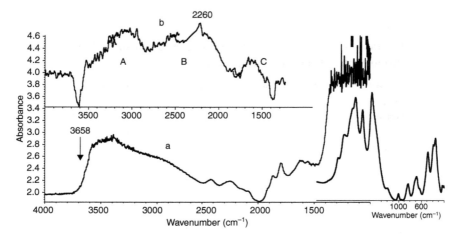

FIGURE 9.7 (a) FT-IR spectrum of "solid phosphoric acid." (b) Subtraction spectrum after pivalonitrile adsorption at r.t. on "solid phosphoric acid"

On the contrary, dehydroxylation of ionic oxides is easier, and results in the generation of Lewis acid sites, which are sufficiently stable to stay as such in relatively mild conditions. The strongest Lewis acidic oxides in normal conditions are alumina and gallia (and silica after very strong pretreatments) that is, oxides of elements at the limit of the metallic character. The same elements also give rise to halides characterized by an even stronger Lewis acidity.

9.4.2 Density of Lewis Sites

Lewis acid sites can adsorb water as any other molecules having some amount of basicity. As discussed in the case of alcohol adsorption [127], the coordination on Lewis sites of basic but dissociable molecules activates them toward dissociation if a sufficiently basic site is located nearby. This condition can be fulfilled for ionic oxides, thus molecular coordination of water can be precursor for its dissociation, and the overall process is, in principle, reversible (Scheme 9.2).

The reactions (both dissociative and nondissociative adsorption of water) are highly exothermic and their equilibrium is shifted toward left by heating and lowering water vapor partial pressure. On the other hand, at least for ionic oxides, it is clear that bases stronger than water can also displace water from Lewis sites. This has been found, for example, for ammonia [128]. Consequently, the number of Lewis acid sites actually available depends on the degree of dehydroxylation of the surface that depends primarily on the temperature, on the composition of the environment and on the ionicity–covalency of the oxide (Scheme 9.2).

As for example, microcalorimetric experiments, like those reported in Figure 9.2, clearly show that the number of Lewis sites with medium strength available on titania-anatase is higher than on alumina [40,44,46], although few very strong Lewis sites are only present on alumina. In general, it can be deduced

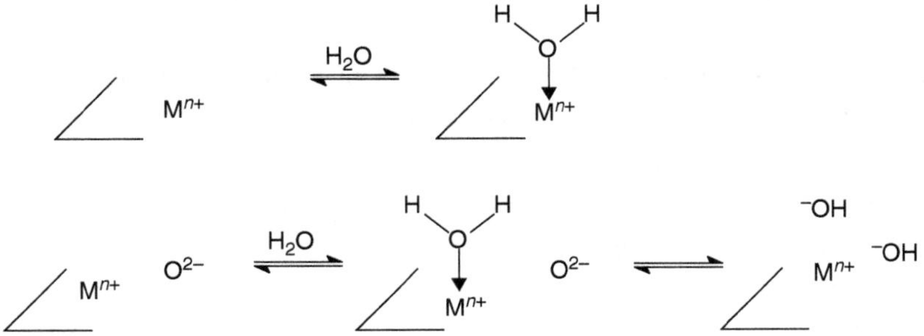

SCHEME 9.2 Molecular (up) and dissociative (down) adsorption of water on metal oxides

that, the stronger the Lewis acidity, the fewer available sites. This is because the hydroxylation degree is higher, in the same conditions. This is somehow valid principle at least for oxides having a predominantly acid character. The oxides of the nonmetals would (in theory) have the strongest Lewis acidity: in fact, they are characterized by the smallest "cationic" sizes. These sites would, however, be so strong that they cannot be formed at all in usual conditions, like in the case of silica [123].

9.4.3 Lewis Acidity and Basicity

Experimental studies show that covalent oxides in usual conditions present neither Lewis acidity nor significant basicity. This can be evidenced by the lack of any significant chemisorption of probe molecules with acidic or electrophylic character. In particular, they are unable to chemisorb carbon dioxide in the form of carbonates or sulfur dioxide in the form of sulfites.

In agreement with the discussion, surface Lewis acidity and basicity (the latter associated to surface anions) have the same origin: they appear because of the ionicity of the element–oxygen bond. Thus, Lewis acidity is generally associated to basicity. Surface basic sites are, in the case of metal oxides, either oxide or hydroxide ions. Their basic strength depends upon the extent of their coordinative unsaturation and, inversely, on the Lewis acid strength of the cations. This means that, essentially, the stronger the Lewis acid sites, the weaker the basic sites and vice versa (assumed ionicity of the bond). This is experimentally, more or less, true (see Table 9.6). Thus, the characterization of the acidity allows information on the basicity too. This allows to overcome in some extent the difficulties in the characterization of the surface basicity mentioned earlier. On the other hand, basic and Lewis acidic sites mostly work synergically, and this means that acido-basicity more than pure basicity properties are relevant.

9.4.4 Brønsted Acidity

According to their nature of "basic oxides," most of ionic metal oxides do not carry sufficiently strong Brønsted acidity to protonate bases such as pyridine or ammonia

TABLE 9.7

Evaluation of the Brønsted Acid Strength of Surface Hydroxy-Groups on Catalytic Materials by Different IR Techniques

Catalysts	Protonation of acetonitrile	Pyridine	NH_3	n-butyl amine	Piperidine
↓ P.A. →	783	912	846		933
pKa →	−10.1	5.25	9.2	10.8	11.2
HZSM5	Yes[a]	Yes	Yes	Yes	Yes
Ferrierite	No	Yes	Yes		
$SO_4^=/TiO_2$	No	Yes	Yes	Yes	Yes
WO_3/TiO_2	No	Yes	Yes	Yes	Yes
$SiO_2–Al_2O_3$	No	Yes	Yes	Yes	Yes
H_3PO_4/SiO_2	No	Yes	Yes	Yes	Yes
WO_3	No	Yes	Yes		
MoO_3	No	Yes	Yes		
V_2O_5	No	Yes	Yes		
Nb_2O_5	No	Yes	Yes		
ALF_3	No	Tr	Yes	Yes	Yes
Cr_2O_3 oxidized	No	No	Yes		
$SiO_2–TiO_2$	No	No	Yes	Yes	Yes
$Al_2O_3–B_2O_3$	No	No	Yes[b]	Yes	Yes
SiO_2 on γ-AL_2O_3	No	No	Yes[b]	Yes	Yes
γ-Al_2O_3	No	No	Yes[b]	Yes	Yes
TiO_2-anatase	No	No	Tr	Tr	Yes
β-Ga_2O_3	No	No	No		
am-SiO_2	No	No	No	No	No
ZrO_2	No	No	No	No	No
Cr_2O_3 reduced	No	No	No	No	No
MgO	No	No	No	No	No

The Brønsted acid strengths decrease from top to bottom. P.A: Proton affinity (kJ/mol); am: amorphous.

[a] Protonated dimeric species are likely formed.

[b] Formation of a "disproportionation" reaction → $NH_4^+ + NH_2^-$ possibly occurs.

at room temperature (Table 9.7). On the other hand, the surface hydroxy groups of the more acidic metal oxides are actually able to make quite strong hydrogen bondings with medium-strong bases (Table 9.8), at room temperature, and to act as Brønsted acidic catalysts at higher temperature. In spite of this, the surface OHs of most ionic oxides have a basic more than an acidic character.

Covalent low-valency nonmetal oxides (such as silica, germania, and boria) also show quite weak Brønsted acidic properties. In these cases, in fact, although the M–OH bond allows covalent polarization of the O–H bond, the negative charge arising from dissociation cannot be delocalized. Strong Brønsted acidity, resulting in protonation of ammonia and pyridine and in cationic oligomerization of olefins,

TABLE 9.8

Evaluation of the Brønsted Acid Strength of Surface Hydroxy-Groups on Catalytic Materials by Different IR Techniques

| Catalysts | $\Delta\nu$OH nitriles | Behaviour with n-C_4H_8 | Polymerization of | | | |
			C_2H_4	C_3H_6	i-C_4H_8	C_4H_6
HZSM5	ABC^a	Polymerization + isomerization	Yes	Yes	Yes	Yes
Ferrierite	ABC	Polymerization				
$SO_4^=/TiO_2$	ABC	Polymerization	Yes	Yes	Yes	Yes
WO_3/TiO_2	ABC	Polymerization	Tr	Yes	Yes	Yes
SiO_2–Al_2O_3	ABC	Polymerization	No	Yes	Yes	Yes
H_3PO_4/SiO_2	ABC	Polymerization	No	No	Yes	Yes
AlF_3	>500				Yes	Yes
Nb_2O_5	~500					
SiO_2–TiO_2	~450	Polym. traces	No	No	Tr	
Al_2O_3–B_2O_3	480–420 330–280	butoxide traces	No	No		Yes
SiO_2 on γ-Al_2O_3	480–420 330–280	$\Delta\nu$OH 200–300	No	No	No	
γ-Al_2O_3	450–400 330–280	$\Delta\nu$OH 200–300	No	No	No	Yes
am-SiO_2	400	$\Delta\nu$OH 150–300	No	No	No	No
TiO_2-anatase	<300		No	No	No	No
MgO	~450		No	No	No	No

The Brønsted acid strength decrease from top to bottom.

^{a} Protonated dimeric species are likely formed.

appears for oxides of elements with valency five or higher (Table 9.8). In these cases, in fact, M=O double bonds exist and allow delocalization of the anionic charge arising from OH's dissociation. Tungsta, molybdena, vanadia, phosphoric anhydride, and niobia as well as sulfate-containing oxides, all present significant to very strong Brønsted acidity. The detection of typically sharp PO–H stretching bands at ~3660 cm^{-1} for samples containing phosphate species and supported P_2O_5 is due to the formation of species like I in Scheme 9.3 while the detection of only broad bands for supported tungsta and for sulfated oxides is due to the formation of H-bonded OH's like species II in Scheme 9.3. These species occur for very high valency metal-oxo centers and to the possibility to delocalize extensively the negative charge arising from dissociation, with the generation of very strong acidity. Sometimes water is directly involved in the structure of the catalyst. This is the case of "solid phosphoric acid" whose IR spectrum is reported in Figure 9.7. The broad band in the region 3700 to 2000 cm^{-1} is due to the OH stretchings of the POH groups and of water molecules with which they interact. The subtraction

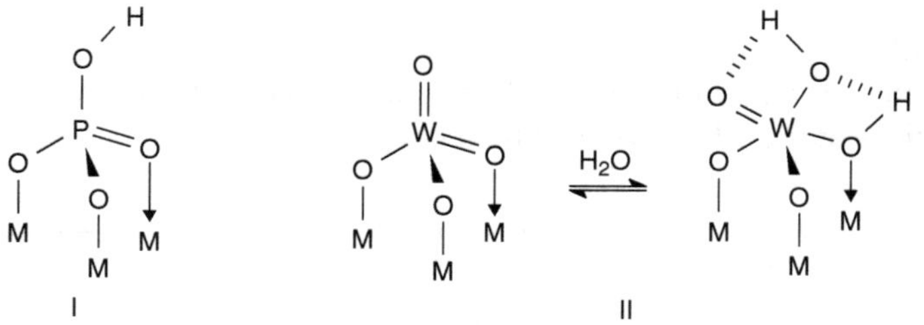

SCHEME 9.3 Structures for Brønsted acidic hydroxy groups on phosphate catalysts (I, not convertible into Lewis sites in usual conditions) and on tungsta and tungstates (II, convertible in Lewis sites by outgassing)

spectrum in Figure 9.7(b), shows the perturbations arising from adsorption of a nitrile probe. An ABC-type spectrum is formed [44] that provides evidence for a strong H-bonding of the probe with some acidic hydroxy group. The negative feature at 3658 cm^{-1} shows that the OHs which actually interact with the probe are typical POH groups free from H-bonding in the activated samples, even if their existence is not easy to be realized, from the unsubtracted spectrum.

The mechanism of formation of strong Brønsted acidity on silica–alumina and protonic zeolites will be discussed later. It results from the association of covalent silica based structures with aluminum in different ways. In any case, in spite of the so-called "acidity generation theories" in the case of mixed oxides [33,129,130], it seems clear today [131] that these effects are not general phenomena.

9.5 Structure-Related Surface Properties of Metal Oxides. CASE STUDIES

As discussed in a previous review [9], acidity and basicity of metal oxides are basically linked to the nature of the element involved, whose valency and atomic size are the main factors generating both the bulk structure and the surface chemistry. This relation is summarized in Table 9.9 for binary oxides and Table 9.10 for mixed oxides. However, the particular preparation of any compound has its own properties in relation to morphology and purity/impurity, which arise from the particular preparation method. In the following case studies, we will take into consideration also some of these aspects.

9.5.1 The Aluminas

9.5.1.1 Structure-related and morphology-related surface properties of aluminas

Aluminum oxide is a polymorphic material [132,133]. The thermodynamically stable phase is α-Al$_2$O$_3$ (corundum, S.G. R-3c, $Z = 6$) where all Al ions are

TABLE 9.9
Tentative Summary of the Acid–Base Properties of Binary Metal Oxides

Element	Oxidation state	Cation size (radius*, Å)	M–O bond nature	Acidity type	Acidity strength	Basicity, nucleophilicity	Examples
Semi-metal	$\geq +5$	Very small ≤ 0.2	Covalent	Brønsted	Medium strong	None	(SO_3); P_2O_5
	$+3$–$+4$	Small ≤ 0.4	Covalent	Brønsted	Medium weak	None	B_2O_3, SiO_2, GeO_2
Metal	High	Small to medium	Largely covalent	Brønsted and	Medium to strong	None	WO_3, MoO_3, CrO_3, Ta_2O_5, Nb_2O_5, V_2O_5
	$+5$–$+7$	0.3–0.7		Lewis			
	Medium	Small 0.35–0.5		Lewis	Strong	Weak	γ-Al_2O_3, β-Ga_2O_3
	$+3$–$+4$	Medium 0.5–0.6	Ionic	Lewis	Medium	Medium weak	TiO_2, Fe_2O_3, Cr_2O_3
		Large 0.7–1.2		Lewis	Medium weak	Medium strong	La_2O_3, SnO_2, ZrO_2, CeO_2, ThO_2 (Bi_2O_3, Sb_2O_3)
	Low	Large to very large	Lewis	Medium to	Strong to	Mgo, Cao, Sro, Bao, CoO,
	$+1$–$+2$	0.7–1.5		very weak	very strong	NiO, CuO, ZnO, (Cu_2O)

Source: Data from Shannon, R.D. and Prewitt, C.T. *Acta Cryptalogr.* 1969, *B25*, 925.

TABLE 9.10

Acid–Base Properties of "Mixed Oxides" as Determined by Experimental Data

Types	Structures-compounds	Brønsted acidity	Lewis acidity	Basicity	M–O bond type	"acidic" component	"basic" component
"True" mixed oxides	Rock-salt solid solutions	No	Weak	Strong			
	Corundum solid solutions	No	Medium	Medium			
	Spinels AB_2O_4	No	Medium–strong	Medium	Ionic	No	Both
	Ilmenites ABO_3	No	Medium–strong	Medium	+ionic		
	Perovskites ABO_3	No	Weak	Strong			
	GeP_2O_7	Yes	No	No	Covalent	Both	No
"Salts"	$(VO)_2 P_2O_7$	Yes	Medium–strong	No	+covalent	Strong	Weak
	$Ti P_2O_7$	Yes	Medium–strong	No	Covalent	Strong	Weak
	Ti-silicalite, TiO_2 in SiO_2	Weak	Strong	No	+ionic	Medium	Weak
	$Mg_3(VO_4)_2$	No	Weak	Medium		Medium	Strong
Supported oxides	V_2O_5–TiO_2, Nb_2O_5–Al_2O_3 "monolayers" or more	Yes	Medium–strong	No-medium	Covalent on ionic	Supported	Support

equivalent in octahedral coordination in a hcp oxide array. α-Al_2O_3 is usually prepared as medium-low surface area powders ($\sim$50 m^2/g) by topotactic decomposition of diaspore α-$AlOOH$ at 750 to 820 K or by calcination of gibbsite $Al(OH)_3$ to near 1300 K, and as very low surface area powders ($\sim$1 to 3 m^2/g) by heating any alumina powder above 1100 to 1300 K. In addition, the powders produced by high-temperature sintering of transition aluminas retain the memory of the lamellar structure of the precursors but with lamellae thicknesses increased significantly by sintering. Corundum powders are applied in catalysis as supports, for example, of silver catalysts for ethylene oxidation to ethylene oxide [134], just because they have low Lewis acidity, low catalytic activity, and, conversely, they are mechanically and thermally very strong. All other alumina polymorphs are metastable, and most of them have a structure, which can be related to that of cubic spinel $MgAl_2O_4$.

γ-Al_2O_3, which is the most used form of alumina, is mostly obtained by decomposition of the boehmite oxyhydroxide γ-$AlOOH$ (giving medium surface area lamellar powders, $\sim$100 m^2/g) or of a poorly crystallized hydrous oxyhydroxide called "pseudobohemite" at 600 to 800 K, giving high surface area materials ($\sim$500 m^2/g). γ-Al_2O_3 is one of the most used materials in any field of technologies. However, the details of its structure are still matter of controversy. It has a cubic structure described by Lippens and de Boer [135] to be a defective spinel (S.G. Fd3m; $Z = 8$), although it can be tetragonally distorted. Being the stoichiometry of the "normal" spinel $MgAl_2O_4$ (with Al ions virtually in octahedral coordination and Mg ions in tetrahedral coordination) the presence of all trivalent cations in γ-Al_2O_3 implies the presence of vacancies in usually occupied tetrahedral or octahedral coordination sites. ^{27}Al NMR spectra not only show that tetrahedral Al is near 25% of all Al ions (Figure 9.6[b]) but also show a small fraction of Al ions that is in coordination five [136], or highly distorted tetrahedral. Soled [137] proposed that the cation charge can be balanced, more than by vacancies, by hydroxy ions at the surface. In fact, γ-Al_2O_3 is always hydroxylated; dehydroxylation occurring only at a temperature where conversion to other alumina forms is obtained. XRD studies using the Rietveld method, performed by Zhou and Snyder [138], suggested that Al^{3+} cations can be in positions different from those of spinels, that is, in trigonal coordination. The possibility of a structure of γ-Al_2O_3, as a "hydrogen-spinel" has been proposed based on IR spectroscopy [139]. Calculations based on the composition HAl_5O_4 have been performed but found that this structure is very unstable [140].

DFT calculations have been performed recently, but did not solve the problem completely. Sohlberg et al. [141,142] arrived to a structure very similar to that proposed by Zhou and Snyder [138], based on spinel but with occupation of extraspinel sites. On the contrary, Digne et al. [143] and Krokidis et al. [144] proposed a structure based on ccp oxide lattice (S.G. $P2_1/m$; $Z = 8$) but different from that of a spinel, with 25% of Al ions in tetrahedral interstices and no structural vacancies. According to these authors, this structure, although unstable with respect to corundum, is more stable than that of the spinel based structures.

Calcination at increasing temperatures gives rise to the sequence γ-$Al_2O_3 \rightarrow$ δ-$Al_2O_3 \rightarrow \theta$-$Al_2O_3 \rightarrow \alpha$-$Al_2O_3$ [145]. According to ^{27}Al MAS-NMR the ratio between tetrahedrally coordinated and octahedrally coordinated aluminum ions increases upon the sequence γ- $\rightarrow$ δ- $\rightarrow$ θ-Al_2O_3. Tetrahedric Al^{3+} is near 25% in γ-Al_2O_3, 30–37% in δ-Al_2O_3 and 50% (in principle) in θ-Al_2O_3.

According to Lippens and de Boer [135] and Wilson and Mc Connell [145], δ-Al_2O_3 is a tetragonal spinel superstructure whose unit cell is constituted by three spinel unit blocs with tetragonal deformation, likely with a partial ordering of Al ions into octahedral sites. It is formed continuously in the range 800 to 900 K. θ-Al_2O_3 is formed above 900 K with simultaneous decrease of the surface area to near $100\ m^2/g$ or less. Its monoclinic structure, which is the same of β-gallia (S.G.: C2/m , Z = 4) can be derived from that of a spinel, with deformation and some ordering of the defects, with half-tetrahedral and half-octahedral Al ions [146]. During the sequence γ-Al_2O_3 $\rightarrow$ δ-Al_2O_3 $\rightarrow$ θ-Al_2O_3 $\rightarrow$ α-Al_2O_3 the lamellar morphology of bohemite is mostly retained but with progressive sintering of the lamellae and disappearance of the slit shaped pores.

η-Al_2O_3 is also considered to be a spinel-derived structure but is obtained by decomposing Beyerite $Al(OH)_3$. Most authors conclude that η-Al_2O_3 corresponds to a defective spinel like γ-Al_2O_3 but with a different distribution of vacancies, namely with more tetrahedrally coordinated (35%) and less octahedrally coordinated Al ions [138,141,142,147,148]. Calcination gives rise to the sequence η-Al_2O_3 $\rightarrow$ θ-Al_2O_3 $\rightarrow$ α-Al_2O_3.

Other metastable forms of alumina, denoted as ρ-Al_2O_3, χ-Al_2O_3, and κ-Al_2O_3 [132,133] also exist and can be obtained from the hydroxides gibbsite and tohdite, but they seem to have less interest in catalysis. Amorphous aluminas [149,150], possibly impure from organic reagents, have also been investigated. They tend to convert into γ-Al_2O_3 upon hydrothermal treatment. Amorphous alumina appear to be quite inactive as a acid catalyst and Al ions there appear to be essentially in octahedral coordination.

Transition aluminas, mostly denoted as γ-Al_2O_3, but actually being frequently a mixture of γ-Al_2O_3, δ-Al_2O_3 and θ-Al_2O_3, or of η-Al_2O_3 and θ-Al_2O_3, have wide application in adsorption and heterogeneous catalysis as catalysts (e.g., for the Claus process) and as supports (e.g., in hydrodesulfurization). γ-Al_2O_3 is also active as an acidic catalyst. As for example it is very active in the dehydration of alcohols to olefins and to ethers [149,151], in the dehydrochlorination of chlorided alkanes to olefins [152] as well as both in double bond isomerization [153] and in skeletal isomerization of olefins [154].

9.5.1.2 The "active sites" of aluminas

Many studies have been devoted to the surface characterization of transitional aluminas, and several models have been proposed, mostly related to the structure and multiplicity of the surface hydroxy groups. After the work of Peri [155], and of Tsyganenko and Filimonov [156], Knözinger and Ratnasamy [157] reported a very popular model of the different exposed planes of spinel type aluminas. This model has been later modified by Busca et al. [158,159]. These studies have been reviewed by Morterra and Magnacca [160]. More recently, additional investigations have

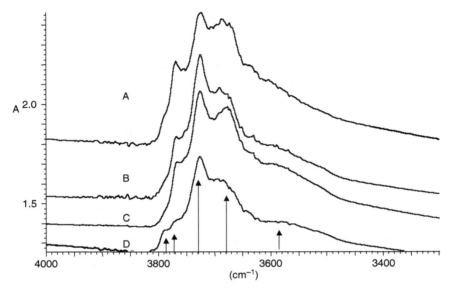

FIGURE 9.8 FT-IR spectra of four commercial γ-Al$_2$O$_3$ samples activated at 773 K by outgassing

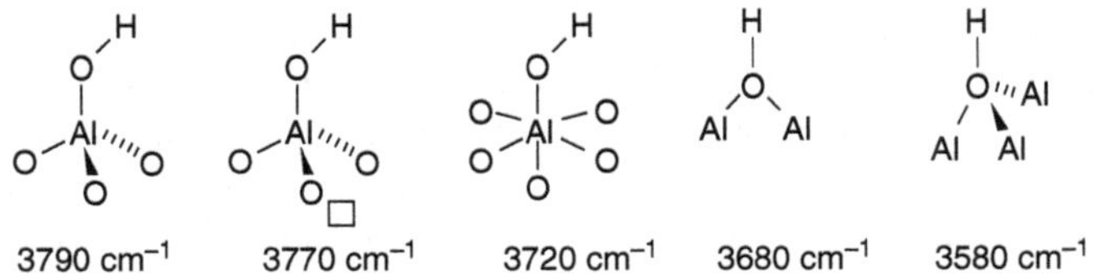

3790 cm^{-1} 3770 cm^{-1} 3720 cm^{-1} 3680 cm^{-1} 3580 cm^{-1}

SCHEME 9.4 Assignment of the OH stretchings of transitional aluminas, following Busca et al. [158,159], based on the comparison with the spectra of nondefective and partially defective spinel aluminates and ferrites. The square represents a vacancy in a normally occupied position of stoichiometric spinels

been published by Tsyganenko and Mardilovich [161] and, based on the theoretical calculations by Fripiat et al. [162] and by Digne et al. [112,143], who attempted to model the interaction of probe molecules. Typical spectra of γ-Al$_2$O$_3$ samples are reported in Figure 9.8. In Scheme 9.4 the assignments of the five main νOH bands of the surface hydroxy groups of transitional aluminas proposed by Busca et al. [159,160] are schematized.

The catalytic activity of transitional aluminas (γ-, η-, δ-, θ-Al$_2$O$_3$ are undoubtedly mostly related to the Lewis acidity of a small number of low coordination surface aluminum ions, as well as to the high ionicity of the surface Al–O bond. The alumina's Lewis sites have been well characterized by adsorption of probes such as pyridine, carbon monoxide, and several bases followed by IR [161], ammonia and amines followed by calorimetry [40,44,46], triphenylphosphine followed by ^{31}P NMR [163], to be the strongest among metal oxides, only

TABLE 9.11

Number (site/nm^2) and Distribution of Lewis Acid Sites on Different Metastable Aluminas as a Function of the Pretreatment Temperature by Using TMP and CO Adsorption (Ref. 107)

	γ-Al$_2$O$_3$			δ-Al$_2$O$_3$			θ-Al$_2$O$_3$		
	Weak	Medium	Strong	Weak	Medium	Strong	Weak	Medium	Strong
673	0.09	0.02	0	0.13	0.02	0	0.07	0.02	0
803	0.22	0.09	0.06	0.14	0.18	0.02	0.21	0.01	0.01
923 TMP	0.31	0.05	0.08	0.31	0.08	0.04	0.28	0.14	0.01
CO			0.78			1.05			1.26

weaker than those of Al halides. Volumetric, TPD and calorimetric experiments allowed also to determine the amount of such very strong Lewis sites present on transitional alumina surfaces, which however depend on the dehydroxylation degree (depending on the activation temperature) and on the peculiar phase and preparation.

Mathonneau [107] recently measured the density of Lewis acid sites on γ-Al$_2$O$_3$, δ-Al$_2$O$_3$, and θ-Al$_2$O$_3$ after outgassing at 673, 803, and 923 K. The site densities reported in Table 9.11 obtained by dosing trimethylphosphine and following adsorption by ^{31}P NMR, are a little smaller than those measured by the same author by dosing the adsorption of CO, and agree roughly with the data reported by Bolis et al. [46] as the result of adsorption of CO. These data are also comparable with those reported by Auroux and Gervasini [40] in their microcalorimetric study concerning ammonia adsorption on a commercial γ-Al$_2$O$_3$ previously outgassed at 673 K for 2 h, and with those arise from adsorption of butylamine (Figure 9.2). The density of the very strong adsorption sites responsible for ammonia adsorption heat of more than 200 kJ/mol is reported to be near 0.1 sites/nm^2 [40]. Additionally, a "continuous heterogeneity" of the acidity of the sites whose total amount (for adsorption heats higher than 100 kJ/mol) is near 0.5 site/nm^2 was found. Taking into account the bulk density of γ-Al$_2$O$_3$, it is easy to calculate that at most one site every 50 to 100 acts as a strong Lewis site on γ-alumina outgassed at 673 to 823 K, the large majority being still hydroxylated or not highly exposed at the surface.

It seems that, although the different alumina "spinel-type" phases react a little differently to outgassing, the density of the strongest Lewis acid sites tends to decrease a little by increasing the historical calcination temperature of the alumina (i.e., upon the sequence $\gamma \rightarrow \delta \rightarrow \theta$, which is also a sequence of decreasing surface area). As a result of this, the number of strongest acid sites per gram significantly decreases in this sequence, although catalyst stability increases.

Characterization studies determined that the transition aluminas also present significant surface basicity. In fact, CO$_2$ adsorption sites on γ-Al$_2$O$_3$ is quite strong although the density of sites for strong adsorption (ΔH$\div$50–180 kJ/mol) is small (0.06 site/nm^2). Apparently, a little more sites (0.2 site/nm^2) can give strong adsorption of hexafluoroisopropanol [48] and even more (>1 site/nm^2) give strong adsorption of SO_2 for aluminas activated at 673 to 773 K [49].

However, the true particular sites of aluminas for most catalytic reactions are very likely the anion–cation couples, which have very high activity and work synergistically. Alcohol adsorption experiments [127,164] allow the characterization of such sites where dissociative adsorption occurs. Mechanistic studies suggest that such cation–anion couples are likely those active in alcohol dehydration [165], in alkylchloride dehydrochlorination [152,166], and in double bond isomerization of olefins [167] over γ-Al_2O_3.

Although most authors correctly attribute to transitional aluminas essentially Lewis acidic properties, several studies show that some of their multiple surface hydroxy groups also have medium strong Brønsted acidity [168]. Actually, as shown in Table 9.7 and Table 9.8, alumina seems to be, among the pure ionic oxides, one of the strongest Brønsted acids. The activity of pure gamma alumina as a good catalyst of skeletal n-butene isomerization to isobutene has been attributed to its medium-strong Brønsted acidity, sufficient to protonate n-butenes at high temperature, producing carbenium ions, but too low to cause much cracking and coking [154].

9.5.1.3 Impurity related surface properties of aluminas

In Figure 9.8, the IR spectra of four γ-Al_2O_3 commercial preparations are reported in the OH stretching region. The characteristics of these samples are reported in Table 9.12. These materials have similar surface areas and are sold as highly "pure"gamma-aluminas. XRD and morphology data do not show significant differences among them. Additionally, qualitative IR investigations (concerning the nature of the surface hydroxy-groups and of the surface Lewis and basic sites)

TABLE 9.12

Characteristics of Four γ-Al_2O_3 Commercial Samples (Figure 9.8), and catalytic activity in n-butene conversion (isobutene main product, the conditions are the same for all catalysts)

Sample	Intensity[a] T_{act} 500°C	I/m^2 T_{act} 500°C	Intensity [a] T_{act} 700°C	I/m^2 T_{act} 700°C	SA m^2/g	PV cc/g	Na[b] ppm	X[c] 1 but
A	60	0.316	67	0.353	190	0.454	<10	37.3
B	55	0.325	48	0.284	169	0.702	<10	25.4
C	48	0.257	23	0.123	187	0.719	25	23.8
D	34	0.173	13	0.066	196	0.734	70	12.2

[a] Arbitrary units.
[b] Chemical analysis.
[c] n-butene conversion.

failed to find significant differences. However, their catalytic activity in n-butene skeletal isomerization is definitely different, being decreasing in the order A > B > C > D. Actually, the activity trend correlates well with the total integrated intensity of the νOH stretching band of the surface hydroxy groups (I/m^2) and, inversely, with the sodium content derived by chemical analysis (Table 9.12). This content is always very low, but differs significantly among the samples. It seems likely the sodium exchanges the protons of the surface hydroxy groups. It has been concluded that the amount of residual sodium, although always low, is determinant for decreasing the number of the active sites for n-butene isomerization on γ-Al$_2$O$_3$, which is believed to be a proton-catalyzed reaction [154].

Sodium content on alumina strongly depends on the preparation method. Aluminas derived from Aluminum metal via alkoxide have Na content ($\leq$40 ppm as Na$_2$O) generally about ten times lower than those derived from Bauxite via the Bayer process. Sodium impurities decrease not only the number of active sites but also their strength according to induction effects [169,170], so finally decreasing the alumina activity in acid catalyzed reactions. As it will be discussed later, sodium cations are so big that they are unable to enter the cavities of ccp oxide ions of spinel-type structures. For this reason, even when their total concentration is small, they concentrate at the surface and have a relevant poisoning effect. This however, is sometimes beneficial, like when alumina acts as a support. It has been shown that the high dehydrochlorination activity of alumina toward ethylene dichloride is a main cause for loss in selectivity upon ethylene oxychlorination over alumina-supported copper chloride catalysts [166,171]. To decrease such an unwanted support activity, doping is performed with alkali or alkali-earth chlorides, which finally poison the alumina active sites.

In addition, residual chlorine can modify significantly the properties of transitional aluminas. As for example, it has been shown that aluminas derived from ignition of AlCl$_3$ have different spectra of the surface hydroxy groups [172], as well as in a significantly different catalytic behavior, with the enhanced acidity typical of chlorided aluminas [173]. On the other hand, the amount of residual chlorine can also depend strongly on the pretreatment used prior to use.

9.5.2 The "Mixed Oxides" of Silicon and Aluminum: Composition and Structure Properties

Silica forms many different crystalline structures. All the structures which are stable at ambient pressure present tetrahedrally coordinated silicon atom and the structure is associated to a covalent network [174]. In Figure 9.1(d) the structure of α-quartz is shown. On the other hand, silica is also the best-known glass forming material [175], that is, it shows very stable amorphous states, which also consist of a tetrahedral covalent network structure, although disordered.

Crystalline "mixed oxides" of silicon and aluminum exist, called aluminum silicates. They are three polymorphic forms of Al$_2$SiO$_5$ (kyanite, andalusite, and sillimanite) and mullite (whose composition ranges between 3Al$_2$O$_3 \cdot$ 2SiO$_2$ and 2Al$_2$O$_3 \cdot$ SiO$_2$. Silicon is always tetrahedral while Al ions is octahedral in kyanite, half-octahedral and half-tetrahedral in sillimanite, half-octahedral and half-pentacoordinated in andalusite. In mullite, Al is basically octahedral but a variable amount of it occupies also tetrahedral sites. These are Al-rich crystalline materials generally obtained at high temperature as sintered ceramic materials. A spinel-type phase with composition $6Al_2O_3 \cdot SiO_2$, where Si substitutes for Al in tetrahedral coordination, has also been reported as a metastable form [176].

The substitution of aluminum for silicon in a silica covalent network leads to a charge unbalance, which must be compensated by "extra-framework" cations, mostly alkaline. This occurs in the cases of the so-called "stuffed silicas": these materials have structures strictly related to the crystalline forms of silica, but with cations in the interstices to counterbalance the presence of Al ions substituting for Si. This is the case, for example, of Eucriptite ($LaAlSiO_4$, a stuffed β-quartz) or nepheline ($NaAlSiO_4$, a stuffed tridymite). A similar mechanism also occurs in the amorphous networks of glasses [175].

A similar situation is that of zeolites. They are natural framework silicoaluminates where, however, cations (usually alkali or alkali earth) are located in relatively large cavities in the interior of the $[Si_{1-x}Al_xO_2]^{x-}$ negatively charged framework. These cavities are connected by channels that give rise to a variety of microporous structures, which can be penetrated only by sufficiently small molecules, so giving rise to the "molecular sieving" effect [177] as well as the "shape selectivity" effect in catalysis [178,179]. The cations are exchangeable, so zeolites also act as cationic exchangers. The exchange can be performed with ammonium ions which can be later decomposed into gaseous ammonia and a proton. This allows to produce protonic zeolites, which are very strong solid Brønsted acids. On the other hand, protonic zeolites can also be synthesized by using templating agents [180]. In this case, the protons may be residual from the combustion or decomposition of the templating agents. Protonic zeolites are formally crystalline Si–Al mixed oxides or solid solutions of alumina in crystalline silica network, although the protons are necessary for stoichiometry. Their general formula is, in principle, $H_xSi_{1-x}Al_xO_2$. The value of x is generally quite low, the protonic structures becoming unstable when Al content is relatively high, although this depends from the particular zeolite structure. On the other hand, the totally siliceous material ($x = 0$) not always can actually be synthesized. In Figure 9.9, a scheme of the MFI structure is reported with the structures of protonic sites and exchanging cations.

The structural details of the (mostly) amorphous oxides resulting from coprecipitation or co-gelling of Si and Al compounds, that is, the so-called silica–aluminas or alumina-silicas, are still less clear, if at all. Commercial materials are available with any composition starting from pure aluminas up to pure silicas. The silica-rich materials are generally fully amorphous and are called "silica–aluminas." They behave as strongly acidic materials and have been used for some decades (1930 to 1960) as catalysts for catalytic cracking processes, and still find relevant industrial application [1]. IR spectra clearly provide evidence for a tetrahedral silica-based amorphous network. In recent years mesoporous silica–alumina containing ordered relatively big channels have been developed and they could somehow considered like very large pore zeolites. Actually, this

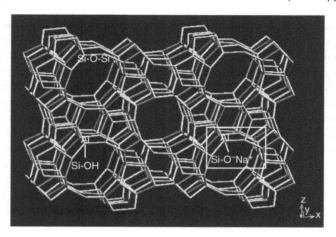

FIGURE 9.9 Structure of a typical zeolite (MFI) with normal oxygen bridges, protonic sites, and exchanging cations

is not true, because these materials are essentially amorphous silica–alumina with nonstructural although sometimes ordered mesopores. ^{27}Al MAS-NMR studies (Figure 9.6c for a typical cracking catalyst) provided evidence for tetrahedral coordination of Al in silica–aluminas, although variable amounts of octahedral Al species as well as pentacoordinated or distorted tetrahedral Al are also present [181–183]. According to Omegna et al. [183], two kinds of octahedrally coordinated Al species actually exist in silica–aluminas. Part of it in fact changes its coordination when adsorption of bases like ammonia occurs, converting into tetrahedral. This is possibly associated to surface aluminum species that, upon the effect of a base, generates a Brønsted acidic site by bridging of a silanol over Al cation.

Alumina-rich materials have been the focus of few investigations. According to Trombetta et al. [184] and Daniell et al. [185] they have the structure of γ-alumina, silica being mostly located at the surface. In the field of catalysis, a number of other materials belonging to the silicon/aluminum mixed oxide system have been considered. The so-called "silicated aluminas," obtained by a "reactive" deposition of silica precursors (such as TEOS) usually onto of the γ-alumina, are a relevant class of materials, being good catalysts for the skeletal isomerization of butenes to isobutene [186,187]. In addition, alumina can be deposed on silica. Finally, it can be mentioned that several commercial aluminas actually contain small amounts of silica essentially to stabilize them against the phase transformation to corundum. However, only silica–aluminas and protonic zeolites appear to have acidic properties significantly different from those of alumina and silica.

9.5.2.1 The Brønsted acidic sites of protonic zeolites

It is unanimously recognized that the bridging hydroxy groups Al–(OH)–Si, which are located in the walls of the zeolitic cavities constitute the strong acidic sites of protonic zeolites. The proton balances the charge defect due to the Al for Si substitution in the framework (Figure 9.9). It has been recently underlined that the bridging OH's are only detected in the interior of the zeolitic cavities, being the corresponding spectroscopic features absent in any nonzeolitic material based on silica and alumina [187] and also on the external surfaces of different zeolites (see later). Thus, the existence of the bridging hydroxy groups Al–(OH)–Si implies the existence of the cavity. In other words, cavity is involved in the site. This means that many theoretical studies based on "clusters" where bridging OH's are present without the cavity, frequently used in theoretical studies [108–111], are possibly unappropriate being models of nonexisting structures.

The bridging hydroxy groups of zeolites are well characterized by the presence, in the IR spectrum, of a quite definite and strong band in the region between 3650 and 3500 cm^{-1}. In Table 9.13, some structural data of the most used protonic zeolites together with the position of the OH stretching of the Brønsted acidic sites, are summarized. These groups are also characterized by an evident ^1H MAS-NMR narrow peak near 4.4 ppm [90,93,94,188,189]. Parallel ^1H NMR and IR studies show that the IR extinction coefficient of the zeolite's bridging OH's is far higher than for silanol groups, and this allowed Kazansky et al. [190] to propose to use the intensity of the IR band to determine the surface acid strength. Kustov [63] discussed the presence of additional broad OH stretching bands in H-zeolite's IR spectra.

Most data agree suggesting that, when the Al content is relatively low, the amount of Brønsted sites in zeolites actually strictly depends on Al concentration, according to the theory. The ratio between catalytically active sites and Al

---

TABLE 9.13

Channel Structure and Application of the Protonic Zeolites Investigated Here

Symbol	Channel structures	νOH IR	δH NMR
H-FER	10-ring channel [001] 4.2 Å × 5.4 Å	3595	4.4
	8-ring channel [010] 3.5 Å × 4.8 Å		
H-MFI	10-ring channel [010] 5.3 Å × 5.6 Å (straight)	3620–3610	4.2–4.4
	10-ring channel [100] 5.1 Å × 5.5 Å (sinusoidal)		
H-BEA	12-ring channel [100] 6.6 Å × 6.7 Å	3608	4.6
	12-ring channel [001] 5.6 Å × 5.6 Å	3605	5.6
H-MOR	12-ring main channel [001] 6.5 Å × 7.0 Å	3605, 3609	4.6
	8-ring compressed channel [001] 2.6 Å × 5.7 Å		
	8-ring side pockets [010] 3.4 Å × 4.8 Å	3588	
H-FAU	12-ring main channels [111] 7.4 Å × 7.4 Å	3625	3.9–3.8
	Sodalite cavities accessed through 6-ring	3553	4.6–4.4
	channels ca 2.7 Å × 2.7		

ions ranges apparently from 80 to 100% for highly siliceous extraframework-species-free zeolites [191,192]. Different opinions seem still to exist on whether the position of the OH stretching band can actually be correlated with the Brønsted acidity of the site and/or what are the factors determining IR OH band and ^{1}H NMR peak positions and Brønsted acidity. The OH stretching band position and width can be influenced by H-bondings through the cavities [193]. On the other hand, some authors suggested that a correlation exists between OH stretching frequency and the Si–O(H)–Al bond angle [194].

The position of the IR band due to bridging OH's, is somehow dependent on the size of the zeolites cavities species, as shown in Table 9.13. In the case of zeolites with more than one type of quite different cavities, splitting of the band of the bridging hydroxy groups can be observed. This can be very clearly observed in the case of H–FAU (H–Y) zeolite, where two well defined although weak bands appear for bridging OH's (Figure 9.10[a]). The high frequency HF band (3627 cm^{-1}) has been assigned to bridging OH's located in the so-called supercage, while the low frequency LF band (3562 cm^{-1}) has been assigned to OH's located near the middle of the six bond rings connecting the sodalite cages [193,195]. Although the cavity where the LF OH group is likely located is small and is accessed through a quite compressed six-ring channel, even large and hindered molecules can interact with this site. This is shown by using 2,2-dimethylpropionitrile (pivalonitrile) as a probe, in Figure 9.10(b)–(d). From Figure 9.10 and Figure 9.11, where corresponding subtraction spectra are reported, is evident that pivalonitrile is able to interact with

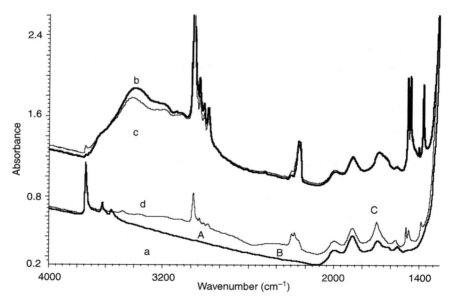

FIGURE 9.10 FT-IR spectra of HY zeolite after activation at 673 K (a), adsorption of pivalonitrile (b), outgassing at r.t. (c), and at 273 K (d)

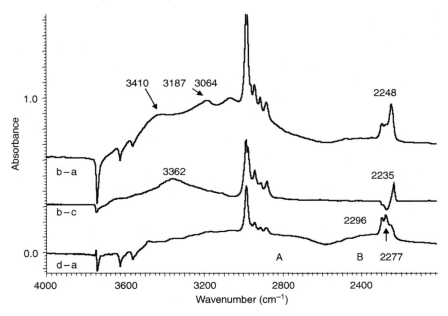

FIGURE 9.11   Subtraction spectra relative to Figure 9.10

both bridging OH families giving rise to "quasi-symmetrical hydrogen bondings" responsible for the so called A,B,C spectrum [196], stable even by outgassing at 273 K. This is because the main cavity is so large to allow the pivalonitrile molecule to rotate and to put the nitrile group (whose dimensions are sufficiently small) in the six-atom ring.

In the case of H-BEA, the band of bridging OH's is a little split [192,197]. The Brønsted acid sites present in the sinusoidal channels (νOH 3608 cm^{-1}) can be distinguished from those located in the bigger ones (νOH 3620 to 3612 cm^{-1}). In fact, most of the band, centered at 3620 to 3612 cm^{-1}, is consumed upon the adsorption of 2,2-dimethyl-propionitrile (pivalonitrile) showing that most of the bridging OH's can be accessible to such a molecule [197,198]. This agrees with the diameter of the main 12-ring channels of BEA, which are sufficiently big to be entered by molecules containing the tert-butyl group. On the other hand, a component can be still seen even in presence of pivalonitrile near 3608 cm^{-1}. This suggests that the residual unperturbed OH's are located in the smaller 10-ring channels that actually should not allow the entrance of pivalonitrile. Kotrel et al. [192] observed that the intensity of the low frequency component only, correlates with the catalytic activity in n-hexane cracking for Na-poisoned H-BEAs.

Several recent studies have been devoted to the characterization and to the loc-alization of the "framework" protonic sites in H-MOR [199–203]. Several authors reported that the OH's in the so-called side pockets and smaller channels are asso-ciated to a band which is located at distinctly lower frequencies (near 3580 cm^{-1} with respect to those located in the main channels. Identification of distinct

Brønsted sites in H-MOR have also been reported based on NMR techniques [96,196].

Studies on the adsorption of hindered nitriles [82,204] allowed to distinguish three families of bridging OH's in H-MOR. The spectra of the activated H-MOR (Figure 9.12[a]) and of the zeolite in contact with propionitrile (Figure 9.12[b]), 2-methylpropionitrile (isobutironitrile, Figure 9.12[c]), 2,2-dimethylpropionitrile (pivalonitrile, Figure 9.12[d]), and 2,2-diphenyl-propionitrile (Figure 9.12[e]) provide evidence for the different access of these molecules and of the different locations of the bridging OH's. Those responsible for a band at 3588 cm^{-1} are available to interact with linear nitriles, but not to branched and aromatic ones: they are likely located inside the side-pockets. Others, interact with linear and mono-branched nitrile (isobutyronitrile), but not with doubly branched pivalonitrile and to aromatic nitriles: they, absorbing at 3609 cm^{-1}, are likely at the intersection between side pockets and main channels. The small dimension of the main channel of mordenite and the rigidity of the pivalonitrile and benzonitrile molecules do not allow the rotation and the bending of the probes, which, consequently, cannot probe such an intersection. A third family interacts with all nitriles investigated except 2,2-diphenyl-propionitrile. They absorb at 3605 cm^{-1} and are thought to stand near the center of the main channels. 2,2-diphenyl-propionitrile is so large a molecule that it cannot enter even the main channels of mordenite. Consequently, this molecule probes the external surface only of H-MOR, where no Bridging OH's are found: however, terminal silanols are mostly located there, together with Lewis sites.

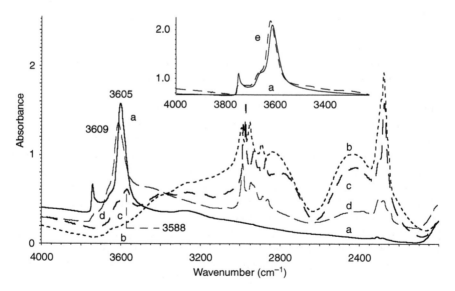

FIGURE 9.12 FTIR spectra of H-MOR zeolite after activation at 673 K (a) and after saturation with nitrile vapors: (b) propionitrile. (c) Isobutironitrile, (d) pivalonitrile, and (e) diphenylpropionitrile

Although the channels of H-ZSM5 are very similar to each other, and the OH's are usually considered to be homogeneous, the benzene-driven access of pivalonitrile [205] suggests that the OH's in the channels intersection, considered to be less stable by theory [111], absorb at a slightly higher frequency (3610 to 3620 cm^{-1}) than those located in the channels 3610 to 3590 cm^{-1} [205]. The adsorption of hindered nitriles also allowed to investigate independently the OH's located in the two channels of H-FER. In fact isobutironitrile enters only the larger channels of H-FER, where the OH's vibrate at 3594 cm^{-1}, leaving free the OH's of the smaller channels that absorb at 3597 cm^{-1} [206].

The effect of the zeolite structure on their acidity has been investigated by several authors with different techniques. Zecchina and coworkers have investigated the hydrogen bonding of different basic probes with different zeolites and other Brønsted acids such as HF and H-Nafion [201,208,209]. They found protonic zeolites to be weaker acids than H-Nafion and stronger acids than HF. On the other hand, it seems that differentiation of the acid strength of protonic zeolites, such as H-MFI, H-MOR, and H-BEA is difficult, although all of them seem to be slightly more acidic than H-FAU. A quite similar trend has been reported by Auroux [39], who summarized the microcalorimetric data concerning ammonia adsorption: according to these data, the trend for the cited zeolites is H-MOR ($\Delta H_{adsNH3} > 160$ kJ/mol) $\geq$ H-ZSM5 $\geq$ H-FER > H-BEA $\geq$ H-FAU ($\Delta H_{adsNH3} \sim 130$ kJ/mol), although the secondary "solvation" effects in the cavities could be even more determinant in this measure. Also, ammonia TPD data reported by Niwa and coworkers [191] give a similar trend: H-MOR > H-MFI > H-BEA > H-FAU. Thibault-Starzyk et al. [80], instead, deduced from the low temperature CO adsorption the trend H-MOR > H-FAU > H-MFI > H-FER, while, using the high temperature protonation of acetonitrile obtained a trend that correlates better with n-hexane cracking, that is, H-MOR > H-FER > H-MFI > H-FAU. Theoretical calculations by Brändle and Sauer [111], in contrast, indicate that deprotonation energy is higher for highly siliceous H-FAU than for H-MOR and H-MFI, thus supporting a higher acidity of H-FAU and a low one for H-MFI.

9.5.2.2 The acidic sites of silica–aluminas

According to several authors in the 1990s [5,210], the active site for silica–alumina and protonic zeolites is the same (i.e., it is constituted by the bridging hydroxy groups bonded to a silicon and an aluminum atom) in contrast to the fact that the catalytic activity of protonic zeolites in proton-catalyzed reactions is by far higher than that of silica–aluminas.

In the 1970s and 1980s already was evident how different are the spectroscopic features of silica–alumina and zeolites [211]. More recently, however, some papers reported the presence of very small bands near 3600 cm^{-1} in the spectra of mesoporous silica–aluminas [212–214], supposed to be due to bridging zeolite-type sites. Theoretical works till the end of the 1990s modeled the active site for zeolites and silica–alumina in the same way, as Al–(OH)–Si bridging hydroxy groups. In contrast, it seems established today the complete absence of the IR band of bridging OH's in the case of all silica–aluminas, either micropor-ous or mesoporous [64,186,187,215–217]. This provides evidence of a different Brønsted active site in silica–aluminas, with respect to H-zeolites, or of a more complex behavior. The IR spectra of silica–aluminas always present a very sharp band near 3747 cm^{-1} certainly due terminal silanols, spectroscopically very sim-ilar to those of pure silicas and of any silica-containing material, with a tail likely due (as on pure silica too) to H-bonded silanols. This in spite of the definitely higher Brønsted acidity of silica–aluminas with respect to silica, detected, for example, by protonation of ammonia, pyridine [64], amines [45] and also by the strong H-bonding with nitriles [187,218]. The strong acidity of silica–alumina has also been demonstrated by microcalorimetric studies of the adsorption of ammonia and pyridine [36] as well as by TPD data of desorption of the same bases and in catalytic activity studies.

In the case of silica–aluminas, the OH stretching band contains; however, two or more components almost unresolved in the activated sample but due to OH's interacting with distinctly different strengths with bases [187,218]. According to Trombetta et al. [215] the Brønsted acidic centers of silica alumina are due to terminal silanols, which are somehow modified by nearest Al cations.

^{1}H MAS NMR provides evidence for a structurally different active site in silica aluminas with respect to zeolites. Proton NMR spectra of activated silica–aluminas in fact usually show a single peak at δ 1–7 1.8 ppm assigned to terminal silanols, with a broader component located near 2.8 ppm attributed to Al–OH [90,182,219], in contrast to the typical bridging OH of zeolites which resonates sharp near 4.2 to 4.4 ppm. According to Kuroda [182] a broad weak signal in the case of silicalumina can be detected near 3.8 ppm could be due more acidic OH's.

Two questions arise from this picture: (i) Why well evident bridging OH's are formed in the case of protonic zeolites, while they are not evident in amorphous and mesoporous silica–aluminas? (ii) How Al ions generate the Brønsted acidity of silica–aluminas? According to Trombetta and Busca [187,215,220], the rigid-ity of the zeolite frameworks and the interaction of the proton with the electric field of the cavity should stabilize the bridging hydroxyl structure only in the case of the internal cavities of zeolites. They also proposed that the bridging of the silanol over a nearest Al ion cannot occur on silica if no bases are present, but when a base increases the polarization of the O–H bonding, the oxygen can bridge over the nearest Al ions and deprotonation becomes easier (Scheme 9.5, up). This is proposed to be the mechanism for Brønsted acidity generation in silica–aluminas. This mechanism has been supported recently by the structural study reported by Skowronska-Ptasinka et al. [221] of a aluminosilsesquioxane complex where bridging Si–OH–Al group exist but the proton is just involved in the protonation of a base (triethylamine), like proposed by Trombetta and Busca (Scheme 9.5, up). To our knowledge, still chemical species containing bridging silanol complexes free from a base have not been prepared.

Based on proton NMR spectroscopy, Heeribout et al. [219] proposed a mechan-ism for the formation-destruction of strong Brønsted acid sites in silica–aluminas by adsorption–desorption of water (Scheme 9.5, middle). According to these authors, in fact, bridging silanol groups exist only in the presence of some water, giving rise to hydroxonium ions. When water is forced to desorb, the bridge opens giving rise to Al–OH species. This mechanism forecasts the generation of an unsaturated silicon species, which seems quite unlikely. The mechanism proposed by Trombetta and Busca [220] has been considered to be compatible with the ^{27}Al MAS NMR experiments of Omegna et al. [183], who proposed a mechanism of Brønsted acidity generation upon adsorption–desorption of water involving flexibly coordinated Al species (reported in Scheme 9.5, down).

More recently, Busco and Ugliengo [222] addressed this problem on the base of computational methods. These authors showed that, on a nonzeolitic structure, the

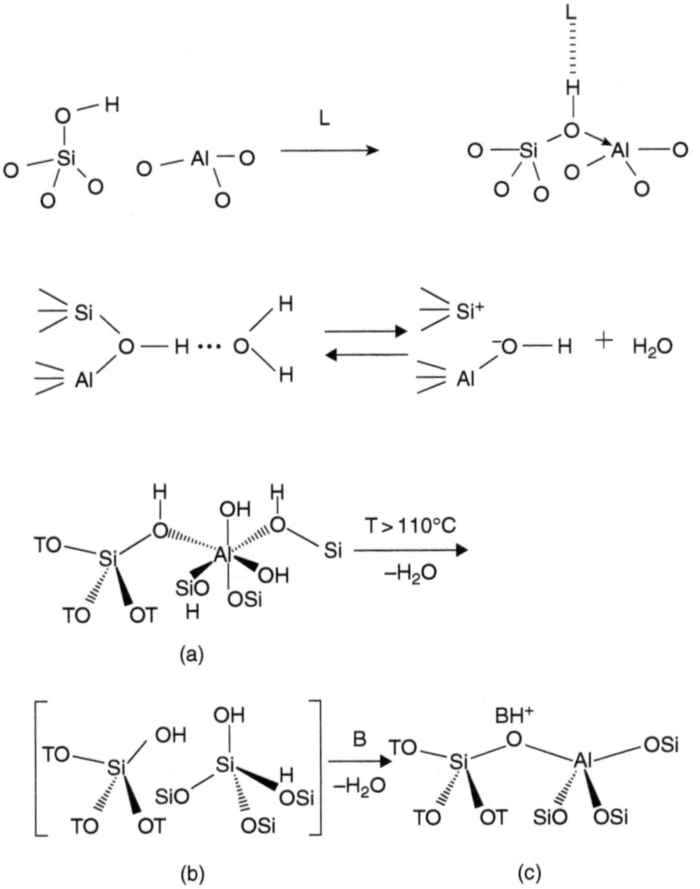

SCHEME 9.5 Mechanisms explaining the Brønsted acidity of silica alumina and the absence, in usual conditions, of bridging hydroxy groups. (a) Trombetta and Busca [220]. L = a base, (b) Heeribout et al. [219], and (c) Omegna et al. [183]

bridging OH structure is unstable toward the internal protonation of a nearby OH group bonded to the Al^{3+} ion and the water molecule is tightly bound to the Lewis site. However, the Bridging site could be formed back by water adsorption. On the other hand, it must also be mentioned that on the surface of silica alumina very strong Lewis acid sites can be detected, for example, by adsorption of pyridine [64], nitriles [187] and CO [189]. They are certainly due to highly uncoordinated Al ions and correspond to the strongest Lewis sites of transitional alumina or perhaps are even stronger, due to the induction effect of the covalent silica matrix. This makes silica–alumina also a very strong catalyst for Lewis acid catalyzed reactions, such as, for example, the dehydrochlorination and the steam reforming of halided hydrocarbons [223,224].

9.5.2.3 The external surface of protonic zeolites

The external surface of protonic zeolites can be relevant in acid catalysis. Several data suggest that nonshape-selective catalysis can occur at these sites, like in the case of alkylaromatics conversions over H-MFI [225,226]. On the other hand, H-zeolites also catalyze reactions of molecules, which do not enter the cavities due to their bigger size. Therefore, the external surface of zeolites is certainly active in acid catalysis. Additionally, the bulk versus surface Si/Al composition of a zeolite could be different and different preparation procedures can allow to modify this ratio [226]. Corma et al. [84] reported data on the accessibility of protonic sites of different zeolites to 2,6-di-ter-butyl-pyridine (DTBP). This molecule has been considered "selective" for Brønsted sites, due to its impossible interaction with Lewis sites for steric hindrance. According to these authors, however, the interpretation of the data is not straightforward, for several reasons such as the presence of different cavities and the big size of the probe itself. Surprisingly, Corma et al. found a complete accessibility of the sites of beta zeolite to DTBP. This contrasts the data of Trombetta et al. [197], who showed that protonic sites exist also on the smaller channels of beta, whose access to DTBP seems very unlikely. In a more recent publication, Farcasiu et al. [227] reported an accessibility of 90% of the protons of H-BEA to DTBP, much higher than the 36% for H-MOR and 31% for H-USY.

The external surfaces of H-FER [215,220] and H-MFI [220,228] have been studied by IR spectroscopy of adsorbed hindered nitriles. The result on H-FER has been confirmed using pyridine [187], lutidine, and aromatic hydrocarbons [83]. In both cases terminal silanols and Lewis acid sites exist at the external surface of the zeolites. Interestingly, the acidity of the external silanol OH's of zeolites can be, however, enhanced with respect to those of silica. The very strong bridging Brønsted acid sites, instead, only are located at the internal surface. Otero Arean et al. [229] obtained similar results on H-MFI using adamantine-carbonitrile as the probe. A similar result has been obtained on H-MOR using diphenylpropionitrile as a probe (Figure 9.12[e]). This molecule interacts with silanol groups but cannot access the cavities where all bridging OH's are located. Figure 9.11(a,b) shows that part of the silanol groups of H-FAU are shifted upon adsorption of pivalonitrile down to 3187 and 3064 cm^{-1}, showing also in this case an acidity definitely higher than those of typical silica's silanols.

Onida et al. [230] recently investigated the acidity of samples of ITQ-2 produced by exfoliation of MCM22 zeolite. They showed that upon exfoliation the band due to bridging OH's decreases strongly in intensity, while the band due to terminal silanols strongly increases. This indicates, that actually bridging sites cannot resist when the cavity disappears and they become exposed at the surface. These authors, however, report the existence on the surface of sites with intermediate acidity between silanols and bridging OH's, they consider not directly detectable in IR spectra. Possibly, they are indistinguishable spectroscopically from terminal silanols, as suggested by Trombetta et al. [187,215,220].

As cited previously, the adsorption of hindered nitriles reveal the presence of Lewis sites at the external surface of H-FER, H-MFI [220], and H-MOR (Figure 9.12[e]), after previous outgassing at 673 to 773 K. Recently, van Bokhoven et al. [231] showed that, in these conditions, tri-coordinated Al species can be detected by in situ XANES at the Al–K-edge on H-MOR and H-BEA. These authors propose these species is at framework position, but they cannot determine if these sites are internal or external. The presence of Lewis acid sites at the external surface of H-FER, H-MFI, and H-MOR may be relevant in catalysis. According to a recent work [232], an environmental-friendly vapor phase synthesis of 2-hydroxyacetophenone (or other 2-hydroxyphenylalkylketones) is possible through the Fries rearrangement of phenylacetates on a commercial H-MFI zeolite. In this case, treatments allowing the poisoning of the external Lewis sites and the formation of defects where few Brønsted sites are accessible with relaxed shape selectivity effects are likely the key to the optimal behavior of the catalyst. According to Trombetta et al. [220], the external surface of protonic zeolites is similar to that of silica alumina. This is reasonable because at the external surface of zeolites, the "rigidity" of the crystal is relaxed and the cavity effects, obviously, do not exist.

9.5.2.4 Extraframework material in protonic zeolites

Zeolite catalysts are actually applied frequently after treatments that tend to increase their stability and also, in case, to further enhance surface acidity and shape selectivity effects. These treatments, like steam dealumination, can cause the decrease of the framework Al content and the release from the framework of aluminum-containing species that contribute in stabilizing the framework, but can also contain additional catalytically active acid sites. Extraframework material (EF) can also arise from the preparation or the activation procedure or by addition of other components by impregnation or ion exchange. The presence of EF gives rise to the presence of strong additional bands in the OH stretching spectrum. In general, bands above 3750 cm^{-1}, and in the region 3730 to 3650 cm^{-1} in protonic zeolites is attributed to OH's on EF materials. Bands at 3780 and 3670 cm^{-1} in H-ZSM5 (Figure 5[a]) are in fact due to EF. Similarly, the detection of octahedral Al ions in ^{27}Al NMR techniques is evidence of EF. Several authors also attribute

Lewis acidity of zeolites to extraframework species, neglecting the evidence of their presence also at the external surface of the framework.

H-Y (H-FAU) zeolites are largely applied for catalytic cracking but in this case they must be stabilized by steam dealumination. The resulting materials are hydrothermally more stable (the so-called Ultrastable Y zeolite, USY). A similar stabilization effect is obtained by introducing in the cavities rare-earth elements such as lanthanum (REY zeolite), which can also create additional active sites. The IR spectrum of the surface OH's of low-Al content H-FAU, where the band of silanols and those of the HF and LF bridging OH's are sharp at 3743, 3627 and 3562 cm^{-1} (Figure 9.13[a]), is compared in Figure 9.13 with those of USY (Figure 9.13[b]), of a H-FAU with Si/Al 5 (Figure 9.13[c]), and of a REY (Figure 9.13[d]) It is evident in all such cases the presence of features due to extraframework species. The characterization of EF in USY and REY has been the object of several studies. According to ^{27}Al 3Q NMR and ^{29}Si MAS NMR studies, van Bokhoven et al. [233] concluded that extraframework octahedral cationic EF Al species causes a perturbation on framework tetrahedral Al ions. In agreement with this, DFT studies suggest that EF Al species would tend to coordinate to oxygen atoms near the framework Al atoms [234]. In this way, an enhancement of the Brønsted acidity of the regular framework sites discussed here can be obtained. However, it seems that EF species possess their own Brønsted acidity having been identified as silica–alumina debris [216] where part of Al in a flexible octahedral environment [183]. According to Menezes et al. [235] MQMAS NMR and IR experiments show that tetrahedral Al species are also formed by steaming in USY and give rise to Lewis acidity, which however, is not completely removed

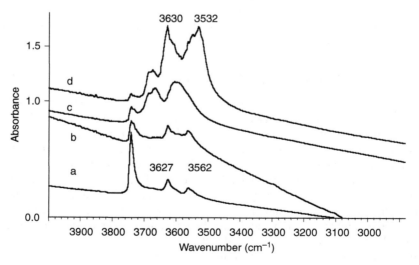

FIGURE 9.13 FT-IR spectra of the hydroxy groups of H-FAU zeolite (Si/Al 30) (a), USY (b), H-FAU zeolite (Si/Al 5) (c), and rare earth-Y (d)

by leaching. Daniell et al. [81] discussed the applicability of different probes to IR characterization of USY.

Hydrothermal treatment producing EF species has been reported to affect positively the isobutene selectivity upon n-butene conversion over H-FER [236] and the activity of H-MOR for light alkane conversion [237]. Some authors believe that EF is released at the external surface of zeolites [238]. The use of hindered nitriles; however, allowed to show that the EF material produced by thermal treatment in H-MOR is in the interior of the side pockets [82]. Similarly, in a sample of H-MFI EF material was found to be located in the interior of the channels [239]. Low temperature CO adsorption experiments on the same sample (Figure 9.5[b,c]) allow to evaluate the Brønsted acid strength of the OH's attributed to EF species. They are responsible for a peak at 3670 cm^{-1}, that shifts to 3470 cm^{-1} by interacting with CO. The shift $\Delta\nu$OH $\approx$ 200 cm^{-1} is lower than that of the bridging OH's ($\Delta\nu$OH $\approx$ 200 cm^{-1}) but bigger than that of the external silanols ($\Delta\nu$OH $\approx$ 150 cm^{-1}), showing that they are actually quite strong.

EF material is considered beneficial also for the selective acylation of 2-methoxy-naphthalene over H-BEA [240]. An IR study on H-BEA [202] suggests that at least two different types of extraframework structures exist in unleached Beta zeolite. One of them is characterized by the presence of H-bonded OH's responsible for a broad absorption in the region 3700 to 3200 cm^{-1} region and for strong Lewis acid sites, characterized by a CN stretching of adsorbed pivalonitrile at 2300 cm^{-1}. These features, decrease by calcination, disappear by acid leaching, and are related to structures that are internal to the zeolite pores. They are identified as Al hydroxo-ions interacting with the internal wall of the zeolite cavities. These species, presumably highly dispersed into the channels, probably do not hinder so much the channels. By heating they likely dehydrate coalescing into bigger particles (those associated to the OH stretching band at 3785 cm^{-1}) which give rise to the improved shape selectivity effect observed in the acylation of 2-methoxy-naphthalene. IR data provide also evidence for the presence of defects in the relatively perfect BEA sample obtained by double acid leaching. In fact, this sample shows a small fraction of terminal silanols that are unaccessible both to pyridine and to pivalonitrile. However, the frequency of these residual sites is 3744 cm^{-1}, that is, nearly intermediate between that of the sites considered to be external (3747 cm^{-1}) and those considered to be internal (3736 cm^{-1}). This suggests that some big holes exist which are however in contact with the external atmosphere by very small channels. Further studies of H-BEA dealumination have been published recently [241]. Tri-coordinated Al ions, as detected by XANES on heat-treated H-MOR and H-BEA [231], could be precursors for EF formation.

9.5.3 Cationic Zeolites and Other Metal-Containing Zeolite-Like Structures

Natural zeolites mostly occur in the cationic form with alkali or alkali-earth ions in the cavities, and water molecules coordinated to them. For example, natural faujasite can have the following formula $Na_{12}Ca_{12}Mg_{11}[Si_{133}Al_{59}O_{384}]$ 260 H_2O [177].

Cation exchanged structure may be produced by cation exchange starting from other cationic forms or from the ammonium or protonic forms. Water can desorb by thermal treatment and outgassing. The quite strong but reversible bonding of water to such cations is an evidence of the significant Lewis acidity of the cations that become more or less naked in the cavities after drying. Alkali and alkali-earth zeolites when dried adsorb quite strongly basic molecules and can also do the same when hydrated, by displacement of water molecules. Cation positions are quite fixed in particular in the dry zeolites, but can change reversibly upon hydration–dehydration cycles or upon adsorption of different molecules [242].

The significant medium Lewis acidity of alkali and alkali-earth zeolites is well-known and is the key feature for the use of these materials as regenerable adsorbants, like for the application of Na–A zeolite for gas drying. In spite of the low PP of alkali and alkali-metal cations, increased however by the loss of ligands in dry zeolites, alkali zeolites adsorb with well detectable hexothermal effects also very weak bases such as CO and nitrogen [243]. IR studies of adsorbed nitriles [244] and of CO [245] on alkali and alkali-metal cations provided evidence of such medium Lewis acidity. CO_2 adsorption experiments also showed the significant basicity of the framework oxygen atoms in the case of NaX, with the formation of carbonate-like species [245]. The mild Lewis acidity of the alkali cations allow them to coordinate different molecules, such as olefinic compounds via π-bonds [244].

The introduction of monovalent and bivalent transition metal cations into zeolites is also possible and introduces in zeolites sites with redox activity. Several of these systems have wide application in catalysis. In particular, Co-zeolites, such as Co-MFI and Co-FER, have been deeply investigated for their activity in the CH_4-SCR reaction [246]. In this case the adsorption of bases such as nitriles and ammonia, followed by IR and by TPD technique, show that they act as medium-strong Lewis acid sites. The current opinion is that these sites are catalytically active for the $DeNO_x$ reaction just when they are "isolated" in the zeolite cavities. A recent investigation provided evidence for the deposition of part of Co ions also at the external surface of the zeolite upon cation exchanging [85] and to their likely nonnegligible catalytic activity [247]. The deposition of Co species at the external cavities can be a reason for only apparent over-exchanging (i.e., production of zeolites with Co^{2+}/Al^{3+} atomic ratios >0.5).

A very popular system is that of Cu-zeolites, mostly Cu–MFI. This is the only well known system that has significant activity in the decomposition of NO to N_2 and O_2 [248]. The zeolites can be prepared by ion exchange and by gas-phase deposition procedures. Multiple ion exchange procedures are usually performed with Cu^{2+} salts and the resulting material can be over-exchanged (Cu^{2+}/Al^{3+} atomic ratios >0.5) possibly because of the autoreduction of Cu^{2+} to Cu^+. Over-exchanged catalysts are the most active. Most authors believe that extralattice oxygen also exist in overexchanged catalysts although some controversy exists on the possible role of dimeric Cu–O–Cu complexes or in monomeric Cu–O species. The $deNO_x$ reaction is certainly a very complex multi-step redox reaction, Lewis acidity being involved in the first step, that is, the coordination of NO on the

Cu center, which however can also involve π-type back-donation effects. Many studies have been published on the interaction of NO and CO on Cu–MFI zeolites. However, investigations of the pure acidity–basicity properties of such systems seem to be lacking. Studies performed with nitriles [249] and ammonia [250] as the probes show that even on over-exchanged materials the true exchange of the zeolitic protons is not complete, zeolitic Brønsted acid sites being still present, and that Cu sites present medium Lewis acidity. The adsorption of hindered nitriles indicates that even in good catalysts a part of Cu ions is actually located at the external surface of the zeolite [249], although their role in catalysis never has been considered. CO_2 adsorption found negligible basicity [250] in spite of the likely presence of extralattice oxygen species.

Fe-zeolites, mainly Fe-ZSM5, have also received much attention in recent years, due in particular to their activity in benzene oxidation by N_2O. They can be prepared starting from Fe-silicalites (i.e., MFI zeolitic structures with substitutional Fe^{3+} ions): by steaming or heat treatment they segregate extraframework Fe-containing species thought to be the active species. Alternatively, they can be prepared by aqueous ions exchange, impregnation, solid-state exchange, or chemical vapor deposition of $FeCl_3$ starting from H-MFI or silicalite. Different types of Fe oxides nanoparticles are formed in the cavities of Fe-MFI but the presence of binuclear Fe oxo-hydroxo complexes seem to be well established today [251]. Both framework Fe ions and extraframework Fe oxide species are responsible for acid sites. Fe^{3+} in a silicalite framework causes, like Al^{3+}, the formation of bridging Brønsted sites (νOH 3630 cm^{-1} in the MFI structure) that are able to protonate ammonia and are found by CO adsorption to be at little less acidic than those of H-ZSM5 [252]. Ammonia TPD show that the total acidity of Fe–silicalite can be even higher than that of H-MFI [253], but possibly extraframework sites contribute with their Lewis acidity. Ammonia TPD and pyridine adsorption experiments showed that strong acidity, both of the Lewis and of the Brønsted type, can be associated to extraframework species containing Fe^{3+} but possibly also Al^{3+} [254].

Another deeply investigated system is that of Ti–MFI (or Ti silicalite) [255]. Only a very small amount of Ti actually enters the silicalite structure in normal tetrahedral sites, although some of them can be bonded to hydroxy groups in open defects [256]. Anatase-like extraframework Ti oxide particles are formed when excess titanium is present upon the preparation. The presence of Ti makes Ti–silicalite less defective than pure silicalite and also changes its room temperature structure from monoclinic to orthorhombic. A recent neutron diffraction study showed that Ti sitting is preferential on the sites where defects are frequently present in pure silicalite [257].

Due to the valency IV of titanium, its substitution for Si does not cause any charge unbalance. Accordingly, no Brønsted acidity is formed on Ti silicalite, as deduced, for example, by the lack of ammonia protonation [73], although an enhancement of the acidity of the silanols does possibly occur (Figure 9.4). Some debate exists on whether framework Ti^{4+} ions can actually act as Lewis acid sites at the gas–solid interface. According to Manoilova et al. [258] CO adsorption experiments detect Lewis acidity of such sites, in contrast to Zecchina et al. [259] that attributed such interaction with CO to extraframework Ti species. The experiment shown in Figure 9.4 indicates that CO at low temperature does not reveal Lewis acidity of a good TS-1 sample. However, an IR study by Astorino et al. [73] showed that some ammonia coordinates strongly on extraframework-species-free Ti silicalite, so supporting a relevant Lewis acidity of framework Ti. These data were confirmed later by Bolis et al. [260] that also reported calorimetric data. Studies using hindered nitriles show that, on well-prepared Ti silicalite, at least part of Lewis acidic Ti sites are present at the external surface [72].

9.5.4 Sulfated and Tungstated Zirconia: True Superacids?

Zirconia is a polymorphic material. It presents three structures that are thermo-dynamically stable in three different temperature ranges. Monoclinic zirconia (baddeleyite, S.G. $P2_1/a$, $Z = 4$) is the room temperature form, tetragonal zirconia (S.G. $P4_2/nmc$, $Z = 2$) is stable above 1200 K while cubic zirconia (S.G. Fm3m, $Z = 4$) is stable above 2400 K. Tetragonal and cubic zirconia, however, may exist as metastable forms at room temperature, mostly if stabilized by dopants such as Yttrium. Frequently, zirconia powders are mixed tetragonal and monoclinic. Several characterization studies have been performed on pure zirconias and showed it is a typical ionic material, characterized by medium Lewis acidity, significant surface basicity and very low Brønsted acidity, if at all. According to Bolis et al. [46], CO adsorption experiments provide evidence for two slightly different types of Lewis acidic Zr^{4+} ions on both monoclinic and tetragonal zirconia. On the other hand, the quality of the sites is very similar in the two phases, but little stronger sites exist on the tetragonal phase. Slightly different concentrations of Lewis sites can be found on the two solids, depending on outgassing temperature. CO_2 adsorption studies [57] reveal significant basicity.

Also, titania is a polymorphic material: the most usual phases are anatase (SG = $I4_1/amd$, $Z = 4$) and rutile (S.G. $P4_2/mnm$, $Z = 2$), the latter being always thermodynamically stable. In addition, titanias are highly ionic oxides with medium-high Lewis acidity, significant basicity and weak Brønsted acidity if at all. Characterization data show that on anatase stronger Lewis acid sites are usually detectable than on rutile [261,262].

Both titania (anatase more than rutile) and, even more, zirconia (tetragonal more than monoclinic), when sulfated or covered with tungsten oxide become very active for some hydrocarbon conversion reactions such as *n*-butane skeletal isomerization [263]. For this reason, a discussion began on whether these materials have to be considered "superacidic." Spectroscopic studies showed that the sulfate ions [264] as well as the tungstate ions [265,266] on ionic oxides in dry conditions, are tetracoordinated with one short S=O and W=O bond (mono-oxo structure) as shown in Scheme 9.3(II). Polymeric forms of tungstate species could also be present [267]. However, in the presence of water the situation changes very much. According to the Lewis acidity of wolframyl species, it is believed that it can react with water and be converted in a hydrated form, as shown in Scheme 9.3. Residual surface OH's of such materials do not present well-defined sharp bands but very broad features, suggesting that the OH's are in fact involved in H-bondings. Also, sulfate species are strongly perturbed by hydration.

The presence of wolframate species on both titania and zirconia causes a little increase of the Lewis acid strength of the residual Lewis sites, an almost full disappearance of the surface anions acting as basic sites and the appearance of a very strong Brønsted acidity [268–271]. Strong Brønsted and Lewis acidity are also found on pure WO_3 [272]. Similar effects have been found for sulfated titanias [261]. The Brønsted acid strength of these materials, measured by the "olefin oligomerization" method (Table 9.8) is superior to that of silica–alumina and comparable to that of protonic zeolites [268].

The case of sulfated zirconia has been investigated in great detail. The very high catalytic activity appears when a certain number of requirements is satisfied: in particular, it must be prepared by an amorphous sulfated precursor calcined at $T \geq 823$ K in order to have tetragonal sulfated phase, and be properly activated [273]. Also, in this case most Lewis acidity and basicity disappear by sulfation, but the residual Lewis sites are a little stronger. However, very strong Brønsted acidity is also formed, responsible for broad IR absorption and for characteristic ^{1}H MAS and broad line NMR signals [94] with larger chemical shifts than zeolites protons. Microcalorimetric ammonia chemisorption studies [274] and ammonia TPD experiments [275] suggested that dry sulfated zirconia surfaces have Brønsted acidity not stronger than protonic zeolites. However, water is needed for high catalytic activity [276] and, according to Katada et al. [277], generates "superacidic" Brønsted centers.

However, reaction mechanism studies revealed that the mechanism of n-alkane isomerization is a very complex one, presenting steps that are certainly not acid-catalyzed and where butene is likely an intermediate [278]. This means that the good performances of such catalysts can be due to other features besides (super)acidity. In particular, hydrogenation–dehydrogenation properties could be beneficial.

UV-Vis studies by Gutierrez-Alejandre et al. [270,279] contributed to underline the possible role of the semiconducting nature of zirconia and titania in these catalytic materials. In the same conditions, insulating materials like aluminas when sulfated or tungstated seem to give less active catalysts [263], in spite of their even stronger Lewis and Brønsted acidity. In the case of tungsta-based catalysts, different opinions have been reported on the possible role of tungsten oxide reduction in generating different Brønsted sites [280,281], or, possibly, playing a role in nonacidic steps. Monoclinic zirconia, in whose bulk Zr ions have coordination seven, and half oxygen ions have coordination three (in contrast to both cubic and tetragonal ZrO_2 where coordination is eight at Zr and four at oxygen) has an optical gap significantly lower than tetragonal zirconia [282], which gives better catalysts; this could explain a different electronic interaction with tungsten oxide species as well as can be in relation to a better behavior in nonacidic catalytic steps, such as those involved in dehydrogenation–hydrogenation of hydrocarbons. In parallel, also rutile has an optical gap significantly lower than anatase (which gives more active catalysts) [283]. The other sulfated oxide that has recently been reported to be superacidic, as deduced by Ar TPD [284,285], and even more active than sulfated zirconia, is sulfated tin oxide, that is, another semiconducting material.

9.5.5 Solid Basic Catalysts: Oxides Containing Alkali, Alkali-Earth, and Rare Earth Cations

The oxides of alkali-earth metals are the strongest solid bases among usually stable metal oxides. They present very high reactivity when treated in vacuum at very high temperatures. These treatments generate highly exposed cations and anions that induce very complex radical-type chemistry, like that producing polymeric CO species from carbon monoxide [68]. In their normal use, however, alkali-earth oxides are partly hydroxylated and carbonated, but still present high surface basicity. Strong bases are also lanthanide oxides such as La_2O_3 and Nd_2O_3 [59], as well as La or other rare-earth containing perovskites [58]. This is essentially due to the very low PP of the corresponding cations, which have large size and relatively low charge. Even stronger bases should be the oxides of alkali metals, whose basic reactivity is so high that they are essentially nonstable in usual conditions, that is, in the presence of water vapor, that is sufficient to convert them into the corresponding hydroxides. On the other hand, similar stability problems occur also with pure alkali-earth oxides. For this reason, basic solid oxide catalysts mostly contain alkali and/or alkali-earth cations in thermally stable mixed oxide phases. Basic zeolites and systems based on oxides (mainly alumina) containing alkali or alkali-earth ions are apparently the most used both for industrial applications and in academic research [286]. Even stronger bases are the "superbases" where alkali metallic phases are produced, which will be not considered here.

The perspectives of the use of basic zeolites as catalysts has been reviewed recently by Davis [287]. Extraframework material free, alkali exchanged zeolites are used as quite mild basic catalysts. As seen earlier, light alkali-metal zeolites, such as Na–X and Na–Y, have a mild Lewis acid behavior and do not appear to have strong basic character [245]. However, heavy-alkali metal zeolites like Cs–Y act actually as basic catalysts, or better as acid–base catalysts, for example, for toluene side chain alkylation. Stronger basic character arises from impregnation of alkali-zeolites with alkali salts, later decomposed to occluded alkali oxide, as evidenced by CO_2 adsorption microcalorimetry. The characterization of such materials is still quite poor.

Several studies have been devoted to the characterization of metal oxides carriers impregnated by alkali and alkali-earth metal ions. Doping of most metal oxides with large-size cations results in their accumulation at the surface. In fact, such cations cannot penetrate for their sizes into the vacancy sites of the close packing oxide arrays of most oxide carriers such as for example, alumina and titania. The alkali and alkali-metal cations remain exposed at the surface where their weak Lewis acidity (corresponding to the strong acidity of the oxide anions) is well detectable by conventional probes. This has been shown using IR spectroscopy of adsorbed ammonia and pyridine, for example, for the systems $K_2O–TiO_2$ [288],

$SrO–TiO_2$ [289] and $CaO–Al_2O_3$ [290]. Such surface layers adsorb very strongly carbon dioxide so that bulk carbonate particles can form like in the case of the $BaO–Al_2O_3$ system [291]. Similarly, TPD studies show that the temperature for CO_2 desorption from zirconia grows from below 473 to 673–873 K upon KNO_3 doping giving rise to $K_2O–ZrO_2$ [292].

The ternary mixed oxides of such large cations with trivalent elements produce as thermodynamically stable compounds β-alumina phases, like for $BaAl_9O_{14.5}$, $BaAl_{12}O_{19}$, and $BaAl_{14}O_{22}$. Such solids are constituted by spinel type alumina blocks separated by alkali or alkali-earth oxide planes where these large cation can give rise to cationic conductivity [293]. On these highly crystalline solids, such as for Ba-β-alumina [294] and for La-β-alumina, the surface apparently only exposes the large low valency cations, well detectable by adsorbing bases, and very basic oxygen species that adsorb CO_2 in the form of carbonates. The decomposition temperature of surface carbonates on Ba-β-alumina ($BaAl_{12}O_{19}$) has been followed by IR under outgassing, and is compared with the same experiment as performed with MgO (Figure 9.14). Actually, $BaAl_{12}O_{19}$ appears to by far more basic than pure alumina and most pure oxides but less basic than pure alkali-earth oxides.

In addition, perovskite type phases [58] form when small cations and large cations combine in a mixed oxide. The surface, however, appears to be largely dominated by the large cations and basic oxide anions, like in the cases of $BaTiO_3$ [295], $SrTiO_3$ [289] and several La perovskites [296]. Low temperature CO adsorption studies on sintered $LaCrO_3$ powders allow to detect both La^{3+} and Cr^{3+} coordinatively unsaturated cations, but they are thought to be exposed on different faces [68].

Mg-aluminates produced by decomposition of hydrothalcites are also very popular basic catalysts today. Hydrothalcite itself has the formula $Mg_6Al_2(OH)_{16}CO_3 \cdot 4H_2O$. Its thermal decomposition gives rise to a mixed oxide whose virtual composition is $5 MgO \cdot MgAl_2O_4$, although these phase give rise

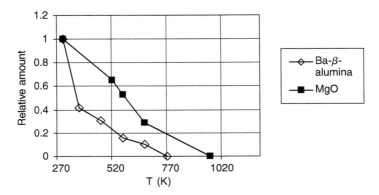

FIGURE 9.14 Relative amount of residual carbonate species (measured from the integrated intensities of the IR bands) upon outgassing at increasing temperatures on two basic catalysts

to partial solid solutions depending on decomposition temperature. Such materials find several applications in industrial heterogeneous catalysis today [297,298]. Stoichiometric $MgAl_2O_4$, is essentially a normal spinel phase with tetrahedrally coordinated Mg and octahedrally coordinated Al. Actually, due to partial inversion of the spinel structure, low-coordination Al cations typical of spinel-type aluminas can be detected at the surface and produce a small density of very strong Lewis acid sites [299]. The surface basicity of $MgAl_2O_4$, has been evaluated to be very similar to that of γ-Al_2O_3 by microcalorimetry of adsorbed hexafluoro-isopropanol [299]. Excess Mg ions resulting in the segregation of large amounts of MgO causes the predominance of the basic character of MgO. Low temperature CO adsorption experiments reveal the presence of cations with medium-weak Lewis acidity attributed to Mg^{2+} ions [300], although ammonia adsorption experiments [301,302], provide evidence for sites with higher strength, identified as Al^{3+} ions. CO_2 adsorption forms surface carbonates that are stable up to near 773 K.

9.6 CONCLUDING REMARKS

In this chapter, a brief and incomplete overview of the acid–base properties of oxide materials used in catalysis and adsorption technologies has been done. While the most general and fundamental aspects of solid acidity and basicity are quite well established and unanimously recognized, many questions are still opened in several particular systems of practical interest. Some of them have been underlined earlier. The most classical experimental techniques briefly described here allow to have a quite complete picture of the materials under study, but still the knowledge needs to be refined in many cases. Theoretical and computational methods are in rapid progress and allow now to go deeper in the understanding of the experimental results.

REFERENCES

1. Tanabe, K. and Hölderich, W.F. Industrial application of solid acid–base catalysts. *Appl. Catal. A: Gen.* 1999, *181*, 399–434.
2. Calatayud, M., Markovits, A., Menetrey, M., Mguig, B., and Minot, C. Adsorption on perfect and reduced surfaces of metal oxides. *Catal. Today* 2003, *85*, 125–143.
3. Lavaste, V., Watts, J.F., Chehimi, M.M., and Lowe, C. Surface characterisation of components used in coil coating primers. *Int. J. Adhes. Adhes.* 2000, *20*, 1–10.
4. Horn, R.G. Surface forces and their action in ceramic materials. *J. Am. Ceram. Soc.* 1990, *73*, 1117–1135.
5. Fraissard, J. and Petrakis, L., Eds. *Acidity and Basicity of Solids: Theory, Assessment and Utility*, NATO ASI Series, Kluwer Academic Publishers, Dordrecht, 1994.
6. Corma, A. Inorganic solid acids and their use in acid-catalyzed hydrocarbon reactions. *Chem. Rev.* 1995, *95*, 559–614.
7. Farneth, W.E. and Gorte, R.J. Methods for characterizing zeolite acidity. *Chem. Rev.* 1995, *95*, 615–635.

8. Zecchina, A., Lamberti C., and Bordiga, S. Surface acidity and basicity: general concepts. *Catal. Today* 1998, *41*, 169–177.
9. Busca, G. The surface acidity of solid oxides and its characterization by IR spectroscopic methods. An attempt at systematization. *Phys. Chem. Chem. Phys.* 1999, *1*, 723–726.
10. Brønsted, J.N. Some remarks on the concept of acids and bases. *Rec. Trav. Chim. Pays-Bas* 1923, *42*, 718–728.
11. Lowry, T.M. *Chim. Ind.* (London) 42, 43 (1923); *Trans. Faraday Soc.* 1924, *20*, 13.
12. Lewis, G.N. *Valency and Structure of Atoms and Molecules*, John Wiley & Sons, New York, 1923.
13. Pearson, R.G. Hard and soft acids and bases. *J. Am. Chem. Soc.* 1963, *85*, 3533–3539.
14. Pearson, R.G. The principle of maximum hardness. *Acc. Chem. Res.* 1993, *26*, 250–255.
15. Corma, A. and Garcia, H. Lewis acids: from conventional homogeneous to green homogeneous and heterogeneous catalysis. *Chem. Rev.* 2003, *103*, 4307–4365.
16. Corma, A., Llopis, F., Viruela P., and Zicovich-Wilson, C. Acid softness and hardness in large pore zeolites as a determinant parameter to control selectivity in orbital controlled reactions. *J. Am. Chem. Soc.* 1994, *116*, 134–142.
17. Papakondylis, A.A. and Sautet, P. *Ab inizio* study of the α-MoO_3 solid and study of the adsorption of H_2O and CO molecules on its (100) surface. *J. Phys. Chem.* 1996, *100*, 10681–10688.
18. Busca, G., Ramis, G., and Lorenzelli, V. FT–IR study of the surface properties of polycrystalline vanadia. *J. Mol. Catal.* 1989, *50*, 231–240.
19. Bartmess, J.E. and Mc Iver, R.T. In *Gas Phase Ion Chemistry*, Aue, D.H. and Bowers, M.T., Eds. Academic Press, New York, 1979, Vol. 2, p. 87.
20. Bellamy, L.J., Hallam H.E., and Williams, R.E. *Trans. Faraday Soc.* 1958, *54*, 1120.
21. Paukshtis, E.A. and Yurchenko, E.N. Study of the acid–base properties of heterogenous catalysts by infrared spectroscopy. *Russ. Chem. Rev. Engl. Transl.* 1983, *52*, 242–258.
22. Pazé, C., Bordiga, S., Lamberti, C., Salvalaggio, M., Zecchina, A., and Bellussi, G. Acidic properties of H-β zeolite as probed by bases with proton affinity in the 118–204 kcal mol^{-1} range: a FTIR investigation. *J. Phys. Chem.* B 1997, *101*, 4740–4751.
23. Teunissen, E.H., van Santen, R.A., Janssen, A.P.J., and van Duijneveldt, F.B. Ammonium in zeolites: coordination and solvation effects. *J. Phys. Chem.* 1993, *97*, 203–210.
24. Onida, B., Monelli, B., Borello, L., Fiorilli, S., Geobaldo, F., and Garrone, E. IR evidence that secondary interactions may camper H-bonding at protonic sites in zeolites. *J. Phys. Chem.* B 2002, *106*, 10518–10522.
25. Bolis, V., Broyer, M., Barbaglia, A., Busco, C., Foddanu, G.M., and Ugliengo, P. Van der Waals interactions on acidic centers in zeolites nanocavities: a calorimetric and computer modelling study. *J. Mol. Catal. A: Chem.* 2003, 204–205, 561–569.
26. Gorte, R.J. What do we know about the acidity of solid acids? *Catal. Lett.* 1999, *62*, 1–13.

27. Camia, M., Gherardi, P., Gubitosa, G., Petrera M., and Pernicone, N. In *Proceedings of the 8th ICC, Berline, 1984*, Verlag Chemie, Weinheim, 1984, Vol. 4, p. 747.

28. Spitz, R.N., Barton, J.E., Barteau, M.A., Stanley, R.H., and Sleight, A.W. Characterization of the surface acid–base properties of metal oxides by titration/displacement reactions. *J. Phys. Chem.* 1986, *90*, 4067–4075.

29. Fierro, J.L.G., Ed. *Spectroscopic Characterization of Heterogeneous Catalysts*. Elsevier, Amsterdam, 1990.

30. Imelik, B. and Vedrine, J.C., Ed. *Catalyst Characterisation, Physical Techniques for Solid Materials*. Plenum Press, New York, 1994.

31. Sun, C. and Berg, J.C. A review of the different techniques for solid surface acid–base characterization. *Adv. Colloids Interface Sci.* 2003, *105*, 151–175.

32. Benesi, H.A. and Winquist, B.H.C. Surface acidity of solid catalysts. *Adv. Catal.* 1978, *27*, 97–182.

33. Tanabe, K., Misono, M., Ono, Y., and Hattori, H. *New Solid Acids and Bases, their Catalytic Properties*. Elsevier, Amsterdam, 1989.

34. Deeba, M. and Hall, W.K. The measurement of catalyst acidity II. Chemisorption studies. *Zeit. Phys. Chem. Neue Folge* 1985, *144*, 85–103.

35. Stockenhuber, M. and Lercher, J.A. Characterization and removal of extra lattice species in faujasites. *Micropor. Mater.* 1995, *3*, 457–465.

36. Cardona-Martinez, N. and Dumesic, J.A. Application of adsorption microcalorimetry to the study of heterogeneous catalysis. *Adv. Catal.* 1992, *38*, 149–244.

37. Auroux, A. Acidity characterization by microcalorimetry and relationship with reactivity. *Top. Catal.* 1997, *4*, 71–89.

38. Solinas, V. and Ferino, I. Microcalorimetric characterization of acidic–basic catalysts. *Catal. Today* 1998, *41*, 179–189.

39. Auroux, A. Microcalorimetry methods to study the acidity and reactivity of zeolites, pillared clays and mesoporous materials. *Top. Catal.* 2002, *19*, 205–213.

40. Auroux A. and Gervasini, A. Microcalorimetric study of the acidity and basicity of metal oxide surfaces. *J. Phys. Chem.* 1990, *94*, 6371–6379.

41. Cutrufello, M.G., Ferino, I., Monaci, R., Rombi, E., and Solinas, V. Acid–base properties of zirconium cerium and lanthanum oxides by calorimetric and catalytic investigation. *Top. Catal.* 2002, *19*, 225–240.

42. Gallardo Amores, J.M., Sanchez Escribano, V., Ramis, G., and Busca, G. An FT–IR study of ammonia adsorption and oxidation over anatase-supported metal oxide. *Appl. Catal. B: Environ.* 1997, *13*, 45–58.

43. Busca, G. and Lorenzelli, V. Infrared spectroscopic identification of species arising from reactive adsorption of carbon oxides on metal oxide surfaces. *Mater. Chem.* 1982, *7*, 89–126.

44. Ramis, G., Rossi, P.F., Busca, G., Lorenzelli, V., La Ginestra, A., and Patrono, P. Phosphoric acid on oxide carriers. Part 2. Surface acidity and reactivity towards olefins. *Langmuir* 1989, *5*, 917–922.

45. Ramis, G. and Busca, G. FT–IR spectra of adsorbed *n*-butyl-amine. *J. Mol. Struct.* 1989, *193*, 93–100.

46. Bolis, V., Cerrato, G., Magnacca, G., and Morterra, C. Surface acidity of metal oxides. Combined microcalorimetric and IR-spectroscopic studies of variously dehydrated systems. *Thermochim. Acta* 1998, *312*, 63–77.

47. Rossi, P.F., Busca, G., Lorenzelli, V., Lion, M., and Lavalley, J.C. Characterization of the surface basicity of oxides by means of microcalorimetry and FT–IR spectroscopy of adsorbed hexafluoroisopropanol. *J. Catal.* 1988, *109*, 378–386.

48. Rossi, P.F., Busca, G., Lorenzelli, V., Saur, O., and Lavalley, J.C. Microcalorimetric and FT–IR spectroscopic study of the adsorption of isopropyl alcohol and hexafluoro-isopropyl alcohol on titanium dioxide. *Langmuir* 1987, *3*, 52–58.

49. Petre, A.L., Perdigon-Melon, J.A., Gervasini A., and Auroux, A. Acid–base properties of alumina-supported M_2O_3 (M = B, Ga, In) catalysts. *Top. Catal.* 2002, *19*, 271–281.

50. Zajac, J., Dutrartre, R., Jones D.J., and Rozières, J. Determination of surface acidity of powdered porous materials based on ammonia chemisorption: comparison of flow microcalorimetry with bath volumetric method and temperature programmed desorption. *Thermochim. Acta* 2001, *379*, 123–130.

51. Tombacz, E., Szekeres, M., and Klumpp, E. Interfacial acid–base reactions of aluminum oxide dispersed in aqueous electrolyte solutions. 2. Calorimetric study on ionization of surface sites. *Langmuir* 2001, *17*, 1420–1425.

52. Niwa, M. and Katada, N. Measurements of acidic property of zeolites by temperature programmed desorption of ammonia. *Catal. Surv. Jpn.* 1997, *1*, 215–226.

53. Costa, C., Lopes, J.M., Lemos F., and Ramoa Ribeiro F. Activity–acidity relationship in zeolite Y. Part 2. Determination of the acid strength distribution by temperature programmed desorption of ammonia. *J. Mol. Catal. A: Chem.* 1999, *144*, 221–231.

54. Essayem, N., Frety, R., Coudurier G., and Vedrine, J. Ammonia adsorption–desorption over the strong solid acid catalyst $H_3PW_{12}O_{40}$ and its Cs^+ and NH_4^+ salts. *J. Chem. Soc., Faraday Trans.* 1997, *93*, 3243–3248.

55. Stevens, R.W., Chuang S.S.C., and Davis, B.H. Temperature programmed desorption / decomposition with simultaneous DRIFTS analysis: adsorbed pyridine on sulphated ZrO_2 and Pt-promoted sulphated ZrO_2. *Thermochim. Acta* 2003, *407*, 61–71.

56. Matsuhashi, H. and Arata, K. Temperature programmed desorption of argon for evaluation of surface acidity of solid superacids. *Chem. Commun.* 2000, 387–388.

57. Pokrovskki, K., Jung, K.T., and Bell, A.T. Investigation of CO and CO_2 adsorption on tetragonal and monoclinic zirconia. *Langmuir* 2001, *17*, 4297–4303.

58. Peña, M.A. and Fierro, J.L.G. Chemical structure and performance of perovskite oxides. *Chem. Rev.* 2001, *101*, 1981–2017.

59. Kus, S., Otremba, M., and Taniewski, M. The catalytic performance in oxidative coupling of methane and the surface basicity of La_2O_3, Nd_2O_3, ZrO_2 and Nb_2O_5. *Fuel* 2003, *82*, 1331–1338.

60. Abee, M.W. and Cox, D.F. BF_3 adsorption on α-Cr_2O_3 (10–12): probing the Lewis basicity of surface oxygen anions. *J. Phys. Chem.* B 2001, *105*, 8375–8380.

61. Lercher, J.A., Grundling C., and Eder-Mirth, G. Infrared studies of the surface acidity of oxides and zeolites using adsorbed probe molecules. *Catal. Today* 1996, *27*, 353–376.

62. Knözinger, H. In *Handbook of Heterogeneous Catalysis*. Ertl, G., Knözinger, H., and Weitkamp, J. Eds. VCH, Weinheim, 1997, Vol. 2, p. 707.

63. Kustov, L.M. New trends in IR spectroscopic characterization of acidic and basic sites in zeolites and oxide catalysts. *Top. Catal.* 1997, *4*, 131–144.

64. Busca, G. Spectroscopic characterization of the acid properties of metal oxide catalysts. *Catal. Today* 1998, *41*, 191–206.

65. Lavalley, J.C. Infrared spectrometric studies of the surface basicity of metal oxides and zeolites using adsorbed probe molecules. *Catal. Today* 1997, *27*, 377–401.

66. Dines, T.J., MacGregor, L.D., and Rochester, C.H. The surface acidity of oxides probed by IR spectroscopy of adsorbed diazines. *Phys. Chem. Chem. Phys.* 2001, *3*, 2676–2685.

67. Knözinger, H. and Huber, S. IR spectroscopy of small and weakly interacting molecular probes for acidic and basic zeolites. *J. Chem. Soc., Faraday Trans.* 1998, *94*, 2047–2059.

68. Zecchina, A., Scarano, D, Bordiga, S., Spoto, G., and Lamberti, C. Surface structures of oxides and halides and their relationships to catalytic properties. *Adv. Catal.* 2001, *46*, 265–397.

69. Hadjiivanov K.I. and Vayssilov, G.N. Characterization of oxide surfaces and zeolites by carbon monoxide as an IR probe molecule. *Adv. Catal.* 2002, *47*, 307–511.

70. Otero Areán, C., Rodríguez Delgado, M., Manoilova, O.V., Turnes Palomino, G., Tsyganenko A.A., and Garrone E. Linkage isomerism of carbonyl coordination complexes formed upon CO adsorption on the zeolite Li-ZSM-5: variable-temperature FTIR studies. *Chem. Phys. Lett.* 2002, *362*, 109–113.

71. Storozhev, P. Yu., Otero Areán, C., Garrone, E., Ugliengo, P., Ermoshin V.A., and Tsyganenko, A. A. FTIR spectroscopic and *ab initio* evidence for an amphipathic character of CO bonding with silanol groups. *Chem. Phys. Lett.* 2003, *374*, 439–445.

72. Armaroli, T., Bevilacqua, M., Trombetta, M., Milella, F., Gutièrrez Alejandre, A., Ramirez Solis, J., Notari, B., Willey, R.J., and Busca, G. A study of the external sites of MFI-type zeolitic materials through the FT–IR investigation of the adsorption of nitriles. *Appl. Catal. A: Gen.* 2001, *216*, 59–71.

73. Astorino, E., Peri, J., Willey, R.J., and Busca, G. Spectroscopic characterization of silicalite-1 and titanium silicalite-1. *J. Catal.* 1995, *157*, 482–500.

74. Platon, A. and Thomson, W.J. Quantitative Bronsted/Lewis ratios using DRIFTS. *Ind. Eng. Chem. Res.* 2003, *42*, 5988–5992.

75. Ferraris, G., De Rossi, S., Gazzoli, D., Pettiti, I., Valigi, M., Magnaccia, G., and Morterra, C. WO_x/ZrO_2 catalysts: Part 3. Surface coverage as investigated by low temperature CO adsorption: FT–IR and volumetric studies. *Appl. Catal. A: Gen.* 2003, *240*, 119–128.

76. Morterra C. and Cerrato, G. Titrating surface acidity of sulfated zirconia catalysts: is the adsorption of pyridine a suitable probe? *Phys. Chem. Chem. Phys.* 1999, *1*, 2825–2831.

77. Morterra, C., Cerrato, G., Pinna, F., and Meligrana, G. Limits in the use of pyridine adsorption, as an analytical tool to test the surface acidity of oxidic systems. The case of sulfated zirconia catalysts. *Top. Catal.* 2001, *15*, 53–61.

78. Busca, G., Lietti, L., Ramis, G., and Berti, F. Chemical and mechanistic aspects of the selective catalytic reduction of NO_x by ammonia over oxide catalysts: a review. *Appl. Catal. B: Environ.* 1998, *18*, 1–36.

79. Jug, K., Homann, T., and Bredow, T. Reaction mechanism of the selective catalytic reduction of NO with NH_3 to N_2 and H_2O. *J. Phys. Chem.* A 2004, *108*, 2966–2971

80. Thibault-Starzyk, F., Travet, A., Saussey, J., and Lavalley, J.C. Correlation between activity and acidity on zeolites: a high temperature infrared study of adsorbed acetonitrile. *Top.Catal.* 1998, *6*, 111–118.

81. Daniell, W., Topsøe, N.Y., and Knözinger, H. An FTIR study of the surface acidity of USY zeolites: comparison of CO, CD_3CN and C_5H_5N probe molecules. *Langmuir* 2001, *17*, 6233–6239.

82. Bevilacqua, M., and Busca, G. A study of the localization and accessibility of Brønsted and Lewis acid sites of H-mordenite through the FT–IR spectroscopy of adsorbed branched nitriles. *Catal. Commun.* 2002, *3*, 497–502.

83. Armaroli, A., Bevilacqua, M., Trombetta, M., Gutièrrez Alejandre, A., Ramirez, J., and Busca, G. A Study of the adsorption of aromatic hydrocarbons and 2,6-lutidine on H-FER and H-ZSM5 zeolites. *Appl. Catal. A: Gen.* 2001, *220*, 181–190.

84. Corma, A., Fornés, V., Forni, L., Marquez, F., Martinez-Triguero, J., and Moscotti, D. 2,6-di-tert-butyl-pyridine as a probe molecule to measure external acidity of zeolites. *J. Catal.* 1998, *179*, 451–458.

85. Montanari, T., Bevilacqua, M., Resini, C., Busca, G. UV-Vis and FT–IR study of the nature and location of the active sites of partially exchanged Co–H zeolites. *J. Phys. Chem. B* 2004, *108*, 2120–2127.

86. Thomasson, P., Tyagi, O.S., and Knözinger, H. Characterization of the basicity of modified MgO catalysts. *Appl. Catal. A: Gen.* 1999, *181*, 181–188.

87. Nesterenko, N., Lima, E., Graffin, P., Charles de menorval, L., Laspéras, M., Tichit, D., and Fajula, F. Probing the basicity of oxide surfaces by FTIR spectroscopy of isocyanic acid generated *in situ* by thermal decomposition of nitromethane. *New J. Chem.* 1999, *23*, 665–666.

88. Sahibed.-Dine, A., Aboulayt, A., Bensitel, M., and Mohammed Saad, A.B., Daturi, M., and Lavalley, J.C. IR study of CS_2 adsorption on metal oxides: relation with their surface oxygen basicity and mobility. *J. Mol. Catal. A: Chem.* 2000, *162*, 125–134.

89. Hasan, M.A, Zaki, M.I., and Pasupulety, L. A spectroscopic investigation of isopropanol and methylbutynol as infrared reactive probes for base sites on polycrystalline metal oxide surfaces. *J. Mol. Catal. A: Chem.* 2002, *178*, 125–137.

90. Brunner, E. Characterization of solid acids by spectroscopy. *Catal. Today* 1997, *38*, 361–376.

91. Haw, J.F. and Teng Xu. NMR studies of solid acidity. *Adv. Catal.* 1998, *42*, 115–180.

92. Haw, J.F. Zeolite acid strength and reaction mechanisms in catalysis. *Phys. Chem. Chem. Phys.* 2002, *4*, 5431–5441.

93. Heeribout, L., Semmer, V., Batamack, P., Doremiueux-Morin, C., and Fraissard, J. Brønsted acid strength of zeolites studied by ^1H NMR: scaling, influence of defects. *Micropor. Meospor. Mater.* 1998, *21*, 565–570.

94. Semmer, V., Batamack, P., Doremiueux-Morin, C., and Fraissard, J. NMR studies of the Bronsted acidity of solids. Application to superacidic solids. *Top. Catal.* 1998, *6*, 119–125.

95. Kentgens, A.P.M., Iuga, D., Kalwei, M., and Koller, H. Direct observation of Bronsted acidic sites in dehydrated zeolite H-ZSM5 using DFS-enhanced ^{27}Al MQMAS NMR spectroscopy. *J. Am. Chem. Soc.* 2001, *123*, 2925–2926.

96. Ganapathy, S., Kumar, R., Delevoye, L., and Amoureux, J.P. Identification of distinct Brønsted acidic sites in zeolite mordenite by proton localization and [[27]Al]-^{1}H READPOR NMR spectroscopy. *Chem. Commun.* 2003, 2076–2077.

97. Zuzaniuk, V. and Prins, R. Synthesis and characterization of silica-supported transition-metal phosphides as HDN catalysts. *J. Catal.* 2003, *219*, 85–96.

98. Forni, L., Fornasari, G., Tosi, C., Trifirò, F., Vaccari, A., Dumeignil F., and Grimblot, J. Non-conventional sol–gel synthesis for the production of boron–alumina catalyst applied to the vapour phase Beckmann rearrangement. *Appl. Catal. A: Gen.* 2003, *248*, 47–57.

99. Gore, K.U., Abraham, A., Hegde, S.G., Kumar, R., Amoureux, J.-P., and Ganapathy, S. ^{29}Si and ^{27}Al MAS/3Q-MAS NMR studies of high silica USY zeolites. *J. Phys. Chem.* B 2002, *106*, 6115–6120.

100. Eckert, H. and Wachs, I.E. Solid-state vanadium-51 NMR structural studies on supported vanadium(V) oxide catalysts: vanadium oxide surface layers on alumina and titania supports. *J. Phys. Chem.* 1989, *93*, 6796–6805.

101. Gorte, R.J. and White, D. Interactions of chemical species with acid sites in zeolites. *Top. Catal.* 1997, *4*, 57–69.

102. Haw, J.F. *In situ* NMR of heterogeneous catalysis: new methods and opportunities. *Top. Catal.* 1999, *8*, 81–86.

103. Lima, E., Charles de Menorval, L., Tichit, D., Lasperas, M., Graffin, P., and Fajula, F. Characterization of acid–base properties of oxide surfaces by ^{13}C CP/MAS NMR using adsorption of nitromethane. *J. Phys. Chem.* B 2003, *107*, 4070–4073.

104. Earl, W.L., Fritz, P.O., Gibson, A.A.V., and Lunsford, J.H. A solid-state NMR study of acid sites in zeolite Y using ammonia and trimethylamine as probe molecules. *J. Phys. Chem.* 1987, *91*, 2091–2095.

105. Kao, Hsien-Ming, Yu, Chun-Yu, and Yeh, Ming-Chu. Detection of the inhomogeneity of Brønsted acidity in H-mordenite and H-β zeolites: a comparative NMR study using trimethylphosphine and trimethylphosphine oxide as ^{31}P NMR probe. *Micropor. Mesopor. Mater.* 2002, *53*, 1–12.

106. Osegovic, J.P. and Drago, S.R. A solid acidity scale based on the ^{31}P MAS NMR shift of chemisorbed triethylphopshine oxide. *J. Catal.* 1999, *182*, 1–4.

107. Mathonneau, E. Characterisation par RMN et Infrasound de molecules sonde adsorbées de l'acidité et de la basicité d'alumines de transition. Thesis, University of Caen, France, 2003.

108. Bromley, S.T, Catlow, C.R.A., and Maschmeyer, Th. Computational modelling of active sites in heterogeneous catalysis. *CATTECH* 2003, *7*, 164–175.

109. Nicholas, J.B. Density functional theory studies of zeolite structure, acidity and reactivity. *Top. Catal.* 1997, *4*, 157–171.

110. Bates S.P. and van Santen, R.A. The molecular basis of zeolite catalysis: a review of theoretical simulations. *Adv. Catal.* 1998, *42*, 1–114.

111. Brändle, M. and Sauer, J. Acidity differences between inorganic solids induced by their framework structure. A combined quantum mechanics/molecular mechanics *ab initio* study on zeolites. *J. Am. Chem. Soc.* 1998, *120*, 1556–1570.

112. Digne, M., Sautet, P., Raybaud, P., and Euzen, P., and Toulhoat, H. Hydroxyl groups on γ-alumina surfaces: a DFT study. *J. Catal.* 2002, *211*, 1–5.
113. Karlesen, E.J., Nygren, M.A., and Pettersson, L.G.M. Comparative study on structures and energetics of NO_x, SO_x and CO_x adsorption on alkaline earth metal oxides. *J. Phys. Chem. B* 2003, *107*, 7795–7802.
114. Forni, L. Standard reaction tests for microporous catalysts characterization. *Catal. Today* 1998, *41*, 221–228.
115. Käßner, P. and Baerns, M. Comparative characterization of acidity of metal oxide catalysts for the oxidative coupling of methane by different methods. *Appl. Catal. A: Gen.* 1996, *139*, 107–129.
116. Gervasini, A., Fenyvesi, J., and Auroux, A. Study of the acidic character of modified metal oxide surfaces using the test of isopropanol decomposition. *Catal. Lett.* 1997, *43*, 219–228.
117. Martin, D. and Duprez, D. Evaluation of acid–base surface properties of several oxides and supported metal catalysts by means of model reactions. *J. Mol. Catal. A: Chem.* 1997, *118*, 113–128.
118. Trombetta, M., Busca, G., Lenarda, M., Storaro, L., Ganzerla, R., Piovesan, L., Jimenez Lopez, A., Alcantara-Rodrìguez, M., and Rodríguez-Castellón, E. Solid acid catalysts from clays. Evaluation of surface acidity of mono- and bi-pillared smectites by FT–IR spectroscopy measurements, NH_3-TPD and catalytic tests. *Appl. Catal. A: Gen.* 2000, *193*, 55–69.
119. Matsuhashi, H., Oikawa, M., and Arata, K. Formation of superbase sites on alkaline earth metal oxides by doping of alkali metals. *Langmuir* 2000, *16*, 8201–8205.
120. Essayem, N., Ben Taarit, Y., Feche, C., Gayraud, P.Y., Saopaly, G., and Naccache, C. Comparative study of *n*-pentane isomerization over solid acid catalysts, heteropolyacid, sulphated zirconia, and mordenite: dependence on hydrogen and pèlatinum addition. *J. Catal.* 2003, *219*, 97–106.
121. De Vos, D.E., Ernst, S., Perego, C., O'Connor C.T., and Stocker, M. Standard reaction of the International Zeolite Association for acidity characterization: ethylbenzene disproportionation over LaNaY. *Micropor. Mesopor. Mater.* 2002, *56*, 185–192.
122. Hart, M.P. and Brown, D.R. Surface acidities and catalytic activities of acid-activated clays. *J. Mol. Catal. A: Chem.* *212*, 315–321.
123. Morrow, B.A. and Cody, I.A. Infrared studies of reactions on oxide surfaces. 5. Lewis acid sites on dehydroxylated silica. *J. Phys. Chem.* 1976, *80*, 1995–1998.
124. Low, M.J.D. and Matsushita, K. Infrared spectra of the surface species produced by reactions of ammonia with germania gel. *J. Phys. Chem.* 1969, *73*, 908–910.
125. Ramis, G., Busca, G., Lorenzelli, V., La Ginestra, A., Galli, P., and Massicci, M.A. Surface acidity of the layered pyrophosphates of quadrivalent Ti, Zr, Ge and Sn and their activity in some acid-catalyzed reactions. *J. Chem. Soc., Dalton Trans.* 1988, 881–886.
126. Marrone, M., Montanari, T., Busca, G., Conzatti, L., Costa, G., Castellano, M., and Turturro, A. An FT–IR study of the reaction of tri-ethoxysilane (TES) and of bis [3-triethoxysilylpropyl-]tetrasulphane (TESPT) with the surface of amorphous silica. *J. Phys. Chem. B* 2004, *108*, 3563–3572

127. Busca, G., Rossi, P.F., Lorenzelli, V., Benaissa, M., Travet, J., and Lavalley, J.C. Microcalorimetric and FT–IR spectroscopic studies of methanol adsorption on Al_2O_3. *J. Phys. Chem.* 1985, *89*, 5433–5439.
128. Ramis, G., Busca, G., Bregani, F., and Forzatti, P. FT–IR study of the adsorption and coadsorption of NO, NO_2 and ammonia on vanadia–titania and mechanism of the SCR reaction. *Appl. Catal.* 1990, *64*, 259–278.
129. Thomas, C.L. Chemistry of cracking catalysts. *Ind. Eng. Chem.* 1949, *41*, 2564–2573.
130. Kung, H.H. Transition *Metal Oxides: Surface Chemistry and Catalysis*. Elsevier, Amsterdam, 1989.
131. Shangwei Hu, Willey, R.J., and Notari, B. An investigation on the catalytic properties of titania–silica materials. *J. Catal.* 2003, *220*, 240–248.
132. Wefers, K. and Misra, C. *Oxides and Hydroxides of Aluminium*, Alcoa: Pittsburgh, PA, 1987.
133. Oberlander, R.K. Aluminas for catalysts. In *Applied Industrial Catalysis*, Leach, B.E., Ed. Academic Press, Orlando, 1983, Vol. 3, pp. 64–112.
134. Berty, J.M. Ethylene oxide synthesis. In *Applied Industrial Catalysis*, Leach, B.E., Ed. Academic Press, Orlando, 1983, Vol. 1, pp. 207–238.
135. Lippens B.C. and de Boer, J.H. Study of phase transformations during calcination of aluminum hydroxides by selected area electron diffraction. *Acta Crystallogr.* 1964, *17*, 1312–1321.
136. Chen, F.R., Davis J.G., and Fripiat, J.J. Aluminum coordination and Lewis acidity in transition aluminas. *J. Catal.* 1992, *133*, 263–278.
137. Soled, S. γ-Al_2O_3 viewed as a defect oxyhydroxide. *J. Catal.* 1983, *81*, 252–257.
138. Zhou. R.S. and Snyder, R.L. Structures and transformation mechanisms of the η, γ and θ transition aluminas. *Acta Crystallogr.* B 1991, *47*, 617–630.
139. Tsyganenko, A.A., Smirnov, K.S., RzhevskiJ, A.M., and Mardilovich, P.P. Infrared spectroscopic evidence for the structural OH groups of spèinel alumina modifications. *Mater. Chem. Phys.* 1990, *26*, 35–46.
140. Wolverton C. and Hass, K.C. Phase stability and structure of spinel-based transition aluminas. *Phys. Rev. B* 2001, *63*, 24102.
141. Sohlberg, K., Pantelides T.S., and Pennycook, S.J. Surface reconstruction and the difference in surface acidity between γ- and η-alumina. *J. Am. Chem. Soc.* 2001, *123*, 26–29.
142. Sohlberg, K., Pennycook, S.J., and Pantelides T.S. The bulk and surface structure of γ-alumina. *Chem. Eng. Commun.* 2000, *181*, 107–135.
143. Digne, M. Ph.D. thesis, Ecole Normale Superieure de Lyon; Sautet, P. Personal communications.
144. Krokidis, X., Raybaud, P., Gobichon, A.-E., Rebours, B., Euzen, P., and Toulhoat, H. Theoretical study of the dehydration process of boehmite to γ-alumina. *J. Phys. Chem. B* 2001, *105*, 5121–5130.
145. Wilson, S.J. and Mc Connell, J.D.C. A kinetic study of the system γ-AlOOH/Al_2O_3. *J. Solid State Chem.* 1980, *34*, 315–322.
146. Hyde B.G. and Andersson, S. *Inorganic Crystal Structures*. John Wiley & Sons New York, 1989, p. 156.
147. John, C.S., Alma, N.C.M. and Hays, G.R. Characterization of transitional alumina by solid-state magic angle spinning aluminium NMR. *Appl. Catal.* 1983, *6*, 341–346.

148. Pechartroman, C., Sobrados, I., Iglesias, J.E., Gonzales Carreno, T., and Sanz, J. Thermal evolution of transitional aluminas followed by NMR and IR spectroscopies. *J. Phys. Chem. B* 1999, *103*, 6160–6170.

149. Abbattista, F., Delmastro, S., Gozzelino, G., Mazza, D., Vallino, M., Busca, G., Lorenzelli, V., and Ramis, G. Surface characterization of amorphous alumina and of its crystallization products. *J. Catal.* 1989, *117*, 42–51.

150. Cocke, D.L., Johnson, E.D., and Merrill, R.P. Planar models for alumina based catalysts. *Catal. Rev. Sci. Eng.* 1984, *26*, 163–231.

151. Pines, H. and Manassen, J. The mechanism of dehydration of alcohols over alumina catalysts. *Adv. Catal.* 1966, *16*, 49–94.

152. Pistarino, C., Finocchio, E., Romezzano, G., Brichese, F., DiFelice, R., Busca, G., and Baldi, M. A study of the catalytic dehydrochlorination of 2-chloropropane in oxidizing conditions. *Ind. Eng. Chem. Res.* 2000, *39*, 2752–2760.

153. Hong, Y., Chen, F.R., and Fripiat, J.J. *Catal. Lett.* 1993, *17*, 187.

154. Trombetta, M., Busca, G., Rossini, S., Piccoli, V., and Cornaro, U. FT–IR studies on light olefin skeletal isomerization catalysis. Part I: the interaction of C4 olefins with pure γ-alumina. *J. Catal.* 1997, *168*, 334–348.

155. Peri, J.B. Infrared and gravimetry study of the surface hydration of γ-alumina. *J. Phys. Chem.* 1965, *69*, 211–220.

156. Tsyganenko, A. and Filimonov, V.N. Infrared spectra of surface hydroxyl groups and crystalline structure of oxides. *Spectrosc. Lett.* 1972, *5*, 477–487.

157. Knözinger, H. and Ratnasamy, P. Catalytic aluminas: surface models and characterization of surface sites. *Catal. Rev. Sci. Eng.* 1978, *17*, 31–70.

158. Busca, G., Lorenzelli, V., Ramis, G., and Willey, R.J. Surface sites on high-area spinel-type metal oxides. *Langmuir* 1993, *9*, 1492–1499.

159. Busca, G., Lorenzelli, V., Sanchez Escribano, V., and Guidetti, R. FT–IR study of the surface properties of the spinels $NiAl_2O_4$ and $CoAl_2O_4$ in relation to those of transitional aluminas. *J. Catal.* 1991, *131*, 167–177.

160. Morterra, C. and Magnacca, G. A case study: surface chemistry and surface structure of catalytic aluminas, as studied by vibrational spectroscopy of adsorbed species. *Catal. Today* 1996, *27*, 497–532.

161. Tsyganenko, A.A. and Mardilovich, P.P. Structure of alumina surfaces. *J. Chem. Soc., Faraday Trans.* 1996, *92*, 4843–4852.

162. Fripiat, J., Alvarez, L., Sanchez Sanchez, S., Martinez Morades, E., Saniger, J., and Sanchez, N. Simulation of the infrared spectra of transition aluminas from direct measurement of Al coordination and molecular dynamics. *Appl. Catal. A: Gen.* 2001, *215*, 91–100.

163. Guillaume, D., Gautier, S., Despujol, J., Alario, F., and Beccat, P. Characterization of acid sites on γ-alumina and chlorinated γ-alumina by ^{31}P NMR of adsorbed trimethylphosphine. *Catal. Lett.* 1997, *4*, 213–218.

164. Knözinger, H., Krietenbrink, H., Müller, H.D., and Schulz, W. Co-operative effects in surface chemical and heterogeneously-catalyzed reactions on metal oxides. ICC, London, 1976, Proceedings, pp. 183–190.

165. Busca, G. Infrared studies of the reactive adsorption of organic molecules over metal oxides and of the mechanisms of their heterogeneously catalyzed oxidation. *Catal. Today* 1996, *27*, 457–496.

166. Carmello, D., Finocchio, E., Marsella, A., Cremaschi, B., Leofanti, G., Padovan, M., and Busca, G. An FT–IR and reactor study of the

dehydrochlorination activity of $CuCl_2/\gamma$-Al_2O_3 based oxychlorination catalysts. *J. Catal.* 2000, *191*, 354–363.

167. Busca, G., Finocchio, E., Lorenzelli, V., Trombetta, M., and Rossini, S.A. Infrared study of olefin allylic activation on magnesium ferrite and alumina catalysts. *J. Chem. Soc., Faraday Trans.* 1996, *92*, 4687–4693.

168. Tretyakov, N.E. and Filimonov, V.N. *Kinet. Catal.* 1972, *13*, 815.

169. Lietti, L., Forzatti, P., Ramis, G., Busca, G., and Bregani, F. Potassium doping of vanadia–titania De-NO_xing catalysts: surface characterization and reactivity study. *Appl. Catal. B: Environ.* 1993, *3*, 13–36.

170. Mohammed Saad, A.B., Ivanov, V.A., Lavalley, J.C., Nortier, P., and Luck, F. Comparative study of the effects of sodium impurity and amorphisation on the Lewis acidity of γ-alumina. *Appl. Catal. A: Gen.* 1993, *94*, 71–83.

171. Finocchio, E., Rossi, N., Busca, G., Padovan, M., Leofanti, G., Cremaschi, B., Marsella, A., and Carmello, D. Characterization and catalytic activity of $CuCl_2$–Al_2O_3 ethylene oxychlorination catalysts. *J. Catal.* 1998, *179*, 606–618.

172. Lavalley, J.C., Benaissa, M., Busca, G., and Lorenzelli, V. FT–IR study of the effect of pretreatment on the surface properties of alumina produce by flame hydrolysis of $AlCl_3$. *Appl. Catal.* 1986, *24*, 249–255.

173. Cornet, D., Goupil, J.-M., Szabo, G., Poirier, J.-L., and Clet, G. Alkylation of isobutane by ethylene catalyzed by chlorided alumina: influence of experimental conditions. *Appl. Catal. A: Gen.* 1996, *141*, 193–205.

174. Griffen, D.T. *Silicate Crystal Chemistry.* Oxford University Press, Oxford, 1992.

175. Vogel, W. *Chemistry of glass.* The American Ceramic Society, Columbus, 1985.

176. Okada K. and Otsuka, N. Characterization of the spinel phase from SiO_2–Al_2O_3 xerogels and the formation process of mullite. *J. Am. Ceram. Soc.* 1986, *69*, 652–656.

177. Barrer, R.M. *Zeolites and Clay Minerals as Sorbents and Molecular Sieves.* Academic Press, New York, 1978.

178. Chen, N.Y., Garwood, W.E., and Dwyer, F.G. *Shape Selective Catalysis in Industrial Applications*, 2nd ed. Dekker, New York, 1996.

179. Guisnet, M. and Gilson, J.P., Eds. *Zeolites for Cleaner Technologies.* Imperial College Press: London, 2002.

180. Beck, L.W. and Davis, M.E. Alkylammonium polycations as structure-directing agents in MFI zeolite synthesis. *Micropor. Mesopor. Mater.* 1998, *22*, 107–114.

181. Feng, Xiaobing, Lee, Jae Sung, Lee, Jun Won, Lee, Jecong Yong, Wei, Di, and Haller, G.L. Effect of pore size of mesoporous molecular sieves (MCM-41) on Al stability and acidity. *Chem. Eng. J.* 1996, *64*, 255–263.

182. Kuroda, Y., Mori T., and Yoshikawa, Y. Improvement in the surface acidity of Al_2O_3–SiO_2 due to a high Al dispersion. *Chem. Commun.* 2001, (11), 1006–1007.

183. Omegna, A., Van Bekhoven, J.A., and Prins, R. Flexible aluminium coordination in alumino silicates. Structure of zeolite H-USY and amorphous silica–alumina. *J. Phys. Chem. B* 2003, *107*, 8854–8860.

184. Trombetta, M., Busca, G., and Willey, R.J. Characterization of silica-containing aluminum hydroxide and oxide aerogels. *J. Colloids Interface Sci.* 1997, *190*, 416–426.

185. Daniell, W., Schubert, U., Glöckler, R., Meyer, A., Noweck, K., and Knözinger, H. Enhanced surface acidity in mixed alumina-silicas: a low temperature FT–IR study. *Appl. Catal. A: Gen.* 2000, *196*, 247–260.

186. Finocchio, E., Busca, G., Rossini, S., Cornaro, U., Piccoli,V., and Miglio, R. FT–IR characterization of silicated aluminas, active olefin skeletal isomerization catalysts. *Catal. Today* 1997, *33*, 335–352.

187. Trombetta, M., Busca, G., Rossini, S., Piccoli, V., Cornaro, U., Guercio, A., Catani, R., and Wippey, R.J. FT–IR studies on light olefin skeletal isomerization catalysis. Part III: Surface acidity and activity of amorphous and crystalline catalysts belonging to the SiO_2–Al_2O_3 system. *J. Catal.* 1998, *179*, 581–596.

188. Hunger, M. Brønsted acid sites in zeolites characterized by multi-nuclear solid-state NMR spectroscopy. *Catal. Rev. Sci. Eng.* 1997, *39*, 345–393.

189. Pazé, C., Zecchina, A., Spera, S., Spano, G., and Rivetti, F. Acetonitrile as probe molecule for an integrated ^{1}H NMR and FTIR study of zeolitic Brønsted acidity: interaction with zeolites H-ferrierite and H-beta. *Phys. Chem. Chem. Phys.* 2000, *2*, 5756–5760.

190. Kazansky, V.B., Seryk, A.I., Semmer-Herledan, V., and Fraissard, J. Intensities of OH stretching bands as a measure of the intrinsic acidity of bridging hydroxyl groups in zeolites. *Phys. Chem. Chem. Phys.* 2003, *5*, 966–969.

191. Miyamoto, Y., Takada N., and Niwa, M. Acidity of β-zeolite with different Si/Al ratio as measured by temperature programmed desorption of ammonia. *Micropor. Mesopor. Mater.* 2000, *40*, 271–278.

192. Kotrel, S., Rosynek, M.P., and Lunsford, J.H. Quantification of acid sites in H.ZSM5, H-β and H-Y zeolites. *J. Catal.* 1999, *182*, 278–281.

193. Eichler, U., Brändle M., and Sauer, J. Predicting absolute and site specific acidities for zeolite catalysts by a combined quantum mechanics/ interatomic potential function approach. *J. Phys. Chem. B* 1997, *101*, 10035–10050.

194. Schroder, K.P., Sauer, J., Leslie, M., Catlow, C.R.A., and Thomas, J.M. Bridging hydroxyl groups in zeolitic catalysts: a computer simulation of their structure, vibrational properties and acidity in protonated faujasites (H–Y zeolites). *Chem. Phys. Lett.* 1992, *188*, 320–325.

195. Lonyi, F., Valyon, J., and Pal-Borbély, G. A DRIFT spectroscopic study of the N_2 adsorption and acidity of H-faujasite, *Micropor. Mesopor. Mater.* 2003, *66*, 273–282.

196. Pelmenschikov, A.G. and van Santen, R.A. Acetonitrile-d3 as a probe of Lewis and Bronsted acidity of zeolites. *J. Phys. Chem.* 1993, *97*, 10071–11074.

197. Trombetta, M., Busca, G., Storaro, L., Lenarda, M., Casagrande, M., and Zambon, A. Surface acidity modifications induced by thermal treatments and acid leaching on microcrystalline H-BEA zeolite. AFT–IR, XRD and MAS–NMR study. *Phys. Chem. Chem. Phys.* 2000, *2*, 3529–3537.

198. Busca, G., Bevilacqua, M., Armaroli, T., and Trombetta, M. FT–IR studies of internal, external and extraframework sites of FER, MFI, BEA and MOR type protonic zeolite materials. *Stud. Surf. Sci. Catal.* 2002, *142*, 975–982.

199. Maache, M., Janin, A., Lavalley J.C., and Benazzi, E. FT-infrared study of Brønsted acidity of H-mordenites: heterogeneity and effect of dealumination. *Zeolites* 1995, *15*, 507–516.

200. Bordiga, S., Lamberti, C., Geobaldo, F., Zecchina, A., Turnes Palomino, G., and Otero Arean C. Fourier-transform infrared study of CO adsorbed at 77 K on H-mordenite and alkali-metal-exchanged mordenites. *Langmuir* 1995, *11*, 527–533.

201. Datka, J., Gil, B., and Weglarski, J. Heterogeneity of OH groups in mordenites: IR studies of benzene and carbon monoxide sorption and NMR studies. *Micropor. Mesopor. Mater.* 1998, *21*, 75–79.

202. Martucci, A., Cruciani, G., Alberti, A., Ritter, C., Ciambelli, P., and Rapacciuolo, M. Location of Brønsted sites in D-mordenites by neutron powder diffraction. *Micropor. Mesopor. Mater.* 2000, *35–36*, 405–412.

203. Kao, H.M., Yu, C.Y., and Yeh, M.C. Detection of the inhomogeneity of Brønsted acidity in H-mordenite and H-β-zeolites: a comparative NMR study using trimethylphosphine and trimethyl-phosphine oxide as ^{31}P NMR probes. *Micropor. Mesopor. Mater.* 2002, *53*, 1–12.

204. Bevilacqua, M., Gutièrrez Alejandre, A., Resini, C., Casagrande, M., Ramirez, J., and Busca, G. An FT–IR study of the accessibility of the protonic sites of H-mordenites. *Phys. Chem. Chem. Phys.* 2002, *4*, 4575–4583.

205. Armaroli, T., Trombetta, M., Gutièrrez Alejandre, A., Ramirez Solis, J., and Busca, G. FT–IR study of the interaction of some branched aliphatic molecules with the external and the internal sites of H-ZSM5 zeolite. *Phys. Chem. Chem. Phys.* 2000, *2*, 3341–3348.

206. Montanari, T., Bevilacqua, M., and Busca G. unpublished results.

207. Zecchina, A., Bordiga, S., Spoto, G., Scarano, D., Spanò, G., and Geobaldo, F. IR spectroscopy of neutral and ionic hydrogen-bonded complexes formed upon interaction of CH_3OH, C_2H_5OH, $(CH_3)_2O$, $(C_2H_5)_2O$ and C_4H_8O with H-Y, H-ZSM-5 and H-mordenite: comparison with analogous adducts formed on the H-Nafion superacidic membrane. *J. Chem. Soc., Faraday Trans.* 1996, *92*, 4863–4875.

208. Zecchina, A., Geobaldo, F., Spoto, G., Bordiga, S., Ricchiardi, G., Buzzoni, R., and Petrini, G. FTIR Investigation of the formation of neutral and ionic hydrogen-bonded complexes by interaction of H-ZSM-5 and H-mordenite with CH_3CN and H_2O: comparison with the H-NAFION superacidic system. *J. Phys. Chem.* 1996, *100*, 16584–16599.

209. Pazé, C., Bordiga, S., Lamberti, C., Salvataggio, M., Zecchina, A., and Bellussi, G. Acidic properties of H-β-zeolite as probed by bases with proton affinity in the 118–204 kcal mol^{-1} range: a FTIR investigation. *J. Phys. Chem. B* 1997, *101*, 4740–4751.

210. Zamaraev, K.I. and Thomas, J.M. Structural and mechanistic aspects of the dehydration of isomeric butyl alcohols over porous aluminosilicate acid catalysts. *Adv. Catal.* 1996, *41*, 335–358.

211. Boehm, H.-P. and Knözinger, H. Nature and estimation of functional groups on solid surfaces. *Catal. Sci. Technol.* 1983, *4*, 39–208.

212. Morin, S., Ayrault, P., El Mouahid, S., Gnep, N.S., and Guisnet, M. Particular selectivity of *m*-xylene isomerization over MCM-41 mesoporous aluminosilicates. *Appl. Catal. A: Gen.* 1997, *159*, 317–331.

213. Corma, A., Grande, M.S., Gonzalez-Alfonso, V., and Orchilles, A.V. Cracking activity and hydrothermal stability of MCM-41 and its comparison with amorphous silica-alumina and a USY zeolite. *J. Catal.* 1996, *159*, 375–382.

214. Di Renzo, F., Chiche, B., Fajula, F., Viale, S., and Garrone, E. Mesoporous MCM-41 alumino-silicates as model silica–alumina catalysts: spectroscopic characterization of the acidity. *Stud. Surf. Sci. Catal.* 1996, *101*, 851–860.

215. Trombetta, M., Busca, G., Lenarda, M., Storaro, L., and Pavan, M. An investigation of the surface acidity of mesoporous Al-containing MCM41 and of the

external surface of ferrierite through pivalonitrile adsorption. *Appl. Catal. A: Gen.* 1999, *182*, 225–235.

216. Cairon , O., Chevreau, T., and Lavalley, J.C. Brønsted acidity of extraframework debris in steamed Y zeolites from the FTIR study of CO adsorption. *J. Chem. Soc., Faraday Trans.* 1994, (19), 3039–3047.

217. Bonelli, B., Onida, B., Chen, J.D., Galarneau, A., DiRenzo, F., Fajula, F., and Garrone, E. Spectroscopic characterization of the acidic sites of Al-rich microporous micelle-templated silicates. *Micropor. Mesopor. Mater.* 2004, *67*, 95–106.

218. Scokart, P.O. and Rouxhet, P.G. Comparison of the acid–base properties of various oxides and chemically treated oxides. *J. Colloid Interface Sci.* 1982, *86*, 96–104.

219. Heeribout, L., Vincent, R., Batamack, P., Dorémieux-Morin, C., and Fraissard, J. Bronsted acidity of amorphous silica-alumina studied by ^{1}H NMR. *Catal. Lett.* 1998, *53*, 23–31.

220. Trombetta, M. and Busca, G. On the characterization of the external acid sites of ferrierite and other zeolites. *J. Catal.* 1999, *187*, 521–523.

221. Skowronska-Ptasinka, M.D., Duchateau, R., van Santen, R.A., and Yap, G.P.A. Steric factors affecting the Brønsted acidity of aluminosilsesquioxanes. *Eur. J. Inorg. Chem.* 2001, 133–137.

222. Busco, C. Ph.D. thesis, University of Turin, 2003; Ugliengo, P. Personal communication.

223. Pistarino, C., Finocchio, E., Larrubia, M.A., Serra, B., Braggio, S., Busca, G., and Baldi, M. A study of the dehydrochlorination of 1,2-dichloropropane over silica-alumina catalysts. *Ind. Eng. Chem. Res.* 2001, *40*, 3262–3269.

224. Finocchio, E., Pistarino, C., Dellepiane, S., Serra, B., Braggio, S., Baldi, M., and Busca, G. Studies on the catalytic dechlorination and abatement of chlorided VOC: the cases of 2-chloropropane, 1,2-dichloropropane and trichloroethylene. *Catal. Today* 2002, *75*, 263–267.

225. Paparatto, G., Moretti, E., Leofanti, G., and Gatti, F. *J. Catal.* 1987, *105*, 227.

226. Kunieda, T., Kim J.-H., and Niwa, M. Source of selectivity of *p*-xylene formation in the toluene disproportionation over HZSM-5. Zeolites. *J. Catal.* 1999, *188*, 431–433.

227. Farcasiu, D., Leu, R., and Corma, A. Evaluation of the accessible acid sites on solids by ^{15}N NMR spectroscopy with 2,6-di-tert-butylpyridine as base. *J. Phys. Chem. B* 2002, *106*, 928–932.

228. Trombetta, M., Armaroli, T., Gutiérrez-Alejandre, A., Ramírez, J., and Busca, G. A FT–IR study of the internal and external surfaces of HZSM5 zeolite. *Appl. Catal. A: Gen.* 2000, *192*, 125–136.

229. Otero Arean, C., Escalona Platero, E., Penarroya Mentruit, M., Rodriguez Delgado, M., Llabbres i Xamena, F.X., Garcia Raso, A., and Morterra, C. The combined use of acetonitrile and adamantine-carbonitrile as IR spectroscopic probes to discriminate between external and internal surfaces of medium pore zeolites. *Mesopor. Micropor. Mater.* 2000, *34*, 55–60.

230. Onida, B., Borello, L., Monelli, B., Geobaldo, F., and Garrone, E. IR study of the acidità of ITQ-2, an "all-surface" zeolitic system. *J. Catal.* 2003, *214*, 191–199.

231. Van Bokhoven, J.A., van der Eerden, A.M.J., and Koningsberger, D.C. Three coordinate Aluminum in zeolites observed with *in situ* x-ray absorption near

edge spectroscopy at the Al K-edge: flexibility of Al coordinations in zeolites. *J. Am. Chem. Soc.* 2003, *125*, 7435–7442.

232. Borzatta, V., Busca, G., Poluzzi, E., Rossetti, V., Trombetta M., and Vaccari, A. As to the reasons of the high activity of a commercial pentasil-type zeolite in the vapor-phase fries rearrangement. *Appl. Catal. A: Gen.* 2004, *257*, 85–95.

233. van Bokhoven, J.A., Roest, A.L., Koningsberger, D.C., Miller, J.T., Nachtegaal, G.H., and Kentgens, P.M. Changes in structural and electronic properties of the zeolite framework induced by extraframework Al and La in H-USY and La(x)NaY: a ^{29}Si and ^{27}Al MAS NMR and ^{27}Al MQMAS NMR study. *J. Phys. Chem. B* 2000, *104*, 6743–6754.

234. Behring, D.L., Ramirez-Solis, A., and Mota, C.J.A. A density functional theory based approach to extraframework Al species in zeolites. *J. Phys. Chem. B* 2003, *107*, 4342–4347.

235. Menezes, S.M.C, Camorim, V.L., Lam, Y.L., San Gil, R.A.S., Bailly, A., and Amoureux, J.P. Characterization of extraframework of steamed and washed faujasite by MQMAS NMR and IR measurements. *Appl. Catal. A: Gen.* 2001, *207*, 367–377.

236. Mériaudeau, P. and Naccache, C. Skeletal isomerization of *n*-butenes catalyzed by medium-pore zeolites and aluminosilicates. *Adv. Catal.* 1999, *44*, 505–543.

237. Van Bokhoven, J.A., Tromp, M., Koningsberger, D.C., Miller, J.T., Pieterse, J.A.Z., Lercher, J.A., Williams, B.A., and Kung, H.H. An explanation of the enhanced activity for light alkane conversion in mildly steam dealuminated mordenite: the dominant role of adsorption. *J. Catal.* 2001, *202*, 129–140.

238. Remy, M.J., Stanica, D., Poncelet, G., Feijen, E.J.P., and Grobet, P. Dealuminated H-Y zeolites: relation between physicochemical properties and catalytic activity in heptane and decane isomerization. *J. Phys. Chem.* 1996, *100*, 12440–12447.

239. Trombetta, M., Armaroli, T., Gutièrrez Alejandre, A., Gonzalez, H., Ramirez Solis, J., and Busca, G. Conversion and hydroconversion of hydrocarbons on zeolite-based catalysts: an FT–IR study. *Catal. Today* 2001, *65*, 285–292.

240. Casagrande, M., Storaro, L., Lenarda, M., and Ganzerla, R. Highly selective Friedel–Crafts acylation of 2-methoxynaphthlene catalyzed by H-BEA zeolite. *Appl. Catal. A: Gen.* 2000, *20*, 263–270.

241. Marques, J.P., Gener, I., Ayrault, P., Bordado, J.C., Lopes, J.M., Ramoa Ribeiro, F., and Guisnet, M. Infrared spectroscopic study of the acid properties of dealuminated BEA zeolite. *Micropor. Mesopor. Mater.* 2003, *60*, 251–262.

242. Lim, K.H. and Grey, C.P. Characterization of extraframework cation position in zeolites NaX and NaY with very fast ^{23}Na MAS and multiple quantum MAS NMR spectroscopy. *J. Am. Chem. Soc.* 2000, *122*, 9768–9780.

243. Savitz, S., Myers, A.L., and Gorte, R.J. A calorimetric investigation of CO, N_2 and O_2 in alkali exchanged MFI. *Micropor. Mesopor. Mater.* 2000, *37*, 33–40.

244. Armaroli, T., Finocchio, E., Busca, G., and Rossini, S. A FT–IR study of the adsorption of C5 olefinic compounds on NaX zeolite. *Vibrat. Spectrosc.* 1999, *20*, 85–94.

245. Martra, G., Ocule, R., Marchese, L., Centi, G., and Coluccia, S. Alkali and alkaline earth exchanged faujasites: strength of Lewis base and acid centers and

cation site occupancy in Na- and Ba-Y and Na- and Ba-X zeolites. *Catal. Today* 2002, *73*, 83–93.

246. Resini, C., Montanari, T., Nappi, L., Bagnasco, G., Turco, M., Busca, G., Bregani, F., Notaro M., and Rocchini, G. Selective catalytic reduction of NO_x by methane over Co–H–MFI AND Co–H–FER zeolite catalysts: characterisation and catalytic activity. *J. Catal.* 2003, *214*, 179–190.

247. Bagnasco, G., Turco, M., Resini, C., Montanari, T., Bevilacqua, M., and Busca, G. On the role of external cobalt sites in NO oxidation and reduction by CH_4 over Co–H–MFI zeolite. *J. Catal.*, 2004, *225*, 536–540.

248. Yahiro, H. and Iwamoto, M. Copper ion exchanged zeolite catalysts in de-NO_x reactions. *Appl. Catal. A: Gen.* 2001, *222*, 163–181.

249. Montanari, T., Bevilacqua, M., Resini, C., Busca, G., Pirone, R., and Ruoppolo, G. A spectroscopic study of the nature and accessibility of protonic and cationic sites in H- and partially exchanged Cu- and Co-MFI zeolites. submitted.

250. Hadjiivanov, K., Klissurski, D., Ramis, G., and Busca, G. Fourier transform IR study of NO_x adsorption on CuZSM5 DeNO$_x$ catalysts. *Appl. Catal. B: Environ.* 1996, *7*, 251–267.

251. Battiston, A.A., Bitter, J.H., Heijboer, W.M., de Groot, F.M.F., and Koningsberger, D.C. Reactivity of Fe binuclear complexes in overexchanged Fe/SM5, studied by *in situ* XAFS spectroscopy. *J. Catal.* 2003, *215*, 279–293.

252. Zecchina, A., Geobaldo, F., Lamberti, C., Bordiga, S., Turnes Palomino, G., and Otero Aréan, C. *Catal. Lett.* 1996, *42*, 25–29.

253. Katada, N., Miyamoto, T., Begum, H.A., Naito, N., Niwa, M., Matsumoto, A., and Tsutsumi, K. Strong acidity of MFI-type ferrisilicate determined by TPD of ammonia. *J. Phys. Chem. B* 2000, *104*, 5511–5518.

254. Meloni, D., Monaci, R., Solinas, V., Berlier, G., Bordiga, S., Rossetti, I., Oliva, C., and Forni, L. Activity and deactivation of Fe–MFI catalysts for benzene hydroxylation to phenol by N_2O. *J. Catal.* 2003, *214*, 169–178.

255. Notari, B. Mesoporous crystalline titanosilicates. *Adv. Catal.* 1996, *41*, 253–334.

256. Lamberti, C., Bordiga, S., Arduino, D., Zecchina, A., Geobaldo, F., Spanò, G., Genoni, F., Petrini, G., Carati, A., Villain, F., and Vlaic, G. Evidence of the presence of two different framework Ti(IV) species in Ti-silicalite-1 *in vacuo* conditions: an EXAFS and a photoluminescence study. *J. Phys. Chem. B* 1998, *102*, 6382–6390.

257. Lamberti, C., Bordiga, S., Zecchina, A., Artioli, G., Marra, G., and Spanò, G. Ti location in the MFI framework of Ti silicalite: a neutron powder diffraction study. *J. Am. Chem. Soc.* 2001, *123*, 2204–2212.

258. Manoilova, O.V., Dakka, J., Sheldon, R.A., and Tsyganenko, A.A. *Stud. Surf. Sci. Catal.* 1995, *94*, 163.

259. Zecchina, A., Spoto, G., Bordiga, S., Padovan, M., Leofanti, G., and Petrini, G. *Stud. Surf. Sci. Catal.* 1991, *65*, 671.

260. Bolis, V., Bordiga, S., Lamberti, C., Zecchina, A., Carati, A., Rivetti, F., Spanò, G., and Petrini, G. A calorimetric, IR, XANES and EXAFS study of the adsorption of NH_3 on Ti-silicalite as a function of the sample pre-treatment. *Micropor. Mesopor. Mater.* 1999, *30*, 67–76.

261. Busca, G., Saussey, H., Saur, O., Lavalley, J.C., and Lorenzelli, V. FT–IR characterization of the surface acidity of different TiO_2 anatase preparations. *Appl. Catal.* 1985, *14*, 245–260.

262. Ferretto, L. and Glisenti, A. Surface acidity and basicity of a rutile powder. *Chem. Mater.* 2003, *15*, 1181–1188.

263. Arata, K. Solid superacids. *Adv. Catal.* 1990, *37*, 165–211.

264. Saur, O., Bensitel, M., Mohammed Saad, A.B., Lavalley, J.C., Tripp, C.P., and Morrow, B.A. The structure and stability of sulfated alumina and titania. *J. Catal.* 1986, *99*, 104–110.

265. Busca, G. and Lavalley, J.C. Use of overtone bands to monitor the state of catalyst active phases during infrared studies of adsorption and catalytic reaction. *Spectrochim. Acta* 1986, *42* A, 443–445.

266. Busca, G. Differentiation of mono-oxo and poly-oxo and of monomeric and polymeric vanadate, molybdate and wolframate species in metal oxide catalysts by IR and Raman spectroscopies. *J. Raman Spectrosc.* 2002, *33*, 348–358.

267. Scheithauer, M., Grasselli, R.K., and Knozinger, H. Genesis and structure of WO_x/ZrO_2 solid acid catalysts. *Langmuir* 1998, *14*, 3019–3029.

268. Ramis, G., Busca, G., and Lorenzelli, V. Surface acidity of solid acids and superacids: a FT–IR study of the behaviour of titania doped with phosphoric, sulphuric, tungstic and molybdic acids. *Stud. Surf. Sci. Catal.* 1989, *48*, 777–786.

269. Ramis, G., Busca, G., Cristiani, C., Lietti, L., Forzatti, P., and Bregani, F. Characterization of tungsta-titania catalysts. *Langmuir* 1992, *8*, 1744–1749.

270. Gutierrez-Alejandre, A., Castillo, P., Ramirez, J., Ramis, G., and Busca, G. Redox and acid reactivity of wolframyl centers on oxide carriers: Bronsted, Lewis and redox sites. *Appl. Catal. A: Gen.* 2001, *216*, 181–194.

271. Calabro, D.C., Vartuli, J.C., and Santiesteban, J.G. The characterization of tungsten oxide modified zirconia supports for dual functional catalysis. *Top. Catal.* 2002, *18*, 231–242.

272. Ramis, G., Cristiani, C., Elmi, A.S., Villa, P.L., and Busca, G. Characterization of the surface properties of polycrystalline tungsten trioxide. *J. Mol. Catal.* 1990, *61*, 319–331.

273. Bolis, V., Magnacca, G., Cerrato, G., and Morterra, C. Effect of sulfation on the acid–base properties of teragonal zirconia. A calorimetric and IR spectroscopic study. *Top. Catal.* 2002, *19*, 259–269.

274. Yaluris, G., Larson, R.B., Kobe, J.M., Gonzales, M.R., Fogash, K.B., and Dumesic, J.A. Selective poisoning and deactivation of acid sites on sulfated zirconia catalysts for *n*-butane isomerization. *J. Catal.* 1996, *158*, 336–342.

275. Barthos, R., Onyestyak, Gy. and Valyon, J. An IR, FR and TPD study on the acidity of H-ZSM5, sulphated zirconia and sulphated zirconia–titania using ammonia as the probe molecule. *J. Phys. Chem.* B 2000, *104*, 7311–7319.

276. Dijs, I.J., Geus, J.W., and Jenneskens, L.W. Effect of size and extent of sulphation of bulk and silica-supported ZrO_2 on catalytic activity in gas- and liquid-phase reactions. *J. Phys. Chem.* B 2003, *107*, 13403–13413.

277. Katada, N., Endo, J., Notsu, K., Yasunobu, N., Naito, N., and Niwa, M. Superacidity and catalytic activity of sulfated zirconia. *J. Phys. Chem.* B 2000, *104*, 10321–10328.

278. Hammache, S. and Goodwin, J.G. Characteristics of the active sites on sulphated zirconia for *n*-butane isomerization. *J. Catal.* 2003, *218*, 258–266.

279. Gutierrez-Alejandre, A., Ramirez, J., and Busca, G. The electronic structure of oxide supported tungsten oxide catalysts as studied by UV spectroscopy. *Catal. Lett.* 1998, *56*, 29–33.

280. Baertsch, C.D., Komala, K.T., Chua, Y.-H., and Iglesia E. Genesis of Brønsted acid sites during dehydration of 2-butanol on tungsten oxide catalysts. *J. Catal.* 2002, *205*, 44–57.

281. Kuba, S., Che, M., Grasselli, R.K., and Knozinger, H. Evidence of the formation of W^{5+} centers and OH groups upon hydrogen reduction of platinum-promoted tungstated zirconia catalysts. *J. Phys. Chem.* B 2003, *107*, 3459–3462.

282. Fernandez-Lopez, E., Sanchez-Escribano, V., Panizza, M., Carnasciali, M.M., and Busca, G. Vibrational and electronic spectroscopic properties of zirconia powders. *J. Mater. Chem.* 2001, *11*, 1891–1897.

283. Gallardo Amores, J.M., Sanchez Escribano, V., and Busca, G. Anatase crystal growth and phase transformation to rutile in high-area TiO_2, MoO_3–TiO_2 and related catalytic materials. *J. Mater. Chem.* 1995, *5*, 1245–1249.

284. Matasuhashi, H., Miyazaki, H., Kawamura, Y., Nakamura, H., and Arata, K. Preparation of a solid superacid of sulphated tin oxide with acidità higher than that of sulphated zirconia and its applications to aldol condensation and benzoylation. *Chem. Mater.* 2001, *13*, 3038–3042.

285. Matasuhashi, H., Tanaka, T., and Arata, K. Measurement of heat of argon adsorption for the evaluation of relative acid strength of some sulphated metal oxides and H-type zeolites. *J. Phys. Chem.* B 2001, *105*, 9669–9671.

286. Weitkamp, J., Hunger, M., and Rymsa, U. Base catalysis on microporous and mesoporous materials: recent progress and perspectives. *Micropor. Mesopor. Mater.* 2001, *48*, 255–270.

287. Davis, R.J. New perspectives on basic zeolites as catalysts and catalyst supports. *J. Catal.* 2003, *216*, 396–405.

288. Busca, G. and Ramis, G. FT–IR study of the surface properties of K_2O–TiO_2. *Appl. Surf. Sci.* 1986, *27*, 114–126.

289. Gallardo Amores, J.M., Sanchez Escribano, V., Daturi, M., and Busca, G. Preparation, characterization and surface structure of coprecipitated high-area Sr_xTiO_{2+x} ($0 \leq x \leq 1$) powders. *J. Mater. Chem.* 1996, *6*, 879–886.

290. Morterra, C., Magnacca, G., Cerrato, G., Del Favero, N., Filippi, F., and Folonari, C.V. X-ray diffraction, high-resolution electron microscopy and Fourier transform infrared study of Ca-doped Al_2O_3. *J. Chem. Soc., Faraday Trans.* 1993, *89*, 135–150.

291. Prinetto, F., Ghiotti, G., Nova, I., Lietti, L., Tronconi, E., and Forzatti, P. FT–IR and TPD investigation of the NO_x storage properties of BaO/Al_2O_3 and Pt-BaO/Al_2O_3 catalysts. *J. Phys. Chem.* B 2001, *105*, 12732–12735.

292. Wang, Y., Huang, W.Y., Chun, Y., Xia, J.R., and Zhu, J.H. Dispersion of potassium nitrate and the resulting strong basicity on zirconia. *Chem. Mater.* 2001, *13*, 670–677.

293. Bellotto, M., Busca, G., Cristiani, C., and Groppi, G. FT–IR skeletal powder spectra of Ba-ß-aluminas with compositions $BaAl_9O_{14.5}$, $BaAl_{12}O_{19}$ and $BaAl_{14}O_{22}$ and of Ba-ferrite $BaFe_{12}O_{19}$. *J. Solid State Chem.* 1995, *117*, 8–15.

294. Busca, G., Cristiani, C., Forzatti, P., and Groppi, G. Surface characterization of Ba-β-alumina. *Catal. Lett.* 1995, *31*, 65–74.

295. Busca, G., Buscaglia, V., Leoni, M., and Nanni, P. Solid state and surface spectroscopic characterization of $BaTiO_3$ fine powders. *Chem. Mater.* 1994, *6*, 955–961.

296. Daturi, M., Busca, G., and Willey, R.J. Surface and structure characterization of some perovskite-type powders to be used as combustion catalysts. *Chem. Mater.* 1995, *7*, 2115–2126.

297. Vaccari, A. Clays and catalysis: a promising future. *Appl. Clay Sci.* 1999, *14*, 161–198.

298. Tichit, D. and Coq, B. Catalysis by hydrotalcites and related materials. *CATTECH* 2003, *7*, 206–217.

299. Rossi, P.F., Busca, G., Lorenzelli, V., Waquif, M., Saur, O., and Lavalley, J.C. Surface basicity of mixed oxides: stoichiometric and non-stoichiometric magnesium and zinc aluminates. *Langmuir* 1991, *7*, 2677–2681.

300. Kannan, S., Kishore, D., Hadjiivcanov, K., and Knozinger, H. FT–IR study of low temperature CO adsorption on MgAl hydrothalcite and its calcined forms. *Langmuir* 2003, *19*, 5742–5747.

301. Trombetta, M., Ramis, G., Busca, G., Montanari, B., and Vaccai, A. Ammonia adsorption and oxidation on Cu/Mg/Al mixed oxide catalysts prepared *via* hydrotalcite-type precursors. *Langmuir* 1997, *13*, 4628–4637.

302. Prinetto, F., Ghiotti, G., Durand, R., and Tichit, D. Investigation of acid–base properties of catalysts obtained from layered double hydroxides. *J. Phys. Chem. B* 2000, *104*, 11117–11126.

10 Optical Basicity: A Scale of Acidity/Basicity of Solids and Its Application to Oxidation Catalysis

E. Bordes-Richard
Laboratoire de Catalyse de Lille, Cité Scientifique,
Villeneuve d'Ascq, France

P. Courtine
Département de Génie Chimique,
Université de Technologie de Compiègne, Compiègne, France

CONTENTS

10.1 INTRODUCTION

Solid oxides used in catalytic reactions belong to the whole periodic table. Their role in catalysis depends mostly on their acid–base and redox properties. Schematically, acidic (basic) oxides of the right (left) part of the periodic table are involved in cracking, isomerization, alkylation, etc., of hydrocarbons, and are used as supports of metals or of other oxides. Redox catalysts are mainly found among the transition metal oxides and are more particularly involved in mild or total oxidation of hydrocarbons or of other molecules (e.g., alcohols). If parameters describing the catalytic activity are numerous and well-known, selectivity is always related to kinetics, for example, expressed as a ratio of the rates of one process compared with another. Finding parameters that are able to account for selectivity in mild oxidation has been one of our concerns, but until recently our attempts failed because of the lack of parameters accounting for the solid-state properties of transition metal oxides. Indeed, there are few answers when one tries to understand why only vanadium oxide supported on titania is able to selectively transform o-xylene into phthalic anhydride, or why Mo_3VO_{11} is selective in the oxidation of acrolein to acrylic acid. Similarly, the fact that, for example, n-butane is selectively oxidized to butadiene, maleic anhydride, or CO_2, if $CoMoO_4$, $(VO)_2P_2O_7$, or $LaCrO_3$, are used as respective catalysts, is not explained. This means that well-defined properties of the catalyst are required to get selectivity in a given type of reaction. In these reactions, the surface lattice oxygens (O^{2-}) of the metallic oxide are directly responsible for the selective formation of the required product. The commonly used two-stepped scheme proposed by Mars and van Krevelen [1] describes this participation:

$$[R–C–H] + 2KO \rightarrow [R–C–O] + H_2O + 2K \qquad (10.1)$$

$$2K + O_2 \rightarrow 2KO \qquad (10.2)$$

The first step is the transfer of surface lattice oxygen O to the molecules of the products (RCO and H_2O) leaving the catalyst in its reduced form K. K is regenerated as KO by gaseous dioxygen which is, generally co-fed with the reactant [R–C–H] in usual reactors. Therefore, the properties of O^{2-} species linked to metallic cations determine the catalytic properties, and particularly the selectivity to products. Among them, a key parameter is the nucleophilicity as noticed by Haber [2], who showed that nucleophilic (O^{2-}) and electrophilic (O_2^-, O_2^{2-}) oxygen species were responsible for mild and total oxidation, respectively. However, there was no scale of nucleophilicity/electrophilicity or, more generally, of basicity/acidity utilizable for transition metal oxides until recently. To account for selectivity as a general concept that allows the classification of reactions, it would be necessary to have available parameters accounting for a similar type of property in the gas (liquid) phase (organic chemical reaction) and in the solid phase (mineral catalyst). Moreover, thermodynamic parameters would be more appropriate than kinetic parameters, which depend strongly on experimental conditions: the initial state would be the reactant facing the oxidized KO form and the final state would be the required product facing the reduced form K of the "selective" catalyst. Since oxidation catalysis proceeds by an exchange of electrons that accompanies the exchange of O species, we have thought that the *ionization energy* of the organic molecules could be useful to account for the gas phase reaction. Finding a parameter of the same type for solid oxides was not so easy. Modern theories of reactivity propose to consider the oxidizing power, acidity, basicity, and reducing power, as steps along a same continuum, instead of distinct phenomena [3]. Among parameters, Zhang's scale of electronegativity [4], optical electronegativity and optical basicity [5,6], electronic polarizability [7], Racah parameter [8], and ionic–covalent parameter (ICP) [9,10], were used recently to quantitatively describe the acid (base) or redox behavior of oxides in various applications. General correlations of catalytic *activity*, with, for example, Me–O bond strength, electronegativity, or oxygen partial charge, are numerous [11–14]. Auroux and Gervasini [15,16] used some of these parameters in order to predict heats of adsorption and acid strength. Differential heats of adsorption of probe molecules such as NH_3 or CO_2 as a function of coverage have been worked out for different simple oxides. Also, relationships between the charge/radius ratio or the percentage of ionic character and average heats of CO_2 or ammonia adsorption, respectively, were obtained. Recently, Idriss and Seebauer [17] showed that the rates of oxidative dehydrogenation (ODH) of ethanol to acetaldehyde and of benzaldehyde esterification are correlated with the oxygen electronic polarizability α_O.

Successful correlations with experimental catalytic selectivity are far less numerous. For example, Vinek et al. [18] showed on a series of oxides that the x-ray photoelectron spectroscopy (XPS) binding energy of O1s is related to the electron pair donating (EPD) strength of oxygen, and hence to selective dehydrogenation (on more basic oxides, high EPD of O^{2-}) or selective dehydration (on more acidic oxides, low EPD of O^{2-}) of alcohols. This EPD/EPA approach was later used [19] in oxidation reactions and it was shown that selectivity increased with acid strength.

We have an ambitious goal: to classify catalytic oxides known to be selective in different oxidation reactions, in order to further use the resulting classification as a predictive trend. Among the parameters mentioned earlier, optical basicity Λ as well as ICP seemed to be the most appropriate parameters, because both take into account not only the type (ionic to covalent) of Me–O bond but also the extent (through polarizability) of the negative charge borne by oxygen. Duffy et al.

[20–22] started from the principle that a quantitative scale of basicity could be provided if the magnitude of the electron density of the oxygen atoms was known. The word "optical" means that optical spectroscopy was the origin of the measurements performed by using suitable probes. Metal ions (Tl^+, Pb^{2+}, Bi^{3+}) were inserted into the oxidic material whose acidity had to be determined. The expansion of their d outer orbital (nephelauxetic effect) upon coordination to the O^{2-} Lewis base depends on the polarization state of these O^{2-} and the shift of the UV frequency is related to Λ. This method was particularly useful in the case of molten oxidic systems (metallurgical slags) and glasses, and Λ could be successfully correlated with several properties of oxidic media in the solid or liquid state: refractive index [6,23], viscosity [24], redox equilibria [25–27] and acid–base neutralization reaction between oxyanion species [28]. However, the method is not applicable when transition metal oxides are concerned because of the presence of the own d orbitals of metals. Lebouteiller and Courtine [29] plotted optical basicity against ICP values and were able to set-up a new quantitative scale of optical basicity (or better said: optical acidity) of all types of cations related to the same O^{2-} base. This scale has been recently used by Lenglet [30,31], who found correlations with the Racah parameter of transition metal oxides. Catalytic applications to selective oxidation of various reactants were proposed by Moriceau et al. [32–34] by plotting optical basicity of the corresponding catalysts against the potential ionization difference, $\Delta I = I_R - I_P$, of the gas phase, and linear correlations were obtained. Calibrations of these lines with the most well-known oxidation reactions, and within a reasonable limit of error, were carried out. These correlations allow to classify reactions and also to a priori determine the optimum optical basicity that would be required for a catalyst to be selective in the chosen reaction.

This chapter aims to summarize most of the findings using optical basicity scale. First, the concepts of optical basicity and of ICP are recalled and some examples given. Then, the choice of ionization potential as a means to account for a "thermodynamic" selectivity is justified. The linear correlations obtained when plotting ΔI against the optical basicity are then discussed. Finally, the thermodynamic nature of optical basicity is discussed and attempts to account for the linearity of correlations between ΔI and Λ are proposed.

10.2 SCALE OF OPTICAL BASICITY OF SOLID OXIDES

10.2.1 After Duffy

As first proposed by Duffy et al. [5,6,20–22], the so-called optical basicity characterized by Λ allows to classify oxides as a scale of acidity that is referred to the same O^{2-} base. It accounts for physical and chemical behavior of phases and gives indications, for example, on the structural modifications in glasses, the effective electronic charge carried by ions in an oxide, the M–O bond lengths, or on redox equilibria in melted glasses. The optical basicity is built on Lewis acidity concept and is particularly adapted to the study of nonaqueous nonprotonated systems. This parameter, first obtained from UV spectroscopic measurements, is a "measure" of the donating electron power of O^{2-} to the M^{n+} cation in MO_y oxide. Indeed the polarizing power of M^{n+} influences the capacity of O^{2-} ions to give a part of their electronic charge to other cations in the solid phase, and as such it depends on M coordination. Therefore, Λ characterizes the electron acceptor power of the M^{n+} cation as well as the electron donor power of O^{2-}, and the result is that it characterizes the acidity of the whole solid versus the same O^{2-} base. For example, Λ MgO = 0.78 ($\Lambda^{VI}Mg^{2+}$ = 0.78, in which valence and coordination of $^{VI}Mg^{2+}$ are VI and +2). The reference chosen by Duffy et al. is CaO for which Λ = 1.0.

Duffy's scale is satisfactory for most s–p oxides but not for transition metal oxides. Examples are WO_3, MoO_3, and V_2O_5, the optical basicity Λ of which was found to be >1.0, which is the value of basic CaO [6,7]. The main reason is that the optical basicity of transition metal oxides cannot be directly measured because the own d orbitals of cations are involved and their metal–oxygen bonds have an ionic–covalent character. This problem was side-stepped by Portier et al. [9,10], who succeeded in building a similar scale, but based on the new ICP parameter. The ICP, which is an adimensional number, accounts for the influence of the covalence of metal–oxygen bond on the acid strength of the metallic cation, and is calculated by Equation (10.3):

$$ICP = \log P - 1.38\chi + 2.07 \qquad (10.3)$$

where P is the polarizing power of the cation ($P = z/r^2$, z = formal charge, r = Shannon ionic radius [35] with $rO^{2-} = 1.40$ Å). The electronegativity χ (in Pauling-type scale) is calculated by (10.4):

$$\chi = 0.274z - 0.15zr - 0.01r + 1 + c \qquad (10.4)$$

in which c is a correcting term, depending on each cation. For a given cation, the value of ICP depends on its valence, its coordination, and eventually its spin state (low or high) in the solid [9,10].

By plotting Λ values determined for simple oxides by Duffy et al. [5,6] against ICP of cations, Lebouteiller and Courtine [29] showed that cations are distributed among straight lines depending upon their electronic configuration. Five lines for s–p, d^0, d^1–d^9, d^{10}, and $d^{10}s^2$ configurations were obtained, the equations of which are presented in Table 10.1. As ICP does, the theoretical optical basicity Λ_{th} obtained from these correlations depends on the valence, coordination, and spin of the cation(s) [32]. When several valences and coordinations exist, the basicity is obtained as follows, for example, for Fe_3O_4, which crystallizes as inverse spinel:

$$\Lambda_{th} Fe_3O_4 = 3\Lambda^{IV}Fe^{3+} + 2\Lambda^{VI}Fe^{2+} + 3\Lambda^{IV}Fe^{3+} = 0.785$$

with $\Lambda^{IV}Fe^{3+} = 0.66$, $\Lambda^{IV}Fe^{2+} = 0.76$; $\Lambda^{IV}Fe^{3+} = 0.77$; $\Lambda^{VI}Fe^{2+} = 1.00$ (high spin configurations). In the case of two polymorphs, the value of Λ_{th} is different

TABLE 10.1
ICP Plotted Against Optical Basicity of Cations

Electronic configuration	a	b	Reliability factor R^2
s–p	−0.672	1.451	0.989
d^0	+2.982	−1.322	0.999
d^1–d^9	−0.644	1.144	0.991
d^{10}	−0.729	1.390	0.999
$d^{10}s^2$	−13.844	17.134	0.896

Note: Slope a and intercept b of linear relationships ICP $= a\Lambda + b$
Lebouteiller and Courtine [29].

only if the coordination of cation(s) is different. This is the case of α and γ forms of Al_2O_3. The latter is a defective spinel, with $\Lambda_{th}\ \gamma$-$Al_2O_3 = 0.50$ while $\Lambda\alpha$-Al_2O_3 ($\Lambda = 0.60$), is less acidic. Similarly, the acidity of GeO_2-quartz ($\Lambda_{th}\ ^{IV}Ge^{4+} = 0.54$) is higher than that of GeO_2-rutile ($\Lambda_{th}\ ^{VI}Ge^{4+} = 0.66$). Conversely, the anatase and rutile forms of TiO_2, the band gap of which is slightly different (3.2 and 3.0 eV, respectively) cannot be distinguished by the value of Λ_{th} because of VI coordination in both oxides.

Figure 10.1 shows the example of Λ vs ICP lines for d^0 and d^1–d^9 configuration, which gather most transition metal cations that are commonly used in catalytic oxidation. The most acidic cations lie on the left of the figure. Redox couples involve transition metal cations which may be both in d^0/d^{1-9} configuration (first group), such as W^{6+}/W^{5+} or W^{6+}/W^{4+}, Mo^{6+}/Mo^{5+}, V^{5+}/V^{4+} etc., or both in (second group) such as Fe^{3+}/Fe^{2+} or Mn^{4+}/Mn^{3+}, or in d^9/d^{10} such as Cu^{2+}/Cu^+ (third group). The acidity/basicity of a same element depends mostly on its configuration at the highest valence and on its coordination. Considering cations of the first group (d^0 line) with the same valence, acidity decreases with decreasing coordination ($\Lambda_{th}\ ^{VI}W^{6+} = 0.51$, $\Lambda_{th}\ ^{IV}W^{6+} = 0.54$). For the same coordination, acidity generally decreases with decreasing valence ($^{VI}V^{5+} > {}^{VI}V^{4+}$; $^{VI}Mo^{6+} > {}^{VI}Mo^{5+}$). However, the case of group VIa M cations ($M = Cr, Mo, W$) is special because $\Lambda_{th,}\ ^{VI}M^{6+} < \Lambda_{th,}\ ^{VI}M^{4+} < \Lambda_{th,}\ ^{VI}M^{5+}$, that is, M^{5+} is less acidic than M^{4+}. Moreover, when comparing V and Mo oxides which are among the most used in selective mild oxidation, it is striking that Mo^{5+} is far more basic than Mo^{6+} ($\Delta\Lambda = \Lambda\ ^{VI}Mo^{5+} - {}^{VI}Mo^{6+} = 1.17 - 0.52 = 0.65$) while $\Lambda\Delta = 0.68 - 0.63 = 0.05$ only for V^{4+}/V^{5+} couple. For redox couples lying on the same d^1–d^9 line (group 2), cations in their higher valence state are generally less acidic than in their lower state ($\Lambda^{VI}Mn^{4+} = 0.88$; $\Lambda\ ^{VI}Mn^{3+} = 0.81$; this was the case of $\Lambda_{th}^{VI}Mo^{4+} < \Lambda_{th}\ ^{VI}Mo^{5+}$, just seen), and the lowest coordination corresponds to more acidic cations (in high spin configuration, $\Lambda_{th}\ ^{VI}Fe^{3+} = 0.88$, $\Lambda_{th}\ ^{IV}Fe^{3+} = 0.66$). Therefore, the value of Λ_{th} is meaningful as it reflects most structural characteristics.

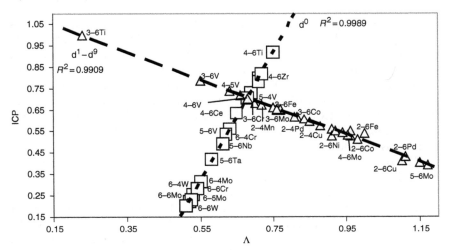

FIGURE 10.1 Linear relationships between ICP and optical basicity Λ_{th} for transition metal cations. Cations M are labeled as z-CNM, where z = valence and CN = coordination (N.B. $***$Suppressed because not seen on the figure)

10.2.2 Mixed Oxides

By using the optical basicity Λ_{th} of cations in their appropriate coordination, valence, and spin, it is possible to calculate the theoretical optical basicity of mixed oxides, oxysalts, or of any oxygen-containing solid, if the stoichiometry is known. Given a $M_i^{z_i+}{}_{xi}O_n^{2-}$ oxide or mixture, Λ_{th} is calculated by the linear combination of stoichiometry x_i, valence z_i, and Λ_i of the i cations (n = oxygen stoichiometry), according to:

$$\Lambda = \frac{1}{2n} \sum_i x_i z_i \Lambda_i \qquad (10.5)$$

or $\Lambda_{th} = (ax\Lambda_A + by\Lambda_B)/2n$ for $A_x^{a+}B_y^{b+}O_n^{2-}$. Using data from [ICP, Λ] correlations, the optical basicity of, for example, $(VO)_2P_2O_7$ is ($\Lambda^{VI}V^{4+} = 0.68$, $\Lambda^{IV}P^{5+} = 0.33$):

$$\Lambda_{th} V_2P_2O_9 = (2 \times 4 \times 0.68 + 2 \times 5 \times 0.33)/18 = 0.486$$

Such Λ_{th} values can be calculated for any oxygen-containing phase, including hydrated/anhydrous/cation exchanged zeolites (proton being considered as any cation, $\Lambda^{I}H^+ = 0.40$), or supported oxides. In the latter case, if the support has only a textural effect consisting of the dispersion of the active phase, it may be neglected. But if its action is also synergistic or leads to bifunctionality its Λ_{th} must be included in the calculation. In the case of, for example, the V_2O_5/TiO_2

system in which synergistic effects are well-known [36,37], the role of TiO_2 is prominent. For $0.1V_2O_5/TiO_2$ and V_2O_5/TiO_2, $\Lambda = 0.72$ and 0.68, respectively (Λ $^{VI}Ti^{4+} = 0.75$). Another interest of the correlation is to account for the variation of acidity (Λ) upon modification by another cation or by a promoter. For example, $H_4PMo_{11}VO_{40}$ is a strong acid as $\Lambda_{th} = 0.51$. By replacing all or some of the protons by cations the basicity of O^{2-} is modified in Λ_{th} ranging from 0.511 ($H_3Zn_{0.5}$) to 0.532 ($HCe_{0.75}$) [32]. In the case of $0.05V_2O_5/TiO_2$ example, doping by 5% K [38] results in higher $\Lambda_{th} = 0.743$ (instead of 0.737) and then in higher basicity, as expected.

The optical basicity of mixtures like the so-called multicomponent oxides claimed in patents is also calculable. However, it is necessary to make assumptions as far as the valence and coordination of cations are concerned, on which, moreover, the oxygen stoichiometry depends. For example, in the case of $P_{1.5}Mo_{12}V_{0.5}Bi_{0.3}As_{0.4}Cu_{0.3}Cs_{1.4}O_y$, which has been claimed selective in the oxidation of isobutane to methacrylic acid [39], the oxygen stoichiometry can be calculated by assuming the most stable valence and usual coordination for all elements ($y = 43.45$), so that Λ_{th} is 0.547.

Figure 10.2 and Figure 10.3 present a decreasing (from the left to the right) acidity scale of some known catalysts and of some oxides often used as supports. As expected, zeolites are the most acidic compounds, followed by heteropolyacids and then by WO_3, MoO_3, $(VO)_2P_2O_7$, V_2O_5, etc. More basic (e.g., Bi_2MoO_6, Cu_2O) solids are found on the right. Clays such as chrysotile, $Mg_3(OH)_4Si_2O_5$ ($\Lambda_{th} = 0.562$), may be considered as basic [40] as compared with zeolites because of the contribution of Mg (Λ $^{VI}Mg^{2+} = 0.78$) to Si (Λ $^{IV}Si^{4+} = 0.48$), but they are less basic than most oxides used in selective oxidation. Silica is the most acidic oxide in the support series, followed by γ-Al_2O_3, α-Al_2O_3, etc., and MgO is the most basic. The strongest bases are found in the K_2O–Cs_2O series ($\Lambda_{th} = 1.40$ to 1.7) [29].

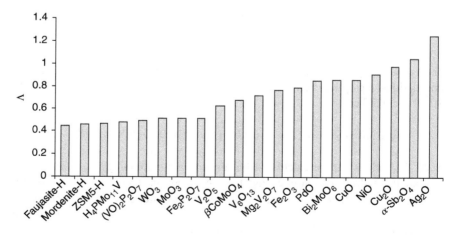

FIGURE 10.2   Scale of acidity/basicity (Λ_{th}) of some catalysts

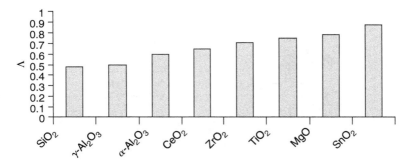

FIGURE 10.3   Scale of acidity/basicity (Λ_{th}) of some oxidic supports

10.2.3 Surface Evaluation of Optical Basicity

The former considerations on optical basicity Λ are driven for bulk but not for surface of solids, which, obviously, is concerned when dealing not only with catalysis, but also with adhesion, corrosion, etc. As it is well-known, the surface compared with the bulk may have very different properties and the main problem to solve is to know to what extent. First, we have considered the cases where the crystal structure of the catalyst is anisotropic so that their layered morphology is directly obtained by preparation. Mild oxidation being structure-sensitive reactions, it is known that by changing the crystal morphology the selectivity is modified, as shown on MoO_3 and V_2O_5 [41,42], or $(VO)_2P_2O_7$ [43–46]. Therefore we have assumed that the known selective reaction proceeds on crystals with their natural habitus, for example, $\{010\}$ V_2O_5 for o-xylene to phthalic anhydride, or $\{100\}(VO)_2P_2O_7$ for n-butane to maleic anhydride, or $\{010\}Bi_2MoO_6$ for propene to acrolein. In such cases we have considered that $\Lambda_{surf} \sim \Lambda_{th}$. The same remark holds for monolayers of active phases on supports and Λ_{th} of the monolayer/support system is used directly. In the case of "bulky" compounds such as NiO or Co_3O_4, the structures of which are cubic, we propose a correction on Λ_{th} by considering the results obtained by Iguchi and Nakatsugawa [47] on NiO. These authors have found that the polarizability of oxygen αO^{2-} is increased on the surface as compared with the bulk of NiO. By using a relation driven by Duffy [16] between Λ and αO^{2-} the mean estimation of Λ_{surf} is $\Lambda_{surf} = \Lambda_{th} + 0.12$, which is assumed to give the maximum of deviation of Λ when a cubic solid is dealt with [32].

A more accurate way is to consider XPS data, which provides a way to correlate surface to bulk properties. Delamar [48] found a linear correlation between IEPS and (DO + DM), where DO is the difference between the binding energy (BE) of O1s in the oxide and that in MgO taken as reference, and DM is the difference of BE of cation and of metal atom, respectively. We also found a linear correlation by plotting Λ_{th} (or Λ_{surf} when appropriate) against (DO + DM) [32]. Another try is to use the BE O1s data [49,50]. Plotting it against Λ results in the following

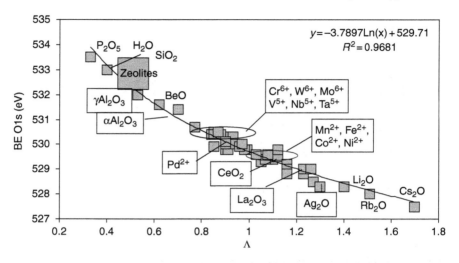

FIGURE 10.4 Logarithmic relationship between the BE O1s and optical basicity Λ_{th} of oxides. Range of BE O1s and Λ_{th} for oxides (represented by cations, e.g. Cr^{6+} for CrO_3) is shown by ellipses

logarithmic equation (Figure 10.4):

$$BE\ O1s = -3.79 Ln\ \Lambda + 529.7 \tag{10.6}$$

in which the intercept value, 529.7 eV, is the BE O1s found experimentally for CaO (taken by Duffy as reference, $\Lambda = 1$). However, the Auger parameter, which is related to refraction index [51], would be more suitable to go further because BE O1s is known to vary, sometimes widely, with the real surface state of the examined material. Since photoelectrons emerge from a thin surface layer of ~20 to 30Å thick, the O1s line may be distorted because of the contribution of OH, H_2O, and CO_3^{2-} groups [52]. In this context, however, there is a general correlation between BE O1s and Λ of the simple oxides (which could be refined by taking into account their structure), from the most acidic (P_2O_5) to the most basic (Cs_2O) oxides. As such, it gathers metal and transition metal oxides, including zeolites and most supports. Because of the small range of BE O1s when considering d^0 (mean BE ~ 530.2 eV) and d^1–d^9 (mean BE ~ 529.5 eV) oxides, it is quite difficult, today, to use the correlation reliably when dealing only with a series of closely related oxides such as V, Mo, W, Nb oxides, unless accurate experimental data are considered.

As a partial conclusion, we have made available a parameter, Λ_{th}, which varies with stoichiometry and with valence, coordination, spin state of cation(s) in the oxide structure, and which is calculable for any oxygen-containing solid. Scales of bulk and of surface acidity/basicity are proposed by using optical basicity as a suitable parameter, which allows to rank oxides, and which can be used for several applications particularly in catalysis as seen further.

10.3 IONIZATION POTENTIAL AND SELECTIVITY

Selectivity to a given product is a kinetic parameter generally understood as a ratio of rates, and which depends, therefore, on experiments. Here, we are looking for a "thermodynamic selectivity" that could be used to rank selective oxidation reactions, and further be compared with the acidity/basicity of catalysts. Free enthalpies of formation or ionization potential are suitable thermodynamic parameters for the reactant and the required product to compare the oxidation reactions. Free enthalpies per mole of oxygen were used by Bordes [53] in the following manner: the $\Delta_r G_{MOX}$ of Mild OXidation (e.g., butene to butadiene, to furan, to maleic anhydride and to other intermediates) reactions was compared with the $\Delta_r G_{TOX}$ of Total OXidation of the same reactant (e.g., butene to CO_2). The difference was plotted against the initial selectivity of these reactions, which resulted in a linear correlation.

More recently, we have chosen the ionization potential of molecules as a representative parameter to be correlated to acid–base properties in several reactions. This parameter has already been used by Ai [54] to compare reactions, as well as by Richardson [55] who has correlated the degree of ionization of aromatics determined from ESR spectroscopy data with the electron affinity of cations in cationic zeolites. The ionization potential I of a C_n (n = number of carbons) hydrocarbon depends on n and on the type (linear or branched) of the isomer considered. The I potential decreases when n increases, because the greater the n, the higher is the HOMO orbital, the weaker is the ionization energy, and then the more reactive is the molecule. This is verified with a series of C_2–C_4 paraffins and olefins, respectively (Figure 10.5[a]). When oxygen is present in a molecule, I increases with the amount of O in the same series (saturated C–C or unsaturated C=C compounds). For example, $I = 9.95$ and 10.52 eV for propanal and propanoic acid, while $I = 10.1$ and 10.6 eV for acrolein and acrylic acid, respectively.

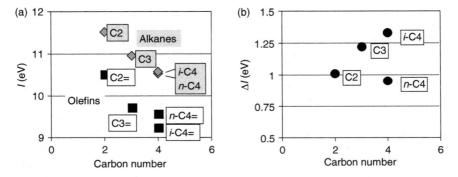

FIGURE 10.5   Effect of the number n of carbons. (a) On the ionization potential of C_2–C_4 hydrocarbons, C_n = indicates olefin, (b) In the case of ODH, on the difference of ionization potential ΔI between reactant and product; C_n indicates ODH of C_n alkane (e.g., C_2 for ethane to ethene)

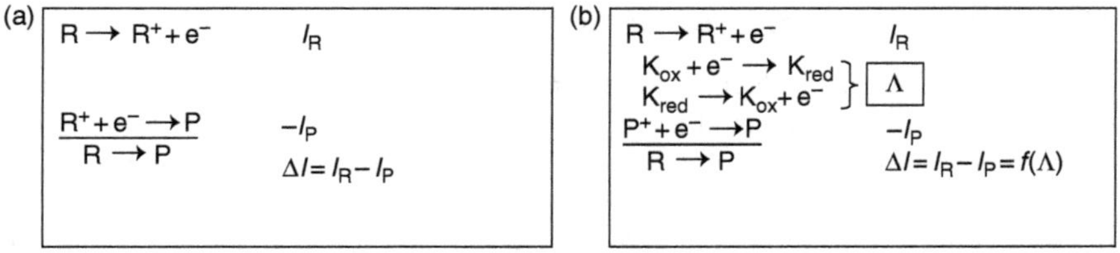

(a)
$$R \rightarrow R^+ + e^- \qquad I_R$$

$$\frac{R^+ + e^- \rightarrow P}{R \rightarrow P} \qquad \begin{array}{l} -I_P \\ \Delta I = I_R - I_P \end{array}$$

(b)
$$R \rightarrow R^+ + e^- \qquad I_R$$
$$\left.\begin{array}{l} K_{ox} + e^- \rightarrow K_{red} \\ K_{red} \rightarrow K_{ox} + e^- \end{array}\right\} \boxed{\Lambda}$$
$$\frac{P^+ + e^- \rightarrow P}{R \rightarrow P} \qquad \begin{array}{l} -I_P \\ \Delta I = I_R - I_P = f(\Lambda) \end{array}$$

SCHEME 10.1   Comparison of ΔI for (a) noncatalytic and (b) catalytic oxidation (one-electron redox)

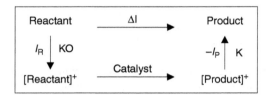

SCHEME 10.2   Thermodynamic cycle applied to catalytic oxidation

To account for selectivity, we proposed to use the difference of ionization potential between the reactant and the product that is required (which is not CO_2 when mild oxidation is considered). When starting from a same reactant, the deepness of oxidation is accounted for by this difference, $\Delta I = (I_{Reactant} - I_{Product})$. For example, $\Delta I = 0.37$ and 0.87 eV for propene oxidation to acrolein and to acrylic acid, respectively. ΔI represents the difference of electron donor power when the reactant becomes the product (Scheme 10.1[a]). The main difference with noncatalytic reactions is the exchange of electrons which proceeds through the catalyst (Scheme 10.1[b]), and thus we assume that the electron donor power (Λ) of a selective catalyst must fit the difference of electron acceptor power (ΔI) of the gas (liquid) phase reaction. A simple thermodynamic cycle may be drawn because ionization energy is a state function (Scheme 10.2). The interest of thermodynamic relationships has been put in evidence, for example, by Hodnett et al. [56,57] who used enthalpies of formation of reactant and of product to account for the difficulty to get a high selectivity in mild oxidation of alkanes. However enthalpies are not state functions, contrary to ionization energies that allow to predict what type of catalyst (characterized by Λ) could fit a given reaction (characterized by ΔI). ΔI is not only the net exchange of electrons during the reaction, but it also depends on the structure of the molecules of reactant and of product. Figure 10.5(b) shows what happens when ΔI is plotted against n in the case of ODH of C_2–C_4 alkanes, during which two electrons are exchanged per molecule. It is seen that ΔIs are different and that ΔI increases with n, including for ODH of i-butane to i-butene, but decreases for n-butane to but-1-ene (ΔI very close for but-2-enes). Now, considering the (more or less) selective catalysts described in the literature [58,59] for ODH reactions, it is striking that catalysts based on vanadium oxide (or vanadates) are rather more selective for C_2 and n-C_4 (ΔI =1.01 to

0.95 eV), while catalysts based on molybdenum oxide (or molybdates) are rather more selective for C_3 and i-C_4 ($\Delta I = 1.24$ to 1.27 eV) oxidations.

ΔIs have been calculated for a series of mild and total oxidation reactions involving molecules with $n = 1$–8. ΔIs range from 0.15 to 2.01 eV in the case of mild oxidation, and up to 5.3 eV when combustion of phenol is considered [33,34]. When the number of carbons differs during the reaction (e.g., benzene oxidation to maleic anhydride), the following correction is applied:

$$\Delta I = (I_R - I_P)n_P/n_R \qquad (10.7)$$

where n_R and n_P are the carbon number of reactant and product, respectively. In the case of ammoxidation reaction, ammoniac is considered as a reactant (the other being oxygen on which the correlations are based), and ΔI is calculated by:

$$\Delta I = [(I_R + I_{NH3})/2 - I_P]n_P/n_R \qquad (10.8)$$

Sometimes the reaction is not selective enough (or better said, the fully selective catalysts have not yet been found), so when two P_1 and P_2 products are equally obtained the following equation can be used:

$$\Delta I = [I_R - (I_{P1} + I_{P2})/2]n_P/n_R \qquad (10.9)$$

This is the case, for example, of propene, which is generally found in noticeable amounts besides acrylic acid or acrolein in the oxidation of propane.

As a partial conclusion, by considering a "thermodynamic selectivity," which is the difference between ionization energies of reactant and product, various reactions of oxidation can be ranked. ΔI represents the electron acceptor power of the gas (liquid) phase reaction and is the analog of Λ, which represents the electron donor power of the selective solid catalyst.

10.4 CORRELATION BETWEEN IONIZATION ENERGY AND OPTICAL BASICITY

10.4.1 Correlations between Λ and Experimental Values of Selectivity

Although the aim of this chapter is to show how a thermodynamic relationship between Λ and ΔI allows to predict the type of catalysts needed for a reaction, it is worth recalling that Λ can be correlated with experimental parameters related to catalysis, or values of selectivity, provided the same reaction is studied [33]. For example, by using data proposed by Matsuura [60], the heat of adsorption ΔH_{ads} for a series of catalysts of oxidation of 1-butene to butadiene, or the Mössbauer quadruple shift values for Fe^{3+}-containing catalysts of propene ammoxidation, could be related to the Λ value of the respective catalysts [33]. In a study of the ODH of isobutyric acid to methacrylic acid on polyoxometallates $H_{4-zx}M_xPMo_{11}VO_{40}$ (z = valence of M) [61–63], in which M was varied, selectivities from 57 to 77 mol% were obtained at isoconversion (90 mol%). These experimental values of selectivity were correlated to the calculated Λ_{th} values. Indeed, by replacing all or some of the protons of $H_4PMo_{11}VO_{40}$, which is a strong acid ($\Lambda_{th} = 0.51$), by various cations, the basicity of O^{2-} is modified in a Λ_{th} range of 0.511 ($H_3Zn_{0.5}$) to 0.532 ($HCe_{0.75}$). By plotting selectivity versus Λ, a volcano curve was obtained, the maximum of which corresponds to $H_2CuPMo_{11}VO_{40}$ ($\Lambda_{th} = 0.519$). The electron donor power of the active oxygens in this compound is assumed to be optimum to ensure both the activation of the reactant (a basic oxygen is needed to abstract H from H–(C–) in isobutyric acid) and the desorption of the product (a more basic site is needed to allow desorption of the more basic C $=$ C of methacrylic acid) [33]. Another highly selective catalyst for the same reaction is $Fe_2P_2O_7$ which is the main active phase found at the steady state [64,65] and it is remarkable that its optical basicity ($\Lambda_{th} = 0.52$) is very close to that of the other selective catalyst, $H_2CuPMo_{11}VO_{40}$.

10.4.2 Linear [ΔI, Λ] Correlations: Alkanes and Alkyl-Aromatics

Correlations [ΔI, Λ] were first drawn for partial oxidation of alkanes and of unsaturated hydrocarbons. As ΔI, difference of ionization energy, may be negative or positive according to whether the product is more or less stable than the reactant, the absolute value $|\Delta I|$ (further written ΔI) was used to be plotted against Λ in order to compare the various reactions starting from C_1 (methane) to C_8 (o-xylene) reactants. Both ODH and MOX reactions were examined, the idea being that the mechanism is more or less similar, at least for the activation step. Let us recall that, used as a predictive trend, such correlations cannot be based on experimental values of selectivities, as it is obvious that the catalytic properties of the catalyst depend strongly on not only its preparation but also its operating parameters. Therefore, for well-known reactions performed industrially like, for example, propene, n-butane or benzene, o-xylene oxidations to acrolein, maleic anhydride and phthalic anhydride [66–69], respectively, the optical basicity was calculated for the "well-known" selective catalyst, that is, for bismuth molybdates, vanadyl pyrophosphate or $0.7V_2O_5$–$0.3MoO_3$, and V_2O_5/TiO_2, respectively. A first calibration curve was then drawn. In the case of ODH or MOX of alkanes, highly "selective" catalysts are not yet known and a variety of formula taken from academic or patent literature can be used. Examples were chosen to get a range of Λ_{th} and then a new, optimized, calibration line was obtained. An example is isobutane oxidation to methacrylic acid ($\Delta I = 0.42$ eV, Equation [10.7]) for which the formula $P_{1.5}Mo_{12}V_{0.5}Bi_{0.3}As_{0.4}Cu_{0.3}Cs_{1.4}O_y$ ($\Lambda_{th} = 0.54$, vide supra) has been claimed selective [37]. In another patent [70], methacrolein is found besides methacrylic acid ($\Delta I = 0.565$, Equation [10.10]) for the formula $PMo_{12}VAg_{0.1}As_{0.25}KO_y$, thus $y = 41.92$ and $\Lambda_{th} = 0.53$. Obviously in these cases the actual valence state and coordination of cations at the steady state is not known. This point is particularly important each time molybdenum is involved, as compared with vanadium. As already mentioned, for ^{VI}Mo, there is a large difference of Λ_{th} between Mo^{6+} (0.52) and Mo^{5+} (1.17), whose Λ_{th} is higher than that of Mo^{4+} (0.96) (Figure 10.1).

Each reaction/catalyst couple being characterized by [ΔI, Λ_{th} (or Λ_{surf})], it has been noticed that the 52 catalysts ($\Lambda = 0.45$ to 0.95) in 26 reactions ($\Delta I = 0.15$ to 2.01 eV) are distributed between two straight lines with reliability factors R^2 close to 0.94 to 0.93. Slopes are positive or negative when oxidation of paraffinic bonds or of unsaturated bonds is concerned, respectively, as explained in Section 10.4.2.1.

10.4.2.1 The Mild Oxidation of Paraffins Line

The Mild Oxidation of Paraffins (MOPs) line (positive slope) gathers 29 [ΔI, Λ_{th}] couples and is attributed to reactions involving paraffinic bonds (Figure 10.6) [33]. Alkyl-aromatics such as toluene, ethyl-benzene, or *o*-xylene (oxidized into benzoic acid, styrene, or phthalic anhydride, respectively) also fit this MOP line because only the methyl (ethyl) group(s) are oxidised to acid or anhydride. Some O-containing reactants (ODH of isobutyric acid, MOX of acrolein and methacrolein) fit the line for similar reasons (C–C → C=C, or –(O=)C–H → –(O=)C–OH). However, this is not the case of alcohols (vide infra). Another advantage of thermodynamics, which considers only the initial and the final stages and not the way the transformation happens, is that the points representing methane coupling (to ethylene) performed, for example, on Li_2O/MgO, the mechanism of which is known to involve radicals (at least partly) [71,72], are situated close to the MOP line. Similarly, the [ΔI, Λ] of MOX of methane to synthesis gas

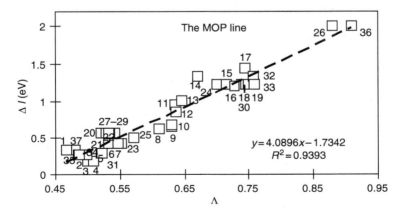

FIGURE 10.6 The MOP line. Mild oxidation and ODH of C_1 to C_8 paraffins and alkyl–aromatics. Linear correlation between ΔI and Λ of catalysts. Several catalysts are represented by their Λ value for the same reaction. Reproduced by permission of the PCCP Owner Societies

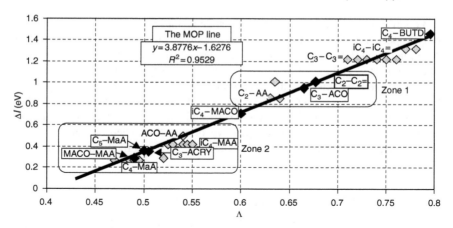

FIGURE 10.7 The MOP line restricted to MOX and ODH of alkanes. Comparison between optimal values of Λ for some examples (BUTD = butadiene, (M)ACO = (metha)crolein, ACRY = acrylic acid, MAA = methacrylic acid, MaA = maleic anhydride)

($\Delta I = 2.27$ eV, Equation [10.8]) also fits the line if NiO is considered to be the actual selective phase ($\Lambda_{surf} = 1.03$).

It is interesting to examine more closely the reaction/catalyst couples in the light of the known experimental characteristics, the more so because highly selective catalysts are not yet known in alkane oxidation (Figure 10.7). The exception is $(VO)_2P_2O_7$ ($\Lambda = 0.486$). Known as the selective catalyst for n-butane oxidation to maleic anhydride (MaA) [64,66,73], its acidity is close from that of catalysts like aluminum phosphate (ethyl-benzene to styrene) [74], or from heteropolyacids like $H_4PMo_{11}VO_{40}$. By adding promoters such as Fe^{3+} or Co^{2+} its selectivity to MaA is known to increase [66,75,76]. The addition of these dopants, which are less acidic ($\Lambda = 0.77$ and 0.98, respectively) than V^{4+} (0.68), brings the representative Λ values closer to the calibration line. The effect occurs with more acidic V^{5+} (0.63), which must be present in small amounts on the surface [73]. Pentane oxidation to MaA on $(VO)_2P_2O_7$ also fits the line. When considering the oxidation of $C_2–C_4$ alkanes, ΔIs of MOX are far lower than ΔIs of ODH, which means that the optimum optical basicity of catalysts is also different. Less acidic cations are needed for ODH, from an optimal $\Lambda_{opt} = 0.68$ for ethane to ethylene to $\Lambda_{opt} = 0.78$ for i-butane to i-butene. Operating conditions like the alkane-to-oxygen ratio, which is generally greater than one, render the feed more reducing. This is in accordance with the fact that the acidity of d^0 cations is lower (at iso-coordination) when reduced ($d^1–d^9$) than when oxidized (see Figure 10.1). In MOX of alkanes, the observed formation of the olefin in substantial amounts besides the acid (e.g., propane to propene and to acrolein or to acrylic acid) may also be explained by this reducing action of the reactant on the cation. Actually, it would be necessary to know the mean valence state and coordination of the cations *at the steady state* to get a more realistic value of optical basicity. A good example is given by $AgMo_3P_2O_{14}$ studied by Costentin et al. [77], in ODH of propane, who measured the Mo^{6+}/Mo^{5+} ratio by XPS of the most selective catalyst and found it equal to 2/1. Calculating Λ of $AgMo_3P_2O_{14}$ gives 0.514 (too acidic), but for $AgMo_2^{5+}Mo^{6+}P_2O_{13.5}$ the value of Λ is 0.735, which fits the MOP line. Other examples may be found in Reference 33.

Two zones are worthwhile to be examined (Figure 10.7). Beginning by ethane oxidation (zone 1), the ΔIs of both MOX and ODH lie in the same range (0.86 to 0.94 eV), which is not the case of C_3 or C_4 alkanes. Although it depends on experimental conditions and thus on the reducing power of the mixture, ethylene and acetic acid are often obtained together which is accounted for by the correlation ($\Lambda_{opt} = 0.64$ to 0.66). Today the best catalysts are based on MoVO system (V/Mo < 1) in which the reduced Mo_5O_{14} oxide is identified [78–83]. Its calculated Λ_{th} is 0.752, but if MoO_3 ($\Lambda_{th} = 0.52$), which is always detected besides Mo_5O_{14} is taken into account, $\Lambda_{th} = 0.65$ for $Mo_5O_{14}/MoO_3 = 1/1$. If that type of catalysts is promoted with Pd, its Λ_{th} increases a little ($\Lambda_{th}{}^{VI}Pd^{2+} = 1.11$), while it decreases upon addition of Nb^{5+} ($\Lambda_{th}{}^{VI}Nb^{5+} = 0.61$). As already mentioned, the actual Λ_{th} depends on the relative amounts of cations and of their valence at the steady state. Catalysts based on V_2O_5/TiO_2 ($\Lambda_{th} = 0.69$) were also studied for the same reaction [84]. To account for the oxidation of o-xylene to phthalic anhydride which proceeds on the same system, it is necessary to use less vanadium as compared with TiO_2, so that $\Lambda_{th} = 0.737$ for a theoretical formula like $0.05V_2O_5/TiO_2$, value which can be increased by considering all V^{4+} ($\Lambda_{th} = 0.68$) and thus $\Lambda_{th} = 0.743$ better fits the MOP line. MOX of propane to acrolein ($\Delta I = 0.95$) with $\Lambda_{opt} \approx 0.67$ lies also in this zone.

In the second zone, three MOX reactions involving C_3 and i-C_4 (direct oxidation of propane to acrylic acid, isobutane to methacrylic acid and methacrolein to methacrylic acid), are in the same ΔI range (0.30 to 0.42 eV). The optimum Λ_{th} are ~0.49 to 0.55. It is striking that patents claim for the same type of multicomponent Mo-based oxides, the Λ of which is very close from that of polyoxomolybdates, often claimed for MOX of isobutane. Recently, new systems were proposed, beginning by MoVO doped with both acidic (e.g., Al, $\Lambda_{th}{}^{VI}Al^{3+} = 0.60$) and more basic cations like Te ($\Lambda_{th}{}^{VI}Te^{6+} = 1.173$, $\Lambda_{th}{}^{VI}Te^{4+} = 0.751$). For example, by adding Al and Te to the raw formula $Mo_6^{6+}V_2^{4+}O_{22}$ (hypothesis), this material ($\Lambda_{th} = 0.549$) becomes less and less acidic along $Mo_6^{6+}V_2^{4+}Al^{3+}O_{22.5} < Mo_6^{6+}V_2^{4+}Al^{3+}Te_{0.5}^{4+}O_{23.5} < Mo_6^{6+}V_2^{4+}Al^{3+}Te_{0.5}^{6+}O_{24}$ ($\Lambda_{th} = 0.552 < 0.578 < 0.589$). Catalysts for MOX of acrolein to acrylic acid ($\Delta I = 0.50$ eV) must be slightly less acidic ($\Lambda_{th} = 0.57$). As far as propylene is concerned as a reactant, $\Delta I = 0.37$ eV for its oxidation to acrolein lies in the same range. However, because propene is not an alkane, the basicity of the catalysts must be higher, as discussed further (see Section 10.4.2.2, Figure 10.8).

10.4.2.2 The Mild Oxidation of Olefins Line

The Mild Oxidation of Olefins (MOOs) gathers MOX and ODH reactions of C_2–C_4 olefins to epoxides, aldehydes, and acids or anhydrides (Figure 10.8). Epoxidation of ethylene proceeds on alumina supported silver which, here, is supposed to undergo oxidation to Ag_2O ($\Lambda_{th} = 1.25$) during the course of the reaction.

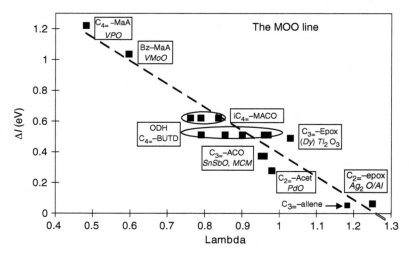

FIGURE 10.8 The MOO line. Mild oxidation and ODH of C_2 to C_6 olefins and aromatics. Linear correlation between ΔI and Λ of catalysts. Several catalysts are represented by their Λ value for the same reaction (BUTD = butadiene, (M)ACO = (metha)crolein, ACRY = acrylic acid, MAA = methacrylic acid, MaA = maleic anhydride; Epox = epoxide, Acet = acetaldehyde, MCM = multicomponent molybdates)

Heterogeneous catalysts for epoxidation of propylene are not very selective. In the past, we tried to use thallium oxide doped with dysprosium [85] and we decided to take this point ($\Lambda_{th} = 1.25$) into account because it lies not too far from the calibrated line. Benzene and but-1-ene oxidations to maleic anhydride on $(V_{0.7}Mo_{0.3})_2O_5$ and $VOPO_4/(VO)_2P_2O_7$ [86], respectively, fit the line. According to the MOO line, more basic catalyst should be used to oxidize propene ($\Lambda_{opt} = 0.93$) than propane ($\Lambda_{opt} = 0.67$) to acrolein. Several catalysts have been reported for MOX of propene to acrolein (Sb_2O_4/SnO_2, Cu_2O, USb_3O_{10}, and bismuth molybdates) [87–89] whose Λ range from 0.71 ($Bi_2Mo_3O_{12}$) to 0.97 (USb_3O_{10}, Cu_2O) while $\Lambda_{th} = 0.96$ for Sb_2O_4/SnO_2. Pure bismuth molybdates therefore do not seem to fit the line although they are very well-known catalysts of this reaction. Because their rate of reoxidation (by air) is far higher than their rate of reduction (by propene), the presence of Mo^{5+} is difficult to check but this cation is known to form temporarily. If it is assumed that $\frac{2}{3}$ of Mo is Mo^{5+} at the steady state, the obtained Λ_{th} is 0.998 ($\Lambda_{opt} = 1.0$). When multicomponent molybdates are used, the role of cations such as Co^{2+}, Ni^{2+}, Bi^{3+}, etc., is to bring more basicity to acidic Mo^{6+} so that a better fit may be obtained. Several experiments [66,89,90] have shown that Bi^{3+} in bismuth molybdates could be responsible for the oxidation step of propene to allene ($\Delta I = 0.04$ eV) and Λ_{th} is roughly in accordance since Λ_{th} $^{VIII}Bi^{3+} = 1.18$. As it is common to also find acrylic acid besides acrolein, Equation (10.8) gives $\Delta I = 0.62$ eV, which is the same value than for MOX of isobutene to methacrolein. The same Mo-based multicomponent oxides are generally used for both reactions, which is accounted for by the correlation. According to the MOO line, the optimum Λ_{opt} of a catalyst for the direct MOX of propylene to acrylic acid ($\Delta I = 0.87$ eV) should be 0.67. Industrially, the reaction is performed in two reactors, catalysts based on molybdate type ensuring propylene oxidation to acrolein ($\Delta I = 0.37$, line MOO, $\Lambda_{opt} = 1.0$) and for example, Mo_3VO_{11} [91] ($\Lambda_{th} = 0.54$) ensuring oxidation of acrolein to acrylic acid ($\Delta I = 0.50$ eV, line MOP, $\Lambda_{opt} = 0.54$).

If the standard error is small on ΔI (± 0.02 eV), it is certainly high on Λ because the value of Λ on surface or at the steady state cannot be ascertained. Moreover, several solids are tried for ODH of alkanes, as already mentioned, with little success as far as selectivity is concerned. For sure, however, the correlations show that a catalyst for ODH must be less acidic than for MOX of alkanes. It must be emphasized that these correlations are given for a "theoretical selectivity" but do not apply to activity parameters, as far as different reactions are compared. Up to now, and to give an example, our attempts failed in correlating optical basicity of oxides with the turnover frequency measured for various oxidation reactions by using the very thorough work presented by Wachs et al. [92].

10.4.3 Other Mild Oxidation Reactions

Following the same method, $[\Delta I, \Lambda]$ couples were plotted for the mild oxidation of C_1 to C_6 alcohols to aldehydes or ketones (Table 10.2). The positive slope of the obtained MOA line shown in Figure 10.8 is smaller than that found for MOX of paraffinic bonds (see Table 10.5). The ammoxidation of hydrocarbons was also examined but the literature is scarce and examples are not numerous enough. In the case of paraffins and alkyl-aromatics, data are available mainly for propane and toluene. A $[\Delta I, \Lambda]$ AP line was tentatively drawn to represent this type of reaction [34], where ΔI was calculated by Equation (10.9). The large contribution of the ionization energy of ammoniac to ΔI, which is high, is responsible for the small positive slope, and the reliability coefficient is poor also because of the chosen reactions, which are not numerous enough. However, slope and intercept of AP line are gathered in Table 10.5, together with those of other lines.

10.4.4 Combustion of Paraffins and of Olefins

Table 10.3 and Table 10.4 gather $[\Delta I, \Lambda]$ couples obtained for the total oxidation of paraffins (TOP line) and of toluene (C_1–C_7), and that of olefins (TOO line) (C_2H_4 to i-C_6H_{12}) and of phenol. The slope of the TOP line is positive, whereas it is negative for the TOO (Figure 10.5), as it was observed for MOX of paraffins and MOX of olefins, respectively. To replace noble metals that are the most active catalysts, several perovskite-type oxides are commonly used in these reactions of combustion, which become more and more important for pollution abatement (combustion, soot particulates, Volatile Organic Compounds). These solids have a versatile ABO_3 structure, well-known to accommodate nearly all types of cations A and B, the coordination of which may change according to stoichiometry.

TABLE 10.2
Mild Oxidation of Alcohols (the MOA Line)

Λ_{th}	ΔI (eV)	Reactant	Product	Catalyst	Ref.
0.55	0.02	Methanol	Formaldehyde	$2MoO_3–Cr_2Mo_3O_{12}$	1
0.56	0.02	Methanol	Formaldehyde	$4MoO_3–Cr_2Mo_3O_{12}–Fe_2Mo_3O_{12}$	2
0.57	0.02	Methanol	Formaldehyde	$0.5Fe_2(MoO_4)_3–MoO_3$	3
0.625	0.23	Ethanol	Acetaldehyde	$0.08MoO_3–0.25TiO_2–0.8Al_2O_3$	4
0.65	0.23	Ethanol	Acetaldehyde	V_4O_9	5
0.74	0.23	Ethanol	Acetaldehyde	Mg_5CeO_7	6
0.845	0.38	Butanol-2	Butanone-2	$9SnO_2–MoO_3$	7
0.92	0.44	Propanol-2	Acetone	Mn_3O_4	8
0.92	0.41	Allylic alcohol	Acrolein	Mn_3O_4	9
0.98	0.60	Cyclohexanol	Cyclohexanone	$2CuO–ZnO–0.4Al_2O_3$	10
1.04	0.60	Cyclohexanol	Cyclohexanol	$CuO–CoO$	11

TABLE 10.3
Total Oxidation of Paraffins (the TOP Line)

Λ_{th}	ΔI (eV)	Reactant	Catalyst	1
0.908	1.23	Methane	$La_{0.5}^{3+}Sr_{0.5}^{2+}Co^{2+}O_3$	2
0.853	1.23	Methane	$La_{0.2}^{3+}Sr_{0.8}^{2+}Co^{2+}O_3$	3
0.902	1.23	Methane	$La_{0.2}^{3+}Sr_{0.8}^{2+}Co_{0.5}^{2+}Co_{0.5}^{3+}O_3$	4
0.952	2.82	Propane	$La_{0.2}^{3+}Sr_{0.8}^{2+}Co^{3+}O_3$	5
0.957	2.82	Propane	$La_{0.5}^{3+}Sr_{0.5}^{2+}Co_{0.5}^{2+}Co_{0.5}^{3+}O_3$	6
0.995	3.24	Butane	$La_{0.5}^{3+}Sr_{0.5}Co_{1-y}^{3+}Fe_y^{3+}O_3 \; y = 0.2$	7
0.987	3.24	Butane	$y = 0.4$	8
0.979	3.24	Butane	$y = 0.6$	9
0.971	3.24	Butane	$y = 0.8$	10
0.963	3.24	Butane	$y = 1.0$	11
1.00	3.643	Hexane	$\gamma\text{-}MnO_2$	12
1.027	4.95	Toluene	$Y^{3+}Ba_2^{2+}Cu_3O_{6+x}$	13

10.4.4.1 Total oxidation (combustion) of paraffins (the TOP line)

The most used catalysts are perovskite-type oxides containing B = Fe, Co, and Mn cations, all of which exhibits various valences, and which may form solid solutions together [93,94]. Cation A is mainly represented by lanthanum, which can be partially substituted, most often by Sr, to increase defects and catalytic activity (here, activity may be confounded with selectivity). The optical basicity varies widely upon substitution of A (e.g., La^{3+} substituted by more basic Sr^{2+}) and slightly less upon substitution of B. Studying n-butane combustion on $La_{0.2}Sr_{0.8}Co_{1-y}Fe_{1-y}O_3$ catalyst, Yamazoe et al. [95] have shown that activity increases along $y = 1 < 0.6 < 0.8 < 0.4$ which is nearly parallel to the calculated Λ_{surf}, increasing as $y = 1 < 0.8 < 0.6 < 0.4$ (Table 10.3). For the same reaction, Seiyama et al. [96] found that $La_{0.8}Sr_{0.2}Co_{0.4}Fe_{0.6}O_3$ ($\Lambda_{surf} = 1.23$) is more active than $La_{0.6}Sr_{0.4}CoO_3$ ($\Lambda_{surf} = 1.0$), itself known as the most active composition in the $La_{1-x}Sr_xCoO_3$ system [97]. When the B site composition was fixed as in $La_{1-x}Sr_xCo_{0.4}Fe_{0.6}O_3$ and x was varied, the activity was found to increase as $x = 1.0 < 0.8 < 0.6 < 0 < 0.4 < 0.2$, again nearly parallel to Λ_{surf} increasing along $x = 1.0 < 0.8 < 0.6 < 0.4 < 0.2 < 0$. Then we may conclude that activities roughly follow the same trend than optical basicity Λ_{surf}. The cation-deficient perovskites such as $La_{1-x}Sr_xMnO_3$, are known to act as "suprafacial" catalysts [88–99] because their structure does not contain enough ion vacancies to allow lattice oxygen to be mobile enough, whereas in Co- and Fe-perovskites, considered to be "intrafacial catalysts," oxide ions are highly mobile [88,89,100–102]. The oxide ion vacancies that are created by charge compensation are able to accommodate extra-oxygen in the lattice, which leads to the variation of the oxygen composition depending on temperature and oxygen partial pressure. This phenomenon cannot be directly accounted for by optical basicity (unless the exact stoichiometry, valence, coordination, spin state at the steady state are known), and may explain the few discrepancies observed in the above parallelism between the activities and Λ_{surf} in the 0.97 to 1.32 range.

These and other examples of combustion catalysts are shown to fit the TOP line (Figure 10.9). Another type of perovskite-related are $YBa_2Cu_3O_{6+y}$ catalysts (known for their superconductivity). When $y = 1$ ($\Lambda_{surf} = 1.027$), the catalyst is highly selective in the combustion of toluene at all oxygen pressures that were investigated [90].

10.4.4.2 Catalytic combustion of olefins and other unsaturated compounds (the TOO line)

Perovskites and spinels are also claimed to be active in the combustion of olefinic compounds. Examples including other oxides are given in Table 10.4 and Figure 10.10 (the TOO line). Among numerous catalyst to eliminate Diesel engine soots, the spinel-type $Cu_{1-x}K_xFe_2O_4$ ($x = 0$ to 0.2) catalyst was proposed by Shangguan et al. [103] who found that the surface composition of potassium was quite different from the overall composition of their sample and increased with the doping amount. They showed that all K was not dissolved in the spinel lattice and segregated on surface. The figurative point for K_2O (Figure 10.10) is a measure of the deviation from the point C_2 on the TOO line representing ideally the catalytic combustion of pure olefinic bonds of the $\sigma-\pi$ type, which

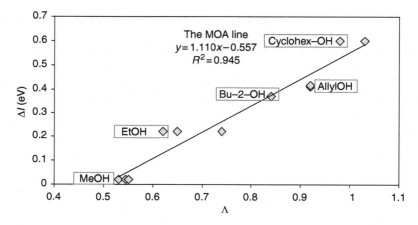

FIGURE 10.9 The MOA line. Mild oxidation of C_1 to C_6 alcohols. Linear correlation between ΔI and Λ of catalysts. Several catalysts are represented by their Λ value for the same reaction. Me = methyl; Et = ethyl, etc.

TABLE 10.4
Total Oxidation of Olefins (the TOO Line)[a]

Λ_{th}	ΔI (eV)	Reactant	Catalyst	Ref. TOO
0.65	5.30	C_6H_5OH	$CuO(10\%)/Al_2O_3$	1
0.75	5.04	C_9H_{12}	SnO_2-V_2O_5	2
1.038	3.27	C_2H_4	$La_{0.8}Sr_{0.2}MnO_3$	3
1.135	3.27	C_2H_4	$La_{1.2}Sr_{0.8}Co^{3+}O_3$	4
1.075	3.27	C_2H_4	$LaMnO_3$	5
1.012	4.04	C_3H_6	$La_{0.85}Ca_{0.15}Co^{3+}O_3$	6
1.042	4.04	C_3H_6	$La_{0.85}Sr_{0.15}CoO_3$	7
0.965	4.04	C_3H_6	Co_3O_4	8
0.849	4.53	i-C_4H_8	$La_{0.2}Sr_{0.8}Co^{3+}O_3$	9
0.904	4.53	i-C_4H_8	$La_{0.5}Sr_{0.5}Co^{3+}O_3$	10
0.908	4.53	i-C_4H_8	$La_{0.8}Sr_{0.2}Fe^{3+}O_3$	11
1.317	2.51	C_1[a]	La_2CuO_4	12
1.50	1.52	C_2[b]	$La_2CuO_4 + K_2O$	13

[a] *Handbook of Physical Chemistry*, 76th ed., CRC Press (1966).
[b] C_2 is considered here as a main constituent of soot with σ–π bonds

are the main constituent of soots. Other examples of catalysts of combustion of VOC are also provided. Phenol and pseudocumene C_9H_{12}, often considered as VOC representative, were studied [104] using CuO/Al_2O_3 and SnO_2–V_2O_5 catalysts.

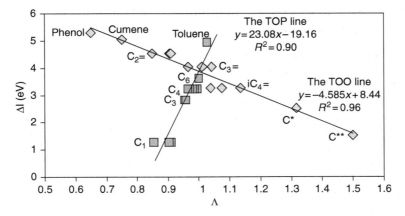

FIGURE 10.10 The TOP and TOO lines. Total oxidation of C_1 to C_9 hydrocarbons, of VOC and carbon. Linear correlations between ΔI and Λ of catalysts for paraffinic C–C or C–OH bonds and for C=C bonds. Several catalysts are represented for the same reaction. Same labels as in Figure 10.5

TABLE 10.5
Linear Relationships between ΔI and Λ Drawn for Mild Oxidation and Total Oxidation Of Various Reactants

Reactions	ΔI range (eV)	Λ range	$\Delta I = p\Lambda + m$		Reliability factor R^2
			p	m	
Paraffinic bonds mild oxid. MOP	0.18–2.00	0.48–0.91	4.09	−1.73	0.94
Olefinic bonds mild oxid. MOO	0.06–1.22	0.48–1.25	−1.51	+1.90	0.93
Paraffinic bonds ammox. AP	0.11–0.34	0.49–0.93	0.58	−0.15	0.81
Alcohol mild oxidation. MOA	0.02–0.60	0.53–1.03	1.11	−0.56	0.95
Paraffinic bonds, total oxid. TOP	1.26–4.90	0.87–1.07	23.08	−19.16	0.90
Olefinic bonds, total oxid. TOO	3.20–4.54	0.85–0.93	−4.585	+8.44	0.96

10.5 THEORETICAL ASPECTS

The linear correlations between Λ and ΔI have distinct slopes and intercepts (Table 10.5), which seem to depend on: (i) the relative electron donor power or (Lewis-related) acidity/basicity of the reactant and (ii) on the extent of the considered oxidation reactions. Although it is not (yet) possible to explain all of these findings, we shall try to bring more light, on the one hand, by considering thermodynamics of redox systems to account for the particular nature of optical basicity, which looks like an intensive parameter, and on the other hand, by using the charge transfer theory to account for the linear nature of $[\Delta I, \Lambda]$ correlations.

10.5.1 Attempts to Give a Thermodynamic Definition of Optical Basicity

It is worth recalling that in selective oxidation reactions, whatever their goal (partial or total oxidation), the same type of mechanism has been proposed by Mars and van Krevelen for mild oxidation as well as by Voorhoeve [105] ("intrafacial" mechanism) for combustion. Even though the "suprafacial" mechanism is considered in certain cases, it is obvious that the basicity of surface oxygen species bound to, for example a perovskite, is higher than that on vanadium oxide. Therefore, as surface lattice oxygen participates to the reaction and is inserted in the oxygenated product, it is legitimate to try to find a parameter to account for its "quality." Meanwhile the cation undergoes reduction, which is not taken into account in the above work because, as a catalyst, it is supposed to come back to its initial, oxidized state, by means of gaseous dioxygen. However, there are applications to electrochemistry in which optical basicity is used as a parameter useful to predict redox equilibria in particular media, like in molten silicates [21,22,26,27], and is related to standard electrode potentials $E°$ in aqueous solution [106].

Duffy et al. [6,27,106] have studied the bulk redox equilibria of many cationic couples such as Fe^{3+}/Fe^{2+}, Sn^{4+}/Sn^{2+}, As^{5+}/As^{3+}, etc., taking place in molten alkaline silicates (at 1400°C) and containing various concentrations of Li^+, Na^+, or K^+. In the presence of oxygen, and applied to, for example, Fe^{3+}/Fe^{2+}, the following equation represents the combination of two half-equations according to:

$$4Fe^{3+} + 2O^{2-} = 4Fe^{2+} + O_2 \qquad (10.10)$$

In alkali iron silicate glasses, $\ln([Fe^{2+}]/[Fe^{3+}])$ was found to decrease with increasing basicity along $Li_2O < Na_2O < K_2O$ (activities confounded with concentrations). Three distinct $\ln([Fe^{2+}]/[Fe^{3+}]) = f(\Lambda_{exp})$ curves (Λ_{exp} is the experimental optical basicity) depending on the alkali oxide were drawn, and were combined in a single relationship according to Equation (10.11):

$$\ln[Fe^{2+}]/[Fe^{3+}] = 7.51 - 15.33\Lambda, \quad \text{or} \quad \Lambda = 0.489 - 0.065 \ln[Fe^{2+}]/[Fe^{3+}]$$
$$(10.11)$$

This equation looks similar to the well-known Nernst equation for one-electron redox Equation (10.12):

$$E = E° + (RT/F) \ln[Fe^{3+}]/[Fe^{2+}],$$
$$\text{where } E = -\Delta_r G_T/F \text{ and } E° = -\Delta_r G°_{298}/F (F = \text{Faraday}) \quad (10.12)$$

In other words, Equation (10.11) is related to the free enthalpy ΔG of the redox system as if Λ would take part in a thermodynamic state function of the system.

In order to check this hypothesis, let us combine Equations (10.11) and (10.12), in which E and $E° = -0.77$ V refer to the equilibrium potential of Fe^{2+}/Fe^{3+}

redox (in molten silicate medium, $F/RT = 6.9378$ at 1673 K) and at 25°C (in aqueous medium), respectively. It gives Equation [10.13]:

$$\ln[Fe^{2+}]/[Fe^{3+}] = -6.9378(0.77 + E) = 7.51 - 15.33\Lambda, \quad \text{and}$$

$$\Lambda = 0.838 + 0.452E \tag{10.13}$$

Using the experimental data provided by Duffy and choosing the value $\ln[Fe^{2+}]/[Fe^{3+}] = -1$, for instance, gives $\Lambda_{exp} \sim 0.55$ (after Figure 10.7[b] in Reference 6). On the other hand, the value of E can be calculated from the Nernst equation by means of the chosen value, $\ln[Fe^{2+}]/[Fe^{3+}] = -1$, and thus $E = -0.77 + 1/6.9378 = -0.625$ V. Using Equation (10.13), we find $\Lambda = 0.838 + [0.452(-0.625)] = 0.55$, that is, as expected, the same numerical value than Λ_{exp} in Reference 6.

We assume therefore that Λ is a dimensionless coefficient that is tightly associated to $\Delta_r G_T$ state function. Using Equation (10.11) and replacing $\ln([Fe^{2+}]/[Fe^{3+}])$ by $(E° - E)F/RT$, which is equal to $(-\Delta_r G_T° + \Delta_r G_T)/RT$, leads to the expression:

$$\Lambda = a \cdot \Delta_r G_T + b \tag{10.14}$$

where $a = 1/15.33RT$ and $b = -(1/15.33RT)(\Delta_r G° + 7.51RT)$ for Fe^{3+}/Fe^{2+} redox system. Other values of a and b may be found in Reference 105.

Taking into account the definition of the free enthalpy state function, and more particularly its consequences on solid-state properties [107], it is not surprising that Λ_{th} has been already related to one of the intensive variables, such as refractive index, viscosity, etc., as recalled in the Introduction.

10.5.2 Attempts to Account for Linear Relationships Between Ionization Potential of Molecules and Optical Basicity of Catalysts

According to Richardson [55], who has applied the charge transfer theory to catalysis on cationic zeolites, the number of cation radicals N^+ coming from the activation of N molecules is related to the ionization potential of the molecule, to the electron affinity of the cation in the oxide, and to the dissociation energy W of the excited state of the charge transfer complex (all in eV), according to:

$$N^+ = N \exp[-(I - A - W)/kT] \tag{10.15}$$

($T =$ temperature and $k =$ Boltzman constant). Following the thermodynamic cycle presented in Scheme 10.2, this relationship can be applied to the [reactant molecule-KO] and [product-K] couples. The reactant molecule, R, is adsorbed on KO (oxidized form) and the product, P, desorbs from K (reduced form of the catalyst). In the following, I_R and I_P are the ionization potential of reactant

R and product P, A_{KO}, and A_K are the electron affinity of cation in KO and K, and W_{R-KO} and W_{P-K} are the dissociation energy of the excited state of the charge transfer complexes corresponding to R on KO and P on K, respectively [33].

The following equations hold for reactant R and product P, respectively:

$$(-I_R + A_{KO} + W_{R-KO})/kT = \ln(N_R^+/N_R),$$

or

$$I_R = (A_{KO} + W_{R-KO}) - kT \ln(N_R^+/N_R)$$

and

$$(-I_P + A_K + W_{P-K})/kT = \ln(N_P^+/N_P),$$

or

$$I_P = (A_K + W_{P-K}) - kT \ln(N_P^+/N_P)$$

Then

$$\Delta I = |I_P - I_R| = (A_K + W_{P-K}) - (A_{KO} + W_{R-KO})$$
$$+ kT[\ln(N_R^+/N_R) - \ln(N_P^+/N_P)]$$
$$= (A_K - A_{KO}) + (W_{P-K} - W_{R-KO}) + kT[\ln(N_R^+/N_P^+) + \ln(N_P/N_R)]$$

Let us consider that the N_P/N_R ratio represents the conversion of R to P. At the steady state, this ratio is constant, and $(A_K - A_{KO})$ and $(W_{P-K} - W_{R-KO})$ also are constant. Then:

$$\Delta I = (A_K - A_{KO}) + (W_{P-K} - W_{R-KO}) + kT[\ln(N_P/N_R) + \ln(N_R^+/N_P^+)]$$

or : $\Delta I = \alpha + \beta \ln(N_R^+/N_P^+)$

where $\alpha = [(A_K - A_{KO}) + (W_{P-K} - W_{R-KO})] + kT \ln(N_P/N_R)$, Λ and $\beta = kT$.

$$(10.16)$$

Let us assume that, in the temperature range (350 to 550°C) corresponding to that generally used in mild oxidation, α and β vary little with T. If, furthermore, the molecular N_R^+/N_P^+ ratio is supposed to be close from the $[N_R^+]/[N_P^+]$ concentration ratio, then it should be proportional to the [KO]/[K] redox ratio, so that relation (10.16) becomes:

$$\Delta I = \alpha' + \beta' \ln [KO]/[K] \tag{10.17}$$

Using a generalized equation similar to Equation (10.11) by which Λ was related to $\ln[Fe^{2+}]/[Fe^{3+}]$ gives:

$$\ln[M^{z+}]/[M^{(z+n)+}] = a'\Lambda_{th} + b' \tag{10.18}$$

in which a' and b' are constant and depend on the $M^{z+}/M^{(z+n)}+$ redox. By combining Equations (10.17) and (10.18), where [KO]/[K] replaces $M^{z+}/M^{(z+n)+}$, the following expression (10.19) is obtained:

$$\Delta I = \alpha' - \beta'[a' \cdot \Lambda + b'], \quad \text{and then:}$$
$$\Delta I = m + p\Lambda \tag{10.19}$$

where $p = -\beta'a'$ and $m = \alpha' - \beta b'$. Therefore, Equation (10.19) accounts for the linear equations gathered in Table 10.5, if p and m are supposed to be constant in the considered range of temperature ($\sim$300 to 550°C for MOX, 700 to 1000°C for combustion).

10.6 CONCLUSION

This chapter aimed at finding out whether theoretical or empirical parameters are able (i) to account for the mean acidity/basicity or electron acceptor/donor power of the solid catalyst, (ii) to account for the mean acidity/basicity or electron acceptor/donor power of the reactant-to-product selective reaction, in order to, (iii) combine them to account for the selective action of the catalyst suited for the reaction, and, later, to, (iv) forecast the optimum acidity/basicity of a (new) catalyst in a given (new) reaction. In our opinion, this ambitious goal is quite well reached by plotting the optical basicity against the variation of ionization potential during the reaction. After recalling the concept and applications of Duffy's optical basicity of a solid oxide and its extension to transition metal oxides, catalysts and supports have been classified by means of Λ_{th}, from acidic (including zeolites) to basic oxides. As Λ_{th} depends on valence, coordination and spin state of the cation(s), this parameter really accounts for structural characteristics of the solid oxide. The way the acidity is varied by adding dopants or other oxides is easily accounted for by Λ_{th}. In the case of multicomponent catalysts (e.g., molybdates) of known stoichiometry, an evaluation of Λ_{th} is obtained, provided the valence and coordination of cations is "estimated". The evaluation of the *surface* optical basicity has been tentatively addressed by proposing a standard correction for bulky solids, and by a relationship between the O1s binding energy BE and Λ_{th}.

A simple criterion, which has the same nature as Λ_{th} of the catalyst, was used to represent the electron donor power during reaction: the absolute value of the potential ionization difference, $\Delta I = |I_R - I_P|$, weighed by the ratio n_P/n_R of carbon in product and reactant molecules respectively. The linear correlations obtained between ΔI and Λ_{th} show that their slope is related to the electron donor power of the reactant, positive when C–C (alkanes, alkyl-aromatics) or C–H (alcohols) bonds are to be transformed, and negative when C=C bonds are concerned. The intercept depends on the extent of oxidation, and its absolute value increases from mild to total oxidation, respectively. A main difficulty is the actual state of cations at the *steady state*, but each time accurate experiments allow determining the mean valence state, the calculated Λ_{th} fits well the correlations. These lines may be used as a *predictive trend*, and allow, for example, to precise that more basic catalysts are needed for alkane ODH than for its mild oxidation to oxygenated compound. As a variety of solids have been catalytically experienced in literature, it would be worthwhile to consider far more examples than what is proposed here to refine the relationships observed. Finally, theoretical considerations are proposed to tentatively account for these linear relationships. Optical basicity would be closely related to the free enthalpy, and, as an intensive thermodynamic parameter, it is normal that it could be related to several characteristic properties, including now catalytic properties.

REFERENCES

1. Mars, J. and van Krevelen, D.W. Oxidations carried out by means of vanadium oxide catalysts. *Chem. Eng. Sci., Special Suppl.* 1954, *3*, 41–57.
2. Haber, J. Catalysis by transition metal oxides. In *Solid State Chemistry in Catalysis*; Grasselli, R.K. and Brazdil, J.F., Eds. ACS Symposium Series, Washington: ACS, 1985, Vol. 279, pp. 3–21.
3. Noguera, C. *Physique et chimie des surfaces d'oxydes*. Paris: Eyrolles, 1995, 219 pp.
4. Zhang, Y. Electronegativities of elements in valence states and their applications. *Inorg. Chem.* 1982, *21*, 3886–3893.
5. Duffy, J.A. *Bonding, Energy Levels and Bands in Inorganic Solids*. Longman Scientific & Technical: New York, 1990, 249 pp.
6. Duffy, J.A. A review of optical basicity and its applications to oxidic systems. *Geochim. Cosmochim. Acta* 1993, *57*, 3961–3970.
7. Dimitrov, V. and Sakka, S. Electronic oxide polarizability and optical basicity of simple oxides. I. *J. Appl. Phys.* 1996, *79*, 1736–1740.
8. Lenglet, M. Spectroscopic study of the chemical bond in 3d transition metal oxides. Correlation with the ionic–covalent parameter. *Trends Chem. Phys.* 1997, *6*, 121–543.
9. Portier, J., Campet, G., Etourneau, J., Shastry, M.C.R., and Tanguy, B. A simple approach to materials design: role played by an ionic-covalent parameter based on polarizing power and electronegativity. *J. Alloys Compd.* 1994, *209*, 59–64.

10. Portier, J., Campet, G., Etourneau, J., and Tanguy, B. A simple model for the estimation of electronegativities of cations in different electronic states and coordinations. *J. Alloys Compd.* 1994, *209*, 285–289.

11. Tanaka, K.-I. and Ozaki, A. Acid–base properties and catalytic activity of solid surfaces. *J. Catal.* 1967, *8*, 1–7.

12. Klier, K. Oxidation–reduction potentials and their relation to the catalytic activity of transition metal oxides. *J. Catal.* 1967, *8*, 14–21.

13. Sokolovskii, V.D. Donor–acceptor concept of covalent interaction of substrate with catalysts. *React. Kinet. Catal. Lett.* 1982, *20*, 79–86.

14. Pankratiev, Yu.D. Correlation between oxygen binding energy and catalytic activity of oxides. *React. Kinet. Catal. Lett.* 1982, *20*, 255–259.

15. Auroux, A. and Gervasini, A. Microcalorimetric study of the acidity and basicity of metal oxide surfaces. *J. Phys. Chem.* 1990, *94*, 6371–6379.

16. Auroux, A. From strong solid acids to strong solid bases: characterization of the acid/base centers of catalyst surfaces by adsorption microcalorimetry. In *Proceedings of the NATAS 29th Annual Conference on Thermal Analysis and Applications*, Kociba, K.J. Ed., St Louis, Missouri, Sept. 24–26, 2001, pp. 53–58. B and K Publishing.

17. Idriss, H. and Seebauer, E.G. Effect of oxygen electronic polarisability on catalytic reactions over oxides. *Catal. Lett.* 2000, *66*, 139–145.

18. Vinek, H., Noller, H., Ebel, M., and Schwarz, K. X-ray photoelectron spectroscopy and heterogeneous catalysis, with elimination reactions as an example. *J. Chem. Soc., Faraday Trans. I.* 1977, *73*, 734–746.

19. Noller, H. and Vinek, H. Coordination chemical approach to catalytic oxidation reactions. *J. Mol. Catal.* 1989, *51*, 285–294.

20. Duffy, J.A. The refractivity and optical basicity of glass. *J. Non-Cryst. Solids* 1986, *86*, 149–160.

21. Duffy, J.A. and Ingram, M.D. Establishment of an optical scale for Lewis basicity in inorganic oxyacids, molten salts and glasses. *J. Am. Chem. Soc.* 1971, *93*, 6448–6454.

22. Duffy, J.A. Optical basicity: a practical acid-base theory for oxides and oxyanions. *J. Chem. Education* 1996, *73*, 1138–1142.

23. Iwamoto, N., Makino, Y., and Kasahara, S. Correlation between refraction basicity and theoretical optical basicity. Part II. Lead monoxide–silicon dioxide, calcium oxide–aluminum oxide–silicon dioxide, and potassium oxide–titanium dioxide–silicon dioxide glasses. *J. Non-Cryst. Solids* 1984, *68*, 379–397.

24. Mills, K.C. Viscosities of Molten Slags. National Physics Laboratory, Teddington/Middlesex, U.K. NPL report DMM(A) 1992, 116 pp.

25. Jeddeloh, G. The redox equilibrium in silicate melts. *Phys. Chem. Glass.* 1984, *25*, 163–164.

26. Baucke, F.G.K. and Duffy, J.A. Oxidation states of metal ions in glass melts. *Phys. Chem. Glasses* 1994, *35*, 17–21.

27. Baucke, F.G.K. and Duffy, J.A. The effect of basicity on redox equilibria in molten glasses. *Phys. Chem. Glasses* 1991, *32*, 211–218.

28. Dent-Glasser, L.S. and Duffy, J.A. Analysis and prediction of acid–base reactions between oxides and oxysalts using the optical basicity concept. *J. Chem. Soc., Dalton. Trans.* 1987, *20*, 2323–2328.

29. Lebouteiller, A. and Courtine, P. Improvement of a bulk optical basicity table for oxidic systems. *J. Solid State Chem.* 1998, *137*, 94–103.

30. Lenglet M. Spectroscopic study of the chemical bond in 3d transition metal oxides. Correlation with the ionic–covalent parameter. *Trends Chem. Phys.* 1997, *6*, 121–142.

31. Lenglet M. Ligand field spectroscopy and chemical bonding in Cr^{3+}-, Fe^{3+}-, Co^{2+}- and Ni^{2+}-containing oxidic solids. *Mater. Res. Bull.* 2000, *35*, 531–543.

32. Moriceau, P., Lebouteiller, A., Bordes, E., and Courtine, P. A new concept related to selectivity in mild oxidation catalysis of hydrocarbons: the optical basicity of catalyst oxygen. *Phys. Chem. Chem. Phys.* 1999, *1*, 5735–5744.

33. Moriceau, P., Taouk, B., Bordes, E., and Courtine, P. A common concept accounting for selectivity in mild and total oxidation reactions: the optical basicity of catalysts. In *Proceedings of the 12th International Congress on Catalysis*, Granada (Spain), July 17–24, 2000; *Studies in Surface Science Catalysis Series*, Corma, A., Melo, F.V., Mendioroz, S. and Fierro, J.L.G. Eds. 2000, Vol. 130B, pp. 1811–1816.

34. Moriceau, P., Taouk, B., Bordes, E., and Courtine, P. Correlations between the optical basicity of catalysts and their selectivity in oxidation of alcohols, ammoxidation and combustion of hydrocarbons. *Catal. Today* 2000, *61*, 197–201.

35. Shannon, R.D. Revised effective ionic radii and systematic studies of interatomic distances in halides and chalcogenides. *Acta Crystallogr.* A 1976, *A32*, 751–767.

36. Courtine, P. and Bordes, E. Mode of arrangement of components in mixed vanadia catalysts and its bearing for oxidation catalysis. *Appl. Catal. A: Gen.* 1997, *157*, 45–65.

37. Bordes, E. and Courtine, P. Synergistic effects in multicomponent catalysts for selective oxidation. In *Proceedings of the 3rd World Congress on Oxidation Catalysis*, San Diego, CA, Sept. 21–26, 1996; *Studies in Surface Science Catalysis Series*. Grasselli, R.K., Oyama, S.T., Gaffney, A.M., and Lyons, J.E. Eds., *1997*, Vol. 110, pp. 177–184.

38. Grzybowska, B., Gressel, I., Samson, K., Wcislo, K., Stoch, J., Mikolajczyk, M., and Dautzenberg, F. Effect of the potassium promoter content in V_2O_5/TiO_2 catalysts on their physicochemical and catalytic properties in oxidative dehydrogenation of propane. *Polish J. Chem.* 2001, *75*, 1513–1519.

39. Sumitomo Chemicals, JP 09,020,700-A, 1995.

40. R. Burch, Ed. Pillared clays. *Catal. Today* 1988, *2*. Special issue, 367 pp.

41. Volta, J.C. and Portefaix, J.L. Structure sensitivity of mild oxidation reactions on oxide catalysts. A review. *Appl. Catal.* 1985, *18*, 1–32.

42. Volta, J.C. and Tatibouet, J.M. Structure sensitivity of oxide in mild oxidation of propylene molybdenum(VI). *J. Catal.* 1985, *93*, 467–480.

43. Bordes, E. Crystallochemistry of V–P–O phases and application to catalysis. *Catal. Today* 1987, *1*, 499–526.

44. Bordes, E. Reactivity and crystal chemistry of V–P–O phases related to C_4-hydrocarbon catalytic oxidation. *Catal. Today* 1987, *3*, 163–174.

45. Higarashi, H., Tsuji, K., Okuhara, T., and Misono, M. Effects of consecutive oxidation on the production of maleic anhydride in butane oxidation over four kinds of well-characterized vanadyl pyrophosphates. *J. Phys. Chem.* 1993, *97*, 7065–7071.

46. Bordes, E., Ziolkowski, J., and Courtine, P. A dynamic model of the oxidation of *n*-butane and 1-butene on various crystalline faces of $(VO)_2P_2O_7$. In *Proceedings of the New Developments in Selective Oxidation, Studies in Surface Science Catalysis Series*, Centi, G. and Trifirò, F. Eds. 1990, Vol. 55, pp. 625–633.

47. Iguchi, E. and Nakatsugawa, H. Application of a polarizable point-ion shell model to a two-dimensional periodic structure: the NiO (001) surface. *Phys. Rev. B: Condens. Matter* 1995, *51*, 10956–10964.

48. Delamar, M. Correlation between the isoelectric point of solid surfaces of metal oxides and X-ray photoelectron spectroscopy chemical shifts. *J. Electron. Spectrosc. Relat. Phenom.* 1990, *53*, c11–c14.

49. http://www.lasurface.com

50. http://www.srdata.nist.gov

51. Dhanavantri, C. and Karekar, R.N. An empirical generalized relationship between the change in Auger parameter and the refractive indices of optical dielectric materials. *J. Electron Spectrosc. Relat. Phenom.* 1993, *61*, 357–365.

52. Nefedov, V.I. , Gati, D. , Dzhurinskii, B.F., Sergushin, N.P., and Salyn, Ya.V. An X-ray photoelectron spectroscopic study of certain oxides. *Russ. J. Inorg. Chem.* 1975, *20*, 1279–1283.

53. Bordes, E. Propriétés structurales des phases catalytiques à l'état stationnaire, Doctorat d'Etat, Compiègne (France), 1979.

54. Ai, M. Oxidation of propane to acrylic acid. *Catal. Today* 1992, *13*, 679–684.

55. Richardson J.T. ESR studies of cationic oxidation sites in faujasite. *J. Catal.* 1967, *9*, 172–177.

56. Batiot C. and Hodnett, B.K. The role of reactant and product bond energies in determining limitations to selective catalytic oxidations. *Appl. Catal.* 1996, *137*, 179–91.

57. Batiot, C., Cassidy, F.E., Doyle, A.M., and Hodnett, B.K. Selectivity of active sites on oxide catalysts. In *Proceedings of the 3rd World Congress on Oxidation Catalysis*, San Diego, CA, Sept. 21–26, 1996; *Studies in Surface Science Catalysis Series*, 1997, Vol. 110. 1097–1106.

58. Kung, H.H. *Transition Metal Oxides: Surface Chemistry and Catalysis*, Studies in Surface Science Catalysis, 1987, Vol. 45.

59. Bettahar, M.M., Costentin, G., Savary, L., and Lavalley, J.C. On the partial oxidation of propane and propylene on mixed metal oxide catalysts. *Appl. Catal. A: Gen.* 1996, *145*, 1–48.

60. Matsuura, I. The nature of the active sites on olefin oxidation catalysts. In *Proceedings of the 6th International Congress on Catalysis*, London: Vol. B1, 1976, p. 819.

61. Desquilles, C., Bartoli, M.J., Bordes, E., Hecquet, G., and Courtine, P. Oxidative dehydrogenation of isobutyric acid by heteropolycompounds. Effect of alkali-containing supports. *Erdoel Erdgas Kohle* 1993, *109*, 130–133.

62. Bartoli, M.J., Monceaux, L., Bordes, E., Hecquet, G., and Courtine, P. Eds. Stabilization of heteropoly acids by various supports. In *New Developments in Selective Oxidation by Heterogeneous Catalysis*, Louvain-la-Neuve (Belgium), Apr. 8–10, 1991; *Studies in Surface Science Catalysis Series*, Ruiz, P., and Delmon, B., Eds. Amsterdam: Elsevier Science, Vol. 72, 1992, pp. 81–90.

63. Desquilles, C., Bartoli, M.J., Bordes, E., Courtine, P., and Hecquet, G. Oxidative dehydrogenation of isobutyric acid by heteropoly compounds — effect of alkali containing supports. In *Proceedings of the DGMK-Conference on Selective Oxidations in Petrochemistry*, Erdoel Erdgas Kohle, Vol. 9204, 1992, 69–79.

64. Belkouch, J., Taouk, B., Monceaux, L., Bordes, E., Courtine, P., and Hecquet, G. Cesium promotion of iron phosphate catalysts in the oxidative dehydrogenation of isobutyric acid to methacrylic acid. *Stud. Surf. Sci. Catal.* 1994, *82*, 819–828.

65. Millet, J.M.M. and Védrine, J. Role of cesium in iron phosphates used in isobutyric acid oxidative dehydrogenation. *Appl. Catal.* 1991, *76*, 209–219.

66. Grasselli, Robert K. and Burrington, James D. Selective oxidation and ammoxidation of propylene by heterogeneous catalysis. *Adv. Catal.* 1981, *30*, 133–163.

67. Cavani, F., Trifirò, F., and Arpentinier, P. *The Catalytic Technology of Oxidation*. Paris: Technip, 2001. Vol. 1. 324 pp.

68. Bielański, A. and Najbar, M. V_2O_5-MoO_3 catalysts for benzene oxidation. *Appl. Catal. A: Gen.* 1997, *157*, 223–261.

69. Hodnett, B.K. Vanadium—Phosphorus oxide catalysts for the selective oxidation of C4 hydrocarbons to maleic anhydride. *Catal. Rev.-Sci. Eng.* 1985, *27*, 373–424.

70. Asahi Chemical Industry. JP 02,042,032, 1990.

71. Pitchai, P. and Klier, K. Partial oxidation of methane. *Catal. Rev.-Sci. Eng.* 1986, *28*, 13–88.

72. Baerns, M. and Ross, J.R.H. Catalytic chemistry of methane conversion. In *Perspectives in Catalysis*, Thomas, J.M. and Zamaraev, K.I. Eds. London: Blackwell Science Publication, 1993, pp. 315–335.

73. Bordes, E. Nature of the active and selective sites in vanadyl pyrophosphate, catalyst of oxidation of *n*-butane, butene and *n*-pentane in maleic anhydride. *Catal. Today* 1993, *16*, 27–38.

74. Bautista, F.M., Campelo, J.M., Garcia, A., Luna, D., Marinas, J.M., and Quiros, R.A. Gas-phase catalytic oxydehydrogenation of ethylbenzene on $AlPO_4$ catalysts. In *Proceedings of the New Developments in Selective Oxidation II*. Benalmádena (Spain), Sept. 20–24, 1993; *Studies in Surface Science Catalysis Series*, Cortès-Corberan, V. and Vic Bellón, S. Eds., Vol. 82, 1994, pp. 759–768.

75. Hodnett, B.K. and Delmon, B. Influence of cobalt on the textural, redox and catalytic properties of stoichiometric vanadium phosphate. *Appl. Catal.* 1983, *6*, 245–259.

76. Sananes-Schulz, M.T., Tuel, A., Hutchings, G.J., and Volta, J.C. The V^{4+}/V^{5+} balance as a criterion of selection of vanadium phosphorus oxide catalysts for *n*-butane oxidation to maleic anhydride: a proposal to explain the role of Co and Fe dopants. *J. Catal.* 1997, *166*, 388–392.

77. Costentin, G., Lavalley, J.C., and Studer, F. Mo oxidation state of Cd, Fe, and Ag catalysts under propane mild oxidation reaction conditions. *J. Catal.* 2001, *200*, 360–369.

78. Merzouki, M., Bordes, E., Taouk, B., Monceaux, L., and Courtine, P. Catalytic properties of promoted vanadium oxide in the oxidation of ethane in acetic acid. In *Proceedings of the New Development in Selective Oxidation by Heterogeneous Catalysis. Louvain-la-Neuve (Belgium), Apr. 8–10, 1991; Studies in Surface*

Science Catalysis Series, Ruiz, P., and Delmon, B., Eds.: Vol. 72, 1992, pp. 165–179.

79. Merzouki, M., Bordes, E., Taouk, B., Monceaux, L., and Courtine, P. Correlation between catalytic and structural properties of modified molybdenum and vanadium oxides in the oxidation of ethane to acetic acid or ethylene. In *Proceedings of the 10th International Congress on Catalysis*. Budapest (Hungary), July 19–24, 1992; *New Frontiers in Catalysis*, Guczi, L., Solymosi, F. and Tétényi, P. Eds. 1993, pp. 753–764.

80. Werner, H. , Timpe, O., Herein, D., Uchida, Y., Pfänder, N., Wild, U., and Schlögl, R. Relevance of a glassy nanocrystalline state of Mo_4VO_{14} for its action as selective oxidation catalysts. *Catal. Lett.* 1997, *44*, 153–163.

81. Karim, K., Al-Hazmi, M., and Khan, A. to Sabic, U.S. 6,060,421, 2000.

82. Roussel, M., Bouchard, M., Bordes-Richard, E., and Karim, K. Oxidation of ethane to acetic acid and ethylene by MoVNbO catalysts. *Catal. Today* 2005, *99*, 77–87.

83. Linke, D., Wolf, D., Baerns, M., Timpe, O., Schloegl, R., Zeyss, S., and Dingerdissen, U. Catalytic partial oxidation of ethane to acetic acid over $MoV_{0.25}Nb_{0.12}Pd_{0.0005}O_x$. *J. Catal.* 2002, *205*, 16–31.

84. Tessier, L., Bordes, E., and Gubelmann-Bonneau, M. Active species on vanadium containing catalysts for the selective oxidation of ethane to acetic acid. *Catal. Today* 1995, *24*, 335–340.

85. Askander, J. Ph.D. Propriétés physicochimiques et catalytiques de l'oxyde de thallium Tl_2O_3. Compiègne, 1982.

86. Bordes, E. and Courtine, P. Some selectivity criteria in mild oxidation catalysis: V–P–O phases in butene oxidation to maleic anhydride. *J. Catal.* 1979, *57*, 237–252.

87. Figueras, F., Forissier, M., Lacharme, J.P., and Portefaix, J.L. Catalytic oxidation of propene over antimony–tin–oxygen mixed oxides. *Appl. Catal.* 1985, *19*, 21–32.

88. Grasselli, R.K. and Suresh, D.D. Aspects of structure and activity in uranium–antimony oxide acrylonitrile catalysts. *J. Catal.* 1972, *25*, 273–291.

89. Grasselli, R.K. Selectivity and activity factors in bismuth–molybdate oxidation catalysts. *Appl. Catal.* 1985, *15*, 127–139.

90. Grzybowska, B., Haber, J., and Janas, J. Interaction of allyl iodide with molybdate catalysts for the selective oxidation of hydrocarbons. *J. Catal.* 1977, *49*, 150–163.

91. Tichý, J. Oxidation of acrolein to acrylic acid over vanadium–molybdenum oxide catalysts. *Appl. Catal. A: Gen.* 1997, *157*, 363–385.

92. Briand, Laura E., Hirt, Andrew M., and Wachs, Israel E. Quantitative determination of the number of surface active sites and the turnover frequencies for methanol oxidation over metal oxide catalysts: application to bulk metal molybdates and pure metal oxide catalysts. *J. Catal.* 2001, *202*, 268–278.

93. Tejuca, L.G. and Fierro, J.L.G., Eds. *Properties and Applications of Perovskite-Type Oxides*; *Chemical Industries Series*. Dekker, New-York. Vol. 50, 1993, 382 pp.

94. Fierro, J.L.G. Composition and structure of perovskite surfaces. In *Properties and Application of Perovskite Type Oxides*, Tejuea, L and Fierro, J.L.G. Eds. *Chemical Industries* Dekker, New-York. Vol. 50, 1993, pp. 195–214.

95. Yamazoe, N. and Teraoka, T. Oxidation catalysis of perovskites — relationships to bulk structure and composition (valency, defect, etc.). In *Perovskites*, Misono, M. and Lombardo, E.A. Eds. *Catal. Today*, Vol. 8, 1990, 175–199.

96. Seiyama, T., Yamazoe, N., and Eguchi, K. Characterization and activity of some mixed metal oxide catalysts. *Ind. Eng. Chem., Prod. Res. Dev.* 1985, *24*, 19–27.

97. Teraoka, Y., Farukawa, S., Yamazoe, N., and Seiyama, T. Oxygen-sorptive and catalytic properties of defect perovskite-type $La_{1-x}Sr_xCoO_{3-}$. *Nippon Kagaku Kaishi* 1985, (8), 1529–1534.

98. Zhang, H.M., Yamazoe, N., and Teraoka, Y. Effects of B site partial substitutions of perovskite-type lanthanum strontium cobalt oxide ($La_{0.6}Sr_{0.4}CoO_3$) on oxygen desorption. *J. Mater. Sci. Lett.* 1989, *8*, 995–996.

99. Nitadori, T., Kurikara, S., and Misono, M. Catalytic properties of lanthanum manganese mixed oxides ($La_{1-x}A'_xMnO_3$(A' = Sr, Ce, Hf)). *J. Catal.* 1986, *98*, 221–228.

100. Teraoka, T., Zhang, K.M., Okamoto, K., and Yamazoe, Y. Mixed ionic-electronic conductivity of lanthanum strontium cobalt iron oxide ($La_{1-x}Sr_xCo_{1-y}FeyO_{3-}$) perovskite-type oxides. *Mater. Res. Bull.* 1988, *23*, 51–58.

101. Nakamura, T., Misono, M., and Yoneda, Y. Catalytic properties of perovskite-type mixed oxides, $La_{1-x}Sr_xCoO_3$. *Bull. Chem. Soc. Jpn.* 1982, *55*, 394–399.

102. Nitadori, T. and Misono, M. Catalytic properties of $La_{1-x}A'_xFeO_3$ (A' = Sr, Ce) and lanthanum cerium cobalt oxide ($La_{1-x}Ce_xCoO_3$). *J. Catal.* 1985, *93*, 459–466.

103. Shangguan, W.F., Teraoka, Y., and Kagawa, S. Promotion effect of potassium on the catalytic property of $CuFe_2O_4$ for the simultaneous removal of NO_x and diesel soot particulate. *Appl. Catal. B: Environ.* 1998, *16*, 149–154.

104. Spivey, J.J. Complete catalytic oxidation of volatile organics. *Ind. Eng. Chem. Res.* 1987, *26*, 2165–2180.

105. Voorhoeve, R.J.H. Perovskite-related oxides as oxidation-reduction catalysts. *Adv. Mater. Catal.* 1977, 129–180.

106. Baucke, F.G.K. and Duffy, J.A. Redox equilibria and corrosion in molten silicates: relationship with electrode potentials in aqueous solution. *J. Phys. Chem.* 1995, *99*, 9189–9193.

107. Courtine, P. Thermodynamic and structural aspects of interfacial effects in mild oxidation catalysts. In *Solid State Chemistry in Catalysis*. Grasselli, R.K. and Brazdil, J.F. Eds. ACS. Symposium Series, Vol. 279, 1985, p. 37–56.

11 Investigation of the Nature and Number of Surface Active Sites of Supported and Bulk Metal Oxide Catalysts through Methanol Chemisorption

L.E. Briand

Centro de Investigación y Desarrollo en Ciencias
Aplicadas-Dr. Jorge J. Ronco, CONICET — Univ. Nacional de
La Plata, La Plata, Buenos Aires-Argentina

CONTENTS

11.1 INTRODUCTION

One of the most fundamental questions of all catalytic studies about bulk catalysts
is the true surface composition and its difference from the bulk composition. This
question has been the driving force of many research efforts in order to develop
only one methodology to describe the nature, properties, and surface density of
the active surface sites of oxide-based catalytic materials. However, the scientific
literature shows many studies that correlate the catalytic activity with the bulk
properties assuming (erroneously) that the composition, structure, and properties
of the bulk and the surface are similar.

X-ray photoelectron spectroscopy (XPS) is a valuable tool that measures the
concentration of atoms in a near surface region of a sample. The photoelectrons that
escape from the solid, and are subsequently detected during the analysis, originate
from a narrow mean free path of 1 to 4 nm for kinetic energies in the range 15 to
1000 eV [1]. Therefore, only about 30% of the signal arises from the outermost
sample layer.

More recently, low-energy ion scattering spectroscopy (LEISS) has been used
to elucidate the surface coverage of the monolayer supported and bulk vanadium
and molybdenum oxide-based catalysts under *in situ* thermal treatments [2]. In
LEISS, a beam of ions (typically, He^+) with a certain energy (0.1 to 10 keV)
scatters elastically from the outermost layer of atoms since the ions that penetrates
the solid are neutralized [1].

A reactive molecule only interacts with the exterior surface of a catalyst and
may be more reliable for the determination of the true surface composition, number
of active surface sites and redox, acid–base properties than a physical spectroscopic
technique. The following sections summarize the most relevant investigations

concerning the use of probe molecules to investigate the surface of oxide materials from the 1970 to the present.

11.2   MAPPING THE SURFACE WITH PROBE MOLECULES: PAST AND PRESENT

11.2.1   Redox Sites Selective Molecular Probes

Transition metal oxide-based materials are widely applied as catalytic materials for heterogeneous and homogeneous selective oxidation processes. Although many oxidants, such as hydrogen peroxide and iodosylarenes, are also used in these processes, there is no doubt that molecular oxygen is the most common. Therefore, it is not surprising that O_2 was applied as a probe molecule to investigate the surface of vanadium pentoxide in the early 1950 [3].

Dyrek [4] combined molecular oxygen chemisorption, electron paramagnetic resonance (EPR), and magnetic susceptibility measurements to investigate the surface coordination of partially reduced V_2O_5. This study focused on the observation that VO_4 coordinated surface V^{4+} sites adsorb oxygen as paramagnetic O_2^- and O^- species, whereas VO_6 coordinated V^{4+} sites form diamagnetic O^{2-} adsorbed ions, assuming the following mechanism:

$$2V^{4+}_{(s)} + \tfrac{1}{2}O_{2(g)} \rightarrow 2V^{5+}_{(s)} + O^{2-}_{(ads)} \tag{11.1}$$

The paramagnetic species yield EPR signals at room temperature while diamagnetic species can only be detected at $-195°C$. The authors concluded that, oxygen chemisorption performed between 250 and 350°C formed diamagnetic O^{2-} species, which suggests the presence of distorted VO_6 coordinated species, most of which originated from the oxidation of V^{3+} species during oxygen adsorption. However, the high temperature range of oxygen chemisorption reoxidizes both surface and bulk vanadium sites. Therefore, it is not clear if the conclusion of this study can be only ascribed to the surface of the material.

Dyrek's observations led to an improvement of the technique known as the low-temperature oxygen chemisorption (LTOC) method [5,6]. The technique involves the prereduction of the material at 500°C under a hydrogen atmosphere prior to the chemisorption of oxygen at temperatures as low as $-78°C$. The reduction led to the formation of coordinatively unsaturated sites (CUSs), which would be able to chemisorb oxygen at a monolayer level. However, again it is not clear how deep is the reduction (surface and bulk) or the final oxidation state of the molybdenum oxide species. Moreover, the reduction with hydrogen should mimic the reduction under reaction conditions in order to make any correlation between the amount of oxygen adsorbed and the catalytic activity of the material. Although, the results obtained about the surface active phase might be completely different from the actual surface under reaction conditions, the LTOC method was applied over transition metal oxide supported catalysts and sulfide phases in an attempt to determine the monolayer surface coverage as well as the hydrodesulfurization and oxidation activity [5–8].

Moreover, the application of oxygen chemisorption techniques assumes that reduced surface transition metal species are responsible for the catalytic activity [4]. However, detailed investigation through successive cycles of reactant adsorption and temperature programmed surface reaction (TPSR) without reoxidation of the surface showed the deactivation of molybdenum and vanadium oxide-based catalysts upon reduction [9]. These results clearly indicate that oxidized surface metal oxide species are the active surface sites to be investigated.

This observation led Murakami and coworkers to develop a new method to measure the concentration of V=O species on oxidized vanadium oxide-based catalysts [10,11]. The method known as rectangular pulse technique involves the reaction of NO and NH_3 with surface V=O species to produce molecular nitrogen according to the following stoichiometric reaction:

$$NO + NH_3 + V{=}O \rightarrow N_2 + H_2O + V{-}OH \qquad (11.2)$$

In the second step, V–OH species are reoxidized with gaseous O_2 or through the diffusion of bulk V = O species,

$$2V{-}OH \xrightarrow{\text{gaseous } O_2 \text{ or bulk } V{=}O} 2V{=}O + H_2O \qquad (11.3)$$

The authors were able to distinguish between the amount of nitrogen catalyzed by surface V=O species and the contribution of reduced vanadium sites and bulk species (generated in the secondary process) through the deconvolution of the initial signal and the tailing portion of a graph of N_2 produced on successive pulses of the reactants mixture.

Although, this method is questionable to measure the density of surface active sites when the reactants are not NO–NH_3, there is no doubt that this technique is the right choice to investigate the surface of those materials active on selective catalytic reduction (SCR) of nitric oxide with ammonia. In fact, 10 years later, Dumesic and coworkers [12,13] upgraded the use of NO–NH_3 as probe molecules in their investigation of a series of supported vanadium on titania catalysts during DeNO$_x$ process. The authors performed TPSR (online mass spectrometric analysis of desorbed products upon heating) and *in situ* infrared (IR) analysis over the catalysts with preadsorbed ammonia exposed to either NO, O_2, or NO + O_2 mixture.

The combination of spectroscopic tools and the use of the actual reactants allowed identifying for the first time, the nature of the surface active sites and the intermediate reactive species of vanadium-based catalysts during DeNO$_x$ reaction. In fact, the authors observed that the catalytic activity was related to the ammonia adsorbed as $V^{5+}{-}ONH_3H^+$ species over oxidized $V^{+5}{-}OH$ that further interact with vanadyl species $V^{5+}{=}O$ to form a $V^{5+}{-}ONH_3H^+{-}O{-}V^{4+}$ complex that reacts with NO.

This observation evidences that the NO–NH_3 probe mixture titrates both $V^{5+}{=}O$ and $V^{5+}{-}OH$ active surface sites and that two vanadium sites are involved in the reaction, which demonstrates that reaction (11.2) was not correct. Another important observation of the authors is about the interaction of the catalysts with H_2 during temperature programmed reduction (TPR) analysis. It is well known that molecular hydrogen is widely used as a probe to measure the reducibility of a catalytic material. The combination of TPR and *in situ* IR analysis demonstrated that monolayer supported vanadium species on TiO_2 agglomerates during reduction causing the breakage of the monolayer and exposing the Ti–OH species of the oxide support [12]. In contrast, an O_2 atmosphere reverses this phenomenon spreading the vanadium oxide layer over the oxide support.

This is direct evidence of the structural changes that a probe molecule might cause on the nature of the surface active sites. Moreover, erroneous conclusions might be obtained when the surface properties are correlated with the catalytic behavior of the material. There is no doubt, however, that this problem is avoided when the surface active sites are studied with the actual reactant of the process under investigation.

Sleight and coworkers [14–17] performed pioneering work on this idea in the 1980, when studying bulk MoO_3 and iron molybdate catalysts. The interest in metal molybdates and MoO_3-metal molybdate mixtures began in 1926 when the high efficiency of the $MoO_3/Fe_2(MoO_4)_3$ system to produce formaldehyde from methanol in excess oxygen was first discovered [18]. During the following years, much research was devoted to understanding the catalytic behavior and the nature of the active surface sites of molybdenum-based catalytic systems. Sleight and coworkers performed spectroscopic studies on methanol adsorbed on pure MoO_3 and $MoO_3/Fe_2(MoO_4)_3$ with TPSR spectroscopy in order to obtain more insights into the number of active surface sites and the nature of surface intermediates during methanol selective oxidation. The authors demonstrated that methanol is adsorbed as undissociated molecules and methoxy species only on a fully oxidized surface. Surface methoxy, CH_3O_{ads}, species are the intermediate species for formaldehyde production and the breaking of the carbon–hydrogen bond is the rate-limiting step in the reaction of the surface methoxy groups over molybdate catalysts.

Further kinetic studies of methanol selective oxidation to formaldehyde over ferric and ferrous molybdate; bismuth, chromium, aluminum, and heteropoly molybdates, demonstrated that a wide range of metal molybdates are active and selective for this reaction and that the same mechanism of methanol adsorption and reaction occurs [19].

The major technical contribution of their work was the determination of the number of surface active sites through the quantification of the methanol molecules adsorbed at room temperature. A decade later, this pioneering work inspired the development of a novel chemisorption technique to determine the density of surface active sites of supported and bulk oxide catalysts that will be discussed in the following sections.

Oyama and coworkers [20] reported another example of the use of the actual reactant as a surface probe. The author used ozone to investigate the number of surface sites of supported manganese oxides that are active in the decomposition of ozone to oxygen. The technique involves the adsorption of ozone at

$-40°C$ and quantification of the oxygen produced during temperature programmed reaction.

11.2.2 Acid–Base Sites Selective Molecular Probes

Probe molecules such as, pyridine, ammonia, carbon monoxide, piperidine, and *n*-butylamine are used to investigate acid–base surface properties due to their capability to distinguish between surface Lewis and Brønsted acid sites of oxide materials. These molecules possess the ability to interact as distinct adsorbed surface species according to the nature of the acid site.

A detailed review of the species generated upon adsorption of a series of basic probe molecules on metal oxide catalysts and their characteristic IR bands was published by Busca [21]. Moreover, the characterization of a series of binary and ternary oxides are also presented.

Recently, more complex heterocyclic nitrogen bases such as *N*-methylpyrrole, 2-chloropyridine, quinoline, pyrazine, pyrimidine, and pyridazine have been tested as reliable probes to obtain information about the surface acidity of SiO_2, TiO_2, ZrO_2, SiO_2–Al_2O_3, H-mordenite, and sepiolite [22–25].

N-methylpyrrole (Scheme a) and 2-chloropyridine (Scheme b) interact with an oxide material through the following adsorption mechanisms: hydrogen-bonding interactions with hydroxylated sites (weak Brønsted acid sites), transferring electrons to Lewis acid sites, or are protonated only by strong Brønsted acid sites [22,23]. Each of these interactions is also associated with a characteristic IR band, which allows distinguishing between weak and strong Brønsted acid sites. This is an advantage compared with conventional NH_3 and pyridine probes that are protonated even by weak Brønsted acid sites. The behavior is attributed to the lower pK_a values of the conjugated acids of *N*-methylpyrrole and 2-chloropyridine than NH_3 and pyridine.

An interesting method that uses inorganic probes in the liquid media was proposed by Jaroniec and coworkers [26]. This method involves the investigation of the luminescence spectra of hydrated uranyl groups UO_2^{2+} adsorbed on SiO_2 and Al_2O_3 nanoparticles dispersed in aqueous media. Uranyl groups are sensitive to the energetic heterogeneity of the surface sites generating different adsorption complexes. Therefore, the electronic perturbation of the molecular ions results in the broadening of their emission spectra under the simultaneous excitation of all the adsorption complexes. The application of a selective laser excitation (480 to 520 nm) allows obtaining each elementary luminescence spectrum that corresponds to a single type of adsorption complex. The authors demonstrated that uranyl groups chemisorb on hydroxylated surface sites (silanol groups Si–OH and Al–OH) through two different adsorption complexes. The bimodal distribution of the adsorption energy indicates the presence of two energetically different M–OH sites.

Fluoride ions F^- have also been used as a probe to assess the nature and the number of surface sites of FeOOH (goethite) in aqueous media [27]. IR and nuclear magnetic resonance studies showed that fluoride ions replace surface OH groups according to the reactions,

$$FeOH^{-1/2} + H^+_{(aq)} + F^-_{(aq)} \rightarrow FeF^{-1/2} + H_2O_{(l)} \qquad (11.4)$$

$$Fe_2OH^0 + H^+_{(aq)} + F^-_{(aq)} \rightarrow Fe_2F^0 + H_2O_{(l)} \qquad (11.5)$$

This inorganic probe allows the establishing of the coordination of the surface sites since fluoride ions exchange singly coordinated OH groups at low F^- concentrations (reaction [11.4]) and doubly coordinated OH groups at high concentration (reaction [11.5]).

Acid molecules such as CO_2, SO_2, pyrrole, chloroform, $(CF_3)_3COH$, $(CF_3)_2CHOH$, CF_3CH_2OH, H_2S, CH_3SH, and boric acid trimethyl ether $B(OCH_3)_3$ are used to characterize surface basic sites. Those molecules possess the ability of adsorbing on the electron donor sites (basic sites) such as hydroxyl groups and oxygen centers (O^{2-}). A detailed discussion of IR spectrometric investigations of the surface basicity of metal oxides and zeolites using adsorbed probe molecules was reviewed by Lavalley [28].

11.2.3 Super-Acid Sites of Heteropoly-Oxo Compounds

Keggin and Wells-Dawson heteropoly-oxo acids possess the structures $X^{n+}M_{12}O_{40}H_{(3,4)}$, and $(X^{n+})_2M_{18}O_{62}H_{(16-2n)}$, respectively, where X^{n+} represents a central atom (silicon [IV], phosphorous [V], arsenic [V], sulfur [VI], or fluorine) surrounded by a cage of M atoms, such as, tungsten (VI), molybdenum (VI), vanadium (V), or a mixture of elements [29]. These complex structures are increasingly being applied as solid super-acid catalysts in eco-friendly chemical processes. The use of solid, bulk, or supported heteropoly acid (HPA) to replace conventional liquid acids (HCl, HF, H_2SO_4), allows recovering and reusing the catalyst, which diminishes undesirable liquid wastes [29].

A variety of methods to study the acidity of heteropoly compounds are reported in the literature in line with the study of the catalytic applications of the HPAs. These methods, mainly applied on Keggin type structures, evaluate the acid strength and the amount of molecules adsorbed with probe molecules (ammonia, pyridine) different from actual reactants. The studies that used actual reactant molecules, such as alcohols, determine the total amount of adsorption sites rather than the number of active acid sites at a certain temperature of reaction. In fact, the studies reported in the literature focused on the intermediate adsorbed species rather than on the *in situ* identification and quantification of the products of reaction and the relationship between that information and the density of active sites. Moreover, a detailed review of the literature of the past 10 years shows almost no studies of the acidity of Wells–Dawson structures with probe molecules [30].

The acid strength of solid HPA compounds (mainly of the Keggin type structure) is measured through the heat of adsorption and the temperature of desorption of ammonia [31]. Additionally, Misono [32] studied the surface acidity of phospho-tungstic cesium salts through the IR analysis of CO adsorbed at 110 K. The interaction of the probe molecule with the HPA generates three different signals depending on physisorbed or chemisorbed species.

The activity toward isopropanol dehydration is also used to characterize the acidity of HPA compounds since the product distribution depends on the nature of surface active sites (and bulk active sites considering the pseudo-liquid behavior of heteropoly compounds). Strong surface Brønsted (H^+) and Lewis acid sites catalyze the dehydration of isopropanol to propylene (di-isopropyl ether for weak Lewis acid sites) and redox/basic sites lead to the dehydrogenation of the alcohol to acetone.

Photoacoustic Fourier transform infrared (FTIR) spectroscopy studies performed by Moffat and coworkers [33] on C_1 to C_4 alcohols adsorbed over phospho-tungstic Keggin acid $H_3PW_{12}O_{40}$, demonstrated the formation of alkoxyl intermediates. The authors observed the chemisorption of 3–4 isopropanol molecules per Keggin anion after 5 to 10 h of exposure to the alcohol at 25°C. The alcohol initially adsorbs as a protonated species i-$C_3H_7OH_2^+$ and forms isopropoxyl species upon evacuation at about 50°C.

The kinetics of adsorption and the amount of polar (alcohols) and nonpolar molecules, Keggin type compounds, was extensively studied through IR spectroscopy and thermogravimetric analysis [16,31,32,34,35]. Misono and coworkers [31] established that the rate of alcohol adsorption depends on the size of the probe molecule, and that the amount of adsorbed molecules in $H_3PW_{12}O_{40}$ is an integral multiple of the number of protons. The observation of the diffusion coefficients and adsorption of polar molecules into the bulk structure of the HPAs proved unequivocally the "pseudo-liquid phase" behavior of those compounds.

Bielański et al. [34,35] performed a detailed study of methanol adsorption–desorption on dehydrated $H_4SiW_{12}O_{40}$ through IR spectroscopy and thermogravimetric analysis in order to obtain evidences on the mechanism of methyl-tert-butyl ether (MTBE) synthesis. The authors demonstrated that up to 16 molecules of methanol are adsorbed per Keggin unit at 18°C. The formation of dimethyl ether and water were observed upon heating the catalyst up to 100°C although methanol is not completely desorbed out of the structure even at 250°C.

11.3 Designing a Tailored Molecular Probe for a Specific Application

The previous sections summarize a variety of probe molecules and methods to explore the surface of oxide materials. In general, these methods possess the following limitations:

1. The nature and the number of surface active sites are determined with a different probe molecule than the actual reactant.
2. The oxide catalysts are reduced with hydrogen and the reduction stoichiometry of the catalysts is usually not known.
3. The temperature of the chemisorption is far from the actual reaction conditions.
4. Both surface and bulk sites are involved in the reduction–oxidation measurements.

The aim of the investigations discussed here have been the discovery of only one "miraculous" molecular probe that can give information about the active surface sites of any oxide catalysts regardless of the catalytic application of the material. This is impossible if we consider that, the surface of a catalyst is dynamic under reaction conditions and, therefore, might be significantly modified by a specific reactant. The "holy grail" of the chemical probe molecules does not exist. Nevertheless, it is possible to design a tailored molecular probe for a specific application. The process of tailoring a probe molecule to measure the density of active surface sites is described as follows:

1. The investigation of the surface reaction mechanism of the adsorption-reaction processes in order to identify the reactive surface intermediate species.
2. The knowledge of temperature and partial pressure that allows formation of a monolayer of surface reactive species on the oxide surface.
3. The measurement of the amount of reactive surface intermediate species in order to quantitatively determine the density of surface active sites (sites per unit surface area).
4. The number of surface atoms occupied by one reactant molecule.

The following sections describe the development of a chemisorption method that applies the dissociative chemisorption of methanol to quantitatively determine the density of active surface sites of metal oxide catalysts for methanol selective oxidation.

11.3.1 Surface Reaction Mechanism of Methanol Adsorption-Reaction on Oxide Surfaces

The spectroscopic investigations performed by Sleight and coworkers [14–17] demonstrated that methanol chemisorption on MoO_3 at room temperature results in a combination of molecular and dissociative adsorption mechanisms. The first mechanism can be considered as a physical adsorption since the methanol molecules adsorb intact on the surface. Dissociative adsorption is a chemisorption process that involves the formation of metal–methoxy ($M–OCH_3$) groups. Further,

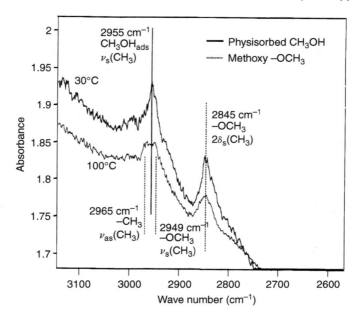

FIGURE 11.1   IR spectra of methanol adsorbed on monolayer supported molybdenum oxide on Al_2O_3 (18% MoO_3/Al_2O_3) at 30 and 100°C (From Briand, L.E., Farneth, W.E., and Wachs, I.E. *Catal. Today* 2000, *62*, 219–229. With permission.)

IR studies carried out by Groff indicated that at 100°C molecularly adsorbed methanol species are volatile while the surface methoxy groups remain intact on the MoO_3 surface [36].

More recently, Wachs and coworkers [37–40] performed a detailed investigation of the methanol adsorption mechanism on a series of supported and bulk oxide catalysts through *in situ* IR analysis. The authors demonstrated that methanol adsorption is greatly influenced by the temperature and the concentration of methanol in the gas phase. The *in situ* IR spectra of methanol adsorbed on monolayer supported molybdenum oxide on Al_2O_3 (18% MoO_3/Al_2O_3) at 30 and 100°C, are presented in Figure 11.1. Methanol adsorbed on an oxide surface at low temperature generates two bands in the C–H stretching region (2700 to 3000 cm^{-1}). The bands at 2845 and 2955 cm^{-1} are assigned to the symmetric bend ($2\delta_s$) of the CH_3 unit in the adsorbed surface methoxy species and the symmetric stretch (v_s) of the CH_3 unit in the intact methanol molecule, respectively. Upon heating the sample to 100°C, the band at 2845 cm^{-1} does not change; however, the signal at 2955 cm^{-1} decreases in intensity and splits into two new bands at 2949 and 2965 cm^{-1} (assigned to the symmetric (v_s) and asymmetric stretch (v_{as}) of the CH_3 unit in the surface methoxy species) [37,38].

Further studies showed that methanol adsorbs as surface methoxy species and methanol-like species strongly coordinated to surface metal cations (Lewis acid

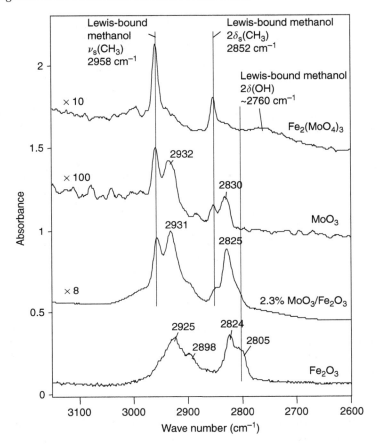

FIGURE 11.2 IR spectra of methanol adsorbed over Fe_2O_3, MoO_3, and monolayer suppor-ted molybdenum over iron oxide and bulk iron molybdate at 110°C (From Burcham, L.J., Briand, L.E., and Wachs, I.E. *Langmuir* 2001, *17*, 6164–6174 and 6175–6184. With permission.)

sites) at temperatures equal to or above 100°C. Figure 11.2 shows a typical IR spectra of methanol adsorbed over Fe_2O_3, MoO_3, and monolayer surface coverage of supported molybdenum over iron oxide and bulk iron molybdate at 110°C [39]. The spectra exhibit bands at ∼2960/2935 and ∼2852 cm^{-1} that correspond to intact methanol-like species, and ∼2850/2835 and ∼2930 cm^{-1} assigned to –OCH_3 species. The authors observed that the basic materials such as Fe_2O_3, NiO, ZnO_2, ZrO_2, and Bi_2O_3 do not adsorb molecular methanol as Lewis-bounded surface species, but favor the formation of surface methoxy species. In general, monolayer coverage of supported molybdenum oxide and bulk metal molybdate catalysts showed comparable intensities of both types of surface methoxylated species.

   Figure 11.3 presents the *in situ* IR spectra of methanol oxidation over monolayer supported vanadium on titania 5% V_2O_5/TiO_2 as a function of

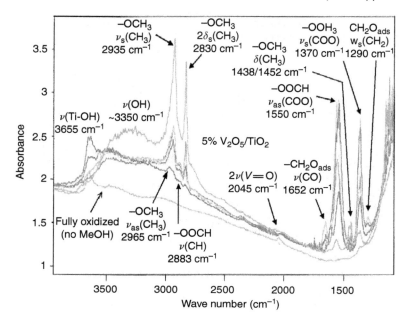

FIGURE 11.3 *In situ* IR spectra of methanol adsorbed over 5% V_2O_5/TiO_2 during methanol oxidation as a function of temperature (lowest temperature on top: 100, 200, 225, 250, 275, and 300°C) (From Burcham, L.J. and Wachs, I.E. *Catal. Today* 1999, *49*, 467–484. With permission.)

temperature [40]. The spectra show that surface methoxylated species react toward formaldehyde or are further oxidized toward formate species HCOO– (1550 and 1370 cm^{-1}) upon increasing the catalyst temperature.

Wachs and Burcham [37] concluded that methanol chemisorption proceeds by two different pathways. The first pathway adsorbs methanol associatively to produce an intact Lewis-bound adsorbed surface molecular methanol species that remains stable to relatively high temperatures under vacuum (at least 100 to 200°C). The second pathway dissociatively adsorbs methanol to form surface methoxy species ($-OCH_3$) and surface hydroxyls. In both cases, the oxygen of the methoxyl group is coordinated to a surface Lewis acid cation, while in the dissociated case the methanol alcoholic proton must coordinate to a basic surface oxygen anion or to a surface hydroxyl (producing either surface hydroxyls or water, respectively). The authors observed that all surface hydroxyls present on the initial oxide catalysts are titrated upon monolayer surface coverage of the methoxylated surface species. Moreover, Lewis-bound water is rarely observed even at saturation and any residual water on the metal oxide surfaces present prior to methanol exposure is displaced by the adsorbed methoxylated surface species. Surface hydroxyls are generally not present in cases where only dissociated species are formed, indicating their condensation and evolution from the surface as gaseous. The authors suggested that, the absence of Lewis-bound water on the catalysts indicates that the high temperature of methanol chemisorption ($\sim 100°C$) forces the water equilibrium toward the vapor phase and allows for the assumption of complete water loss.

11.3.2 Saturation of the Oxide Surface with Reactive Surface Intermediate Species

More detailed information about the nature and the stability of the surface intermediate species formed during methanol adsorption is obtained from TPSR experiments. The difference in the nature of the adsorbed species is also apparent from the desorption profiles from the 18% MoO_3/Al_2O_3 catalyst shown in Figure 11.4 and Figure 11.5 [38]. The products and desorption temperatures (T_{max}) after saturation of the surface with methanol at $23°C$ are similar to the results previously reported in the literature for bulk MoO_3 [16]. The first desorption peak starting at $50°C$ ($T_{max} \sim 100°C$) is due to pure methanol ($m/e = 32$). The shape and broadness of the signal can be ascribed to the presence of physisorbed (molecular methanol) and chemisorbed methoxylated species on the catalyst surface that result in CH_3OH formation and desorption. Formaldehyde ($m/e = 30$) desorbs as a broad signal with $T_{max} \sim 250°C$. The signal with $m/e = 28$ corresponds both to a cracking product of methanol in the mass spectrometer and formaldehyde. Both surface methoxy formation and decomposition generate water that desorbs in a wide temperature range. Exposed acid sites of the alumina support

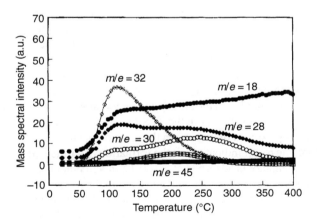

FIGURE 11.4 TPSR of methanol adsorbed over 18% MoO_3/Al_2O_3 catalyst at $23°C$: methanol ($m/e = 32$), formaldehyde ($m/e = 30$), water ($m/e = 18$), dimethyl ether ($m/e = 45$), and product of methanol and formaldehyde cracking in the mass spectrometer ($m/e = 28$) (From Briand, L.E., Farneth, W.E., and Wachs, I.E. *Catal. Today* 2000, *62*, 219–229. With permission.)

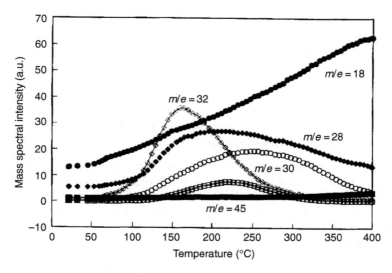

Figure 11.5 TPSR spectra of methanol adsorbed over 18% MoO_3/Al_2O_3 catalyst at 100°C (From Briand, L.E., Farneth, W.E., and Wachs, I.E. *Catal. Today* 2000, *62*, 219–229. With permission.)

and acidic surface sites associated with the MoO_x monolayer lead to methanol dehydration–condensation and formation of a small amount of dimethyl ether.

The TPSR spectra from methanol adsorbed at 100 and 23°C are similar, but some important differences in the T_{max}, shape of methanol signal, and the area of the CH_2O peak, can be observed (Figure 11.5). For adsorption at 100°C, methanol begins to desorb at 100°C and the peak maximum is shifted to 150°C, which is associated with a higher stability of the species formed from methanol adsorbed at 100°C. According to the literature, surface methoxy species are also able to recombine with adsorbed hydrogen atoms ($OCH_{3(ads)} + H_{ads} \rightarrow CH_3OH$) and desorb as methanol at the temperatures of H_2CO formation [16]. The T_{max} for formaldehyde is not modified compared with the adsorption temperature; however, the integral of the signal is considerably greater in the TPSR of methanol chemisorbed at 100°C. Formaldehyde is produced only through the reaction of surface methoxylated species; therefore, the result indicates that the formation of surface methoxy species is favored when methanol is adsorbed at ~100°C, in agreement with the IR studies.

The partial pressure of methanol in contact with the oxide surface also needs to be adjusted in order to saturate the surface of the catalyst with a monolayer of surface methoxy species. Controlled methanol adsorption on the 18% MoO_3/Al_2O_3 catalyst was performed by exposing it to different methanol partial pressures until saturation of the surface was reached [38]. Figure 11.6 shows methanol adsorption at 23 and 100°C in a microbalance as a function of the exposure in units of Langmuir ($1L = 10^{-6}$ torr sec). The chemisorption isotherms fit the Langmuir adsorption model, as already reported for bulk MoO_3 [16]. Regardless of the temperature,

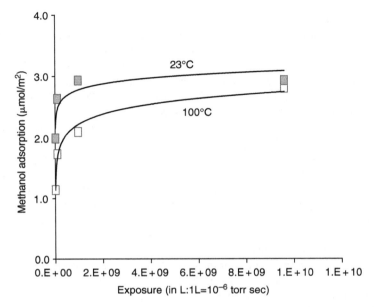

FIGURE 11.6 Methanol adsorption over 18% MoO_3/Al_2O_3 at 23 and 100°C as a function of increasing partial pressures of methanol (From Briand, L.E., Farneth, W.E., and Wachs, I.E. *Catal. Today* 2000, *62*, 219–229. With permission.)

the surface of the catalyst is saturated when exposed to methanol partial pressures above 1000 mtorr. The difference in methanol uptake with the temperature of the adsorption is related to the nature of the species involved in the process. The values of micromols of methanol adsorbed at different partial pressures assume molecular methanol regardless of the temperature of adsorption.

At low temperature, methanol is mainly molecularly physisorbed, therefore, methanol molecular weight should be considered in order to calculate molar adsorption from the weight gain value. However, methanol adsorbs as surface methoxylated species at ~100°C as discussed earlier.

11.3.3 Measurement of the Amount of Surface Reactive Intermediate Species for the Determination of the Density of Surface Active Sites

Surface methoxy groups are the intermediate reaction species in the production of partially oxygenated reaction products (formaldehyde, methyl formate, methylal, etc.) during methanol selective oxidation on catalysts containing transition metal oxides (oxides of vanadium, molybdenum, chromium, etc.). Therefore, the knowledge of the amount of surface methoxy species formed during methanol chemisorption is the key for the determination of the number of surface active sites available for methanol selective oxidation. The results described in the previous sections demonstrated that, in general, an oxide surface in contact with 1000 mtorr of methanol vapor at $\sim 100°C$ is suitable to be covered with a stable monolayer of surface methoxylated species. Although this statement is valid for a wide series of supported and bulk oxide catalysts, it does not apply to heteropoly-oxo acids and some bulk metal oxides that are highly reactive even at $100°C$. This observation will be further discussed in the following sections.

There is no doubt that the temperature for the formation of a stable monolayer of the surface intermediate species must be tuned for each material through the investigation of the adsorbed surface species as a function of the temperature.

The quantification of the adsorbed intermediate species has been performed through many different ways: the determination of the amount of water evolved during the adsorption, weight change of the material with a microbalance, directly measuring the methoxylated surface species through IR spectroscopy or quantifying the products desorbed during TPSR.

Farneth et al. [41] performed adsorption–desorption studies with a high vacuum microbalance/temperature programmed reaction apparatus coupled with a mass spectrometer. Chemisorption was performed exposing the MoO_3 sample to 10, 10^2, 10^3, and 10^4 mtorr of methanol vapor. The amount of adsorbed methanol was measured gravimetrically as the stable weight after pumping out up to 10^{-8} torr at room temperature. The authors also used 3A-sieves to trap the water produced during the adsorption of methanol, ethanol, isopropanol, and tert-butanol on MoO_3 [16]. The alcohols dissociatively chemisorb as surface alkoxy and hydroxyl species, and hydroxyls recombine and desorb as water. The stoichiometry of the adsorption allows for the quantitative adsorption of the alcohol adsorbed,

$$MoO_3 + 2HOR \rightarrow 2(RO)-MoO_2 + H_2O\uparrow \qquad (11.6)$$

Methanol chemisorption experiments have also been carried out at atmospheric pressure in a specially adapted thermal gravimetric analyzer (TGA) microbalance coupled with a PC for temperature and weight monitoring [38]. The system allowed a controlled flow of high purity gases: air for pretreatment, helium and a mixture of 2000 ppm methanol in helium for adsorption experiments.

Several attempts to quantify the amount of adsorbed methanol species through IR spectroscopy were performed over CeO_2 [42], MgO [43], MoO_3 [35], SiO_2 [44], and ZrO_2 [45]. Wachs and coworkers [37,39,40] performed an extensive investigation of the *in situ* spectroscopic quantification of surface methoxylated species before and during methanol reaction over oxide catalysts. It was determined from the integrated molar extinction coefficient (IMEC) calibration of the IR bands at 2830 to 2850 cm^{-1} that correspond to C–H vibration of the methoxylated species. The calibration was achieved by quantitatively adsorbing 1 μmol of methanol onto the catalysts at $110°C$ under vacuum and integrating the IR signals (in transmission mode) of the initial doses. The authors also designed a novel *in situ* cell that operates as a plug-flow reactor and allows measurement of the steady-state concentrations of surface methoxylated species under methanol oxidation reaction conditions [46]. This IR transmission cell possesses a fixed-bed configuration that forces a convective gas flow through the catalyst bed and is coupled with an online gas chromatograph for products analysis. This new experimental methodology has been coined "operando spectroscopy" by Bañares and Wachs [47].

The authors determined the adsorption and the surface reaction kinetic parameters of methanol reaction over monolayer supported vanadium, molybdenum, chromium, rhenium catalysts, and bulk iron molybdate.

More recently, Briand and coworkers [48] determined the amount of active surface acid sites for WO_3, monolayer supported tungsten on titania, phospho-tungstic Keggin, and Wells–Dawson HPAs ($H_3PW_{12}O_{40}$ and $H_6P_2W_{18}O_{62}$) through TPSR spectroscopy. Isopropanol was chemisorbed and the number of active acid sites measured by the amount of propylene generated during the TPSR was determined.

11.4 NUMBER OF SURFACE ACTIVE SITES AND TURNOVER FREQUENCY: WHY ARE THESE PARAMETERS RELEVANT?

As previously stated, the number of surface intermediate species formed during the chemisorption of a probe molecule is directly related to the number of surface active sites available for the reaction of that molecule. Therefore, the amount of surface intermediate species is expressed as the number of accessible surface active sites per unit of surface area (N_s).

This parameter is relevant for the calculation of the turnover frequency (TOF) since it allows normalizing the reaction rate per active surface site (TOF = molecules converted to products per second per active surface site). This is the "true catalytic activity" since it is independent of the contact time of the reactant stream and mass, surface area, and amount of surface active sites of the catalytic material in the reactor. This concept is critical since the TOF allows comparing the reactivity of different metal oxide-based catalytic systems exclusively as a function of their surface properties.

Section 11.4.1 to Section 11.4.4 discuss the stoichiometry of adsorption, kinetic parameters, reactivity, and surface morphology of several catalytic materials that have been obtained through the application of methanol chemisorption to determine N_s and TOF values.

11.4.1 Stoichiometry of Methanol Adsorption and "Ligand Effect" of Monolayer Supported Metal Oxide-Based Catalysts

Monolayer supported metal oxide catalysts possess a two-dimensional overlayer of an active metal oxide that is molecularly dispersed over a high surface area support. Usually, all the metal atoms deposited onto the oxide support are considered as the number of surface active sites. However, the application of methanol chemisorption as surface intermediate methoxy species on monolayer supported molybdenum,

TABLE 11.1

Surface Molybdenum Oxide Structures and Methanol Chemisorption Data

Catalyst	Mo molecular structure	Mo surface density (μmol/m^2)	N_s (μmol methoxy/m^2)	Methoxy adsorbed per Mo atom
3% MoO$_3$/ZrO$_2$	$O_h + T_d$ (polym.)	5.31	1.25	0.24
6% MoO$_3$/Nb$_2$O$_5$	O_h (polym.)	7.64	2.10	0.28
6% MoO$_3$/TiO$_2$	O_h (polym.)	7.64	3.07	0.40
2% MoO$_3$/MnO	Not determined	7.50	1.14	0.15
3% MoO$_3$/Cr$_2$O$_3$	Not determined	7.42	2.87	0.39
18% MoO$_3$/Al$_2$O$_3$	$O_h + T_d$ (polym.)	6.97	2.78	0.40
4% MoO$_3$/NiO	Not determined	7.32	1.47	0.20
5% MoO$_3$/SiO$_2$	T_d (isolated)	0.70	0.95	1.35

O_h: octahedral coordination; T_d: tetrahedral coordination; polym.: polymerized.

Source: From Briand, L.E., Farneth, W.E., and Wachs, I.E. *Catal. Today* 2000, *62*, 219–229. With permission.

vanadium, chromium, niobium, rhenium, and tungsten demonstrated that there exists a steric limitation of ~0.3 methoxylated surface species (both Lewis-bound CH_3OH_{ads} and methoxy –$OCH_{3,ads}$ species) [38,40].

Table 11.1 shows some results of the investigation performed on a series of monolayer supported molybdenum oxide catalysts [38]. The table summarizes the surface molybdenum oxide molecular structures, surface concentration of Mo atoms, surface methoxy concentration expressed as the number of accessible surface active sites per unit surface area (N_s), and the concentration of surface methoxy species adsorbed per surface molybdenum atom. The amount of adsorbed intermediate species was determined through a microbalance and *in situ* IR techniques with similar results.

The surface Mo coverage of the catalysts corresponds to a monolayer of surface Mo oxide species (~7 μmol of Mo/m^2). At monolayer surface coverage, the active phase is 100% dispersed on the oxide supports, which avoids bulk MoO$_3$ formation and minimizes the adsorption of methanol on exposed oxide support sites. The surface molybdenum oxide species are polymerized; MoO$_5$/MoO$_6$ coordinated on Nb$_2$O$_5$ and TiO$_2$ and possess a mixture of MoO$_5$/MoO$_6$ and MoO$_4$ coordinations on ZrO$_2$ and Al$_2$O$_3$ under dehydrated conditions. Silica is not completely covered with molybdenum due to the low density and reactivity of the silica surface hydroxyls, which accounts for the low-Mo surface density and the isolation of the surface molybdenum species on the silica support.

The number of surface active sites, N_s, does not depend on the surface molybdenum oxide coordination and is lower than the number of molybdenum atoms dispersed on the support. Furthermore, surface methoxy chemisorption stoichiometry is ~ 1 CH_3O_{ads} per 3 to 4 Mo_s for a polymerized molybdenum oxide structure and ~ 1 CH_3O_{ads} per Mo_s for isolated molybdenum oxide species. These results reflect the presence of lateral interactions in the surface methoxy overlayer and demonstrate that, even when every exposed Mo atom could be an active redox site, not all of them are simultaneously available under reaction conditions.

The specific activity of monolayer supported oxide catalysts considering the actual density of surface active sites is a factor of ~ 3 higher than the TOFs obtained with the amount of dispersed metal atoms. This observation is expected since less than half of the exposed metal oxide atoms can simultaneously participate as "active surface sites" during steady-state reaction due to the steric hindrance in methanol adsorption.

Nevertheless, the authors observed similar trends in the "ligand effect" with the new reevaluated TOFs, as they found previously. This effect is a variation of the TOF values in the selective oxidation of methanol when changing the electronegativity of the oxide support. In general, the low electronegativity of the metal cation of the oxide support ($Zr \sim Ce > Ti > Nb > Al \gg Si$) results in a high TOF of the catalyst. The authors obtained more insights into the "ligand effect" determining the adsorption equilibrium constant, K_{ads}, and the kinetic rate constant for the surface decomposition of intermediate species, k_{rds}, of the selective oxidation of methanol toward formaldehyde according to,

$$CH_3OH + S-O-S' \rightarrow CH_3O-S + S'-O-H \quad K_{ads} \tag{11.7}$$

$$CH_3O-S + O_{ads} \rightarrow CH_2O + S-OH \qquad k_{rds} \tag{11.8}$$

where S, S' are metal cations, $S = Mo$, $S' = $ support or Mo.

These parameters were calculated through the fractional methoxy surface coverage and TOF values obtained with a specially designed in situ IR cell reactor during steady-state methanol selective oxidation [46]. This in situ cell, previously described, allows simultaneous determining of the number of adsorbed methoxylated species and the reaction rate at low-methanol conversion.

The authors observed that the surface coverage decreases with increasing temperature due to the increase on the reactivity of the surface intermediate species (typical results are presented in Figure 11.7). In fact, the authors demonstrated that the surface decomposition rate constant, k_{rds}, is directly influenced by the nature of the oxide support and accounts for the "ligand effect."

11.4.2 Surface Morphology of Bulk Metal Oxide Catalysts

Badlani et al. [49,50] performed the most detailed study on the reactivity and the surface properties of bulk metal oxides. The authors investigated the number

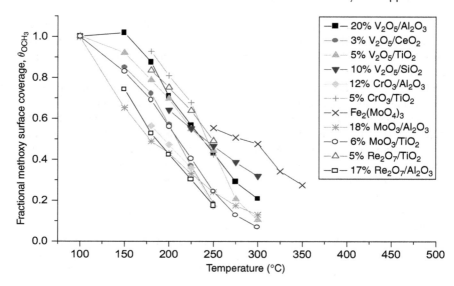

FIGURE 11.7   Fractional surface coverage of adsorbed methoxy species determined with *in situ* IR during methanol selective oxidation as a function of temperature of reaction (From Burcham, L.J., Badlani, M., and Wachs, I.E. *J. Catal.* 2001, *203*, 104–121. With permission.)

and nature of the surface active sites, and the TOF for methanol selective oxidation of one-component bulk metal oxides through methanol chemisorption and temperature programmed reaction in a microbalance.

Table 11.2 and Figure 11.8 present the N_s values and the dependency of the redox TOF with the temperature of surface methoxy decomposition, respectively. The redox TOF involves the reaction rate toward redox products (formaldehyde, dimethoxy methane, and methyl formate) per surface active site. Methanol was adsorbed as surface methoxylated species at 100°C (similar to monolayer supported oxide catalysts) over transition metal oxides with the exception of Rh_2O_3, PdO, PtO, CuO, Ag_2O, and Sb_2O_3, which required a lower temperature of adsorption (50°C) due to their high reactivity. Table 11.2 shows that MoO_3, V_2O_5, and ZnO possess lower values of N_s (~0.6 μmol/m^2) compared with the average ~3 to 4 $OCH_{3,ads}$ μmol/m^2 of most of the bulk metal oxides. This observation was attributed to the anisotropic (platelet) surface morphology of MoO_3, V_2O_5, and ZnO and the preferential adsorption of methanol on the edge sites of the crystals.

The authors demonstrated for the first time in the literature that the redox TOF is not related with the enthalpy of formation, the metal–oxygen bond strength, the isotopic dioxygen exchange rate, or the temperature of reduction of metal oxides, as has many times been stated in the literature. However, the redox TOF possesses a strong inverse relationship with the temperature of surface methoxy decomposition (see Figure 11.8), which again evidences that the decomposition of adsorbed intermediate species is the rate-determining step in methanol selective oxidation.

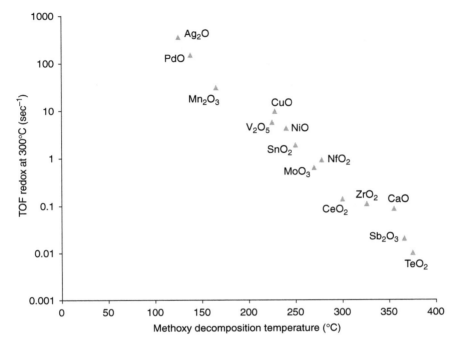

FIGURE 11.8 Redox TOF of bulk metal oxides toward methanol selective oxidation as a function of the temperature of surface methoxy decomposition (From Badlani, M. and Wachs, I.E. *Catal Lett*. 2001, *75*, 137–149. With permission.)

11.4.3 Surface Composition of Bulk Metal Molybdate and Vanadate Catalysts

The number and nature of the active surface sites and the catalytic activity of bulk metal molybdates and vanadates were also investigated through methanol chemisorption [51–53]. These materials proved to be equally or more active and stable than the industrial catalyst $MoO_3/Fe_2(MoO_4)_3$ in formaldehyde production [54–56].

The significant difference between the TOF and selectivity of bulk metal molybdates and vanadates compared with pure metal oxides was a key factor in uncovering the true surface composition of those bulk catalysts. Table 11.3 and Table 11.4 show the number of surface active sites, redox TOF, and selectivity toward methanol selective oxidation products of bulk metal vanadates and the corresponding metal oxide, respectively. Similar results were obtained for bulk metal molybdates. Bulk metal vanadates possess a high selectivity to formaldehyde with some selectivity to dimethoxy methane (nickel vanadate), dimethyl ether (niobium, chromium, and aluminum vanadates), methyl formate (magnesium, chromium, and copper vanadates), and CO_2 (niobium and silver vanadates).

TABLE 11.2
Number of Surface Active Sites for Methanol Selective Oxidation of Bulk Metal Oxides

Catalyst	N_s (μmol/m^2)	Catalyst	N_s (μmol/m^2)
MgO	22.5	Co_3O_4	2.8
CaO	5.4	Rh_2O_3	3.5
SrO	4.3	NiO	6.5
BaO	3.8	PdO	4.3
Y_2O_3	4.9	PtO	3.1
La_2O_3	34.1	CuO	3.6
TiO_2	3.7	Ag_2O	5.2
ZrO_2	1.1	ZnO	0.3
HfO_2	2.6	Al_2O_3	5.6
CeO_2	4.2	SiO_2	0.2
V_2O_5	0.7	Ga_2O_3	4.1
Nb_2O_5	2.6	In_2O_3	2.7
Ta_2O_5	4.6	SnO_2	1.6
Cr_2O_3	12.4	P_2O_5	3.6
MoO_3	0.8	Sb_2O_3	4.9
WO_3	2.3	Bi_2O_3	2.1
Mn_2O_3	1.6	TeO_2	4.1
Fe_2O_3	3.7		

Source: From Badlani, M. and Wachs, I.E. Catal. Lett. 2001, 75, 137–149. With permission.

Turnover frequencies and selectivity results of methanol oxidation over pure metal oxide catalysts were obtained at different temperatures (typically 300°C) in order to maintain low methanol conversions. The surface redox/acid sites of bulk V_2O_5 lead to a high selectivity to formaldehyde, along with dimethoxy methane and dimethyl ether. The redox/basic character of the surface active sites of MgO, NiO, MnO, Cr_2O_3, CoO, and ZnO yield formaldehyde and CO_2. The surface acidic sites of Al_2O_3, Nb_2O_5, and Fe_2O_3 catalyze methanol dehydration to dimethyl ether. Formaldehyde is the only product of the methanol oxidation reaction over CuO and Ag_2O.

The observation that bulk metal vanadates posses a high selectivity toward formaldehyde strongly suggests that the surface of bulk metal vanadates is composed of vanadium oxide sites with redox properties that cover the counter-cation sites (Mg, Ni, Mn, Cr, Co, Zn, Al, Nb, Fe, Cu, and Ag) thus, inhibit methanol total oxidation. The TOF values of the bulk metal vanadates are similar indicating that there is no significant influence of the specific nature of the metal oxide counter-cation on the catalytic behavior. Moreover, there are significant differences between the TOF values of bulk metal vanadates and pure metal oxides,

TABLE 11.3

TOF and Selectivity of Bulk Metal Vanadates Toward Methanol Oxidation at Low Methanol Conversion

Catalyst	N_s (μmol/m^2)	Redox TOF[a] (sec^{-1})	Selectivity[b] (%)
Mg$_3$(VO$_4$)$_2$	0.6	1.43	100.0
NbVO$_5$	1.6	3.1	87.9
CrVO$_4$	0.7	14.4	99.3
Mn$_3$(VO$_4$)$_2$	3.9	0.4	100.0
AlVO$_4$	2.7	2.2	98.0
AgVO$_3$	21.1	1.6	92.5
Ni$_3$(VO$_4$)$_2$	0.4	4.3	100.0
Co$_3$(VO$_4$)$_2$	2.4	2.1	100.0
Cu$_3$(VO$_4$)$_2$	1.0	6.2	100.0
FeVO$_4$	2.0	4.0	100.0
Zn$_3$(VO$_4$)$_2$	4.3	0.2	100.0

[a] TOF based on selective oxidation products (formaldehyde, methyl formate, and dimethoxy methane) at 300°C.

[b] Selectivity toward methanol partial oxidation products (formaldehyde, methyl formate, and dimethoxy methane) at 300°C.

Source: From Briand, L.E., Jehng, J.-M., Cornaglia, L.M., Hirt, A.M., and Wachs, I.E. *Catal. Today* 2003, 78, 257–268. With permission.

which give more evidence that the metal cations are not exposed on the surface of bulk metal vanadates.

Studies of thermal and reaction induced spreading of MoO_3 and V_2O_5 over metal oxides demonstrated that molybdenum and vanadium spontaneously migrate at 400 to 500°C and react with methanol forming a mobile metal–methoxy complex at 200 to 250°C [57]. This property of molybdenum and vanadium oxide species suggests that the surface enrichment of bulk metal molybdates and vanadates is produced during calcination at 500°C and increases due to the reaction environment (6% methanol/oxygen–helium at 380°C). The thermal treatment of the precursors and the nature of the reactant play an important role in the formation of the catalyst surface. The surface enrichment of bulk metal molybdates and vanadates was further confirmed through XPS and more recently, with *in situ* ion scattering spectroscopy LEISS [58].

11.4.4 Monolayer Supported versus Bulk Oxide Catalysts: Which is More Active?

The catalytic activity of bulk and supported catalysts has been a matter of controversy in the catalytic field. The higher catalytic activity of supported than bulk

TABLE 11.4

TOF and Selectivity of Pure Metal Oxide Catalysts Toward Methanol Oxidation at Low Conversion

Catalyst (reaction temperature, °C)	Redox TOF[a] (sec^{-1})	Selectivity (%)			
		FA	CO$_2$	DME	Others[b]
MgO (300)	0.02	60.0	40.0	—	—
Nb$_2$O$_5$ (300)	0.0	—	—	100.0	—
Cr$_2$O$_3$ (290)	7.1	35.7	59.6	—	4.7
MnO (300)	31.0	79.5	20.5	—	—
Al$_2$O$_3$ (300)	0.0	—	—	100.0	—
Ag$_2$O (300)	359.0	100.0	—	—	—
NiO (300)	6.4	82.6	17.4	—	—
CoO (270)	24.7	61.2	34.2	—	4.7
CuO (330)	15.9	100.0	—	—	—
Fe$_2$O$_3$ (300)	3.3	57.9	—	36.4	5.7
ZnO (380)	18.2	32.7	31.5	—	35.7
V$_2$O$_5$ (300)	9.8	79.0	—	10.5	10.4

[a] TOF based on selective oxidation products (formaldehyde, methyl formate, and dimethoxy methane).

[b] Other products are methyl formate and dimethoxy methane.

Source: From Briand, L.E., Jehng, J.-M., Cornaglia, L.M., Hirt, A.M., and Wachs, I.E. Catal. Today 2003, 78, 257–268. With permission.

catalysts has been repeatedly reported in the literature. However, this conclusion was made based on the conversion of a reactant toward products instead of the specific activities due to the uncertainty regarding the number of surface active sites of bulk catalysts [59–61].

The investigation of Briand et al. [52,53] on the methanol TOFs of bulk metal molybdate and vanadate catalysts overcomes this problem since methanol chemisorption enables the quantification of surface active sites and the calculation of the activity per surface active site. In order to compare the catalytic activity of bulk metal molybdates and monolayer molybdenum oxide supported catalysts, the reaction rate and TOF values of supported catalysts were extrapolated to 380°C considering that the activation energy for methanol reaction over supported molybdenum oxide catalysts is typically 20 kcal/mol. Figure 11.9 compares the TOFs toward selective oxidation products of monolayer molybdenum oxide supported catalysts and bulk metal molybdates with the same components [Zr(MoO$_4$)$_2$ and 3% MoO$_3$/ZrO$_2$, MnMoO$_4$ and 2% MoO$_3$/MnO, NiMoO$_4$ and 4% MoO$_3$/NiO, Al$_2$(MoO$_4$)$_3$, and 18% MoO$_3$/Al$_2$O$_3$, and Cr$_2$(MoO$_4$)$_3$ and 3% MoO$_3$/Cr$_2$O$_3$]. The surprising similarity of the TOF of bulk and supported molybdenum oxide-based catalysts gives further proof that the surface composition of these systems

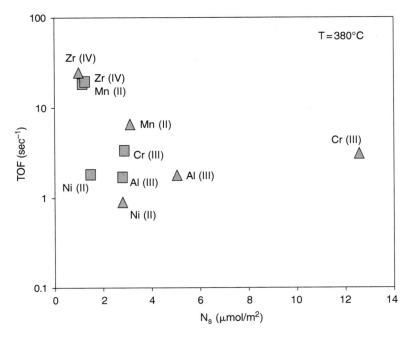

FIGURE 11.9 Comparison of the TOFs toward methanol selective oxidation products of monolayer supported molybdenum (square symbols) and bulk metal molybdate (triangles) catalysts at 380°C (From Briand, L.E., Hirt, A.M., and Wachs, I.E. *J. Catal.* 2001, *202*, 268–278. With permission.)

is alike although it is not possible to determine the surface molybdenum oxide structures.

Similarly, Figure 11.10 compares the TOF values of bulk metal vanadates with the corresponding supported vanadium oxide monolayer catalysts. The TOF values of $Ni_3(VO_4)_2$ and 1% V_2O_5/NiO, $CrVO_4$ and 5% V_2O_5/Cr_2O_3, $Mn_3(VO_4)_2$ and 1% V_2O_5/MnO, $FeVO_4$ and 4% V_2O_5/Fe_2O_3, $Co_3(VO_4)_2$ and 3% V_2O_5/Co_3O_4, and $AlVO_4$ and 20% V_2O_5/Al_2O_3 reveal that bulk metal vanadates and supported vanadium oxide monolayer catalysts are rather similar. This behavior is in agreement with the methanol oxidation selectivity results presented earlier, that both catalytic systems are composed of surface vanadium oxide species.

The TOF values of bulk metal molybdates were extrapolated to 300°C in order to compare their values with the corresponding bulk metal vanadate catalysts. Figure 11.11 shows that, in general, bulk metal vanadates possess one order of magnitude higher TOF values ($\sim$2 to 14 sec^{-1}) than their corresponding bulk metal molybdates ($\sim$0.1 sec^{-1}) for methanol selective oxidation. The TOF values of pure V_2O_5 (9.8 sec^{-1}) and MoO_3 (0.6 sec^{-1}) crystals are also presented to demonstrate that the difference in the activity of bulk metal molybdates and vanadates is based on the nature of the surface species (in this case, VO_x versus MoO_x). The results

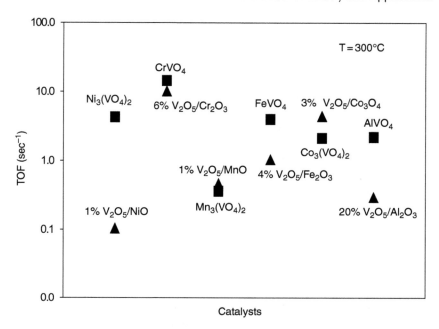

FIGURE 11.10 Comparison of the TOFs towards methanol selective oxidation products of monolayer supported vanadium (▲ symbols) and bulk metal vanadate catalysts (■) at 300°C (From Briand, L.E., Jehng, J.-M., Cornaglia, L.M., Hirt, A.M., and Wachs, I.E. *Catal. Today* 2003, *78*, 257–268. With permission.)

clearly show that surface vanadium oxide sites are more active than surface molybdenum oxide sites regardless of the bulk structure, bulk coordination, or density of surface active sites.

11.5 ISOPROPANOL AS A PROBE MOLECULE: REDOX VERSUS ACID PROPERTIES OF OXIDE CATALYSTS

The previous sections discussed the applications of methanol chemisorption as a tool to describe the surfaces of oxide catalysts. Other molecules, such as isopropanol and formic acid, have also been used as chemical probes to measure the number of surface active sites of metal oxide catalysts but to a lesser extent. The ability of isopropanol to distinguish between surface redox and acid sites and the observation that it adsorbs as a stable monolayer of surface isopropoxy species over oxide materials, also makes this molecule a suitable surface chemical probe.

As was discussed previously, strong Brønsted (H^+) and Lewis acid sites catalyze the dehydration of isopropanol to propylene (di-isopropyl ether for weak Lewis acid sites) and redox/basic sites lead to the dehydrogenation of the alcohol

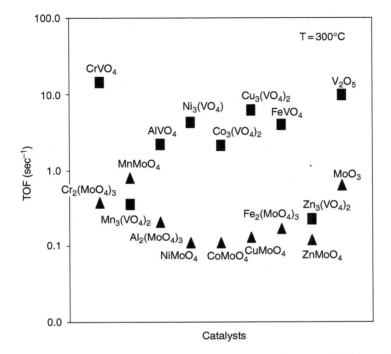

FIGURE 11.11 Comparison of the TOFs toward methanol selective oxidation products of V_2O_5, MoO_3, bulk metal vanadates, and molybdates at 300°C (From Briand, L.E., Jehng, J.-M., Cornaglia, L.M., Hirt, A.M., and Wachs, I.E. *Catal. Today* 2003, *78*, 257–268. With permission.)

to acetone. The process of isopropanol dehydration to propylene involves the adsorption of the alcohol through the OH group and a β-hydrogen, followed by the abstraction of the OH group to produce an intermediate carbocation species and the abstraction of the β-hydrogen to yield the corresponding olefin [21,33,62,63].

11.5.1 Isopropanol Adsorption over One-Component Bulk Metal Oxides

Analogous to previous work on methanol chemisorption, Kulkarni et al. [64] used isopropanol to determine the number of surface active sites (N_s) and TOFs over a series of bulk metal oxides. Most of the oxides possessed an active surface sites density of ∼2 to 4 μmol/m^2 although precious metal oxides had a higher N_s value (∼5.5 μmol/m^2). The knowledge of the N_s and the reaction rates toward redox (acetone) and acid (propylene and di-isopropyl ether) allowed for determination the redox and acid TOFs. The TOF for acidic oxides (TiO_2, ZrO_2, HfO_2, Nb_2O_5, Ta_2O_5, WO_3, Al_2O_3, Ga_2O_3, etc.) varied by eight-orders of magnitude (from 10^1 to 10^{-7} sec^{-1}) and TOFs for redox catalysts (MgO, CaO, BaO,Y$_2$O$_3$, La$_2$O$_3$, TiO$_2$, CeO$_2$, V$_2$O$_5$, Mn$_2$O$_3$, etc.) varied by six-orders of magnitude (from 10^2 to 10^{-4} sec^{-1}).

As for methanol selective oxidation, the authors did not observe a correlation between the TOF and bulk properties of metal oxides such as enthalpy of formation or temperature of reduction. However, the redox TOF showed a strong inverse trend with the temperature of surface decomposition of adsorbed isopropoxy species (Figure 11.12).

Interestingly, most of the catalysts exhibit 100% selectivity to either redox or acid products regardless of the TOF, even though this parameter varied by many orders of magnitude as discussed earlier (Figure 11.13). This observation dismissed the belief, frequently stated in the literature, that extremely active metal oxides lead to over-oxidation.

Fein et al. [65] also investigated the TOFs of bulk metal oxides toward formic acid oxidation through the dissociative chemisorption of the HCOOH to surface formate species HCOO–M. The authors obtained similar structure–activity relationships as observed for methanol and isopropanol.

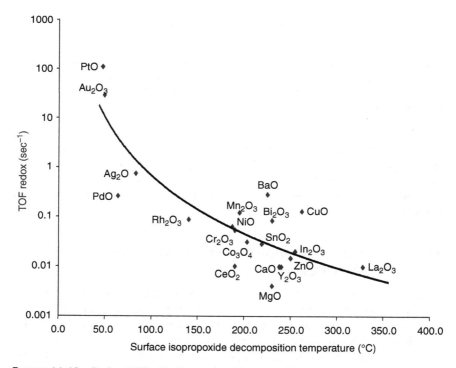

FIGURE 11.12 Redox TOF of bulk metal oxides toward isopropanol selective oxidation as a function of the temperature of surface isopropoxy decomposition (From Kulkarni, D. and Wachs, I.E. *Appl. Catal.* 2002, *237*, 121–137. With permission.)

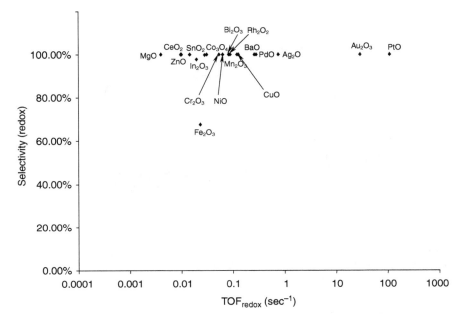

FIGURE 11.13 Redox selectivity of bulk metal oxides as a function of the redox TOF toward isopropanol selective oxidation (From Kulkarni, D. and Wachs, I.E. *Appl. Catal.* 2002, *237*, 121–137. With permission.)

11.5.2 Super-Acid Properties of Complex Heteropoly-Oxo Metallates

Briand and coworkers [48] used *in situ* isopropanol chemisorption and TPSR to determine the nature, amount, and acid strength of the active sites of bulk WO_3, monolayer supported tungsten oxide over titania, and phospho-tungstic Wells-Dawson and Keggin type HPAs. Heteropoly compounds are highly reactive materials and, therefore, only isopropanol chemisorption at room temperature leads to adsorbed isopropoxy species and avoids surface reaction and molecular adsorption. These species dehydrate toward propylene (no redox products were observed) over the investigated catalysts. The amount of active sites (N_s) was determined as the amount of propylene desorbed during the temperature programmed reaction analysis (see Figure 11.14 for a typical TPSR analysis).

Table 11.5 shows the number of surface active sites and the temperature of reaction of adsorbed isopropoxy species over dehydrated bulk tungsten oxide WO_3, monolayer supported tungsten oxide over titania, and phospho-tungstic Keggin type acid $H_3PW_{12}O_{40}$. The number of available surface sites for isopropanol adsorption of bulk tungsten trioxide and monolayer supported tungsten oxide catalysts is orders of magnitude (0.9 μmol/m^2) lower than the HPAs (8 to 19 μmol/m^2). In the particular case of WO_3 and monolayer supported tungsten oxide catalysts, the adsorption of alcohol occurs exclusively at the surface. In contrast to bulk WO_3

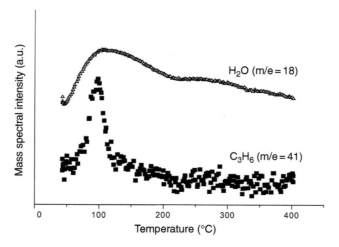

FIGURE 11.14 TPSR of isopropanol adsorbed over phospho-tungstic heteropoly acid $H_6P_2W_{18}O_{62}$ at $40°C$ (From Gambaro, L.A. and Briand, L.E. *Appl. Catal.* 2004, *264*, 151–159. With permission.)

TABLE 11.5

Maximum Number of Active Sites for Isopropanol Adsorption of WO_3, Monolayer Supported Tungsten Oxide on TiO_2, Phospho-Tungstic Keggin and Wells-Dawson Acids

Catalyst	S_{BET} (m²/g)	Temperature of adsorption (°C)	N_s (μmol/m²)	Temperature of isopropoxy reaction (°C)
WO_3	1.7	110	0.9	156
3% WO_3/TiO_2	37.3	70	0.9	140
$H_6P_2W_{18}O_{62}$	2.8	40	8.1	96
$H_3PW_{12}O_{40}$	1.9	40	19.8	112

Source: Gambaro, L.A. and Briand, L.E. *Appl. Catal.* 2004, *264*, 151–159. With permission.

---

and supported tungsten oxide species, the structure of the HPAs allows the adsorption of alcohol on the external surface as well as the bulk of the solid. This unique property known as "pseudo-liquid phase behavior" is attributed to the wide distance between the Wells-Dawson units $P_2W_{18}O_{62}^{6-}$ in the crystal structure that allows the adsorption of polar molecules inside the solid bulk structure [31,32]. This property explains the higher N_s of the HPAs rather than conventional catalysts where all the active sites are on the surface. Briand et al. attributed the higher temperature of surface reaction over bulk tungsten trioxide and monolayer supported tungsten oxide catalysts than HPAs, to the differences in the acidity of the active sites.

A central phosphorous atom PO_4 surrounded by a cage of tungsten atoms compose the structure of Wells–Dawson and Keggin type HPAs. Similar to monolayer supported tungsten oxide catalysts, each tungsten atom is composed of WO_6 octahedral units with one terminal double bonded oxygen and are linked together through W–O–W bonds. However, many studies demonstrated that solid HPAs are Brønsted acids and are stronger than conventional solid acids such as SiO_2–Al_2O_3, sulfated zirconia SO_4^{2-}/ZrO_2, titania SO_4^{2-}/TiO_2, etc. [31,32].

The lower temperature of surface reaction of isopropoxy species over dehydrated Wells-Dawson (96°C) than dehydrated Keggin (112°C) HPA is a measurement of the higher acid strength of the phospho-tungstic Wells–Dawson acid in the anhydrous form.

Interestingly, this investigation allowed distinguishing between the number of active and adsorption acid sites measuring *in situ* the amount of propylene produced during the adsorption and the amount of propylene during TPSR analysis. This application gives similar information that the *in situ* IR reactor cell developed by Burcham et al. that was discussed previously. The authors observed that the quantity of propylene produced during the adsorption is proportional to the amount of active acid sites. Moreover, the quantity of isopropanol adsorbed without reaction is desorbed as propylene during the TPSR experiment. This amount is proportional to the number of adsorption sites that are occupied by intermediate alkoxy species at a certain temperature. The addition of the number of active and adsorption sites is the maximum amount of sites available for isopropanol adsorption/reaction of the Wells-Dawson HPA. Table 11.6 shows the number of acid sites (N_s), the fraction of active sites, and the fraction of sites covered with a stable layer of intermediate species at different temperatures. The maximum amount of adsorbed isopropoxy species is achieved through isopropanol adsorption at 40°C that avoids further reaction toward propylene. Under these conditions, the maximum number of active acid sites is suitable to be determined as the amount of propylene desorbed during temperature programmed reaction experiments.

11.5.2.1 The influence of the degree of hydration on the amount of accessible active acid sites of complex Wells–Dawson structures

Briand and coworkers [48] also addressed the role of the degree of hydration on the catalytic activity of HPAs, which is a matter of controversy in the literature. In fact, the nature of the acid sites is greatly modified upon dehydration although the primary structure remains unaltered up to 600°C. The fully hydrated acid possesses large protonated water clusters $H^+(H_2O)_n$, which degrade to $H_5O_2^+$ at 200°C and finally to isolated protonic sites when the acid is completely dehydrated.

According to previous theoretical studies, the isolated acid sites trapped between Wells-Dawson units are inaccessible to the reagents causing the drop of the catalytic activity of the acid. The deactivation of the acid upon full dehydration

TABLE 11.6
Maximum Number of Active Sites and Adsorption Sites for Isopropanol Reaction.

Temperature (°C)	Ns_1 [a] ($\mu mol/m^2$)	Ns_2 [b] ($\mu mol/m^2$)	θ^c_{active}	$\theta^d_{adsorption}$
40	8.1	8.0	0.0	1.0
70	7.5	2.3	0.7	0.3
100	6.5	0.0	1.0	0.0

Fraction of active sites and adsorption sites at different temperatures of dehydrated phospho-tungstic Wells-Dawson acid $H_6P_2W_{18}O_{62}$.

[a] Maximum number of active sites for adsorption/reaction of isopropanol calculated from the addition of the amount of propylene produced during chemisorption and TPSR experiments.

[b] Number of adsorption sites calculated from propylene desorption during TPSR.

[c] Fraction of active sites $= (Ns_1 - Ns_2)/Ns_1$.

[d] Fraction of adsorption sites $= Ns_2/Ns_1$.

Source: From Gambaro, L.A. and Briand, L.E. *Appl. Catal.* 2004, 264, 151–159. With permission.

was observed during methanol reaction and the synthesis of MTBE in the gas phase at temperatures above 200°C [30,66].

Isopropanol adsorption (at 40°C) and TPSR analysis was performed on the fully hydrated acid and after *in situ* calcination at 70, 115, and 320°C in order to obtain more insights on the role of the degree of hydration on the catalytic activity of the phospho-tungstic Wells-Dawson type acid.

The TPSR analysis, after saturation of the fully hydrated acid with successive pulses of isopropanol, showed the desorption of molecular alcohol at 77°C, propylene at 107°C, CO with three maxima at ~120, 177, and 334°C, and water. The analysis after dehydration at 70, 115, and 320°C is similar although no CO was detected. The desorption of molecular isopropanol only on the hydrated acid (no molecular alcohol desorbs upon complete dehydration at 400°C) indicated that the alcohol dissolves in an aqueous layer surrounding the primary structure of the $P_2W_{18}O_{62}^{6-}$ heteropoly anion.

The production of CO during alcohols desorption experiments is attributed to the readsorption and further oxidation of primary products (e.g., the readsorption and further oxidation of formaldehyde to CO_x during methanol chemisorption and TPSR over oxide catalysts). Similarly, the production of CO could be attributed to the oxidation of propylene that is unable to desorb out of the surface due to the presence of an aqueous layer.

Table 11.7 presents the amount of isopropanol molecularly absorbed and propylene produced as a function of the degree of hydration of the Wells-Dawson acid. The results give more evidences that the hydration water influences both molecular absorption of the alcohol and propylene production.

TABLE 11.7

Calcination Temperatures, Degree of Hydration, Amount of Methanol Molecularly Absorbed, Amount of Propylene produced during TPSR Analysis and Temperature of Surface Reaction of Isopropoxy Species towards Propylene.

Calcination temperature (°C)	Molecular formula	Degree of hydration[a]	Alcohol absorbed[b] (μmol/g)	Propylene[c] (μmol/g)	Temperature of isopropoxy reaction (°C)
No calcination	$H_6P_2W_{18}O_{62} \cdot 24H_2O$	100%	13.2	51.8	108
70	$H_6P_2W_{18}O_{62} \cdot 7H_2O$	29%	8.7	50.0	94
115	$H_6P_2W_{18}O_{62} \cdot 2H_2O$	8%	10.3	10.9	105
320	$H_6P_2W_{18}O_{62}$	Dehydrated	9.0	25.4	96
400	$H_6P_2W_{18}O_{62}$	Dehydrated	0.0	24.7	96

[a] The degree of hydration was calculated considering that the fully hydrated acid contains 24 molecules of water (100%) and remains 7 and 2 water molecules upon calcination at 70 and 115°C, respectively. Complete dehydration is achieved upon calcination at 400°C.

[b] Amount of molecular isopropanol desorbed during TPSR analysis per unit weight of Wells-Dawson acid with the corresponding degree of hydration ($H_6P_2W_{18}O_{62} \cdot xH_2O$).

[c] Amount of propylene desorbed during TPSR analysis per unit weight of Wells-Dawson acid with the corresponding degree of hydration ($H_6P_2W_{18}O_{62} \cdot xH_2O$).

Source: From Gambaro, L.A. and Briand, L.E. Appl. Catal. 2004, 264, 151–159. With permission.

The amount of intermediate isopropoxy species toward propylene is directly influenced by the hydration water since fully hydrated Wells-Dawson acid produces twice the amount of propylene (52 μmol/g) than the anhydrous acid (25 μmol/g). Moreover, a continuous drop of the amount of propylene is observed with the decrease in the degree of hydration of the Wells-Dawson structure.

This observation evidences that the loss of water leads to the shortening of the distance between the Wells-Dawson units and the decrease of the available active sites for isopropanol chemisorption.

The observation that the temperature of surface reaction of isopropoxy species is ~96°C regardless of the degree of hydration of the Wells-Dawson acid shows that, in general, the nature of the acid Lewis sites (most numerous) are not modify by water. This statement cannot be applied to Brønsted acid sites (protonic sites), whose nature is modified by the degree of hydration according to previous investigations [66].

11.6 FINAL REMARKS: THE FUTURE ON SURFACE SCIENCE THROUGH MOLECULAR PROBES

The investigations discussed earlier clearly demonstrate that the reactant itself should be used as a probe molecule in order to obtain reliable surface structure–catalytic activity correlations. A controlled chemisorption of the reactant to cover the surface with intermediates species to products is the key for the determination of the number of surface active sites. TPSR, which allows to identify and quantify the products, therefore, is unique to determine the nature and number of the active sites. Moreover, the analysis would be useful to determine the density of redox, acid, or basic sites individually, through the quantification of the products selectively catalyzed by each type of active site.

The catalytic activity per surface active site (TOF) toward a specific reaction is the right parameter to obtain reliable surface structure–activity correlations. The knowledge of the TOFs values dismissed the believes that the catalytic activity is influenced by bulk properties and that monolayer supported oxide catalysts are more active than bulk oxide catalysts. There is no doubt that the specific activity would contribute to design more active and selective catalytic materials at a molecular level.

ACKNOWLEDGMENTS

The author acknowledges the Consejo Nacional de Investigaciones Científicas y Técnicas CONICET (Argentina) and the National Science Foundation (NSF, USA) for the international scientific collaboration program (Res. No 0060). To CONICET for the grant PEI No 6132 for young scientists and the Agencia Nacional de Promoción Científica y Tecnológica (SECyT-Argentina) for the grant PICT 14-12161.

REFERENCES

1. Niemantsverdriet, J.W. *Spectroscopy in Catalysis: An Introduction*. VCH Publishers, Germany, 1995, pp. 37–55.
2. (a) Briand, L.E., Tkachenko, O.P., Guraya, M., Wachs, I.E., and Grünert, W. Methodological aspects in the surface analysis of supported molybdena catalysts. *Surf. Interface Anal.* 2004, *36*, 238–245. (b) Briand, L.E., Tkachenko, O.P., Guraya, M., Gao, X., Wachs, I.E., and Grünert, W. Surface analytical studies of supported vanadium oxide monolayer catalysts. *J. Phys. Chem. B* 2004, *108*, 4823–4830.
3. Clark, H. and Berets, D.J. *Advances in Catalysis*. Academic Press, New York, 1957, Vol. IX, p. 204.
4. Dyrek, K. Chemisorption of oxygen on partially reduced V_2O_5 studied by EPR. *Bull. Acad. Pol. Sci.* 1974, *22*, 605–612.
5. Parekh, B.S. and Weller, S.W. Specific surface area of molybdena in reduced supported catalysts. *J. Catal.* 1977, *47*, 100–108.
6. Chary, K.V.R., Rama Rao, B., and Subrahmanyam, V.S. Characterization of supported vanadium oxide catalysts by a low-temperature oxygen chemisorption technique. III. The V_2O_5/ZrO_2 system. *Appl. Catal.* 1991, *74*, 1–13.
7. Nag, N.K., Prasada Rao, K.S., Chary, K.V.R., Rama Rao, B., and Subrahmanyam, V.S. Characterization of γ-alumina supported tungsten sulfide hydroprocessing catalysts. I. Low-temperature oxygen chemisorption. *Appl. Catal.* 1988, *41*, 165–176.

8. Nag, N.K., Chary, K.V.R., Rama Rao, B., and Subrahmanyam, V.S. Characterization of supported vanadium oxide catalysts by low temperature oxygen chemisorption. II. The V_2O_5/SiO_2 system. *Appl. Catal.* 1987, *31*, 73–85.
9. Wachs, I.E., Jehng, J-M., and Ueda, W. Determination of the chemical nature of active surface sites present on bulk mixed metal oxide catalysts. *J. Phys. Chem. B*, 2005, *109*, 2275–2284.
10. Miyamoto, A., Yamazaki, Y., Inomata, M., and Murakami, Y. Determination of the number of V=O species on the surface of vanadium oxide catalysts. 1. Unsupported V_2O_5 and V_2O_5/TiO_2 treated with an ammoniacal solution. *J. Phys. Chem.* 1981, *85*, 2366–2372.
11. Inomata, M., Miyamoto, A., and Murakami, Y. Determination of the number of V=O species on the surface of vanadium oxide catalysts. 2. V_2O_5/TiO_2 catalysts. *J. Phys. Chem.* 1981, *85*, 2372–2377.
12. Topsøe, N.-Y., Topsøe, H., and Dumesic, J.A. Vanadia/titania catalysts for selective catalytic reduction (SCR) of nitric oxide by ammonia. I. Combined temperature programmed *in situ* FTIR and on-line mass spectroscopy studies. *J. Catal.* 1995, *151*, 226–240.
13. Topsøe, N.-Y., Dumesic, J.A., and Topsøe, H. Vanadia/titania catalysts for selective catalytic reduction of nitric oxide by ammonia. II. Studies of active sites and formulation of catalytic cycles. *J. Catal.* 1995, *151*, 241–252.
14. Cheng, W.-H., Chowdhry, U., Ferretti, A., Firment, L.E., Groff, R.P., Machiels, C.J., McCarron, E.M., Ohuchi, F., Staley, R.H., and Sleight, A.W. Heterogeneous catalysis. In *Proceedings of the 2nd Symposium of IUCCP of Department Chemistry, Texas A&M Univeristy, 1984*, Shapiro, B.L., Ed. Texas A&M University Press, College Station, TX, 1984, pp. 165–181.
15. Machiels, C.J., Cheng, W.H., Chowdhry, U., Farneth, W.E., Hong, F., McCarron, M.E., and Sleight, A.W. The effect of the structure of molybdenum oxides on the selective oxidation of methanol. *Appl. Catal.* 1986, *25*, 249–256.
16. Farneth, W.E., Staley, R.H., and Sleight, A.W. Stoichiometry and structural effects in alcohol chemisorption/temperature-programmed desorption on MoO_3. *J. Am. Chem. Soc.* 1986, *108*, 2327–2332.
17. Holstein, W.L. and Machiels, C.J. Inhibition of methanol oxidation by water vapor-effect on measured kinetics and relevance to the mechanism. *J. Catal.* 1996, *162*, 118–124.
18. Adkins, H. and Peterson, W.R. The oxidation of methanol with air over iron, molybdenum, and iron-molybdenum oxides. *J. Am. Chem. Soc.* 1931, *53*, 1512–1518.
19. Machiels, C.J. and Sleight, A.W. In *Proceedings of the 4th International Conference on the Chemistry and Uses of Molybdenum*, Ann Arbor, MI; Barry, H.F. and Mitchell, P.C.H., Eds. Climax Molybdenum Co., 1982, p. 411.
20. Radhakrishnan, R. and Oyama, S.T. Reactant-probe method for estimating active site number in catalysis. *J. Catal.* 2001, *204*, 516–519.
21. Busca, G. Spectroscopic characterization of the acid properties of metal oxide catalysts. *Catal. Today* 1998, *41*, 191–206.
22. Dines, T.J., MacGregor, L.D., and Rochester, C.H. IR spectroscopy of *N*-methylpyrrole adsorbed on oxides. A probe of surface acidity. *J. Colloid Interface Sci.* 2002, *245*, 221–229.
23. Dines, T.J., MacGregor, L.D., and Rochester, C.H. Adsorption of 2-chloropyridine on oxides-an infrared spectroscopic study. *Spectrosc. Acta Part A* 2003, *59*, 3205–3217.

24. Dines, T.J., MacGregor, L.D., and Rochester, C.H. The surface acidity of oxides probed by IR spectroscopy of adsorbed diazines. *Phys. Chem. Chem. Phys.* 2001, *3*, 2676–2685.

25. Dines, T.J., MacGregor, L.D., and Rochester, C.H. IR spectroscopic investigation of the interaction of quinoline with acid sites on oxide surfaces. *Langmuir* 2002, *18*, 2300–2308.

26. Glinka, Y.D., Jaroniec, C.P., and Jaroniec, M. Studies of surface properties of disperse silica and alumina by luminescence measurements and nitrogen adsorption. *J. Colloid Interface Sci.* 1998, *201*, 210–219.

27. Hiemstra, T. and Van Riemsdijk, W.H. Fluoride adsorption on goethite in relation to different types of surface sites. *J. Colloid Interface Sci.* 2000, *225*, 94–104.

28. Lavalley, J.C. Infrared spectrometric studies of the surface basicity of metal oxides and zeolites using probe molecules. *Catal. Today* 1996, *27*, 377–401.

29. Pope, M.T. *Heteropoly and Isopoly Oxometalates*. Springer-Verlag, Berlin, 1983, pp. 58–90.

30. Briand, L.E., Baronetti, G.T., and Thomas, H.J. The state of the art on Wells–Dawson heteropoly-compounds. A review of their properties and applications. *Appl. Catal. A: Gen.* 2003, *256*, 37–50 and other papers included in this special issue dedicated to heteropoly-oxo compounds.

31. Okuhara, T., Mizuno, N., and Misono, M. Catalytic chemistry of heteropoly compounds. In *Advances in Catalysis*, Eley, D.D., Haag, W.O., and Gates, B., Eds. Academic Press Inc., New York, 1996, Vol. 41, pp. 113–252.

32. Misono, M. Unique acid catalysts of heteropoly compounds (heteropolyoxometalates) in the solid state. *Chem. Commun.* 2001, *14*, 1141–1152.

33. Highfiled, J.G. and Moffat, J.B. Characterization of sorbed intermediates and implications for the mechanism of chain growth in the conversion of methanol and ethanol to hydrocarbons over 12-tungstophosphoric acid using infrared photoacoustic spectroscopy. *J. Catal.* 1986, *98*, 245–258.

34. Bielański, A., Datka, J., Gil, B., Malecka-Lubańska, A., and Micek-Ilnicka, A. Sorption of methanol on tungstosilicic acid. *Phys. Chem. Chem. Phys.* 1999, *1*, 2355–2360.

35. Bielański, A., Malecka-Lubańska, A., Micek-Ilnicka, A., and Poźniczek, J. The role of protons in acid base type reactions on heteropolyacid catalysts: Gas phase MTBE synthesis on $H_4SiW_{12}O_{40}$. *Top. Catal.* 2000, *11–12*, 43–53.

36. Croff, R.P. An infrared study of methanol and ammonia adsorption on molybdenum trioxide. *J. Catal.* 1984, *86*, 215–218.

37. Burcham, L.J. The Origin of the Ligand Effect in Supported and Bulk Metal Oxide Catalysts: In Situ Infrared, Raman, and Kinetic Studies During Methanol Oxidation. Ph.D. thesis, Lehigh University, Bethlehem, PA, 1999.

38. Briand, L.E., Farneth, W.E., and Wachs, I.E. Quantitative determination of the number of active sites and turnover frequencies for methanol oxidation over metal oxide catalysts. I. Fundamentals of the methanol chemisorption technique and application to monolayer supported molybdenum oxide catalysts. *Catal. Today* 2000, *62*, 219–229.

39. (a) Burcham, L.J., Briand, L.E., and Wachs, I.E. Quantification of active sites for the determination of methanol oxidation turn-over frequencies using methanol chemisorption and in situ infrared techniques. I. Supported metal oxide catalysts. *Langmuir* 2001, *17*, 6164–6174. (b) Burcham, L.J., Briand, L.E., and Wachs, I.E.

Quantification of active sites for the determination of methanol oxidation turnover frequencies using methanol chemisorption and in situ infrared techniques. 2. Balk metal oxide 43 catalysts. *Langmuir* 2001, *17*, 6175–6184.

40. Burcham, L.J. and Wachs, I.E. The origin of the support effect in supported metal metal oxides catalysts: in situ infrared and kinetic studies during methanol oxidation. *Catal. Today* 1999, *49*, 467–484.

41. Farneth, W.E., Ohuchi, F., Staley, R.H., Chowdhry, U., and Sleight, A.W. Mechanism of partial oxidation of over molybdenum (VI) oxide as studied by temperature programmed desorption. *J. Phys. Chem.* 1985, *89*, 2493–2497.

42. (a) Badri, A., Binet, C., and Lavalley, J.-C. Use of methanol as an IR molecular probe to study the surface of polycrystalline ceria. *J. Chem. Soc., Faraday Trans.* 1997, *93*, 1159–1168. (b) Binet, C., Daturi, M., and Lavalley, J.-C. IR study of polycrystalline ceria in oxidized and reduced states. *Catal. Today* 1999, *50*, 207–225.

43. Spitz, R.N., Barton, J.E., Barteau, M.A., Staley, R.H., and Sleight, A.W. Characterization of the surface acid-base properties of metal oxides by titration/ displacement reactions. *J. Phys. Chem.* 1986, *90*, 4067–4075.

44. (a) Borello, E., Zecchina, A., and Morterra, C. Infrared study of methanol adsoprtion on aerosil. I. Chemisorption at room temperature. *J. Phys. Chem.* 1967, *71*, 2938–2945. (b) Borello, E., Zecchina, A., Morterra, C., and Ghiotti, G. Infrared study of methanol adsoprtion on aerosil. II. Physical adsorption at room temperature. *J. Phys. Chem.* 1967, *71*, 2945–2951.

45. (a) Ouyang, F., Kondo, J.N., Maruya, K., and Domen, K. Site conversion of methoxy species on ZrO_2. *J. Phys. Chem. B* 1997, *101*, 4867–4869. (b) Ouyang, F., Kondo, J.N., Maruya, K., and Domen, K. IR study on migration of $^{18}OCH_3$ species on ZrO_2. *Catal. Lett.* 1998, *50*, 179–181.

46. Burcham, L.J., Badlani, M., and Wachs, I.E. The origin of the ligand effect in metal oxide catalysts: novel fixed-bed in situ infrared and kinetic studies during methanol oxidation. *J. Catal.* 2001, *203*, 104–121.

47. Bañares, M.A. and Wachs, I.E. Molecular structures of suported metal oxide catalysts under different environments. *J. Raman Spectrosc.* 2002, *33*, 359–380.

48. Gambaro, L.A. and Briand, L.E. *In-situ* quantification of the active acid sites of $H_6P_2W_{18}O_{62}.nH_2O$ heteropoly-acid through chemisorption and temperature programmed surface reaction of isopropanol. *Appl. Catal.* 2004, *264*, 151–159.

49. Badlani, M. Methanol: A Smart Chemical Probe Molecule. Master thesis, Lehigh University, Bethlehem, PA, 2000.

50. Badlani, M. and Wachs, I.E. Methanol a smart chemical probe molecule. *Catal. Lett.* 2001, *75*, 137–149.

51. Briand, L.E. and Wachs, I.E. Quantitative determination of the number of active surface sites and the turnover frequencies for methanol oxidation over metal oxide catalysts. In *Studies in Surface Science and Catalysis*. Elsevier, Amsterdam, 2000, Vol. 103, pp. 305–310.

52. Briand, L.E., Hirt, A.M., and Wachs, I.E. Quantitative determination of the number of active sites and turnover frequencies for methanol oxidation over metal oxide catalysts. *J. Catal.* 2001, *202*, 268–278.

53. Briand, L.E., Jehng, J.-M., Cornaglia, L.M., Hirt, A.M., and Wachs, I.E. Quantitative determination of the number of surface active sites and the turnover frequencies for metanol oxidation over bulk metal vanadates. *Catal. Today* 2003, *78*, 257–268.

54. Wachs, I.E. and Briand, L.E. *In situ* Formation of Iron-Molybdate Catalysts for Methanol Oxidation to Formaldehyde. U.S. patent No. 6,331,503 B1, December 18, 2001.

55. Wachs, I.E. and Briand, L.E. Metal Molybdate/Iron Molybdate Dual Catalysts Bed System and Process Using The Same for Methanol Oxidation to Formaldehyde. U.S. patent No. 6,518,463, February 11, 2002.

56. Wachs, I.E. and Briand, L.E. In situ Formation of Metal-Molybdate Catalysts for Methanol Oxidation to Formaldehyde. U.S. patent No. 6,624,332 B2, September 23, 2003.

57. Wang, C.-B., Cai, Y., and Wachs, I.E. Reaction-induced spreading of metal oxides onto surfaces of oxide supports during alcohol oxidation: phenomenon, nature, and mechanism. *Langmuir* 1999, *15*, 1223–1235.

58. Gruenent, W., Briand, L., Tkachenko, O. P., Tolkatchev, N.N. and Wachs, I.E. Surface composition of supported and bulk mixed metal oxide catalyst: New insight from ion scattering spectroscopic studies. Proceedings of the American Chemical Society 228th National Meeting, Philadelphia, PA (August 22-26, 2004).

59. Yang, T.-J. and Lunsford, J.H. Partial oxidation of methanol to formaldehyde over molybdenum oxide on silica. *J. Catal.* 1987, *103*, 55–64.

60. Klissurski, D., Pesheva, Y., Abadjieva, N., Mitov, I., Filkova, D., and Petrov, L. Multicomponent oxide catalysts for the oxidation of methanol to formaldehyde. *Appl. Catal.* 1991, *77*, 55–66.

61. Niwa, M., Yamada, H., and Murakami, Y. Activity for the oxidation of methanol of a molibdena monolayer, supported on tin oxide. *J. Catal.* 1992, *134*, 331–339.

62. Gervasini, A., Fenyvesi, J., and Auroux, A. Study of acidic character of of modified metal oxide surfaces using the test of isopropanol decomposition. *Catal. Lett.* 1997, *43*, 219–218.

63. Youssef, A.M., Khalil, L.B., and Girgis, B.S. Decomposition of isopropanol on magnesium oxide/silica in relation to texture, acidity and chemical composition. *Appl. Catal.* 1992, *81*, 1–13.

64. Kulkarni, D. and Wachs, I.E. Isopropanol oxidation by pure metal oxide catalysts: number of active surface sites and turnover frequencies. *Appl. Catal.* 2002, *237*, 121–137.

65. Fein, D.E. and Wachs, I.E. Quantitative determination of the catalytic activity of bulk metal oxides for formic acid oxidation. *J. Catal.* 2002, *210*, 241–254.

66. Sambeth, J.E., Baronetti, G.T., and Thomas, H.J. A theoretical–experimental study of Wells–Dawson acid. An explanation of their catalytic activity. *J. Mol. Catal. A: Chem.* 2003, *191*, 35–43.

67. Miyamoto, A., Hattori, A., and Murakami, Y. The concentration of V=O species on oxidized vanadium oxide catalysts. *J. Solid State Chem.* 1983, *47*, 373–375.

12 Combinatorial Approaches to Design Complex Metal Oxides

Ferdi Schüth
MPI für Kohlenforschung, Mülheim, Germany

Stephan A. Schunk
hte Aktiengesellschaft, Heidelberg, Germany

CONTENTS

12.1 INTRODUCTION

Combinatorial or high-throughput approaches toward the synthesis and performance evaluation of solids are gaining increasing interest over the last ten years [1–3]. While initially the work focused primarily on optical and electronic properties, recently attention has shifted more toward catalysis. Requirements for both types of applications, however, are rather different. While for optical and electronic properties most often a perfect bulk structure is desired, where

391

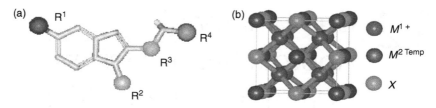

FIGURE 12.1   (a) Principle of combinatorial chemistry of molecules. For a common scaffold, substituents R^1–R^4 are varied independently, resulting in a multitude of different molecules. (b) Concept for a transfer to solids. In a solid, such as a spinel, the different constituents are varied independently, with temperature being an additional synthesis parameter. However, well-defined entities are not necessarily formed in this case

grain boundaries and other defects may mask the bulk behavior, in catalysis it is often the defect structure that determines the performance of a material. Synthesis procedures for both types of applications therefore may differ strongly. Whatever the desired application is, however, it is always mandatory that the synthesis of the material is highly reproducible. Since samples are often prepared in minute amounts in a high-throughput program, full analysis is almost impossible. If a "hit" is identified in a later screen, one has to be able to reproduce the sample and to scale-up the synthesis.

The concepts of combinatorial chemistry were originally developed for the synthesis of small organic molecules (Figure 12.1[a]). If a family of molecules has a common core structure, normally termed "scaffold," bearing different functional groups, which can be varied independently of each other, then a large number of different molecules can be generated based on this common scaffold, following combinatorial principles. A similar building concept can be used for a linear backbone molecule with backbones consisting of variable building blocks, such as oligo- or polypeptides. In theory, one may envisage a transfer of these concepts to solid-state chemistry, and specifically to the synthesis of oxides (Figure 12.1[b]). Many oxide structures are highly flexible with respect to the occupation of the cation sites, spinels being a very prominent example for such flexibility. One may then envisage synthesis strategies where the A and B cations in a spinel are combinatorially varied, resulting in the formation of a library of different spinels. As an additional parameter, one may also take into account the anion sites, if one moves away from pure oxides, or one could use treatment temperature as an additional parameter. However, for solids, the situation is much more difficult than for molecules. While for well-defined reaction steps, the product of an organic synthetic sequence is generally also well defined; this is not necessarily true for solids. The particle sizes may differ, the crystallinity can be dependent on treatment temperature, certain phases may need different temperature to form at different levels of composition, different precursors will have different reactivities, and so on. A simple combinatorial synthesis of complex oxides is therefore not straightforward, and the situation is made even more complicated by the fact that a synthesis planning for solids synthesis is only in its infancy [4], that is, we

TABLE 12.1
Differences in High-Throughput Experimentation for Molecular Entities in Pharmaceutical Research and Inorganic Solids

Organic molecular entities	Inorganic functional materials
Discrete molecular entities	Three-dimensional structures, often multielement composites, potentially metastable materials
Finite number of "active centers," often well characterized, can in many cases be modeled, good understanding of chemistry on a molecular level	Only for few materials basic understanding of type and function of active centers
Purities of over 85% achievable via combinatorial synthesis	Pure substances often without effect, especially multicomponent systems with defect structures show largest wealth
Highly developed screening procedures for biological activity available	Characterization of numerous often independent parameters challenging
Descriptors for library diversity well developed	Basis development work for descriptors still required
Good availability of synthetic building blocks and methods	Adaptation of complete "unit-operations" for automated synthesis required

typically cannot predict, what solid structure we will obtain if certain precursors are mixed. Therefore, the approach in the high-throughput synthesis of complex oxides is typically just the opposite as in organic and medicinal chemistry: while in the latter case a defined molecule is the target of the synthesis, in solid state chemistry usually precursors are mixed and one later analyzes which structure was obtained or what performance the material has, irrespective of whether it is a phase pure material or a mixture of different phases. Table 12.1 compares the characteristics of high-throughput synthesis of molecules with that of solids.

12.2 STRATEGIES IN HIGH-THROUGHPUT EXPERIMENTATION

A general division in Stage I and Stage II screening is useful for distinction, as both screening modes have complementary goals and, respectively, features that need to be realized and thus also taken into account in choosing the method of synthesis used for the preparation of the libraries. Stage I screening is usually employed for the screening of large material libraries for a certain target application. The focus of the screening lies on a large variety of chemistries that are intended to be screened for the target applications, in some cases the test conditions may even be far from a later technical application. Generally, the goal of a Stage I screening is to test several ten- up to hundred thousand materials per year and test unit. This is different for Stage II screening: in Stage II screening the focus of the investigation is to obtain results that are close to the ones obtained with conventional test procedures. Here, the throughput usually ranges from several hundred up to several thousand samples per test unit and year. As the distinctive properties of the materials, such as formation and aging behavior, are an important target of the investigation, usually a number of test conditions is applied so that a longer investigation time is justified. One has to keep in mind that the testing of materials with regard to useful properties is in most cases the bottleneck of the screening process regardless whether it is Stage I or Stage II screening. The natural consequence of this division of time needed for the different ways for screening leads, as indicated above, to an adaptation of the synthetic technique to an appropriate speed in order to supply the test units with materials of interest while having acceptable synthetic accuracy and reproducibility.

The above-described factors explain why the reports on synthetic efforts are rather scarce and usually coupled to the respective test techniques. Still it is worthwhile to take a closer look at the different approaches taken to synthesize materials with target properties.

12.2.1 Substrate-Based Approaches

Most approaches described in the literature for the synthesis of complex oxides deal with parallel or fast sequential synthesis by the aid of robotic systems. The variability of synthetic procedures employed ranges from gas phase deposition methods pioneered by the initial work of Hanak [5,6] and other authors who used the basic principle and refined the synthetic technique with regard to the deposition features and the chemistries employed [7–9]. Typical deposition techniques use several source materials and spatial resolution is mostly achieved via masking techniques. An essential feature of these synthetic approaches is the fact that a plurality of compounds is generated on a single substrate so that the result of the synthetic procedure is a substrate-bound library, where the position of each library member "encodes" the synthetic information, as composition or other synthetic steps that the material has been subjected to.

Deposition techniques such as sputtering, chemical vapor deposition, or epitaxy methods are especially useful for the synthesis of "dense" and defect free solids. On the other hand, for a large range of applications, such as catalysis, usually hydrothermal, sol–gel, or solution-based methodologies are employed as a synthetic concept for the generation of substrate-bound libraries. Different than for the deposition techniques discussed before, the target solid materials can in many cases represent metastable modifications, which still are highly interesting for a certain range of applications, in particular catalysis. Several groups reported the use of sol–gel, hydrothermal, or solvent-based chemistries for the generation of substrate-bound libraries of mixed oxides, oxide mixtures, and even mixed metal compounds [10–13]. A typical characteristic of all of the examples of substrate-bound libraries that are cited here is the main application focus on Stage I screening. So far, the authors are not aware of any substrate-bound library approaches that could be exploited for Stage II screening as well.

12.2.2 Substrate-Free Approaches

Substrate free synthesis plays a very important role in the generation of materials for Stage I and Stage II screening approaches. The essential difference lies in the fact that here no common substrate is the essential part of the whole library, but that the different materials are synthesized in parallel or fast sequential fashion via automated procedures and are obtained as separate materials, which can be individually handled and manipulated. Good examples for the parallel or sequential synthesis of materials are found in the catalysis community, although only in few cases bulk oxide catalysts have been synthesized. Mostly supported materials were generated: the group of Senkan reported the successful synthesis of impregnated catalysts by using alumina extrudates as support materials [14,15]. The single supports of a size of 4 mm × 1 mm were impregnated with Pt/Pd/In as active metals using aqueous precursor solutions, which were dispensed by an automated liquid handling system. After the impregnation step, drying and calcination were performed. Impregnation and co-precipitation techniques for larger scales are also described by Baerns and coworkers [16–18]. The relevant scales and testing procedures are in a range that can surely be called "close to conventional" and allow conclusions to the catalyst behavior on a larger scale. Hoffmann et al. [19] report the automated synthesis of Au-based catalysts via pH-controlled co-precipitation and equilibrium adsorption techniques. From a preparative point of view, these techniques require a high level of sophistication and the complexity of automating these techniques has to be adapted to these requirements. The point that makes this preparative work extremely important is the fact that a lot of the techniques required to prepare this type of catalysts have high technical relevance and mimic unit operations used for the synthesis of certain catalyst types on an industrial scale [20].

If one compares the approach of non-substrate bound parallel or fast sequential synthesis with the substrate bound approaches it becomes obvious that the focus of the nonsubstrate bound approaches is clearly directed more toward the technical goals of the final application. Usually the degree of parallelization for the respective test unit in Stage II screening is also moderate (6-, 16- up to 48-fold [17,19]). Library handling of nonsubstrate bound libraries is completely different and the idea of unified substrate handling does not play a role. The identification of the different samples usually follows the rules of conventional sample identification, although in many cases higher degrees of sophistication such as bar coding of single vessels for electronic identification will prove to be a useful tool. Even automated solutions for sequential or parallel sample handling can be employed as the handling of single samples can also become a bottleneck and therefore has to be analyzed with respect to its time limitation within the workflow envisaged.

In general, the conditions of the synthetic protocol will roughly determine the types of materials that can be expected as a result of the synthesis effort. For instance, hydrothermal conditions in the presence of amines and ions such as silicon or aluminum, which prefer tetrahedral coordination by oxygen, will favor the formation of zeolite type materials, while high temperature synthesis of suitable precursors with ions in specific ranges of ionic radii and favorable stoichiometry may induce the formation of spinels or spinel-type materials. However, exact planning with respect to the structure on the atomic level, such as it is known in medicinal chemistry for small molecular entities, is not possible for solids yet. Therefore, combinatorial approaches to produce oxide libraries are generally based more on a preparative approach, which consists of mixing precursors for oxides by different techniques and then treating these precursors to generate the oxide materials of different nature.

In the following, the different methods to create libraries of oxide materials will be discussed. We place only little emphasis on the synthesis of supported catalysts or modified zeolites by ion-exchange, wet impregnation, or other methods. Such procedures usually rely on custom manufactured; prefabricated, typically oxidic support materials and oxides are not necessarily formed. To a much larger extent than in the carrier beads, which are used in the split-and-pool approach (see Section 12.3.6), the support material plays an essential role in the final materials, be it for high dispersion of a noble metal compound or by providing catalytically active sites as in a zeolite.

12.3 SPECIFIC TECHNIQUES

12.3.1 Sputter or Evaporation Procedures

Sputter processes using either compositional spreads [5,6,21] or masking techniques [22] were the first methods employed to create complex libraries of oxides (Figure 12.2). In the generation of compositional spreads, gradients of different

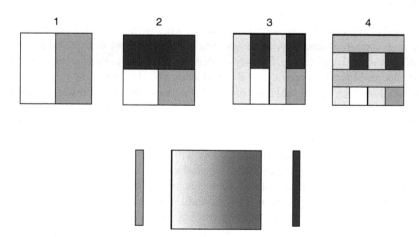

FIGURE 12.2   Creation of sputtered libraries by a masking technique in four steps, resulting in spatially defined compounds (top) or by coevaporation of two components, resulting in a compositional spread (bottom)

materials are produced, which result in the formation of gradients in oxide composition after annealing. The result is not a materials library in the sense that at each position a defined compound is formed, but the nature of the materials formed may gradually vary or may show abrupt changes, depending on the phase diagram of a system. After analysis of properties of a point in such a compositional spread, the composition and the nature of the material formed at the position of a "hit" needs to be analyzed. This is different if masks are used; then each element in an array has a defined composition. However, this situation does not necessarily imply the formation of a defined compound; therefore, a careful analysis may be necessary. Nevertheless, it will be substantially easier to resynthesize a "hit" found in a library prepared with masks than in a compositional spread.

One of the most impressive examples for such a library generated by successive use of masks was described earlier on during the development of high-throughput research in materials science. Danielson et al. [23] used an evaporation technique to synthesize a library of over 25,000 different potential luminescent materials (Figure 12.3). In this case, first electron beam evaporation of different oxide precursors was used to deposit various oxides in different film thickness on substrates (film thickness ranged from 100 to 400 nm for the matrix oxides and 0 to 25 nm for the rare earth dopants), while masks controlled the position where specific compositions were formed. Subsequently, mixing of the deposited layers was induced by oxidative thermal processing. In the formed oxide library, novel luminescent materials were identified. Following this synthetic procedure, not only the variation in composition is a process parameter, but also the conditions of the oxidative thermal processing, since not all materials will respond in a similar manner to certain temperatures and partial pressures. However, with sufficiently high temperature at least homogenization of the constituents over the different deposited layers will occur. As previously mentioned, it would be easier to resynthesize a "hit" than with a compositional spread. This was demonstrated using an example from the library shown in Figure 12.3. A white light emitting material had been discovered in a compositional range where it had not been expected. This region was therefore mapped in more detail and eventually a $SrCe_2O_4$ was identified as the source of this emission, which could be an interesting luminescent material.

12.3.2 Hydrothermal Reactions

Hydrothermal reactions are especially important in the synthesis of one particular class of catalytically relevant oxides, that is, in the synthesis of zeolites and related compounds. For such materials, high-throughput approaches appear especially interesting for three main reasons: (i) the structural and compositional diversity of zeolitic materials can be very large, (ii) the rational design of zeolitic structures is only in its infancy and large parameter spaces have to be investigated empirically to discover novel materials, and (iii) synthesis times normally are rather long on the order of days, which means that synthesis is most often indeed the bottleneck.

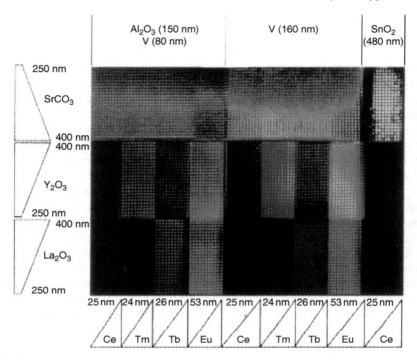

FIGURE 12.3 Library with 25,000 members created by sputtering with masks, viewed under illumination with UV-light. The library members are mostly fluorescent, which is seen (Reproduced from Danielson, E., Devenney, M., Giaquinta, D.M., Golden, J.H., Haushalter, R.C., McFarland, E.W., Poojary, D.M., Reaves, C.M., Weinberg, W.H., and Wu, X.D. *Science* 1998, *279*, 837–839. With kind permission.)

Thus, hydrothermal reactions were among the first to be studied using parallel synthesis technology [24,25]. Akporiaye et al. [24] designed a macroscopic autoclave block, in which multiples of 100 samples could be processed and exposed to hydrothermal conditions simultaneously, while the rest of the processing was done in a more conventional way, that is, X-ray diffraction (XRD) analysis and so on. Klein et al. [25] went one step further in miniaturizing the wells and also automating the analysis: they used a miniaturized 37-well reactor array, which only contained sufficient amounts of reagents for the synthesis of minute amounts of zeolite on the submilligram scale. This created the additional problem of sample analysis, for which a special type of diffractometer, the general area detector diffraction (GADDS) system (Bruker-AXS), was used. In addition, for semiquantitative analysis of the composition of the solids a micro-X-ray fluorescence system was employed [26]. With these systems, it was possible to reproduce the synthesis conditions for the synthesis of TS-1, a titanosilicate of the MFI family. However, it had not been used for the discovery of novel zeolites. Subsequently the SINTEF team developed automated methods for the analysis of the synthesis products resulting from high-throughput hydrothermal synthesis. Akporiaye et al. [27] were able to identify several different metallophosphate phases, such as, for instance, of the AFI and CHA topology. However, as yet unidentified structures were also observed, presumably layered materials. The software they had developed allowed the solution of a serious problem, that is, the automatic detection and deconvolution of phase mixtures, which notoriously form in hydrothermal synthesis starting with arbitrary mixtures.

Another parallelized autoclave for the hydrothermal synthesis of solids was introduced by Bein and coworkers [28]. In this system, the products are recovered by a customized centrifugation device and product identification is done via XRD using an automated XY-translation stage. The system was used to study the influence of different template molecules, notably of metallocenium ions, on the formation of aluminophosphates.

In a slightly different setup, which makes use of the so-called dry gel conversion, Zhang et al. [29] investigated conditions under which the silicoaluminophosphate SAPO-34 is formed. However, this system seems to be more limited in scope, since only a fraction of the known crystalline microporous materials is accessible via the dry gel conversion route. Another special synthesis procedure was developed by the group of Gavalas, who describe the parallel synthesis of zeolite films [30]. The synthesis mixtures are formed in each well of a 21-well setup, then the substrate is placed on top of each well and the system turned around to expose the substrate to the synthesis mixture. Also in this case, product identification proceeds via XRD. Ideally, a microdiffractometer should be used for analysis as the authors point out, but due to the unavailability of such a system, analysis was performed by conventional X-ray diffraction, covering all but one sample by lead foil.

12.3.3 Sol–Gel Synthesis

Sol–gel synthesis is a synthetic procedure where first a sol is formed from suitable precursor species, and this sol is subsequently converted to a container-spanning gel, the hydrogel. After drying, which is normally accompanied by some shrinkage, a porous solid, the so-called xerogel, is obtained. The sol–gel synthesis lends itself ideally to parallelization and miniaturization, since a major fraction of the solution volume is converted to the solid and normally there are no complex isolation steps for the solid involved. This makes it possible to only use rather small wells and still produce reasonable amounts of solid for subsequent performance evaluation. Figure 12.4 shows an example of such a library produced by the sol–gel method by Maier and coworkers [26]. With respect to catalytic applications, the sol–gel route has the advantage that typically highly porous materials are obtained, which provide sufficient surface area to allow a detectable reaction rate even for small amounts of sample. There are, however, some disadvantages associated with the sol–gel synthesis: (i) some of the precursors are susceptible to hydrolysis so that the library synthesis would ideally be performed under inert atmosphere, that is, in a glove-box or using some other means to prevent hydrolysis; (ii) not all materials

FIGURE 12.4   Library of oxides synthesized by a sol–gel method. The library is synthesized on a substrate which can directly be analyzed in a subsequent screen, so that removal of the solids from the wells is not necessary (Reproduced from Scheidtmann, J., Weiß, P.A., and Maier, W.F. *Appl. Catal. A: Gen.* 2001, *222*, 79–89. With kind permission.)

are accessible via the sol–gel route, because suitable precursors are not available for all elements; (iii) the product can be difficult to remove from the container, especially if small amounts are prepared, and thus often the performance analysis is done on the same substrate as the synthesis.

An apt example for a sol–gel synthesis in very small quantities in a highly miniaturized system has been given by Reddington et al. [11], who have converted an ink-jet printer to be able to deliver precursors for oxide materials. These precursors were "printed" on graphite paper and the formed materials were evaluated with respect to their performance as anode catalysts in direct methanol fuel cells.

12.3.4 Precipitation Procedures

For most catalytic materials in the form of bulk oxides, precipitation is the pathway via which the oxides themselves or their precursors are generated [31]. A precipitation reaction is characterized by high volumes of solution that need to be handled, even when relatively concentrated solutions are being used. This makes such reactions not very amenable to automated synthesis in high-throughput mode. Consequently, there are very few attempts described in the literature to use this technique in a high-throughput program.

Since the handling of solids in an automated fashion is challenging, working in suspension as long as possible can be a suitable workaround procedure. This concept was the basis for the synthesis of Haruta-type gold catalysts via a co-precipitation route developed in our laboratory [19]. The method used a Gilson automatic dispensing system, which was additionally equipped with a rack for wide test tubes on a shaker tray. Precipitation was carried out in the test tubes by first filling in the support metal and the gold precursor solution, then adding the precipitation agent by means of the dispenser. The action of the shaker prevented settling of the precipitate, so that the solid formed could be transferred with a wide bore syringe of the dispenser system to the filtration module, where the solid was separated from the mother liquor and washed. Using this technique, it was possible to increase the level of reproducibility in the synthesis of supported gold catalysts. It should be noted, though, that the method cannot be generalized. Depending on the nature of the precipitate formed, it may not be possible to maintain the solid in suspension, which would make separation difficult. Even worse, a fractionation in smaller, well-suspended particles and heavier particles that settle may occur, which would seriously affect the reproducibility of the synthesis.

Separation of the precipitated solid may be achieved by centrifugation, a step that can be automated, although no application example has been published for the synthesis of catalysts. However, an automated centrifugation system capable of handling rather large solution volumes is rather expensive and requires high maintenance. In such cases, one has to carefully consider the alternative of synthesizing precipitated solid oxides in an only semiautomated or fully manual manner. General methods for the high-throughput synthesis of precipitated oxides is certainly still a major challenge, in spite of the technical importance of such materials.

12.3.5 Activated Carbon Route

An alternative high-throughput method for the synthesis of bulk oxides was recently developed in our laboratories. It is based on the exotemplating effect of the pore system of activated carbon [32]. It has been found, that high surface area oxides can be generated, if suitable activated carbon is impregnated with highly concentrated solutions of metal oxide precursors that decompose easily and without residue. Calcination of the impregnated carbon at temperatures of typically 500°C leads to the formation of the oxides [33]. The method was found to be suitable not only to generate binary oxides, but also to produce defined ternary oxides, such as spinels or perowskites, if the carbon is impregnated with suitable precursors in the correct stoichiometry. The mass-based yield amounts to several 10% of the mass of the carbon used as the exotemplate, which means that sizable amounts of oxides can be synthesized with relatively small amounts of carbon.

While the method is suitable for the production of various defined oxide phases, such as series of different spinels with different degree of substitution of cations in the structure, alternatively random compositions of mixed oxides can be synthesized to cover a wider range of different solids. The practical implementation of such a method is as follows: measured amounts of activated carbon (volumetric measurement is sufficient, the deviations in weight are on the order of 1% for about 100 mg) are placed in small quartz containers (in our setup a 7×11 matrix is used; however, any other format is also possible). A tray with these containers is placed in the working area of a dispensing system. The dispensing system takes the required volumes from stock solution containers with concentrated solutions and premixes the solutions that are supposed to be impregnated in the carbon in one container in a rack with Eppendorf vials. Since the amount of solution needed to fully imbibe the carbon is known, all carbon samples are impregnated with the same total solution volume. The mixed solutions are then impregnated in the carbon, the tray is covered with gauze to prevent the carbon from spilling if the combustion reaction is too vigorous, and then the carbon is burned off in a furnace at temperatures around 500°C. If stock solutions are prepared and a pipetting program has been written, it is possible to synthesize 77 different oxide materials with a minimum of manual labor.

The pipetting program is standardized, and if random libraries are desired, a computer program generates random compositions of the oxides to be synthesized, converts the target compositions to solution volumes, and transfers all values to a disk, which is then used to control the movements of the dispensing system.

The materials generated by this procedure have been found to be attractive potential catalysts in several applications. With rather small libraries of only a few hundred compounds, it was possible to find a low-temperature active CO-oxidation catalyst, which was superior to various noble metal-based systems [34] and a promising material for the development of a NO_x storage catalyst [35].

12.3.6 Split-and-Pool Synthesis

Since only little information on inorganic split-and-pool synthesis is available in the literature [36–38], and since this method is not simply a parallelization or automatization of known synthetic approaches, this concept is discussed here in detail.

One crucial feature of the split-and-pool synthesis as known in medicinal chemistry is the attachment of the desired compounds to a carrier particle or their confinement within a container. Both the carrier or the container are typically tagged so that the sequence of steps they have undergone can be backtracked and thus the nature of the compound attached to the carrier or present in the container can be established after a subsequent screening. The most often used combinatorial principle in organic synthesis is nowadays the concept of "one bead one compound" [39–41]. If this could be transferred to the world of materials science, a powerful synthetic methodology would become available.

Figure 12.5 illustrates the principle of the method developed for the combinatorial synthesis of organic molecules. In a first step, the beads to which the molecules should be attached are divided into sub-batches — into three batches in the example. After the first three reactions, a different group is attached to beads from each batch. The beads are then pooled and divided again into sub-batches. Each batch is exposed to a second reaction step. After this step, altogether nine compounds have been generated. Pooling, splitting, and reacting in a third step gives 27 different compounds attached to the different beads. One can easily see that the number of different compounds increases dramatically, if the number of

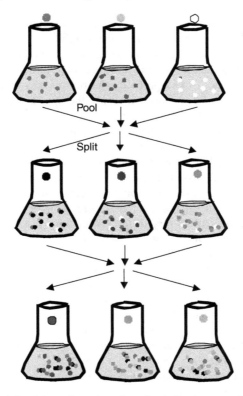

FIGURE 12.5   Principle of the split-and-pool synthesis for the example of three steps with three different reactions in each step. This results in 3 different compounds after the first step, 9 different compounds after the second step, and 27 different compounds after the final step. For each of the three reaction vessels in the final step, deconvolution of 9 different compounds is necessary, since the last reagent added is known for the final step

batches into which the beads are divided or the number of splitting and pooling steps are increased. There is, however, a price that has to be paid for the high efficiency: after the reaction sequence, each bead has only one compound attached to it, but it is not known which of the possible compounds this is. Since the analysis of minute amounts of the attached compounds is very difficult, different encoding strategies have been developed to backtrack the sequence of reaction steps after a "hit" has been identified [42]. These strategies can make use of chemical tags, where additional reactions, yielding easy-to-analyze compounds, are carried out simultaneously with each reaction step. However, small containers can also be used to carry a lesser number of beads through the different steps. If the containers are tagged, for instance with radiofrequency labels as in the IRORI-approach [43–45], backtracking is also possible.

Several problems arise if one attempts to transfer the split-and-pool concept to the synthesis of oxides and other inorganic solids. In organic synthesis, strong covalent bonds are formed which are not easily cleaved in subsequent steps. For inorganic materials, redissolution of an oxide precipitated on a bead may occur in the following step, thus leading to contamination of the other beads in the batch. In addition, suitable strategies for encoding are more difficult to implement than in medicinal chemistry. This is probably the reason why it took several years before first attempts had been made to apply the split-and-pool method to the combinatorial synthesis of inorganic solids.

The basis of the single-bead concept for inorganic materials is the use of single shaped bodies as the carrier of the material to be generated. The advantages of using such carriers shall therefore be briefly discussed before the synthetic methods available will be addressed. The carriers may, in principle, be of any shape, but usually spherical particles are applied for convenience. In accordance with approaches known from organic synthesis, such spherical particles are called "beads," although they fulfill very different functions in comparison with their application in combinatorial setups in organic chemistry and biochemistry [46].

After having gone through the desired sequence of synthesis steps, each bead represents one catalyst as a member of a library of solid catalysts. It may consist of a nonporous material, such as α-Al$_2$O$_3$ Steatit, or of typical porous support materials, such as γ-Al$_2$O$_3$, SiO$_2$, or TiO$_2$, depending on the application focus of the material. These beads can be subjected to different synthesis procedures and sequences, such as impregnation, coating, etc. In addition, full mixed metal oxide catalysts can also be formed to spherical particles via different shaping techniques like tabletting or the like.

Using single beads for high-throughput experimentation has a number of advantages, especially when it comes to the field of catalysis. First, such beads are comparable to well-known fixed bed catalysts and the synthetic pathways can be the same as for conventional materials, which may facilitate scale-up procedures. A number of common preparation procedures for these beads are available and can be carried out in standard laboratory environments. The second main advantage is that each bead is a single entity that can be handled independently from other beads or the final reactor configuration. Starting from masterbatches, a large diversity of materials can easily be prepared. Furthermore, different beads may be treated individually, for instance, subjected to different preparation steps or pretreatments such as calcination or steaming, rather than handling the complete library, as necessary for substrate-bound thin or thick film catalysts. These characteristics allow the use of synthesis procedures different from the parallel approach. The potential use of *ex situ* synthesis procedures may, furthermore, present a significant advantage especially in micro-chemical systems, as an *in situ* preparation of the catalysts in the reactor could cause problems with contamination or thermal and chemical stability of the reactor material. From an inorganic synthetic viewpoint, it is interesting to see how far the "bead principle" can be exploited for the synthesis of complete combinatorial libraries.

If we compare the chemistries involved for the creation of organic libraries and libraries of inorganic materials, one can conclude that for the synthesis of organic libraries a series of well-defined synthetic steps leads to libraries of defined chemical entities, whereas for inorganic materials libraries, a series of well-defined synthetic steps leads to libraries of complex materials, which are characterized by the synthetic pathway employed to synthesize the respective library member. However, the single synthetic operation and the overall pathway may not have the same effect over the whole library of complex materials. The synthetic possibilities with regard to inorganic split-and-pool synthesis offer a wide range of perspectives for variation. Purity of compounds, potential aging of precursor solutions, and sequence and history of steps during addition of compounds offer a range of synthetic parameters that can be varied and used to create diverse libraries.

As discussed earlier, among the different methodologies for the efficient synthesis of libraries having a high degree of chemical diversity, the split-and-pool principle is by far superior to other synthetic methodologies. This makes the split-and-pool method an ideal tool for Stage I synthetic efforts. With a relatively small number of operations, compared with standard parallel synthesis procedures, a wide variation in the properties of the library members can be achieved. In principle, the synthetic steps of splitting and pooling in the creation of inorganic materials libraries are identical to the ones applied in organic chemistry. The particular steps that are used, however, certainly differ due to the different demands arising from the different chemistries required for inorganic solid state chemistry.

The synthetic approach to the synthesis of inorganic materials via the split-and-pool methodology is closely connected to the question of synthesizing materials on or within physical entities that will carry the library member throughout the synthetic process. If physical entities are employed as "carriers" that show sufficient chemical inertness toward the screening process, a separation of the final library compound may not be necessary at the end of the synthetic steps. In some cases, it may even be desirable to employ "carriers" that may in fact become an integral functional part of the desired material, so that the separation of the created compound is not an issue of interest any longer.

Generally, solution-based approaches for the generation of inorganic split-and-pool libraries have substantial advantages over approaches where solid phases are introduced as chemical sources during the different synthetic steps. Solution chemistry offers potentially a wide range of synthetic opportunities that can be exploited not only for the purpose of parallel synthesis, but also for synthetic steps for split-and-pool library creation. Redissolution of compounds deposited in previous steps in later steps, however, is always a possibility that should be kept in mind, since this may create problems.

For a large range of applications impregnation or equilibrium adsorption methods followed by subsequent drying or calcination steps for solvent removal can be employed as a synthetic approach to obtain large libraries of materials. Carriers that can be used are available or can be customized. A wide range of chemistries can be performed from a synthetic point of view based on these "carrier" materials. Usually a spherical shape is preferred as it eases transfer operations and minimizes potential damages during transfer. Incipient wetness impregnation was used to create a library of about 1000 members containing the five elements Mo, Ni, Co, Bi, and Fe on one out of four concentration levels each [37]. In each step, a batch of beads was impregnated with a solution corresponding to 80% of the water uptake volume of the respective amount of carrier. Between successive impregnation steps, the batches were dried and calcined at 400°C, which decreases the potential for cross-contamination. A smaller, ten-member noble metal-based library had previously been reported in the literature, which was also created by impregnation followed by calcinations in three steps [38].

Alternative to impregnation procedures coating procedures are another possibility for the application of different chemical compositions to a "carrier"-body. A sequential procedure can lead to an onion like structure of the deposited components, which may not in all cases be desirable; however, for some applications one may also exploit this concept. In combination with impregnation procedures, coating procedures may prove to be most powerful and efficient.

If bead-based methods are used with the beads being modified while freely floating in solutions, a tagging scheme has to be developed. There are different schemes reported in the literature. The most generally useful concept seems to be micro-XRF. Other than in organic chemistry, where tagging is necessary due to the minute amounts of very similar molecules that cannot be analyzed, it was found that the chemical composition itself was a suitable tag in this case. Schunk and coworkers described the analysis if the Bi, Mo, Co, Ni, and Fe library mentioned above. Figure 12.6 shows the readout of chemical analysis of several beads from such a library using micro-XRF. The two higher loading levels can be reliably assessed, while the lower concentrations could not be discriminated. Also in the publication by Mallouk and coworkers [38], micro-XRF has been used as one of the techniques to determine the identity of a bead. Both studies demonstrate that in principle a split-and-pool library can be analyzed using micro-XRF after a "hit" has been identified.

In the study by Mallouk and coworkers [38] an additional method is reported, which can be used for tagging, that is, attachment of a fluorescent label. It was shown that dyes could be attached and read out again, so, in principle, this is a method for tagging. However, the applicability is probably limited, because only a limited number of dyes can be detected simultaneously and the dyes would not survive high temperature treatment steps.

For "carrier"-free synthetic approaches, techniques employing different vessels, like cans, or alternatively "containers" that are able to absorb and contain considerable amounts of solutions with regard to their own volume, are a central component of the synthetic effort. Starting from solutions these can be "split" using vessels or cans, creating a library with members of different identity by either adding components or by altering the solution by any form of physical treatment (heat, evaporation of solvent, or the like). The pooling step in many cases becomes absurd if all the solutions of the first split step are reunited. Still if only some "carriers" or parts of the solutions from the "carriers" are united during the pooling step, this step may very well make sense from a synthetic point of view and can be usefully employed, although in many cases the number of pooling steps may be smaller than the number of splitting steps. Via addition of further components, the complexity of the mixture is increased and the final performance

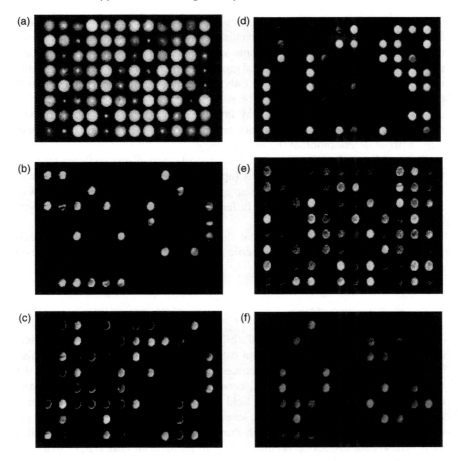

FIGURE 12.6 Optical photograph (upper left corner) and XRF maps for different elements of a solid state library created by a split-and-pool synthesis. (b) Ni, (c) Fe, (d) Bi, (e) Mo, and (f) (Co) (Reproduced from Newsam, J., Schunk, S.A., and Klein, J.)

of a precipitation step P leads to solid library members. The elegance of the procedure lies in the fact that encoding is eased if "carriers" in the form of vessels or cans are employed, which can be equipped with tags. For many applications where solutions are employed to perform split-and-pool synthesis, a shaping step of the resulting library members will be mandatory. Obtaining a powder as a product leaves a range of methods for shaping open, which can be adapted according to the demands of the screening procedure employed.

Another possible option for the application of the split-and-pool principle for obtaining "carrier"-free libraries in the form of powders is the use of combustible "carriers" that can contain solutions. A variety of organic polymers is available for this purpose, the ones that proved to be of highest value were acrylates or hydrophilized polyolefins, suitable for the needs of an approach based on aqueous chemistries or other highly polar solvents. In this case, the similarity with regard to organic synthetic approaches seems very high, from a point of view of the polymer properties, discrete functional groups that bind to the entities added during the synthetic steps are obsolete, the only compatibility that is demanded is that the carrier is capable of absorbing the solvent. As many steps in the generation of functional inorganic materials may involve a thermal treatment of the given material, such absorbent materials are preferred that tolerate thermal treatment steps without deterioration of the shaped entity.

From a chemical point of view, inorganic solid "carrier"-materials are far from being inert toward catalytic reactions (apart from selected exceptions) and will in many cases become an active component as such via the synthetic pathway followed. This is also an important differentiation with regard to the traditional organic split-and-pool synthesis.

It should be mentioned here that the efficient screening of the split-and-pool libraries goes along with specific reactor formats, which are described elsewhere [47].

12.4 SUMMARY AND OUTLOOK

The preceding discussion shows that the high-throughput synthesis of oxidic materials is at the transition to become an established technology. However, it also becomes clear that pronounced differences exist between the conventional and the high-throughput synthesis of oxides, conceptually as well as with respect to the experimental procedures. Since the field is developing rapidly, substantial progress can be expected over the next several years. There are areas where so far no satisfactory solutions for the high-throughput/automated synthesis of oxides exist. These deficiencies lie predominantly in the field of bulk oxides made by precipitation, spray drying, melt processes, etc. However, before such techniques are automated, it is sensible to ask whether the effect would be worth the effort. If the synthesis is not the time limiting step in the overall workflow, it may be advisable to continue with manual synthesis of such materials and rather focus on "de-bottlenecking" of other steps. On the other hand, automation of the synthesis with the often resulting higher degree of reproducibility is of value in itself. Therefore, developments will and should continue in the different laboratories active in this field, and the coming years will show how much high-throughput techniques can penetrate the materials science of oxidic solids.

REFERENCES

1. Newsam, J.M. and Schüth, F. Combinatorial approaches as a component of high-throughput experimentation (HTE) in catalysis research. *Biotechnol. Bioeng.* 1999, *61*, 203–216.

2. Jandeleit, B., Schaefer, D.J., Powers, T.S., Turner, H.W., and Weinberg, W.H. Combinatorial materials science and catalysis. *Angew. Chem. Int. Ed.* 1999, *38*, 2495–2532.

3. Senkan, S. Combinatorial heterogeneous catalysis — A new path in an old field. *Angew. Chem. Int. Ed.* 2001, *40*, 312–329.

4. Schön, J.C. and Jansen, M. First step towards planning of syntheses in solid-state chemistry: determination of promising structure candidates by global optimization. *Angew. Chem. Int. Ed.* 1999, *35*, 1287–1304.

5. Hanak, J.J. Multiple-sample-concept in materials research — Synthesis, compositional analysis and testing of entire multicomponent systems. *J. Mater. Sci.* 1970, *5*, 964–971.

6. Hanak, J.J. Compositional determination of RF co-sputtered multicomponent systems. *J. Vac. Sci. Technol.* 1971, *8*, 172–177.

7. Schultz, P.G. and Xiang, X.D. Combinatorial approaches to materials science. *Curr. Opin. Solid State Mater. Sci.* 1998, *3*, 153–158.

8. Liu, D.R. and Schultz, P.G. Generating new molecular function: a lesson from nature. *Angew. Chem. Int. Ed.* 1999, *38*, 36–54.

9. McFarland, E.W. and Weinberg, W.H. Combinatorial approaches to materials discovery. *Trends Biotechnol.* 1999, *17*, 107–115.

10. Sun, X.D., Wang, K.A., Yoo, Y., Wallace-Freedman, W.G., Gao, C., and Xiang, X.D. Solution-phase synthesis of luminescent materials libraries. *Adv. Mater.* 1997, *9*, 1046–1050.

11. Reddington, E., Sapienza, A., Gurau, B., Viswanathan, R., Sarangapani, S., Smotkin, E.S., and Mallouk, T.E. Combinatorial electrochemistry: a highly parallel, optical screening method for discovery of better electrocatalysts. *Science* 1998, *280*, 1735–1737.

12. Rantala, J.T., Kololuma, T., and Kivimaki, L. Combinatorial methods in sol–gel technology. *Proc. SPIE.* 2000, *3941*, 11–18.

13. Liu, Y., Cong, P., Doolen, R.D., Turner, H.W., and Weinberg, W.H. High-throughput synthesis and screening of V–Al–Nb and Cr–Al–Nb oxide libraries for ethane oxidative dehydrogenation to ethylene. *Catal. Today* 2000, *61*, 87–92.

14. Senkan, S.M. and Öztürk, S. Discovery and optimization of heterogeneous catalysts by using combinatorial chemistry. *Angew. Chem. Int. Ed.* 1999, *38*, 791–795.

15. Senkan, S.M., Krantz, K., and Öztürk, S. High-throughput testing of heterogeneous catalyst libraries using array microreactors and mass spectrometry. *Angew. Chem. Int. Ed.* 1999, *38*, 2794–2799.

16. Rodemerck, U., Ignaszewski, P., Lucas, M., Claus, P., and Baerns, M. Synthesis and fast catalytic testing of catalyst libraries for oxidation reactions conducted in parallel. *Chem. Ing. Tech.* 1999, *71*, 873–877.

17. Buyevskaya, O., Wolf, D., and Baerns, M. Ethylene and propene by oxidative dehydrogenation of ethane and propane — "Performance of rare-earth oxide-based catalysts and development of redox-type catalytic materials by combinatorial methods." *Catal. Today* 2000, *62*, 91–99.

18. Buyeskaya, O.V., Brückner, A., Kondratenko, E.V., Wolf, D., and Baerns, M. Fundamental and combinatorial approaches in the search for optimisation of catalytic materials for the oxidative dehydrogenation of propane to propene. *Catal. Today* 2001, *67*, 369–378.

19. Hoffmann, C., Wolf, A., and Schüth, F. Parallel synthesis and testing of catalysts under nearly conventional testing conditions. *Angew. Chem. Int. Ed.* 1999, *38*, 2800–2803.

20. Stiles, A.B. *Catalyst Manufacture-Laboratory and Commercial Preparations.* Marcel Dekker: New York, 1983.

21. van Dover, R.B., Schneemeyer, L.D., and Fleming, R.M. Discovery of a useful thin-film dielectric using a composition-spread approach. *Nature* 1998, *392*, 162–164.

22. Danielson, E., Golden, J.H., McFarland, E.W., Reaves, C.M., Weinberg, W.H., and Wu, X.D. A combinatorial approach to the discovery and optimization of luminescent materials. *Nature* 1997, *389*, 944–948.

23. Danielson, E., Devenney, M., Giaquinta, D.M., Golden, J.H., Haushalter, R.C., McFarland, E.W., Poojary, D.M., Reaves, C.M., Weinberg, W.H., and Wu, X.D. A rare-earth phosphor containing one-dimensional chains identified through combinatorial methods. *Science* 1998, *279*, 837–839.

24. Akporiaye, D.E., Dahl, I.M., Karlsson, A., and Wendelbo, R. Combinatorial approach to the hydrothermal synthesis of zeolites. *Angew. Chem. Int. Ed.* 1998, *37*, 609–611.

25. Klein, J., Lehmann, C.W., Schmidt, H.-W., and Maier, W.F. Combinatorial material libraries on the microgram scale with an example of hydrothermal synthesis. *Angew. Chem. Int. Ed.* 1998, *37*, 3369–3372.

26. Scheidtmann, J., Weiß, P.A., and Maier, W.F. Hunting for better catalysts and materials-combinatorial chemistry and high-throughput technology. *Appl. Catal. A: Gen.* 2001, *222*, 79–89.

27. Akporiaye, D., Dahl, I., Karlsson, A., Plassen, M., Wendelbo, R., Bem, D.S., Broach, R.W., Lewis, G.J., Miller, J., and Moscoso, J. Combinatorial chemistry — The emperor's new clothes? *Micropor. Mesopor. Mater.* 2001, *48*, 367–373.

28. Choi, K., Gardner, D., Hilbrandt, N., and Bein, T. Combinatorial methods for the synthesis of aluminophosphate molecular sieves. *Angew. Chem. Int. Ed.* 1999, *38*, 2891.

29. Zhang, L.X., Yao, J.F., Zeng, C.F., and Xu, N.P. Combinatorial synthesis of SAPO-34 via vapor-phase transport. *Chem. Commun.* 2003, 2232–2233.

30. Lai, R., Kang, B.S., and Gavalas, G.R. Parallel synthesis of ZSM-5 zeolite films from clear organic-free solutions. *Angew. Chem. Int. Ed.* 2001, *40*, 408–411.

31. Schüth, F. and Unger, K. Preparation of catalysts by precipitation and coprecipitation and by precipitation from organic solvents. In *Handbook of Heterogeneous Catalysis*; Ertl, G., Knözinger, H., and Weitkamp, J., Eds., Wiley-VCH: Heidelberg, 1997, p. 72.

32. Schüth, F. Endo- and exotemplating to create high surface area inorganic materials. *Angew. Chem. Int. Ed.* 2003, *42*, 3604–3622.

33. Schwickardi, M., Johann, T., Schmidt, W., and Schüth, F. High-surface-area oxides obtained by an activated carbon route. *Chem. Mater.* 2002, *14*, 3913–3919.

34. Johann, T., Brenner, A., Busch, O., Marlow, F., Schunk, S., and Schüth, F. Real-time photoacoustic parallel detection of products from catalyst libraries. *Angew. Chem. Int. Ed.* 2002, *41*, 2966–2968.

35. Busch, O., Hoffmann, C., Johann, T., Schmidt, H.-W., Strehlau, W., and Schüth, F. Application of a new color detection based method for the fast parallel screening of DeNO$_x$ catalysts. *J. Am. Chem. Soc.* 2002, *124*, 13527–13532.

36. Newsam, J., Schunk, S.A., and Klein, J. Process for Producing a Multiplicity of Building Blocks of a Material's Library. DE 100 59 890 and WO 02/43860 to hte Aktiengesellschaft.
37. Klein, J., Zech, T., Newsam, J.M., and Schunk, S.A. Application of a novel Split & pool-principle for the fully combinatorial synthesis of functional inorganic materials. *Appl. Catal. A: Gen.* 2003, *254*, 121–131.
38. Sun, Y., Chan, B.C., Ramnarayanan, R., Leventry, W.M., and Mallouk, T.E. Split-pool method for synthesis of solid state material combinatorial library. *J. Comb. Chem.* 2002, *4*, 569–575.
39. Furka, A., Sebestyen, F., Asgedom, M., and Dibo, G. Abstract. In *Proceedings of the 14th International Congress Biochemistry*, Vol. 5, Prague, 1988. p. 47.
40. Furka, A., Sebestyen, F., Asgedom, M., and Dibo, G. General method for rapid synthesis of multicomponent peptide mixtures. *Int. J. Pept. Prot. Res.* 1991, *37*, 487–493.
41. Balkenhohl, F., von dem Bussche-Hünnefeld, C., Lansky, A., and Zechel, C. Combinatorial synthesis of small organic molecules. *Angew. Chem. Int. Ed.* 1996, *108*, 2289–2337.
42. Yan, B. Single bead analysis in combinatorial chemistry. *Curr. Opin. Chem. Biol.* 2002, *6*, 328–332.
43. http://www.discoverypartners.com (accessed December 2003).
44. Nicolaou, K.C., Xiao, X.Y., Parandoosh, Z., Senyei, A., and Nova, M.P. Radio-frequency encoded combinatorial chemistry. *Angew. Chem. Int. Ed.* 1995, *34*, 2289–2291.
45. Moran, E.J., Sarshar, S., Cargill, G.F., Shahbaz, J.M., Lio, A., Mjalli, A.M.M., and Armstrong, R.W. Radio frequency tag encoded combinatorial library method for the discovery of tripeptide-substituted cinnamic acid inhibitors of the protein tyrosine phosphatase PTP1B. *J. Am. Chem. Soc.* 1995, *117*, 10787–10788.
46. Klein, J., Laus, O., Newsam, J.M., Strasser, A., Sundermann, A., Vietze, U., Zech, T., and Schunk, S.A. First use of the Split & Pool-principle for the synthesis of inorganic solids. *Petrol. Chem. Div. Prepr.* 2002, *47*, 254.
47. Zech, T., Klein, J., Schunk, S.A., Johann, T., Schüth, F., Kleditzsch, S., and Deutschmann, O. Miniaturized reactor concepts and advanced analytics for primary screening in high-throughput experimentation. In *High Throughput Analysis: A Tool for Combinatorial Materials Science*; Potyrailo, R.A. and Amis, E.J., Eds. Kluwer Academic/Plenum Publishers: Dordrecht, 2003.

13 Propane Selective Oxidation to Propene and Oxygenates on Metal Oxides

E.K. Novakova
CenTACat, Queen's Unversity Belfast David Keir Building
Belfast, Ireland, UK

J.C. Védrine
Laboratoire de Physico-Chimie des Surfaces, Ecole Nationale
Supérieure de Chimie de Paris, Paris, France

CONTENTS

13.1 INTRODUCTION

The evolution of the modern society is directly related and even dependent on the availability of vast number and variety of chemicals, used in the chemical, polymer, food, textile, etc., industries. Preparation of the majority of these compounds involves or directly originates from petrochemicals. However, due to the limited and continuously and inexorably diminishing reserves of petroleum worldwide, alternative methods for making these chemicals need to be found. One alternative is the utilization of natural gas, composed mainly not only of methane but also nonnegligible amounts of C_2–C_4 alkanes, depending on the resource location. The estimations show that the reserves of natural gas are unlimited and currently its main application is as fuel.

Despite the apparent attractiveness of the conversion of natural gas into high value chemicals as a cheap raw material, the nature of the alkanes (i.e., inertness in chemical reaction at mild temperatures) is a definite obstacle. The way to overcome this problem is by finding a catalyst, which is able to activate the alkane under mild conditions and selectively transform it into a desired product. Many different reactions have been studied and within them, particularly, dehydrogenation, oxidative dehydrogenation (ODH), and partial oxidation.

Oxidation of light alkanes is an advantageous method of producing oxygen-containing compounds due to the low cost of the starting material and fewer side products because of the low reactivity of the alkane. Oxidation of n-butane to maleic anhydride over vanadium phosphorous oxides is a good example of a successful catalytic and industrially applied process. Currently, the efforts are concentrated on the oxidation of propane to acrylic acid, which proves to be more difficult due to the even lower reactivity of propane compared with n-butane. As a result, higher temperatures are required; however, at these temperatures the unselective reactions are also enhanced and, in particular, the total combustion. In addition, acrylic acid is less thermally stable than maleic anhydride and readily oxidizes further under these conditions.

Several types of catalysts have been proposed to be active and selective for the propane oxidation to acrylic acid, however, none show performance challenging the existing industrial method of production in a two-step process starting from propene. The most promising catalysts are the mixed transition metal oxides, containing two or more oxides of Mo, V, Te, or Sb, and Nb. The extensive characterization and catalytic testing of these materials reveal the mechanism of interaction between catalyst surface and reactants and products. Two very complex, in terms of steric and atomic organization, phases, designated as M1 and M2, have been identified as essential for the selective conversion of propane via propene and acrolein to acrylic acid in the Mo–V–Te(Sb)–Nb–O_x catalyst. M1 contributes to the activity and the selectivity, while M2 improves only the selectivity. The absence of any of these phases and the presence of single oxide phase promotes unselective reaction routes such as the formation of acetic acid via acetone and total combustion. Reaction mechanism has been identified and two major routes were proposed. The first route proceeds via propene and then acrolein as intermediates to acrylic acid.

The second route proceeds via isopropanol and then acetone, which is oxidatively cleaved into acetic acid and CO_2.

Lewis acid site of medium strength was found to favor the first route mechanism to acrylic acid, whereas Brønsted acid sites was found to favor the second route to acetic acid, particularly for strong acidity and in the presence of water.

13.2 ACRYLIC ACID — INDUSTRIAL APPLICATION AND PRODUCTION

Oxidation processes occupy a prominent place in both the science of catalysis and the catalysis-based modern chemical industry. Selective oxidation, ODH, and ammoxidation represent about 11% in value of catalysts consumed by the chemical industry and more than 60% in products quantity [1]. Table 13.1 presents the production capacity of the major industrial (amm)oxidation processes [2].

Typically, the feedstocks used for the industrial production of these products are olefins, aromatic compounds, or molecules that already contain oxygen. These starting materials have high value and constitute a large portion of the cost of production. The only process where a cheaper alternative is found is the production of maleic anhydride from n-butane. Light alkanes are highly abundant as they constitute the major fraction of the natural gas, cheap and currently, mainly used as a fuel. In addition, alkanes are more environmental friendly as they are less reactive and produce less partial oxidation products. The main obstacle for the wider application of alkanes in industry is the alkane difficult and selective activation and its extensive complete oxidation. Therefore, it is critical to find catalysts, which can activate the alkane in a selective manner at reasonable temperatures.

The success of the butane-based maleic anhydride process occurred at the cost of a dramatic drop of molar selectivity (at most 65–67% compared with 75–77% when starting from benzene) and a drop of productivity by nearly 20%. Compared with the benzene route, butane oxidation gives molar yields per pass at about 55% instead of 75% in the cyclic process [2].

Acrylic acid is a widely used material in the chemical industry and the demand for it and its derivatives increases every year. The growth of acrylic acid production between 1995 and 2000 is found to be ~6.4% per year in USA (Figure 13.1) and

TABLE 13.1
Production Capacities (million tonnes per year)

Product	Formaldehyde	Acrylonitrile	Maleic anhydride	Phthalic anhydride	Acrylic acid	Methyl methacrylate	Ethylene oxide
World production	6.7	4.5	0.65	3.3	2.0	2.2	11.0

Source: Hecquet, G. *Plenary Lecture in the 2nd European Congress* on *Catalysis*, EUROPACAT II, Maastricht, 1995. With permission.

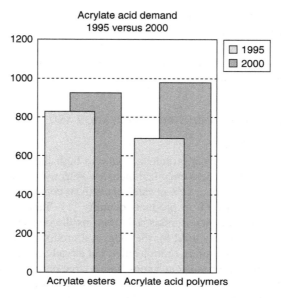

Figure 13.1 Acrylic acid and esters demand in USA in 1995 versus 2000 (From US Outlook: Acrylic Acid and Derivatives with Forecasts to 2006 and 2011. Freedonia Industry Study #1488, 2001.)

it is projected another 4.2% annual increase throughout 2005 [3,4]. The interest in the utilization of acrylic acid and its derivatives is due to the highly reactive double bond and conjugated to it carboxyl functional group. Acrylic acid is mainly used as superabsorbent polymers in the hygiene industry (e.g., baby diapers, sanitary care products, etc.) and as detergent polymers for water treatment. A wide range of acrylate esters are also produced and are mainly used as paints, coating materials, adhesive, and plastics.

Currently, the main capacity for acrylic acid production is in the United States (38%), followed by Western Europe (28%), Japan (14%), the Asia/Pacific region (16%), and the rest in smaller factories in Mexico, Brazil, and Eastern Europe [3]. The main manufacturers are BASF, Celanese, Dow Chemical, Rohm and Haas, Americal Acryl, and Stockhausen. The commercial process for making acrylic acid is a two-step process, starting from propene via acrolein as follows [5]:

Step 1:

$$H_2C = CH - CH_3 + O_2 \longrightarrow H_2C = CH - CHO + H_2O$$
$$(\Delta H = -333 \text{ kJ/mol})$$

Step 2:

$$H_2C = CH - CHO + \tfrac{1}{2}O_2 \longrightarrow H_2C = CH - COOH$$
$$(\Delta H = -265 \text{ kJ/mol})$$

Several multicomponent metal oxide catalysts, developed for this process, have achieved excellent product selectivity with a high conversion of propene: Mo–Bi–Fe–Co–M–K–O (M = V or W) used for the first step can attain >90% acrolein yields [6,7] while for the second step Mo–V–Cu-based oxides can lead to >97% acrylic acid yields [8,9], giving, in theory, an overall acrylic acid yield from propene of 87%. In addition to the compositional differences in the catalysts for the two-step process, there is also a difference in the optimal reaction temperatures: 320–330°C for the first step and 210–255°C for the second step. One has to keep in mind that propene and oxygen can form an explosive mixture and therefore, certain limitations in the feed composition (propene: oxygen (air): steam) exist. In addition, the acrylic acid easily dimerises at temperatures above 90°C, meaning that the reactor effluent should be quickly quenched after the second catalyst bed to temperatures below this critical value.

Alternatively, acrylic acid can also be produced from a one-step oxidation of propene:

$$H_2C = CH–CH_3 + 1.5O_2 \rightarrow H_2C = CH–COOH + H_2O$$

$$(\Delta H = -598 \text{ kJ/mol})$$

The catalyst used for this process is again based on complex mixed oxides, for example, Nippon Shokubai's Mo–W–Te–Sn–Co–O or Nb–W–Co– Ni–Bi–Fe–Mn–Si–Zr–O catalysts giving 65 and 73% yield, respectively [10,11]. The temperature of the reaction lies between 325 and 350°C in order to afford economic propene conversion, but under such conditions, significant total oxidation takes place. Gas-phase homogeneous reaction also occurs, which implies that the size of the catalyst bed, the void volume, and the shape of the reactor are important. These are the reasons why the one-step process yields much less acrylic acid than the two-step process, making the latter the commercially preferred route.

13.3 CHARACTERISTICS OF PROPANE OXIDATION REACTION

As mentioned earlier, the current scientific interest lies in finding cheaper alternatives for the production of acrylic acid. Propene price accounts for 50% or more of the total manufacturing cost of acrylic acid [12]. Furthermore, the price of propene tends to increase (15% in 2001), resulting in increase of the price of acrylic acid by 5% [4]. Such economic alternative is the one-step oxidation of propane to acrylic acid. This reaction has been extensively studied over the past 20 years and yet no catalyst, active and selective enough to make it commercially viable, has been found. Propane requires significantly higher temperatures to achieve reasonable conversion, but under these conditions, great number of reactions take place (Scheme 13.1):

1. Oxidative dehydrogenation (ODH) of propane to propene
2. Selective oxidation reactions to carbonyls (acrolein, propanal, acetone) and carboxylic acids (acrylic and propionic acids)

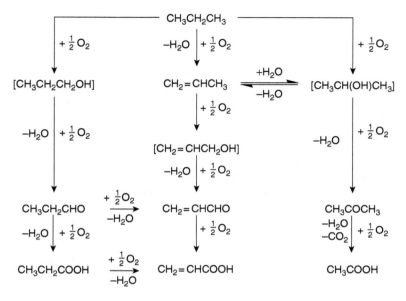

SCHEME 13.1 Main oxygenated products of the partial oxidation of propane; compound in brackets usually not observed

3. Oxidative breaking of C–C bonds giving acetic acid, acetaldehyde, methanol, and formaldehyde
4. Cracking reactions to methane and ethylene
5. Total oxidation reaction to CO and CO_2

There are three main routes: the first one, leading to acrylic acid, is the desired route and goes via propene and acrolein. Allyl alcohol is usually not observed as it is shown to be highly reactive and exist only as surface adsorbed species. The second route leading to acetic acid is a commonly observed side reaction via acetone. The route going to propionic acid is a minor route and rarely reported. The products and the intermediates from all three routes can be further oxidized to CO_x.

Very different catalytic materials have been tested in the selective oxidation of propane, which includes vanadium phosphorous oxide (VPO) catalysts, industrially applied for n-butane oxidation to maleic anhydride. The other catalysts systems include Keggin structure heteropolyoxometallic compounds (HPCs) and multicomponent mixed oxides (MMOs). These materials have different structure and properties, but have something in common: they contain reducible metal oxo species.

The most widely accepted mechanism describing the interaction between a hydrocarbon (R), molecular oxygen, and reducible metal oxide surface is the Mars and van Krevelen mechanism [13]. In this mechanism, the cationic active site oxidizes the reactant R creating lattice oxygen O_L vacancy □, and being reduced yields the product P. The catalyst is then reoxidized by gas-phase oxygen in a subsequent step, according to the following scheme, where p represents the partial pressure and θ the surface coverage by the hydrocarbon:

$$R_{(g)} + O_L \xrightarrow{k_{red}} P + \square \quad \text{and} \quad \square + (\tfrac{1}{2})O_2 \xrightarrow{k_{ox}} O_L$$

$$\text{with } r_{red} = k_{red}\, p_R \theta \quad \text{and} \quad r_{ox} = k_{ox}\, p'_{O_2}(1 - \theta)$$

A selective oxidation reaction involves one (or several) H atoms(s) abstraction from the hydrocarbon molecule, one (or several) O atom(s) insertion from the lattice and several electron transfers. The ease with which the hydrocarbon can be activated, strongly depends on its nature. While it is relatively easy to activate olefins due to their relatively high reactivity induced by the C=C bond, it is much more difficult for less reactive alkanes. Propane oxidation to acrylic acid requires 4 lattice oxygen ions, the abstraction of 4 hydrogen atoms from the substrate, the insertion of 2 oxygen atoms, and 8 electron transfers. This means that surface cations are reduced and dehydroxylation occurs creating lattice oxygen vacancies, which are further replenished by molecular oxygen that reoxidizes the cations by electron transfers at the surface or in the bulk of the solid. This implies good lattice oxygen mobility and electron conductivity, that is, redox properties and acid–base properties.

Various types of surface oxygen species such as O_{2ads}, O_{ads}, O_3^-, O_2^-, O^-, $O^\bullet$, lattice O^{2-} anions, etc., may exist and are usually classified, following Bielanski and Haber's suggestions [14,15] as electrophilic and nucleophilic: 1. $M^{\delta+} = O^{\delta-}$ (nucleophilic)$\rightarrow O^{2-}$ 2. $M=O \rightarrow O^-$ and 3. $M^{\delta-} = O^{\delta+}$ (electrophilic) $\rightarrow O$. Each case corresponds to specific properties:

1. Nucleophilic oxygen — participates in the activation of a C–H bond in α-position to the double bond of alkenes as in the case of propene and is considered selective in case of alkane activation.
2. Weakly polarized oxygen — favors predominantly homolytic fragment-ation of C–H bond in the coordination sphere of the acceptor metal ion and transfer of the hydrogen to the oxygen ion in -2 oxidation state causing the metal–oxygen bonds to break with a transfer of an electron toward the metal.
3. Electrophilic oxygen — leads to direct attack on a double bond (oxidative breaking) and unselective activation in the case of alkanes.

The transition metal oxides are nonstoichiometric compounds whose compos-itions depend on the surrounding gas-phase environment. The formation of point defects or alteration of the linkage between polyhedra under catalytic reaction conditions, because of lattice oxygen incorporation in the adsorbed hydrocarbon, leads to the reorganization of the surface and to the formation of extended defects in crystallographic shear planes. This has been demonstrated by Gai [16] for VPO catalysts for n-butane oxidation to maleic anhydride. It follows that the oxide surface is changing during catalytic reaction and can be considered as "living and in breathing movement." It also corresponds to the frequent suggestion that the catalyst uppermost surface is rather labile and even amorphous [17,18].

The highly electronegative character of the oxygen determines ionic bonding with metals and covalent with nonmetals. Oxides of metals at high oxidation state such as V^{5+} or Mo^{6+} are characterized by a covalent metal to oxygen bond and behave as acidic oxides, whereas the same elements at lower oxidation state have more ionic character and behave as basic oxides. The acid–base characteristics of the oxide have a major effect on the activation of the reactants, the relative rates of competitive reaction pathways, and the rate of adsorption and desorption of the reactants and products. These characteristics may change under reaction conditions depending on the oxidation state of the catalyst surface.

On a basic oxide, one expects the allylic species, formed by the first H abstraction, to be anionic, and π-bonded to a Lewis acid site. It is then susceptible to an attack at the α-hydrogen by neighboring lattice oxygen. The allyl species can be side-on or end-on bonded depending on the metal. The former may undergo nucleophilic attack and give a ketone or an aldehyde, while the latter may dimerise.

On an acidic oxide, the allylic species exist as carbocations giving an unsaturated ketone $R-CO-CH=CH_2$ or aldehyde $R-CH=CH-CHO$. Another possibility is that, the hydrocarbon is attacked by a Brønsted acid site and forms an alkoxide intermediate, which gives a saturated ketone $R-CO-CH_2-CH_3$. It is clear that controlling the acid–base properties of the catalyst surface will affect the first C–H activation and first reaction intermediate, and thus the overall reaction scheme.

There are a number of key factors ensuring the selective oxidation of light alkanes as summarized by Cavani and Trifiro [19]:

1. *Activation of oxygen and alkane*:
   a. The role of adsorbed oxygen species.
   b. The importance of the mode of alkane adsorption.
2. *Relative reactivity of the reactant and the product(s)*:
   a. The mechanism of activation of the C–H bond (heterolytic versus homolytic activation).
   b. The role of the stability of the products.
3. *The mechanism of the transformation of the reactant*:
   a. The importance of nondesorption of reaction intermediates.
   b. The importance of the relative ratio between the intermediate olefin (oxi)dehydrogenation and oxygen insertion in affecting the selectivity to oxygenated products.
   c. The role of the nature of the reaction intermediate (hydrocarbon fragment) in determining the direction of oxidative transformation.
   d. The contribution of homogeneous reactions, especially for processes, which require temperatures higher than 400 to 450°C.
   e. The effect of coadsorbates in facilitating the dissociative adsorption of saturated hydrocarbons.

The authors also suggest a number of key properties of the bulk and the surface of catalysts as fundamental tools for controlling the catalytic performance and products distribution:

1. *The surface of the catalyst:* The nature of the active sites and how the surface is affected by the bulk features, specifically:
   a. The density of the active sites — "site isolation" theory.
   b. The role of surface acidity/basicity.
   c. The need of intrinsic surface polyfunctionality.
2. *The structure of the catalyst:*
   a. The redox properties of the metal in transition metal oxide-based catalysts, in terms of reducibility and reoxidizability of the active sites, and the metal–oxygen bond strength.
   b. The reactivity of specific crystal phases in the different transformations, which constitute the reaction network.
   c. The role of structural defects in favoring the mobility of ionic species in the bulk.
   d. The importance of cooperative effects of different phases in obtaining catalysts with improved performances.
   e. The importance of the interaction between the support and the active phase in modifying the catalytic properties of the latter.

Most of these aspects may be applied, in general, to hydrocarbon oxidation while some of them are specific for alkane transformation. Later in this chapter, we are going to focus on the effect of redox and more acid–base properties of three different catalyst systems on the selective oxidation of propane to acrylic acid.

13.4 CATALYSTS FOR SELECTIVE OXIDATION OF PROPANE

Considering the nature of the reactants, the interactions between reactants in gas phase, reactants with catalyst, and the subsequent reactions on the catalyst surface, three main classes of catalysts have been proposed for propane partial oxidation:

1. Catalysts based on reducible metal oxides (typically transition metal oxides) where heterogeneous, redox-type of mechanism is operating.
2. Catalysts based on nonreducible metal oxides or systems, which do not reduce under the reaction conditions, for which the mechanism is initiated on the catalyst and then transferred into the gas phase or in close proximity to the catalyst surface.
3. Noble metal-based catalysts, which are usually considered as nonselective oxidation systems (i.e., used for combustion reactions), but which under particular reaction conditions may become selective catalysts.

The first group of catalysts is the most promising for partial oxidation of propane and from those, the best results are achieved using VPO, HPCs, and especially mixed oxides.

13.4.1 Vanadium Phosphorous Oxides Catalysts

Usually, catalysis research for a new reaction starts with investigating systems effective for similar reactions. Therefore, the most obvious choice is the vanadium phosphorous oxide (VPO) catalyst, which is successfully implemented in the industry for n-butane oxidation. The reported maleic anhydride selectivity varies from 45 to 65% with n-butane conversion of 65% [20,21]. VPO is also well known to catalyze selectively O- and N-insertion reaction on aliphatics, methyl aromatics, and methyl heteroaromatics [22,23].

Despite the relatively simple catalyst composition, many crystalline phases exist, classified as V^{5+} phases: hydrated or pure $VOPO_4$ (e.g., α_I-, α_{II}-, β-, γ-, δ-$VOPO_4$ (.$2H_2O$)) and V^{4+} phases: hydrogenated, hydrated or pure $VOHPO_4 \cdot 2H_2O$, $VOHPO_4 \cdot 0.5H_2O$, $VO(H_2PO_4)_2$, $(VO)_2P_2O_7$, $VO(PO_3)_2$, etc., [24]. Among these phases, $VOHPO_4 \cdot 0.5H_2O$ is particularly important since it is the precursor of the $(V^{4+}O)_2P_2O_7$ crystallized phase, usually observed in the final, activated VPO catalyst. V^{5+} oxides also exist as isolated species at the surface of the $(VO)_2P_2O_7$, as shown by spin echo mapping of ^{31}P nuclear magnetic resonance (NMR) peak [24], in connection with some defects (electron vacancies p^+) formed on its (100) crystalline face:

$$V^{5+} \rightarrow V^{4+} + p^+$$

such an electron vacancy can be filled by one neighboring O^{2-} anion:

$$O^{2-} + p^+ \rightarrow O^-$$

which globally gives:

$$V^{5+} + O^{2-} \rightarrow V^{4+} + O^-$$

The V^{4+}/O^- couple is then believed to be the center for n-butane oxidation according to the equation:

$$C_4H_{10} + O^- \rightarrow C_4H_9^\bullet + O^{2-} + H^+$$

$C_4H_9^\bullet$ radicals are, then, the starting reacting entities for further O-insertion. The importance of the domains of V^{5+} located on the (100) face is well documented [25,26]. The active V^{4+}/V^{5+} pair is observed for different microcrystalline and amorphous $V^{5+}OPO_4$ phases dispersed on crystalline $(VO)_2P_2O_7$ phase, making it difficult to deduct if the latter phase is the active phase or it is a combination of crystalline $(VO)_2P_2O_7$ and amorphous $V^{5+}PO$ phases for n-butane oxidation. A mature, working VPO catalyst always shows an average vanadium oxidation state higher than 4.0, proving the importance of the redox properties of the vanadium and the presence of both oxidation states.

The VPO catalyst acidity is also very important as H-abstraction occurs on Lewis acid sites (i.e., V^{4+}) and the *n*-butane C–H bond cleavage results from the interaction between Lewis and Brønsted acid sites (P–OH). Brønsted acidity is reported to favor the stabilization of reaction intermediates, the stabilization of adsorbed oxygen species or in the generation of organic surface species, involved in oxygen activation and transport [27].

Addition of steam to the reaction feed proves to be crucial for obtaining high yields of maleic anhydride by modifying the physico-chemistry of the VPO catalyst: improved overall crystallinity, decreased ratio of V^{5+}/V^{4+}, enhanced Brønsted acidity, but decreased acidic strength. It has been determined that the rate of maleic anhydride formation is proportional to the decay of V^{5+} species [28].

Propane is a homologue of *n*-butane with one less CH_2 group and therefore, the same mechanism of catalyst operation may apply. Unfortunately, VPO catalysts are not particularly effective in propane selective oxidation ($<15\%$ yield to acrylic acid) due to lesser reactivity of propane (higher reaction temperatures) and lesser stability of acrylic acid compared with maleic anhydride (enhanced total combustion) [29]. At the higher reaction temperatures, coke is also formed in some cases on the catalyst surface [30]. Typical catalytic results for propane partial oxidation over VPO catalysts, reported in the literature, are summarized in Table 13.2.

TABLE 13.2

Propane Oxidation in Presence of Stream Over VPO Catalysts

Catalyst	Rxn T (°C)	X_{C3} (%)	S_{AA} (%)	S_{AcA} (%)	Y_{AA} (%)	Reference
VPO (P:V = 1.15atomic)	390	36	18	—	6.5	
Te-VPO	390	32	28	—	9	29
(P:V:Te = 1.15:1:0.1)						
VPO (P:V = 1.2)	350	45	17	17	7.5	31
VPO (P:V = 1.04) [a]	300	24	7.5 ($C_3^=$)	—	trace	32
VPO (P:V = 1) as:						
(VO$_2$)P$_2$O$_7$	400	6	4 (Acr)	10 [b]	—	33
VOPO$_4$	400	8	4 (Acr)	25	—	
VPO (P:V = 1.1)	420	46	32	N/A	14.7	34
VPO bulk (P:V = 0.9)	300	15	25	22	3.7	30
16.7%VPO/TiO$_2$-SiO$_2$	300	22	39	22	8.5	
VPO (P:V = 1)	400	23	48	—	11.2	35
Zr-VPO	340	18	81	—	14.8	
(P:V:Zr = 1.5:1:0.5)						
0.01% Ce-VPO	390	28	68	N/A	18.8	36

X_{C3}: propane conversion; S_{AA}/Y_{AA}: selectivity/yield of acrylic acid, S_{AcA}: selectivity to acetic acid; N/A: data are not available.

[a] No steam is introduced.

[b] Total yield of acetic and propionic acids.

The initial work on propane oxidation over VPO was done by Ai [29] who found that there is an optimum P:V ratio of 1.15 over a wide range of propane conversions, where high selectivities to acrylic acid are obtained. The data in Table 13.2 are for catalysts with compositions close to this optimum (P:V = 0.9 to 1.2 atomic ratio); however, the results are widely spread (acrylic acid yields from 0 to 15%). Variations in the catalyst preparation such as the source of vanadium (oxides and salts), solvent (organic or aqueous), type of reducing agent, etc., and the heat treatment can yield a very active or completely inactive material. This is a problem, which has been often met in n-butane oxidation reaction.

Adding small amounts of dopants, such as Te, Zr, or Ce, significantly improve the yields by making the reaction more selective. For example, introducing Te into the Ai's VPO catalyst increases acrylic acid yield from 6.5 to 9% and in the case of adding Zr, the improvement is from 11.2 to 14.8%. The Ce-doped (0.01%) VPO catalyst shows the highest yield reported till date of 18.8%.

In one of the examples in Table 13.2, catalyst improvement is achieved by introducing amorphous TiO_2–SiO_2 xerogel as a matrix for the active VPO catalyst. The material embodies high thermal stability, inertness in propane oxidation, high surface area, and ability to store and transfer oxygen [30]. The results show that acrylic acid yields increase from 3.7 to 8.5%.

For the majority of catalysts listed in Table 13.2, there are three products of the reaction: acrylic acid, CO, and CO_2. Occasionally, some of the intermediates in the acrylic acid route (see Scheme 13.1), for example, propene ($C_3^=$) and acrolein (Acr) are observed. When no steam is added to the feed (entry 4), the only product observed is propene, indicating the importance of water to the reaction. In other cases, acetic acid is also detected in comparable with acrylic acid amounts [30,31]. As mentioned earlier, acetic acid is formed via a parallel route through isopropanol and acetone. The acetone undergoes C–C bond cleavage, which is usually acid-catalyzed process. However, the examples where this product is observed, have P:V ratio between 0.9 and 1.2 [P:V = 1 is the stoichiometric ratio for $(VO)_2P_2O_7$]. Excess P would induce more Brønsted acidity (P:V > 1), which can explain the formation of the acetic acid in this case, but it is difficult to see why at ratio 0.9 acetic acid is formed and for the catalyst with P:V = 1.15 (entry 1, Table 13.2) it is not.

13.4.2 Heteropolyoxometallic Compounds

The Heteropolyoxometallic compounds (HPCs) are a class of inorganic metal–oxygen cluster compounds and can be formed through a polymerization of metal oxoanions around central cations in aqueous solutions at low pH. Unlike other catalyst systems, the preparation of HPCs does not involve a calcination step and the catalytic activity and thermal stability depends entirely on the formation of specific structure. A number of cage-like structures (Keggin, Dawson, Anderson, etc., type) are known and well characterized by x-ray diffraction (XRD) and infrared (IR) techniques. The most studied Keggin structure with a general formula for a unit cell of $[X^{n+}M_{12}O_{40}]^{(8-n)-}$ has T_d symmetry and is composed of a central

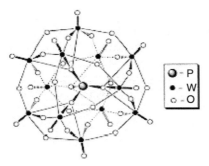

Figure 13.2 Schematic representation of HPCs with Keggin structure

heteroatom tetrahedron (XO_4; X = Si, P, As, B, or Ge) surrounded by 12 peripheral or "addenda" atom octahedra (MO_6; M=Mo, W, or V). The peripheral MO_6 units form four trimetallic clusters (Figure 13.2). Four edge-shared O atoms hold the trimetallic clusters together — one linking to the central heteroatom and three bridging peripheral MO_6 octahedra. The trimetallic clusters are linked to each other by corner-shared oxygen atoms and each metal atom has a single terminal $M=O_t$ ligand. The Keggin unit is anionic, and its negative charge is balanced by cations in the solid state: protons (typically coordinated with water as $[H_5O_2]^+$ or $[H_9O_4]^+$), in which case the complex is acidic (identified as "heteropoly acid") or alkali metals and ammonium derivatives.

Only few types of addenda atoms have been observed in Keggin structures; most typical are Mo and W due to the similar cation size and their ability to accept electron density from the 2p orbitals of the terminal oxygen atoms [37]. The particular radii of Mo^{6+} and W^{6+} (0.073 and 0.074 nm, respectively) allow them to fit into the anionic oxygen framework of the Keggin structure. However, many cations have radii very close to 0.074 nm, but they lack the strong ability to accept electron density via $p\pi-d\pi$ interactions with terminal oxygen atoms. This is important since, the Keggin structure gains stability by elongation of the X–O–M bond lengths, which separates the highly oxidized metal addenda from one another in the otherwise rigid oxygen framework. This also results in contraction of the $M=O_t$ bond due to $p\pi-d\pi$ interaction making the terminal oxygen anions less basic. This effect is stronger with tungsten than with molybdenum, hence the terminal oxygen atoms have more negative charge with molybdenum addenda. As a result, they bind protons more strongly and are weaker acids [38]. Such a property will be used in Part 4 to tune acid-strength properties of HPCs by substituting W for Mo in the Keggin structure.

Based on the above information, Keggin-type HPCs demonstrate activity as acid (both Brønsted and Lewis) and redox (Mo^{6+} and V^{5+}) catalysts. The latter property was initially recognized in 1975 for the oxidation of methacrolein to methacrylic acid using $(NH_4)_2HPMo_{12}O_{40}$ and the process was industrialized in 1982 [39]. The acidic properties of the Keggin structure are utilized in the hydrations of isobutene and 1-nbutene to butyl alcohols in solutions (brought into industry in 1985 and 1986, respectively) [39]. Other applications are continuously discovered as oxidation catalyst in alcohols dehydrogenation and allylic ODH of aldehydes, acids, ketones, and nitriles (e.g., isobutene to methacrylonitrile, acrolein to acrylic acid, phenol to hydroquinone, etc.) and as acid catalyst for isomerization, etherification, and alkylation reactions (e.g., isopropanol to propene, 2-methylpropene, and methanol to methyl tert-butyl ether, esterification of acetic acid with ethylene, etc.).

The first attempts at the use of heteropoly compounds as catalysts for the partial oxidation of propane can be dated back to the early 1980s with the research conducted by Rohm and Haas company [40] (Table 13.3). Since then, many and very different HPCs are studied for this reaction, but none can achieve acrylic acid yields higher than 15%.

Keggin-type HPCs ($H_3PMO_{12}O_{40}$) are not very thermally stable and begin decomposing at temperatures below 400°C. This is unfortunate as propane requires high temperature for its activation and under the conditions of the reaction (340 to 380°C), the conversion is limited. Ueda et al. [41,42] performed the reaction at 340°C and found that $H_3PMO_{12}O_{40}$ is practically inactive. This is also attributed to the low surface area of these catalysts (3.8 m^2/g). In order to enhance the surface area, that is, propane conversion, the catalyst can be treated with pyridine (12.8 m^2/g) as previously suggested in the literature [48]. The treated catalyst shows significantly different results with main products being acrylic and acetic acids. The overall yield is still low, mainly due to the low propane conversion, but from a scientific point of view, these results are significant, because they show that neutralizing the Brønsted sites (pyridine selectively blocks these sites) and keeping the catalyst in a highly reduced state enhance the catalyst performance selectively.

Introducing steam to the reactants feed further improves the performance, similarly to VPO catalysts [49–51]. The mechanism by which water affects the reactivity of the Keggin structure has not been elucidated in the literature, but several studies point to the thermal stabilization and reconstructive effects of water on the Keggin structure. The reconstruction effect of steam has been observed for $H_4SiMo_{12}O_{40}$ on SiO_2 after dry calcination at temperatures of up to 600°C, well above the decomposition temperature of about 300°C [52]. Ilkenhans et al. [53] have also implicated water in the reconstruction of $H_4PVMo_{11}O_{40}$ catalysts during the oxidation of isobutyric acid.

Substituting Mo^{6+} addenda atoms for V^{5+} (i.e., effectively reducing the catalyst) also leads to improvement of the catalytic performance [42], but in a nonlinear way. There is a maximum at one V atom. Further, neutralization of the Brønsted acidity by exchanging the protons with different metal cations (Cs/Fe, Sb, Nb, etc.) brings the acrylic acid yield up to 13% (entries 3 to 6 in Table 13.3). This is a better way of reducing the acidity than with pyridine, because the latter tends to elude from the catalyst into the reaction effluent with time and the metal cations (Cs, in particular) significantly improves the thermal stability ($\approx$450°C) and the surface area.

The best results (15% acrylic acid yield) are obtained for supported HPCs catalysts on high surface area materials such as SiO_2. The catalyst is again

TABLE 13.3
Propane Oxidation over HPCs Catalysts

Catalyst	Rxn T (°C)	X_{C3} (%)	S_{AA} (%)	S_{AcA} (%)	Y_{AA} (%)	Reference
$H_3PMo_{12}O_{40}$		0.9	73[a]	10.6	trace	
$(NH_4)_3PMo_{12}O_{40}$	340	4.5	6	26.2	0.3	41,42
$(PyH)_3PMo_{12}O_{40}$		7.5	29	15.3	2.2	
$H_{1.26}Cs_{2.5}Fe_{0.08}PVMo_{11}O_{40}$	380	47	28	11	13	43,44
$H_{3-n}Sb_nP_1Mo_{12}O_{40}$	340	10	19	N/A	2	40
$Nb(pyr)-PMo_{11}VO_{40}$	380	21	49	23[b]	10.3	45
30 wt% HPA on $Cs_3PMo_{12}O_{40}$:						
$H_2(VO)_{0.5}PMo_{12}O_{40}$	<400	44.8	22.7	28.3	10.2	46
$H_2(VO)_{0.5}(2,2\text{-bipy})PMo_{12}O_{40}$		50.4	21.5	20.5	10.8	
(20%) $H_xCu_{0.6}Cr_{0.6}PMo_{10}V_2As_{0.6}O_{40}/SiO_2$	390	38	39	19	14.8	47

Abbreviations as in Table 13.2.
[a] Selectivity to propene.
[b] Also 15% selectivity to maleic acid.

Mo–V-Keggin-type HPCs with large number of metal cations. It is worth noting that the reaction temperature is 390°C, at which the HPCs is partially decomposed [47]. The authors speculate that during the gradual decomposition, the number of the coordinated oxygen atoms decreases, leading to a change in addenda atoms environment. As a result, specific active species with low coordinated oxygen number and aberrant polyhedron structure could form, facilitating the oxygen insertion process during reaction. The partly destroyed and at the same time partially reduced HPCs (Cu acts as a reducing agent) increases the amount of oxygen scarcity, which is believed to maintain a good redox cycle for the selective oxidation of alkanes.

The product distribution in propane oxidation over HPCs is wider compared with VPO. In all cases, propene, acrylic and acetic acids, and CO_x are also observed. The acrylic and acetic acids are formed in significant amounts, but their ratio varies depending on the catalyst composition (treatment) and reaction conditions: AA/AcA ratio >1 for the catalysts with "neutralized" Brønsted acidity such as Py-treated $H_3PMo_{12}O_{40}$ and $Nb(pyr)-PMo_{11}VO_{40}$ (AA/AcA $= 1.89$ and 2.13, respectively) and $H_{1.26}Cs_{2.5}Fe_{0.08}PVMo_{11}O_{40}$ (AA/AcA $= 2.54$) and AA/AcA ratio <1 for the "acidic" catalysts such as: $H_3PMo_{12}O_{40}$ and $(NH_4)_3PMo_{12}O_{40}$ (AA/AcA $= 0.12$ and 0.23, respectively), and $H_2(VO)_{0.5}PMo_{12}O_{40}$ (AA/AcA $= 0.8$). The exchange of the protons with ammonia is not a very effective method to reduce acidity as ammonia easily desorbs from the catalyst at elevated (reaction) temperatures. Ueda and Suzuki [41] demonstrated this by measuring the ratio between IR absorption intensities of ν [P–O] (1064 cm^{-1}; designated as I_{P-O}) and ν [Mo=O] (962 cm^{-1}; $I_{Mo=O}$) of H-, NH$_3$-, and Py-form $H_3PMo_{12}O_{40}$ after reaction at 340°C and found $I_P-O/I_{Mo=O} = 0.9$; 0.8, and 0.3, respectively. This clearly shows that NH$_3$-exchanged HPCs has been transformed to H$^+$-HPCs under the reaction conditions, while pyridine has remained on the catalyst reducing acidity (P–O) and oxidation state (Mo=O) of the catalyst. All these data clearly show that Brønsted acidity of HPCs, known to be strong, favors the acetone and acetic acid route at the expense of the propene and acrylic acid route. This aspect is discussed above in Section 13.4.

13.4.3 Multicomponent Mixed Metal Oxides

Reducible mixed metal oxides (MMOs) catalysts are currently the most promising catalytic materials for the selective oxidation of propane to acrylic acid. They consist of two or more mixed oxides of transition metals, most typically based on Mo and V oxides. In general, MMO do not have well-defined structure as HPCs; in fact, they are mixture of multiple crystallized and amorphous phases. Mixed oxides are typically prepared via calcination at high temperature, which gives them an excellent thermal stability.

Historically, the application of MMO as catalysts for propane oxidation to acrylic acid began in the late 1970s with Mo–V–Nb mixed oxides, previously reported as a catalyst for ethane oxidation [54]. The results of propane oxidation over this catalyst show that propane could be activated at 300°C, but producing only acetic acid, acetaldehyde, and carbon oxides. However, the possibility to activate

TABLE 13.4
Propane Oxidation over MMO Catalysts

Catalyst	Rxn T (°C)	X_{C3} (%)	S_{AA} (%)	S_{AcA} (%)	Y_{AA} (%)	Reference
$Bi\,Mo_{12}V_5Nb_{0.5}SbKO_x$	400	18	28	16	5	55
$Mo_1V_{0.3}Te_{0.23}Nb_{0.12}O_n$	380	84	63	N/A	53	56
$Mo_1V_{0.3}Te_{0.23}Nb_{0.12}O_n$	380	80	60	N/A	48	57, 58
$Mo_1V_{0.3}Te_{0.23}Nb_{0.12}O_n$	390	71	59	Small	42	59
$Mo_1V_{0.44}Te_{0.10}O_n$	380	36.2	46.6	16.5	17	60
$Mo_1V_{0.25}Te_{0.11}Nb_{0.12}O_n$	380	33.4	62.4	7.3	21	60
$Mo_1V_{0.3}Sb_{0.16}Nb_{0.05}O_n$	380	50	32	N/A	16	61
$Mo_1V_{0.3}Sb_{0-0.1}Nb_{0.1}O_n$	400	39–40	43–42	N/A	17	62
$Mo_1V_{0.3}Sb_{0.25}Nb_{0.12}K_{0.013}O_n$	420	39	64	N/A	25	63
$Mo_1V_{0.18}Sb_{0.15}K_{0.005}O_n$	380	27	31	11	8.4	64
$Ni_1Mo_{1.51}Te_{0.01}P_{0.02}O_x$	460	12	23	—	3	65, 66
Te–Ni–Mo–O	420	40	41	20	16	67

Abbreviations as in Table 13.2.

saturated hydrocarbons at relatively low temperatures encouraged the research of these catalysts. Table 13.4 lists examples of successful MMO catalysts for the oxidation of propane.

The most effective, from the listed catalysts till date, for propane oxidation to acrylic acid is Mo–V–Te–Nb mixed oxides, patented by Ushikubo et al. [56,57] and Lin and Linsen [59], which give more than 40% yield of acrylic acid. The catalyst with the same elemental composition appears to be very active and selective for propane ammoxidation reaction (58% yield of acrylonitrile at 89% propane conversion) [68]. This indicates that propane oxidation and propane ammoxidation share some fundamental reaction steps and active crystalline phases.

In fact, in the recent years, two phases have been identified as responsible for the superior catalytic behavior of Mo–V–Te–Nb–O over other catalysts [69]. They have been named M1 and M2 phases, which stands for orthorhombic and hexagonal, respectively, type structures containing all four metallic elements. The M2 phase (Figure 13.3[b]) has been shown to have hexagonal bronze structure where Mo, V, and Nb cations are octahedrally coordinated to oxygen atoms forming a MO_6 lattice (M=Mo, V, and Nb) with hexagonal channels occupied by tellurium cations and oxygen anions [70,71]. The structure of M1 phase is orthorhombic and exhibits more complex network of MO_6 octahedra forming not only hexagonal channels, but also pentagonal and heptagonal ones. The pentagonal channels are occupied by M cations, the hexagonal ones by Te cations and O anions, and the heptagonal channels are empty (Figure 13.3[a]). Such structural arrangement has already been described for $Cs_{0.7}(Nb_{2.7}W_{2.3})O_{14}$ system [72].

The XRD patterns of these two phases are well established (Figure 13.4), indexed, and the unit cell parameters calculated: M1 phase has $a = 2.1207$ nm;

(a)

(b)

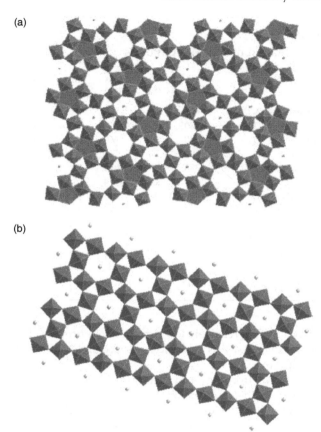

FIGURE 13.3 Projection of the average structures of (a) M1 and (b) M2 phases along [001] direction showing the arrangement of MO_6 octahedra (M=Mo,V, and Nb) and Te (small dots inside the hexagonal channels); the oxygen atoms between Te cations in the channel are not presented for clarity. The sites in the pentagonal channels are occupied by M cations (From Millet, J.M.M., Roussel, H., Pigamo, A., Dubois, J.L., and Jumas, J.C. *Appl. Catal. A: Gen.* 2001, *232*, 77–92. With permission.)

$b = 2.683$ nm, and $c = 0.8047$ nm and M2 has $a = 2.87$ nm and $c = 0.403$ nm [73]. Usually, the XRD pattern of an active catalyst for propane oxidation to acrylic acid is a mixture of the patterns of both phases. Each of these phases show catalytic activity as shown by Baca et al. [74] but does not achieve the performance of a catalyst containing both phases (Table 13.5). Pure M1 is almost as selective as the mixture, but less active, while pure M2 is almost inactive. This suggests that M1 phase is active and selective phase of the efficient mixed phases Mo–V–Te–Nb oxide catalysts, M2 seems to have only some positive effect on the activity of M1.

The morphology of the M1 and M2 phases is determined using SEM (Figure 13.5) [75]. M1 phase forms needles while M2 crystallizes in platelets.

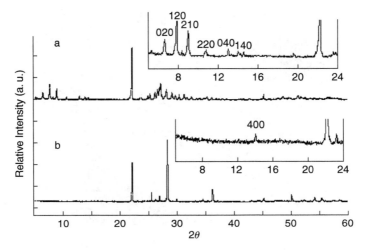

FIGURE 13.4 XRD patterns of single phase catalysts: (a) M1 and (b) M2; enlarged patterns between 5 and 24° with indexation based on [69] (Data taken from Baca, M., Pigamo, A., Dubois, J.L., and Millet, J.M.M. *Top. Catal.* 2003, *23*, 39–46. With permission.)

TABLE 13.5

Catalytic Performance of Catalysts Containing M1, M2, and M1 + M2 phases

Catalyst	T (°C)	X_{C3}	S_{AA}	S_{AcA}	$S_{C_3^=}$	S_{COx}
M1	380	27.9	52	6	9	33
M2	410	0.6	22	—	39	50
M1 + M2	380	34.9	53	4	7	36

Source: From Baca, M., Pigamo, A., Dubois, J.L., and Millet, J.M.M. *Top. Catal.* 2003, 23, 39–46. With permission.

These results are in accordance with other studies [76]. The catalytic activity and selectivity of Mo–V–Te–Nb–O_x increase by grinding, which strongly suggests that the basal plane (side face of the needles) of M1 phase comprises the key catalytic sites, since grinding develops it and that intimate mixture of both M1 and M2 phases is favorable to the reaction.

Despite the large number of techniques employed to characterize these two phases, little is known about how they catalyze the propane transformation into acrylic acid and other products. Normally, propane activation should start with abstraction of methylene H, which is widely believed to occur on $V^{5+} =O$ center on the catalyst surface (M1 phase). This vanadyl group favors radical-type H-abstraction through its resonance form: $V^{5+} =O \leftrightarrow {}^{4+}V^\bullet{-}O^\bullet$ to form secondary propyl radical (see Figure 13.6 right) [75]. Other authors also proposed this first

(a) Orthohombic phase

(b) Pseudo-hexagonal phase

FIGURE 13.5 SEM image of the orthorhombic M1 phase and hexagonal M2 phase (From Grasselli, R.K., Burrington, J.D., Buttrey, D.J., DeSanto, P., Jr., Lugmair, C.G., Volpe, A.F., Jr., and Weingand, T. *Top. Catal.* 2003, 23, 5–22. With permission.)

step for V–Sb–O$_x$ [77] and for *n*-butane activation on VPO [78]. A second H from the methyl group should then be abstracted in order to form the olefin. This, according to Grasselli and coworkers [75,79] occurs on the Te center, adjacent to the V centers. Te is found to be tetravalent in the bulk of the catalyst, but there is some evidence (XPS) that it exists as Te^{6+} on the surface. Without desorption, the π-electrons of the propene molecule can coordinate to an adjacent Mo^{6+} center, where O is inserted via another H abstraction on neighboring Te to form allyl radical. If the Te is originally in 6+ oxidation state, then this second H abstraction on the same center can go without regeneration (Te^{4+} has lone pair of electrons), but if it is tellurium is at 4+ oxidation state as the bulk, reoxidation is required between the H abstractions. The Nb^{5+} centers stabilize the primary active centers and

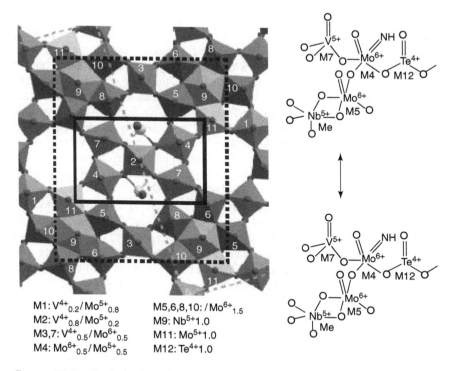

M1: $V^{4+}_{0.2}/Mo^{5+}_{0.8}$ M5,6,8,10: / $Mo^{6+}_{1.5}$

M2: $V^{4+}_{0.8}/Mo^{5+}_{0.2}$ M9: $Nb^{5+}1.0$

M3,7: $V^{4+}_{0.5}/Mo^{6+}_{0.5}$ M11: $Mo^{5+}1.0$

M4: $Mo^{6+}_{0.5}/Mo^{5+}_{0.5}$ M12: $Te^{4+}1.0$

FIGURE 13.6 Catalytically active center of $Mo_{7.5}V_{1.5}TeNbO_{29}$ (M1) in [001] projection on the left and Chem Draw of the active site on the right (From Grasselli, R.K., Burrington, J.D., Buttrey, D.J., DeSanto, P., Jr., Lugmair, C.G., Volpe, A.F., Jr., and Weingand, T. *Top. Catal.* 2003, *23*, 5–22. With permission.)

isolate them structurally from each other. Grasselli and coworkers [80] proposed this mechanism based on structural considerations of M1 phase (Figure 13.6), catalytic performance and previously reported findings. However, one must keep in mind that the M1 phase proposed there ($Mo_{7.5}V_{1.5}TeNbO_{29}$; $a = 2.1134$ nm, $b = 2.6658$ nm, and $c = 0.4015$ nm)) has different stoichiometry and unit cell parameters from those reported by Millet et al. [79] ($Te_2M_{20}O_{57}$ or $(TeO)_2M_{20}O_{56}$; M=Mo, V, Nb; $a = 2.1207$ nm, $b = 2.683$ nm, and $c = 0.8047$ nm).

The second most active and selective mixed oxide catalyst for the propane oxidation to acrylic acid is based again on $Mo–V–Te–Nb–O_x$, in which Te is substituted by Sb [61]. Test results on this catalyst suggested that Sb-based catalyst is less active and selective than its Te analog, but the overall performance was still quite good (16% acrylic acid yield). Takahashi et al. [63] proposed further improvements to this catalyst system by introducing potassium into the catalyst, resulting in raise of the yield to 25%.

$Mo–V–Sb–Nb–O_x$ also has the same two M1 and M2 phases as the Te form [81]. Under hydrothermal preparation, rod-shape crystallites are formed comparable with M1 phase of $Mo–V–Te–Nb–O_x$. Similarly to the latter catalyst, grinding the Mo–V–Sb–Nb–O_x improved the yield to acrylic acid by 13 times, indicating again that the cross-section of the rods is the active for the propane oxidation and that intimate mixtures are more favorable.

Antimony such as Te^{4+} presents a lone pair of electrons and should be able to accommodate the same type of coordination (inside the hexagonal channels of M1 and M2). However, Aouine et al. [70] proposed that Sb also partially occupies the MO_6 sites, which was not possible for Te due to steric hindrance. Antimony exists as Sb^{3+} and Sb^{5+}, and the substitution of Te^{4+} with Sb should involve either an equal amount of cations with both valences or a modification of the valence states of the other elements in the total composition of the phases. The characterization of this catalyst with XANES and EXAFS shows that M1 phase contain predominantly Sb^{5+} ($\approx 89\%$), while M2 has mainly Sb^{3+} ($\approx 63\%$) [82].

Previously, it was reported that M2 is isomorphous to $Sb_4^{3+}Mo_{10}^{5+}O_{31}$, while M1 shows similarities in the unit cell parameters to $Sb_2^{5+}Mo_{10}^{5.2+}O_{31}$ ($a = 2.1207$ and 2.023 nm; $b = 2.6831$ and 0.809 nm; and $c = 0.8047$ and 0.717 nm, respectively) [70]. The discrepancies in the b parameter indicate that M1 has a structure that is more complex and could be a result of both high degree of substitution of Mo cations by smaller V cations and a possible ordering of the metal cations on the sites. It also suggested a possible intergrowth of M1 and M2 structures leading to this more complex organization of M1 phase.

Acetic acid is again a major by-product of the propane oxidation reaction over Mo–V–(Te/Sb)–Nb–O_x. It is observed in all cases (Table 13.4); however, in some of the original publications, no numeric values are reported. Baca et al. [74] made a relation between acrylic/acetic acid formation and the two active M1 and M2 phases (see Table 13.5). The positive effect of M2 phase on the selectivity to acrylic acid in M1+M2 catalyst could only be explained with the lower affinity of this phase toward acetone/acetic acid formation compared with M1 phase. The transformation of propene to acetone proceeds through a hydration-oxidation mechanism as postulated by Luo et al. [83]. For this mechanism to take place, acidic sites are required. Therefore, the acidity of M1 and M2 phases were compared by pyridine thermodesorption followed by Fourier transform infrared (FTIR) [75]. The bands at 1608, 1487 and 1450 cm^{-1} belonging to pyridine adsorbed on Lewis acid sites are observed in the spectra of both phases, but not those at 1638 and 1535 cm^{-1} typical for Brønsted acid sites. The Lewis acid sites are found to be of medium strength in both phases. Figure 13.7 shows the variation of the area of the band at 1450 cm^{-1} as a function of the desorbing temperature.

It is very clear then that M2 has very few acid sites compared with M1, which is in agreement with superior activity (propane conversion) and selectivity to acrylic acid of the former phase. On the other hand, the authors suggest that the medium Lewis acid sites of M1 may account for the high efficiency in propane activation.

13.4.4　Zeolites and Molecular Sieves

Reducible transition metal modified zeolites and molecular sieves, designated as Zeo&MS, find their application in the selective oxidation of alkanes. This is due

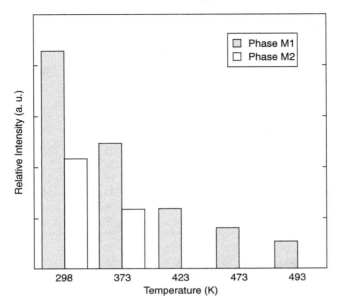

FIGURE 13.7   Relative variation of the area of Lewis ($1450 \ cm^{-1}$) pyridine band as a function of temperature of the desorption of M1 and M2 phases (From Baca, M., Pigamo, A., Dubois, J.L., and Millet, J.M.M. *Top. Catal.* 2003, *23*, 39–46. With permission.)

to the uniform structure of the zeolite/molecular sieve, where the transition metals are anchored inside and outside the framework in an isolated manner. In addition, in some cases, the zeolite structure stabilizes an unusual coordination of the metal cations to the framework. Very wide distribution of results is obtained depending on the metal nature, the position in the framework (inside or extra), and the nature of the matrix (Table 13.6). However, a common feature is the tendency to form alkenes and not acrylic acid, which is presumably due to the high isolation of the active reducible metal cations (it was shown in Section 13.2 how complex is the transformation of propane to oxygenates and the importance of synergy of different metal cations).

Propene is mainly formed over V-substituted molecular sieves, in particular, AlPO-5 and silicalites. The mechanism of its formation is not very clear. However, there are several experimentally established aspects of successful ODH catalyst: V has to be incorporated inside the framework as tetrahedral $V^{5+}O_4$ species [87,89]. Still, there is no evidence of the actual existence of P–O–V and Al–O–V bonds [89] and seems that if there is any real substitution, it occurs at low extent. Furthermore, low vanadium content catalysts (1V/u.c. or $\approx$2 wt% V) show better performance, which is explained by extra-framework V_2O_5 species (less active) in the case of high loadings. This is experimentally demonstrated by preparing a VAPO-5 catalyst, where vanadium is added during the hydrothermal preparation of the molecular sieve and another, where V is impregnated onto AlPO-5 (entries 2 and 3 in Table 13.6) [85]. Both show similar conversions, but

TABLE 13.6
Propane Oxidation over Zeo&MS Materials

Catalyst	Rxn T (°C)	X_{C3} (%)	S_{CO_x} (%)	S_{AcA} (%)	$S_{C_3^=}$ (%)	Reference
VAPO-5	480	16	20	—	80	84
(0.38 wt% V) VAPO-5	540	31	47	—	53	85
(2.22 wt% V) V_2O_5-AlPO-5	540	37	66	—	34	
VAPO-5			28	—	72	
VO-SAPO-5	400	5[a]	39	—	30	86
CoAPO-5			51	—	19	
V-Silicalite-1 (Si/V=237)	500	20	36	12[b]	52	87
VO–H–Beta; Si/Al = 25; V:Al = 0.61	350	1.6	58	21	21	83
	450	13	79	8	11	
(0.42 wt%) Mn-ZSM −5; Si/Al = 50; Mn/Al=0.2	330	7	59	39[c]	2	88
	370	11	79	17	4	
(0.43 wt%) Fe-ZSM −5; Si/Al=50; Fe/Al=0.2	330	9	65	30[c]	5	88
	330[d]	15	0	28	70	

Abbreviations as in Table 13.2.
[a] Fixed conversion of 5%.
[b] Selectivity to aromatics.
[c] Total selectivity to oxygenates.
[d] N_2O as oxidant at difference to air in all other cases.

a big difference in propene selectivity (36% improvement with six times smaller V loading). It is found that extraction of the uncoordinated vanadium (V_2O_5) by ammonium acetate significantly improves the performance. The same is also true for silicalite-1-based catalysts [87]. There, the effect of Brønsted acidity is also discussed. Pure silicalite has only weak acid sites due to the silanol groups; however, the addition of vanadium creates moderate Brønsted acidity because of the presence of V–OH groups. It is found, when compared with other matrixes, that the increase in the strength of Brønsted acid sites increases the rate of transformation of propene into aromatics, however, a moderate acidity is found to be beneficial to the selectivity of propene. Finally, the authors suggest that the channel structure has a marginal role on the reactivity of propane in presence of oxygen.

Impregnating $(VO)SO_4$ onto SAPO-5 or incorporating (fully) Co into the framework of AlPO-5 gives much less selective catalysts. VO-SAPO-5 has better selectivity toward propene, which has been assigned to partial incorporation of V into the framework [86].

Oxygenates (different from carbon oxides) are observed on transition metal modified zeolites, such as Beta, ZSM-5, and Y. The main products over VO–H–Beta catalyst are acetic acid, propene, and CO_x [83]. No acrolein or acrylic acid is formed. The authors assigned this behavior to the acidity this catalyst exhibits, as most of the reaction steps are acid driven: hydration of the olefin to the corresponding alcohol followed by ODH to ketones.

Propane oxidation over Mn (Fe)-ZSM-5 gives higher yields toward oxygenates at lower temperatures. However, the product distribution is significantly different: on the M-ZSM-5, oxygen-containing products are C_1 (methanol), C_2 (ethanol and acetaldehyde), and C_3 (isopropanol — precursor of acetic acid) indicative of large degree of C–C bond cleavages, probably of propene, on this stronger acidic zeolite (Si/Al = 50 compared with 25 for the VO–H–Beta). One still has to keep in mind that independent of the metal and conditions, the main product of the propane oxidation remains carbon oxides (>60%). A way of improvement is by using a different oxidant such as N_2O [88]. In this case, the main product becomes propene and oxygenates. At 330°C, no carbon oxides are formed at all. It is believed that iron–zeolite complexes decompose N_2O to release so-called α-oxygen and also O^- species. α-oxygen may be easily inserted into the hydrocarbon molecule resulting in oxygenated products, while O^- species are reported to facilitate ODH.

Despite the obvious improvement in the oxidation ability of the metal modified zeolites compared with molecular sieves, the main products of this reaction are carbon oxides due to the high acidity. A recent study on effect of Brønsted acidity on the propane selective oxidation using alkaline earth exchanged zeolite Y reveals that the number of Brønsted acid sites is directly related to acetone formation [90]. It must be marked that this catalyst does not have a reducible metal and the mechanism of propane transformation is based on electrostatic interactions of the propane molecule with the framework oxygen, followed by formation of charge-transfer complex: $[C_3H_8^+O_2^-]_{ads}$. When the framework negative charge is compensated by cation with low electronegativity, the charge on the oxygen atoms may be high enough to create basic properties. The average lattice oxygen charge of alkaline earth exchanged zeolite Y lies between -0.21 and -0.224. The basicity of the framework oxygen increases with the increase in cation radius: Mg < Ca < Sr < Ba, opposite to their Brønsted acidity. It was found that the amount of propane absorbed increases with increase in the basicity of the framework. The reaction products observed by *in situ* IR are isopropylhydroperoxide and acetone formed as shown below:

$$CH_3CH_2CH_3 + O_2 \rightarrow [C_3H_8^+O_2^-]_{ads} \rightarrow CH_3C(OOH)HCH_3$$
$$\rightarrow CH_3COCH_3 + H_2O$$

The formation of peroxide (oxygenates precursor) is believed to occur on the cation site. The ratio between isopropylhydroperoxide and acetone is highest for BaY and decreases for SrY, followed by CaY and MgY, opposite to their Brønsted acidity BaY<SrY<CaY<MgY. Furthermore, the products ratio continuously decreases with time, which is probably due to the additional formation of acid sites via hydrolysis of the *in situ* produced water:

$$M^{2+} + H_2O_{ads} + Si–O–Al \leftrightarrow M(OH)^+ + Si–O(H^+)–Al$$

This indicates that the peroxide decomposition to acetone and water occurs on the Brønsted sites. Therefore, reducing the Brønsted acidity would stabilize the peroxide intermediate and under conditions that are more suitable could be converted into more desirable products.

13.5 CASE STUDY: PROPANE OXIDATION OVER MIXED OXIDES, HPCs, AND ZEOLITES

This section presents a summary and comparison of four different projects carried out at Leverhulme Centre for Innovative Catalysis, University of Liverpool, UK on propane (amm)oxidation over Mo–V–Sb–Nb mixed oxides, V- and W-modified Keggin structure HPCs and Ga exchanged H-ZSM-5. These three examples represent the main groups of catalytic materials for propane oxidation to acrylic acid (see Section 13.3). The results, which will be discussed here, on the effect of the different redox and acid–base properties on the reaction, aim at bringing further insight into the catalytic transformation of propane. Introduction of steam or ammonia to the propane:oxygen mixture over some of the catalysts is demonstrated to be a crucial parameter for more selective reaction.

13.5.1 Experimental

13.5.1.1 Catalyst preparation

1. Mo–V–Sb–Nb–O$_x$: A Mo$_1$V$_{0.3}$Sb$_{0.25}$Nb$_{0.08}$O$_n$ mixed oxide sample was prepared according to the patent literature [91] by a co-precipitation method, based on the preparation of slurry containing the salts of the four metals (Aldrich): ammonium molybdate, ammonium metavanadate, antimony trioxide, and niobium oxalate in water. The suspension was refluxed at 90°C for 12 h, then concentrated and dried at 120°C. Varying the calcination conditions and further *in situ* activation prior reaction was the method chosen to modify the properties of the catalyst. The same batch of raw material is used in all heat-treatments ensuring reliable results. The patent calcination procedure (600°C/ flow of air/2 h) has been extended by calcining the sample at 500 or 600°C under different atmospheres: nitrogen, 2% NH$_3$/air, 2% H$_2$O/air, and dry air for 2 h. The catalyst was further activated at 500°C under oxidative (20% O$_2$ in He) or inert (He) atmospheres for 1 h *in situ* in the reactor before starting the reaction.

2. HPCs: The HPCs with chemical formula Cs$_{2.5}$H$_{0.5+x}$PV$_x$Mo$_{12-x}$O$_{40}$, Cs$_{2.5}$H$_{0.5}$PMo$_{12-x}$W$_x$O$_{40}$, Cs$_{2.5}$H$_{1.5}$PV$_1$Mo$_{11-x}$W$_x$O$_{40}$, and Cs$_{2.5}$H$_{1.5-3x}$M$_x$PV$_1$Mo$_{11}$O$_{40}$ were prepared according to the following procedures: the acid forms (H$_{3+x}$PV$_x$Mo$_{12-x}$O$_{40}$, H$_3$PMo$_{12-x}$W$_x$O$_{40}$, H$_4$PV$_1$Mo$_{11-x}$W$_x$O$_{40}$, and H$_4$PVMo$_{11}$O$_{40}$) were prepared by adding calculated amounts of V$_2$O$_5$ and H$_2$WO$_4$, MoO$_3$, and H$_3$PO$_4$ in water. The solutions were stirred, heated up to 80°C and maintained at this temperature under reflux for 24 h. The solid samples were obtained by evaporating the solution to dryness at 50°C without filtration.

The Cs$_{2.5}$ salts for H$_{3+x}$PV$_x$Mo$_{12-x}$O$_{40}$, H$_3$PMo$_{12-x}$W$_x$O$_{40}$, and H$_4$PV$_1$Mo$_{11-x}$W$_x$O$_{40}$ samples were prepared by precipitating the aqueous solution of each acid with stoichiometric amount of Cs$_2$CO$_3$ and then evaporated to dryness at 50°C without filtration.

The $Cs_{2.5}H_{1.5-3x}Ga_xPV_1Mo_{11}O_{40}$ samples were precipitated by adding a mixture of the desired amount of $Ga(NO_3)_3$ and Cs_2CO_3 to the $H_4PVMo_{11}O_{40}$ acid sample. The solid samples were obtained by evaporating the solution to dryness at 50°C without filtration. Prior to reaction, the samples were calcined for 1 h at 400°C under the O_2 flow.

3. Ga-ZSM-5: NH_4-form ZSM-5 (Zeolyst) was calcined in air at 500°C to yield H-form ZSM-5. A known amount of the latter was then stirred with 0.05 M aqueous solution of $Ga(NO_3)_3$ for 2 to 4 h at 100°C (volume of exchange solution to mass of zeolite = 10 cm^3/g). The solid was recovered by filtration and dried at 150°C. The Ga/H-ZSM-5 materials were then calcined at 500°C in air to yield Ga_2O_3/H-ZSM-5. Catalyst activation was done by three successive H_2–O_2 treatments at 550°C for 6 h (1 h in each atmosphere).

13.5.1.2 Catalyst characterization

The catalysts were characterized using elemental, bulk, and surface area analyses. XRD, TPR, XPS, and ESR techniques were used for mixed oxides; XRD, FTIR, ^{31}P MAS NMR, and TGA-DSC techniques for HPCs and EDX-TEM, ^{27}Al and ^{29}Si MAS NMR, FTIR (of hydroxyl groups and lattice vibrational bands) techniques for Ga/H-ZSM-5. The elemental composition of the catalysts was determined by either acrylic acid or ICP (Spectro CIROS 120 SOP). The surface area measurements were performed with a Micromeritics ASAP2000 apparatus using nitrogen as probe. The XRD characterization was carried out at room temperature using a D-5000 Siemens diffractometer with a Co anticathode between 10 and 90° 2-Theta angle. The TPR characterization was performed in Micromeritics TPO/TPR 2900 apparatus from room temperature to 1100°C (5°C/min) under 5% H_2 in Ar. XPS analyses were done using a VG Scientific spectrometer with Mg anode as a source of nonmonochromatized x-ray radiation. All binding energy values were determined by using the C_{1s} line as a reference at 284.5 eV. ESR spectra were recorded on Varian E109 spectrometer at X-band frequencies (9.145 GHz) and 100 kHz field modulation at room or −173°C (flow of cooled N_2 by liquid nitrogen).

FTIR analyses were done on Nicolet Nexus FTIR Raman spectrometer at room temperature in diffusion–reflectance mode (2 wt% HPC in KBr). Solid-state NMR analyses were carried out on a Bruker 400 MHz spectrometer. TGA-DSC analyses were performed on Setaram TGA-DSC 111 apparatus by heating the samples up to 600°C in N_2 at heating rate of 5°C/min. The mass changes and heat flow were recorded against the temperature. HR-TEM and EDX analyses were performed using a JEOL 2010 microscope, operating at 200 kV.

13.5.1.3 Catalytic testing

Catalytic experiments were performed in a continuous flow fixed bed microreactor using SiC diluted catalyst to obtain predetermined total volume of catalyst bed (i.e., GHSV, h^{-1}). The feed composition comprises propane, oxygen, and, in some cases, steam or ammonia. In the case of mixed oxide catalyst and Ga-ZSM-5, the condensable products (acrylic and acetic acids, acrolein, acetone, acrylonitrile, acetonitrile, etc.) were collected in a system of two successive traps at room temperature and the gases were analyzed on-line (propane, propene, carbon oxides, etc.). For HPCs catalyst, all products were analyzed on-line. The GC analyzes were performed with a Varian chromatograph, using three columns: Haysep D (CO_2, H_2O, and hydrocarbons) and MolSieve 13X (O_2, CO, and CH_4) packed columns set in series and a FFAP capillary column, connected in parallel to the other two, to analyze the oxygenates and propane.

13.5.2 Results

13.5.2.1 Mo–V–Sb–Nb mixed oxides [92]

The catalytic performance of transition metals mixed oxides is determined not only by the properties of the individual metal cations, but also by their interaction during preparation and heat treatment. Depending on the calcination procedure and the activation prior to reaction, catalysts with very different crystalline structures are obtained (Table 13.7). No effect of calcination temperature is observed on the XRD patterns and, therefore, the discussion will be focused only for the catalysts calcined at 500°C. For catalysts calcined in air with or without additives, very similar XRD patterns are found with MoO_3 being the major phase together with a phase of type $(M_xM_yMo_{1-x-y})_5O_{14}$, with $x = V$ and $y = Nb$. It has been shown in the literature that the former enhances the activity of the catalyst for partial oxidation reactions, but has a negative effect on the selectivity to oxygenated product [93]. The lack of any effect of water or ammonia in air on the crystalline structure suggests that either higher concentration levels of additives are required or the effect is insignificant in the oxidative environment.

The catalyst calcined in nitrogen (activated in inert atmosphere), however, has a very different XRD pattern than the earlier one; no single oxide was identified, except for trace of molybdenum suboxide. The major phase is $Sb_4Mo_{10}O_{31}$ again together with strong presence of a mixed oxide phase of the type of $(M_xM_yMo_{1-x-y})_5O_{14}$, with $x = V$ and $y = Nb$.

As discussed in Section 13.3.3, $Sb_4Mo_{10}O_{31}$ is isomorphous to M2 phase determined for Mo–V–Te–Nb–O_x. The other mixed oxide phase observed in our study, $(M_xM_yMo_{1-x-y})_5O_{14}$ is very interesting and represent the basis of M1 phase. Unfortunately, we could not run the XRD at 2-Theta angles lower than 10° where the most characteristic peaks of M1 are situated to verify if actually the phase, we have observed, is the important M1 phase. M1 structure is a mixed metal molybdenum bronze with features similar to Mo_5O_{14} (designated as θ-molybdenum oxide) and $Mo_{17}O_{47}$ (χ-molybdenum oxide) [80]. Mo_5O_{14} orthorhombic framework is built by MoO_6 polyhedra and MoO_7 pentagonal bipyramids mutually connected by sharing corners and edges, forming hexagonal, pentagonal, triangular, and square channels (Figure 13.8) [93,94]. The lattice parameters of the tetragonal sub-cell are found equal to $a = b = 2.2989$ nm and $c = 0.3938$ nm, that is, u.c. volume of 2.081 nm^3. This phase is observed to be metastable [94] and to decompose to a mixture of $Mo_{17}O_{47}$ and MoO_3 and

TABLE 13.7

Characterisation and Testing Results for Mo–V–Sb–Nb–O$_x$, Calcined and Activated under Different Conditions

Calcination at 500°	Activation at 500°	Main XRD phases	SA m²/g	Catalytic testing[a]				
				X_{C_3}	S_{AA}	$S_{C_3^=}$	S_{AcA}	S_{CO_x}
Air			9.1	5.5	24.2	8.3	29.3	36.5
H$_2$O/air		MoO$_3$ ($M_xM_yMo_{1-x-y}$)$_5$O$_{14}$	7.6	5.1	19.9	9.1	32.9	45.8
NH$_3$/air	O$_2$/He	($M_xM_yMo_{1-x-y}$)$_5$O$_{14}$	9.3	6.1	22.3	8.2	26.9	41.9
N$_2$		Sb$_2$Mo$_{10}$O$_{31}$ ($M_xM_yMo_{1-x-y}$)$_5$O$_{14}$	6.1	7.5	32.1	11.1	33.3	22.4
N$_2$	He	Sb$_4$Mo$_{10}$O$_{31}$ ($M_xM_yMo_{1-x-y}$)$_5$O$_{14}$	6.1	11.4	54.1	12.9	17.7	13.8

Abbreviations as in Table 13.2.

[a] 0.7 g catalyst diluted in SiC (total volume 1.2 cm³); GHSV = 1500 h^{-1}; C$_3$H$_8$:O$_2$:H$_2$O = 52:13:35 vol%, 400°C.

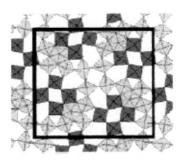

FIGURE 13.8 Crystal structure of Mo_5O_{14} (From Dieterle, M., Mestl, G., Jager, J., Uchida, Y., Hibst, H., and Schlögl, R. *J. Mol. Catal. A: Chem.* 2000, *174*, 169–185. With permission.)

further to MoO_2 and MoO_3, but stabilizes by a partial substitution of transition metals such as V, Nb, and W for Mo or their incorporation into the pentagonal channels. The ternary phase diagram of $(M_xMo_{1-x})_5O_{14}$ with M = V, or Nb was established by Ekström and Nygren [95,96]. They found that the pure θ phase was obtained in the range $x = 0.06$–0.11 for V [95] and $x = 0.05$–0.15 for Nb [96], and in mixtures for x up to 0.2 for V and up to 0.4 for Nb. The a and b parameters were observed to decrease and the c parameter to increase with increasing V substitution for Mo, resulting in a negligible u.c. volume change, although V^{5+} radius is smaller than Mo^{6+} in six coordination (0.054 against 0.059 nm). For Nb, the a and b parameters are similar while the c parameter increased with increase in Nb content [47], resulting in a slight increase in the u.c. volume up to 2.100 to 2.105 nm^3 (radius of $Nb^{5+} = 0.064$ nm in six coordination). In our case, the calculations from XRD patterns gave $a = b = 2.2810$ nm and $c = 0.4042$ nm, that is, u.c. volume of 2.103 nm^3 for air calcined samples and $a = b = 2.2898$ nm and $c = 0.4025$ nm, that is, u.c. volume of 2.110 nm^3. These data indicate that both Nb and V cations substitute Mo in the Mo_5O_{14} structure. Therefore, the complex phase we observe $(M_xM_yMo_{1-x-y})_5O_{14}$, with $x = $ V and $y = $ Nb is probably analogous to the M1 phase.

The *in situ* activation of the already calcined catalysts does not lead to any modifications in the XRD patterns except for the N_2 calcined sample activated in 20% O_2/He flow. The major $Sb_4Mo_{10}O_{31}$ phase is transformed into more oxidized $Sb_2Mo_{10}O_{31}$ phase (similar to M1; see Section 13.3.3) under these oxidative conditions.

The calcination in different atmospheres clearly shows that the catalyst readily changes oxidation state [97]. The TPR measurements (Figure 13.9) demonstrate great degree of reduction with loss of four oxidation states (stoichiometry of reduction is 1 mol catalyst reducible species to 2 mol H_2). At difference to XRD, the TPR shows an effect of the additive to air calcination: addition of NH_3 into air partially reduces the catalyst compared with the dry air calcination, whereas water acts as an oxidant. For nitrogen-calcined sample, the hydrogen uptake is smaller than NH_3/air-calcined catalyst, which is indicative of further reduction.

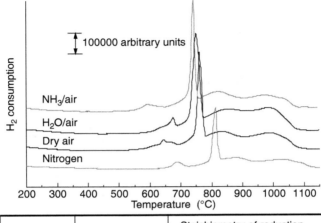

Calcinationat 500°C	Total moles of H$_2$ consumed	Stoichiometry of reduction	
		1cat : 1H$_2$ (%)	1cat : 2H$_2$ (%)
Air	1.45×10^{-4}	172	86
H$_2$O/air	1.02×10^{-4}	192	96
NH$_3$/air	1.62×10^{-4}	120	60
N$_2$	0.90×10^{-4}	106	53

FIGURE 13.9   TPR results for Mo–V–Sb–Nb–O$_x$ calcined under different conditions

XPS clearly shows the presence of some V^{4+} (binding energy of 516.2 eV, V$_2$O$_5$: 516.8 eV, and VO$_2$: 515.9 eV [98]) and Sb^{3+} (539.3 eV; Sb$_2$O$_3$: 539.4 eV) in the catalysts calcined in nitrogen and only traces of V^{4+} (516.6 eV) in the catalyst calcined in air.

ESR shows presence of bulk Mo^{5+}, paramagnetic d^1 cations, with a broad band centred at g$_\perp = 1.9634$ and g$_\parallel = 1.9072$ in both cases of air- and nitrogen-calcined catalysts (Figure 13.10). No difference was observed when the analyses were performed at RT in air and $-173°C$ in vacuum, where the electron transfer between adjacent metal paramagnetic species is diminished, and the paramagnetic O$_2$ from the atmosphere does not interact with the surface of the catalyst, which should result in ESR peak broadening beyond detection. This is a good indication that the paramagnetic species are in the bulk and not on the surface, which is in agreement with the XPS data. Rough calculations indicate that about 15 and 20% of the molybdenum in the nitrogen- and air-calcined catalysts, respectively is Mo^{5+}. In addition to the broad Mo^{5+} signal, a superimposed eight peak hyperfine structure is observed with the characteristic for V^{4+} hyperfine coupling ($I = 7/2$). Even more, expanding the spectra shows that there are two V^{4+} species in different environments and after simulation of the

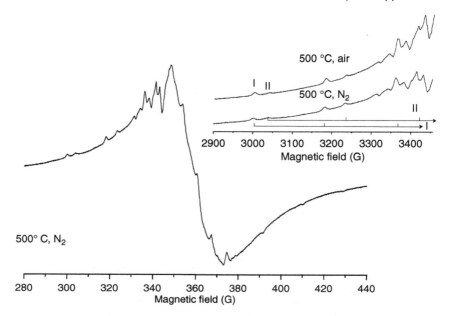

FIGURE 13.10 ESR at $-173°C$ of Mo–V–Sb–Nb–O$_x$ sample calcined in N$_2$; expansion: the two types V^{4+} species

nitrogen-calcined sample spectrum, it was found that there is 4:1 split in the V^{4+} cation environment.

The acidity of the samples was determined by ammonia TPD. Two calcination conditions are investigated: (a) calcination atmosphere — air versus nitrogen at 600°C and (b) calcination temperature — 500 versus 600°C in nitrogen. The patterns obtained are shown in Figure 13.11 with the dashed line designating the fitted Gaussian curves. As could be seen, three different peaks are obtained: the first is between 150 and 220°C and is attributed to weak acidic sites; the second peak is between 250 and 330°C and is due to medium acidic sites; and the third peak at around 420°C is due to relatively strong acidic sites.

TPD data after normalization to the same weight of catalyst are shown in Table 13.8. In order to support the TPD results, acid–base titration was performed with a standardized solution of NH$_4$OH. The results from these titrations are comparable with those from TPD experiments and show $-$lg[NH$_3$] $= 5.8$ and 6.7 for air- and nitrogen-calcined catalyst, respectively. This indicates very low acidity of Lewis type (IR does not show characteristic bands for Brønsted acidity).

Catalytic testing results for the range of air-calcined catalysts show similar performance in terms of both conversion and product selectivities, which is in agreement with their XRD patterns (Table 13.7). Propane conversion seems to be proportional to the surface area, which is between 7.6 and 9.3 m^2/g for H$_2$O/air and NH$_3$/air samples, respectively. Air and NH$_3$/air-calcined catalysts are the most active; however, CO$_x$ and acetic acid formation prevail over acrylic acid. As

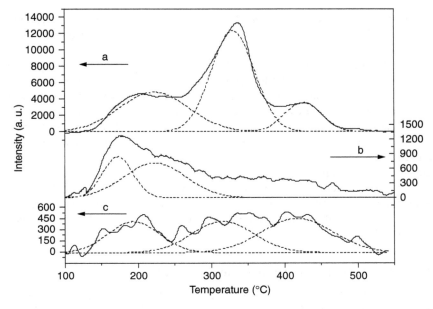

FIGURE 13.11 NH$_3$-TPD patterns of various catalysts: (a) air calcined at 600°C; (b) nitrogen calcined at 500°C, and (c) nitrogen calcined at 600°C

TABLE 13.8
Results from TPD and Titration

Catalyst	NH$_3$-TPD results				H$^+$ titration	
	NH$_3$ moles[a]	$-lg[NH_3]$[a]	NH$_3$ moles[a]	$-lg[NH_3]$[a]	$-lg[NH_3]$[a]	$-lg[NH_3]$[a]
600°C/Air	15×10^{-7}	5.8	11×10^{-7}	5.9	5.5	5.4
600°C/N$_2$	1.1×10^{-7}	7	0.9×10^{-7}	7.1	6.5	6.6
500°C/N$_2$	1.8×10^{-7}	6.7	1.4×10^{-7}	6.9	6.4	6.5

[a] For repeated experiment.

---

mentioned previously, the main crystalline phase for these catalysts is MoO$_3$ followed by the $(M_xM_yMo_{1-x-y})_5O_{14}$, the analog of M1. Therefore, the observed activity and selectivity is probably due to the presence of $(M_xM_yMo_{1-x-y})_5O_{14}$ phase, but the lack of the selective M2 phase and the presence of the unselective MoO$_3$ resulted in the high selectivities toward CO$_x$.

N$_2$-calcined catalyst activated in O$_2$/He shows higher activity despite of the lower surface area (6.1 m^2/g) and higher selectivities toward all valuable products, particularly acrylic acid. The CO$_x$ formation is halved compared with the air-calcined samples. The comparison with the crystalline structure suggests that both $Sb_2Mo_{10}O_{31}$ and $(M_xM_yMo_{1-x-y})_5O_{14}$ contribute to the catalytic performance. However, under the oxygen-limiting conditions, the conversion is still low.

Further improvement is observed for the N_2-calcined and He-activated catalyst, where a maximum in both propane conversion and acrylic acid is reached (54 molC% at 11% conversion). This is a material in which both active and selective combination of phases is achieved: $(M_xM_yMo_{1-x-y})_5O_{14}$ (M1 analog) and $Sb_4Mo_{10}O_{31}$ (M2 isomorph). This is also the only case, where the selectivity to acrylic acid prevails by several times over one of acetic acid, which, based on the literature, could be also due to the lower acidity of nitrogen-calcined catalyst than the air-calcined catalyst.

Therefore, the combined presence of (i) correct crystalline structure, (ii) partial reduction of the transition metal oxides, and (iii) low Lewis acidity, are crucial elements for successful propane oxidation to acrylic acid over Mo–V–Sb–Nb mixed oxide catalyst.

13.5.2.2 Keggin-type heteropolyoxo compounds

Contrary to the Mo–V–Sb–Nb mixed oxides catalyst, where selective propane activation depends largely on creating the right crystalline structure, the Keggin structure HPCs [99] catalyst ($H_3PMo_{12}O_{40}$) is characterized by well-defined atomic arrangement and with predetermined physico-chemical properties. One way, chosen by us, to vary the redox and acid–base properties is by substituting Mo addenda atoms with V^{5+} and W^{6+} cations. This ensures that the latter cations are fully incorporated into the Keggin structure. Another approach, taken here, is by partially exchanging constitutional protons with a transition metal such as Ga^{3+}, known to be a dehydrogenating element, without modifying the Keggin structure. In order to improve the surface area ($H_{3+x}PMo_{12-x}M_xO_{40}$ range between 3 and 8 m^2/g) and the thermal stability (decomposition around 350°C for M=V and 450°C for M=W), in some cases, the protons are partially exchanged with Cs^+. Moffat [38] attributes the dramatic increase in the surface area (order of 100 m^2/g) to the opening of channels in the crystalline structure by the separation of terminal oxygen atoms (breaking H-bonds) of HPCs molecules. Another implication of replacing the protons (typically coordinated with water as $[H_5O_2]^+$ or $[H_9O_4]^+$) with Cs is the reduced Brønsted acidity of the catalysts.

FTIR of the heteropoly compounds demonstrates the degree of incorporation of V and W into the Keggin structure by the level at which the positions of the bands characteristic of the Keggin structure are affected: $\nu_{as}(P–O_a)$ 1080–1060 cm^{-1}, $\nu_{as}(Mo–O_d)$ 990–960 cm^{-1}, $\nu_{as}(Mo–O_b–Mo)$ 900–870 cm^{-1}, and $\nu_{as}(Mo–O_c–Mo)$ 810–760 cm^{-1} (O_a refers to O atom common to PO_4 tetrahedron and a trimetallic edge-shared MoO_6 octahedra, the Mo_3O_{13} triplet; O_b refers to the O atom connecting two Mo_3O_{13} units by corner sharing; O_c refers to the O atom connecting two MoO_6 octahedra inside a Mo_3O_{13} unit by edge sharing and O_d refers to the terminal O atom). Typical IR data of nondehydrated samples with the positions of the bands are given in Table 13.9. The addition of V^{5+} in $H_{3+x}PV_xMo_{12-x}O_{40}$

TABLE 13.9
FT-IR Data in cm^{-1} of HPCs Before Catalytic Reaction

Heteropoly compounds	ν_{as} (P–O$_a$)	ν_{as} (Mo–O$_d$)	ν_{as} (Mo–O$_b$–Mo)	ν_{as} (Mo–O$_c$–Mo)
$H_3PMo_{12}O_{40}$ commercial	1065	962	870	789
$H_3PMo_{12}O_{40}$	1064	962	868	789
$H_4PVMo_{11}O_{40}$	1063	[a]996, 962	867	785
$H_5PV_2Mo_{10}O_{40}$	1060	[a]996, 961	862	781
$H_6PV_3Mo_9O_{40}$	1059	[a]995, 961	864	783
$Cs_{2.5}H_{0.5}PMo_{12}O_{40}$	1064	962	868	789
$Cs_{2.5}H_{0.5}PMo_{11}W_1O_{40}$	1064	970	868	800
$Cs_{2.5}H_{0.5}PMo_9W_3O_{40}$	1065	973	871	807
$Cs_{2.5}H_{0.5}PMo_6W_6O_{40}$	1066	974	871	806
$Cs_{2.5}H_{1.5}PVMo_{11}W_0O_{40}$	1063, [a]1036	[a]997, 967	867	800
$Cs_{2.5}H_{1.5}PVMo_{10}W_1O_{40}$	1063	[a]997, 968	867	794
$Cs_{2.5}H_{1.5}PVMo_9W_2O_{40}$	1064	970	869	796
$Cs_{2.5}H_{1.5}PVMo_8W_3O_{40}$	1065	972	870	808
$Cs_{2.5}H_{1.5}PVMo_7W_4O_{40}$	1066	972	871	792
$Cs_{2.5}H_{1.5}PVMo_5W_6O_{40}$	1068	974	874	795
$Cs_{2.5}H_{1.5}M_0PVMo_{11}O_{40}$	[a]1076, 1060	[a]997, 970	868	802
$Cs_{2.5}H_{1.26}Ga_{0.08}PVMo_{11}O_{40}$	[a]1077, 1061	[a]997, 969	868	803
$Cs_{2.5}H_{1.02}Ga_{0.16}PVMo_{11}O_{40}$	[a]1076, 1062	[a]997, 970	868	803
$Cs_{2.5}H_{0.54}Ga_{0.32}PVMo_{11}O_{40}$	[a]1076, 1061	[a]997, 965	865	796
$Cs_{2.5}H_{0.06}Ga_{0.48}PVMo_{11}O_{40}$	[a]1076, 1061	[a]997, 965	865	795

[a] Indicates presence of a shoulder.

compounds shifts the main bands to lower frequencies, which shows that vanadium is incorporated inside the Keggin unit. Moreover, no evidence of free vanadium and molybdenum oxides (characteristic bands at 1024 and 996 cm^{-1}, respectively) is found.

For $Cs_{2.5}H_{0.5}PMo_{12-x}W_xO_{40}$ and $Cs_{2.5}H_{1.5}PV_1Mo_{11-x}W_xO_{40}$ samples, the introduction of W^{6+} lead to a general increase of the frequencies of the main bands, for example, for $Cs_{2.5}H_{1.5}PV_1Mo_{11-x}W_xO_{40}$ samples ν_{as}(P–O$_a$) increases from 1062 to 1068 cm^{-1}, ν_{as}(Mo–O$_b$–Mo) from 866 to 874 cm^{-1}, and ν_{as}(Mo–O$_d$) from 964 to 974 cm^{-1}, while ν_{as}(Mo–O$_c$–Mo) does not change appreciably suggesting that incorporation of the W^{6+} is achieved with, as a consequence, the strengthening of the W–O bonds and thus increased Brønsted acidity, since the protons as Keggin anion compensating charge are freer.

FTIR data of $Cs_{2.5}H_{1.5-3x}Ga_xPV_1Mo_{11}O_{40}$ samples show no difference when the transition metal is exchanged in the place of protons.

^{31}P -MAS NMR study was done for some of the HPCs samples and shows some interesting features (Table 13.10). Introduction of 1V atom in the Keggin

TABLE 13.10

^{31}P-MAS NMR Chemical Shifts for $Cs_{2.5}H_{1.5-3x}M_xPVMo_{11}O_{40}$ Catalysts Series before Catalytic Reaction

Heteropoly compounds	Chemical shift (ppm)
$H_3PMo_{12}O_{40}$	-3.7
$H_4PVMo_{11}O_{40}$	-3.1 (18%), -3.7 (82%)
$Cs_{2.5}H_{0.5}PV_0Mo_{12}O_{40}$	-4.5
$Cs_{2.5}H_{2.5}PV_1Mo_{11}O_{40}$	-2.6 (3%), -3.5 (53%), -4.2 (44%)
$Cs_{2.5}H_{1.5}PVMo_0Mo_{11}O_{40}$	-2.4 (3%), -3.5 (49%), -4.3 (48%)
$Cs_{2.5}H_{1.26}Ga_{0.08}PVMo_{11}O_{40}$	-2.6 (3%), -3.6 (67%), $-n4.4$ (30%)
$Cs_{2.5}H_{0.54}Ga_{0.32}PVMo_{11}O_{40}$	-2.4 (5%), -3.5 (67%), -4.2 (28%)
$Cs_{2.5}H_{0.06}Ga_{0.48}PVMo_{11}O_{40}$	-2.5 (2.5%), -3.5 (49%), -4.3 (48%)

structure results in the appearance of two signals, indicating either the presence of a mixture of heteropolyoxo compounds or the presence of the same heteropolyoxo compound but in two different hydrated states. In general, partial exchange of protons by $Cs_{2.5}$ results in the shift of the signal from -3.7 ppm for $H_3PMo_{12}O_{40}$ to -4.5 ppm, while for $H_4PVMo_{11}O_{40}$ the peak is shifted to -4.2 ppm with a small presence of a peak at -3.5 ppm. Moreover, $Cs_3PMo_{12}O_{40}$ and $Cs_4PV_1Mo_{11}O_{40}$ give the same signal as $Cs_{2.5}H_{0.5}PMo_{12}O_{40}$ salt at -4.5 ppm. Similarly, $Cs_{2.5}H_{1.5-3x}Ga_xPVMo_{11}O_{40}$ compounds show comparable results indicating that a mixture of the $Cs_{2.5}$, Cs_3 or Cs_4 heteropolyoxo salt and the free acid form coexist. The addition of Ga^{3+} in the position of protons does not appear to modify the Keggin structure.

These NMR data indicate that the $Cs_{2.5}$ salts are in fact composed of the acid form H_3 or H_4 laying on the Cs_3 or Cs_4 salt, rather than a pure $Cs_{2.5}$ phase, whatever they contain V or Ga in their structure or in exchangeable position.

The catalytic results for different HPCs at 5% propane conversion, but without water in the feed, are presented in Table 13.11. The first entry gives the results for unmodified $H_3PMo_{12}O_{40}$ obtained at 368°C. The main product is propene with major side products being CO_x ($S_{CO_2} > S_{CO}$). Traces of acrolein are also detected, but no acrylic acid. Introducing V_x^{5+} to the H-form HPCs (V_1, V_2, and V_3) means increasing the number of protons $(3 + x)$ and therefore Brønsted acidity of the catalyst. The catalytic results do not follow simple pattern with respect to the V substitution; however, the beneficial effect on the formation of oxygenates is apparent. The preferred oxygenated product is acetic acid, which is formed by stepwise oxygen insertion and C–C bond cleavage (acid-catalyzed, see Section 13.3). V_2 gives maximum performance in terms of propane activation (lowest reaction temperature required for 5% conversion) and selectivity to acrylic and acetic acids at difference to Ueda, who observed this maximum for V_1 [42]. CO_x selectivities increase with the increase of H^+ content in the catalyst.

TABLE 13.11

Catalytic Results for V- and W-Substituted (Cs/H)-HPC Catalysts

| HPCs | M (x/K.U.) | H+/ K.U.a | T for 5% X_{C_3} (°C) | Catalytic testing^b | | | | |
				S_{CO_x}	$S_{C_3^=}$	S_{AcA}	S_{AA}	S_{Acr}
$H_3PMo_{12}O_{40}$		2.3 (3)	368	24.8	73.8	—	—	1.4
	1	3.9 (4)	382	36.0	57.3	2.6	—	4.2
$H_{3+x}PMo_{12-x}V_xO_{40}$	2	5.7 (5)	304	44.3	20.0	24.9	9.1	1.7
	3	8.3 (6)	343	63.2	19.8	15.6	—	15
$Cs_{2.5}H_{0.5}PMo_{12}O_{40}$		1 (0.5)	413	91.8	8.2	—	—	—
$Cs_{2.5}H_{0.5+x}PMo_{12-x}V_xO_{40}$	1	1.6 (1.5)	292	20.1	79.9	—	—	—
$Cs_{2.5}H_{0.5}PMo_{12-x}W_xO_{40}$	1	1.6 (0.5)	369	46.5	37.2	9.2	1.3	5.8
	6	1.7 (0.5)	354	44.3	33.8	13.0	2.1	6.9
$Cs_{2.5}H_{1.5}PV_1Mo_{11}O_{40}$		1.6 (1.5)	292	20.1	79.9	—	—	—
	1	2.5 (1.5)	343	31.4	65.9	2.2	—	0.4
$Cs_{2.5}H_{1.5}PV_1Mo_{11-x}W_xO_{40}$	2	2.6 (1.5)	347	42.3	35.4	13.2	3.7	5.3
	4	2.8 (1.5)	333	59.9	24.1	13.4	—	2.6
	0.16	3.0 (1.0)	367	46	48.8	—	—	5.2
$Cs_{2.5}H_{1.5-3x}Ga_xPVMo_{11}O_{40}$	0.32	3.1 (1.2^c)	370	50.1	41.8	4.5	—	3.7
	0.48	4.2 (1^c)	365	41.0	51.8	3.7	—	2.9

Abbreviations as in Table 13.2.

^{a} Experimental values for $H^+/K.U.$ measured by TGA-DSC (200–450°C); in brackets are given calculated values from chemical formula.

^{b} 0.6 to 0.75 g catalyst; $C_3:O_2:He=2:1:2$ vol; total flow 7.5 to 30 cm^3 min^{-1}.

^{c} Calculated values assuming $Ga(OH)_2^+$ as exchangeable cation instead of Ga^{3+}.

It is obvious that protons play a significant role in catalytic performance and in activation of propane and therefore similar study was performed with Cs substituted HPCs samples ($Cs_{2.5}H_{0.5}PMo_{12}O_{40}$) [100,101]. Surprisingly, the results for unmodified catalyst show, mainly, formation of CO_2 and some propene. Introduction of one V atom, change over the ratio between these two products. This could be explained with the higher reaction temperature required to achieve 5% conversion for the former: 413°C compared with 368°C for H-form HPCs and 292°C for V_1 Cs-HPC. No oxygenates are detected, which might imply a necessity for stronger acidity. It is clear that a certain amount of acidity is necessary for the activation of propane and also for the formation of oxygenated products.

Introduction of W ($x = 1$ or 6) as addenda atoms in Cs-exchanged HPCs leads to different catalytic behavior of the catalyst. Propane activation occurs at higher temperatures with W-modified HPCs than with V-modified ones, but the formation of oxygenates, especially acetic acid is enhanced. As discussed earlier, the addenda atoms (W more than Mo) accept electron density from the terminal oxygen atoms (O_t) via $p\pi$–$d\pi$ interactions, making the solid more acidic. Hence, the O_t has more negative charge with molybdenum addenda and bonds more strongly with H^+ and/or Cs^+, resulting in a weaker acidity [38]. Experimental evidence for the enhanced acidity with W-substituted HPCs, is found in H^+ content per Keggin Unit, determined as weight loss due to dehydroxylation, which in all cases is between two and three times higher than the theoretical number of protons, deduced from their chemical composition.

Based on these results, the simultaneous presence of both substitutes (V and W) on $Cs_{2.5}H_{0.5}PMo_{12}O_{40}$ should improve propane activation and dehydrogenation via V substitution, while W will be beneficial for the formation of oxygenates (acetic acid) via the enhanced Brønsted acidity [101]. The results for $Cs_{2.5}H_{1.5}PV_1Mo_{11-x}W_xO_{40}$ ($x = 1$, 2, and 4) indicate the predominant formation of propene for W_0 and W_1, while the formation of oxygenates (acetic acid) and particularly CO_x is enhanced for W_2 and W_4 catalysts. These show that introducing one vanadium atom into HPCs with Mo and W addenda improves the activation of propane (lower reaction temperatures) but does not have effect on the selectivity at higher W loading.

For $Cs_{2.5}H_{1.5-3x}Ga_xPV_1Mo_{11}O_{40}$ samples, the amount of Ga^{3+} influences the products distribution. As mentioned earlier, the Ga-free catalyst ($Cs_{2.5}H_{1.5}PV_1Mo_{11}O_{40}$) gives predominantly propene and some CO_x. Proton exchanging with Ga leads to the formation of not only some oxygenated but also large amounts of CO_x. This can be explained with an increase in the Brønsted acidity at high Ga^{3+} loadings due to the formation of $Ga(OH)^{2+}$ or even $Ga(OH)_2^+$ cations rather than Ga^{3+} for proton exchange on the surface.

Addition of steam to the propane–oxygen feed proves to have beneficial effect on the formation of oxygenates as demonstrated for three of the previously discussed catalysts (Table 13.12). For $Cs_{2.5}H_{1.5}PVMo_{11}O_{40}$ catalyst, presence of steam leads to a dramatic increase in acetic acid selectivity (from 24 to 90% at the expense of propene) at very little loss in conversion. Acrylic acid is also observed. Similar changes are found for the second and the third catalyst ($Cs_{2.5}H_{1.5}PVMo_9W_2O_{40}$ and $Cs_{2.5}H_{0.54}Ga_{0.32}PVMo_{11}O_{40}$) with doubling the selectivities to the acids at the expense of CO_x and propene. Moreover, $Cs_{2.5}H_{0.54}Ga_{0.32}PVMo_{11}O_{40}$ catalyst shows the highest conversion and selectivity to oxygenates, especially to acrylic acid. These results suggest different roles of steam than those already discussed, namely reconstruction effect on the Keggin structure. The reaction temperature used here is only 340°C where the Keggin structure is intact. We suggest two possible effects: enhanced hydration of propene to 2-propanol and acetic acid in presence of steam as observed over zeolites and also partial blocking of the protons of the HPCs via H-bonding, which facilitates the desorption of the partial oxygenated products as observed with MMO and decreased Brønsted acid strength, thus CO_x formation.

13.5.2.3 Ga exchanged H-ZSM-5

Ga exchanged H-ZSM-5 is a proven catalyst for aromatization of light alkanes (C_3 and C_4) and is applied in the CYCLAR process [102]. They act in this reaction as bifunctional catalysts with gallium species promoting the activation and dehydrogenation of the alkane and acid centers, where the polymerization

TABLE 13.12
Effects of Water in the Feed on Catalytic Results for V, W-Substituted and Ga-Exchanged Cs-HPCs

HPCs	$C_3:O_2:H_2O^a$	X_{C_3}	S_{COx}	$S_{C_3^=}$	S_{AcA}	S_{AA}	S_{Acr}
$Cs_{2.5}H_{1.5}PVMo_{11}O_{40}$	2:1:0	8.2	26.2	45.6	24.0	—	4.2
	2:1:1	6.7	2.3	3.2	90.1	4.0	0.4
$Cs_{2.5}H_{1.5}PVMo_9W_2O_{40}$	2:1:0	4.7	34.4	31.0	32.1	5.5	7.0
	2:1:1	3.7	2.3	3.5	83.4	10.3	0.5
$Cs_{2.5}H_{0.54}Ga_{0.32}PVMo_{11}O_{40}$	2:1:0	14.4	25.8	17.2	52.6	1.1	3.3
	2:1:1	13	1.7	1.4	85.6	10.9	0.5

^{a} 340°C, 0.6 g catalyst in SiC (1.1 cm^3 total volume), GHSV=818 h^{-1}.

to aromatics occurs. However, in presence of oxygen these catalysts show oxidation activity. Ga-ZSM-5 similarly to the HPCs, have ordered atomic and steric structure, and also strong Brønsted acidity determined by the ratio of Si/Al. Ion-exchange method, used for this study, is repeated as many times as necessary to obtain desired Ga loading. As a result of this, Ga-species can be deposited outside the intracrystalline volume of the zeolite as GaO(OH), which after calcination transforms into Ga_2O_3 small crystallites (2 to 3 nm in size) In presence of hydrogen, the latter crystallites are reduced to Ga_2O, which migrates into the channels of the zeolite and is "anchored" by cationic exchange of protons by Ga cationic species, as shown by HR-TEM, EDX and ^{29}Si and ^{27}Al MAS-NMR studies [103]. Such H_2–O_2 cycles lead to substantial increase in Ga dispersion. In oxidizing environment, Ga is present as Ga^+ and GaO^+, which in the presence of steam (introduced or reaction by-product) may be converted to $GaOH^{2+}$ or $Ga(OH)_2^+$. Either one or all of these three species are believed to be the active sites for propane activation [104]. However, during propane oxidation, the acidity of the zeolite and the Ga Brønsted acidity proves to be far too high to obtain any oxygenated products apart of CO_x (Table 13.13) even under oxygen-limited conditions. The synergy between the Ga species and the zeolite acid sites appears to be poor. In addition, the presence of external Ga_2O_3 also contributes to unselective oxidation.

Addition of a base, such as ammonia, to propane–oxygen mixture (propane ammoxidation over Ga/H-ZSM-5) has a significant effect by enhancing the formation of valuable products (propene, acetonitrile, and acrylonitrile) over total combustion (Table 13.14) [105]. However, acetonitrile (C_2 molecule) prevails over acrylonitrile (C_3 molecule) in a similar manner as acetic prevails over acrylic acid in propane oxidation over HPCs. The increase in Si/Al ratio, meaning the reduction of the number of Brønsted acid sites, benefits the selectivity to propene and acrylonitrile at the expense of acetonitrile. CO_x selectivities are not sensitive either to change in acidity or Ga content, which could suggest that those are

TABLE 13.13

Catalytic Results for Ga-ZSM-5 in Propane Oxidation

Catalyst	$C_3:O_2{}^a$	X_{C_3}	S_{CO_x}	$S_{C_3^=}$	$S_{Ace+AcA}$	S_{Acr+AA}	S_{MA}	S_{Arom}
Si/Al = 18	1:11	34.4	93.2	1.1	1.5	0.2	—	1.8
Si/Ga = 33	2:1	10.3	75.6	10.6	1.7	2.0	Tr	7.9
Si/Al = 150	1:11	15.0	96.6	1.2	—	0.1	—	0.3
Si/Ga=38								

Abbreviations as in Table 13.2; MA: methacrylic acid.

[a] 400°C, GHSV = 1500 h^{-1}.

TABLE 13.14

Catalytic Results for Ga-ZSM-5 in Propane Ammoxidation

Catalyst		Catalytic testing[a]					
Si/Al	Ga, wt%	X_{C_3}	$S_{C_3^=}$	S_{AcCN}	S_{ACN}	S_{COx}	S_{HCN}
27	0.4	23	29	41	1	25	1.6
	1.6	41	25	47	1.4	23	2
150	0.4	13	36	25	7	23	2
	1.6	18	32	26	10	26	3

Abbreviations as in Table 13.2: AcCN; acetonitrile, ACN: acrylonitrile.

[a] 475°C, $C_3:O_2:NH_3$ =1:2:1, GHSV = 1600 h^{-1}.

formed on the external Ga_2O_3 as for propane oxidation. It has been experimentally established that multiple H_2–O_2 treatments reduce the carbon oxides formation.

13.6 DISCUSSION AND CONCLUSIONS

It has been shown that the presence of particular structure, redox properties, and the levels and type of acidity determine the catalytic behavior for each of the three types of catalysts. One or another reaction pathway becomes more favorable depending on these properties [106–108]. Scheme 13.2 schematizes possible reaction pathways for propane oxidation:

Route I leads to acrylic acid via intermediate formation of propene and acrolein. This route is preferred for mixed oxides, where $(M_xM_yMo_{1-x-y})_5O_{14}$, with $x =$ V and $y =$ Nb (M1 analogu) and $Sb_4Mo_{10}O_{31}$ (M2 isomorph) phases are simultaneously present. Our results also suggest that propane is activated and oxidized to acrylic acid on $(M_xM_yMo_{1-x-y})_5O_{14}$ (as observed for air-calcined catalysts) and

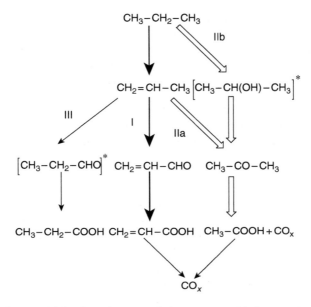

SCHEME 13.2   Reaction network for propane oxidation reaction

further facilitated by $Sb_4Mo_{10}O_{31}$ phase. When the pure $Sb_4Mo_{10}O_{31}$ phase was prepared and tested, less than 1% conversion was obtained with 20% selectivity to acrylic acid.

These phases are characterized by partial reduction of all Mo, V, and Sb cations and low Lewis acidity. The same combination of physico-chemical properties for $Cs_{2.5}H_{1.5}PV_1Mo_{11-x}W_xO_{40}$, and $Cs_{2.5}H_{1.5-3x}Ga_xPV_1Mo_{11}O_{40}$ (reduced acidity due to presence of Cs/Cs–Ga/Cs–W and partial reduction via substitution of Mo^{6+} for V^{5+}) ensured the best selectivity to acrylic acid among all HPCs catalyst presented here. The presence of water is particularly beneficial for the formation of oxygenated products over any type of catalyst. It appears that steam improves the formation of valuable products via (a) modifying the catalyst, for example, reconstructing the Keggin structure of the HPCs, decreasing the strong Brønsted acid catalyst surface, and facilitating of product desorption and (b) changing the kinetics of the reaction such propane activation [109].

Route II leads to acetic acid and can proceed via Route IIa: propene, which is oxidized to acetone and further to acetic acid or Route IIb: oxidation of propane or hydration of propene to isopropanol, further oxidized to acetone and acetic acid. Isopropanol has not been detected in any of the studies; however, it is believed to be too reactive and to exist only as a surface adsorbed species. Route II has been the major route for catalyst with medium to strong Brønsted acidity (HPCs) and also for literature reported zeolites (see VO–H–Beta and M–Y). It is also major for the air-calcined mixed oxide catalyst, where MoO_3 is the main phase. The latter is highly oxidized with medium strength Lewis acidity, which in presence of steam can transform into Brønsted type acid. Route IIa is preferred for catalysts with strong dehydrogenation abilities such as V-containing HPCs and more reduced mixed oxides and in absence of water, while route IIb is the preferred route in presence of water, because of the enhanced hydration of propane and propene. This is especially valid for mixed oxides, where acetic acid is considered a primary product together with propene.

Route III is observed only for mixed oxides and lead to formation of thermally stable propionic acid. It is a minor route (selectivities of less than 5%) and chemically not very probable because propene is hydro-oxidized to propanol against the Markovnikov rule.

The products from all routes can be over-oxidized to CO_x in presence of strong Brønsted acidity, as observed for some of the HPCs catalysts and all Ga-ZSM-5 (combined zeolite and Ga acidity) tested in propane oxidation reaction. In the latter case, presence of the surface Ga_2O_3 additionally catalyzes propane to CO_x. Blocking the Brønsted acid sites with water/ammonia or exchanging the protons with reducible metals in the case of Ga exchanged HPCs ($Cs_{2.5}H_{1.5-3x}Ga_xPVMo_{11}O_{40}$) reduces both C–C bond cleavage (AA/AcA and ACN/AcN ratios increase) and complete oxidation.

The catalysts described here are singular examples of three major groups of materials under investigation for propane (amm)oxidation. They do not exhibit outstanding catalytic performance and it has not been the aim of the research projects carried out at LCIC. However, summarizing the results, the way it is done here, gives fundamental insight on the process and incentives for further reaction and catalyst development. Each of these three types of catalysts can activate propane under relatively mild reaction conditions (300 to 400°C) and with careful tuning of the catalyst properties, it should be possible to achieve reasonable selectivities to chemically valuable products. The intensifying flow of research publications on this subject is clear evidence for the potentials of these materials.

In conclusion, there are some requirements for the selective propane oxidation reaction:

- Redox elements, particularly V, are needed. If not, acetone would be the major product as for H–Y or Ca–Y zeolites.
- Lewis acidity with a relatively mild strength is a key parameter to initiate the formation of oxygenates, but to avoid route II becoming major.
- Brønsted acidity is important to activate propane but should be controlled efficiently to avoid over-oxidation to CO_x and routes IIa and IIb to become major. Partial neutralization by Cs, Ga, or pyridinium ions is compulsory to orientate the reaction toward acrylic acid.
- For propane oxidation to acrylic acid, mixed metal oxides appear to be the most promising catalysts in presence of steam. The presence of two phases such as M1 and M2 turn out to be determining.

Acknowledgment

We thankfully acknowledge Dr. Izabela Nowak for the preparation, characterization, and testing results of propane (amm)oxidation on Ga-ZSM5 catalysts.

References

1. Delmon, B. The future of industrial oxidation catalysis spurred by fundamental advances. *Stud. Surf. Sci. Catal.* 1997, *110*, 43–59.
2. Hecquet, G. *Plenary Lecture in the 2nd European Congress on Catalysis.* EUROPACAT II, Maastricht, 1995.
3. US Outlook: Acrylic Acid and Derivatives with Forecasts to 2006 and 2011. Freedonia Industry study #1488, 2001.
4. Chemical Profile: Acrylic Acid, Commercial report published by The Innovation group in *Chemical Market Reporter.* Schnell Publishing Company, New York, 2002.
5. Nojiri, N., Sakai, Y., and Watanabe, Y. Two catalytic technologies of much influence on progress in chemical process-development in Japan. *Catal. Rev.-Sci. Eng.* 1995, *37*, 145–178.
6. Ohara, T., Ueshima, M., and Yanagisawa, I. JP patent 47-42241B, 1972.
7. Umemura, S., Odan, K., Asada, H., Nakamura, Y., and Tsuruoka, M. JP patent 58-49535B, 1983.
8. Ogawa, M. JP patent 62-34742B, 1987.
9. Wada, M., Ninomiya, M., Yanagisawa, I., and Ohara, T. JP patent 53-6127B, 1978.
10. Kurata, N., Matsumoto, T., Ohara, T., and Oda, K. JP patent 42-9805B, 1967.
11. Nagai, I., Yanagisawa, I., Ninomiya, M., and Ohara, T. JP patent 58-17172B, 1983.
12. Lin, M.M. Complex metal-oxide catalysts for selective oxidation of propane and derivatives: I. Catalysts preparation and application in propane selective oxidation to acrylic acid. *Appl. Catal. A: Gen.* 2003, *250*, 305–318.
13. Mars, P. and van Krevelen, D.W. Oxidations carried out by means of vanadium oxide catalysts. *Eng. Sci.* 1954, *3* (Special Suppl.), 41–59.
14. Bielanski, A. and Haber, J. *Oxygen in Catalysis.* Marcel Dekker, New York, 1991.
15. Haber, J. Molecular mechanism of heterogeneous oxidation — Organic and solid state chemists' views. *Stud. Surf. Sci. Catal.* 1997, *110*, 1–17.
16. Gai, P.L. Environmental high resolution electron microscopy of gas-catalyst reactions. *Top. Catal.* 1999, *8*, 97–113.
17. Védrine J.C. Main description of the V_2O_5-WO_3/TiO_2 EUROCAT oxide SCR standard catalysts and of their support. *Catal. Today* 2000, *56*, 455–460; Védrine, J.C. The role of redox, acid–base and collective properties and of crystalline state of heterogeneous catalysts in the selective oxidation of propane. *Top.Catal.* 2002, *21*, 97–106.
18. Hutchings, G.J., Bartley, J.K., Webster, J.M., López-Sanchez, J.A., Gilbert, D.J., Kiely, C.J., Carley, A.F., Howdle, S.W., Sajip, S., Caldarelli, S., Rhodes, C., Volta, J.C., and Poliakoff, M. Amorphous vanadium phosphate catalysts from supercritical antisolvent precipitation. *J. Catal.* 2001, *197*, 232–235.

19. Cavani, F. and Trifiro, F. Selective oxidation of light alkanes: interaction between the catalyst and the gas phase on different classes of catalytic materials. *Catal. Today* 1999, *51*, 561–580.

20. Abon, M., Bere, K., Tuel, A., and Delichere, P. Evolution of a VPO catalyst in *n*-butane oxidation reaction during the activation time. *J. Catal.* 1995, *156*, 28–36.

21. Ruitenbeek, M., van Dillen, A.J., Barbon, A., van Faassen, E.E., Koningsberger, D.C., and Geus, J.W. The selective oxidation of *n*-butane to maleic anhydride: comparison of bulk and supported V–P–O catalysts. *Catal. Lett.* 1998, *55*, 133–139.

22. Centi, G. Vanadyl pyrophosphate — A critical overview. *Catal. Today* 1993, *16*, 5–26.

23. Martin, A. and Lücke, B. Ammoxidation and oxidation of substituted methyl aromatics on vanadium-containing catalysts. *Catal. Today* 2000, *57*, 61–70.

24. Volta, J.C. Vanadium phosphorus oxides, a reference catalyst for mild oxidation of light alkanes: a review. *C.R. Acad. Sci. Paris, Serie IIc, Chem.* 2000, *3*, 717–723.

25. Volta, J.C., Béré, K., Zhang, Y.J., and Olivier, R. V–P–O catalysts in *n*-butane oxidation to maleic-anhydride — Study using an *in-situ* Raman cell. In *ACS Symposium Series. Catalytic Selective Oxidation*, Oyama, T. and Hightower J.W., Eds., 1993, *523*, 217–230.

26. Zhang, Y., Sneeden, R., and Volta, J.C. On the nature of the active sites of the VPO catalysts for *n*-butane oxidation to maleic anhydride. *Catal. Today* 1993, *16*, 39–49.

27. Centi, G., Colinelli, G., and Busca, G. Modification of the surface pathways in alkane oxidation by selective doping of Broensted acid sites of vanadyl pyrophosphate. *J. Phys. Chem.* 1990, *94*, 6813–6819.

28. Coulston, G., Bare, S., Kung, H., Birkeland, K., Bethke, G., Harlow, R., Herron, N., and Lee, P. The kinetic significance of V^{5+} in *n*-butane oxidation catalyzed by vanadium phosphates. *Science* 1997, *275*, 191–193.

29. Ai, M. Oxidation of propane to acrylic acid on V_2O_5–P_2O_5-based catalysts. *J. Catal.* 1986, *101*, 389–395.

30. Han, Y., Wang, H., Cheng, H., Lin, R., and Deng, J. Dispersed vanadium phosphorus oxide on titania-silica xerogels: highly active for selective oxidation of propane. *New J. Chem.* 1998, 22(11), 1175–1178.

31. Ai, M. Oxidation of propane over V_2O_5–P_2O_5-based catalysts at relatively low temperatures. *Catal. Today* 1998, *42*, 297–301.

32. Centi, G. and Trifiro, F. Functionalization of paraffinic hydrocarbons by heterogeneous vapour-phase oxidation. III. Conversion of the C_1–C_7 alkane series. *Catal. Today* 1988, *3*, 151–162.

33. Takita, Y., Yamashita, H., and Moritaka, K. Selective partial oxidation of propane over metal phosphate catalysts. *Chem. Lett.* 1989, (10), 1733–1736.

34. Wang, Z., Wei, W., Liu, G., Mao, G., and Kuang, D. Oxidation of propane to acrylic acid on V-P-O catalysts. *Shiyou Xuebao, Shiyou Jiagong* 1998, *14*, 21–26.

35. Han, Y., Wang, H., Cheng, H. and Deng, J. V-Zr-P oxide catalysts for highly selective oxidation of propane to acrylic acid. *Chem. Commun.* 1999, (6), 521–522.

36. Cheng, H., Han, Y., and Wang, H. Shiyou. Direct oxidation of propane to acrylic acid over Ce-doped V–P–O catalyst. *Huagong* 1999, *28*, 803–807.

37. Pope, M.T. *Heteropoly and Isopoly Oxometallates.* Springer-Verlag, New York, 1983.
38. Moffat, J.B. Cation-anion effects in heteropoly compounds with Keggin structures. *Polyhedron* 1986, *5*, 261–269.
39. Misono, M. Heterogeneous catalysis by heteropoly compounds of molybdenum and tungsten. *Catal. Rev. Sci. Eng.* 1987, *29*, 269–321.
40. Krieger, H. and Kirch, L. Process for the Production of Unsaturated Acids. U.S. patent 4,260,822, April 7, 1981.
41. Ueda, W. and Suzuki, Y. Partial oxidation of propane to acrylic acid over reduced heteropolymolybdate catalysts. *Chem. Soc. Jpn. Chem. Lett.* 1995, (7), 541–542.
42. Li, W., Oshikara, K., and Ueda, W. Catalytic performance for propane selective oxidation and surface properties of 12-molybdophosphoric acid treated with pyridine. *Appl. Catal. A: Gen.* 1999, *182*, 357–363.
43. Mizuno, N., Tateishi, M., and Iwamoto, M. Pronounced catalytic activity of $Fe_{0.08}Cs_{2.5}H_{1.26}PVMo_{11}O_{40}$ for direct oxidation of propane into acrylic acid. *Appl. Catal. A: Gen.* 1995, *128*, L165–L170.
44. Mizuno, N., Suh, D., Han, W., and Kudo, T. Catalytic performance of $Cs_{2.5}Fe_{0.08}H_{1.26}PVMo_{11}O_{40}$ for direct oxidation of lower alkanes. *J. Mol. Catal. A: Chem.* 1996, *114*, 309–317.
45. Holles, J.H., Dillon, C.J., Labinger, J.A., and Davis, M.E. A substrate-versatile catalyst for the selective oxidation of light alkanes: I. Reactivity. *J. Catal.* 2003, *218*, 42–53.
46. Lyons, J., Volpe, A., Ellis, P., and Karmakar, S. Conversion of alkanes to unsaturated carboxylic acids over heteroploy acids supported on polyoxometallate salts. U.S. patent 5,990,348, November 23, 1999.
47. Jiang, H-S., Mao, X., Xie, S-J., and Zhong, B-K. Partially reduced heteropoly compound catalysts for the selective oxidation of propane. *J. Mol. Catal. A: Chem.* 2002, *185*, 143–149.
48. Ueshima, M., Tuneki, H., and Shimizu, N. Heteropoly acid as oxidation catalyst. Oxidation of molybdovanadophosphoric acid and methacrolein. *Hyoumen* 1986, *24*, 582–594.
49. Howard, M.J., Sunley, G.J., Poole, A.D., Watt, R.J., and Sharma, B.K. New acetyls technologies from BP chemicals. *Stud. Surf. Sci. Catal.* 1999, *121*, 61–68.
50. Davis, M.E., Dillon, C.J., Holes, J.H., and Labinger, J. A new catalyst for the selective oxidation of butane and propane. *Angew. Chem. Int. Ed.* 2002, *41*, 858–860.
51. Rocchiccioli-Deltcheff, C., Amirouche, M., Che, M., Tatibouët, J., and Fournier, M. Structure and catalytic properties of silica-supported polyoxomolybdates. 1. Mo/SiO_2 catalysts prepared from hexamolybdate. *J. Catal.* 1990, *125*, 292–310.
52. Sugino, T., Kido, A., Azuma, N., Ueno, A., and Udagawa, Y. Partial oxidation of methane on silica-supported silicomolybdic acid catalysts in an excess amount of water vapor. *J. Catal.* 2000, *190*, 118–127.
53. Ilkenhans, T., Siegert, H., and Schlögl, R. The mechanism of the synthesis in connection with assignments for a solid reaction cycle of the HPA catalyst during catalytic reactions. *Catal. Today* 1996, *32*, 337–347.
54. Thorsteinson, E.M., Wilson, T.P., Young, F.G., and Kasai, P.H. Oxidative dehydrogenation of ethane over catalysts containing mixed oxides of molybdenum and vanadium. *J. Catal.* 1978, *52*, 116–132.

55. Bartek, J., Ebner, A., and Brazdil, J. Process for oxidation of propane. U.S. patent 5,198,580, March 30, 1993.
56. Ushikubo, T. and Oshima, K. Production of alpha, beta-unsaturated carboxyl acid. JP patent 10,036,311, February 10, 1998.
57. Ushikubo, T., Nakamura, H., Koyasu, Y., and Wajiki, S. Method for producing an unsaturated carboxylic acid. U.S. patent 5,380,933, January 10, 1995.
58. Ushikubo, T. Activation of propane and butanes over niobium- and tantalum-based oxide catalysts. *Catal. Today* 2003, *78*, 79–84.
59. Lin, M. and Linsen, M. A process for preparing a multi-metal oxide catalyst. EP patent 962,253 A2, December 08, 1999.
60. Vitry, D., Moriwaka, Y., Dubois, J.L., and Ueda, W. Mo–V–Te–(Nb)–O mixed metal oxides prepared by hydrothermal synthesis for catalytic selective oxidations of propane and propene to acrylic acid. *Appl. Catal. A: Gen.* 2003, *251*, 411–424.
61. Ushikubo, T., Koyasu, Y., Nakamura, H., and Wajiki, S. Production of acrylic acid. JP patent 10,045,664, May 12, 1998.
62. Al-Saeedi, J.N., Guliants, V.V., Guerrero-Perez, O., and Banares, M.A. Bulk structure and catalytic properties of mixed Mo–V–Sb–Nb oxides for selective propane oxidation to acrylic acid. *J. Catal.* 2003, *215*, 108–115.
63. Takahashi, M., To, S., and Hirose, S. Production of acrylic acid. JP patent 10,120,617, May 12, 1998.
64. Botella, P., Concepcion, P., Lopez-Nieto, J.M., and Solsona, B. Effect of potassium doping on the catalytic behavior of Mo–V–Sb mixed oxide catalysts in the oxidation of propane to acrylic acid. *Catal. Lett.* 2003, *89*, 249–253.
65. Mazzocchia, C., Tempesti, E., Anouchinsky, R., and Kaddouri, A. Catalyseur et procédé d'oxydation ménagée et selective d'alcanes. FR. patent 2,693,384, July 10, 1994.
66. Kaddouri, A., Mazzocchia, C., and Tempesti, E. The synthesis of acrolein and acrylic acid by direct propane oxidation with Ni–Mo–Te–P–O catalysts. *Appl. Catal. A: Gen.* 1999, *180*, 271–275.
67. Fujikawa, N., Wakui, K., Tomita, K., Ooue, N., and Ueda, W. Selective oxidation of propane over nickel molybdate modified with telluromolybdate. *Catal. Today* 2001, *71*, 83–88.
68. Hatano, M. and Kayo, A. Process for producing nitriles. EP patent 0,318,295 B1, May 31, 1992.
69. Ushikubo, T., Oshima, K., Kayou, A., and Hatano, M. Ammoxidation of propane over Mo–V–Nb–Te mixed oxide catalysts. *Stud. Surf. Sci. Catal.* 1997, *112*, 473–480.
70. Aouine, M., Dubois, J.L., and Millet, J.M.M. Crystal chemistry and phase composition of the MoVTeNbO catalysts for the ammoxidation of propane. *Chem. Commun.* 2001, (13), 1180–1181.
71. Magneli, A. Studies on the hexagonal tungsten bronzes of potassium, rubidium and cesium. *Acta Chem. Scand.* 1953, *7*, 315–324.
72. Lundberg, M. and Sundberg, M. New complex structures in the cesium-niobium-tungsten- oxide system revealed by HREM. *Ultramicroscopy* 1993, *52*, 429–435.
73. Stevenson, C. The crystal structure of orthorhombic antimony trioxide, Sb_2O_3. *Acta Crystallogr. B* 1974, *30*, 458–461.
74. Baca, M., Pigamo, A., Dubois, J.L., and Millet, J.M.M. Propane oxidation on MoVTeNbO mixed oxide catalysts: study of the phase composition of active and selective catalysts. *Top. Catal.* 2003, *23*, 39–46.

75. Grasselli, R.K., Burrington, J.D., Buttrey, D.J., DeSanto, P., Jr., Lugmair, C.G., Volpe, A.F., Jr., and Weingand, T. Multifunctionality of active centers in (amm)oxidation catalysts: from $Bi-Mo-O_x$ to $Mo-V-Nb-(Te, Sb)-O_x$. *Top. Catal.* 2003, *23*, 5–22.
76. Oshikara, K., Hisano, T., and Ueda, W. Catalytic oxidative activation of light alkanes over Mo–V-based oxides having controlled surface. *Top. Catal.* 2001, *15*, 153–160.
77. Andersson, A., Andersson, S.L.T., Centi, G., Grasselli, R.K., Sanati, M., and Trifiro, F. Direct propane ammoxidation to acrylonitrile — Kinetics and nature of the active phase. *Stud. Surf. Sci. Catal.* 1992, *75*, 691–705.
78. Bluhm, H., Haevecker, M., Kleimenov, E., Knop-Gericke, A., Liskowski, A., Schlogl, R., and Su, D. *In situ* surface analysis in selective oxidation catalysis: *n*-butane conversion over VPP. *Top. Catal.* 2003, *23*, 99–107.
79. Millet, J.M.M., Roussel, H., Pigamo, A., Dubois, J.L., and Jumas, J.C. Characterization of tellurium in MoVTeNbO catalysts for propane oxidation or ammoxidation. *Appl. Catal. A: Gen.* 2001, *232*, 77–92.
80. DeSanto, P., Jr., Buttrey, D.J., Grasselli, R.K., Lugmair, C.G., Volpe, A.F., Toby, B.H., and Vogt, T. Structural characterization of the orthorhombic phase *M1* in MoVNbTeO propane ammoxidation catalyst. *Top. Catal.* 2003, *23*, 23–38.
81. Ueda, W. and Oshikara, T. Selective oxidation of light alkanes over hydrothermally synthesized Mo–V–M–O (M = Al, Ga, Bi, Sb, and Te) oxide catalysts. *Appl. Catal. A: Gen.* 2000, *200*, 135–143.
82. Millet, J.M.M., Baca, M., Pigamo, A., Vitry, D., Ueda, W., and Dubois, J.L. Study of the valence state and coordination of antimony in MoVSbO catalysts determined by XANES and EXAFS. *Appl. Catal. A: Gen.* 2003, *244*, 359–370.
83. Luo, L., Labinger, J.A., and Davis, M.E. Comparison of reaction pathways for the partial oxidation of propane over vanadyl ion-exchanged zeolite beta and $Mo_1V_{0.3}Te_{0.23}Nb_{0.12}O_x$. *J. Catal.* 2001, *200*, 222–231.
84. Salem, G.F., Besecker, C.J., Kenzig, S.M., Kowlaski, W.J., and Cirjak, L.M. Oxidizing paraffin hydrocarbons. U.S. patent 5,306,858, April 26, 1994.
85. Concepcion, P., Lopez-Nieto, J.M., and Perez-Pariente, J. The selective oxidative dehydrogenation of propane on vanadium aluminophosphate catalysts. *Catal. Lett.* 1993, *19*, 333–337.
86. Okamoto, M., Luo, L., Labinger, J.A., and Davis, M.E. Oxydehydrogenation of propane over vanadyl ion-containing VAPO-5 and CoAPO-5. *J. Catal.* 2000, *192*, 128–136.
87. Centi, G. and Trifiro, F. Catalytic behavior of V-containing zeolites in the transformation of propane in the presence of oxygen. *Appl. Catal. A: Gen.* 1996, *143*, 3–16.
88. Nowinska, K., Waclaw, A., and Izbinska, A. Propane oxydehydrogenation over transition metal modified zeolite ZSM-5. *Appl. Catal. A: Gen.* 2003, *243*, 225–236.
89. Blasco, T., Concepcion, P., Lopez-Nieto, J.M., and Perez-Pariente, J. Preparation, characterization, and catalytic properties of VAPO-5 for the oxydehydrogenation of propane. *J. Catal.* 1995, *152*, 1–17.
90. Xu, J., Mojet, B.L., van Ommen, J.G., and Lefferts, L. Propane selective oxidation on alkaline earth exchanged zeolite Y: Room temperature *in situ* IR study. *Phys. Chem. Chem. Phys.* 2003, *5*, 4407–4413.

91. To, S., Takahashi, M., and Hirose, S. Production of catalyst for producing acrylic acid. JP patent 10,137,585-A, May 26, 1996.

92. Novakova E.K. Ph.D. thesis, *Catalytic Studies of Propane Oxidation to Acrylic Acid on Mo–V–Sb–Nb Mixed Oxides Catalysts.* University of Liverpool, 2002.

93. Dieterle, M., Mestl, G., Jager, J., Uchida, Y., Hibst, H., and Schlögl, R. Mixed molybdenum oxide based partial oxidation catalyst: 2. Combined X-ray diffraction, electron microscopy and Raman investigation of the phase stability of $(MoVW)_5O_{14}$-type oxides *J. Mol. Catal. A: Chem.* 2000, *174*, 169–185.

94. Kihlborg, L. Studies on molybdenum oxides. *Acta Chem. Scand.* 1959, *13*, 954–962; The crystal structure of $Mo_{17}O_{47}$. *Acta Chem. Scand.* 1960, *14*, 1612–1622; Stabilization of tunnel structure of Mo_5O_{14} by partial metal atom substitution. *Acta Chem. Scand.* 1969, *23*, 1834.

95. Ekström, T. and Nygren, M. Ternary phases with Mo_5O_{14} type of structure. 1. Study of molybdenum–vanadium–oxygen system. *Acta Chem. Scand.* 1972, *26*, 1827–1835.

96. Ekström, T. and Nygren, M., Ternary phases with Mo_5O_{14} type of structure. 2. Study of molybdenum–niobium–oxygen and molybdenum–tantalum–oxygen systems. *Acta Chem. Scand.* 1972, *26*, 1836–1842.

97. Védrine, J.C., Novakova, E.K., and Derouane, E.G. Recent developments in the selective oxidation of propane to acrylic and acetic acids. *Catal. Today* 2003, *81*, 247–262.

98. Moulder, J.F. and Strickle, W.F. *Handbook of X-Ray Photospectroscopy.* Perkin-Elmer, USA, 1992.

99. Dimitratos, N. Ph.D. thesis, *Catalytic Studies of Heteropoly Compounds on Propane Oxidation.* University of Liverpool, March 2003.

100. Dimitratos, N. and Védrine, J.C. Properties of $Cs_{2.5}$ salts of transition metal M substituted Keggin-type $M_{1-x}PV_1Mo_{11-x}O_{40}$ heteropolyoxometallates in propane oxidation. *Appl. Catal. A: Gen.* 2003, *256*, 251–263.

101. Dimitratos, N. and Védrine, J.C. Selective oxidation of propane on $Cs_{2.5}H_{1.5}PV_1W_xMo_{1-x}O_{40}$ heteropolyoxometallate compounds. In *Catalysis in Application*, Jackson, S.D., Hargreaves J.S.J., and Lennon, D., Eds. British Royal Society of Chemistry, Cambridge, UK, 2003, pp. 145–152.

102. Gregory, R. and Colombos, A.J. Chemical Process making Aromatic Hydrocarbons over Gallium Catalyst. U.S. patent 4,056,575, November 1, 1977; Bulford, S.N. and Davis, E.E. Process for aromatizing C_3–C_8 hydrocarbon feedstocks using a gallium containing catalyst supported on certain silicas. U.S. patent 4,157,356, June 5, 1979.

103. Nowak, I., Quartararo, J., Derouane, E.G., and Védrine, J.C. Effect of H_2–O_2 pre-treatments on the state of gallium in Ga/H-ZSM-5 propane aromatisation catalysts. *Appl. Catal. A: Gen.* 2003, *251*, 107–120.

104. Derouane, E.G., Abd Hamid, S.B., Ivanova, I.I., Blom, N., and Højlumd-Nielsen, P.E. Thermodynamic and mechanistic studies of initial stages in propane aromatisation over Ga-modified H-ZSM-5 catalysts. *J. Mol. Catal.* 1994, *86*, 371–400.

105. Pal, P., Quartararo, J., Abd Hamid, S.B., Derouane, E.G., Védrine, J.C., Magusin, P.C.M.M., and Anderson, B.G. A MAS-NMR and DRIFT study of the Ga-species in Ga/H-ZSM5 catalysts and their effect on propane ammoxidation, *Canad. J. Chem.* in press 2005.

106. Novakova, E.K., Védrine, J.C., and Derouane, E.G. Propane oxidation on Mo–V–Sb–Nb mixed-oxide catalysts, 1. Kinetic and mechanistic studies. *J. Catal.* 2002, *211*, 226–234.
107. Novakova, E.K., Védrine, J.C., and Derouane, E.G. Propane oxidation on Mo–V–Sb–Nb mixed-oxide catalysts, 2. Influence of catalyst activation methods on the reaction mechanism. *J. Catal.* 2002, *211*, 235–243.
108. Dimitratos, N. and Védrine, J.C. Role of acid and redox properties on propane oxidative dehydrogenation over polyoxometallates. *Catal. Today* 2003, *81*, 561–571.
109. Novakova, E.K., Derouane, E.G., and Védrine, J.C. Effect of water on the partial oxidation of propane to acrylic and acetic acids on Mo–V–Sb–Nb mixed oxides. *Catal. Lett.* 2002, *83*, 177–182.

14 Methane Oxidation on Metal Oxides

R.M. Navarro, M.A. Peña, and J.L.G. Fierro
Institute of Catalysis and Petrochemistry, CSIC, Madrid, Spain

Contents

14.1 Introduction

The transformation of methane, the major component of natural gas, to more valuable products has been a challenging task because methane is extremely difficult to activate. In the methane molecule, consisting of a single C-atom surrounded by four H-atoms (CH$_4$), the sp^3 hybridization of the atomic orbitals of carbon makes the carbon–hydrogen bonds very strong. A barrier to converting CH$_4$ into useful chemicals is that the products are less thermodynamically stable than the

CH_4 reactant and hence are further converted to undesired products, namely carbon oxides. There are a number of potential routes to oxygenates such as methanol (CH_3OH) and formaldehyde (HCHO) and all have been studied with a view to find ways to simplify the conventional technology to produce the chemicals.

Over the past decade much effort has been devoted to develop effective processes and technology for optimum utilization of the abundant natural gas, whose main constituent is methane, for the production of chemicals and energy. The methods for converting methane into value-added products include: (i) the energy efficient conversion of methane to syngas based on partial oxidation at very low contact times [1–4], or the coupling of partial oxidation with endothermic steam and CO_2 reforming reactions [5,6]; (ii) the oxidative coupling of methane to ethane and ethene [7–9]; (iii) methane reforming for CO-free hydrogen production for fuel cells [10–12]; and (iv) methane homologation to higher hydrocarbons [13–15]. Most of these processes are conducted in the presence of metal oxide catalysts under specific reaction conditions. In regard to these advances, further research is required before any of the processes mentioned can become industrially viable.

Natural gas is also used to generate power during its combustion in gas-turbines [15]. Currently, most power plants employ the thermal combustion of natural gas with a flame temperature range of 1773 to 2273 K, under which the formation of nitrogen oxides (NO_x), resulting from the nitrogen and oxygen present in the air, is thermodynamically favored [15]. Such NO_x emissions can be minimized, or even inhibited, by the use of catalytic combustion in gas-fired power plants and furnaces [15,16]. Although catalytic combustion has several advantages over conventional flame combustion, several issues need to be addressed prior to large-scale implementation. The design of a suitable catalyst remains a classic concern in the catalytic combustion process. Much effort has been directed toward the design of robust and active catalysts and the development of innovative technology for catalytic combustion in power generation units. The catalyst system used in combustion generally consists of an active component and a support/substrate. The large number of catalyst systems for the combustion process makes it difficult to categorize them systematically. Greater complexity is introduced because certain components that are themselves active as combustion catalysts may be used as supports/substrates in other catalytic systems (e.g., perovskites and hexaaluminates) [15–23]. In this chapter, attention is only focused on the different metal oxide catalysts (single metal oxides, perovskites, doped metal oxides, and hexaaluminates).

Other chemical routes for the direct conversion of methane into valuable chemicals, that is, CH_3OH and HCHO, involve partial oxidation under specific reaction conditions [24–28]. As a general rule, these conversion processes use fuel-rich mixtures with the oxidant to minimize the extent of combustion reactions, which yield unwanted carbon oxides. Under these conditions, purely gas-phase oxidation reactions require high temperatures, which are detrimental for the control of selectivity of the desired products. Among the plethora of catalysts employed for this purpose, metal oxides — most of them transition metal oxides — are prominent. These catalyst systems are considered in this chapter.

Among the conversion processes, the direct conversion of CH_4 to CH_3OH and HCHO offers many advantages over the currently available two-step syngas technology process. Although great effort has been taken over the past ten years, till date no direct process has been reported to compete with the well-established syngas route, in which steam reforming, that is, the first step in the production of syngas — accounts for almost 60% of the total cost of methanol production. The direct one-step conversion methods appear to be more expensive than steam reforming, and hence none of them has as yet proved to be as scalable as steam reforming.

14.2 CH$_4$ ACTIVATION ON METAL OXIDES

The activation of C–H bond in methane is a crucial first step in its combustion for power and heat generation. Once the first bond has been broken, sequential oxidation reactions to CO_2 and H_2O are relatively easy. A basic understanding of the activation of C–H bonds in methane is of vital importance since it permits one to assess the influence of catalytic and process parameters on the rate and efficiency of its catalytic combustion [29]. Several types of catalysts — including nonreducible oxides, reducible oxides, and metals — are capable of oxidizing methane with varying efficiencies. It is harder for C–H bonds to be activated in methane than in other hydrocarbons, due to the weaker adsorption of methane on oxides or on oxidized metal surfaces [30]. Strong adsorption of a saturated hydrocarbon is a prerequisite for combustion, but factors other than the strength of the C–H bond also affect the rate of combustion.

The combustion of methane on oxides requires relatively high temperatures. Under these circumstances, evidence from work on methane-coupling catalysts suggests that gas-phase contributions to combustion may be important [31]. For example, there is evidence that methyl radicals can be observed in the gas phase surrounding the oxides at temperature as low as 773 K [32]. The initial activation of the C–H bond is, however, a surface-catalyzed process which may involve homolytic or heterolytic breakage. At present, there is no general agreement about the true mechanism involved. Since the C–H bond in methane is only very weakly acidic ($pK_a = 46$), it follows that surface sites able to deprotonate methane must be strongly basic. Such sites would be expected to adsorb CO_2 strongly, and hence they might also be expected to become rapidly self-poisoned. Indeed, evidence from H–D isotopic exchange reactions, which are thought to occur by homolytic splitting of C–H and C–D bonds, show that even small traces of CO_2 cause almost complete poisoning of this reaction.

On the other hand, several authors have proposed that C–H bond activation occurs through deprotonation, although the evidence in support of such a proposal is not entirely clear. For example, Choudhary and Rane [33] have compared the activity of several oxides for methane conversion with the acid/base properties of the catalysts but failed to find a simple correlation. They proposed that both acidic and basic sites would be important and rationalized the results by suggesting the occurrence of an acid–base pair as the active site, which comprises of a metal cation (the acidic site) and an oxide anion (the basic site). Deprotonation would lead to the formation of an OH(ads) and a CH_3^- ion attached to a metal cation.

14.3 METHANE COMBUSTION

14.3.1 Single Metal Oxides

Transition metal oxides, such as copper, cobalt, manganese, and chromium, are known to be active combustion catalysts [34–39]. Dispersed on a porous substrate, these metal oxides are more active than the respective bulk oxides. This trend is very clear on alumina-supported CuO, in which methane combustion is much higher than on bulk CuO [34]. In this system, the copper oxide loading has considerable influence on the structure, and therefore on the performance, of the supported CuO phase. Thus, at low metal-oxide contents, the CuO phase is highly dispersed and the Cu–O bond is basically ionic. However, at high CuO contents the metal results in the formation of more covalent oxide. One weakness of the CuO catalyst is its low stability. In general, aging at high temperatures has a detrimental effect on combustion activity due to the solid-state reaction between the CuO and the Al_2O_3 substrate. This drawback can be circumvented by using spinel-type (AB_2O_4) supports [40]. Thus, CuO does not interact on the surface of zinc-aluminate (spinel-type oxide), which renders the catalyst very stable for high temperature operation. Nevertheless, the dispersion of CuO is very low on the conventionally prepared zinc aluminate as a consequence of its low specific area. Zinc aluminate can be prepared by the sol–gel method, with a high specific area, and the catalyst resulting from the deposition of CuO on this sol–gel zinc aluminate becomes highly active for methane combustion [41]. Consistent with the above method, another work dealing with the structure of supported CuO catalysts on the methane oxidation revealed that methane oxidation activity decreases upon increasing Cu-loading [42]. This behavior can be interpreted assuming that isolated Cu surface species are more active than crystalline CuO, and that higher Cu contents result in the formation of poorly dispersed CuO crystallites, which in turn are less active in methane combustion.

Busca and coworkers [43–45] carried out an exhaustive study on the combustion of C_3 hydrocarbons on metal-oxide catalysts. While the authors observed propene and traces of ethylene at incomplete propane conversion on Mn_3O_4 in the absence of oxygen, only CO_2 was detected in a highly oxidizing atmosphere and for complete propane conversion [43]. The authors proposed a propane combustion mechanism on Mn_3O_4 involving propene and acrolein as intermediate species. Subsequent work [44] revealed that the pure Fe_2O_3 catalyst exhibits lower propane and propene combustion activity as compared with the values for manganese oxide catalysts. The results of a recent investigation involving methane combustion on MnO_x-based catalysts by Zaki et al. [46] suggested that acidic supports such as silica–alumina are more suited for methane oxidation. MgO, which is a relatively thermally stable metal oxide, has been used as a catalyst support for noble metals [47]. Berg and Jaras [48] observed that MgO itself shows some activity for methane combustion; a conversion of 10% for methane was approached at a much lower temperature in the presence of MgO than in case of an empty reactor tube. High temperatures and O_2/CH_4 ratios were found to favor selectivity for CO_2. CoO–MgO solid solution (calcined at 1173 to 1673 K) was found to be an active and thermally stable catalyst for high temperature combustion of dilute methane at high space velocities [49]. The apparent activation energy for the methane combustion reaction decreased with increasing Co/Mg ratio, and increased with increasing catalyst calcination temperature.

Uranium oxide catalysts show a considerable degree of activity for lower alkane combustion [50]. The dispersion of uranium oxide on a silica substrate and the addition of Cr were found to enhance the activity of the uranium oxide catalysts. Employing high surface area, $LaAlO_3$, as a support for several single metal-oxide catalysts, McCarthy et al. [51] observed the following sequence for methane oxidation activity: $Co_3O_4 > CuO > NiO > Mn_2O_3 > Cr_2O_3$. In a recent investigation into methane combustion on single metal oxides at temperatures in the range of 573 to 873 K, it was found that Cr_2O_3, Co_3O_4, and Mn_3O_4 exhibited comparable activities whereas Fe_2O_3 was less active [52].

14.3.2 Double Oxides

Perovskite-type oxides with a general formula of ABO_3 are particularly promising for the combustion of methane due to their good stability, at least below 1300 K, and high catalytic activity [53,54]. Arai et al. [55] studied the catalytic combustion of methane over $LaBO_3$ (B = transition element) and Sr-substituted $La_{1-x}Sr_xBO_3$ $(0 < x < 0.4)$ perovskites. In particular, Sr-substituted manganites, cobaltites, and ferrites exhibited very high oxidation activities that were comparable with the activity of a Pt/Al_2O_3 combustion catalyst [55]. High catalytic activities for methane combustion over perovskite oxides were also reported by McCarthy and Wise [56]. An interesting characteristic of these perovskites is the possibility of varying the dimensions of the unit cell by substitution of the A ion, and thereby the covalency of the B–O bond in the ABO_3 structure. Moreover, partial substitution at the A-site can strongly affect catalytic activity due to stabilization of the unusual oxidation states of the B component and due to the simultaneous formation of structural defects. Structural defects are responsible not only for part of the catalytic activity, but also for oxygen mobility within the crystal lattice, due to the nonstoichiometry created by substituting the A-site. Overviews of these aspects, as well as the implications for methane combustion, can be found in our previous reviews [53,54,57].

In a recent contribution, Falcon et al. [58] used oxygen-deficient $SrFeO_{3-\delta}$ $(0.02 < \delta < 0.26)$ perovskite oxides for methane oxidation. Combustion was investigated in the temperature range between 600 and 1000 K. At moderate temperatures, a significantly higher catalytic activity was observed for the most oxygen-deficient sample, which was correlated with the oxygen desorption ability exhibited by this material, containing a large number of oxygen vacancies in the crystal structure (Figure 14.1). All the samples exhibited an S-shaped profile for CH_4 conversion as a function of the reaction temperature, with ignition

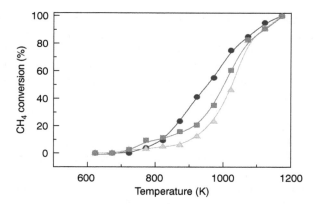

FIGURE 14.1   Temperature dependence of CH_4 conversion on $SrFeO_{3-\delta}$. ($\bullet$), $\delta = 0.26$, ($\blacksquare$), $\delta = 0.09$, and ($\blacktriangle$), $\delta = 0.02$. Reaction conditions: catalyst weight = 0.20 g, CH_4 concentration = 22.9 to 34.4% (molar), feed ratio $N_2:CH_4:O_2$ =4:2:1 (molar), and atmospheric pressure (Reproduced with permission from American Chemical Society)

temperatures in the 600 to 800 K range. At 850 K, the reaction rate reached the highest value (99.5 mmol/s.g) for the $SrFeO_{2.74}$ sample, decreased for $SrFeO_{2.91}$ (72.6 mmol/s.g) and even more so for the $SrFeO_{2.98}$ counterpart. The reaction rates appeared to be associated with the relative ease with which oxygen became activated on the O-deficient materials.

On perovskite oxides, the combustion of methane is assumed to occur through an intrafacial mechanism, in which the adsorbed oxygen is partly consumed and regenerated along a continuous cycle. Thus, $SrFeO_{3-\delta}$ offers an interesting model for examining the feasibility of inserting O^{2-} ions into the lattice from gas-phase O_2. The O_2 temperature-programmed desorption (TPD) profiles shown in Figure 14.2 convincingly demonstrate the presence of two types of oxygen (α- and β-oxygen) [53,54,59–61] in such systems. The low temperature O_2 TPD peak (α-oxygen) is accommodated in the O^{2-} vacancies generated by inserting Sr^{II} ions into a $SrFeO_{3-\delta}$ lattice. This form is believed to be more active and reacts with CH_4 at lower temperatures than the β-oxygen. Thus, the combustion of methane proceeds according to a redox mechanism; several kinetic models have been proposed to account for the reaction rate of this mechanism [53,54,62,63]. The rates are considered to be first order in CH_4 partial pressure, while the order in oxygen may vary from 0 to 0.5. For a simple first-order kinetic model, the reaction rate is:

$$r_{CH_4} = k \cdot P_{CH_4} \tag{14.1}$$

where k is the specific rate constant and P_{CH_4} is the methane partial pressure. By combining the design equation for a plug-flow reactor and the Arrhenius equation, the following relationship can be established:

$$\ln[-\ln(1-x)] = \ln A - E_a/RT \tag{14.2}$$

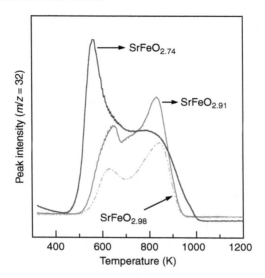

FIGURE 14.2 TPD profiles for $SrFeO_{3-\delta}$ oxides (Reproduced with permission from American Chemical Society)

where x is CH_4 conversion, A is the preexponential factor, T is the reaction temperature in K, E_a is the apparent activation energy, and R is the gas constant. According to Equation (14.2) apparent activation energies of 90, 73, and 115 kJ/mol have been calculated for $SrFeO_{2.74}$, $SrFeO_{2.91}$, and $SrFeO_{2.98}$ samples, respectively. The E_a values for the first two samples are similar to those reported by Belessi et al. [64] for La–Ce–Fe–O oxides, in which activity was explained in terms of the ability of the $LaFeO_{3-\delta}$ phase to uphold very large amounts of oxygen in the O-vacancies of its structure. The slightly higher activation energy measured for $SrFeO_{2.98}$ indicates a lower reactivity of this system.

Monolith perovskites prepared from ultradispersed powders of mixed oxides of rare earths (La–Ce or Dy–Y) and transition metals (Ni, Fe, and Mn) have recently been used in methane combustion [65]. These preshaped structures were seen to be active and selective in the target reaction over a wide range of temperatures. The scale of the specific activity and apparent activation energy for monoliths paralleled that found for powdered samples. The catalyst decreased the ignition temperature down to 200 K to achieve a 10% methane conversion, and enhanced selectivity to CO_2 with respect to the uncatalyzed process.

14.3.3 Doped Metal Oxides

The doping of metal oxides has been exploited in several investigations with a view to prepare better combustion catalysts. Choudhary et al. [66] observed a sharp increase in methane combustion activity after doping zirconia with transition metals such as Mn, Co, Cr, and Fe. These authors demonstrated the involvement of lattice oxygen atoms observed by pulse experiments and confirmed that the redox mechanism was active in the catalytic process. The reactivity of the lattice oxygen in the zirconia-modified catalysts was found to be strongly enhanced due to the doping with the transition metals. This increase in the reactivity of the lattice oxygen was surmised to be due to the creation of crystal defects and an increase in the mobility of lattice oxygen. The investigation revealed that the transition metal-doped zirconia catalysts had higher activity than perovskite catalysts [67] and were comparable with supported noble metal catalysts. Mn-doped zirconia was found to be an excellent combustion catalyst. Thus, the influence of the calcination temperature and the Mn/Zr molar ratio on its surface and catalytic properties for the combustion of methane [68] were studied in some detail. The activity for methane combustion reached a maximum, for Mn/Zr ratios, of around 0.25. An increase in the calcination temperature resulted in a decrease in the surface area; however, the methane combustion activity was found to pass through a maximum with increasing temperature, and the highest methane combustion activity was observed at 873 K. The study revealed that cubic ZrO_2 was more important than the monoclinic ZrO_2 phase for high methane combustion activity. Milt et al. [69] investigated methane oxidation on cobalt-containing zirconia, La-doped zirconia, and La_2O_3 catalysts. Cobalt supported on La-doped zirconia prepared by atomic layer epitaxy exhibited very high activity toward methane combustion relative to the catalysts prepared by wet impregnation and those supported on La_2O_3. The doping of zirconia with La was found to have a positive effect on the stability of the catalyst.

The unique property of oxygen storage capacity exhibited by ceria [70,71] make it an excellent component in combustion catalyst systems. Many studies addressing methane oxidation on ceria-based systems (modified with transition metals [72], noble metals [73], and metal oxides [74–76]) can be found in the literature. Zamar et al. [77] observed that methane combustion was favored upon introduction of Hf and Zr into ceria due to the formation of a defective fluorite-structured oxide. Terribile et al. [74] investigated lower hydrocarbon combustion on a series of ceria–zirconia mixed oxides. Zirconia was found to promote both stability and combustion activity. The introduction of MnO_x and CuO into the ceria–zirconia lattice was found to further enhance the activity of methane combustion. Kundakovic and Flytzani-Stephanopoulos [72] doped ceria with La and Zr to control its crystal size and reduction properties. The methane oxidation activity was found to be higher on modified ceria due to the greater reducibility of the doped samples. The addition of CuO and Ag_2O to the modified ceria samples was found to promote methane oxidation activity [72].

14.3.4 Hexaaluminate Materials

Hexaaluminates exhibit an extremely high thermal stability. This property makes them one of the most promising candidates for high temperature catalytic combustion applications [78]. Since the pioneering work by Arai and coworkers [79,80], these materials have been extensively investigated for catalytic combustion. Hexaaluminates can be represented by the general formula $AAl_{12}O_{19}$, where A is an alkaline or alkaline-earth metal. The high thermal stability of hexaaluminates is derived from their lamellar structure, which consists of stacked spinel blocks separated by a monolayer of oxides. The ionic radius and valence of the A-site cations determines the type of crystal structure for the hexaaluminate (β-alumina or magnetoplumbite).

An ideal catalyst for high temperature combustion should possess high thermal stability as well as high catalytic activity. However, these two characteristics seldom go hand in hand. Similarly, although the unsubstituted hexaaluminates have high thermal stability, they exhibit very low catalytic activity. Considerable effort has been invested in attempting to dope these materials ($A_{1-x}A'_xB_yAl_{12-y}O_{19}$) in order to achieve higher catalytic activity, while preserving the hexaaluminate crystal structure and thereby the inherent high thermal stability [79–84]. It should be noted that the crystal structure of hexaaluminates can be partially substituted. Machida et al. [79] observed that among the various cation-substituted Ba-hexaaluminates ($BaMAl_{11}O_{19-\delta}$; $M = Cr, Mn, Fe, Co,$ and Ni), the Mn-substituted compounds were most active for methane combustion. The temperature required for 10% conversion of the feed was 813 K for the Mn-substituted hexaaluminate, as opposed to 983 K for the unsubstituted compound. The high catalytic activity was related to the ability of Mn to switch between the Mn^{2+} and Mn^{3+} valence states.

The effect of the concentration of the B substituent (value of x) was investigated in some detail [79,83,84]. For the $BaMn_xAl_{12-x}O_{19-\delta}$ system, Machida et al. [79] reported the presence of only a single phase of the hexaaluminate at low additions of Mn, while the addition of Mn above optimum level resulted in the segregation of $BaAl_2O_4$. A more comprehensive study on the same system revealed [85] that at low Mn concentrations (up to $x = 1$) it was preferentially located in the tetrahedral Al sites as Mn^{2+}. However, at higher concentrations, it was found to enter as Mn^{3+} at the octahedral Al sites. The latter process was found to decrease the surface area of the hexaaluminate.

Eguchi and coworkers [86] investigated the influence of the A-site cation (A: La, Pr, Sm, and Nd) on $AMnAl_{11}O_{19-\delta}$. Both the specific area and the catalytic activity increased with increasing ionic radius of the lanthanides. La was found to have the most beneficial effect on catalytic methane combustion activity. The activities of the La-based catalysts were further enhanced upon substitution of Al with Mn and Cu. Interestingly, the activity of these compounds was found to be greater than the B-site substituted Ba-hexaaluminate catalysts. Based on this, the authors concluded that the A-site cations could exert a significant influence on the oxidation state of the catalytically active B ions.

14.4 Oxidative Coupling of Methane

Many metal oxides are able to perform the oxidative coupling reaction of methane molecules. Methane can also be directly converted with oxygen via oxidative coupling (OCM) into ethane and ethylene according to the following reaction:

$$2CH_4 + \tfrac{1}{2}O_2 \rightarrow C_2H_6(C_2H_4) + H_2O \tag{14.3}$$

Since the pioneering work of Keller and Bhasin [87] concerning the use of reducible metal oxides supported on γ-Al_2O_3 and that of Ito and Lunsford [88] on the use of nonreducible metal oxides (Li–MgO), the OCM reaction to C_2H_4 and C_2H_6 (C_2 hydrocarbons) has been widely investigated.

14.4.1 Catalytic Systems

Sofranko et al. [89] at ARCO evaluated many transition metal oxides as oxidative coupling catalysts of CH_4 into C_{2+} hydrocarbons. The reactions were performed under an alternating feed mode in which the oxidized catalyst was placed in contact with CH_4 in the absence of O_2 to form coupling products, and the reduced catalyst was reoxidized with air in a separate step. Redox oxides of Bi, Pb, Sb, Sn, Mn, Ge, and In supported on a silica carrier were found to be active catalysts in the coupling reaction, with 10 to 50% selectivity to C_{2+} hydrocarbons. On MnO_x/SiO_2 catalysts, those authors also observed that C_2H_6 was the initial coupling product, formed by dimerization of CH_3 radicals. Subsequently, C_2H_6 is oxidatively dehydrogenated to C_2H_4. Work carried out by Warren [90] at Union Carbide showed that catalysts containing Ba or Sr, a nonbasic metal oxide, and chlorine on the catalyst and in the feed gas afford very high C_2H_4/C_2H_6 ratios and furthermore are very stable. Acid supports appeared to be essential to reach high C_2H_4/C_2H_6 ratios. Chlorine, added in the form of ppm of ethyl chloride to the reaction feed, was found to improve catalyst performance in three directions: (i) it increased the rate of CH_4 conversion, probably by increasing the number of active sites on the catalysts, (ii) it decreased the rate of CO_2 formation, and (iii) it was responsible for the high C_2H_4/C_2H_6 ratios. These results were interpreted by assuming that chlorine interacts with the support from which chlorine radicals are released into the gas phase, where they initiate free radical reactions, resulting in the dehydrogenation of C_2H_6 into C_2H_4.

After publication of the above works, the number of contributions on the OCM reaction expanded greatly. Many metal oxide catalysts have been tested for this reaction. The catalysts can be grouped as: (i) oxides of groups IV and V metals, (ii) oxides of group III metals, (iii) oxides of group II metals, (iv) oxides of group I metals, (v) lanthanide-based oxides, and (vi) transition metal oxides. The work already done in OCM reaction with the (i)–(vi) categories of catalysts up to 1993 has been reviewed and discussed in detail by Krylov [91] and more recently by Lunsford [92] and by Baerns et al. [93].

Lanthanum oxide has often been used as a major constituent of patented OCM formulations. Lacombe et al. [94] proposed that on La_2O_3 catalysts CO_2 is formed on its surface, involving a slow step of methyl radical oxidation, while CO issues from C_{2+} hydrocarbons. Two types of active sites have been identified on the La_2O_3 surface. Low coordination metal sites localized on step edges are related to the total oxidation pathway, while basic sites associated with oxygen vacancies enable the oxygen activation that leads to the initial methane activation. Proper modification of the lanthana surface by specific additives inhibits CO_2 formation. Thus, Sarkany et al. [95] reported the lowest temperature for methane coupling, which in turn produced high space time yields of C_{2+} hydrocarbons and was achieved by sulfate doping of the basic Sr-doped La_2O_3 catalyst to inhibit poisoning by CO_2. With increasing SrO contents in the Sr-doped La_2O_3 catalyst, surface segregation of strontium oxide occurs [96] and the diffusive exchange of oxygen anions and anion vacancies between the bulk and the outer surface is limited. Therefore, the adsorption of gaseous oxygen and the transformation of adsorbed oxygen into lattice oxygen species is restrained, resulting in a loss of selectivity.

The diffusion mechanism and the rate of diffusion have recently been examined by Nagy et al. [97] using silver as a coupling catalyst. They reported experimental evidence that a strongly bound, Lewis basic, oxygen species intercalated in the silver crystal structure is formed because of diffusional processes. This specie, referred to as O_γ, acts as a catalytic active site for the direct dehydrogenation of methane. The activation energy for methane coupling over the silver catalyst of 138 kJ/mol almost coincides with the value of 140 kJ/mol for oxygen diffusion in silver measured under similar conditions. This correlation between the diffusion kinetics of bulk-dissolved oxygen and the kinetics of the reaction of the OCM to C_{2+} hydrocarbons suggests that the reaction is limited by the formation of O_γ species via surface segregation of bulk-dissolved oxygen. Moreover, catalysis over fresh silver catalysts indicated an initially preferential oxidation of CH_4 to complete oxidation products.

14.4.2 Influence of the Oxidant on the Coupling Reaction

In most studies dealing with the OCM reaction, molecular oxygen has been used as the oxidant. Because the coupling reaction is a complex network of heterogeneous and homogeneous reactions, other oxidants such as nitrous oxide were also used. Thus, Yamamoto et al. [98] found that nitrous oxide is an efficient oxidant for the conversion of CH_4 to C_2H_6 and C_2H_4 over a Li/MgO catalyst. In addition, the rate of oxygen insertion into the oxide lattice and the regeneration of active sites were much slower with N_2O than with O_2. This means that the rate-limiting step is oxygen incorporation rather than the activation of CH_4 at the catalyst surface. Those authors also derived a kinetic model, based on competitive surface reactions, in addition to homogeneous reactions, that accounts for conversion as a function of reagent concentrations and temperature and selectivity to C_2H_6 and C_2H_4. At comparable levels of CH_4 conversion O_2 is always a less selective oxidant than N_2O, not because of the participation of homogeneous reactions, but because of surface reactions involving reactive oxygen species of the O^- type.

14.4.3 Prospects and Opportunities

Since many catalytic systems have been tested for their performance and no further enhancements are envisaged OCM catalysis has reached its maturity. Selectivities of up to 85% for C_2 hydrocarbons have been obtained at methane conversions in the order of 10 to 15%. The general trend observed is that increasing the methane conversion, which is usually achieved by decreasing the CH_4/O_2 ratio in the feed, leads to a drop in selectivity. The highest C_2 yields fall around 22% under single-pass reactor operation. High temperatures (>1000 K) and basic metal-oxide type catalysts are key ingredients for the generation of methyl radicals while avoiding deep oxidation to CO_x, which occurs through methoxy species preferentially formed at low temperatures. The mechanism that first involves the formation of methyl radicals and then their recombination, was discovered by Lunsford and coworkers [32] and subsequently confirmed by high vacuum transient experiments [99]. Subsequently, it was shown that the electronic properties of the solid catalyst play a major role [100]. Apart from this general trend, sustained research efforts for the development of better oxides are essential to make the OCM reaction commercially feasible. Supported rare-earth oxide catalysts promoted by alkaline-earth oxide (particularly SrO) are highly active, selective, and thermally/hydrothermally stable for coupling reaction, and hence show great promise for use in the OCM reaction.

14.5 METHANE CONVERSION INTO C_1 OXYGENATES

The direct conversion of methane into CH_3OH and HCHO involves partial oxidation under specific reaction conditions [24–28]. This reaction uses fuel-rich mixtures with the oxidant to minimize the extent of combustion reactions. Under these conditions, purely gas-phase oxidation reactions require high temperatures, which are detrimental to the control of selectivity of the desired products. Accordingly, in the last 15 years considerable efforts have been made to develop active and selective catalysts for the partial oxidation of methane [101–103]. In addition, high selectivities for methanol have already been obtained by working under noncatalytic reaction conditions [25,104]. Thermodynamic and kinetic studies reveal that the rate-limiting step of the partial oxidation of methane is the first H-abstraction from the C–H bond forming methyl radicals. Thus, initiators and sensitizers have been examined with a view to decreasing the energy barrier of this H-abstraction. In particular, nitrogen oxides have been used to promote gas-phase reactions with methane [28,105–107]. The combination of both the homogeneous activation of CH_4/O_2 reaction mixtures by small amounts of nitrogen oxides (0.39 vol%) and the heterogeneous reaction on a V_2O_5/SiO_2 catalyst resulted in a high yield of CH_3OH and HCHO (up to 15.3%).

14.5.1 Silica

Silica and silica-supported oxide catalysts exhibit greater functionality in the partial oxidation of methane [108,109]. Silica itself has measurable activity for CH_4 conversion to HCHO, although at lower levels of activity than most other catalysts. Kasztelan and Moffat [110] have shown that up to 4.5% of CH_4 co-fed with O_2 can be converted at 866 K at a relatively high contact time and with low selectivity (8%). Kastanas et al. [111] noted that the Vycor or quartz walls of the reactor tubing had discernable activity for HCHO formation. Coupled products (mainly C_2H_6)

were also observed at the highest reaction temperature (above 893 K), pointing to an important production of CH_3 radical. At the same temperature, alumina and magnesia produced only CO and CO_2, confirming the specificity of silica for the partial oxidation. Remarkably better results were obtained by Guliev et al. [112] using silica at a rather high space velocity (6000 h^{-1}) and they obtained much higher CH_4 conversion at 873 K, with reasonable selectivity. The texture of the silicas and the need to maintain the specific surface area as high as possible were seen to have a strong effect on performance.

The particular reactivity of bare SiO_2 for the production of HCHO is a matter of debate and has not yet been completely rationalized. Parmaliana et al. [113] pointed out that the performance of the silica surface in CH_4 partial oxidation is controlled by the preparation method. For several commercial SiO_2 samples, the following reactivity trend has been established, based on the preparation method: precipitation > sol–gel > pyrolysis. The activity of such silicas has been correlated with the density of surface sites stabilized under steady-state conditions acting as O_2 activation centers [114], and the reaction rate was the same for all the silicas when expressed as TOF (turnover frequency). Klier and coworkers [115] reported the activity data for the partial oxidation of CH_4 by O_2 to form HCHO and C_2 hydrocarbons over fumed Cabosil and silica gel at temperatures ranging from 903 to 1953 K under ambient pressure. They observed that short residence times enhanced HCHO (and C_2 hydrocarbon) selectivity, suggesting that HCHO did not originate from methyl radicals, but rather from methoxy complexes formed upon direct chemisorption.

An explanation concerning the mechanism of methane activation over silica came from the surface chemistry of reactive silicas. The available spectroscopic information suggests that the activity of silicas probably comes from the presence of unusual reaction sites. As shown in Scheme 14.1, a surface structure designated as R can be modified by O_2 chemisorption at room temperature to give the OR structure.

This scheme summarizes the properties of the sites, that is, of a pair of silicon radicals associated with two anomalously reactive oxygen atoms — but provides little information concerning the geometry, apart from the requirement that the two silicon atoms must be closely spaced. Dissociative chemisorption of several molecules — that is, H_2O, CH_3OH, O_2, etc. — takes place on dehydroxylated silica [116] as compared with nonpretreated surfaces, and this can be primarily associated with the presence of Si defects, as depicted by structure R.

Another possible activation mechanism of the CH_4 molecule resides in the iron impurities still present in bare silicas. Kobayashi et al. [117] reported that doping fumed silica with Fe^{3+} ions strongly enhanced HCHO productivity and that the

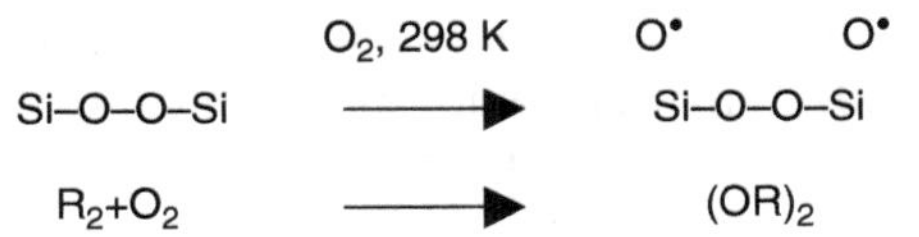

$$Si-O-O-Si \quad \xrightarrow{O_2,\ 298\ K} \quad \overset{O^\bullet \quad \ O^\bullet}{Si-O-O-Si}$$

$$R_2 + O_2 \quad \longrightarrow \quad (OR)_2$$

SCHEME 14.1

Fe loading, its dispersion, and the coordination of the Fe^{3+} ions with oxygen were critical parameters governing the partial oxidation reaction. More recently, Otsuka et al. [118] has claimed that the Fe^{3+} sites in several oxide systems enhance the formation of peroxide (O_2^{2-}) species, which are responsible for the selective activation of the CH_4 molecule. These works provide sufficient arguments that iron impurities might play some role, in addition to defective silicon sites, in the partial oxidation reaction. In a very recent contribution, Parmaliana et al. [119] used the electron spin resonance technique to monitor the presence of isolated Fe^{3+} centers and small iron oxide clusters at the surface of bare silicas. Such sites are able to chemisorb oxygen in an irreversible way when the samples are pretreated under reaction conditions in a redox process within the Fe^{3+}/Fe^{2+} couple, involving O-lattice exchange, which can in turn participate in the CH_4 oxidation reaction by a Mars-van Krevelen mechanism. The promotion of silica surfaces by deliberately adding Fe^{3+} ions enhances the partial oxidation of CH_4 due to the creation of more isolated (or in small oxide clusters) Fe^{3+} sites at the surface. However, if large ferric oxide particles are developed, total combustion of methane becomes the dominant reaction.

14.5.2 Redox Oxides

14.5.2.1 Molybdenum oxide

Molybdenum-containing catalysts belong to one of the categories of oxides widely used for the direct partial oxidation of CH_4 into C_1-oxygenates. For silica-supported MoO_3 catalysts, Spencer and Pereira [120] reported that HCHO selectivity was correlated with the molybdenum oxide structures. Many studies have attempted to elucidate these structures [110,121,122] in the search for connections between molybdenum dispersion and procedures for Mo incorporation into the silica substrate. For molybdenum densities below 1.5 Mo/nm^2, the formation of molybdates is thought to be induced by the presence of Na or Ca impurities on the silica surface [123]. Upon increasing the molybdenum loadings, polymeric molybdenum oxide species become dominant until the onset of bulk MoO_3 formation. The involvement of the different Mo species in MoO_3/SiO_2 catalysts prepared with varying surface concentrations of Mo atoms in the performance for methane partial oxidation has been investigated [121]. For MoO_3/SiO_2 catalysts, the increase in the reaction temperature resulted not only in an increase in CH_4 conversion, as expected, but also in a drop in selectivity toward partial oxidation products. A similar trend was observed when increasing the CH_4 residence time but maintaining the reaction temperature. HCHO decreased monotonically with CH_4 conversion, indicating that HCHO is a primary reaction product. This conclusion was supported by the results obtained with a tracer isotopic labeling technique using $^{18}O_2$ [124]. From these experiments, it was concluded that the oxygen is incorporated into HCHO from the lattice molybdenum oxide and not from the gas-phase molecular O_2, whereas CO_2 comes from CH_4. The role of molecular oxygen is to restore the oxidation state of molybdenum while the catalyst works

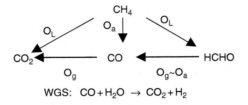

SCHEME 14.2 Reaction pathway for the oxidation of methane into HCHO and carbon oxides on silica-supported molybdenum oxide catalysts

in a redox Mars-van Krevelen cycle. This suggests that the different activities of the molybdenum species must be determined by their different abilities to release and insert oxygen into the methane molecule, which depends on the flexibility of the surface structures to accommodate the generated anion vacancy. Moreover, CO_2 can also be produced from CO oxidation with the unreacted oxygen and also from the water gas-shift reaction of CO. In sum, the reaction pathway, as depicted in Scheme 14.2, is that lattice oxygen yields HCHO and CO_2, while gas-phase molecular oxygen is mainly responsible for the oxidation of HCHO into CO.

A different interpretation of the role played by the different molybdenum oxide species deposited on the surface of silica was proposed for the oxidation of CH_4 into HCHO. Thus, Suzuki et al. [125] concluded that well-dispersed molybdenum oxide clusters on the silica surface are the active species for the reaction. By contrast, Ozkan and coworkers [126,127] proposed that the Mo=O sites present in the MoO_3 crystals are catalytically active while Mo–O–Mo bonds accelerate the deep oxidation of methane. However, the reactivity of supported molybdenum oxide differs from bulk MoO_3, and no direct extrapolation can be made [103].

Zeolites, instead of silica, have also been used as carriers to prepare molybdenum oxide catalysts for CH_4 oxidation. Since the specific activity of the monomeric molybdenum species is high, a few attempts have been made to deposit highly dispersed molybdenum oxide species on HY [128] and HZSM-5 [129] zeolites. Highly dispersed MoO_3 species have been obtained by different methods [128]: (i) a solid-state reaction between $MoCl_5$ and the hydroxyl groups of the zeolite, (ii) adsorption-decomposition of the $Mo(CO)_6$ complex, and (iii) conventional impregnation with ammonium heptamolybdate, followed by thermal decomposition under very low pressure for long periods of time. For the HZSM-5 zeolite [129], molybdenum was incorporated by a different procedure, consisting of the anchorage of the $MoO_2(acac)_2$ (acac = acetylacetonate) complex onto the surface of the zeolite. The isolated oxo species of molybdenum in the Mo^{VI} oxidation state with a moderately high dispersion led to moderate activity at temperatures of around 850 K and selectivities to HCHO below 15%.

14.5.2.2 Vanadium oxide

Vanadium oxide dispersed on a silica substrate has been widely investigated in the selective oxidation of methane into CH_3OH and HCHO [103,120,130–133]. Using

V_2O_5/SiO_2 catalysts, Spencer and Pereira [120] reported HCHO selectivities in excess of 70% at very low conversion levels but these decreased dramatically upon increasing conversion. A sequential reaction path, where HCHO is formed from CH_4, CO from HCHO, and CO_2 from CO, was proposed. This mechanism contrasts with MoO_3/SiO_2 catalysts, in which CO_2 is formed in a parallel way either completely [120] or partially [134] from CH_4. The selectivity-conversion data on V_2O_5/SiO_2 catalysts revealed a marked dependence on the reaction temperature, in contrast with that plotted on a common curve for the MoO_3/SiO_2 systems.

A tracer $^{18}O_2$ isotopic analysis confirmed that partial oxidation proceeds via a Mars–van Krevelen mechanism in which the lattice oxygen is incorporated into the HCHO while the consumed oxygen is restored by O_2 from the gas phase [124]. Koranne et al. [131] showed that O-exchange between the gas-phase oxygen and lattice oxygen of V_2O_5/SiO_2 increased significantly in the presence of CH_4, pointing to a redox reaction as being responsible for making the surface more active for exchange. It was speculated that the oxygen associated with highly dispersed tetrahedral surface vanadia would be involved in a primary oxidation reaction, whereas the oxygen associated with the bulk-V_2O_5 and with the support would be involved in secondary oxidation reactions of HCHO.

Faraldos et al. [103,132] reported further insight into how the catalytic activity of silica-supported V_2O_5 catalysts can be correlated with the fundamental structural and surface parameters of the supported oxide phases. Methane conversion increased with V-loading up to a surface density of 1 V/nm^2 and then decreased slightly. These results show that catalytic activity is related to the appearance of dispersed vanadia on the silica surface. Temperature-programmed *in situ* Raman spectroscopy data revealed that the V_2O_5 coverage on silica has no effect on its structure below its monolayer capacity [133]. However, the interaction among the surface vanadia species under a reducing environment becomes increasingly important. This interaction presumably operates through the sharing of oxygen sites, thus facilitating reduction. Because the probability of this interaction increases with coverage, it is expected that under reaction conditions the catalysts with higher vanadia coverage would have a greater capacity to release oxygen.

The behavior of a very low surface area V_2O_5/SiO_2 catalyst in the conversion of methane in both the absence and presence of NO in the feed has been reported recently [107,134]. Methane conversion started at temperatures above 900 K when CH_4+O_2 mixtures were used in the feed stream, but this was strongly enhanced when small amounts of NO were added (up to 0.39%). The effect of the concentration of NO in the feed on the yield of C_1-oxygenates is depicted in Figure 14.3. The production of C_1-oxygenates increased with increasing NO concentration values, reaching a yield to C_1-oxygenates as high as 7% for 0.39% of NO in the reactants. On the other hand, any increase in the NO concentration afforded very important increases in CH_3OH production. This trend suggests that HCHO is mainly related to the heterogeneous process on the catalyst, while CH_3OH can be produced through gas-phase reactions. In order to gain a better view of the effect of NO addition in the reaction scheme, Figure 14.4 shows the selectivity-conversion plot. In the absence of NO, the selectivity to C_1-oxygenates decreased very fast.

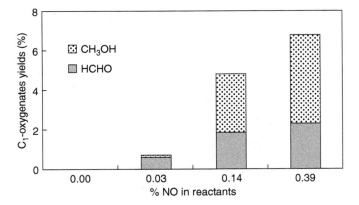

FIGURE 14.3 Yield to HCHO and CH_3OH products at different NO concentration levels. Reaction conditions: temperature = 883 K, CH_4/O_2 =2 (molar), catalyst weight = 0.20 g, and atmospheric pressure (Reproduced with permission from Springer)

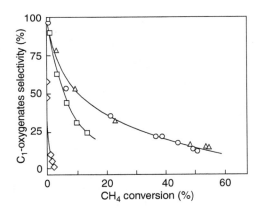

FIGURE 14.4 Selectivity to C_1-oxygenates versus CH_4 conversion at different NO concentration levels in the feed: (◇) 0.0%, (□) 0.03%, (△) 0.14%, and (○) 0.39% (Reproduced with permission from Springer)

Small amounts of NO (0.03%) in the reactants afforded a significant increase in C_1-oxygenate selectivity, which was doubled under 0.14% of NO and was best at 0.39% of NO. The selectivity to C_1-oxygenates remained above 20% at ~30% CH_4 conversion, which is far better than most results reported in the literature. In addition, the selectivity to CH_3OH produced at atmospheric pressure was still observable ($S_{MeOH} \sim 2\%$) at CH_4 conversions above 50%.

Barbero et al. [107] described the effects of both CH_4/O_2 ratio and temperature on the conversion of methane and on the yield values to C_1-oxygenates (Figure 14.5) using a low specific area V_2O_5/SiO_2 catalyst combined with the addition of the NO radical initiator to the feed stream. The conversion of methane

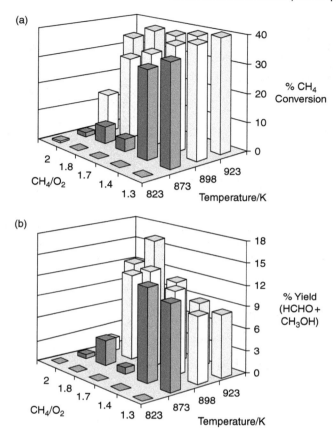

FIGURE 14.5 Effect of temperature and of CH_4/O_2 ratio on the CH_4 conversion (a), and on the yield of C_1-oxygenates (b). Reaction conditions: catalyst weight = 0.30 g, CH_4 concentration = 22.9 to 34.4% (molar), and 1.0 NO/N_2 (Reproduced by permission from Royal Society of Chemistry)

increased with temperature at all oxygen partial pressures (Figure 14.5[a]). Similarly, the yield of C_1-oxygenates increased with the oxygen partial pressure at the lower temperatures; however, it decreased at the higher temperatures (Figure 14.5[b]). The combined effect of both temperature and oxygen partial pressure afforded a yield that reached a maximum value at high temperatures (923 K and low partial pressure of oxygen: $CH_4/O_2 = 1.8$ M). The yield of C_1-oxygenates reached 16% at atmospheric pressure, which to our knowledge is the highest value reported. The strong effect of NO appears to be due to the chain propagation of the radical reactions close to the catalyst bed. These trends are predicted by the reaction models developed by Tabata et al. [106]. The profiles in Figure 14.5(b) clearly underline the relevance of both the oxygen partial pressure and the reaction temperature for a given NO concentration in the feed stream. The radical-due reactions are strongly affected by the reaction conditions. In particular, the methyl–methylperoxo ($CH_3^\bullet/CHO_2^\bullet$) radical equilibrium depends

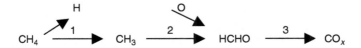

SCHEME 14.3   Simple reaction scheme for CH_4 oxidation on metal oxides

on the concentration of oxygen and on temperature and is shifted to O-containing radicals as temperature decreases. The results presented in Figure 14.5 show how important these two parameters are in the reaction conditions used. An appropriate combination of both temperature and oxygen partial pressure affords very high yields to C_1-oxygenates.

14.5.2.3   Other redox oxides

Although molybdenum and vanadium oxides have been widely used for the partial oxidation of methane into C_1-oxygenates, other metal oxides have also been investigated with the idea in mind. A large number of metal oxides (MO_x) have been tested by Otsuka and Hatano [135] with a view to correlating activity with the electronegativity of the respective cation. The authors proposed a simple, conceptual reaction scheme for CH_4 oxidation (Scheme 14.3) involving H-abstraction in a first step, followed by O-insertion in a second step. Obviously, this requires quite different sites. From this sequential scheme it may be inferred that a compromise oxide with sites having a little of the character needed for steps 1 and 2 might be the best. The highest activity for CH_4 conversion was maximum for Ga_2O_3 and Bi_2O_3, which lay exactly on the middle of the electronegativity scale.

In addition, maximizing the HCHO yield implies inhibition of step 3, and selectivity to HCHO improved for the most strongly electronegative oxides, such as those of P, W, and B. Based on these arguments, Otsuka and Hatano [135] proposed a binary oxide mixture of Be and B supported on silica. Working with this catalyst at 873 K, they achieved a 1% HCHO yield with a CH_4 conversion of 2.8%, which in turn is consistent with the data found with vanadium and molybdenum oxides.

Recently, Zhang et al. [136] reported that antimonium oxides in either Sb_2O_3 or Sb_2O_5 highly dispersed on a silica substrate are selective for the partial oxidation of methane into HCHO. SbO_x/SiO_2 catalysts with SbO_x loadings up to 20 wt% show good HCHO selectivity, even at temperatures as high as 923 K, and the more oxidized Sb_2O_5/SiO_2 catalysts are more selective than the reduced Sb_2O_3 counterparts. HCHO selectivity up to 41% was obtained for the Sb_2O_5/SiO_2 catalyst at 873 K, although this decreased to 18% when increasing the reaction temperature up to 923 K.

14.6   CONCLUSIONS AND OUTLOOK

Metal oxides are useful catalysts for several reactions of methane conversion. In all methane reactions, C–H bond activation is a crucial first step in the formation of partially oxidized or combustion products. Once the first bond has been broken, the resulting CH_3 fragment is highly reactive, and if there is enough oxygen in the environment, reactions to CO_2 and H_2O are relatively easy ("downhill reactions"). As a general rule, the combustion of methane is a relatively easy reaction that can be achieved on many metal oxide catalysts, typically redox oxide catalysts in either simple or mixed systems. The nature and type of the metal oxides used is, however, much more demanding if partial oxidation compounds — for example, dimerization and C_1-oxygenates — are the products targeted. Generally, these later conversion processes use fuel-rich mixtures with the oxidant to minimize the extent of combustion reactions, which yield unwanted carbon oxides. Under these conditions, oxidation reactions require high temperatures, which are detrimental for the control of selectivity to the desired products. The OCM is usually conducted over basic metal oxides, of which rare-earth oxide promoted by alkaline-earth metal oxides are highly active, selective, and thermally stable catalysts and hence show great promise for use in the development of the process. Notwithstanding, considerable efforts have been devoted to developing active and selective catalysts but neither the achieved yield of C_1-oxygenates nor the complete mechanism of the reactions has been clarified. The selective O-insertion into CH_3 or other fragments resulting from the first H-abstraction of the CH_4 molecule, is usually conducted on redox oxides of the MoO_3 and V_2O_5 type. On these systems, the key to maximizing catalytic performances is to keep isolated metal oxide structures on a silica substrate in a slightly reduced state. The presence of these partially reduced oxides allows the reduction–oxidation cycles of catalytic surfaces to proceed more rapidly and smoothly. Additionally, gas-phase radical initiators — that is, nitrogen oxides — strongly enhance catalyst performance. In sum, although much progress has been made along the past 15 years in the three directions of methane conversion on metal oxides addressed in this chapter, sustained efforts for the development of better catalysts and catalytic processes are essential to render these processes commercially feasible.

REFERENCES

1. Choudhary, V.R., Mamman, A.S., and Sansare, S.D. Selective oxidation of methane to CO and H_2 over Ni/MgO at low-temperatures. *Chem. Int. Ed. Engl.* 1992, *31*, 1189–1190.
2. Hickman, D.A. and Schmidt, L.D. Synthesis gas-formation by direct oxidation of methane over Pt moboliths. *J. Catal.* 1992, *138*, 267–282.
3. Hickman, D.A. and Schmidt, L.D. Production of syngas by direct catalytic oxidation of methane. *Science* 1993, *259*, 343–346.
4. Peña, M.A., Gomez, J.P., and Fierro, J.L.G. New catalytic routes for syngas and hydrogen production. *Appl. Catal. A: Gen.* 1996, *144*, 7–57.
5. Ashcroft, A.T., Cheetham, A.K., and Green, M.L.H. Partial oxidation of methane to synthesis gas using carbon dioxide. *Nature* 1991, *352*, 225–226.
6. Choudhary, V.R., Mamman, A.S., and Uphade, B.S. Steam and oxysteam reforming of methane to syngas over $Co_xNi_{1-x}O$ supported on MgO precoated SA-5205. *AIChE J.* 2001, *47*, 1632–1638.

7. Lee, S. and Oyama, S.T. Oxidative coupling of methane to higher hydrocarbons. *Catal. Rev. Sci. Eng.* 1988, *30*, 249–280.

8. Choudhary, V.R., Uphade, B.S., and Mulla, S.A.R. Oxidative coupling of methane over a Sr-promoted La_2O_3 catalyst supported on a low surface area porous catalyst carrier. *Ind. Eng. Chem. Res.* 1997, *36*, 3594–3601.

9. Lunsford, J.H. The catalytic oxidative coupling of methane. *Angew. Chem. Int. Ed. Engl.* 1995, *34*, 970–980.

10. Aiello, R., Fiscus, J.E., Loye, Z., and Amiridis, M.D. Hydrogen production via the direct cracking of methane over Ni/SiO_2: catalyst deactivation and regeneration. *Appl. Catal. A: Gen.* 2000, *192*, 227–234.

11. Choudhary, T.V., Sivadinarayana, C., Chusuei, C., Klinghoffer, A., and Goodman, D.W. Hydrogen production via catalytic decomposition of methane. *J. Catal.* 2001, *199*, 9–18.

12. Belgued, M., Pareja, P., Amariglio, A., and Amariglio, H. Conversion of methane into higher hydrocarbons on platinum. *Nature* 1991, *352*, 789–790.

13. Koerts, T., Deelen, M.J., and van Santen, R.A. Hydrocarbon formation from methane by a low-temperature 2-step reaction sequence. *J. Catal.* 1992, *138*, 101–114.

14. Koranne, M.M., Zajac, G.W., and Goodman, D.W. Direct conversion of methane to higher hydrocarbons via an oxygen-free, low temperature route. *Catal. Lett.* 1995, *30*, 219–234.

15. Zwinkels, M.F.M., Jaras, S.G., and Menon, P.G. Catalytic materials for high-temperature combustion. *Catal. Rev. Sci. Eng.* 1993, *35*, 319–358.

16. Zwinkels, M.F.M., Jaras, S.G., and Menon, P.G. In *Structured Catalysts and Reactors*, Cybulski, A. and Moulijn, J.A., Eds. Marcel Dekker, New York, 1998.

17. Arai, H. and Fukuzawa, H. Research and development on high temperature catalytic combustion. *Catal. Today* 1995, *26*, 217–221.

18. Trimm, D.L. Research and development on high temperature catalytic combustion. *Catal. Today* 1995, *26*, 231–238.

19. Klvana, D., Chaouki, J., Guy, C., and Kirchnerova, J. Catalytic combustion: new catalysts for new technologies. *Combust. Sci. Technol.* 1996, *121*, 51–65.

20. DallaBetta R.A., Schlatter, J.C., Yee, D.K., Loffler, D.G., and Shoji, T.R.L. Catalytic combustion technology to achieve ultra low NO_x emissions: Catalyst design and performance characteristics. *Catal. Today* 1995, *26*, 329–335.

21. Farrauto, R.J. and Heck, R.M. Environmental catalysis into the 21st century. *Catal. Today* 2000, *55*, 179–187.

22. Forzatti, P. and Groppi, G. Catalytic combustion for the production of energy. *Catal. Today* 1999, *54*, 165–180.

23. Ali, A., Chin, Y.H., and Resasco, D.E. Redispersion of Pd on acidic supports and loss of methane combustion activity during the selective reduction of NO by CH_4. *Catal. Lett.* 1998, *56*, 111–117.

24. Arutyunov, V.S., Vasevich, V.Y., Vedeneev, V.I., Parfenov, Y.V., and Sokolov, O.V. Dependence of the kinetics of gas phase methane oxidation at high pressures on the concentration of oxygen and on temperature. *Russ. Chem. Bull.* 1996, *45*, 45–48.

25. Gesser, H.D. and Hunter, N.R. A review of C-1 conversion chemistry. *Catal. Today* 1998, *42*, 183–189.

26. Brown, M.J. Direct oxidation of methane to methanol. *Appl. Catal.* 1988, *45*, 168–169.

27. Wang, Y. and Otsuka, K. Catalytic oxidation of methane to methanol with H_2O_2 gas-mixture at atmospheric pressure. *J. Catal.* 1995, *155*, 256–267.

28. Tabata, K., Teng, Y., Takemoto, T., Suzuki, E., Bañares, M.A., Peña, M.A., and Fierro, J.L.G. Homogeneous activation of methane by oxygen and nitrogen oxides. *Catal. Rev. Sci. Eng.* 2002, *44*, 1–56.

29. Burch, R., Crittle, D.J., and Hayes, M.J. C–H bond activation in hydrocarbon oxidation on heterogeneous catalysts. *Catal. Today* 1999, *47*, 229–234.

30. Burch, R. and Hayes, M.J. H-bond activation in hydrocarbon oxidation on solid catalysts. *J. Mol. Catal. A: Chem.* 1995, *100*, 13–33.

31. Campbell, K.D., Morales, E., and Lunsford, J.H. Gas-phase coupling of methyl radicals during the catalytic partial oxidation of methane. *J. Am. Chem. Soc.* 1987, *109*, 7900–7901.

32. Ito, T., Wang, J.X., Lin, C.H., and Lunsford, J.H. Oxidative dimerization of methane over a lithium-promoted magnesium oxide catalyst. *J. Am. Chem. Soc.* 1985, *107*, 5062–5068.

33. Choudhary, V.R. and Rane, V.H. Acidity–basicity of rare-earth oxides and their catalytic activity in oxidative coupling of methane to C_2-hydrocarbons. *J. Catal.* 1991, *130*, 411–422.

34. Marion, M.C., Garbowski, E., and Primet, M. Physicochemical properties of copper oxide loaded alumina in methane combustion. *J. Chem. Soc., Faraday Trans. I* 1990, *86*, 3027–3032.

35. Zaki, M.I., Hasan, M.A., Pasupulety, L., Fouad, N.E., and Knözinger, H. Bulk and surface characteristics of pure and alkalized Mn_2O_3: TG, IR, XRD, XPS, specific adsorption and redox catalytic studies. *New J. Chem.* 1999, *23*, 875–882.

36. McCarthy, J.G., Chang, Y.F., Wong, V.L., and Johansson, M.E. Kinetics of high temperature methane combustion by metal oxide catalysts. *Prepr. Am. Chem. Soc., Div. Petro. Chem.* 1997, *42*, 158–161.

37. Arnone, S., Bagnasco, G., Busca, G., Lisi, L., Russo, G., and Turco, M. Catalytic combustion of methane over transition metal oxides. *Stud. Surf. Sci. Catal.* 1998, *119*, 65–70.

38. Finnochio, E., Willey, R.J., Busca, G., and Lorenzelli, V. FTIR studies on the selective oxidation and combustion of light hydrocarbons at metal oxide surfaces. 3. Comparison of the oxidation of C_3 organic compounds over Co_3O_4, $MgCr_2O_4$ and CuO. *J. Chem. Soc., Faraday Trans. I* 1997, *93*, 175–180.

39. Xia, B., Duan, L., Xie, Y., and Tang, Y. Preparation of Y_2O_3 stabilised ZrO_2. *Mater. Sci. Technol.* 1999, *15*, 755–760.

40. Artizzu, P., Garbowski, E., Primet, M., Brulle, Y., and Saint-Just, J. Catalytic combustion of methane on aluminate-supported copper oxide. *Catal. Today* 1999, *47*, 83–99.

41. Guilhaume, N. and Primet, M. Catalytic combustion of methane — Copper oxide supported on high specific area spinels synthesized by a sol–gel process. *J. Chem. Soc., Faraday Trans. I* 1994, *90*, 1541–1545.

42. Park, P.W. and Ledford, J.S. The influence of surface structure on the catalytic activity of alumina supported copper oxide catalysts — Oxidation of carbon monoxide and methane. *Appl. Catal. B: Environ.* 1998, *15*, 221–231.

43. Baldi, M., Finocchio, E., Milellea, F., and Busca, G. Characterization of manganese and iron oxides as combustion catalysts for propane and propene. *Appl. Catal. B: Environ.* 1998, *16*, L175–L182.

44. Baldi, M., Escribano, V.S., Amores, J.M.G., Milella, F., and Busca, G. Characterization of manganese and iron oxides as combustion catalysts for propane and propene. *Appl. Catal. B: Environ.* 1998, *17*, L175–L182.
45. Finocchio, E., Busca, G., Lorenzelli, V., and Escribano, V.S. FTIR studies on the selective oxidation and combustion of light hydrocarbons at metal oxide surfaces. 2. Propane and propene oxidation on Co_3O_4. *J. Chem. Soc., Faraday Trans. I* 1996, *92*, 1587–1593.
46. Zaki, M.I., Hasan, M.A., Pasupulety, L., Fouad, N.E., and Knozinger, H. CO and CH_4 total oxidation over manganese oxide supported on ZrO_2, TiO_2,TiO_2–Al_2O_3 and SiO_2–Al_2O_3 catalysts. *New J. Chem.* 1999, *23*, 1197–1202.
47. Matsuura, I., Hashimoto, Y., Takayasu, O., Nitta, K., and Yoshida, Y. Heat-stable ultefine single-crystal magnesium oxide and its character as a support material for high temperature combustion catalysts. *Appl. Catal.* 1991, *74*, 273–280.
48. Berg, M. and Jaras, S. Control of methane emissions from oil/gas-fired furnaces by high temperature combustion of methane over thermally stable CoO–MgO catalyst. *Appl. Catal. A: Gen.* 1994, *114*, 227–241.
49. Choudhary, V.R., Mamman, A.S., Pataskar, S.G., and Banerjee, S. Control of methane emissions from oil/gas-fired furnaces by high temperature combustion of methane over thermally stable CoO–MgO catalyst. *Prepr. Am. Chem. Soc. Fuel Chem. Div.* 2001, *46*, 83–87.
50. Taylor, S.H. and O'Leary, S.R. A study of uranium oxide based catalysts for the oxidative destruction of short chain alkanes. *Appl. Catal. B: Environ.* 2000, *25*, 137–149.
51. McCarthy, J.G., Chang, Y.F., Wong, V.L., and Johansson, M.E. Kinetics of high temperature methane combustion by metal oxide catalysts. *Prepr. Am. Chem. Soc. Div. Petrol. Chem.* 1997, *42*, 158–164.
52. Arnone, S., Bagnasco, G., Busca, G., Lisi, L., Russo, G., and Turco, M. Catalytic combustion of methane over transition metal oxides. *Stud. Surf. Sci. Catal.* 1998, *119*, 65–70.
53. Tejuca, L.G., Fierro, J.L.G., and Tascón, J.M.D. Structure and reactivity of perovskite-type oxides. *Adv. Catal.* 1989, *36*, 237–328.
54. Peña, M.A. and Fierro, J.L.G. Chemical structures and performance of perovskite oxides. *Chem. Rev.* 2001, *101*, 1981–2017.
55. Arai, H., Yamada, T., Eguchi, K., and Seiyama, T. Catalytic combustion of methane over various perovskite-type oxides. *Appl. Catal.* 1986, *26*, 265–276.
56. McCarthy, J.G. Kinetics of PdO combustion catalysis. *Catal. Today* 1990, *26*, 283–293.
57. Tejuca, L.J. and Fierro, J.L.G., Eds. *Properties and Applications of Perovskite Type Oxides.* Marcel Dekker: New York, 1993.
58. Falcon, H., Barbero, J.A., Alonso, J.A., Martinez-Lope, M.J., and Fierro, J.L.G. $SrFeO_{3-\delta}$ perovskite oxides: Chemical features and performance for methane combustion. *Chem. Mater.* 2002, *14*, 2325–2333.
59. Seiyama, T., Yamazoe, N., and Eguchi, K. Characterization and activity of some mixed metal-oxide catalysts. *Ind. Eng. Chem. Prod. Res. Dev.* 1985, *24*, 19–27.
60. Shimizu, T. Partial oxidation of hydrocarbons and oxygenated compounds on perovskite oxides. *Catal. Rev. Sci. Eng.* 1992, *34*, 355.
61. Marchetti, L. and Forni, L. Catalytic combustion of methane over perovskites. *Appl. Catal. B: Environ.* 1998, *15*, 179–187.

62. Ladavos, A.K. and Pomonis, P.J. Catalytic combustion of methane on $La_{2-x}Sr_xNiO_{4-\lambda}$ ($x = 0.00–1.50$) perovskites prepared via the nitrate and citrate routes. *J. Chem. Soc., Faraday Trans. I* 1992, *88*, 2557–2562.

63. Saracco, G., Geobaldo, F., and Baldi, G. Methane combustion on Mg-doped $LaMnO_3$ perovskite catalysts. *Appl. Catal. B: Environ.* 1999, *20*, 277–288.

64. Belessi, V.C., Ladavos, A.K., and Pomonis, P.J. Methane combustion on La–Sr–Ce–Fe–O mixed oxides: bifunctional synergistic action of $SrFeO_{3-x}$ and CeO_x phases. *Appl. Catal. B: Environ.* 2001, *31*, 183–194.

65. Ciambelli, P., Palma, V., Tikhov, S.F., Sadykov, S.V., Isupova, L.A., and Lisi, L. Catalytic activity of powder and monolith perovskites in methane combustion. *Catal. Today* 1999, *47*, 199–207.

66. Choudhary, V.R., Uphade, B.S., Pataskar, S.G., and Keshavraja, A. Low-temperature complete combustion of methane over Mn-, Co-, and Fe-stabilized ZrO_2. *Angew. Chem. Int. Ed. Engl.* 1996, *35*, 2393–2395.

67. Choudhary, V.R., Uphade, B.S., Pataskar, S.G., and Thite, G.A. Low-temperature total oxidation of methane over Ag-doped $LaMO_3$ perovskite oxides. *J. Chem. Soc., Chem. Commun.* 1996, (6), 1021–1022.

68. Choudhary, V.R., Uphade, B.S., and Pataskar, S.G. Low temperature complete combustion of dilute methane over Mn-doped ZrO_2 catalysts: Factors influencing the reactivity of lattice oxygen and methane combustion activity of the catalyst. *Appl. Catal. A: Gen.* 2002, *227*, 29–41.

69. Milt, V.G., Ulla, M.A., and Lombardo, E.A. Cobalt-containing catalysts for the high-temperature combustion of methane. *Catal. Lett.* 2000, *65*, 67–73.

70. Zhang, Y., Andersson, S., and Muhammed, M. Nanophase catalytic oxides. 1. Synthesis of doped cerium oxides as oxygen storage promoters. *Appl. Catal. B: Environ.* 1995, *6*, 325–337.

71. Trovarelli, A. Catalytic properties of ceria and CeO_2-containing materials. *Catal. Rev. Sci. Eng.* 1996, *38*, 439–520.

72. Kundakovic, L. and Flytzani-Stephanopoulos, M. Cu- and Ag-modified cerium oxide catalysts for methane oxidation. *J. Catal.* 1998, *179*, 203–221.

73. Bozo, C., Guilhaume, N., Garbowski, E., and Primet, M. Combustion of methane on $CeO_2–ZrO_2$ based catalysts. *Catal. Today* 2000, *59*, 33–45.

74. Terribile, D., Trovarelli, A., de Leitenburg, C., Primavera, A., and Giuliano, G. Catalytic combustion of hydrocarbons with Mn and Cu-doped ceria-zirconia solid solutions. *Catal. Today* 1999, *47*, 133–140.

75. Palmqvist, A.E.C., Johansson, E.M., Jaras, S.G., and Muhammed, M. Total oxidation of methane over doped nanophase cerium oxides. *Catal. Lett.* 1998, *56*, 69–75.

76. O'Connell, O. and Morris, M.A. New ceria-based catalysts for pollution abatement. *Catal. Today* 2000, *59*, 387–393.

77. Zamar, F., Trovarelli, A., de Leitenburg, C., and Dolcetti, G. CeO_2-based solid solutions with the fluorite structure as novel and effective catalysts for methane combustion. *J. Chem. Soc., Chem. Commun.* 1995, (9) 965–966.

78. Daturi, M., Busca, G., Groppi, G., and Forzatti, P. Preparation and characterisation of $SrTi_{1-x-y}Zr_xMn_yO_3$ solid solution powders in relation to their use in combustion. *Appl. Catal. B: Environ.* 1997, *12*, 325–337.

79. Machida, M., Eguchi, K., and Arai, H. Catalytic properties of of α-$BaCrAl_{11}O_{19}$, α-$BaMnAl_{11}O_{19}$, α-$BaFeAl_{11}O_{19}$, α-$BaCoAl_{11}O_{19}$, and

α-BaNiAl$_{11}$O$_{19}$ for high temperature catalytic combustion. *J. Catal.* 1989, *120*, 377–386.

80. Machida, M., Eguchi, K., and Arai, H. Effect of structural modification on the catalytic property of Mn-substituted hexaaluminates. *J. Catal.* 1990, *123*, 477–485.

81. Bellotto, M., Artioli, G., Cristiani, C., Forzatti, P., and Groppi, G. On the crystal structure and cation valence of Mn in Mn-substituted Ba-beta-Al$_2$O$_3$. *J. Catal.* 1998, *179*, 597–605.

82. Jang, B.W.L., Nelson, R.M., Spivey, J.J., Ocal, M., Oukaci, R., and Marcelin, G. Catalytic oxidation of methane over hexaaluminates and hexaaluminate-supported Pd catalysts. *Catal. Today* 1999, *47*, 103–113.

83. Artizzu-Duart, P., Brulle, Y., Gaillard, F., Garbowski, E., Guilhaume, N., and Primet, M. Catalytic combustion of methane over copper- and manganese-substituted barium hexaaluminates. *Catal. Today* 1999, *54*, 181–190.

84. Artizzu-Duart, P., Millet, J.M., Guilhaume, N., Garbowski, E., and Primet, M. Catalytic combustion of methane on substituted barium hexaaluminates. *Catal. Today* 2000, *59*, 163–177.

85. Widjaja, H., Sekizawa, K., and Eguchi, K. Catalytic combustion of methane over Pd supported on metal oxides. *Chem. Lett.* 1998, (6), 481–482.

86. Widjaja, H., Sekizawa, K., Eguchi, K., and Arai, H. Oxidation of methane over Pd/mixed oxides for catalytic combustion. *Catal. Today* 1999, *47*, 95–101.

87. Keller, G.E. and Bhasin, M.M. Synthesis of ethylene via oxidative coupling of methane. 1. Determination of active catalysts. *J. Catal.* 1982, *73*, 9–19.

88. Ito, T. and Lunsford, J.H. Synthesis of ethylene and ethane by partial oxidation of methane over lithium-doped magnesium oxide. *Nature* 1985, *314*, 721–722.

89. Sofranko, J.A., Leonard, J.J., and Jones, C.A. The oxidative conversion of methane to higher hydrocarbons. *J. Catal.* 1987, *103*, 302–310.

90. Warren, B.K. The role of chlorine in chlorine-promoted methane coupling catalysts. *Catal. Today* 1992, *13*, 311–320.

91. Krylov, O.V. Catalytic reactions of partial methane oxidation. *Catal. Today* 1993, *18*, 209–302.

92. Lunsford, J.H. Catalytic conversion of methane to more useful chemicals and fuels: a challenge for the 21st century. *Catal. Today* 2000, *63*, 165–174.

93. Buyevskaya, O.V., Wolf, D., and Baerns, M. Ethylene and propene by oxidative dehydrogenation of ethane and propane — Performance of rare-earth oxide-based catalysts and development of redox-type catalytic materials by combinatorial methods. *Catal. Today* 2000, *62*, 91–99.

94. Lacombe, S., Geantet, C., and Mirodatos, C. Oxidative coupling of methane over lanthana catalysts. 1. Identification and role of specific active sites. *J. Catal.* 1995, *151*, 439–452.

95. Sarkany, J., Sun, Q., Di Cosimo, J.I., Herman, R.G., and Klier, K. In *Methane and Alkane Conversion Chemistry*, Bhasin, M.M. and Slocum, D.W., Eds. Plenum Press, New York, 1995, p. 31.

96. Borchert, H. and Baerns, M. The effect of oxygen-anion conductivity of metal oxide-doped lanthanum oxide catalysts on hydrocarbon selectivity in the oxidative coupling of methane. *J. Catal.* 1997, *168*, 315–320.

97. Nagy, A.J., Mestl, G., and Schlogl, R. The role of subsurface oxygen in the silver-catalyzed, oxidative coupling of methane. *J. Catal.* 1999, *188*, 58–68.

98. Yamamoto, H., Chu, H.Y., Xu, M., Shi, C., and Lunsford, J.H. Oxidative coupling of methane over a Li^+/MgO catalyst using N_2O as an oxidant. *J. Catal.* 1993, *142*, 325–336.

99. Buyevskaya, O.V., Rothaemel, M., Zanthoff, H.W., and Baerns, M. Transient studies on reaction steps in the oxidative coupling of methane over catalytic surfaces of MgO and Sm_2O_3. *J. Catal.* 1994, *146*, 346–257.

100. Islam, M.S., Ilett, D.J., and Parker, S.C. Surface structures and oxygen hole formation on the La_2O_3 catalyst — A computer-simulation study. *J. Phys. Chem.* 1994, *98*, 9637–9641.

101. Baldwin, T.R., Burch, R., Squire, G.D., and Tsang, S.C. Influence of homogeneous gas-phase reactions in the partial oxidation of methane to methanol and formaldehyde in the presence of oxide catalysts. *Appl. Catal.* 1991, *74*, 137–152.

102. Bañares, M.A. and Fierro, J.L.G. In *Catalytic Selective Oxidation*, Oyama S.T. and Hightower, J.W., Eds. American Chemical Society, Washington, 1992, p. 354.

103. Faraldos, M., Bañares, M.A., Anderson, J.A., Hu, H., Wachs, I.E., and Fierro, J.L.G. Comparison of silica-supported MoO_3 and V_2O_5 catalysts in the selective partial oxidation of methane. *J. Catal.* 1996, *160*, 214–221.

104. Hunter, N.R., Gesser, H.D., Morton, L.A., and Yarlagadda, P.S. Methanol formation at high pressure by the catalyzed oxidation of natural gas and by the sensitized oxidation of methane. *Appl. Catal.* 1990, *57*, 45–54.

105. Otsuka, K., Takahashi, R. and Yamanaka, I. Oxygenates from light alkanes catalyzed by NO_x in the gas phase. *J. Catal.* 1999, *185*, 182–191.

106. Tabata, K., Teng, Y., Yamaguchi, Y., Sakurai, H., and Suzuki, E. Experimental verification of theoretically calculated transition barriers of the reactions in a gaseous selective oxidation of CH_4–O_2–NO_2. *J. Phys. Chem. A* 2000, *104*, 2648–2654.

107. Barbero, J.A., Alvarez, M.C., Bañares, M.A., Peña, M.A., and Fierro, J.L.G. Breakthrought in C_1-oxygenates production via direct methane oxidation. *Chem. Commun.* 2002, 1184–1185.

108. Herman, R.C., Sun, Q., Shi, C., Klier, K., Wang, C.B., Hu, H., Wachs, I.E., and Bhasin, M.M. Development of active oxide catalysts for the direct oxidation of methane to formaldehyde. *Catal. Today* 1997, *37*, 1–14.

109. Arena, F., Giordano, N., and Parmaliana, A. Working mechanism of oxide catalysts in the partial oxidation of methane to formaldehyde. 2. Redox properties and reactivity of SiO_2, MoO_3/SiO_2, V_2O_5/SiO_2, TiO_2, and V_2O_5/TiO_2 systems. *J. Catal.* 1997, *167*, 66–76.

110. Kasztelan, S. and Moffat, J.B. Partial oxidation of methane by oxygen over silica. *J. Chem. Soc., Chem. Commun.* 1987, *21*, 1663–1664.

111. Kastanas, G.N., Tsigdinos, G.A., and Schwank, J. Selective oxidation of methane over Vycor glass, quartz glass and various silica, magnesia and alumina surfaces. *Appl. Catal.* 1988, *44*, 33–59.

112. Guliev, I.A., Mamedov, A.Kh., and Aliev, V.S. *Azerbaijan Khim. Zhur.* 1985, 35.

113. Parmaliana, A., Sokolovskii, V., Miceli, D., Arena, F., and Giordano, N. On the nature of the catalytic activity of silica-based oxide catalysts in the partial oxidation of methane to formaldehyde with O_2. *J. Catal.* 1994, *148*, 514–523.

114. Vikulov, K., Martra, G., Coluccia, S., Miceli, D., Arena, F., Parmaliana, A., and Paukshits, E. FTIR spectroscopic investigation of the active sites on different

types of silica catalysts for methane partial oxidation to formaldehyde. *Catal. Lett.* 1996, *37*, 235–239.

115. Sun, Q., Herman, R.G., and Klier, K. Selective oxidation of methane with air over silica catalysts. *Catal. Lett.* 1992, *16*, 251–261.

116. Low, M.J.D. Reactive silica. 17. The nature of the reaction center. *J. Catal.* 1987, *103*, 496–501.

117. Kobayashi, T., Guilhaume, N., Miki, J., Kitamura, N., and Haruta, M. Oxidation of methane to formaldehyde over Fe/SiO$_2$ and Sn-W mixed oxides. *Catal. Today* 1996, *32*, 171–175.

118. Otsuka, K., Yamanaka, I., and Wang, Y. Reductive activation of oxygen for partial oxidation of light alkanes. *Stud. Surf. Sci. Catal.* 1998, *119*, 15–24.

119. Parmaliana, A., Arena, F., Frusteri, F., Martínez-Arias, A., López-Granados, M., and Fierro, J.L.G. Effect of Fe-addition on the catalytic activity of silicas in the partial oxidation of methane to formaldehyde. *Appl. Catal. A: Gen.* 2002, *226*, 163–174.

120. Spencer, N.D. and Pereira, C.J. V$_2$O$_5$–SiO$_2$-catalyzed methane partial oxidation with molecular oxygen. *J. Catal.* 1989, *116*, 399–406.

121. Bañares, M.A. and Fierro, J.L.G. Selective oxidation of methane to formaldehyde on supported molybdate catalysts. *Catal. Lett.* 1993, *17*, 205–211.

122. Marchi, A.J., Lede, E.J., Requejo, F.G., Renteria, M., Irusta, S., Lombardo, E.A., and Miro, E.E. Laser Raman spectroscopy (LRS) and time differential perturbed angular correlation (TDPAC) study of surface species on Mo/SiO$_2$ and Mo,Na/SiO$_2$. Their role in the partial oxidation of methane. *Catal. Lett.* 1997, *48*, 47–54.

123. Bañares, M.A., Spencer, N.D., Jones, M.D., and Wachs, I.E. Effect of alkali-metal cations on the structure of Mo(VI)/SiO$_2$ catalysts and its relevance to the selective oxidation of methane and methanol. *J. Catal.* 1994, *146*, 204–210.

124. Bañares, M.A., Rodriguez, I., Guerrero, A., and Fierro, J.L.G. Mechanistic aspects of the relective oxidation of methane to C$_1$ oxygenates over MoO$_3$/SiO$_2$ catalysts. In *Proceedings of the 10th International Congress on Catalysis, Budapest, 1992*, Guczi, L., Solymosi, F., and Tetenyi, P., Eds. Akademiai Kiado, Budapest, 1993, Vol. B, p. 1131.

125. Suzuki, K., Hayakawa, T., Shimizu, M., and Takehira, K. Partial oxidation of methane over silica-supported molybdenum oxide catalysts. *Catal. Lett.* 1995, *30*, 159–169.

126. Smith, M.R. and Ozkan, U.S. The partial oxidation of methane to formaldehyde — Role of different crystal planes of MoO$_3$. *J. Catal.* 1993, *141*, 124–139.

127. Smith, M.R., Zhang, L., Driscoll, S.A., and Ozkan, U.S. Effect of surface species on activity and selectivity of MoO$_3$/SiO$_2$ catalysts in partial oxidation of methane to formaldehyde. *Catal. Lett.* 1993, *19*, 1–15.

128. Bañares, M.A., Pawelec, B., and Fierro, J.L.G. Direct conversion of methane to C$_1$-oxygenates over MoO$_3$-USY zeolites. *Zeolites* 1992, *12*, 882–888.

129. Antiñolo, A., Cañizares, P., Carrillo, F., Fernandez-Baeza, J., Funez, F.J., de Lucas, A., Otero, A., Rodriguez, L., and Valverde, J.L. A grafted methane partial oxidation catalyst from MoO$_2$(acac)$_2$ and HZSM-5 zeolite. *Appl. Catal. A: Gen.* 2000, *193*, 139–146.

130. Karthereuser, B., Hodnett, B.K., Zanthoff, H., and Baerns, M. Transient experiments on the selective oxidation of methane to formaldehyde over V_2O_5/SiO_2 studied in the temporal analysis of products reactor. *Catal. Lett.* 1993, *21*, 209–214.

131. Koranne, M.M., Goodwin, J.G. Jr., and Marcelin, G. Partial oxidation of methane over silica-supported and alumina-supported vanadia catalysts. *J. Catal.* 1994, *148*, 388–391.

132. Faraldos, M., Bañares, M.A., Anderson, J.A., and Fierro, J.L.G. In *Methane and Alkane Conversion Chemistry*, Bhasin, M.M. and Slocum, D.W., Eds. Plenum Press, New York, 1995, p. 241.

133. Bañares, M.A., Cardoso, J.H., Agulló-Rueda, F., Bueno, J.M.C., and Fierro, J.L.G. Dynamic states of V-oxide species: Reducibility and performance for methane oxidation on V_2O_5/SiO_2 catalysts as a function of coverage. *Catal. Lett.* 2000, *64*, 1191–1196.

134. Bañares, M.A., Cardoso, J.H., Hutchings, G.J., Bueno, J.M.C., and Fierro, J.L.G. Selective oxidation of methane to methanol and formaldehyde over V_2O_5/SiO_2 catalysts. Role of NO in the gas phase. *Catal. Lett.* 1998, *56*, 149–153.

135. Otsuka, K. and Hatano, M. The catalysts for the synthesis of formaldehyde by partial oxidation of methane. *J. Catal.* 1987, *108*, 252–255.

136. Zhang, H., Ying, P., Zhang, J., Liang, C., Feng, Z., and Li, C. SbO_x/SiO_2 catalysts for the selective oxidation of methane to formaldehyde using molecular oxygen as oxidant. *Stud. Surf. Sci. Catal.* 2004, *147*, 547–552.

15 Oxidative Dehydrogenation (ODH) of Alkanes over Metal Oxide Catalysts

G. Deo, M. Cherian, and T.V.M. Rao

Department of Chemical Engineering, Indian Institute of Technology, Kanpur, India

CONTENTS

15.1 INTRODUCTION

Conversion of lower alkanes (ethane, propane, and butanes) to organic compounds of industrial importance is a daunting task. This is due to the poor reactivity of the C–H bond. Furthermore, the industrially important product(s) is usually more reactive than the alkane itself. Selective catalytic oxidation using oxygen is one of the simplest ways of converting alkanes, especially the lower alkanes, into useful intermediates of the petrochemical industry. Since it is selective under normal conditions and in the presence of air, thermodynamics suggests that undesirable CO_2 is the only stable compound formed; though at higher temperatures CO, which is also undesirable, is also stable [1]. Consequently, all organic materials in the presence of air are only metastable intermediates to carbon oxides. The challenge is,

491

thus, to obtain significant yields of the desired product or products by controlling the operating conditions and catalyst. Of these different factors, the role of the catalyst is paramount and a proper design of a catalyst will assist in converting alkanes to useful products.

Oxidative dehydrogenation (ODH) or oxydehydrogenation is a specific type of selective oxidation process by which alkenes are formed and is a frequently quoted alternative to the dehydrogenation of alkanes process. The dehydrogenation reaction is one of the processes currently used for alkene production, but is limited by thermodynamic, catalyst deactivation, and energy considerations. The ODH reaction is expected to overcome these limitations. However, the ODH reaction possesses disadvantages due to presence of oxygen, which increases the possibility of side reactions occurring and the possibility of an explosive mixture of hydrocarbon and oxygen being present.

During the ODH reaction, a multistep process occurs with a consecutive abstraction of two hydrogen atoms from the alkane molecule and desorption of water. As a result a new C–C bond is formed. This new C–C bond may occur by intermolecular removal of two hydrogen atoms (a new C–C bond between the alkanes) or by intramolecular removal of two hydrogen atoms (a π C–C bond) [2]. Consequently, the C/H ratio of the product is higher. Only the intramolecular removal of two hydrogen atoms from the alkanes that produces alkenes will be considered here. The alkenes that are produced are of significantly higher value compared with the light alkane feedstock and is used in the polymers, detergents, paints, adhesives, and other industries, which form the basis of the ODH reaction study.

During the formation of alkenes by ODH over a metal oxide catalyst, the first step is the adsorption of the basic alkane molecule, which is possible on acid sites of the catalyst surface. Consequently, the acid–base nature of the catalyst surface is important. The first hydrogen atom is removed from the weakest C–H bond [3], which is immaterial for ethane since all C–H bonds are similar. The weakest C–H bond is the C–H bond of the secondary carbon atom for propane and *n*-butane; and is the C–H bond of the tertiary carbon atom for isobutane. The removal of the second hydrogen atom is from the carbon atom that is adjacent to the carbon atom where the first hydrogen atom is removed and the alkene molecule or its immediate precursor is formed [4]. The alkene molecule formed should ideally be desorbed into the gas phase without undergoing further adsorption. Again, the acid–base nature of the catalyst is important since it will affect the readsorption phenomena. For the ODH reaction to be successful, the hydrogen atoms formed previously should combine with the oxidant, usually oxygen, with the formation of water. There are, however, a few studies that suggest that complete combination of hydrogen with the oxygen is not achieved and hydrogen is observed in the outlet of the reactor [5]. The formation of water is what makes the reaction practically irreversible and exothermic. However, the formation of carbon oxides (CO and CO_2) is also possible, which has to be limited. The adsorption of alkane, the reaction of alkane to alkene with no formation of side products, and desorption of alkenes with no readsorption are all features that are to be considered for a proper design of a catalyst.

Metal oxide surfaces can react with gases or solutions; they behave as catalyst or as support for catalyst. Over the years there has been a strong interest in understanding the behavior of metal oxide catalysts as single metal oxide or mostly, as mixed metal oxides [6]. The electronic effects responsible for the reactant becoming the product on a given metal oxide catalyst depends on the nature, strength, and energy of the Me–O bonds, either isolated or bridged, and on their structural arrangement. The mutual adaptation of the molecules (reactants and products) and of the solid determines the selectivity factor and the activity [7]. Transfer of electrons between reacting molecules and solid oxide catalysts is assumed to be a complex phenomenon, however, it can be influenced either by activation of adsorbing molecule or through generation of defects, introduction of dopants, or deposition of the active phase on a support to form oxide monolayer type catalysts [8].

The use of metal oxide catalysts for the selective oxidation of hydrocarbons in general and ODH in particular is well documented. These metal oxides are essentially of two types: reducible and nonreducible. The reducible metal oxide catalysts (e.g., transition metal oxides) undergo the ODH reaction by a redox mechanism and the nonreducible metal oxide catalysts (e.g., rare-earth metal oxides, alkali, and alkaline-earth oxides) used for the ODH reaction are based on the activation by molecular oxygen. The different metal oxide catalysts consist of bulk metal oxides and supported metal oxides, though in several cases efforts have been made to support bulk metal oxide catalysts. Supported metal oxide catalysts are not discussed in detail in this chapter. Several reactor configurations have also been tested and show promise, namely, the fixed-bed reactors, circulating/fluidized-bed reactors, and membrane reactors [9]. Future studies will emphasis on adapting these reactor configurations for macrolevel design [10].

In this chapter, the emphasis is on the bulk metal oxide catalysts that is to be used for the ODH of alkanes. The alkanes considered are ethane, propane, n-butane, and isobutane. The purpose of this discussion is to provide the reader with the present knowledge regarding the different types of metal oxide catalysts used for the ODH of alkanes. In Section 15.2, a brief outline of the mechanisms involved in the ODH reaction is discussed. Section 15.3 discusses bulk metal oxide catalysts used for the ODH reaction. Finally, Section 15.4 concludes this chapter. It is expected that the reader will be able to use this information for future design and engineering of metal oxide catalysts for the ODH of ethane, propane, or butanes or a mixture of these.

15.2 MECHANISMS OF THE ODH REACTIONS

The conversion of alkanes in presence of oxygen comprises of dehydrogenation, oxygenation, and combustion reactions. The product distribution depends on the nature of the catalyst, the nature of the alkane [3], and the operating conditions. Invariably it is noticed that with an increasing alkane conversion the selectivity to carbon oxides increases at the expense of the selectivity to alkenes.

Several studies have been performed to elucidate the mechanisms of the ODH process so as to better design the catalyst that will impart a higher alkene yield. For example, a consecutive reaction network was proposed for ODH of propane over Ni–Co–molybdate catalyst in which the propene is produced by the oxidation of propane, while carbon monoxide is produced by the successive oxidation of propene and carbon dioxide by further oxidation of carbon monoxide [11]. Baldi et al. [12] suggested that direct oxidation of propane to CO_2 occurs under excess oxygen conditions. It appears that the major pathway for CO_x production is from further oxidation of propene formed and minor contribution comes from direct oxidation of propane. It is believed that the parallel reactions have a common surface intermediate formed by activation of the alkane [13]. Morales and Lunsford [14] proposed that two routes exist for the formation of propene over lithium-promoted magnesium oxide catalysts. The alkyl radical generated on the surface of the catalyst can either desorb into the gas phase and form alkene by reacting with oxygen or via a surface alkoxide ion. The latter route is supported by the observation of the isopropoxide as an intermediate species in propene formation [12]. According to Burch and Swarnakar [13], the relative contributions from these two routes for alkene formation are dependent on the reaction temperature and the nature of the catalyst. It was suggested that at higher temperatures there is a greater tendency for alkyl radicals to desorb on nonreducible catalysts, whereas, the tendency to form alkoxide ions may be greater for reducible catalysts, which may also lead to complete combustion [3]. Michaels et al. [15] suggested that the selective ODH of propane to propene and the formation of CO and CO_2 are partly parallel reactions, the first being zero order in oxygen and the second being half order. Furthermore, it was suggested that selective oxidation occurs by reaction with lattice oxygen, while total oxidation occurs with the participation of dissociatively adsorbed oxygen. Results of pulse experiments, however, suggested the involvement of lattice oxygen in both propene and CO_x formation during the ODH of propane reaction, which must be taken into account [14]. Similarly, for ethane and butanes ODH, different reaction pathways have been discussed as given later.

Thus, it emerges that there are six different pathways involved in the ODH reaction scheme: r_1 to r_6. These are shown in Scheme 15.1. Scheme 15.1 shows that propene may be formed as a primary product; CO may be formed as a primary and secondary product; and, CO_2 may be formed as a primary, secondary, and tertiary product. In the scheme, no changes in the catalyst are incorporated and only the state of the gas-phase carbon containing molecules are considered. Furthermore, for the ODH of n-butane the formation of butadiene also needs to be considered. The different pathways shown do not necessarily contain a single route by which the particular reactant is converted to the product. For example, r_1 may contain two routes for converting propane to propene. The objective of the catalyst is to maximize the amount of the desirable product, namely the alkenes, that is, maximize r_1.

Mars–van Krevelen type redox mechanism is widely suggested for the ODH of light alkanes over reducible metal oxides at low temperatures. During this type of reaction, the light alkanes and intermediate products react with lattice

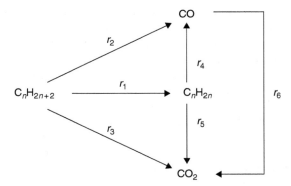

Scheme 15.1   Reaction pathways for the ODH of alkanes. Oxygenated products are not considered

oxygen, reducing the catalyst surface, which is in turn reoxidized by gas-phase O_2. Consequently, the catalyst is undergoing a continuous oxidation–reduction cycle. Under steady-state conditions the catalysts may, however, be partially reduced [3]. The partial reduced nature or the degree of reduction of the metal oxide can play an important role in determining the alkene formation. For example, Creaser et al. [16] observed that propene selectivity in the absence of gas-phase O_2 was superior to steady-state selectivity, at the same propane conversion. Based on this study it was suggested that propene selectivity could be further improved by increasing the degree of reduction of the catalyst [17]. Indeed kinetic and electrical conductivity studies for cesium modified nickel molybdate catalysts demonstrated that the degree of reduction of the catalyst can be modified since the ODH of propane reported smaller kinetic constants for the reduction step obtained with the doped catalyst [11]. The higher concentration of weakly adsorbed oxygen species may account for the lower selectivity in ODH of the undoped nickel molybdate catalysts; the role of cesium is to change the reduction and reoxidation rates of the metal oxide leading to different catalytic behavior. Changing the reducibility of the catalyst by using dopants without changing the nature of the active site is a difficult, if not impossible, task. In addition to changing the reducibility of the catalyst, alkali dopants also change the acid–base characteristics of the catalysts, which will facilitate alkene desorption. The other methods used for achieving a change in the degree of reduction of a specific catalyst is by varying the hydrocarbon/oxygen ratio and using an oxidant other than oxygen. For example, by varying the hydrocarbon/oxygen ratio the degree of reduction was changed for vanadium oxide supported on zirconia catalyst [18]. Studies have revealed that weak oxidants, such as N_2O and CO_2, are also able to modulate the oxidation state or degree of reduction of the metal oxide catalysts during the ODH reaction [19,20]. Under these reducing conditions, the steady-state concentration of active lattice oxygen species is low. Thus, the degree of reduction of the catalyst can indeed be modulated, though this may or may not increase the alkene produced.

The redox mechanism of the ODH for n-butane, isobutane, and ethane has also been studied [21–24]. For the ODH of butane the formation of di-olefins in addition to the olefin is also to be considered. Madeira et al. [22] suggests that lattice oxygen plays a crucial role in butene selectivity and the alkane conversion is limited by the reducibility of the catalyst or by the lattice oxygen availability and mobility. Suppression of total oxidation is attributed to the weakly bound lattice oxygen. The mechanism of the ODH of butane over V–Mg–O was studied by adding intermediate products (butenes, butadiene, and carbon monoxide) and the results suggest that butane adsorption sites are different from those of the products. Furthermore, from n-butane, butadiene, and carbon oxides are formed in parallel with butenes (major route). Additional amounts of butadiene and carbon oxides are produced via consecutive dehydrogenation and deep oxidation steps of butenes. A part of CO is also further oxidized to CO_2 under the reaction conditions studied [25]. A comparative study of thermal cracking with and without oxygen and catalytic n-butane ODH reactions suggest that different initiation steps are involved. In the thermal cracking of n-butane the reaction is initiated by C–C bond scission, whereas, for the thermal cracking of n-butane with oxygen the main initiation step is the removal of the hydrogen atom by molecular oxygen resulting in the formation of C_4H_9 — HO_2 radicals giving butene as the major product. However, butadiene is only obtained via a catalytic pathway, which strongly depends on the reaction conditions [26].

Based on DRIFTS studies of ethane ODH, Karamullaoglu et al. [27] observed the presence of OH, ethyl, vinyl, and formyl species on the surface. As illustrated by the authors, the formation of ethene, CO, CO_2, H_2O, and CH_4 can be described with the help of six steps involving these surface species. Mechanistic studies for the ODH of ethane by Chen et al. [21] suggest that ethene is a primary product and CO_x are secondary products. TAP experiments carried out by Schuurman et al. [28] over Ni suggested that the catalyst irreversibly holds the oxygen species involved in the reaction at 573 K. The irreversibly held oxygen species may be of the O^- type. Furthermore, it was reported that ethane is irreversibly adsorbed and CO originates from a parallel-consecutive scheme.

In general, the alkene formation occurs via a redox mechanism over reducible metal oxides or through a radical mechanism over nonreducible metal oxides. In both cases, breaking of the C–H bond is the major step. For redox mechanism, two possibilities are taken into account: abstraction of hydrogen in the form of a proton and formation of an OH group and a carbanion fragment linked to a metal cation, or abstraction of a hydride ion and formation of an alkoxy group and a metal-hydride type complex. Depending on the type of oxygen participating, the reactions can be electrophilic or nucleophilic oxidation processes. A simple representation for redox mechanism for the formation of an OH group and a carbanion is shown in Scheme 15.2. A similar scheme can be developed for the formation of an alkoxide species and a metal-hydride type complex.

Over nonreducible metal oxides, the proposed mechanism deals with the initial generation of alkyl radicals. In these cases, the contribution of homogeneous reactions becomes fundamental. The alkyl radical species are generated at the

$$\frac{1}{2}O_2 + Z \longrightarrow [O-Z]$$

$$C_nH_{2n+2} + [O-Z] \longrightarrow C_nH^*_{2n+1} + [OH]-Z$$

$$C_nH^*_{2n+1} + [O-Z] \longrightarrow C_nH_{2n} + [OH]-Z$$

$$2[OH]-Z \longrightarrow H_2O + [O-Z] + Z$$

SCHEME 15.2 Proposed Mars–van Krevelen redox mechanism over reducible metal oxide catalysts showing the formation of an OH group and a carbanion, where Z is a vacant site

catalyst surface and they:

- React quickly in the adsorbed state, for instance, by being transformed to the corresponding olefin via β-elimination, or by undergoing nonselective oxidative attack by adsorbed oxygen species.
- React further in close proximity to the catalyst surface.
- Become transferred into the gas phase where the reaction proceeds.

The further evolution of alkyl radical fragments in the gas phase is known to be a function of temperature. In general, the reaction between alkyl radical and O_2 below 603 to 623 K proceeds mainly by a simple reversible addition process with the formation of a peroxide $C_nH_{2n+1}O^*_2$ species that is thermally stable. Under these conditions, the adducts are either converted to oxygenated products, such as aldehydes, that may also be precursors of carbon oxides or are converted directly to CO_2. The oxygen species adsorbed on the catalyst surface would also take part in this reaction. In the case of ethane oxidation over rare-earth oxides and alkaline-earth oxides, the formation of the alkene is preferred with respect to other oxygenated compounds above 653 K. Direct abstraction of H by O_2 may occur, or alternatively, the elementary addition produces an excited alkyl peroxy species, which is quickly decomposed to the alkene and HO^*_2 via an intermediate alkylhydroperoxy species at high temperature [29]. With higher alkanes, however, the high reactivity of the allylic methyl group in the olefins produced causes the selectivity to decrease with increasing temperature.

In the case of propane oxidation, the higher reactivity of the C–H bond makes the contribution of homogeneous reactions more likely than in the case of ethane. Moreover, due to the presence of the activated methyl group in the propene produced, the selectivity for propene is less in these reactions [30]. Propane molecules are activated on the catalyst surface resulting in the formation of propyl radicals that are released into the gas phase. The propyl radicals undergo chain propagation reactions resulting in the products observed.

In the case of *n*-butane, butene may be produced in the gas phase at high temperatures by cracking of the butyl radical, while at low temperatures the butyl radical is preferentially transformed to oxygenated products. Oxygenates can also decompose yielding lower oxygenates such as formaldehyde and acetaldehyde, or are converted to carbon oxides. The lighter oxygenates may also arise from rearrangement and decomposition of the C_4 alkyl peroxy radical. The nature of the products evolving from the alkyl radical species (either oxygenates or olefins) is, therefore, mainly a function of temperature. An increase in the number of carbon atoms makes the mechanism of intramolecular rearrangement of alkyl peroxy radicals and of radical fragmentation more important [29–31].

In the case of isobutane ODH to isobutene over Keggin type heteropoly compounds, the contribution of homogeneous radical process also appears to play a fundamental role [32]. It is reported that the conversion of isobutane increased with an increase in isobutane concentration in the feed. Similar behavior was reported for propane ODH over boria–alumina catalysts [33]. It was assumed that the mechanism begins on the catalyst surface, alkyl radicals are generated, and the reaction is then transferred into the gas phase [29]. Thus, when the reaction mechanism for the ODH of light alkanes includes a contribution of gas-phase reactions, operation at high hydrocarbon concentrations may lead to a considerable increase in the rate of reaction with improvement of both conversion and yields, if the alkene is stable enough to resist consecutive degradation reactions.

Thus, nonreducible metal oxides, such as rare-earth oxides, alkali and alkaline-metal oxides are also capable of activating the hydrocarbon molecules by means of adsorbed oxygen. The reaction continues by chain propagation through alkyl peroxide or hydroxyl peroxide formation. For this type of metal oxide catalyst, contribution from heterogeneous and homogeneous phases are involved. Experimental evidence supports the formation of alkyl peroxide or hydroxyl peroxide intermediates, which propagate the chain mechanism. In general, the main steps involved are shown in Scheme 15.3.

In summary, the alkanes are converted to alkenes, oxygenates, and carbon oxides depending on the active site and process environment. The ability to be oxidized increases with the number of carbon atoms in the hydrocarbon, following the order $C_4 > C_3 > C_2$, which parallels the order in decreasing strength of the weakest C–H bond of the molecule and the number of such bonds in the alkane molecule in the cases where this weakest bond strength is the same. It is observed that the selectivity toward alkenes decrease with the alkane conversion for reducible metal oxides. Despite the effort to generalize the ODH mechanism of light alkanes it is evident that the ODH activity is a complex function of the alkane molecule, active site, and process conditions.

15.3 METAL OXIDE CATALYSTS USED FOR ODH REACTIONS

This section deals with the research studies carried out over binary and mixed metal oxides for ODH during the past few years. Initially, the ODH of ethane is considered followed by the ODH of propane, and finally the ODH of butanes. The ODH of butane is further subdivided into the ODH of n-butane and ODH of isobutane.

15.3.1 ODH of Ethane

Since ethane is the second major component of natural gas, its transformation to ethene at low temperature is of considerable interest. The various catalysts used

$$C_nH_{2n+2} + O_s \longrightarrow C_nH^{\bullet}_{2n+1} + OHs$$

$$C_nH^{\bullet}_{2n+1} + O_2 \longrightarrow C_nH_{2n} + {}^{\bullet}O_2H$$

$$C_nH^{\bullet}_{2n+1} + O_2 \longrightarrow [C_nH_{2n}O] \longrightarrow CO_x$$

or,

$$C_nH_{2n+2} + 2O_2 \longrightarrow C_nH_{2n+1}O^{\bullet}_2 + {}^{\bullet}O_2H$$

$$C_nH_{2n+1} + O^{\bullet}_2 \longrightarrow C_nH_{2n} \cdot {}^{\bullet}O_2H$$

$$C_nH_{2n+1}O^{\bullet}_2 + C_nH_{2n+2} \longrightarrow C_nH_{2n+1}O_2H + C_nH^{\bullet}_{2n+1}$$

$$C_nH_{2n+1}O_2H + C_nH_{2n+1}O^{\bullet}_2 + O_2 \longrightarrow C_nH_{2n}O + C_nH_{2n+2} + {}^{\bullet}O_2H$$

$$C_nH_{2n}O \longrightarrow CO_x$$

$${}^{\bullet}O_2H + {}^{\bullet}O_2H \longrightarrow 2\,OH^{\bullet} + O_2$$

$$C_nH_{2n+2} + OH^{\bullet} \longrightarrow C_nH^{\bullet}_{2n+1} + H_2O$$

$$2C_nH^{\bullet}_{2n+1} \longrightarrow C_{2n}H_{2(2n+1)}$$

SCHEME 15.3 Proposed lower alkane ODH mechanism over nonreducible metal oxide catalysts

for the ODH of ethane, for convenience, are divided into three groups: (i) mixed oxides containing reducible metal oxides, (ii) Nonreducible metal oxides, and (iii) materials containing noble metals.

Catalysts that are reducible are usually active at low temperatures for selective oxidation reactions. The major part of this group comprises of V-based materials. Vanadium–magnesium mixed oxides (V–Mg–O) have been proposed as a selective catalytic system for the ODH of alkanes into the corresponding alkenes [34,35]. The V–Mg–O catalysts are, however, less active for ODH of ethane compared with propane and butane [4]. Then, for the vanadyl pyrophosphate catalyst, the principal product was ethene, which otherwise is selective for oxygenates during ODH of butane and propane [34]. It would appear that no one catalyst would be effective for the alkane ODH reaction in general. V–P–O catalyst possessing a higher P/V ratio of 1.15 is reported to have high ethene selectivities at elevated temperatures. At 700 K, their total selectivity was about 90% [3]. The addition of palladium to the V–P–O catalyst increased the ethane conversion but lowered the ethane selectivity, whereas, the addition of Bi or Zr to V–P–O catalyst improved the catalytic activity and ethene selectivity [36]. It is suggested that the improvement in ethene selectivity was related to the modification of acid–base properties of the catalyst surface. Ethene yields of 18 to 19 mol% and productivity of about 80 $g_{C_2H_4}$ kg_{cat}^{-1} h^{-1} were reported on modified V–P–O catalysts [36]. The ODH of ethane over stoichiometric and nonstoichiometric (mixed) iron phosphate phases with P:Fe ratios of 1.2:1 and 2:1 were reported at temperatures ranging from 673 to 958 K.

The phosphates tend to be unstable under reaction conditions, and generally convert to the quartz-like $FePO_4$ and other phases depending on stoichiometry. Ethene yields that approach 50% over a sample with a P:Fe ratio of 1.2:1 were reported [37].

Vanadium–cobalt substituted aluminophosphate molecular sieve of AEI structure (VCoAPO-18) was found to be active and selective in the ODH of ethane. Its catalytic behavior can be related to the presence of redox (probably related to V^{5+} and Co^{3+}) and acid sites (related to Co^{2+} cations) in addition to its unique structural properties. The conversion and ethene selectivity decreases in the order: VCoAPO-18 $>$ VO_x/CoAPO-18 $>$ CoAPO-18 [38]. At 873 K, the VCoAPO-18 catalyst showed a 50% ethene selectivity at 60% ethane conversion for an ethane/oxygen molar ratio of 4:8. Acid SAPO-34-based microporous catalysts with chabasite structure have been tested for the ODH of ethane in the temperature range of 823 to 973 K. Pure acid and La/Na containing SAPO-34 were catalytically active and a 75 ethene selectivity for 5% ethane conversion and a 60% ethane selectivity for 30% ethane conversion was observed [39].

Molybdenum-based catalysts were also studied for ODH of ethane [21,40,41]. Mixed metal oxide catalyst containing Mo, V, and Nb were reported to be more active (10% conversion at 559 K) and selective (100%) than other binary mixed metal oxides, such as Mo–V, Mo–Mn, and Mo–Ti. For the ODH of ethane at 673 K, a Mo–V–Nb catalyst possessing the chemical formulation of $Mo_1V_{0.6}Nb_{0.12}$ gave 22% conversion and $\sim$60 to 65% selectivity [40]. The studies suggest that the active phase is based on molybdenum and vanadium. The role of niobium is to enhance the intrinsic activity of the molybdenum and vanadium combination and improve the selectivity by inhibiting the total oxidation of ethane to carbon dioxide. $Mo_6V_2Al_1O_x$ mixed oxide catalysts were successfully used for gas-phase ethane oxidation to ethene and acetic acid. The addition of titanium to the $Mo_6V_2Al_1O_x$ oxide catalyst resulted in a marked increase of the activity for the ethane selective oxidation, which was attributed to the morphological change and the increase of surface area of the catalyst particles by the addition of titanium [41]. Synergetic effects were observed for the ODH of ethane on the conversion, ethene yield, and selectivity when Mo–V–O or Ni–V–O catalysts are mechanically mixed with α-Sb_2O_4 [21].

Efforts of using supported bulk metal oxides for the ODH of ethane have also been considered. Lopez et al. [42] reported the ODH of ethane on Al_2O_3 supported MVSb (M = Ni, Co, Bi, and Sn) oxides and their constituents in the temperature range of 673 to 773 K. Under the same reaction conditions, the binary oxides displayed overall activities close to each other. At the same time, they were more active than the ternary catalysts, with the exception of the SnVSb system, which provides nearly the same level of ethane conversion as SnV and VSb oxides. It was noted that the activity of ternary oxides decreases in the sequence SnVSb $>$ CoVSb $>$ BiVSb $>$ NiVSb. The highest ethane selectivity of up to 80% was observed on the alumina supported NiVSb oxide composition. They also observed that C_2H_4, CO, and CO_2 were primary products at short contact times ($<$1.4 sec) and at longer contact times CO_2 was a secondary product.

Rare-earth oxides and alkaline-earth oxides, and in general nonreducible metal oxides, activate ethane only at temperatures higher than 773 to 873 K. These catalysts are based on the activation of molecular oxygen under operating conditions. Besides the high temperatures, the reactant mixtures are often used with compositions that fall inside the flammability region; in the case of more reactive alkanes, the temperatures used are close to the auto ignition temperatures [39,43]. Ethene yields as high as 42% (selectivity > 69%) were obtained on $Sr_xLa_{1.0}Nd_{1.0}O_y$ catalysts at a contact time of ~45 msec (STP) because of the ignition of the reaction mixture and the resultant occurrence of the (oxidative) ethane pyrolysis. Improved yields can be obtained by the optimization of the reaction conditions by varying space velocity, preheat temperature of the feed gas, and ethane/oxygen molar ratio. The highest ethene yield of 56% (selectivity = 71%) is obtained at GHSV = 8.3×10^4 h^{-1} (τ_{STP} = ~45 msec) for an inlet composition of $C_2H_6/O_2/N_2$ = 2/1/1 and a preheat temperature of 873 K corresponding to a maximal reaction temperature of 1213 K. The addition of steam resulted in a decrease of the maximal temperature leading to a decrease in the ethene yield by ~2% [44]. For a perovskite type catalyst with formula $SrCe_{1-x}Yb_xO_{3-0.5x}$, a maximum yield during the ODH of ethane of 49% was obtained at 973 K [43]. The ODH of ethane over $SrCl_2$-promoted REO_x (RE = Ce, Pr, and Tb) samples revealed that the doping of $SrCl_2$ significantly reduced C_2H_4 deep oxidation. The C_2H_4 selectivity and C_2H_6 conversion are enhanced, which is attributed to the promoting activity of lattice O^{2-} [45]. A comparative study over nanoscale Er_2O_3 and 30 mol% $BaCl_2/Er_2O_3$ catalysts prepared by a modified sol–gel method with the large-size counterparts for the ODH of ethane suggested better catalytic performance for nanoscale catalysts than the corresponding large-size ones [46].

Alkaline-earth oxides, such as MgO and CaO, doped with alkali-metal ions, are also found to be active and selective in the ODH of ethane [43]. In these systems, the role of alkali-metal dopants is required for the creation of surface defects that are able to activate molecular oxygen. These metal oxide systems are essentially p-type semiconductors, at least at the temperatures used for lower alkane activation. Promotion by chlorine, either fed directly to the reactor as chlorine containing compounds or added to the catalyst, lowers the reaction temperature considerably and increases the ethene selectivity. TAP studies over Na_2O/CaO, Sm_2O_3/CaO, and Sm_2O_3 catalysts at 923 K using different oxygen partial pressures revealed that a reversible dissociative-adsorption via a molecular precursor provides a good description of the measured transient responses for all catalysts [47]. Furthermore, contact potential difference studies suggest the transformation of adsorbed molecular oxygen species to atomic species [47].

High throughput synthesis for finding out a suitable catalyst for ODH of ethane was also reported [40,48]. Libraries were prepared by doping the Mo–V–Nb oxide system with various dopants, such as Sb, Ca, Li, etc. Improved performance with the catalysts is observed when Sb replaces some of the Nb centers. The performance of V–Al–Nb oxide catalyst library shows that these catalysts are less active in comparison with the Mo–V–Nb oxide library [48]. Furthermore, analysis of the Cr–Al–Nb oxide library suggested a catalyst, $Cr_{0.28}Al_{0.68}Nb_{0.04}O_x$

that is much more active than the most active member in the Mo–V–Nb oxide library. However, the ethene selectivity of the Cr–Al–Nb oxide library (30 to 50%) is lower than that of the Mo–V–Nb oxide library (~90%) [48]. The Ni–Nb–Ta oxide library suggested an optimum composition of $Ni_{0.62}Ta_{0.10}Nb_{0.28}O_x$. At 573 K, ethane conversions of 20.5% and ethene selectivities of 86.2% were obtained [49].

15.3.2 ODH of Propane

The ODH of propane to propene has been studied extensively as the proportional importance of propene in the global market has doubled within the last four years. A large number of catalytic systems are reported over ODH of propane. However, the propene yield has not reached commercial levels. Vanadium-based catalysts are found to be more active whereas molybdenum-based catalysts are more selective. Bulk V_2O_5 was active for ODH of propane only at higher temperatures. However, propene selectivity is considerably low [50]. V–Mg–O catalysts are more active and selective then bulk V_2O_5 since 65% propene selectivity was reported at 10% propane conversion [51,52]. Hew Sam et al. [53] studied the ODH of propane over V–Mg–O catalysts and observed that the propene selectivity at isoconversion decreased as follows: $Mg_2V_2O_7 > MgV_2O_6 > Mg_3V_2O_5$, suggesting an optimum ratio of Mg to V exists. Mamedov and Corberan [3], however, suggested that the cooperation between separate phases present in the V–Mg–O catalyst determines the overall activity and selectivity.

Propane conversions and propene selectivities over rare-earth orthovanadates (La, Pr, Yb, Er, Sm, Ce, Tb, Nd) at 673 K suggests that the activity (conversion) was higher with Er and Sm orthovanadates (~11.5%) with ~29% propene selectivity [54]. Fang et al. [55] studied a series of rare-earth orthovanadate catalysts for the low-temperature ODH of propane. At 773 K and for a propane/oxygen molar ratio of 2:1 Y-doped VO_4 was able to form propene with a selectivity of 49% at 23% conversion [55,56]. Michaels et al. [15] carried out the ODH of propane over Mg–V–Sb oxide catalysts and observed that the propene selectivity decreases from 75 to 5% as the propane conversion increases from 2 to 68%.

V–Mg–O catalysts were compared with V–Zn–O and V–Pb–O catalysts and the activity follows the order of V–Mg–O > V–Pb–O > V–Zn–O. The difference in activity was attributed to the presence of large crystallites with low surface defect concentration [57]. Studies with V–Sm–O-based catalysts prepared by impregnating V_2O_5 with different amounts of Sm_2O_3 reveal that at isoconversion the catalysts containing samarium are slightly more selective than V_2O_5 [58]. A synergy effect between V_2O_5 and Sm_2O_3 was observed at reaction temperatures lower than 723 K, since the propene selectivity is higher than bulk V_2O_5. Doping with molybdenum slightly improved the selectivity. It is suggested that the aggregation state of V-atoms on the surface could have a direct influence on the selectivity of alkenes [59]. Multiphase oxide systems are suggested to perform better by suppressing the side reactions during ODH of propane.

Supported V–Mg–O catalysts have also been considered for the ODH of propane. Gao et al. [60] used γ-Al_2O_3 and TiO_2 to supported 20 wt% V–Mg–O catalysts with an Mg:V atomic ratio ranging from 1:1 to 4:1. It was observed that on both supports, the activity decreases as the Mg/V ratio increases from 1:1 to 4:1, which is partially due to the decrease in the vanadium active component. *In situ* laser Raman spectra revealed that the V–Mg–O species formed on the supports was primarily pyrovanadate and orthovanadate. Correlation of the structure and the reaction data reveals that the nature of the support controls, the V–Mg–O species formed on the surface, and the catalytic performance for the ODH of propane is primarily determined by the concentration of vanadium oxide and the Mg/V atomic ratio of the catalyst. Furthermore, it appears that using TiO_2 as a support for the V–Mg–O system is more favorable for the formation of propene.

Various Nb containing catalysts were also studied for the ODH of propane. Smits and coworkers [61] observed that niobium oxide shows a very high selectivity in this reaction although the conversion was very low. The activity of the Nb containing catalysts was improved without diminishing its selectivity by adding elements, such as V, Cr, or Mo, all of which are reducible transition metals. For a series of V-based catalysts and at isoconversion levels ($\sim$30% conversion), the propene selectivity for 1:1 Nb–V and Sb–V catalysts was $\sim$20 and 40%, respectively. The 1:1 Sb–V and 1:1:5 Nb–V–Si prepared via the nonhydrolytic method offer the best results in terms of propane conversion and propene yield. With the 1:1:5 Nb–V–Si catalysts the propene productivity of 0.23 kg of propene per kilogram of catalyst per hour was achieved at 823 K [62].

Though V–P–O catalysts are nonselective for ODH of propane, vanadium aluminophosphate (VAPO) catalysts displayed higher propene selectivity, especially for samples with low vanadium content. Moreover, ALPO-5 has an initial selectivity of about 100%, which is associated to the specific structure that promotes the accessibility of propane to the dehydrogenation sites, thus contributing to the high selectivity of VAPO catalysts [63]. The vanadium species in the framework of the ALPO-5 was considered as more selective than the polymeric V species. As for their total activity, it appears to be associated with the presence of the vanadium species. Other phosphate containing catalysts have also been tested, such as the manganese pyrophosphate, which shows a propene yield of $\sim$16% at 823 K [64].

The ODH of propane over titanium and vanadium containing zeolites and nonzeolitic catalysts revealed that Ti-silicalite was the most active. The addition of water caused an increase in selectivity, probably due to a competitive adsorption on the active sites. The reaction is proposed to occur on the outer surface of the Ti-silicalite crystallites on Lewis acid sites, and a sulfation of the catalyst, which increases the acidity of these sites, results in a further increase of the catalytic activity. The maximum conversion obtained was 17% with a propene selectivity of up to 74% [65]. Comparison of propane oxidation and ammoxidation over Co-zeolites shows an increase in conversion and propene selectivity during ammoxidation. For a conversion of 14%, 40% propene selectivity was obtained with ammonia, whereas, at 10% conversion the propene selectivity was only 12% with oxygen. The increase in activity and selectivity can be due to the formation of basic sites via ammonia adsorption [38].

Metal molybdates have been proposed as selective catalysts in ODH reactions although they show a catalytic activity lower than V containing catalysts [66,67]. Mitchell and Wass [68] prepared a sample possessing molybdate intercalated in hydrolatcite with 27.3% Mo content and studied it for the ODH of propane. The maximum yield of propene obtained was 22%. Mg-molybdates shows an improved performance, which is explained by a synergic effect between $MgMoO_4$ and MoO_3 phase [66]. Nickel molybdates and modified nickel molybdates are also active for ODH of propane. Specifically, Ba-modified $NiMoO_4$ catalysts show a high selectivity (75%) even at high conversion (41%) [69,70]. However, the catalyst appears to be unstable due to the formation of $BaCO_3$ during reaction. Manganese oxide catalysts impregnated with molybdenum have shown improvements in propene selectivity in comparison with pure MnO_2. These catalysts exhibit catalytic activity and propene yield at temperatures as low as 623 K. At 823 K, however, the highest propene yield was 9% for 30% propane conversion [71,72].

Ce–Ni–O catalysts have been found to be active and selective at relatively low temperatures of about 573 K. For the CeNi1 precipitated catalyst (nNi/nCe (bulk) = 1), the yield of propene amounted to 11% with the selectivity of about 60%. The potassium addition to the Ce–Ni–O impregnated sample led to an increase in the propene selectivity but a decrease in conversion [73] in accordance with the use of alkali dopants discussed previously.

Baerns and coworkers [74] have applied evolutionary algorithms to develop new catalysts for the ODH of propane; 224 mixed oxides based on the elements V, Mg, B, Mo, La, Mn, Fe, and Ga were analyzed. The best yield of 9% at a propene selectivity of 57% was obtained on materials with the concentration of vanadium in the range from 2.6 to 8.3 wt%. For a good catalytic performance, it was suggested that the presence of magnesium oxide is necessary and the presence of Fe, Mo, or Mn is required. From the analysis of V–Mg–O-based systems a propene yield of up to 17% was achieved for $V_{0.22}Mg_{0.47}Mo_{0.11}Ga_{0.2}O_x$ at 773 K [75].

15.3.3 ODH of *n*-Butane

Transition metal oxides and noble metal oxides are active for ODH of n-butane [4,76]. Among them, vanadium-based catalytic materials were reported to possess high activity and selectivity [34,77]. Pure V_2O_5 is active only at sufficiently high temperatures. At 813 K, the conversion and butene selectivity obtained with V_2O_5 was 11 and 7%, respectively [78]. V–Mg–O system is observed to be highly selective (60 to 50%) even up to 40% conversion at 823 K [35]. Since different magnesium vanadate phases exist in the V–Mg–O catalysts [79,80] researchers have tried to establish the active magnesium vanadate phase. Kung and coworkers [52,79] attributed the active component to magnesium orthovanadate ($Mg_3V_2O_8$), whereas, Gao et al. [50] suggested that pyrovanadate was the most selective and reducible phase. The highest selectivity for magnesium pyrovanadate was related to its ability to stabilize V^{4+} ions associated with oxygen vacancies. Butane oxidation studies at 773 K were carried out by Owen and Kung [81] over the orthovanadates of Mg, Zn, Cr, Ni, Cu, Fe, Nd, Sm, and Eu. It was observed that more reducible cations showed lower selectivities. The correlation was interpreted with a model based on the selectivity-determining step of the reaction. It was suggested that the probability that the surface intermediate could form a carbon–oxygen bond depends on the ease of removal of a lattice oxygen in the M–O–V group in the orthovanadate, where M is the metal cation given earlier [81].

Several metal oxides, Mo, W, Cr, and Fe, were studied as promoters for the vanadium–magnesium-based catalysts for the ODH of n-butane [82,83]. It was observed that the presence of molybdenum improves the yield of C_4-olefins. At 773 K, higher selectivities, up to 68% at 12% conversion, were obtained with modified catalysts [82]. Bhattacharya et al. [83] described the effects of Cr_2O_3, MoO_3, and TiO_2 dopants in V–Mg–O catalysts on activity and selectivity. At 823 K, their best catalyst, incorporating Cr_2O_3 and TiO_2 dopants, converted n-butane with a selectivity of 60% to butadiene at 45% conversion. The incorporation of molybdenum was suggested to modify the number of V^{5+} species on the catalyst surface and the reducibility of selective sites [84]. The best catalyst, possessing a Mo-loading of 17.3 wt% as MoO_3 and a bulk Mo/V atomic ratio of 0.6, showed 40% selectivity for 40 to 60% butane conversion at 773 to 823 K. Stern et al. [85] also studied the promoting effect of Mo, B, Al, Ga, Sb, and Sb–P in the catalytic behavior of $Mg_4V_2M_2O_x$. Ga containing catalyst offered the best yield of butene/butadiene (5.2%) under the experimental conditions. The applicability of the nanoscale material to selective oxidation reactions was studied by loading vanadium on nanocrystals of MgO [86]. It was found that nanocrystals supported catalyst gave higher butene selectivity compared with conventionally prepared catalysts. Characterization studies suggest that similar catalyst surface structure exists on both types of catalysts, however, the difference in selectivities arise from the changes in acid–base properties.

In addition to the extensively studied V–Mg–O catalysts, a wide variety of other materials has also been studied for ODH. Several tetravalent metals (Ce, Zr, Sn, Ti) phosphates were evaluated for the ODH of n-butane in the temperature range of 683 to 843 K, at atmospheric pressure. Zirconium and cerium phosphates appeared to be the most active catalysts but titanium phosphate led to the best yields in butenes and butadiene with a maximum of 14% at 803 K. The addition of water to the gas feed was shown to have a negative effect on both the conversion and selectivity to butenes and butadiene with a significant decrease of the yield in butadiene compared with that of butenes [87]. Conductivity studies performed over Ti and Zr phosphates suggested that the alkane activation was related to the p-type semiconducting properties of the solid [88]. The studies carried out with different Ti/P ratio suggest that the Lewis acidity of the catalysts significantly influence the catalytic properties; the greater acidity was correlated with the less selective catalysts. The Brønsted acidity, however, did not appear to be correlated with any catalytic properties in the ODH of n-butane [89].

Catalyst prepared from a mixture of Fe and Zn (Fe/Zn = 5.8) was used for the ODH of n-butane [26]. Butadiene and CO_2 were the main products formed. At 723 K and for a conversion of 19% the butenes yield was 12%. The NiO–MoO_3 system with basic promoters showed interesting activity and improved selectivity for the ODH of n-butane [90]. The basic promoters considered were Li, Na, K, and Cs. Reaction studies reveal that the addition of promoters generally reduces the catalytic activity; however, it induces a high selectivity to butenes and butadiene. The greater the basicity and the amount of promoter ion the higher is the selectivity, and this improvement arises largely by suppression of total oxidation. The combination of activity and selectivity effects leads to improvement of yields of C_4 hydrocarbons by addition of small amounts of alkali promoters up to an optimum loading [91]. Current state-of-the-art catalysts for the ODH of n-butane include mechanical mixtures of $(VO)_2P_2O_7$ and α-Sb_2O_3 [92]. With these catalysts, selectivities as high as 90% (at 10% conversion) are reported at temperatures of 673 K. A maximum conversion of 25% and dehydrogenation-products selectivity of 56% was obtained.

Significant research efforts have also been undertaken to identify proper catalytic systems for ODH of n-butane using high throughput analysis [93–95]. A combinatorial search for novel catalysts was carried out with 950 mixed oxides based on SiO_2, TiO_2, ZrO_2, and Al_2O_3. The best catalyst, $Hf_3Y_3Ti_{94}O_x$, shows a butene selectivity of 75% for 15% conversion at 723 K and atmospheric pressure [96].

15.3.4 ODH of Isobutane

The ODH of isobutane has been hardly studied in comparison of the other lower alkanes despite the industrial importance of this reaction. The reported effective catalysts for this reaction are V–Mg–O based [97]. Some of the other metal oxide catalysts studied for the ODH of isobutene are, heteropoly compounds [98,99], phosphates, pyrophosphates [100,101], and molybdates [102]. The $Ni_2P_2O_7$ catalyst was reported as the most selective with the isobutene selectivity reaching a maximum of 82.2% at 823 K. Pyrophosphates of Ag, Zn, Mg, Cr, Co, Mn, and Sn catalyze the reaction, and the isobutene selectivities are 43.8 to 65.7% at the temperature where the maximum isobutene yield is observed [100]. In contrast, among various pyrophosphates, $V_4(P_2O_7)_3$ was reported as the most active catalyst with an isobutane conversion of 33.5% at 773 K [103]. The rates of isobutene formation over rare-earth phosphates were studied by Takita et al. [101]. The best performance was obtained over $CePO_4$ and $LaPO_4$ (r $\sim$ 30 μ mol min^{-1}) with the selectivity as high as 80% at 773 K. When V–Sb–O and V–Sb–Ni–O catalysts were supported on alumina, their selectivity improved from 12 to 70% even at 42% conversion [104]. The ODH of isobutane over Cr–V–Nb mixed oxide catalysts with different compositions showed a maximum selectivity and yield values as 0.9 and 0.4, respectively, at a fractional conversion level of 0.45 with a $Cr_{0.74}V_{0.19}Nb_{0.07}$ catalyst at 848 K [27]. A decrease in chromium content of the catalyst caused a decrease in the selectivity and activity. ODH of isobutane over La–Ba–Sm oxide catalyst in the temperature range of 723 to 873 K was studied. A conversion of 4% at 723 K, and 30% at 873 K was observed. The selectivity to isobutene of ODH is only about 27% in the whole range of reaction temperatures [105]. To the best of our knowledge, no high throughput analysis for a catalyst has been undertaken for the ODH of isobutane.

TABLE 15.1
Representative Catalytic Data for Ethane ODH Reaction

Catalyst	C_2H_6/O_2	W/Fao, or equivalent	T (K)	X (%)	S (%)	Reference
VMgO	2	$1.5 \, \text{g h mol}^{-1}$	813	5	24	97
VPO	2	$25 \, \text{g h mol}^{-1}$	748	3	72	36
Zr–VPO	2	$25 \, \text{g h mol}^{-1}$	748	10	80	36
Bi–VPO	2	$25 \, \text{g h mol}^{-1}$	748	8	75	36
Fe–Phosphates	1	$3.7 \, \text{g h mol}^{-1}$	923	42	84	37
$Mo_1V_{0.6}Nb_{0.12}$	3	$20.5 \, \text{g h mol}^{-1}$	673	22	61	40
$Mo–V–AlO_x$	3	$7.4 \, \text{g h mol}^{-1}$	613	10	60	21
$Mo_6V_2Al_1Ti_{0.5}O_x$	3	$7.4 \, \text{g h mol}^{-1}$	653	20	54	21
$Mo–V–O + Sb_2O_4$	1.6	$70 \, \text{g h mol}^{-1}$	773	11	31	41
$Ni–V–O + Sb_2O_4$	1.6	$31 \, \text{g h mol}^{-1}$	773	8	27	41
V–Co–Zeolite	0.77	$0.75 \, \text{g h mol}^{-1}$	723	40	20	73
$Cr_{0.78}V_{0.16}Nb_{0.06}$	2	$1.9 \, \text{g h mol}^{-1}$	843	40	20	27
$SrCl_2–Ce/Pr/Tb$	2	$3.8 \, \text{g h mol}^{-1}$	933	73	69	45
				80	71	
				83	76	
$Sr_xLa_1Nd_1O_y$	2	$\tau = 0.045 \, \text{s,}$ $\text{GHSV} = 83{,}000 \, \text{h}^{-1}$	873	78	71	44
$LaSr_{0.02}O_x$	2.5	$0.2 \, \text{g h mol}^{-1}$	1073	50	60	75

15.3.5 Summary

The ODH of ethane, propane, and butanes carried out over representative catalysts are given in Table 15.1 to Table 15.3, respectively. The columns in the tables represent the following: the chemical composition of the catalyst, the alkane to oxygen molar ratio, the contact time or equivalent representations, and the temperature of the reaction are given in columns one, two, three, and four; the alkane conversion and alkene selectivity are given in columns five and six; and the reference of the article is given in column seven.

From Table 15.1 it can be observed that the usual values of the ethane to oxygen molar ratios used are near the stoichiometric value of 2.0 required for ODH and range from 3.0 to 0.77. The temperatures range from 573 to 1073 K and most frequently used temperatures are in the range 900 to 1000 K. The ethane conversions range from 5.2 to 82% and ethene selectivities range from 20 to 94.5%. The catalyst with the highest ethane conversion does not give the highest ethene selectivity. The highest ethene yield, which is the product of the conversion and selectivity, is given by the nonreducible $SrCl_2–Ce/Pr/Tb$ system of Dai et al. [45] at 933 K with an ethane to oxygen ratio of 2.

For the ODH of propane given in Table 15.2, it is observed that the propane to oxygen molar ratios are usually less than the stoichiometric value and

TABLE 15.2

Representative Catalytic Data for Propane ODH Reaction

Catalyst	C_3H_8/O_2	W/Fao, or equivalent	T (K)	X (%)	S (%)	Reference
V_2O_5	2	1.5 g h mol^{-1}	813	22	18	52
$Mg_2V_2O_7$	3.5	6.2 g h mol^{-1}	723	15	50	51
YVO_4	2	1.8 g h mol^{-1}	773	23	49	55
$SmVO_4$	1	3.8 g h mol^{-1}	773	36	30	58
Mo–Sm–VO_4	1	3.8 g h mol^{-1}	773	35	35	58
VNbSi	0.4	10 g h mol^{-1}	773	21	21	62
V–Nb–SbO_2	0.4	10 g h mol^{-1}	773	33	16	62
Mo–V–Te–Nb–O	0.5	205 g h mol^{-1}	653	10	27	41
CrVNbO$_x$	2	1.9 g h mol^{-1}	845	40	20	27
VAPO-5	0.5	90 g h mol^{-1}	823	30	58	63
Ti-Silicalite	0.2	1300 h^{-1} GHSV	823	10	82	65
Ti–BEA	0.2	1300 h^{-1} GHSV	823	17	38	65
Co–BEA	0.75	0.75 g h mol^{-1}	723	6.4	42	38
Ce–Ni–O	0.5	2.2 g h mol^{-1}	573	13	55	73
K–Ni–Ce–O	0.5	2.2 g h mol^{-1}	573	9	72	73
$NiMoO_4$	0.8	1.2 g h mol^{-1}	803	14	67	69
K/Ca/P–$NiMoO_4$	0.8	1.2 g h mol^{-1}	803	14	64–84	69
9%Ba–$NiMoO_4$	1	1.1 g h mol^{-1}	773	41	75	69
Mo–MnO_2	1	0.4 g h mol^{-1}	823	31	29	72
$Mn_2P_2O_7$	4	Not mentioned	823	41	38	64
$V_{0.22}Mg_{0.47}Mo_{0.11}Ga_{0.2}O_x$	3	$\tau = 0.08$ s	773	32.8	53	75

range from 0.2 to 4.0. The temperatures range from 573 to 953 K and the most frequently used temperatures are in the 773 to 873 K region. The propane conversions and propene selectivities range from 5 to 53.5% and 16.1–84%, respectively. As for the case of ethane the catalyst with the highest propane conversion does not give the highest propene selectivity, and the highest propene yield, which is the product of the conversion and selectivity, is given by the 9% Ba–$NiMoO_4$ system of Liu et al. [69] at 773 K with a propane to oxygen ratio of 1. This system, however, has limitation due to the reported instability of the catalyst during the reaction and should be reconfirmed independently. The next two promising catalysts are the VAPO-5 and $V_{0.22}Mg_{0.47}Mo_{0.11}Ga_{0.2}O_x$, at temperatures of 823 and 773 K, respectively, and propane to oxygen ratio of 0.5 and 3.0, respectively.

The summary of the research work done on the ODH of butane and isobutane is given in Table 15.3. The results suggest that for the ODH of n-butane and isobutane, the V–Mg–O catalytic systems are promising in comparison with the other metal oxides studied. The temperature of the reaction for the V–Mg–O catalytic system is in the range of 823 to 873 K with a hydrocarbon to oxygen ratio near 0.5. The

TABLE 15.3
Representative Catalytic Data for Isobutane and *n*-Butane ODH reaction

Catalyst	C_4H_{10}/O_2	W/Fao, or equivalent	T (K)	X (%)	S (%)	Reference
		n-butane				
V_2O_5	2	$1.5\,\mathrm{g\,h\,mol}^{-1}$	813	11	7	78
V–Mg–O	0.5	$10.7\,\mathrm{g\,h\,mol}^{-1}$	773	20	50	35
			823	40	40	
V–Mg–MoO$_3$	0.25	$10.7\,\mathrm{g\,h\,mol}^{-1}$	773	12	68	82
V–Mg–Cr$_2$O$_3$–TiO$_2$	0.5	—	823	45	60	83
Mg$_4$V$_2$Ga$_2$O$_x$	5	$13\,\mathrm{h}^{-1}$ WHSV	773	8	66	85
MgO(nanocrystal)–V$_2$O$_5$	0.5	$7.4\,\mathrm{g\,h\,mol}^{-1}$	773	60	34	86
CeP$_2$O$_7$	1.5	$1000\,\mathrm{h}^{-1}$ WHSV	803	19	56	87
ZrP$_2$O$_7$	1.5	$1000\,\mathrm{h}^{-1}$ WHSV	803	12	61	87
SnP$_2$O$_7$	1.5	$1000\,\mathrm{h}^{-1}$ WHSV	803	19	63	87
TiP$_2$O$_7$	2	$1000\,\mathrm{h}^{-1}$ WHSV	803	25	56	87
Fe/Zn	2	$16\,\mathrm{g\,h\,mol}^{-1}$	623	11	68	26
NiMoO$_4$	0.4	$20\,\mathrm{g\,h\,mol}^{-1}$	798	38	10	91
6%Cs-NiMoO$_4$	0.4	$20\,\mathrm{g\,h\,mol}^{-1}$	798	11	69	91
(VO)$_2$P$_2$O$_7$+Sb$_2$O$_4$	0.08	$2.4\,\mathrm{g\,h\,mol}^{-1}$	773	10	90	92
Hf$_3$Y$_3$Ti$_{94}$O$_x$		$2.24\,\mathrm{g\,h\,mol}^{-1}$	723	15	75	96
		isobutane				
V–Mg–O	0.67	$0.6\,\mathrm{g\,h\,mol}^{-1}$	873	55	50	97
V–Sb–O	1.66	$\tau = 4\,\mathrm{s}$	823	13	11	104
V–Sb–Ni–O	1.66	$\tau = 4\,\mathrm{s}$	823	15	22	104
MgMoO$_4$	3.07	$1.6\,\mathrm{g\,h\,mol}^{-1}$	823	9	81	102
MgO	3.07	$1.6\,\mathrm{g\,h\,mol}^{-1}$	803	6	39	102
MoO$_3$	3.07	$1.6\,\mathrm{g\,h\,mol}^{-1}$	803	4	65	102
LaBaSm	2	$6.2\,\mathrm{g\,h\,mol}^{-1}$	823	20	40	105
CePO$_4$	15	$63.6\,\mathrm{g\,h\,mol}^{-1}$	823	10	86	101
CeP$_2$O$_7$	4.3	$5\,\mathrm{g\,h\,mol}^{-1}$	823	40	51	103
NiP$_2$O$_7$	15	$63.6\,\mathrm{g\,h\,mol}^{-1}$	823	11	83	100
V$_4$(P$_2$O$_7$)$_3$	4.3	$5\,\mathrm{g\,h\,mol}^{-1}$	773	34	47	103
Cr$_{0.74}$V$_{0.19}$Nb$_{0.07}$O$_x$[a]	1	$0.75\,\mathrm{g\,h\,mol}^{-1}$	848	0.45	0.9	27
K$_2$P$_2$W$_{17}$FeO$_{61}$	2	$\tau = 2\,\mathrm{s}$	723	10	60	98

[a] Fractional conversion and selectivity.

temperatures used for the ODH reaction follows the trend of the bond strength of the weakest alkane C–H bond, that is, the temperatures used for ODH of ethane are the highest and the C–H bond is the strongest, and the temperatures used for the ODH of butanes are the lowest and the secondary C–H bond is the weakest. The Cr$_{0.74}$V$_{0.19}$Nb$_{0.07}$ catalyst by Karamullaoglu et al. [27] gives the maximum isobutene yield at an isobutane to oxygen ratio of 1 and temperature of 848 K.

15.4 CONCLUSIONS

Several factors are involved in increasing the alkene yield during lower alkane ODH reaction, which can be achieved by increasing the alkene formation and decreasing the primary and secondary unwanted side reactions. The primary factor is the proper design of a catalyst. Metal oxide-based catalysts are active and selective catalysts for the ODH reaction of lower alkanes. Among the reducible metal oxides vanadium-based metal oxides have been extensively studied and are found to be active, however, the alkene selectivity is low. The reducibility/oxidation state of the V atoms appears to influence the selectivity to ODH products. Furthermore, it is suggested that the activity and selectivity of a given vanadium site depends on the nature of its neighboring atoms, which can be influenced by using additives/dopants. Multiple phases in the oxide system are suggested as more selective by suppressing the side reactions. Chromium- and molybdenum-based oxides are also studied. For the molybdenum-based catalysts, the activities for the ODH reaction are lower than those obtained for the vanadium-based systems, whereas, the alkene selectivities are higher. Combinatorial approach has paved a way to analyze the adequacy of numerous metal oxides for ODH and to screen them based on the relative performance. Interesting results have been obtained for the ODH of propane and ethane. The auto thermal operation using different catalyst for ODH of ethane appears promising; however, the stability of the catalyst and safety factors limits the process. The catalyst commonly used for butane ODH was usually the V–Mg–O based system; however, adequate yields have not been achieved. It is interesting to note that for the same catalytic material, different behavior in activity/selectivity was observed for the C_2 to C_4 hydrocarbons. It appears that the ability of alkanes to be oxidized increases with the number of carbon atoms in the hydrocarbon. Moreover, the orientation of the hydrocarbon with respect to the catalyst active site also influences the activity and selectivity.

REFERENCES

1. Franz, G. and Sheldon, R.A. Oxidation. In *Ullmann's Encyclopedia of Industrial Chemistry*, Elvers, B., Hawkins, S., and Schulz, G., Eds., Vol. A 18, Wiley-VCH Verlag GmbH & Co, KgaA Weinheim, 1991, p. 261–311.
2. Haber, J., Oxidation of hydrocarbons. In *Handbook of Heterogeneous Catalysis*, Vol. 5. Ertl, G., Knozinger, H., and Weitkamp, J., Eds. Wiley, VCH, 1997.
3. Mamedov, E.A. and Cortes Corberan, V. Oxidative dehydrogenation of lower alkanes on vanadium oxide-based catalysts. The present state of the art and outlooks. *Appl. Catal. A: Gen.* 1995, *127*, 1.
4. Kung, H.H. Oxidative dehydrogenation of light alkanes. *Adv. Catal.* 1994, *40*, 1.
5. Ballarini, N., Cavani, F., Cortelli, C., Giunchi, C., Nobili, P., Trifirò, F., Catani, R., and Cornaro, U. Reactivity of V/Nb mixed oxides in the oxide-hydrogenation of propane under co-feed and under redox-decoupling conditions and references therein. *Catal. Today* 2003, *78*, 353.
6. Stone, F.S. Surface processes on oxides and their significance for heterogeneous catalysis. *J. Mol. Catal.* 1990, *59*, 147.

7. Haber, J. Catalysis by transition metal oxides. In *Solid State Chemistry in Catalysis, ACS Symposium Series*, Grasselli, R.K., and Brazdil, J.F., Eds., Vol. 279. Am. Chem. Society, Washington, 1985, p. 3.

8. Haber, J. and Witko, M. Oxidation catalysis — Electronic theory revisited. *J. Catal.* 2003, *216*, 416.

9. Madeira, L.M. and Portela, M.F. Catalytic oxidative dehydrogenation of *n*-butane. *Catal. Rev.* 2002, *44*, 247.

10. Moulijn, J.A., Makkee M., and van Diepen, A.E. *Chemical Process Technology*. Wiley, New York, March 2001.

11. Barsan, M.M. and Thyrion, F.C. Kinetic study of oxidative dehydrogenation of propane over Ni–Co molybdate catalyst. *Catal. Today* 2003, *81*, 159.

12. Baldi, M., Finocchio, E., Pistarino, C., and Busca, G. Evaluation of the mechanism of the oxy-dehydrogenation of propane over manganese oxide. *Appl. Catal. A: Gen.* 1998, *173*, 61.

13. Burch, R. and Swarnakar, R. Oxidative dehydrogenation of ethane on vanadium–molybdenum oxide and vanadium–niobium–molybdenum oxide catalysts. *Appl. Catal.* 1991, *70*, 129.

14. Morales, E. and Lunsford, J.H. Oxidative dehydrogenation of ethane over a lithium-promoted magnesium oxide catalyst. *J. Catal.* 1989, *118*, 255.

15. Michaels, J.N., Stern, D.L., and Grasselli, R.K. Oxydehydrogenation of propane over Mg–V–Sb-oxide catalysts. I. Reaction network. *Catal. Lett.* 1996, *42*, 135.

16. Creaser, D., Anderson, B., Hudgins, R.R., and Silveston, P.L. Transient study of oxidative dehydrogenation of propane. *Appl. Catal. A: Gen.* 1999, *187*, 147.

17. Lee, K.H., Yoon, Y.S., Ueda, W., and Moro-oka, Y. An evidence of active surface MoOx over MgMoO$_4$ for the catalytic oxidative dehydrogenation of propane. *Catal. Lett.* 1997, *46*, 267.

18. Gao, X., Jehng, J.M., and Wachs, I.E. In Situ UV-vis-NIR diffuse reflectance and Raman spectroscopic studies of propane oxidation over ZrO$_2$-supported vanadium oxide catalysts. *J. Catal.* 2002, *209*, 43.

19. Ge, S., Liu, C., Zhang, S., and Li, Z. Effect of carbon dioxide on the reaction performance of oxidative dehydrogenation of *n*-butane over V–Mg–O catalyst. *Chem. Eng. J.* 2003, *94*, 121.

20. Dury, F., Centeno, M.A., Gaigneaux, E.M., and Ruiz, P. Interaction of N$_2$O (as gas dope) with nickel molybdate catalysts during the oxidative dehydrogenation of propane to propylene. *Appl. Catal. A: Gen.* 2003, *247*, 231.

21. Chen, N.F., Oshihara, K., and Ueda, W. Selective oxidation of ethane over hydrothermally synthesized Mo–V–Al–Ti oxide catalyst. *Catal. Today* 2001, *64*, 121.

22. Madeira, L.M., Herrmann, J.M., Disdier, J., Portela, M.F., and Freire, F.G. New evidences of redox mechanism in *n*-butane oxidative dehydrogenation over undoped and Cs-doped nickel molybdates. *Appl. Catal. A: Gen.* 2002, *235*, 1.

23. Agafonov, Yu.A., Nekrasov, N.V., and Gaidai, N.A. Kinetic and mechanistic study of the oxidative dehydrogenation of isobutane over cobalt and nickel molybdates. *Kinet. Catal.* 2001, *42*, 821.

24. Bychkov, V.Yu., Sinev, M.Yu., and Vislovskii, V.P. Thermochemistry and reactivity of lattice oxygen in V–Sb oxide catalysts for the oxidative dehydrogenation of light paraffins. *Kinet. Catal.* 2001, *42*, 574.

25. Lemonidou, A.A. Oxidative dehydrogenation of C_4 hydrocarbons over VMgO catalyst — kinetic investigations. *Appl. Catal. A: Gen.* 2001, *216*, 277.

26. Toledo, J.A., Armendariz, H., and López-Salinas, E. Oxidative dehydrogenation of *n*-butane: a comparative study of thermal and catalytic reaction using Fe–Zn mixed oxides. *Catal. Lett.* 2000, *66*, 19.

27. Karamullaoglu, G., Onen, S., and Dogu, T. Oxidative dehydrogenation of ethane and isobutane with chromium–vanadium–niobium mixed oxide catalysts. *Chem. Eng. Process.* 2002, *41*, 337.

28. Schuurman, Y., Ducarme, V., Chen, T., Li, W., Mirodatos, C., and Martin, G.A. Low temperature oxidative dehydrogenation of ethane over catalysts based on group VIII metals. *Appl. Catal. A: Gen.* 1997, *163*, 227.

29. Cavani, F. and Trifiro, F. Selective oxidation of light alkanes: interaction between the catalyst and the gas phase on different classes of catalytic materials. *Catal. Today* 1999, *51*, 561.

30. Leveles, L., Seshan, K., Lercher, J.A., and Leffert, L. Oxidative conversion of propane over lithium-promoted magnesia catalyst: II. Active site characterization and hydrocarbon activation. *J. Catal.* 2003, *218*, 307.

31. Baerns, M. and Buyevskaya, O. Simple chemical processes based on low molecular-mass alkanes as chemical feedstocks. *Catal. Today* 1998, *45*, 13.

32. Comuzzi, C., Primavera, A., Trovarelli, A., Bini, G., and Cavani, F. Thermal stability and catalytic properties of the Wells–Dawson $K_6P_2W_{18}O_{62} \cdot 10H_2O$ heteropoly compound in the oxidative dehydrogenation of isobutane to isobutene. *Top. Catal.* 1999, *9*, 251.

33. Buyevskaya, O.V., Muller, D., Pitsch, I., and Baerns, M. Selective oxidative conversion of propane to olefins and oxygenates on boria-containing catalysts. *Stud. Surf. Sci. Catal.* 1998, *119*, 671.

34. Michalakos, P.M., Kung, M.C., Jahan, I., and Kung H.H. Selectivity patterns in alkane oxidation over $Mg_3(VO_4)_2$–MgO, $Mg_2V_2O_7$, and $(VO)_2P_2O_7$. *J. Catal.* 1993, *140*, 226.

35. López Nieto, J.M., Soler, J., Concepcion, P., Herguido, J., Menendez, M., and Santamaria, J. Oxidative dehydrogenation of alkanes over V-based catalysts: influence of redox properties on catalytic performance. *J. Catal.* 1999, *185*, 324.

36. Solsona, B., Zazhigalov, V.A., López Nieto, J.M., Bacherikova, I.V., and Diyuk, E.A. Oxidative dehydrogenation of ethane on promoted VPO catalysts. *Appl. Catal. A: Gen.* 2003, *249*, 81.

37. Miller, J.E., Gonzales, M.M., Evans, L., Sault, A.G., Zhang, C., Rao, R., Whitwell, G., Maiti, A., and King-Smith, D. Oxidative dehydrogenation of ethane over iron phosphate catalysts. *Appl. Catal. A: Gen.* 2002, *231*, 281.

38. Bulanek, R., Novoveska, K., and Wichterlova, B. Oxidative dehydrogenation and ammoxidation of ethane and propane over pentasil ring Co-zeolites. *Appl. Catal. A: Gen.* 2002, *235*, 181.

39. Ji, L. and Liu, J. Excellent promotion by lithium of a lanthanum–calcium oxide catalyst for oxidative dehydrogenation of ethane to ethane. *Chem. Commun.* 1996, *10*, 1203.

40. Botella, P., López Nieto, J.M., Dejoz, A., Vázquez, M.I., and Martínez-Arias, A. Mo–V–Nb mixed oxides as catalysts in the selective oxidation of ethane. *Catal. Today* 2003, *78*, 507.

41. Osawa, T., Ruiz, P., and Delmon, B. New results on the oxidative dehydrogenation of ethane to ethylene: promoting catalytic performance of Mo–V- and Ni–V-oxide by α-Sb_2O_4. *Catal. Today* 2000, *61*, 309.

42. Lopez, R.J., Godjayeva, N.S., Corberian, V.C., Fierro, J.L.G., and Mamedov, E.A. Oxidative dehydrogenation of ethane on supported vanadium-containing oxides. *Appl. Catal. A: Gen.* 1995, *124*, 281–296.

43. Velle, O.J., Andersen, A., and Jens, K.J. The oxidative dehydrogenation of ethane by perovskite type catalysts containing oxides of strontium, cerium and ytterbium. *Catal. Today* 1990, *6*, 567.

44. Mulla, S.A.R., Buyevskaya, O.V., and Baerns, M. Autothermal oxidative dehydrogenation of ethane to ethylene using $Sr_xLa_{1.0}Nd_{1.0}O_y$ catalysts as ignitors. *J. Catal.* 2001, *197*, 43.

45. Dai, H.X., Ng, C.F., and Au, C.T. $SrCl_2$-promoted REO_x (RE = Ce, Pr, Tb) catalysts for the selective oxidation of ethane: a study on performance and defect structures for ethene formation. *J. Catal.* 2001, *199*, 177.

46. Zhong, W., Dai, H.X., Ng, C.F., and Au, C.T. A comparison of nanoscale and large-size $BaCl_2$-modified Er_2O_3 catalysts for the selective oxidation of ethane to ethylene. *Appl. Catal. A: Gen.* 2000, *203*, 239.

47. Kondratenko, E.V., Buyevskaya, O., and Baerns, M. Mechanistic insights in the activation of oxygen on oxide catalysts for the oxidative dehydrogenation of ethane from pulse experiments and contact potential difference measurements. *J. Mol. Catal. A: Chem.* 2000, *158*, 199.

48. Cong, P., Dehestani, A., Doolen, R.D., Giaquinta, D.M., Guan, S., Markov, V., Poojary, D., Self, K., Turner, H., and Weinberg, W.H. Combinatorial discovery of oxidative dehydrogenation catalysts within the Mo–V–Nb–O system. *Proc. Natl. Acad. Sci.* 1999, *96*, 11077.

49. Liu, Y., Cong, P., Doolen, R.D., Turner, H.W., and Weinberg, W.H. High-throughput synthesis and screening of V–Al–Nb and Cr–Al–Nb oxide libraries for ethane oxidative dehydrogenation to ethylene. *Catal. Today* 2000, *61*, 87.

50. Gao, X., Ruiz, P., Xin, Q., Guo, X., and Delmon, B. Effect of coexistence of magnesium vanadate phases in the selective oxidation of propane to propene. *J. Catal.* 1994, *148*, 56.

51. Sugiyama, S., Hashimoto, T., Shigemoto, N., and Hayashi, H. Redox behaviors of magnesium vanadate catalysts during the oxidative dehydrogenation of propane. *Catal. Lett.* 2003, *89*, 229.

52. Chaar, M.A., Patel, D., and Kung, H.H. Selective oxidative dehydrogenation of propane over V–Mg–O catalysts. *J. Catal.* 1988, *109*, 463.

53. Siew Hew Sam, D., Soenen, V., and Volta, J.C. Oxidative dehydrogenation of propane over V–Mg–O catalysts. *J. Catal.* 1990, *123*, 417.

54. Corma, A., López Nieto, J.M., Paredes, N., Perez, M., Shen, Y., Cao, H., and Suib, S.L. Oxidative dehydrogenation of propane over supported-vanadium oxide catalysts. *Stud. Surf. Sci. Catal.* 1992, *72*, 213.

55. Fang, Z.M., Hong, Q., Zhou, Z.H., Dai, S.J., Zheng, W., and Wan, H.L. Oxidative dehydrogenation of propane over a series of low-temperature rare earth orthovanadate catalysts prepared by the nitrate method. *Catal. Lett.* 1999, *61*, 39.

56. Zhou, R., Cao, Y., Yan, S., and Fan, K. Rare earth (Y, La, Ce)-promoted V-HMS mesoporous catalysts for oxidative dehydrogenation of propane. *Appl. Catal. A: Gen.* 2002, *236*, 103.

57. Rybarczyk, P., Berndt, H., Radnik, J., Pohl, M.M., Buyevskaya, O., Baerns, M., and Bruckner, A. The structure of active sites in Me–V–O catalysts (Me = Mg, Zn, Pb) and its influence on the catalytic performance in the oxidative dehydrogenation (ODH) of propane. *J. Catal.* 2001, *45*, 202.

58. Barbero, B.P. and Cadus, L.E. Vanadium species: Sm–V–O catalytic system for oxidative dehydrogenation of propane. *Appl. Catal. A: Gen.* 2003, *244*, 235.

59. Barbero, B.P. and Cadus, L.E. Molybdenum role: Mo–Sm–V–O catalytic system for propane oxidative dehydrogenation. *Appl. Catal. A: Gen.* 2003, *252*, 133.

60. Gao, X., Xin, Q., and Guo, X. Support effects on magnesium–vanadium mixed oxides in the oxidative dehydrogenation of propane. *Appl. Catal. A: Gen.* 1994, *114*, 197.

61. Ross, J.R.H., Smits, R.H.H., and Seshan, K. The use of niobia in oxidation catalysis. *Catal. Today* 1993, *16*, 503.

62. Barbieri, F., Cauzzi, D., De Smet, F., Devillers, M., Moggi, P., Predieri, G., and Ruiz, P. Mixed-oxide catalysts involving V, Nb and Si obtained by a non-hydrolytic sol–gel route: preparation and catalytic behaviour in oxidative dehydrogenation of propane. *Catal. Today* 2000, *61*, 353.

63. Blasco, T., Concepción, P., Grotz, P., López Nieto, J.M., and Martínez-Arias, A. On the nature of V and Mg ions in V, Mg-containing AlPO$_4$-5 catalysts. *J. Mol. Catal. A: Chem.* 2000, *162*, 267.

64. Jibril, B.Y., Al-Zahrani, S.M., and Abasaeed, A.E. Propane oxidative dehydrogenation over metal pyrophosphates catalysts. *Catal. Lett.* 2001, *74*, 145.

65. Schuster, W., Niederer, J.P.M., and Hoelderich, W.F. The gas phase oxidative dehydrogenation of propane over TS-1. *Appl. Catal. A: Gen.* 2001, *209*, 131.

66. Lee, K.H., Yoon, Y.S., Ueda, W., and Moro-oka, Y. An evidence of active surface MoO$_x$ over MgMoO$_4$ for the catalytic oxidative dehydrogenation of propane. *Catal. Lett.* 1997, *46*, 267.

67. Cadus, L.E., Gomez, M.F., and Abello, M.C. Synergy effects in the oxidative dehydrogenation of propane over MgMoO$_4$–MoO$_3$ catalysts. *Catal. Lett.* 1997, *43*, 229.

68. Mitchell, P.C.H. and Wass, S.A. Propane dehydrogenation over molybdenum hydrotalcite catalysts. *Appl. Catal. A: Gen.* 2002, *225*, 153.

69. Liu, Y., Wang, J., Zhou, G., Mo Xian, Bi, Y., and Zhen, K. Oxidative dehydrogenation of propane to propene over barium promoted Ni–Mo–O catalyst. *React. Kinet. Catal. Lett.* 2001, *73*, 199.

70. Kaddouri, A., Mazzocchia, C., and Tempesti, E. Propane and isobutane oxidative dehydrogenation with K, Ca and P-doped α- and β-nickel molybdate catalysts. *Appl. Catal. A: Gen.* 1998, *169*, L3.

71. Cadus, L.E. and Ferretti, O. Highly effective molybdena–manganese catalyst for propane oxidative dehydrogenation. *Catal. Lett.* 2000, *69*, 199.

72. Cadus, L.E. and Ferretti, O. Characterization of Mo–MnO catalyst for propane oxidative dehydrogenation. *Appl. Catal. A: Gen.* 2002, *233*, 239.

73. Boizumault-Moriceau, P., Pennequin, A., Grzybowska, B., and Barbaux, Y. Oxidative dehydrogenation of propane on Ni–Ce–O oxide: effect of the preparation method, effect of potassium addition and physical characterization. *Appl. Catal. A: Gen.* 2003, *245*, 55.

74. Buyevskaya, O.V., Wolf, D., and Baerns, M. Ethylene and propene by oxidative dehydrogenation of ethane and propane: performance of rare-earth oxide-based

catalysts and development of redox-type catalytic materials by combinatorial methods. *Catal. Today* 2000, *62*, 91.

75. Buyevskaya, O.V., Brückner, A., Kondratenko, E.V., Wolf, D., and Baerns, M. Fundamental and combinatorial approaches in the search for and optimisation of catalytic materials for the oxidative dehydrogenation of propane to propene. *Catal. Today* 2001, *67*, 369.

76. Grasselli, R.K. Advances and future trends in selective oxidation and ammoxidation catalysis. *Catal. Today* 1999, *49*, 141.

77. López Nieto, J.M., Concepción, P., Dejoz, A., Knozinger, H., Melo, F., and Vázquez, I. Selective oxidation of *n*-butane and butenes over vanadium-containing catalysts. *J. Catal.* 2000, *189*, 147.

78. Owens, L. and Kung, H.H. The effect of loading of vanadia on silica in the oxidation of butane. *J. Catal.* 1993, *144*, 202.

79. Chaar, M.A., Patel, D., Kung, M.C., and Kung, H.H. Selective oxidative dehydrogenation of butane over V–Mg–O catalysts. *J. Catal.* 1987, *105*, 483.

80. Kung, M.C. and Kung, H.H. The effect of potassium in the preparation of magnesium orthovanadate and pyrovanadate on the oxidative dehydrogenation of propane and butane. *J. Catal.* 1992, *134*, 668.

81. Owen, O.S. and Kung, H.H. Effect of cation reducibility on oxidative dehydrogenation of butane on orthovanadates. *J. Mol. Catal.* 1993, *79*, 265.

82. Dejoz, A., Nieto, J.M.L., Marquez, F., and Vázquez, M.I. The role of molybdenum in Mo-doped V–Mg–O catalysts during the oxidative dehydrogenation of *n*-butane. *Appl. Catal. A: Gen.* 1999, *180*, 83.

83. Bhattacharya, D., Bej, S.K., and Rao, M.S. Oxidative dehydrogenation of *n*-butane to butadiene: effect of different promoters on the performance of vanadium–magnesium oxide catalysts. *Appl. Catal. A: Gen.* 1992, *87*, 29.

84. Harding, W.D., Kung, H.H., Kozhevnikov, V.L., and Poeppelmeir, K.R. Phase equilibria and butane oxidation studies of the $MgO–V_2O_5–MoO_3$ system. *J. Catal.* 1993, *144*, 597.

85. Stern, D.L., Michaels, J.N., DeCaul, L., and Grasseli, R.K. Oxydehydrogenation of *n*-butane over promoted Mg–V-oxide based catalysts. *Appl. Catal. A: Gen.* 1997, *153*, 21.

86. Vidal-Michel, R. and Hohn, K.L. Effect of crystal size on the oxidative dehydrogenation of butane on V/MgO catalysts. *J. Catal.* 2004, *221*, 127.

87. Marcu, I.C., Sandulescu, I., and Millet, J.-M.M. Oxidehydrogenation of *n*-butane over tetravalent metal phosphates based catalysts. *Appl. Catal. A: Gen.* 2002, *227*, 309.

88. Marcu, I.C., Millet, J.-M.M., and Herrmann, J.-M. Semiconductive and redox properties of Ti and Zr pyrophosphate catalysts (TiP_2O_7 and ZrP_2O_7). Consequences for the oxidative dehydrogenation of *n*-butane. *Catal. Lett.* 2002, *78*, 273.

89. Pantazidis, A., Auroux, A., Herrmann, J.-M., and Mirodatos, C. Role of acid–base, redox and structural properties of VMgO catalysts in the oxidative dehydrogenation of propane. *Catal. Today* 1996, *32*, 81.

90. Madeira, L.M., Maldonado-Hódar, F.J., Portela, M.F., Freire, F., Martín-Aranda, R.M., and Oliveira, M. Oxidative dehydrogenation of *n*-butane on Cs doped nickel molybdate: kinetics and mechanism. *Appl. Catal. A: Gen.* 1996, *135*, 137.

91. Martín-Aranda, R.M., Portela, M.F., Madeira, L.M., Freire, F., and Oliveira, M. Effect of alkali metal promoters on nickel molybdate catalysts and its relevance to the selective oxidation of butane. *Appl. Catal. A: Gen.* 1995, *127*, 201.

92. Aït-Lachgar-Ben Abdelouahad, K., Roullet, M., Brun, M., Burrows, A., Kiely, C.J., Volta, J.C., and Abon, M. Surface alteration of $(VO)_2P_2O_7$ by α-Sb_2O_4 as a route to control the *n*-butane selective oxidation. *Appl. Catal. A: Gen.* 2001, *210*, 121.

93. Hagemeyer, A., Jandeleit, B., Liu, Y., Poojary, D.M., Turner, H.W., Volpe, A.F., Jr., and Weinberg, W.H. Applications of combinatorial methods in catalysis. *Appl. Catal. A: Gen.* 2001, *221*, 23.

94. Senkan, S. Combinatorial heterogeneous catalysis — A new path in an old field. *Angew. Chem. Int. Ed. Engl.* 2001, *40*, 312.

95. Schüth, F., Busch, O., Hoffmann, C., Johann, T., Kiener, C., Demuth, D., Klein, J., Schunk, S., Strehlau, W., and Zech, T. High-throughput experimentation in oxidation catalysis. *Top. Catal.* 2002, *21*, 55.

96. Urschey, J., Kühnle, and A., Maier, W.F. Combinatorial and conventional development of novel dehydrogenation catalysts. *Appl. Catal. A: Gen.* 2003, *252*, 91.

97. Kung, H.H. and Kung, M.C. Oxidative dehydrogenation of alkanes over vanadium–magnesium-oxides catalysts. *Appl. Catal. A: Gen.* 1997, *157*, 105.

98. Cavani, F., Comuzzi, C., Dolcetti, G., Etienne, E., Finke, R.G., Selleri, G., Trifiro, F., and Trovarelli, A. Oxidative dehydrogenation of isobutane to isobutene: Dawson-type heteropolyoxoanions as stable and selective heterogeneous catalysts. *J. Catal.* 1996, *160*, 317.

99. Comuzzi, C., Primavera, A., Trovarelli, A., Bini, G., and Cavani, F. Thermal stability and catalytic properties of the Wells–Dawson $K_6P_2W_{18}O_{62} \cdot 10H_2O$ heteropoly compound in the oxidative dehydrogenation of isobutane to isobutene. *Top. Catal.* 1999, *9*, 251.

100. Takita, Y., Sano, K., Kuroaki, K., Kawata, N., Nishiguchi, H., Ito, M., and Ishihara, T. Oxidative dehydrogenation of isobutane to isobutene I. Metal phosphate catalysts. *Appl. Catal. A: Gen.* 1998, *167*, 49.

101. Takita, Y., Sano, K., Muraya, T., Nishiguchi, H., Kawata, N., Ito, M., Akbay, T., and Ishihara, T. Oxidative dehydrogenation of isobutane to isobutene II. Rare earth phosphate catalysts. *Appl. Catal. A: Gen.* 1998, *170*, 23.

102. Zhang, Y.J., Rodriguez-Ramos, I., and Guerrero-Ruiz, A. Oxidative dehydrogenation of isobutane over magnesium molybdate catalysts. *Catal. Today* 2000, *61*, 377.

103. Al-Zahrani, S.M., Elbashir, N.O., Abasaeed, A.E., and Abdulwahed, M. Oxidative dehydrogenation of isobutane over pyrophosphates catalytic systems. *Catal. Lett.* 2000, *69*, 65.

104. Vislovskiy, V.P., Shamilov, N.T., Sardarly, A.M., Talyshinskii, R.M., Bychkov, V.Y., Ruiz, P., Cortés Corberán, V., Schay, Z., and Koppany, Zs. Oxidative conversion of isobutane to isobutene over V–Sb–Ni oxide catalysts. *Appl. Catal. A: Gen.* 2003, *250*, 143.

105. Bi, Y.-L. Zhen, K.-J., Valenzuela, R.X., Jia, M.-J., and Cortes Corberan, V. Oxidative dehydrogenation of isobutane over LaBaSm oxide catalyst: influence of the addition of CO_2 in the feed. *Catal. Today* 2000, *61*, 369.

16 Metathesis of Olefins on Metal Oxides

J.L.G. Fierro
Institute of Catalysis and Petrochemistry, CSIC, Madrid, Spain

J.C. Mol
Van't Hoff Institute for Molecular Sciences, University of Amsterdam, Amsterdam, The Netherlands

CONTENTS

16.1 INTRODUCTION

Metathesis reactions are very attractive since by this reaction olefins can be converted into new products via the rupture and reformation of carbon–carbon

517

double bonds (Equation [16.1])

$$2RCH=CHR' \leftrightarrow RCH=CHR + R'CH=CHR' \qquad (16.1)$$

The key step in this process is the reaction between an olefin and a transition metal carbene complex in a $2 + 2$ fashion to generate an unstable metallacyclobutane intermediate. This intermediate can revert either to the starting olefin or open productively to afford a new metal carbene and produce a new olefin. Metathesis reactions open up new industrial routes to important petrochemical intermediates, polymers, special chemicals, and oleochemicals. The metathesis reaction was discovered by Banks and Bailey at Phillips Petroleum Co. when looking at the conversion of olefins into high octane gasoline via olefin–isoparaffin alkylation [1]. Propene molecules were split over supported molybdenum catalysts instead of alkylating the paraffin, and propene was converted into ethene and butene. Since that discovery, industrial applications of the olefin metathesis reaction progressively increased, and other metathesis reactions, such as ring-opening metathesis polymerizations (ROMP) of cyclic olefins, became of great interest. An overview of the current industrial applications of olefin metathesis has recently been published [2]. Based on the existing process for the production of olefins and a limited volume of raw olefins, applications of the olefin metathesis reaction to convert less desirable olefins to more useful ones represent a great potential. This is illustrated by the conversion of low value olefins produced by Fischer–Tropsch synthesis into high value olefins that can be employed in further downstream processes. A technological relevant metathesis reaction is the conversion of low value C_7 α-olefins to internal C_{12} olefins, which can be used as detergent alcohol feedstock [2].

For its relevance, propene is one of the most important olefins. Propene is obtained mainly from naphtha steam cracking as a coproduct with ethene, and also as a coproduct from fluid catalytic cracking (FCC) units at refineries. Relatively small amounts are produced by propane dehydrogenation and by Fischer–Tropsch synthesis. Because of the strong global demand for polypropene, acrylonitrile, oxo alcohol, and acrylic acid products, present propene supply from conventional sources cannot fulfill the market needs. An alternative route to propene is by applying the metathesis reaction for the conversion of a mixture of ethene and 2-butene into propene (Equation [16.2]).

$$CH_2=CH_2 + CH_3CH_2=CH_2CH_3 \leftrightarrow 2CH_3CH_2=CH_2 \qquad (16.2)$$

This process, called OCT (olefin conversion technology), formerly the Phillips Triolefin Process, which utilizes a heterogeneous catalyst system, was originally developed by Phillips Petroleum Co. for the conversion of propene into ethene and butene. The reaction takes place in a fixed-bed reactor over a mixture of the metathesis catalyst WO_3/SiO_2 and the isomerization catalyst MgO at temperatures above 540 K and an overall pressure of approximately 30 bar [3]. 1-Butene present in the feedstock is isomerized to 2-butene as the original 2-butene is consumed in the metathesis reaction.

A large-scale industrial process incorporating olefin metathesis is the Shell higher olefins process (SHOP) for producing linear higher olefin from ethene; in the metathesis step, an alumina-supported molybdate catalyst is used.

The reaction rate is limited by the high activation energy, but it is accelerated in the presence of various metal oxides, such as Mo-, Re-, W-, and Sn-oxide (see [4] and references therein). Besides the above-mentioned heterogeneous metathesis catalysts, other catalyst consisting of ruthenium complexes are very attractive due to their robustness to air, water, and oxygenate high reaction rates and selectivity to target olefins. An extremely active ruthenium-based catalyst is the complex [RuCl$_2$(=CHPh)(H$_2$IPr)(PCy$_3$)] which shows a very high turnover number, TON (TON > 640 000) at room temperature for the metathesis of 1-octene [5]. The high performance of this complex is due to the steric bulk of the ligands and superior electron-donating properties. Therefore, nonproductive metathesis of the internal olefin formed during the reaction is less possible due to this increased steric bulk, which also might slow down catalyst decomposition pathways.

The metathesis reaction of cycloalkenes yields linear unsaturated polymers, so-called polyalkenamers. This ROMP is driven by the release of ring strain in the starting material. Several interesting polymers are commercially produced via the ROMP of different types of unsaturated cyclic monomers such as cyclooctene, norbornene, and dicyclopentadiene, using homogeneous catalyst systems [6]. As an alternative process, the cross-metathesis between a cyclic and an acyclic olefin allows to synthesize certain poly-unsaturated compounds for the special chemical market. Shell [7] developed the FEAST process for the manufacture of hexa-1,5-diene via cross-metathesis of cycloocta-1,5-diene with ethene.

Metathesis reactions of olefins also offer many possibilities in the area of fine chemicals, through major advances in catalyst design. Highly added value products will be commercially synthesized in the near future via metathesis reactions, such as biologically active compounds (pharmaceuticals, insect pheromones, prostaglandins etc.) and advanced polymeric materials in, for example, the oleochemical industry. The metathesis of unsaturated natural fats and oils and their derivatives offers new synthesis routes from cheap feedstocks to valuable new or existing chemical products from renewable resources with high chemo-selectivity, which have good prospects as a contribution to a sustainable chemical industry (see [8,9] and references therein). For example, among other products obtained upon metathesis of alkyl oleates and linoleates are unsaturated dicarboxylic esters, which are potential starting materials for the synthesis of polyesters, polyamides, and macrocyclic compounds. Cross-metathesis of methyl oleate with ethene gives methyl 9-decenoate, a potential starting material in the synthesis of many polymers and copolymers, for example, nylon-10. The metathesis of ethyl oleate results in the formation of 9-octadecene and diethyl 9-octadecene-1,18-dioate. The former can be dimerized and hydrogenated to give 10,11-dioctyleicosane, a lube-oil range hydrocarbon intermediate. On the other hand, diethyl 9-octadecene-1,18-dioate can be converted to civetone, an odorant found in musk perfumes, by a combination of the Dieckmann condensation and a hydrolysis–decarboxylation reaction. However, to commercialize the metathesis of functionalized alkenes, highly active, selective, and stable catalysts are needed that can withstand deactivation from a broad range of organic functional groups.

In the last 15 years, olefin metathesis reached significant importance in the fields of polymer chemistry [2,6,10–14] and organic synthesis [15–21]. This is mainly due to the development of well-defined catalysts based on W, Mo, or Ru, which resemble the active species in terms of oxidation state and ligand coordination sphere. These well-defined catalysts are also stable enough to be characterized by spectroscopic techniques and are allowed to react with an olefin to yield a persistent new alkylidene complex derived from that olefin. However, compounds that are well characterized, and catalyze the metathesis of olefins under specific reaction conditions but differ from the active species involved in the metathesis reaction, are not well-defined catalysts, but are so-called catalyst precursors. In this chapter attention is paid to the understanding of how a metal oxide precursor becomes active for the metathesis reaction, and to define, whenever possible, the nature of the active sites that are responsible for the activation of the olefin and subsequent recombination of the fragments produced after C=C bond splitting of the starting olefin.

16.2 ROLE OF METAL–CARBENES IN METATHESIS REACTIONS

Low oxidation state tungsten carbene complexes are catalyst precursors that played a role in studies concerned with the mechanism of olefin metathesis. For instance, the $(CO)_5W=CPh_2$ complex reacts with certain olefins in a manner consistent with the metallacyclobutane mechanism [22,23]. It was shown that certain strained cyclic olefins are polymerized slowly to give the polymers expected from ROMP when the polymerization is initiated by $(CO)_5W=CPh_2$ or even $(CO)_5W=CPh(OMe)$ [24,25]. These works provided evidence that metal carbenes are involved in olefin metathesis reactions. However, the catalytic ROMP reactions were not conclusive. In spite of the fact that the initial carbene complexes were well characterized, no propagating metal carbene was detected and, therefore, it cannot be concluded that the pentacarbonyl carbene species are responsible for the catalytic reaction. It seems unlikely that Fischer-type W^0 complexes can be propagating species in ROMP reactions in which some catalytic metathesis activity is observed. A small amount of a high oxidation state alkylidene complex is formed upon decomposition of the Fischer-type carbene complex, and it is this high oxidation state species that is responsible for the metathesis activity.

Tantalum complexes are excellent models in understanding how effective well-defined W and Mo catalysts might be designed that mimic metathesis catalysts based on W and Mo [6]. The first example of a stable M=CHR complex is the $Ta(CH-t-Bu)(CH_2-t-Bu)_3$ complex. In this complex, four bulk covalently bound ligands stabilize even an electronically unsaturated specie toward bimolecular decomposition. This compound is sensitive to oxygen, water, and a variety of functionalities, among them are ketones and aldehydes, with which it reacts to yield polymeric $(t-BuCH_2)_3Ta=O$ and the expected olefin [26]. It was established that d^0 alkylidene species were responsible for olefin metathesis and drew attention to the utility of alkoxide ligands in promoting metathesis. Unfortunately, however, the ethylidene and propylidene intermediates that were formed in the metathesis of *cis*-2-pentene, apparently readily rearranged to give ethene and propene, respectively, and therefore could not be observed.

16.3 OLEFIN METATHESIS OVER TRANSITION METAL OXIDES

The most successful metathesis catalysts are those based on rhenium, molybdenum, or tungsten. An account of these catalyst systems is given in Section 16.3.1 to Section 16.3.3.

16.3.1 Rhenium-Based Metathesis Catalysts

Rhenium-based metathesis catalysts are composed of rhenium oxide dispersed over high surface area oxides, such as alumina or silica–alumina. They have high activity and high selectivity at low temperatures (300 to 370 K), while they are also active for the metathesis of substituted olefins when promoted with a tetraalkyl-tin or -lead compound [27]. The structure of the catalyst precursor before and after calcination, as well as the structure of the catalyst during the reaction, has been widely investigated [28]. The most significant parameters governing catalytic activity include the oxidation state of the rhenium ion, the nature and number of the ligands, and the geometry of the active complex.

In a series of Re_2O_7/Al_2O_3 catalysts prepared from an NH_4ReO_4 salt precursor and Re_2O_7 loading from 1.0 to 18 wt%, a remarkable dependence of the rate of propene metathesis on the Re loading was observed. At low Re content, the catalytic activity was extremely low, but it increased exponentially above a Re_2O_7 loading of ~6 wt% Re_2O_7, the highest activity was attained at ~18 wt% Re_2O_7. The catalytic activity is probably only limited by the maximum amount of rhenium oxide that can be anchored on the alumina surface. This behavior can be explained in terms of the structure of the anchored rhenium to the alumina surface. Infrared spectra showed that at very low Re_2O_7 loadings the absorbances of all types of OH groups increase slightly with increasing Re loading. This indicates that ReO_4^- ions probably do not react with OH groups in the first instance, but are adsorbed on coordinatively unsaturated (CUS) Al^{3+} surface sites (Lewis acid sites). This adsorption of ReO_4^- ions on CUS Al^{3+} sites might result in the formation of new OH groups on these adsorbed Re centers, which could explain the slight increase in absorbance at low Re loadings. After this slight increase, all the infrared bands decreased with increasing Re loading, suggesting that ReO_4 groups have replaced the surface OH groups. The consumption of these surface OH groups proceeds in a sequential way. At the lower rhenium loadings, ReO_4^- ions have reacted predominantly with the more basic alumina OH groups, while at higher Re loadings subsequently the neutral and more acidic OH groups have also reacted, the latter resulting in the most active sites. These rhenium centers will be relatively electron-poor and therefore will easily accept the complication of an electron-rich carbon–carbon double bond of the alkene, which is necessary for the propagation of the metathesis reaction [29]. The surface rhenium oxide species is formed up to the monolayer coverage, that is, when all the reactive surface hydroxyls of the support have been titrated. Thus, at an 18 wt% Re_2O_7 loading nearly all surface hydroxyls have reacted with ReO_4^- ions [29]. It is conceivable that the surface rhenium oxide species anchored to neutral and acid OH groups are the precursor of the catalytic active site [30].

Based on these findings, the metathesis activity of a Re_2O_7/Al_2O_3 catalyst can be increased by treating alumina, first, with an acid capable of reacting with its basic OH groups, such as HCl or HF [31]. The activity of the Re_2O_7/Al_2O_3 catalyst, in particular a low loaded catalyst, can also be increased by using higher calcination temperature than that normally employed. The optimum calcination temperature depends on the Re loading of the catalyst and lies between 1100 and 1200 K. Such an increase in activity can be explained in terms of redistribution of ReO_4 groups over the alumina surface, from basic or CUS Al sites to alumina sites previously covered by acid OH groups [32]. In all the cases, selectivity approaches 100% since the possibility of side reactions such as isomerization of the substrate or product olefins is very low under the mild operation conditions of the Re_2O_7/Al_2O_3 catalyst.

16.3.1.1 Effect of the support

The catalytic activity of Re_2O_7/Al_2O_3 catalysts in the region of rhenium loadings below 10 wt% can be greatly improved by using a mixed support, for example, $SiO_2-Al_2O_3$ [33], $Al_2O_3-B_2O_3$ [34], or phosphated alumina [29]. The infrared spectra of a chemisorbed pyridine probe showed that, in the case of Re_2O_7 supported on phosphated alumina (AlPO), at low rhenium loadings the acidic OH groups, bonded to either P or Al, are substituted by ReO_4 groups, resulting in a catalyst with higher activity than the corresponding alumina-supported catalyst [29]. It is also possible that the mixed oxide support better stabilizes the Re species in the oxidation state which is most favorable for the formation of the initial rhenium carbene complex [35].

Some insight in the structure of the rhenium oxide phases present on the $SiO_2-Al_2O_3$ support has been gained by photoelectron spectroscopy. For low rhenium loadings (≤ 3 wt% Re_2O_7), the rhenium oxide supported on $SiO_2-Al_2O_3$ is present predominantly as monomeric species, while at higher loadings three-dimensional clusters are developed. Since no clustering occurs on γ-alumina, it seems likely that the clusters reside on the silica component of the $SiO_2-Al_2O_3$ support. It could be that during calcination the rhenium precursor on the silica part of the $SiO_2-Al_2O_3$ substrate coalesces because of the weak interaction between rhenium oxide and silica. The specific activity (turnover frequency, TOF) of $Re_2O_7/SiO_2-Al_2O_3$ catalysts decreases with increasing rhenium content. The highest activity is obtained for a $SiO_2-Al_2O_3$ support containing approximately 25 wt% Al_2O_3, and coincides with the highest Brønsted acidity of the support. This Brønsted acidity is due to two types of hydroxyl groups: hydroxyl groups attached to a Si atom and bridging hydroxyl groups attached to both a Si and an Al atom.

The latter type of hydroxyl groups is more electron-deficient — more Brønsted acidic — than the silanol groups. At low rhenium loadings, the $[ReO_4]^-$ ions react preferably with the Si–Al bridging surface hydroxyl groups during calcination, resulting in electron-deficient rhenium centers (ReO_4 tetrahedra), the active site precursors. This might explain as to why Re_2O_7 supported on SiO_2–Al_2O_3 is very active at low rhenium loadings. At higher rhenium loadings, the picture is different: the hydroxyl groups attached to a Si atom are also replaced, resulting in inactive rhenium centers of the type Si–O–ReO_3 (or rhenium clusters), as it occurs in Re_2O_7/SiO_2 that are not active in olefin metathesis [36]. The inactivity of Re_2O_7/SiO_2 might be because SiO_2 does not stabilize the necessary intermediate oxidation state of rhenium.

There is an apparent discrepancy in the hypothesis that the $[ReO_4]^-$ ion reacts first with the acidic hydroxyl groups of the silica–alumina support, but first with the Lewis acid and subsequently the basic hydroxyl groups of the alumina support. In fact, the $[ReO_4]^-$ ions prefer to bind to aluminum containing sites, either during impregnation or during calcination.

16.3.1.2 Influence of promoters

The metathesis activity of the rhenium-based catalysts may increase by a factor 5 to 10 by adding a tetraalkyltin promoter [37,38]. Adding a small amount of a tetraalkyltin (R_4Sn, R=Me, Et, or Bu) or a tetraalkyllead (R_4Pb, R=Et, or Bu) compound to a rhenium-based catalyst also results in highly effective catalysts for the metathesis of alkenes containing polar functional groups (e.g., unsaturated carboxylic esters and nitriles) [39]. The active sites generated by R_4Sn may be intrinsically different from those present on the unpromoted catalyst [40]. The nature of the active sites that are considered to be rhenium carbenes depends on the path of formation from the Re_2O_7/Al_2O_3 catalyst precursor and on the activating agent employed, for example, alkene or tetraalkyltin. On Re_2O_7/Al_2O_3 precursors, it is assumed that the generation of the initiating Re carbene occurs via a π-allyl mechanism (see later) [41]. The mechanism by which an alkyl compound generates the active rhenium site is still not well understood, although the partial reduction of Re(VII) species, generation of new initiating metal-alkylidene species (via a double alkylation followed by an α-H-abstraction) and modification of the active site by addition of a metal (Sn, Pb) ligand are principal steps involved in the activation process.

There are a few studies aiming to unravel the nature of the interaction between the tin compound and the surface rhenium species [42–45]. An organorhenium intermediate is proposed to be formed upon treating the Re_2O_7/Al_2O_3 catalyst with an R_4Sn promoter through a single alkylation of a rhenium site [46]. The Raman spectra of Re_2O_7/Al_2O_3 samples before and after addition of $(CH_3)_4Sn$ confirmed the presence of Re–O–Sn bonds [42]. The Raman spectrum of activated Re_2O_7/Al_2O_3 with 1% v/v $(CH_3)_4Sn$ exhibits, apart from rhenium oxide bands at 996, 884, and 338 cm^{-1}, respectively, three new bands at 520, 547, and 777 cm^{-1}. By comparison with Raman spectra of pure $(CH_3)_4Sn$, the bands at

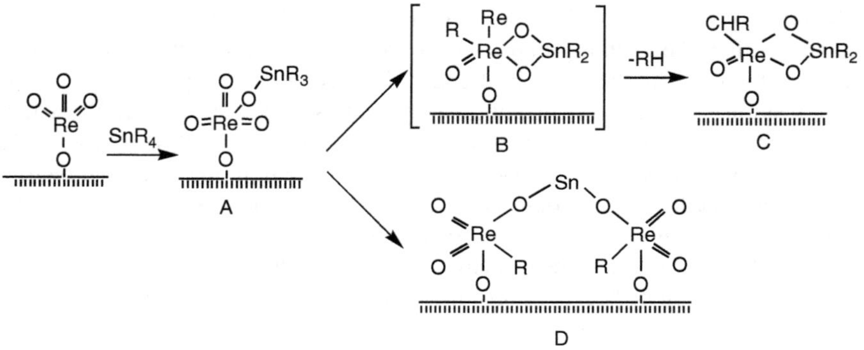

SCHEME 16.1

520 and 547 cm^{-1} may be due to tin–carbon vibrations in $(CH_3)_4Sn$, while the band at 777 cm^{-1} may arise from a new species formed by reaction of $(CH_3)_4Sn$ with the rhenium oxide, which is assumed to contain Re–O–Sn bonds, thereby providing new sites responsible for the metathesis reaction. However, a Raman study of the model compound $(CH_3)_3SnReO_3$ did not show the band at 777 cm^{-1} [47]. Thus, further work is required to confirm the formation of Re–O–Sn bonds.

Based on ^{119}Sn Mössbauer and ^{13}C and ^{119}Sn MAS NMR (nuclear magnetic resonance) spectra recorded with the catalyst system $Re_2O_7/SiO_2–Al_2O_3/R_4Sn$, containing <6 wt% Re_2O_7, a reaction scheme (Scheme 16.1) was proposed [44]. Both ^{119}Sn NMR and Mössbauer spectra accounted for the presence of species A. The R ligand on species A might undergo an electrophilic attack from a remaining OH$^-$ group from the surface with release of an alkane molecule. Mössbauer spectra suggest that higher rhenium loadings favor species D. A double alkylation involving two rhenium species (D) seems highly improbable owing to the low density of such rhenium sites, but cannot be ruled out as the degree of dispersion of $Re_2O_7/SiO_2–Al_2O_3$ decreases strongly above 3 wt% Re_2O_7. If a double alkylation involving only one rhenium site took place, a coordinatively saturated species, B, would be formed. This might undergo an α-H-abstraction, spontaneous, or surface-induced, leading to a rhenium–alkylidene species C. Although an alkylidene ligand in the rhenium coordination sphere could not be spectroscopically identified, chemical reactivity studies suggest its presence in minor amounts. While species A would predominate for $Re_2O_7/SiO_2–Al_2O_3/R_4Sn$ systems, in particular, when R=Et, for the system $Re_2O_7/Al_2O_3/Bu_4Sn$ there would only be species C or D [44,45].

Re_2O_7 supported on silica–alumina and promoted with alkyltin is active in the metathesis of single- and multiple-component fatty acid ester systems [48]. The metathetical transformation of unsaturated esters derived from sunflower oil (viz. alkyl oleates and linoleates) to intermediates for the synthesis of products with medical, cosmetic, polymeric, and other applications, is of particular interest. Single- and multiple-component fatty acid esters derived from sunflower oil easily undergo metathesis in the presence of a 3 wt% $Re_2O_7/SiO_2-Al_2O_3/Bu_4Sn$ catalyst system with high conversions and selectivity above 95% at 293 K. The absence of secondary metathesis products serves as evidence that double-bond migration does not occur under the given reaction conditions. In addition, different reaction products show varying sensitivities toward reaction temperature with the result that maximum yields of specific products can be obtained upon careful selection of the reactor temperature; for example, dicarboxylic esters reach maximum yield at 353 K.

Methyltrioxorhenium, CH_3ReO_3, prepared from Re_2O_7 and $(CH_3)_4Sn$, anchored to a $SiO_2-Al_2O_3$ support is a moderately active catalyst without any additive [49]. Model calculations showed that the active carbene species is formed via hydrogen atom transfer to a Re–O–Si bridging oxygen atom on the support, leading to a methylidene hydroxo derivative [50]:

Recently, a well-defined heterogeneous rhenium–alkylidene complex supported on silica was reported, which brought about the metathesis of methyl oleate with a TON of 900 [51]:

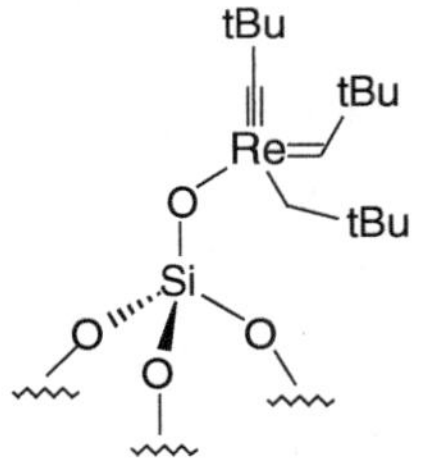

Such a catalyst, however, cannot be regenerated as in the case of promoted supported Re_2O_7 catalysts.

16.3.1.3 Influence of Lewis acidity

There is a broad agreement that carbene species developed on partly oxidized transition metal ions are necessary to bring about the metathesis reaction. The data reported earlier provide support that the environment of the active site does possess acid properties. It has been proposed that there is a relationship between the Brønsted acidity of the rhenium-based catalysts and the metathesis activity [52], and that the activity is not related to their Lewis acidity. Dynamic infrared spectroscopic experiments were conducted by poisoning a Re_2O_7/Al_2O_3 catalyst (8.0 wt% Re) under a propene flow [53]. Ammonia adsorption gave rise to coordinated NH_3 species, characterized by the bands at 1618, 1326, and

$1276 \ cm^{-1}$, but in no case protonated ammonia species were observed (bands at 1680 and $1456 \ cm^{-1}$). Thus, according to the previous work [53], Brønsted acidity does not seem to be a key requirement in the metathesis mechanism, whereas Lewis acidity is necessary for the reaction. However, this conclusion cannot be generalized as other studies provided a relationship between the Brønsted acidity of the Re_2O_7/Al_2O_3 catalysts and their metathesis activity [28].

16.3.1.4 Silicon and germanium promoters

As previously stated, alkyltin and alkyllead enhance the olefin metathesis activity of supported Re_2O_7 catalysts. However, tin- and lead-containing promoters are toxic. Therefore, an attempt has been made to replace these promoters by germanium and silicon compounds, evaluating both their reactivity toward Re_2O_7-based catalysts and the catalytic performance of the resulting systems in the metathesis reaction [54]. The addition of R_4Ge and R_4Si (R=Me, Et, or Bu) to the 3 wt% $Re_2O_7/SiO_2–Al_2O_3$ catalyst results in an enhancement of both activity and selectivity in the metathesis of hex-1-ene. In addition, the promoter decreases side reactions such as double-bond shift in the substrate followed by secondary metathesis reactions. In general, an alkyl germanium compound is a better promoter than silicon compounds. For R=Me, the activity is the lowest. The addition of Me_4Si even deactivates the catalyst. For Si-containing promoters the highest activity was obtained when the temperature of the reaction between promoter and catalyst was 348 K. In the case of Bu_4Ge, a further increase in activation temperature led to a higher amount of released butane, but the catalytic activity did not improve. This suggests that a reaction between unreacted Bu_4Ge and remaining surface OH groups takes place at temperatures above 323 K.

Rhenium oxide phases deposited on a boria-modified $SiO_2–Al_2O_3$ display, in general, higher metathesis activity than the boria-free $Re_2O_7/SiO_2–Al_2O_3$ counterparts. This peculiarity was exploited to perform the metathesis of methyl oleate on a $Re_2O_7/B_2O_3/SiO_2–Al_2O_3$ catalyst [54]. The reactivity of this system is very sensitive to the rhenium oxide loading; the highest specific activity is reached with a loading of 2 wt% Re_2O_7 (at 314 K). Although in the absence of any promoter the catalyst is not active in the metathesis of methyl oleate, it becomes active not only by adding the R_4M promoter (M=Sn, Ge, or Si), but also by incorporating the mono-hydride derivatives $HMBu_3$ (M=Sn, Ge, or Si), although to a lower extent [54]. The *cis/trans* ratio of the diester observed on the R_4M-modified catalysts is about 0.22, which corresponds to the thermodynamic equilibrium value. However, a much higher *cis/trans* ratio overpassing 0.80 is found with the hydride-promoted catalysts, which means that a preference for the *cis* isomer is compared with the thermodynamic value. The change in the distribution of the two isomers is a direct proof of the different environment for the rhenium centers in the active site.

As stated earlier, the interaction between tetraalkyltin and Re_2O_7 supported on $SiO_2–Al_2O_3$ or alumina gives rise to a rhenium alkylidene species. In order to verify whether an alkylidene ligand could be formed via a metathetical exchange reaction, the catalysts promoted with Et_4M (M=Sn, Ge, or Si) were allowed to react with neohexene in a flow reactor. Assuming that an alkylidene complex is formed by reaction between Re_2O_7/support with Et_4Si or Et_4Ge, propene and 4,4-dimethylpent-2-ene should be detected as the products formed first. In addition, self-metathesis of the propene formed would lead to ethene and but-2-ene. Indeed, such products were identified among the reaction products of neohexene metathesis over the Et_4M (M=Sn, Ge, or Si)-promoted rhenium catalysts. In addition, the butene/propene ratios are in agreement with the increasing activity induced by these promoters ($Et_4Si < Et_4Ge < Et_4Sn$). The dihydride Et_2SiH_2 does not give rise to a Re-ethylidene species. Results with Bu_3GeH and Bu_3SnH promoters suggest that no Re-alkylidene complexes are formed with promoters bearing hydride ligands. The carbene species involved in the metathesis reaction would arise from the reaction between olefin and surface hydride species.

16.3.1.5 Catalyst deactivation

Catalyst deactivation is a serious drawback to exploiting alkene metathesis for the production of olefins. There are many routes for the deactivation of a rhenium-based catalyst. Polar compounds, such as H_2O, which might be present as an impurity in the reactants, are catalyst poisons. Other possible routes for the deactivation of rhenium-based catalysts include: (i) reduction of the rhenium below its optimum oxidation state, (ii) adsorption of polymeric by-products on the surface of the catalyst, blocking the active sites, and (iii) reductive elimination of the metallacyclobutane intermediate. Even when the greatest care is taken, deactivation of the rhenium catalyst cannot be avoided. Therefore, reductive elimination of the metallacyclobutane intermediate (to form cyclopropane, or β-elimination to an alkene) is probably the main cause of deactivation and always seems operative [46,55,56].

Despite their relatively fast deactivation, supported rhenium oxide catalysts are sometimes preferred for practical applications because of their low operation temperature. Unpromoted Re_2O_7/Al_2O_3 catalysts can easily be regenerated by oxidation in an air stream at 823 K under atmospheric pressure without any loss of activity.

Although a tetraalkyltin-promoted Re_2O_7/Al_2O_3 catalyst is much more active than an unpromoted catalyst, it deactivates more rapidly. The activity of a deactivated tetraalkyltin-promoted Re_2O_7/Al_2O_3 catalyst can only be partly restored by heating in a stream of oxygen or air, followed by further addition of a promoter. The partial loss of activity is attributed to accumulation of tin on the catalyst surface [57].

16.3.2 Molybdenum-Based Metathesis Catalysts

Supported molybdenum catalysts have received much attention as possible metathesis catalysts because they are widely used already in industrial chemical processes.

16.3.2.1 Molybdenum-oxide catalysts supported on alumina

A MoO_3/Al_2O_3 catalyst precursor is usually prepared by impregnating the support with a solution of a molybdenum salt, such as ammonium heptamolybdate, followed by drying and calcination. After the calcination procedure, the catalyst is still fully oxidized; it shows catalytic activity only after reaction with reactant or hydrogen. H_2O is formed in this reaction, and therefore it can be concluded that the active sites on MoO_3/Al_2O_3 are formed through (slight) reduction [58]. From studies with well-defined supported catalysts obtained by treatment of the support with an organomolybdenum compound, it follows that high catalytic activity attributes to a Mo surface compound in oxidation state +4. The activity of MoO_3/Al_2O_3 for propene metathesis is highly dependent on the molybdenum content. The turnover frequency is very small for low Mo loading (<0.5 atoms/nm^2), but it increases with increasing Mo loading, and then passes through a maximum [59]. The optimal turnover frequency is obtained at 2 Mo atoms/nm^2. The predominant coordination of molybdenum is octahedral, except at low MoO_3 content. The very low activity at low Mo content is attributed to the formation of extremely stable tetrahedrally coordinated Mo^{6+} species that are very difficult to reduce, and which should, consequently, not be particularly active [60].

In the case of molybdenum oxide catalysts, addition of tetramethyltin enhances the activity of the MoO_3/Al_2O_3 catalyst. In propene metathesis at 303 K, the activity is about 20 times higher with, than without a tetramethyltin promoter. The optimal Sn:Mo molar ratio of 0.05 for MoO_3/Al_2O_3 catalysts suggests that the amount of active sites generated by tetramethyltin on the MoO_3/Al_2O_3 surface is not higher than 10% of the total Mo atoms in the catalyst [61].

16.3.2.2 Molybdenum-oxide catalysts supported on silica

The catalytic activity of MoO_3/SiO_2 for propene metathesis gradually increases when the activated catalyst is brought into contact with the reactant (catalyst break-in), illustrating that a reduction of Mo^{6+} is required for the catalyst to become active [62]. The initial activity of this catalyst can also be increased by prereduction in a reducing gas (H_2 or CO) at elevated temperatures. The catalytic activity of MoO_3/SiO_2 also depends on the Mo content; up to approximately 1 Mo ion/nm^2 the turnover frequency increases, together with the formation of a monolayer of tetrahedrally and octahedrally coordinated Mo species with maximum dispersion. At higher molybdenum loading, the activity decreases due to the formation of inactive crystalline MoO_3 [62]. From photoluminescence studies it can be concluded that tetrahedral dioxo Mo^{6+} species, tethered to the silica surface by two oxygen atoms, are reduced to Mo^{4+} species through loss of one oxygen ligand, the latter species being the active site precursor for metathesis [63,64].

Photoreduction of a low loaded calcined MoO_3/SiO_2 catalyst in a CO atmosphere under UV irradiation using a high pressure mercury lamp at room temperature or a laser beam of 308 nm, results in both the selective reduction of Mo^{6+} to Mo^{4+} and the appearance of high catalytic activity in propene metathesis [65,66]. The specific activity of such a catalyst considerably exceeds that of the catalysts activated by thermal reduction in H_2 or CO, due to the conversion of a major part of the original Mo^{6+} ions into coordinatively unsaturated surface Mo^{4+} ions:

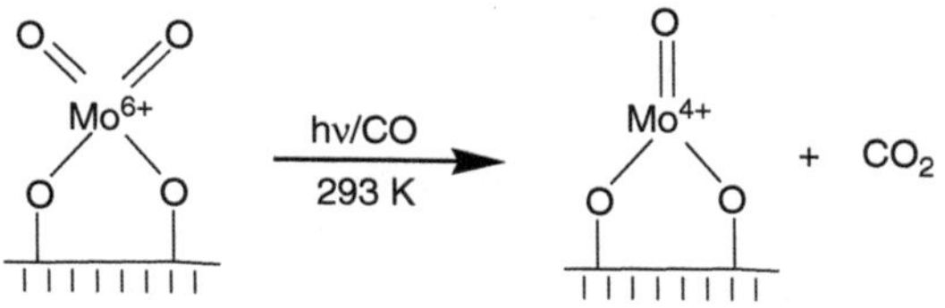

These Mo^{4+} ions can react with alkenes to form active molybdenum carbene complexes [65,67]. Catalysts that are even much more active can be obtained when the photoreduction is followed by the treatment with carbene-generating compounds such as cyclopropane [67,68] at 293 K, followed by evacuation at 623 K. These catalysts are also active for the metathesis of unsaturated esters.

These molybdenum catalysts deactivate faster than the rhenium catalysts due to an intrinsic deactivation mechanism [69].

16.3.3 Tungsten-Based Metathesis Catalysts

Supported WO_3 catalysts are comparable to supported MoO_3 catalysts when considered chemically, but are less active in metathesis. In order to obtain an acceptable activity, much higher reaction temperatures have to be used than for other common catalysts such as Re_2O_7/Al_2O_3 or MoO_3/Al_2O_3. The lower activity of WO_3/Al_2O_3 may be because it is more difficult to reduce than MoO_3/Al_2O_3, owing to the particularly strong interaction between the tungsten and the alumina support.

WO_3/SiO_2 has a high potential for practical applications in metathesis. The high-operation temperatures make it less susceptible to trace quantities of catalyst poisons such as air and water and to coke formation. Recently, this catalyst has extensively been studied in relation to its industrial use for the metathesis of long-chain olefins, such as oct-1-ene and industrial hept-1-ene [70]. The optimum reaction temperature of an 8 wt% WO_3/SiO_2 catalyst was found to be 733 K. The optimum temperature for regeneration in a flow of air is 823 K. A WO_3/SiO_2 catalyst is presently used in the OCT process and also in Phillips's neohexene process [2].

The turnover frequency of WO_3/SiO_2 for propene metathesis is not a constant function of the total transition metal content. Except for very low surface concentrations, the turnover frequency decreases monotonously with the tungsten concentration [63]. Analogous to MoO_3/SiO_2, WO_3/SiO_2 catalysts are composed of a surface phase and crystalline trioxide. Monolayer surface compounds are formed up to concentrations of approximately 1 tungsten atom per nm^2, whereas WO_3 crystals are present at higher concentrations. It has been extensively proven that WO_3 crystals are not active in metathesis and, therefore, the catalytic sites have to be contained in the surface phase. The turnover frequency of the surface phase is independent of the tungsten loading [63].

Studies on propene metathesis showed a pronounced break-in behavior for the WO_3/SiO_2 catalyst. Activation in an inert gas instead of air, which is generally used for activation at 823 K, yields a shorter break-in time and a higher conversion [71]. The break-in time can also be reduced by a controlled treatment with a reducing gas (H_2, CO) at elevated temperatures prior to metathesis; an accompanying color change from yellow to blue indicates reduction of WO_3 to $WO_{2.9}$. At higher activation temperatures in helium (925 K) the break-in disappears completely and the steady-state activity of the catalysts is largely increased [71]. The break-in process is attributed to the formation of a steady-state population of active sites. Trace quantities of acetone and acetaldehyde are formed due to reaction of propene with lattice oxygen from the tungsten oxide, resulting in the initiating tungsten carbene [72]. From a comparison of the activity of catalysts containing 6 wt% WO_3 on different supports for propene metathesis it followed that WO_3/SiO_2 is most active, followed by WO_3 supported on SiO_2–Al_2O_3 and SiO_2. TiO_2 gives the same order of activity as Al_2O_3 at its maximum at 515 K [73].

16.4 OLEFIN METATHESIS OVER NONTRANSITION METAL OXIDES

Nontransition metals supported on inorganic substrates also catalyze olefin metathesis reactions. The system Me_4Sn/Al_2O_3, prepared by deposition of Me_4Sn on a porous alumina support, previously dehydroxylated under vacuum at temperatures above 773 K, is active in the metathesis of light olefins ($C_3^=$–$C_6^=$) at 298 K [74]. Similarly, high specific area SiO_2, activated by evacuation at temperatures above 800 K, brings about the metathesis of lower olefins under photo-irradiation with a 250 W Hg lamp. The reaction occurs on the silica surface free from water molecules and formation of active sites strongly depend on the removal of isolated OH groups; the nature of the active sites is not known [75,76]. Promotion of SiO_2–Al_2O_3 with Bu_4Sn also results in activity for the metathesis reaction [77]. This latter system catalyzes the metathesis of terminal olefins, such as hex-1-ene, oct-1-ene, and dec-1-ene. High conversions into metathesis products were obtained with Bu_4Sn/SiO_2–Al_2O_3 containing $\sim$25 wt% Al_2O_3. The selectivity to primary metathesis products of the starting olefin is a function of its chain length. In the particular case of hex-1-ene, substantial double-bond isomerization was observed, followed by cross-metathesis reactions. The Bu_4Sn/SiO_2–Al_2O_3 system is more active than the system with alumina as support, while under the same experimental conditions a silica-supported catalyst shows no catalytic activity at all. The presence of alumina is a necessary condition for metathesis to take place. Although SiO_2–Al_2O_3 gives the highest activity, there is no correlation between the catalytic

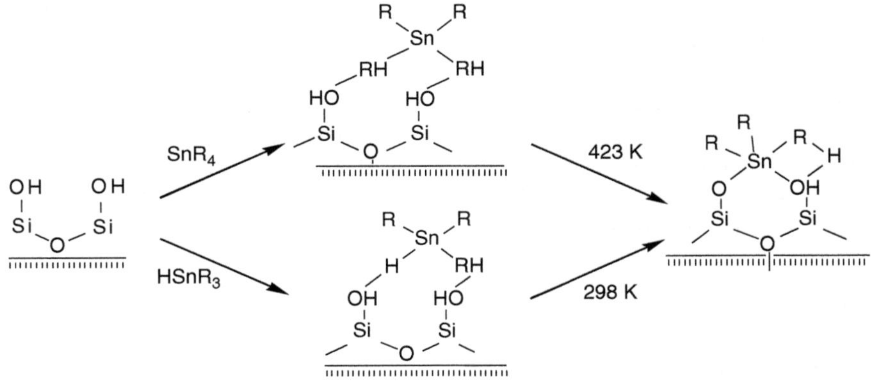

SCHEME 16.2

activity of these oxides and their Brønsted acidity, because for boria–alumina systems the activity decreases when the boria loading increases, that is, when the Brønsted acidity increases.

The anchorage of $(n\text{-Bu})_3\text{SnH}$ on silica–alumina involves the acidic bridging OH groups (Al–(OH)–Si) [78,79]. Such a reaction would lead to a CUS Al site (Scheme 16.2). The same kind of reaction is to be expected for $(n\text{-Bu})_4\text{Sn}$ [78]. On the other hand, when γ-alumina was allowed to react with $(n\text{-Bu})_4\text{Sn}$, the band of hydroxyl groups at 3779 cm^{-1} (basic OH groups) disappeared and simultaneously the ratio of the relative intensities between the bands at 3732 cm^{-1} (neutral OH) and 3688 cm^{-1} (acidic OH) decreased [80]. The findings indicate that basic and neutral OH groups are also involved in the interaction of $(n\text{-Bu})_4\text{Sn}$ with alumina. If the bridging neutral OH groups were directly involved in the formation of AlOSn $(n\text{-Bu})_3$ species, a CUS Al site would also become available: basic OH groups might be responsible for hydrogen bonding with CH_3 groups of the butyl ligands and also in pentaco-ordination to the tin atom. Therefore, the band, due to basic OH groups, would shift to lower wave numbers and could lie under the bands corresponding to neutral or acidic OH groups.

Another point to be considered is the fact that silica–alumina modified with $(n\text{-Bu})_3\text{SnH}$ is almost inactive in the metathesis of dec-4-ene even though the Mössbauer and NMR spectra indicated that $(n\text{-Bu})_3\text{SnH}$ and $(n\text{-Bu})_4\text{Sn}$ lead to the same surface species. The difference lies in the fact that $(n\text{-Bu})_3\text{SnH}$ completely reacts with the surface (through the Sn–H bond) for amounts of Sn between 32 and 80 $\mu\text{mol Sn/g SiO}_2\text{–Al}_2\text{O}_3$. This contrasts with $(n\text{-Bu})_4\text{Sn}$: when starting with 70 $\mu\text{mol Sn/g SiO}_2\text{–Al}_2\text{O}_3$ a maximum of 21 μmol of Sn was grafted on the surface. Moreover, the catalytic activity improved when the ratio of $(n\text{-Bu})_4\text{Sn/SiO}_2\text{–Al}_2\text{O}_3$ was raised from 32 to 85 $\mu\text{mol Sn/g SiO}_2\text{–Al}_2\text{O}_3$ [79]. Thus, some $(n\text{-Bu})_4\text{Sn}$ not involved in the formation of the surface species detected by Mössbauer and ^{119}Sn NMR spectroscopy is responsible for the catalytic activity. Since $(n\text{-Bu})_4\text{Sn}$ alone is not active in metathesis, the active site might arise from some kind of interaction between (n-Bu)$_4$Sn and a CUS Al site. This interaction would be developed under the conditions imposed by the catalytic reaction. However, it is not clear whether the active site involves a carbene species. A few stannene species ([Sn]=CR$_2$) have been reported, for which R must be bulky to prevent association [81]. Their association would be less likely if stannene species were formed from a grafted tin species. If such species were indeed formed in minor amounts, its detection by spectroscopic techniques would be extremely difficult, if not impossible.

16.5 REACTION MECHANISM

16.5.1 Reaction Pathways

Early theories for the effectiveness of the metathesis catalyst were framed in terms of lowering the activation barrier of a surface C$_4$ intermediate transition state [82] by having the substrate orbitals participate in bonding to the reaction transition state and thereby acting as an electron sink to lower the transition state energy. However, work with homogeneous catalysts lead to the proposal of an alternative, two-step model, which suggested that reaction was initiated by the formation of a carbene. In homogeneous phase, including a cocatalyst often provides this. The carbene is then proposed to constitute the active catalytic site by reacting with an alkene to produce a metallocyclic intermediate [22,24,83–86]. This can react via several routes, for example, by reductive elimination to yield cyclopropane [87] or a hydrogen transfer produces an alkene [88]. Both these reactions, of course, destroy the initial carbene active site. Finally, the metallacycle can react via the reverse of its formation pathway to yield an alkene and reform a surface carbene resulting in an overall metathesis reaction since a carbon–carbon double bond has effectively been broken and remade [22,24,83–86]. In fact, in this pathway, complete scission of the double bond is not required (except for the initial carbene synthesis), presumably resulting in a lowering of the reaction activation energy. The carbene can also be consumed by reacting with other carbenes analogously to the polymerization step in Fischer–Tropsch synthesis [89]. This, of course, also effectively constitutes olefin metathesis since carbon–carbon bonds are cleaved and reassembled.

The activity of the molybdenum (and also tungsten and rhenium) catalysts depends on the oxidation state of the metal so that supported metal catalysts are considered to be completely inactive and some higher oxidation state, often thought to be +4, is the most active for the reaction.

The mechanism of metathesis catalysis has been widely investigated and several possible reaction pathways have been advanced. Among the mechanisms proposed, the cyclobutane mechanism [82,90–92], the metallocyclopropane mechanism [93], and the carbene metallacycle mechanism [86,94–96] remain prominent. It is accepted that the olefin metathesis reaction occurs via metallacylcobutane intermediates formed by reacting olefin with the metal–alkylydene complex (M=CHR). Whether this happens on a heterogeneous catalyst, metal–alkylydene sites [M=CHR] should be formed first, from adsorbed olefins. Once [M=CHR] sites are formed on the surface, the olefin metathesis reaction proceeds until the metal-alkylydene species are dismissed by some side reactions. It appears that the olefin metathesis catalysts developed so far are the materials on which [M=CHR] sites are formed from reactions of the adsorbed olefins [97]. Otherwise, if the adsorbed olefins do not yield the [M=CHR] sites on the surface, no active catalyst is obtained. Accordingly, attempts have been made to activate this inactive surface by preparing [M=CHR] sites on the catalyst surface. For instance, a MoO_3 film that is completely inactive in the olefin metathesis reaction and in the olefin isomerization reaction becomes highly active in the olefin metathesis reaction upon condensing the olefin on the inert MoO_3 film at 77 K and then exposing to hydrogen [98]. It is emphasized that the resulting activated surface only catalyzes the olefin metathesis reactions but in no case the isomerization and hydrogenation of olefins. Thus, the Mo=CHR sites are formed on the MoO_3 film from alkyl radicals at 77 K. This key experiment affords some ways to prepare Mo=CHR sites on MoO_x and WO_x. A simple procedure is the activation of supported MoO_x by alkyltin [$Sn(CH_3)_4$] [98], and the other is the preparation of alkyl or carbene radicals on MoO_3 or WO_x films, by reaction of CH_2I_2 with Mg and the reaction of condensed olefins with hydrogen [98,99]. As mentioned earlier, these activation procedures are highly effective in the olefin metathesis reaction while isomerization and hydrogen scrambling reactions of olefins take place at a very limited extent.

Since the metathesis reaction of propene involves either productive or degenerate reactions, they can be followed by using deuterium labeled olefins. Thus, it would be possible to derive a total mechanism for the productive and degenerate metathesis reaction of propene. A complete mechanism for the propene metathesis reaction, where the productive metathesis takes place through alternative formation of Mo=CH_2 and Mo=CH–CH_3 sites, and the degenerate metathesis reaction occurs on either Mo=CH_2 or Mo=CH–CH_3 sites by forming the same alkylydene sites. Either Mo=CH_2 or Mo=CH–CH_3 site predominantly contributes to the degenerate metathesis of propene. In order to solve this uncertainty, isotope-labeled experiments have been designed [100]. The sequences of isotope-labeled analysis derived independently from the four different reactions show that, only the metathesis reaction taking place on the Mo=CH–CH_3 sites retains the conformation of the parent olefins in the product olefins. In other words, the sequences of metallacyclobutane intermediates having two or three substituted alkyl groups are consistent with each other but mono-methyl substituted metallacyclobutane, which is formed on Mo=CH_2 sites, has no preferable stereo conformation so that the conformational feature of parent propene cannot be transferred to the daughter molecules. Based on these experiments, it is concluded that the degenerate metathesis reaction of propene occurs predominantly on Mo=CH–CH_3 sites via the dimethyl substituted metallacyclobutane intermediate and that the reaction on Mo=CH_2 sites is negligible.

The degenerate metathesis of propene occurs 10 (MoO_3/TiO_2 + $Sn(CH_3)_4$) to 27 times (MoO_{3-x}/TiO_2 + $Sn(CH_3)_4$) faster than the productive metathesis reaction. This fact indicates that the methylene group (=CH_2) of propene tends to react and the reaction probability depends strongly on the orientation of propene molecule to Mo=CHR sites (R=H or CH$_3$). If this is the case, degenerate metathesis of propene should be two orders of magnitude faster on Mo=CH–CH$_3$ sites than on Mo=CH$_2$ sites. This mechanism well rationalizes the fact that degenerate metathesis reaction occurs 10 to 30 times on the Mo=CH–CH$_3$ site before the site changes to Mo=CH$_2$ site in accordance with the productive metathesis reaction.

16.5.2 Theoretical Calculations

Several theoretical investigations of olefin metathesis were reported for homogeneous catalysts [101–104]. In the case of the model Mo-alkylidene Schrock-type catalyst [Mo(NH)(CHR)(L)$_2$; R=H, CH$_3$, L=OH, OCH$_3$, OCF$_3$], density functional theory (DFT) calculations showed that the methylcyclobutane intermediates have a trigonal bypiramidal (TBP) or square pyramidal (SP) geometry. The SP geometry of the methylcyclobutane complex with electron-donating L ligands was shown to be more stable than the corresponding TBP geometry, but the opposite trend was predicted with electron-withdrawing L ligands [101]. In contrast to the homogeneous systems, the structures of the surface alkylidene centers are not unambiguously determined. In the case of monomeric centers of supported molybdenum catalysts, a distorted tetrahedral site with one oxo ligand, one alkylidene ligand, and two oxygen bridges with the support surface was proposed [60,69].

DFT was also employed to study the pathway of ethene metathesis proceeding on monomeric Mo(VI) centers of MoO$_3$/Al$_2$O$_3$ catalyst [105,106]. These calculations were carried out with the GAUSSIAN 98 program using the hybrid B3LYP functional. Models of active Mo(VI) sites including alumina cluster of formula Al$_2$(OH)$_6$ were used. Ethene addition to molybdenamethylidene center leading to the TBP molybdacyclobutane as well as the conversion of the TBP molybdocyclobutane to the SP structure was theoretically investigated. The results of these calculations pointed out that the last reaction competes with the decomposition of the TBP molybdocyclobutane to molybdenamethylidene center and ethene. The decomposition of the TBP molybdacyclobutane to molybdenamethylidene center and ethene is a necessary step to continue the catalytic cycle of ethene metathesis. However, this process competes with the exothermic rearrangement of the TBP molybdacyclobutane to the square planar intermediate. Thus, the proceeding of the latter reaction decreases the overall rate of ethene metathesis because of the high energy barrier of the back rearrangement of the SP molybdacyclobutane to the TBP one.

The predicted activation enthalpy of ethene addition to molybdenamethylidene center, according to the B3LYP/LANL2DZ(d) calculations, is about 31 kJ mol^{-1} [107]. The calculated activation Gibbs free energy of this step is high, because of the high negative value of the activation entropy. The decomposition of the TBP molybdacyclobutane to [Mo]=C3H$_2$ molybdenamethylidene surface complex and C1H$_2$=C2H$_2$ is a necessary step to continue the catalytic cycle of ethene metathesis. Because of Cs symmetry structure, the discussion does not require to distinguish between this step and the reverse decomposition.

The calculated activation enthalpy of this decomposition is about 37 kJ mol^{-1} and the activation Gibbs free energy is of the same range. However, the TBP molybdacyclobutane can more easily rearrange to the SP structure because this step has an activation barrier of about 27 kJ mol^{-1} and the similar value for the activation Gibbs free energy. The reverse conversion of the SP molybdacyclobutane to the TBP one has both a much higher activation enthalpy (79 kJ mol^{-1}) and an activation Gibbs free energy (83 kJ mol^{-1}), because the SP molybdacyclobutane is predicted to be more stable than the TBP structure. Therefore, the possible rearrangement of the TBP molybdacyclobutane to the SP geometry decreases the overall rate of ethene metathesis because of the high energy barrier of the reverse rearrangement.

In a recent contribution, Handzlik [108] reported a DFT study of propene metathesis proceeding on monomeric Mo-alkylidene centers of alumina-supported molybdenum catalysts. Based on the applied models of active Mo sites bonded to alumina clusters, including either two or four aluminum atoms, calculations revealed that the mechanism of propene metathesis is more complex than the one usually assumed in microkinetic models. As shown for the homogeneous systems [104] and also for ethene [107], TBP and SP intermediates were distinguished. TBP structure comes from the cycloaddition of the olefin with the Mo-alkylidene center, which undergoes decomposition of the product and a Mo-alkylidene complex, while the SP is formed directly from the TBP structure and this transformation competes with the decomposition of the TBP molybdocyclobutane intermediate to the metathesis product. The activation barriers of propene addition to the Mo-alkylidene centers were found to be higher than that for ethene metathesis, whereas the predicted barrier of ethene addition to the Mo-ethylidene center was lower than its addition to the Mo-methylidene center along the metathesis reaction of ethene.

16.6 SUMMARY

In this chapter, olefin metathesis reactions over heterogeneous catalysts are examined. These reactions, and particularly metathesis of acyclic olefins, are well established in petrochemistry and organic synthesis since, in many cases, high activity and selectivity at low temperatures can be attained. Several ways of modifying the catalyst precursors (supported molybdenum, tungsten, and rhenium) including the addition of an alkyltin promoter, has led to much better performances. The contribution of spectroscopic techniques, organometallic chemistry, and theoretical calculations (DFT methods) provided a better understanding of the nature of the active sites involved in metathesis reactions. It is accepted that olefin metathesis occurs via metallacyclobutane intermediates formed by reacting olefin with the metal–alkylydene complex (M=CHR). Whether this happens on a heterogeneous catalyst, the metal–alkylydene sites [M=CHR] should first be formed from adsorbed olefins, and then the olefin metathesis reaction proceeds until the alkylydene species are dismissed by some side reaction. In any case, further work is required to reveal the exact nature of the active site and how this depends on the nature and acid functionality of the inorganic support substrate.

References

1. Banks, R.L. and Bailey, G.C. Olefin disproportionation — A new catalytic process. *Ind. Eng. Chem. Prod. Res. Develop.* 1964, *3*, 170–174.
2. Mol, J.C. Industrial applications of olefin metathesis. *J. Mol. Catal. A: Chem.* 2004, *213*, 39–45.
3. Parkinson, G. Integration's the word in petrochemicals. *Chem. Eng.* 2001, *108*, 27–35.
4. Bartlett, B., Hossain, M., and Tysoe, W.T. Reaction pathway and stereoselectivity of olefin metathesis at high temperature. *J. Catal.* 1998, *176*, 439–447.
5. Dinger, M.B. and Mol, J.C. High turnover numbers with ruthenium-based metathesis catalysts. *Adv. Synth. Catal.* 2002, *344*, 671–677.
6. Ivin, K.J. and Mol, J.C. *Olefin Metathesis and Metathesis Polymerization.* Academic Press, London, 1997.
7. Chaumont, P. and John, C.S. Olefin disproportionation technology (FEAST). A challenge for process development. *J. Mol. Catal.* 1988, *46*, 317–328.
8. Mol, J.C. Application of olefin metathesis in the oleochemistry: an example of green chemistry. *Green Chem.* 2002, *4*, 5–13.
9. Mol, J.C. Catalytic metathesis of unsaturated fatty acid esters and oils. *Top. Catal.* 2004, *27*, 97–104.
10. Novak, B.M., Risse, W., and Grubbs, R.H. The development of well-defined catalysts for ring-opening olefin metathesis polymerization (ROMP). *Adv. Polym. Sci.* 1992, *102*, 47–72.
11. Schrock, R.R. In *Metathesis Polymerization of Olefins and Polymerization of Alkynes* Imamoglu, Y., Ed. Kluwer, Dordrecht, 1998, Chap. 1, pp. 1–27.
12. Buchmeiser, M.R. Homogeneous metathesis polymerization by well-defined group VI and group VIII transition-metal alkylidenes: fundamentals and applications in the preparation of advanced materials. *Chem. Rev.* 2000, *100*, 1565–1604.
13. Choi, S.H., Gal, Y.S., Jin, S.H., and Kim, H.K. Poly(1,6-heptadiyne)-based materials by metathesis polymerization. *Chem. Rev.* 2000, *100*, 1645–1681.
14. Schrock, R.R. In *Ring-Opening Polimerization*, Brunelle, D.J., Ed. Hanser, Munich, 1993, pp. 129.
15. Grubbs, R.H. and Chang, S. Recent advances in olefin metathesis and its application in organic synthesis. *Tetrahedron* 1998, *54*, 4413–4450.
16. Phillips, A.J. and Abell, A.D. Ring-closing metathesis of nitrogen-containing compounds: applications to heterocycles, alkaloids, and peptidomimetics. *Aldrich Chim. Acta* 1999, *32*, 75–89.
17. Wright, D.L. Application of olefin metathesis to organic synthesis. *Curr. Org. Chem.* 1999, *3*, 211–340.
18. Fürstner, A. Olefin metathesis and beyond. *Angew. Chem. Int. Ed.* 2000, *39*, 3012–3043.
19. Schuster, M. and Blechert, S. Olefin metathesis in organic chemistry. *Angew. Chem. Int. Ed.* 1997, *36*, 2036–2056.
20. Schrock, R.R. and Hoveyda, A.H. Molybdenum and tungsten imido alkylidene complexes as efficient olefin-metathesis catalysts. *Angew. Chem. Int. Ed.* 2003, *42*, 4592–4633.
21. Hoveyda, A.H. and Schrock, R.R. Catalytic asymmetric olefin metathesis. *Chem. Eur. J.* 2001, *7*, 945–950.

22. Casey, C.P., Tuinstra, H.E., and Saemen, M.C. Reactions of $(CO)_5WC(Tol)_2$ with alkenes. A model for structural selectivity in olefin metathesis reaction. *J. Am. Chem. Soc.* 1976, *98*, 608–609.

23. Casey, C.P., Burkhardt, T.J., Bunnell, C.A., and Calabrese, J.C. Synthesis and crystal structure of diphenylcarbene(pentacarbonyl)tungsten(0). *J. Am. Chem. Soc.* 1977, *99*, 2127–2134.

24. Katz, T.J. and Rothchild, R. Mechanism of olefin metathesis of 2,2′-divinylphenyl. *J. Am. Chem. Soc.* 1976, *98*, 2519–2526.

25. Katz, T.J. and McGinnis, J. Metathesis of cyclic and acyclic olefins. *J. Am. Chem. Soc.* 1977, *99*, 1903–1912.

26. Schrock, R.R. Multiple metal–carbon bonds. 5. Reaction of niobium and tantalum neopentylidene complexes with carbonyl function. *J. Am. Chem. Soc.* 1976, *98*, 5399–5400.

27. Mol, J.C. Metathesis of functionalized acyclic olefins. *J. Mol. Catal.* 1991, *65*, 145–162.

28. Mol, J.C. Olefin metathesis over supported rhenium oxide catalysts. *Catal. Today* 1999, *51*, 289–299, erratum *Catal. Today* 1999, *52*, 377.

29. Sibeijn, M., Spronk, R., van Veen, J.A.R., and Mol, J.C. IR studies of Re_2O_7 metathesis catalysts supported on alumina and phosphated alumina. *Catal. Lett.* 1991, *8*, 201–208.

30. Vuurman, M., Stufkens, D.J., Oskam, A., and Wachs, I.E. Structural determination of surface rhenium oxide on various oxide supports (Al_2O_3, ZrO_2, TiO_2, and SiO_2). *J. Mol. Catal.* 1992, *76*, 263–285.

31. Hietala, J., Root, A., and Knuuttila, P. The surface acidity of pure and modified aluminas in Re/Al_2O_3 metathesis catalysts as studied by ^{1}H-NMR spectroscopy and its importance in the ethenolysis of 1,5-cyclooctadiene. *J. Catal.* 1994, *150*, 46–55.

32. Spronk, R., Van Veen, J.A.R., and Mol, J.C. The effect of the calcination temperature on the activity of Re_2O_7/γ-Al_2O_3 catalysts for the metathesis of propene. *J. Catal.* 1993, *144*, 472–483.

33. Xiaoding, Xu and Mol, J.C. Re_2O_7/SiO_2–Al_2O_3-SnR_4 or -PbR_4, a highly active catalyst for the metathesis of functionalized alkenes. *J. Chem. Soc., Chem. Commun.* 1985, 631–633.

34. Xiaoding, Xu, Boelhouwer, C., Benecke, J.I., Vonk, D., and Mol, J.C. $Re_2O_7/Al_2O_3.B_2O_3$ metathesis catalysts. *J. Chem. Soc., Faraday Trans. 1* 1986, *82*, 1945–1953.

35. Sheu, F.C., Hong, C.T., Hwang, W.L., Shih, C.J., Wu, J.C., and Yeh, C.T. Propylene metathesis over $Re_2O_7/Al_2O_3 \cdot B_2O_3$ catalysts. *Catal. Lett.* 1992, *14*, 297–304.

36. Mol, J.C. and Andreini, A. Activity and selectivity of rhenium-based catalysts for alkene metathesis. *J. Mol. Catal.* 1988, *46*, 151–156.

37. Andreini, A., Xu, X., and Mol, J.C. Activity of $Re_2O_7/SiO_2.Al_2O_3$ catalysts for propene metathesis and the influence of alkyltin promoters. *Appl. Catal.* 1986, *27*, 31–40.

38. Fridman, R.A., Nosakova, S.M, Liberov, G., and Bashkirov, A.N. Disproportionation of olefins on aluminum–molybdenum and aluminum–rhenium catalysts, promoted by tin tetraalkyl derivatives. *Izv. Akad. Nauk. SSSR, Ser. Khim.* 1977, *26*, 678–679.

39. Mol, J.C. Metathesis of unsaturated fatty-acid esters and fatty oils. *J. Mol. Catal.* 1994, *90*, 185–199.

40. Moloy, K.G. Evidence for intrinsically distinct active-sites on rhenium-based metathesis catalysts — Norbornene metathesis with Re_2O_7/Al_2O_3. *J. Mol. Catal.* 1994, *91*, 291–302.

41. McCoy, J.R. and Farona, M.F. Olefin metathesis over a Re_2O_7/Al_2O_3 metathesis catalyst — Mechanism for initial metallacarbene formation. *J. Mol. Catal.* 1991, *66*, 51–58.

42. Williams, K.P.J. and Harrison, K. Raman-spectroscopic studies of the effects of tin promotion on a rhenium alumina dismutation catalyst. *J. Chem. Soc., Faraday Trans.* 1990, *86*, 1603–1606.

43. Xu, X., Andreini, A., and Mol, J.C. The role of SnR_4 compounds in the metathesis of alkenes catalyzed by Re_2O_7/γ-Al_2O_3: An ESR study. *J. Mol. Catal.* 1985, *28*, 133–140.

44. Buffon, R., Schuchardt, U., and Abras, A. Mössbauer and solid-state NMR spectroscopic studies of tin-modified rhenium-based metathesis catalysts. *J. Chem. Soc., Faraday Trans.* 1995, *91*, 3511–3517.

45. Buffon, R., Jannini, M.J.D.M., and Abras, A. Effects of the addition of Nb_2O_5 to rhenium-based olefin metathesis catalysts. *J. Mol. Catal. A: Chem.* 1997, *155*, 173–181.

46. Spronk, R., Andreini, A., and Mol, J.C. Deactivation of rhenium-based catalysts for the metathesis of propene. *J. Mol. Catal.* 1991, *65*, 219–235.

47. Mol, J.C. and Wachs, I. Unpublished results.

48. Marvey, B.B., du Plessis, J.A.K., Vosloo, H.C.M., and Mol, J.C. Metathesis of unsaturated fatty acid esters derived from South African sunflower oil in the presence of a 3 wt% Re_2O_7/SiO_2–$Al_2O_3/SnBu_4$ catalyst. *J. Mol. Cat. A: Chem.* 2003, *201*, 297–308.

49. Herrmann, W.A., Wagner, W., Flessner, U.N., Volkhardt, U., and Komber, H. Methyltrioxorhenium as catalyst for olefin metathesis. *Angew. Chem., Int. Ed. Engl.* 1991, *30*, 1636–1638.

50. Morros, L.J., Downs, A.J., Greene, T.M., McGrady, G.S., Herrmann, W.A., Sirsch, P., Scherer, W., and Gropen, O. Matrix photochemistry of methyltrioxorhenium(VII), CH_3ReO_3: formation of the methylidene tautomer $H_2C=Re(O)_2OH$ and its potential relevance to olefin metathesis. *Organometallics* 2001, *20*, 2344–2352.

51. Chabanas, M., Coperet, C., and Basset, J.M. Re-based heterogeneous catalysts for olefin metathesis prepared by surface organometallic chemistry: reactivity and selectivity. *Chem. Eur. J.* 2003, *9*, 971–975.

52. Xu, X.D., Mol, J.C., and Boelhouwer, C. Surface acidity of some Re_2O_7-containing metathesis catalysts — An in situ Fourier Transform infrared study using pyridine adsorption. *J. Chem. Soc., Faraday Trans. I* 1986, *82*, 2707–2718.

53. Schekler-Nahama, F., Clause, O., Commereuc, D., and Saussey, J. Influence of Lewis acidity of rhenium heptoxide supported on alumina catalyst on the catalytic performances in olefin metathesis. *Appl. Catal. A: Gen.* 1998, *167*, 237–245.

54. Buffon, R., Marochio, I.J., Rodella, C.B., and Mol, J.C. Germanium and silicon compounds as promoters for Re_2O_7/SiO_2–Al_2O_3 metathesis catalysts. *J. Mol. Cat. A: Chem.* 2002, *190*, 171–176.

55. Amigues, P., Chauvin, Y., Commereuc, D., Hong, C.T., Lai, C.C., and Lin, Y.H. Metathesis of ethylene–butene mixtures to propylene with rhenium on alumina catalysts. *J. Mol. Catal.* 1991, *65*, 39–50.

56. Moulijn, J.A. and Mol, J.C. Structure and activity of rhenium-based metathesis catalysts. *J. Mol. Catal.* 1988, *46*, 1–14.

57. Spronk, R. and Mol, J.C. Regeneration of rhenium-based catalysts for the metathesis of propene. *Appl. Catal.* 1991, *76*, 143–152.

58. Engelhardt, J., Goldwasser, J., and Hall, W.K. The isomerization and metathesis of n-butenes IV. Reactions of *cis*-2-butene over a molybdena–alumina catalyst of varying extents of reduction. *J. Mol. Catal.* 1982, *15*, 173–185.

59. Thomas, R. and Moulijn, J.A. A comparative study of γ-alumina supported molybdenum and tungsten oxide; relation between metathesis activity and reducibility. *J. Mol. Catal.* 1982, *15*, 157–172.

60. Grünert, W., Stakheev, A.Y., Feldhaus, R., Anders, K., Shipro, E.S., and Minachev, Kh.M. Reduction and metathesis activity of MoO_3/Al_2O_3 catalysts II. The activation of MoO_3/Al_2O_3 catalysts. *J. Catal.* 1992, *135*, 287–299.

61. Handzlik, J. and Ogonowski, J. Activity of molybdena–alumina metathesis catalysts treated with tetramethyltin. *Catal. Lett.* 2002, *83*, 287–290.

62. Thomas, R., Moulijn, J.A., de Beer, V.H.J., and Medema, J. Structure/metathesis-activity relations of silica supported molybdenum and tungsten oxide. *J. Mol. Catal.* 1980, *8*, 161–174.

63. Ono, T., Anpo, M., and Kubokawa, Y. Metathesis activity and structure of MoO_3 highly dispersed on SiO_2. *Chem. Express* 1986, *1*, 181–184.

64. Ono, T., Anpo, M., and Kubokawa, Y. Catalytic activity and structure of molybdenum trioxyde highly dispersed on silica. *J. Phys. Chem.* 1986, *90*, 4780–4784.

65. Shelimov, B.N., Elev, I.V., and Kazansky, V.B. Use of photoreduction for activation of silica–molybdena catalysts for propylene metathesis: comparison with thermal reduction. *J. Catal.* 1986, *98*, 70–81.

66. Mol, J.C. Activity and stability of laser-photoreduced supported molybdenum oxide metathesis catalysts. *Catal. Lett.* 1994, *23*, 113–118.

67. Shelimov, B.N., Elev, I.V., and Kazansky, V.B. Spectroscopic study of formation of active metal-carbene species in photoreduced silica–molybdena catalysts for olefin metathesis. *J. Mol. Catal.* 1988, *46*, 187–200.

68. Elev, I.V., Shelimov, B.N., and Kazansky, V.B. Study of the mechanism of the reaction of metathesis of olefins and the formation of active sites on photoreduced molybdenum–silica catalysts. *Kinet. Kataliz* 1989, *30*, 895–900.

69. Vikulov, K.A., Shelimov, B.N., Kazansky, V.B., and Mol, J.C. The deactivation mechanism of photoreduced and cyclopropane-treated molybdena/silica catalysts for olefin metathesis. *J. Mol. Catal.* 1994, *90*, 61–67.

70. (a) Van Schalkwyk, C., Spamer, A., Moodley, D.J., Dube, T., Reynhardt, J., and Botha, J.M. Application of a WO_3/SiO_2 catalyst in an industrial environment: part I. *Appl. Catal. A: Gen.* 2003, *255*, 121–131. (b) Spamer, A., Dube, T.I., Moodley, D.J., van Schalkwyk, C., and Botha, J.M. Application of a WO_3/SiO_2 catalyst in an industrial environment: part II. *Appl. Catal. A: Gen.* 2003, *255*, 133–142. (c) Factors that could influence the activity of a WO_3/SiO_2 catalyst: part III. *Appl. Catal. A: Gen.* 2003, *255*, 143–152.

71. Andreini, A. and Mol, J.C. Activity of supported tungsten oxide catalysts for the metathesis of propene. *J. Colloid Interface Sci.* 1981, *84*, 57–65.

72. Basrur, A.G., Patwardhan, S.R., and Vyas, S.N. Propene metathesis over silica-supported tungsten oxide catalyst — Catalyst induction mechanism. *J. Catal.* 1991, *127*, 86–95.

73. Andreini, A. and Mol, J.C. Activity of supported tungsten oxide catalysts for the metathesis of propene. *J. Chem. Soc., Faraday Trans. 1* 1985, *81*, 1705–1714.

74. Ahn, H.G., Yamamoto, K., Nakamura, R., and Niiyama, H. A novel metathesis catalyst consisting of non-transition elements — The metathesis of alkenes over tetramethyltin dehydroxylated alumina. *Chem. Lett.* 1992, (3), 503–506.

75. Yoshida, Y., Kimura, K., Inaki, Y., and Hattori, T. Catalytic activity of FSM-16 for photometathesis of propene. *Chem. Commun.* 1997, (1), 129–130.

76. Tanaka, T., Matsuo, S., Maeda, T., Yoshida, H., Funabiki, T., and Yoshida, S. Olefin metathesis over UV-irradiated silica. *Appl. Surf. Sci.* 1997, *121*, 296–300.

77. Jannini, M.J.D.M., Buffon, R., de Wit, A.M., and Mol, J.C. Metathesis of terminal olefins over tin-modified silica–alumina. *J. Mol. Catal. A: Chem.* 1998, *133*, 201–203.

78. Nédez, C., Théolier, A., Lefebvre, F., Choplin, A., Basset, J.M., and Joly, J.F. Surface organometallic chemistry of tin — Reactivity of tetraalkyltin complexes and tributyltin hydride toward silica. *J. Am. Chem. Soc.* 1993, *115*, 722–729.

79. Nédez, C., Lefebvre, F., Choplin, A., Niccolai, G.P., Basset, J.M., and Benazzi, E. Surface organometallic chemistry of tin — reaction of hydridotris(butyl)tin with the surface of partially dehydroxylated aluminas. *J. Am. Chem. Soc.* 1994, *116*, 8638–8646.

80. Buffon, R., Jannin, M.J.D.M., Abras, A., Mol, J.C., de Wit, A.M., and Kellendonk, F.J.A. Olefin metathesis over tin-modified non-transition metal oxides. *J. Mol. Catal. A: Chem.* 1999, *149*, 275–282.

81. Anselme, G., Ranaivonjatovo, H., Escudié, J., Couret, C., and Satgé, J. A stable compound with a formal tin carbon double-bond — The bis(2,4,6-triisopropylphenyl) (fluorophenylidene) stannate. *Organometallics* 1992, *11*, 2748–2750.

82. Adams, C.T. and Brandenberger, S.G. Mechanism of olefin disproportionation. *J. Catal.* 1969, *13*, 360–367.

83. Katz, T.J. and Hersch, W.H. Stereochemistry of olefin metathesis reaction. *Tetrahedron. Lett.* 1977, *6*, 585–588.

84. Tebbe, F.N., Parshall, G.W., and Ovenall, D.W. Titanium-catalyzed olefin metathesis. *J. Am. Chem. Soc.* 1979, *102*, 5074–5075.

85. Wengrovius, J.H., Schrock, R.R., Churchill, M.R., Missert, J.R., and Youngs, W.J. Multiple metal–carbon bonds. 16. Tungsten-oxo alkylidene complexes as olefin metathesis catalysts and the crystal structure of $W(O)(CHCMe_3)(PEt_3)Cl_2$. *J. Am. Chem. Soc.* 1980, *102*, 4515–4516.

86. Howard, T.R., Lee, J.B., and Grubbs, R.H. Titanium metallacarbene–metallacyclobutane reactions: stepwise metathesis. *J. Am. Chem. Soc.* 1980, *102*, 6876–6878.

87. McQuillin, F.J. and Powell, K.C. Stereoselectivity in the carbonyl insertion reaction between tetracarbonyldichlorodirhodium and substituted cyclopropanes. *J. Chem. Soc., Dalton Trans.* 1972, *19*, 2129–2133.

88. Millman, W.S., Crespin, M., Cirillo, A.C., Jr., Abdo, S., and Hall, W.K. Studies of the hydrogen held by solids. XXII. The surface chemistry of reduced molybdena–alumina catalysts. *J. Catal.* 1979, *60*, 404–416.

89. Bartlett, B.F. and Tysoe, W.T. Molybdenum metal catalyzed reaction of ethene. *Catal. Lett.* 1997, *44*, 37–42.

90. Clark, A. and Cook, C. Mechanism of propylene disproportionation. *J. Catal.* 1969, *15*, 420–427.

91. Hughes, W.B. Kinetics and mechanism of homogeneous olefin disproportionation reaction. *J. Am. Chem. Soc.* 1969, *92*, 532–539.
92. Crain, D.L. Mechanism of olefin disproportionation. *J. Catal.* 1969, *13*, 110–118.
93. Biefeld, G.C., Eick, H.A., and Grubbs, R.H. Crystal structure of bis(triphenylphosphine) tetramethyleneplatinum(II). *Inorg. Chem.* 1973, *12*, 2166–2170.
94. Wang, L.P., Millman, W.S., and Tysoe, W.T. The hydrogenation of CO over molybdenum/alumina in the presence of ethylene: coupling of olefin metathesis and CO hydrogenation. *Catal. Lett.* 1988, *1*, 159–167.
95. Kazuta, M. and Tanaka, K. Synthesis of active sites for alkene metathesis reaction on molybdenum oxide films by reaction with alkylidene radicals. *Catal. Lett.* 1988, *1*, 7–10.
96. Vikulov, K.A., Elev, I.V., Shelimov, B.N., and Kazansky, V.B. First IR spectroscopic observation of the stable Mo=CH$_2$ carbene complex on the surface of an active catalyst for olefin metathesis — photoreduced silica–molybdena with chemisorbed cyclopropane. *Catal. Lett.* 1989, *2*, 121–124.
97. Tanaka, K. Catalysts working by self-activation. *Appl. Catal. A: Gen.* 1999, *188*, 37–52.
98. Kazuta, M. and Tanaka, K. Synthesis of active-sites for alkene metathesis reaction on molybdenum- and tungsten-oxide films. *J. Catal.* 1990, *123*, 164–172.
99. Kazuta, M. and Tanaka, K. Preparation of very active molybdenum oxide films for use in the alkene metathesis reaction. *J. Chem. Soc., Chem. Commun.* 1987, (8), 616–617.
100. Tanaka, K., Tanaka, H., Takeo, H., and Matsumura, C. Intermediates for the degenerate and productive metathesis of propene elucidated by the metathesis reaction of (Z)-propene-1-d1. *J. Am. Chem. Soc.* 1987, *109*, 2422–2425.
101. Folga, E. and Ziegler, T. Density functional-study on molybdacyclobutane and its role in olefin metathesis. *Organometallics* 1993, *12*, 325–337.
102. Fox, H.H., Schofield, M.H., and Schrock, R.R. Electronic structure and Mo(VI) alkylidene complexes and an examination of reactive intermediates using the SCF-Xα-SW method. *Organometallics* 1994, *13*, 2804–2815.
103. Aagaard, O.M., Meier, R.J., and Buda, F. Ruthenium-catalyzed olefin metathesis: a quantum molecular dynamics study. *J. Am. Chem. Soc.* 1998, *120*, 7174–7182.
104. Adhart, C., Hinderling, C., Baumann, H., and Chen, P. Mechanistic studies of olefin metathesis by ruthenium carbene complexes using electrospray ionization tandem mass spectrometry. *J. Am. Chem. Soc.* 2000, *122*, 8204–8214.
105. Handzlik, J. and Ogonowski, J. Theoretical study on ethene metathesis proceeding on Mo^{VI} and Mo^{IV} methylidene centers of heterogeneous molybdena–alumina catalysts. *J. Mol. Catal. A: Chem.* 2001, *175*, 215–225.
106. Van Santen, R.A. The cluster approach to molecular heterogeneous catalysis. *J. Mol. Catal. A: Chem.* 1997, *115*, 405–419.
107. Handzlik, J. and Ogonowski, J. DFT study of ethene metathesis proceeding on monomeric Mo^{VI} centres of MoO$_3$/Al$_2$O$_3$ catalyst: the role of the molybdacyclobutane intermediate. *J. Mol. Catal. A: Chem.* 2002, *184*, 371–377.
108. Handzlik, J. Theoretical study of propene metathesis proceeding on monomeric Mo centers of molybdena–alumina catalysts. *J. Catal.* 2003, *220*, 23–34.

17 Applications of Metal Oxides for Volatile Organic Compound Combustion

D.P. Dissanayake

Department of Chemistry, University of Colombo, Colombo, Sri Lanka

CONTENTS

17.1 INTRODUCTION

Volatile organic compounds (VOCs) are emitted into the atmosphere due to various human activities, mainly as unburned fuel from power production, transportation, and solvents from industrial processes. They have direct and indirect injurious effects on human health. Apart from harm caused when exposed to high concentrations, even minute amounts of carcinogenic compounds, such as dioxins and polycyclic aromatic compounds, affect human health [1]. Photochemical smog is an indirect effect of VOCs where nitrogen oxides and VOCs react to produce ozone and peroxy compounds in the presence of sunlight [1]. Other undesirable environmental conditions include ozone depletion and global warming. Halogenated VOCs have been identified as major contributors to ozone depletion. Their high volatility and persistence cause the accumulation of these compounds in the atmosphere, where they act as catalysts for ozone decomposition. Methane, the lightest volatile organic is about 20 times more effective as a green house gas than its complete oxidation product, carbon dioxide [1,2]. The atmospheric lifetime of methane is about 10 years. The increasing incidence of atmospheric methane suggests that sources exceed the sinks by about 10%, which highlights the importance of control of emissions [1].

Abatement of VOCs has attracted the attention of researchers for a long time and more so during the recent past. Established technologies for abatement of VOCs include flame combustion, absorption, condensation, conversion, or catalytic combustion to less harmful compounds. When air is used as the oxidant in high temperature flame combustion of VOCs, significant amounts of nitrogen oxides are formed. In contrast, catalytic combustion is more efficient and can be carried out at low temperatures, which avoids the formation of nitrogen oxides. For example, due to its inert nature, methane requires more severe conditions ($>800°C$) for combustion [3,4]. However, the efficiency of the combustion process can be significantly improved and can be carried out at lower temperatures by employing metal or metal oxide catalysts [4–9], which justifies research on catalysts for VOCs oxidation.

Precious metals are among the highly active catalysts for oxidation of VOCs. Pt and Pd in pure form or supported on high surface area materials, such as alumina, are used in this regard. Three-way catalytic converter capable of catalyzing the oxidation of carbon monoxide, VOCs, and reduction of nitrogen oxides simultaneously is a prime example for the use of precious metals as catalysts [10,11]. Due to high cost and scarcity of precious metals, attention has been focused on metal oxide catalysts for combustion of VOCs. Metal oxides offer another advantage as combustion catalysts, as they have an unlimited potential for modifications. This makes them suitable for applications in highly specific situations, where tailor-made catalysts are required. Supported or unsupported oxides containing single metal or complex formulations containing many metals are used as catalysts for VOCs oxidation.

17.2 VOCs Oxidation Over Metal Oxides

Oxides of transition metals, mainly Cr, Mn, Co, Ni, Fe, Cu, and V are employed in the oxidation of organic compounds. Deep oxidation reactions over these metal oxides are considered to be catalyzed by lattice oxygen. A common feature of these metal oxides is the presence of multiple oxidation states. During catalysis, the metal may be reduced by the hydrocarbon and reoxidized by oxygen. It may cycle between two or more oxidation states thus operating in a redox cycle (Mars–van Krevelen mechanism) [12]. However, the actual mechanism of a working catalyst may involve many steps in a number of consecutive or parallel reactions. Because of its low volatility and low toxicity, MnO_x has received the attention of many researchers.

Manganese oxides assume a wide range of stoichiometries and crystal phases (β-MnO_2, γ-MnO_2, α-Mn_2O_3, γ-Mn_2O_3, α-Mn_3O_4, and Mn_5O_8) where Mn atoms are found in various oxidation states. When heated in air MnO_x undergoes phase transitions. In the temperature range of 500 to 600°C, MnO_2 is converted to Mn_2O_3 and above 890°C to Mn_3O_4 [13]. Mn atoms, depending on the environment, may assume low or high oxidation states. Ability to switch between oxidation states together with defects give rise to the oxygen storage capacity of Mn oxides [14]. Mn oxides are comparatively stable over Al_2O_3 supports due to their low reactivity with Al_2O_3 [15]. Spinel formation ($MnAl_2O_4$), which causes reduction of surface area, resulting low activity, occurs only above $\sim$1000°C [16]. As a result, supported and unsupported MnO_x catalysts are highly stable and have been extensively studied for VOCs oxidation [17,18].

Combustion activity of VOCs over γ-MnO_2 and Pt/TiO_2 was compared by Lahousse et al. [19]. It was found that the activity of γ-MnO_2 for the oxidation of a mixture of VOCs was superior compared with Pt/TiO_2. Moreover, the metal oxide catalyst was found to be less sensitive to the interferences between the VOCs. The presence of water vapor shortened the time required to reach stable activity over γ-MnO_2 catalyst. Carno et al. [20] and Ferrandon et al. [21,22] studied a series of alumina-supported MnO_x catalysts for oxidation of CO, CH_4, C_2H_4, and $C_{10}H_8$. The activity of alumina-supported catalysts depended on the nature of the support as well as on the nature of the phase of Mn oxide. They employed two alumina supports having markedly different surface areas (α-Al_2O_3 5 m^2 g^{-1} and γ-Al_2O_3 50–250 m^2 g^{-1}). Despite the low surface area, Mn oxide supported on α-Al_2O_3 showed high activity for the oxidation of CO. This is attributed to the formation of the Mn_3O_4 phase during the calcination step (600°C) of the Mn/α-Al_2O_3 catalyst. Similar improvement of catalytic activity following thermal treatment of alumina-supported MnO_x has been observed by Tsyrulnikov et al. [23,24]. The Mn_3O_4 phase was not formed over Mn/γ-Al_2O_3 unless the calcination temperature was increased to 1000°C. Also, the Mn/α-Al_2O_3 catalyst displayed high activity for C_2H_4 oxidation. However, when calcined at temperatures below 900°C, CH_4 oxidation activity of Mn/α-Al_2O_3 was found to be lower than that of Mn/γ-Al_2O_3. This is attributed to the availability of a large number of active sites on Mn/γ-Al_2O_3 owing to its high surface area. Compared with the Mn/α-Al_2O_3 catalyst, the surface area of Mn/γ-Al_2O_3 was at least ten times larger. Methane oxidation activity of Mn/α-Al_2O_3 increased when the calcination temperature was increased to 1000°C. As observed by x-ray diffraction (XRD), a complete transformation of Mn_2O_3 phase to Mn_3O_4 was observed following calcination at 1000°C. When calcined at 1000°C, methane oxidation activity over Mn/γ-Al_2O_3 decreased and is attributed to the decrease in surface area of the calcined catalyst. Both catalysts displayed high activity for oxidation of $C_{10}H_8$. It was observed that the temperature for 50% conversion of $C_{10}H_8$ was considerably lower (400°C) than the temperature for 50% conversion of CH_4 (600°C). In their study of propane oxidation over Mn_2O_3 and Mn_3O_4, Finocchio and Busca [25] observed a strong dependence of the catalytic oxidation activity on bulk oxygen mobility. Gallardo-Amores et al. [26] studied MnO_x/TiO_2 catalysts for 2-propanal oxidation. It was found that MnO_x was well dispersed on the surface and part of it was present as Mn^{4+} in bulk TiO_2. The oxidation reaction proceeded by consecutive oxidation of surface adsorbed species. Tahir and Koh [27] studied a series of metal oxides (Mn, Co, Cu, Ce, and Ni) over a SnO_2 support for ethane oxidation. Among the catalysts investigated, MnO_x/SnO_2 and CoO_x/SnO_2 were found to be the most active catalysts. Stability studies of the catalysts showed that MnO_x/SnO_2 is highly stable compared with CoO_x/SnO_2. The low stability arises due to sintering and oxidation state modification of the CoO_x/SnO_2 catalyst. Baldi et. al. [28] studied the activity of Mn_3O_4 for oxidation of C_3 hydrocarbons and oxygenates. They observed propane and traces of ethylene as partial oxidation products. Propane was completely converted to carbon dioxide in a highly oxidizing atmosphere at temperatures above 400°C. A mechanism involving propene and acrolein as intermediates for propane combustion over Mn_3O_4 has been proposed. Incomplete oxidation products, such as propene from propane, CO from propene, acetaldehyde from propanol, propanal, and propanone and carboxylic acids were observed even at high degree of VOCs conversion. The same group reported that the activity of Fe_2O_3 for propane and propene oxidation was low compared with manganese oxide catalysts [29]. Alvarz-Galvan et al. [30] investigated the activity of Al_2O_3-supported MnO_x catalysts (with and without Pd) for formaldehyde and methanol oxidation. The light off temperature decreased from 220 to 90°C when Pd was added to Mn/Al_2O_3. The improved activity of the bimetallic catalyst is attributed to the synergism between supported manganese and palladium phases. This is associated with the ability of the metal oxide (MnO_x) to release oxygen, thereby facilitating the formation of the PdO phase. It is proposed that a fraction of PdO was reduced to Pd and that provided sites for the decomposition of organics.

Oxidation of acetone over Mn oxide/pillared clay catalysts has been investigated by Gandia et al. [31,32]. Complete oxidation of acetone was observed in the temperature range of 337 to 357°C. Further, it was observed that the activity of Al-pillared clays was greater than unpillared clays. Lowest activity was observed with Zr-pillared clays. However, the catalysts synthesized with un-pillared clays were found to be more stable. Lopez et al. [33] studied phenanthrene and isopropanol combustion over Mn–Zr mixed metal oxide catalysts. The high activity of the Mn–Zr catalysts has been attributed to the formation of Mn stabilized tetragonal zerconia solid solution and the presence of Mn(IV) in the structure. It was observed that phenanthrene was completely oxidized over these catalysts and the oxidation of isopropanol produced acetone at low conversions. It was possible to obtain 100% selectivity to carbon dioxide at total conversion of hydrocarbons. Further, it was observed that only 10% of Mn contributed to the formation of the solid solution. Paulis et al. [34] investigated the influence of the surface adsorption–desorption process on the ignition curve of toluene oxidation over Pd/Al_2O_3 and acetone oxidation over Mn_2O_3/Al_2O_3. They reported that deep oxidation occurred by a chain mechanism involving surface adsorbed species. The effect of alkali and acid additives for combustion of methyl–ethyl ketone over Mn_2O_3 has been studied by Gandia et al. [35]. The catalytic activity for deep oxidation improved when Cs^+ or Na^+ was added. Further, it was found that the oxidation is first order with respect to the organic compound and zeroth order with respect to oxygen. Such a situation is possible when the oxygen exchange between gas phase and catalyst is very fast. A study by Zaki et al. [36] has revealed that the use of acidic oxides as supports, such as silica and alumina, has a favorable effect on the oxidation of CH_4 by MnO_x.

Larsson and Andersson [37] investigated a series of catalysts containing Cu, Cu–Ce, Mn, and Mn–Cu supported on Al_2O_3 for oxidation of CO, ethyl acetate, and ethanol. The activity for CO oxidation improved when the alumina support was modified with Ce prior to the deposition of CuO. However, the improvement of activity was not that large for the oxidation of ethyl acetate and ethanol. It was observed that the activity of CuO_x–CeO_2/Al_2O_3 was superior for CO oxidation compared with $CuMn_2O_4$/Al_2O_3. However, the activity for ethyl acetate and ethanol oxidation was higher over $CuMn_2O_4$/Al_2O_3. The phases CuO, $CuAl_2O_4$, and CuO associated with CeO_2 were observed over CuO_x–CeO_2/Al_2O_3. Addition of ceria before copper deposition resulted in a highly dispersed CuO phase and CuO crystallites associated with CeO_2. The enhancement of activity is attributed to high dispersion of CuO and oxygen storage ability of CeO_2. Crystalline $CuMn_2O_4$ and Mn_2O_3 phases were observed over $CuMn_2O_4$/Al_2O_3 catalyst. Solid solutions of CeO–ZrO_2 have also been reported to be active for deep oxidation of methane [38]. Activity of these solid solutions can be further enhanced by doping with MnO_x or CuO_x. Added Mn or Cu dissolves in the solid solution influencing the redox behavior facilitating low temperature reduction of Ce^{4+} and improving the stability for repeated redox cycles. In an investigation of the reducibility of Cu^{2+} over Al_2O_3 and CeO_2/Al_2O_3, Martinez-Arias et al. [39] observed that isolated Cu^{2+} species over an Al_2O_3 support were more easily reduced by CO, compared with clustered Cu^{2+} species. On Cu/CeO_x/Al_2O_3 catalyst, they observed that Cu^{2+} species associated with CeO_x could be reduced at temperatures as low as 200°C. High activity of Cu/CeO_x/Al_2O_3 for CO oxidation is attributed to the easy reduction of Cu^{2+} associated with CeO_x. It has also been reported that the interaction of Cu with Ce improves dispersion of Cu and the activity of Cu containing catalysts [40]. Hutchings et al. [41] have shown that Cu/MnO_x catalysts are active for CO oxidation at ambient temperatures. These studies revealed that the association of Cu with Ce and Mn has a favorable effect on the redox activity of Cu containing catalysts.

Combustion of oxygenated VOCs over CuO catalysts has been studied by Cordi et al. [42]. They proposed that lattice oxygen in CuO is active for deep oxidation and the diffusion of lattice oxygen to the surface is the rate-limiting step. Lattice oxygen is replenished by gas phase oxygen and therefore, the presence of gas phase oxygen is essential for sustained activity. Marion et al. [43] reported that CuO dispersed on Al_2O_3 is active for methane combustion. The activity of the dispersed catalyst is reported to be higher than the activity of CuO in a dilute matrix. They found that low Cu loading on the Al_2O_3 led to better-dispersed ionic CuO while high Cu loading resulted in poorly dispersed catalysts with more covalent character. Catalysts with ionic character were found to be more active in methane combustion. Similar observations have been made by other researchers [44]. A common problem associated with Cu supported on Al_2O_3 catalysts is the formation of $CuAl_2O_4$ phase. When used at high temperatures, CuO reacts with

Al_2O_3 forming $CuAl_2O_4$ [45,46]. This leads to low surface area and therefore, low activity. Several researchers have attempted to circumvent this problem by using $ZnAl_2O_4$ as the catalyst support [47,48]. However, the surface areas of $Cu/ZnAl_2O_4$ catalysts prepared by conventional methods are reported to be low. Catalysts with high surface areas have been obtained by using sol–gel preparation [49]. It has also been reported that the addition of alkali metals improves dispersion of active Cu in Cu containing catalysts [50].

Garbowski et al. [51] studied the activity of Co_3O_4 catalysts supported on Al_2O_3 for catalytic oxidation of methane. They observed that the catalysts were deactivated due to the reaction of the active phase with the support. A stable catalyst could be synthesized using $ZnAl_2O_4$ spinel as the support, where the tetrahedral sites, which could accommodate Co^{2+} are occupied by Zn^{2+}. A study by Finocchio et al. [52] of C_3, C_2, and C_1 oxygenates over Co_3O_4 revealed that the complete oxidation takes place by over oxidation of adsorbed partially oxidized compounds. It is proposed that the oxidation reaction is catalyzed by lattice oxygen anions (O^{2-}) at the oxidized surface. Arnone et al. [53] investigated a series of single metal oxides for complete oxidation of methane. They observed that the activities of Cr, Co, and Mn oxides were comparable and the activity of Fe_2O_3 for methane combustion was low. A study by Perkas et al. [54] revealed that highly dispersed oxides of Co and Fe are more effective for the oxidation of cyclohexane than the zero valent metal. Fe_2O_3 supported on TiO_2 was found to be the most active among the catalysts investigated. The high activity of the catalyst is attributed to the high degree of dispersion and the stabilization of Fe_2O_3 phase over TiO_2 support.

Oxidation of toluene over an alumina supported Cr_2O_3 has been investigated by Younes et al. [55]. They observed that the deep oxidation is catalyzed by a redox mechanism. Further, they found that dissociatively adsorbed oxygen on the surface takes part in the reoxidation of the reduced centers of the catalyst. It was also found that the rate-limiting step is the dissociative adsorption of oxygen. Finocchio et al. [56] compared the activities of Co_3O_4, $MgCr_2O_4$, and CuO for the oxidation of C_3 organics. Among the catalysts, Co_3O_4 and $MgCr_2O_4$ displayed high activities for complete oxidation. They observed that the activity of CuO was quite low at low temperatures and at higher temperatures oxidation over CuO occurred via a redox mechanism. Further, they proposed that the same type of oxygen species are involved in partial and deep oxidation reactions over Co_3O_4 and $MgCr_2O_4$. Wang and Xie [57] in their investigation of Sn–Cr oxide catalysts for CH_4 deep oxidation found that a Sn–Cr catalyst containing equal molar concentrations was the most active. Improved activity of this catalyst is attributed to the presence of high concentrations of active surface oxygen species and high surface area. However, no detailed study is available on the type of oxygen species.

Vanadium oxides are also reported to be active for VOCs oxidation. Ferreira et al. [58] related the oxidation activity of V_2O_5/Al_2O_3 to the content of V^{4+} on the surface of the catalyst. At higher V_2O_5 loadings, high surface concentrations of V^{4+} were observed. Two catalysts containing $VO_x–WO_x/TiO_2$ and MnO_x/TiO_2 have been compared for oxidation of VOCs and chlorinated VOCs [59]. It was found that MnO_x/TiO_2 was more active for VOCs oxidation and $VO_x–WO_x/TiO_2$ was active for chlorinated VOCs oxidation. Further, it was observed that the oxidation activity of MnO_x/TiO_2 decreased when exposed to chlorinated VOCs.

Uranium oxide catalysts have been reported to have high activity for oxidation of volatile organics [60]. Activity for oxidation of short chain linear alkanes improved when Cr was added as a modifier. It is reported that the addition of Cr increases the defects density of U_3O_8 phase. Deep oxidation activity of U_3O_8 was enhanced when 2.6% of water was cofed with VOC [61]. It is proposed that this improvement is due to contributions from the other reaction pathways, such as steam reforming. However, the addition of more water caused the catalytic activity to decrease and similar treatment decreased the VOC oxidation activity of Mn_2O_3.

Ozawa et al. [62] reported that methane combustion activity over PdO/Al_2O_3 catalyst could be improved by addition of Nd_2O_3 and La_2O_3. These metal oxides are capable of generating large quantities of surface oxygen species and are known to be active in oxidative coupling of methane [63–65]. The improved activity is attributed to the stabilization of the PdO phase. Transport of surface oxygen species to the Pd phase may play a key role in the stabilization.

Haber et al. [66] studied V_2O_5, $CuO–Cr_2O_3$ and these oxides supported on γ-Al_2O_3 for oxidation of nitrogen containing VOCs. It was observed that the $CuO–Cr_2O_3/\gamma$-Al_2O_3 catalyst was active for the oxidation of the organic part to CO_2 and CO. However, a large amount of NO_x was formed over this catalyst. Although 100% N_2 selectivity was observed over V_2O_5/γ-Al_2O_3, it produced CO. The difference in catalytic activity/selectivity is attributed to the preferential adsorption of the organic compound on the catalyst surface. A comparative study on the oxidation activity of bulk transition metal oxides has been undertaken by Pradier et al. [67]. Among the metal oxides tested Co_3O_4, Cr_2O_3, CuO, and MnO_2 have been found to be active for n-butane and ethyl acetate oxidation. VOCs oxidation over alumina-supported $Ag–MnO_x$, $Ag–CoO_x$, and $Ag–CeO_x$ catalysts has been examined by Luo et al. [68]. It has been observed that Ag is present as dispersed Ag_2O on the alumina surface and Ag_2O bulk phases. They proposed that hydrogen spill over from Ag facilitated reduction of MnO_x and CoO_x while the oxides of Mn, Co, and Ce promoted oxygen incorporation into the catalyst. Scire et al. [69] proposed that the Ce–O bonds are weakened by Au on an Au/CeO catalyst. The activity of the series $Au/Fe_2O_3 > Ag/Fe_2O_3 > Cu/Fe_2O_3 > Fe_2O_3$ for VOCs oxidation has been explained in terms of the ability of group IB metal to weaken the Fe–O bond [70]. A redox mechanism is proposed for the VOCs oxidation over this catalyst. The catalytic activity of an Au/Fe_2O_3 catalyst was found to be dependent on the pretreatment of the catalyst [71].

The effect of the crystal structure and cation coordination in Ni–Mn oxide catalysts on hydrocarbon oxidation has been investigated by Mehandjiev et al. [72]. They reported that the ilmenite type ($NiMnO_3$) is significantly more active for oxidation of ethyl acetate and benzene, compared with the spinel ($Ni_xMn_{3-x}O_4$) type. Further, it was revealed that ilmenite type $NiMnO_3$ with good crystallinity was the best catalyst for CO oxidation, probably due to face preferential adsorption of CO on the crystalline phases.

17.2.1 Perovskite Oxides in VOCs Oxidation

Perovskites have been extensively studied for VOC oxidation due to their high catalytic activity and stability. Perovskites are represented by the general formula, $ABO_{3\pm\delta}$. The cation A is typically a large metal ion surrounded by 12 oxygen ions and the cation B is a small transition metal ion surrounded by 6 oxygen ions. Both A and B could be partially substituted with other ions. A high degree of electronic and structural defects can be created by substituting the positions of A and B with ions having different valancies. Perovskites are therefore characterized by high ionic conductivity and oxygen mobility [73–79], which give rise to interesting catalytic properties. Experimental [78–86] as well as theoretical [76,77,86–89] investigations on perovskites have yielded valuable information. Many recent reviews discuss catalytic applications of perovskite oxides [73–75].

Oxidation reactions on perovskite oxides have been proposed to be catalyzed by surface oxygen and lattice oxygen [70,81]. At temperatures below $\sim 400°C$, oxidation reactions are mainly catalyzed by surface oxygen and above $400°C$ a major contribution to the reaction comes from lattice oxygen. Therefore, at high temperatures oxygen mobility in the crystal lattice determines the efficiency of oxidation reactions. Combustion activity of perovskites strongly depends on the type of metal ions present in the structure. Mn, Co, Cr, and Fe at "B" position have been extensively studied for combustion reactions. On the "A" site, La is the ion that has been most investigated. Partial substitution of the metal ions at sites A and B drastically modify the catalytic activity of perovskite oxides [91–96].

Methane combustion over $LaMn_{1-x}Mg_xO_3$ and $LaCr_{1-x}Mg_xO_3$, ($x = 0$–0.5) has been investigated by Saracco et al. [97,98]. Authors proposed an Eley–Redial mechanism, where oxygen is dissociatively chemisorbed on active sites and methane reacts from the gas phase. It was also found that the formation of MgO phase in $LaCr_{1-x}Mg_xO_3$ decreased sintering of the perovskite and Mg present in the crystal lattice improved catalytic activity. The improvement of catalytic activity is attributed to the formation of active sites for hydrogen abstraction from methane, which is supposed to be the rate-limiting step in the process of catalytic combustion. Substitution of Mg^{2+} for Cr^{3+} in the perovskite lattice increases the effective charge of the neighboring chromium ions in order for charge compensation. Oxidized chromium reacts with methane generating Cr^{3+} and an oxygen vacancy. Gas phase oxygen replenishes the oxygen vacancies. Kinetic studies suggested that the increased activity was due to the increase in number of active sites and not due to the increased activity per site. This conclusion is based on the fact that Mg doping did not change the activation energy for methane combustion. A similar mechanism has been proposed for methane combustion over $La_{1-x}A_xMnO_3$ where A = Sr, Eu, or Ce [99]. In $LaMnO_3$, a fraction of Mn is already oxidized to Mn^{4+} and charge compensation is established by the creation of metal ion vacancies. Therefore, when a fraction of La^{3+} is substituted by a bivalent ion, the number of metal ion and oxygen vacancies decreases. This results in a decrease in the activity of methane combustion. The reverse happens when a fraction of La^{3+} is substituted by a four valent metal ion. Therefore, methane combustion activity increases when La^{3+} is partly substituted by Ce^{4+}. Presence of Mn^{4+} in Mn-containing perovskites has also been reported by other researchers [100].

Blasin-Aube et al. [101] investigated $La_{0.8}Sr_{0.2}MnO_{3+x}$ type perovskites for combustion of a series of compounds representing alkanes, alkenes, aromatics, and oxygenates. The order of activity varied as propene > methylcyclohexane > toluene > cyclohexane > hexane > benzene > propane. They observed that all the nonoxygenated compounds except cyclohexane could be converted completely to carbon dioxide while benzene was formed when cyclohexane was oxidized. It is proposed that the formation of benzene occurs via successive dehydrogenation of adsorbed cyclohexane. This indicates that cyclohexane is well adsorbed and stays long enough for progressive dehydrogenation. Significant quantities of partially oxygenated byproducts were observed when oxygenated compounds except acetone and propan-2-ol were oxidized. Complete oxidation to CO_2 could be achieved when the temperature was increased above 350°C where all the parent compounds and the byproducts were oxidized. Unlike in partial oxidation, strong adsorption of the organic molecule is preferred for deep oxidation. It is proposed that the rate-limiting step in hydrocarbon oxidation is the breaking of the weakest C–H bond [28,102,103]. This should be manifested as a correlation between the activity and C–H bond enthalpy. Activity, measured as the temperature for 50% conversion of the organic compound is correlated with bond enthalpy for alkanes and aromatics [101]. However, oxygenates lacked such a correlation due to direct reaction of the functional group on the catalyst surface [28,101,104]. Lee et al. [105] studied the effect of K substitution for La in $La_{1-x}K_xMnO_{3+\delta}$ for oxidation of ethane. It was observed that ethane oxidation activity decreased with increasing K content. As the amount of K increased, the ratio of Mn^{4+}/Mn^{3+} at the surface region increased. Authors found that the catalytic activity is correlated with the amount of non-stoichiometric oxygen, characterized by a peak in the temperature programmed oxygen desorption profile of the catalyst at 925 K [105,106]. Similar observations have been made by Yamazoe and Teraoka [107]. Catalytic combustion of C_3 organics over perovskite oxides has been studied by Busca et al. [104]. Authors proposed that lattice oxygen is involved in oxidation reactions. Daturi et al. [108] studied a series of perovskites of composition $SrTi_{1-x-y}Zr_xMn_yO_3$ for methane combustion. It was found that Mn containing catalysts were active for CH_4 combustion. In these catalysts, surface acid sites have been proposed to be involved in C–H bond activation.

Methanol and ethanol combustion activity over Ag modified $La_{0.6}Sr_{0.4}MnO_3$ catalysts has been studied by Wang et al. [109]. Both bulk and surface phases are found to be involved in the catalysis. Authors proposed that the surface oxygen species (O_2^{2-} and O^-) are involved in replenishing lattice oxygen vacancies. Ciambelli et al. [110,111] evaluated the activity of $AFeO_3$, $AMnO_3$ (A = La, Nd, Sm), $LaFe_{1-x}Mg_xO_3$, and $Sm_{1-x}Sr_xMnO_3$ perovskites for CH_4 oxidation. It was observed that the activity of $AFeO_3$ varied in the order of La > Nd > Sm. The same order of activity was observed for the series of $AMnO_3$ perovskites. The activity of $LaFe_{1-x}Mg_xO_3$ decreased with increasing Mg content and the activity of $Sm_{1-x}Sr_xMnO_3$ decreased with increasing Sr content. Authors proposed that, with increasing Sr content, Mn(IV) is reduced to Mn(II) hindering the redox cycle, with an accompanying reduction of catalytic activity. Ponce et al. [96] studied $La_{1-x}Sr_xMnO_3$ for combustion of methane. The catalytic activity was attributed to the stabilization of Mn^{4+} state and the presence of excess oxygen in the perovskite.

Marti et al. [112] studied $La_{0.8}Sr_{0.2}MnO_{3+x}$ perovskites supported on spinels MAl_2O_4 (M = Mg, Ni, Co) for combustion of methane. High activity was observed with a $MgAl_2O_4$ supported catalyst and this is attributed to its high surface area. Zirconia has also been used as supports for oxidation catalysts. Cimino et al. [86] investigated a zirconia supported $LaMnO_3$ catalyst for methane oxidation. The authors found that the catalysts containing crystalline $LaMnO_3$ are highly active for methane oxidation. Catalysts containing low perovskite loadings displayed low activity for methane oxidation.

Methane oxidation kinetics over $LaCr_{1-x}Ni_xO_3$ ($x = 0$ to 1) perovskites has been studied by Stojanovic et al. [113]. The catalytic activity was found to be proportional to x. It was also observed that nickel could be reduced to the metallic form by starving the catalyst of oxygen. A redox process involving surface Ni–O–Ni species in methane oxidation has been proposed.

Ferri and Forni [114] examined a series of perovskites $La_{1-x}Sr_xCoO_3$, $La_{1-x}Eu_xCoO_3$, $La_{1-x}Ce_xCoO_3$, $La_{1-x}Sr_xNiO_3$, and $La_{1-x}Sr_xFeO_3$ for methane oxidation. In these perovskites, a fraction of cobalt exists as Co^{4+} and the charge balance is achieved by the generation of metal ion vacancies. Therefore, the substitution of La by a four valent metal ion (Ce^{4+}) decreases the defect concentration and the substitution by a bivalent metal ion (Sr^{2+} or Eu^{2+}) increases the defects concentration. Methane combustion activity over these catalysts varied accordingly. Forni and Rossetti [115] investigated the activity of $La_{0.9}Ce_{0.1}CoO_{3\pm\delta}$ high surface area material prepared by flame hydrolysis for methane combustion. This method of preparation resulted in a catalyst with high bulk oxygen mobility. The catalyst was found to be significantly active for methane combustion at temperatures as low as 400°C. The activity increased markedly when the temperature was increased. It is proposed that a suprafacial mechanism is operative below 400°C and an intrafacial mechanism (involving bulk oxygen) is operative at high temperatures [107]. In an investigation of methane combustion over the system of $LaCoO_3$–CeO_2–CoO_4, Kirchnerova et al. [116] proposed that phase segregation and phase corporation can occur. Belessi et al. [117] examined methane oxidation activity over a series of catalysts containing La and Fe, where La was partially substituted to give formulae $La_{1-y}Ce_yFeO_3$, $La_{1-x}Sr_xFeO_3$, and $La_{1-x-y}Sr_xCe_yFeO_3$. It was observed that the activity could be significantly enhanced when La was substituted by Sr and Ce. The enhancement of activity is ascribed to the synergistic action of CeO_2 and $SrFeO_{3-x}$ crystal phases. An investigation by Choudhary et al. [118] revealed that the partial substitution of La by Ag enhanced the methane combustion activity of $LaFe_{0.5}Co_{0.5}O_3$ catalysts.

Various preparational techniques have been employed in order to synthesize catalysts possessing improved activity for VOCs oxidation. Highly active perovskite oxides have been prepared by reactive grinding [119]. Total VOCs oxidation has been achieved at high space velocities with a perovskite-based catalytic membrane reactor [120].

17.2.2 Application of Hexaaluminate Materials

Hexaaluminates represent a class of materials that are highly resistant to sintering at high temperature. Therefore, these materials have attracted the attention of researchers who are involved in developing catalysts for high temperature applications. Hexaaluminates have been used as supports as well as the active material in catalytic combustion reactions [121–142]. Hexaaluminates can be represented by the formula $AAl_{12}O_{19}$ where A is an alkaline or alkaline-earth metal. They consist of a lamellar structure and both A cation and Al can be partly substituted by other cations. Incorporation of other cations drastically modify the catalytic activity of these materials. However, such modifications are rather limited compared to the possibilities existing in perovskite type oxides.

Duart et al. studied Cu, Mn [132], and Fe, Mn [133] substituted Ba-hexaaluminates for the combustion of methane. In both cases, the catalytic activity increased with increasing Mn content and directly correlated with the Mn^{3+}/Mn^{2+} ratio. Further, they observed that the Cu-sites were more active for methane oxidation than Mn-sites. However, the Cu substituted catalysts were characterized by low surface area. Many investigations have been carried out on the use of hexaaluminates for high temperature combustion of methane [129,134,135]. Lietti et al. [136] reported that among a series of hexaaluminate catalysts containing Mn, Cr, Co, Ni, and Fe, the catalysts containing Mn and Fe displayed the highest activity for the combustion of methane. Machida et al. [138,139] investigated the effect of surface modifications of hexaaluminate with Mn_3O_4 on the catalytic activity. The authors observed that the catalytic activity of Mn_3O_4/hexaaluminate derived by surface air oxidation is superior compared with the catalysts prepared by conventional evaporation to dryness method. The high activity is attributed to the partial substitution of Mn in the hexaaluminate structure at the surface region. In a study of the catalytic activity for combustion of methane over MgO and Mn substituted hexaaluminate catalysts, Berg and Jaas [140] observed that Mn substituted hexaaluminates are highly active. The temperature for 50% methane conversion is 795°C for MgO while for the hexaaluminate catalyst it is 640°C. Further, they observed that the catalytically initiated homogeneous gas phase reactions contributed significantly to the conversion of methane. Barium hexaaluminates containing CeO_x have also been reported to be active for VOCs combustion. Zarur and Ying [141] prepared a highly active CeO_x deposited nanocrystalline barium hexaaluminate catalyst using microemulsion technique. Total combustion of methane over this catalyst was achieved at 600°C. The authors observed that the catalyst remained stable after operation at temperatures exceeding 1000°C.

17.2.3 Oxidation of Halogen Containing VOCs

Catalysts suitable for the oxidation of halogenated VOCs require high stability. Corrosive action of halides and hydrolyzed products (HCl and HF) tend to destroy the catalytic activity. Formation of volatile or soluble metal chlorides deactivates the catalysts. Metal oxides including the oxides of Mn, Cr, V, and Ce have been extensively investigated in this regard.

MnO_x catalysts for oxidation of chloroaromatics have been investigated by Liu et al. [142]. Highly dispersed MnO_x on TiO_2 support was found to be active for the oxidation of chlorobenzene. Highly dispersed MnO_x over TiO_2 is reported as the precursor for the active phase. The active phase consists of chlorinated manganese oxides (MnO_yCl_z) and is present under working conditions. Padilla et al. [143] studied some commercial chromia-based catalysts for oxidation of chlorinated hydrocarbons. They reported that although highly active, chromium is lost from the catalyst due to attack of chlorine produced during catalytic oxidation. Activity of TiO_2–SiO_2-supported CrO_2 and V_2O_5 for trichloroethane has been compared with the activity over Pt or Pd on the same support [144]. It was revealed that, although the initial activity of the two oxides was superior, they deactivated quickly due to the loss of the active component. Gervasini et al. [145] examined the activities of Cu–Cr/Al_2O_3 and MnO_x for carbon tetrachloride oxidation in the presence and absence of hydrogen supplying compounds, such as n-hexane and toluene. When present in low concentrations, both n-hexane and toluene improved CCl_4 combustion activity. They inhibited the reaction when present at high concentrations. It was also observed that preionization of CCl_4 led to higher conversion. Further, it was found that MnO_x was more active but less selective to CO_2 compared with Cu–Cr/Al_2O_3. Krishnamoorthy and Amiridis [146] investigated o-dichlorobenzene oxidation activity over a V_2O_5/Al_2O_3 catalyst. They reported that the benzene ring remained intact during the adsorption step and partial oxidation products were generated on the catalyst surface. TiO_2–ZrO_2 mixed oxide has been reported to be active for oxidation of dichlorodifluoromethane [147]. Treatment with sulfuric acid further improved the catalytic activity and selectivity to CO_2. Oxidation of allyl iodide on the period IV metal oxides has been studied by Doornkamp et al. [148]. They observed high activity for MnO_2, Co_3O_4, and CuO. The oxidation activity is correlated with the metal oxygen bond strength, where the oxides with weakest metal–oxygen bonds were found to be the most active.

Perovskite oxides have also been investigated for the oxidation of halogenated VOCs. Catalytic decomposition of chlorinated C_1 VOCs has been studied by Sinquin et al. [149]. $LaCoO_3$ and $LaMnO_x$ catalysts were found to be highly active for the said reaction. A mechanism involving hydrolysis of C–Cl and oxidation of C–H has been proposed. CCl_4 reacts mainly by hydrolysis while $CHCl_3$, CH_2Cl_2 and CH_3Cl react in two steps: hydrolysis and oxidation. In an investigation of oxidation of chlorinated C_2 compounds over $LaMnO_{3+\delta}$ perovskites, it was observed that water vapor promoted the oxidation reaction [150]. Further, it was observed that the presence of oxygen is essential in order to preserve the $LaMnO_{3+\delta}$ structure. Oxidation of chloromethane, dichloromethane, and 1-2-dichloroethane over $LaMnO_3$, $LaCoO_3$, and $La_{0.84}Sr_{0.16}Mn_{0.67}Co_{0.33}O_3$ has been studied by Kiebling et al. [151]. In addition to CO_2, H_2O, and HCl, higher chlorinated, C–C coupled and cracked products have been observed. When exposed to chlorinated compounds, a reversible catalyst deactivation has been observed for the other catalysts except $LaCoO_3$.

17.3 Surface Oxygen Species and their Role in VOCs Oxidation

Solid oxide catalysts are capable of generating oxygen species (O_2, O, O_2^-, O_2^{2-}, O_3^-, and O^-) on the surface [12,152–166]. Their reactivity, exchangeability, and efficiency of regeneration play a key role in determining the activity of metal oxides. Electron paramagnetic resonance, x-ray photoelectron spectroscopy, and temperature programmed oxygen desorption techniques have been applied to characterize such oxygen species. Surface oxygen species, such as O^- and O_2^{2-}, have been proposed to be active for hydrogen abstraction from methane during oxidative coupling [63,65,153–156]. In the process, methyl radicals are generated, emanated to the gas phase, and coupled in the gas phase to produce ethane. A fraction of methyl radicals undergoes complete oxidation on the surface of the catalyst. Electrophilic forms of oxygen species weakly adsorbed on metal oxide surfaces are reported to be active for complete oxidation [161,162]. Hydrocarbon molecules adsorbed on metal oxide surface produce carbon oxides via consecutive oxidation steps and such reactions are facilitated when the surface concentration of oxygen is high [154,155]. Therefore, in complete oxidation reactions, adsorbed oxygen species play an important role in oxidizing surface carbon and producing a cleaner surface for continued catalytic activity. In addition, surface oxygen ions are capable of reoxidizing reduced metal ions [107,109]. In a study of oxidative coupling activity of SrO doped Nd_2O_3, Gayko et al. [156] reported that surface oxygen species are converted to lattice oxygen.

Since the other hydrocarbons are more adsorptive on oxide surfaces compared with methane, in principle, a catalyst active for deep oxidation of methane must also be active for the oxidation of other hydrocarbons. However, there is a possibility that the oxidation reaction may not be sustained due to strong adsorption of the hydrocarbon resulting in a surface deprived of oxygen. It may be possible to mitigate this situation by introducing metal oxides having high oxygen affinity as oxygen storing compounds. Supported metal oxide catalysts are often supplemented with such compounds where Mn, Ce, and lanthanide oxides have been the choice of many.

17.4 Carbon Deposition and VOCs Oxidation Activity

Carbon deposition studies have been undertaken primarily on metal catalysts. The mechanisms by which carbon deposits are formed and how they influence the catalytic activity have been extensively investigated [163–170]. Reaction of hydrocarbons over metal surfaces lay down carbon. These carbon deposits contain both C and H. The C/H ratio varies with reaction temperature and the type of the metal [163]. In general, the C/H ratio increases with the increasing temperature.

Carbon deposition has been a major concern during metal catalyzed hydrogenation, dehydrogenation, hydrogenolysis, and oxidation reactions [163–170]. Among the transition metals, Fe, Co, and Ni have been reported to be very prone to carbon deposition [163–167,169,171]. Since oxides of these metals and perovskites containing them are largely employed for oxidation of volatile organics, it is worthwhile examining the chemistry of carbon deposition over these metals in detail. It is generally accepted that the carbon deposition occurs with ease over flat metal surfaces where the metal coordination is high compared with edges and corners with low coordination [163,164,172]. Compared with metals, the corresponding oxides have less propensity for carbon deposition. However, when reduced to the zero valent state, they undergo severe carbon deposition as evidenced by a large number of studies carried out with NiO catalysts [170,171,173–176]. Many researchers have attempted to reduce carbon deposition over NiO-containing catalysts with various modifications. Qing et al. [170] reported that doping of NiO/Al_2O_3 with alkali metals, such as Li, Na, K, or oxides of La, Ce, Y, and Sm improved resistance to carbon deposition during partial oxidation of methane to synthesis gas. These additives also improved the dispersion of Ni and the thermal stability of the catalyst. During syn gas production from methane Battle et al. [177] observed that no carbon was formed on highly active $Ba_3NiRuTaO_9$ perovskite, however, significant carbon formation was observed on a supported Ru metal catalyst. Further, they found that the bulk perovskite structure remained intact. In an investigation of carbon formation during the oxidation of cyclohexane over a series of Al_2O_3 supported catalysts, Hettige et al. [173] reported that the surfaces of MnO_x/Al_2O_3 and CeO/Al_2O_3 catalysts remain relatively free of carbon under catalytic conditions. This is attributed to the high oxygen storage capacity of the MnO_x and CeO_x phases [14]. It is proposed that the surface carbon species are efficiently oxidized over these two catalysts. Therefore, in the synthesis of oxidation catalysts, it is important to maintain a high degree of dispersion together with oxygen storage ability in order to have sustained catalytic activity and resistance to carbon deposition. Attempts have been made to achieve resistance to carbon deposition in many cases by adding Mn, Ce, or lanthanide oxides to the catalysts [14,173,178].

Another important aspect in the reduction of carbon formation is the stabilization of metal oxide phases, which show high resistivity for carbon buildup. Hu and Ruckenstein [174,175] investigated carbon deposition over NiO catalysts during methane oxidation. They reported that CH_4 is completely oxidized to CO_2 over pure NiO. Methane oxidation over NiO/MgO is proposed to occur via two steps. A fast step involving surface oxygen of NiO and a slow step by lattice oxygen bonded to Ni atoms. It was reported that the surface region of the NiO/MgO catalyst consists of a NiO–MgO solid solution, which plays an important role in resisting the reduction of NiO to metallic Ni. The formation of the solid solution results in a high degree of dispersion of NiO and resistance to sintering [175]. Jayathilaka et al. [176] reported that the addition of MgO to $NiO/\alpha\text{-}Al_2O_3$ dramatically reduced carbon deposition while maintaining high activity for complete oxidation. It was observed that the NiO phase was stabilized when MgO was present. It is proposed that the formation of the NiO–MgO solid solution may be the reason for the improved resistance for carbon deposition. Formation of the solid solution prevents the reduction of NiO and aggregation into large particles. Formation of large aggregates has been recognized as a favorable condition for carbon deposition. In general, improved stability of metal oxide phases and thereby resistance to carbon formation can be achieved by using mixed metal oxides.

17.5 Concluding Remarks

Activity of metal oxides for VOCs oxidation is equal or superior to that of precious metal catalysts. Many metal oxides, which are active for VOCs oxidation posses high oxygen storage capacity and high bulk oxygen conductivity. They are capable of fast exchange of oxygen from the gas phase. Many of these metal oxides, including perovskite oxides, are capable of generating surface oxygen species. It has been observed that these oxygen species are involved in oxidation of surface adsorbed organic molecules and in filling lattice oxygen vacancies created during VOCs oxidation. Surface oxygen species play a key role in low temperature oxidation of VOCs. Oxygen storage ability together with fast oxygen exchange of metal oxides facilitates efficient oxidation of surface adsorbed organic molecules and resists build up of carbon on the surface. Metal oxides are also used as supports for precious metal containing catalysts; their role being the efficient supply of oxygen for oxidation reactions. It has been proposed that the presence of precious metals weaken the metal–oxygen bonds in the support oxide, thereby enhancing the supply of oxygen for catalysis.

Catalyst systems based on metals or metal oxides generally suffer from deactivation at temperatures above $\sim$700°C owing to loss of active species and sintering. Catalysts that use oxides of nonmetals especially in the form of complex oxides, such as aluminates, seem more promising. Among them, the catalyst systems based on barium hexaaluminates have received much attention. Metal ions incorporated in the hexaaluminate structure and metal oxides deposited on hexaaluminate crystallites both display high temperature stability. Oxidation of halogenated VOCs, demands highly stable yet highly active catalysts. Although, many metal oxide catalyst systems display promising activities, they suffer from deactivation due to the reaction with the products formed during combustion. This often leads to the loss of the active component from the catalyst.

References

1. Bunce, N. *Environmental Chemistry*. Wuerz Publishing Ltd. 1991, pp. 1–115.
2. Elsom, D.M. *Atmospheric Pollution, A Global Problem*, 2nd ed. Blackwell Publishers: New York, 1992, p. 147.
3. Papaefthimiou, P., Ioannides, T., and Verykios, X.E. Combustion of non-halogenated volatile organic compounds over group (VIII) metal catalysts. *Appl. Catal. A* 1997, *13*, 175–184.

4. Choudhary, T.V., Banerjee, S., and Choudhary, V.R. Catalysts for combustion of methane and lower alkanes. *Appl. Catal. A* 2002, *234*, 1–23.
5. Forni, L. and Rossetti, I. Catalytic combustion of hydrocarbons over perovskites. *Appl. Catal. B: Environ.* 2002, *38*, 29–37.
6. Kummer, J.T. Catalysts for automobile emission control. *Prog. Energy Combust. Sci.* 1980, *6*, 177–199.
7. Subramanian, S., Kudla, R.J., and Chatta, M.S. Removal of methane from compressed natural gas fueled vehicle exhaust. *Ind. Eng. Chem. Res.* 1992, *31*, 2460–2465.
8. Tahir, S.F. and Koh, C.A. Catalytic oxidation for air pollution control. *Environ. Sci. Pollut. Res.* 1996, *3*, 20–23.
9. Papaefthimiou, P., Ioannides, T., and Verykios, X.E. Performance of Pt/TiO$_2$ (W^{6+}) catalysts for combustion of volatile organic compounds (VOCs). *Appl. Catal. B: Environ.* 1998, *15*, 75–92.
10. Koltsakis, G.C. and Stamatelos, A.M. Catalytic automobile aftertreatment. *Prog. Energy Combust. Sci.* 1997, *23*, 1–39.
11. Heck, R.M. and Farrauto, R.J. Automobile exhaust catalysts. *Appl. Catal. A* 2002, *221*, 443–457.
12. Gellings, P.J. and Bouwmeester, H.J.M. Solid state aspects of oxidation catalysts. *Catal. Today* 2000, *58*, 1–53.
13. Reidies, A.H. In *Ullmann's Encyclopedia of Industrial Chemistry*, Vol. A16, 5th ed. Elvers, B., Hawkings, S., Schulz, G., Eds. VCH: New York, 1986, pp. 123–129.
14. Chang, Y. and McCarty, J.G. Novel oxygen storage components for advanced catalysts for emission control in natural gas fueled vehicles. *Catal. Today* 1996, *30*, 163–170.
15. Strohmeier, B.R. and Hercules, D.M. Surface spectroscopic characterization of Mn/Al$_2$O$_3$ catalysts. *J. Phys. Chem.* 1984, *88*, 4922–4929.
16. Ranganathan, T., MacKean, B.E., and Muan, A. The system manganese oxide-alumina in air. *J. Am. Ceram. Soc.* 1962, *45*, 279–281.
17. van de Kleut, D. On the Preparation and Properties of Manganese Oxide Based Combustion Catalysts. Ph.D. thesis, University of Utrecht, The Netherlands, 1994.
18. Kalantar, N.A. and Lindfors, L.E. Catalytic clean-up of emissions from small scale combustion of biofuels. *Fuel* 1998, *77*, 1727–1734.
19. Lahousse, C., Bernier, A., Grange, P., Delmon, B., Papaefthimiou, P., Ioannides, T., and Verykios, X. Evaluation of γ-MnO$_2$ as a VOC removal catalyst: comparison with a noble metal catalyst. *J. Catal.* 1998, *178*, 214–225.
20. Carno, J., Ferrandon, M., Bjornbom, E., and Jaras, S. Mixed manganese oxide/platinum catalysts for total oxidation of model gas from wood boilers. *Appl. Catal. A* 1997, *155*, 265–281.
21. Ferrandon, M., Carno, J., Jaras, S., and Bjornbom, E. Total oxidation catalysts based on manganese or copper oxides and platinum or palladium, I. Characterization. *Appl. Catal. A* 1999, *180*, 141–151.
22. Ferrandon, M., Carno, J., Jaras, S., and Bjornbom, E. Total oxidation catalysts based on manganese or copper oxides and platinum or palladium, II. Activity, hydrothermal stability and sulphur resistance. *Appl. Catal. A* 1999, *180*, 153–161.

23. Tsyrulnikov, P.G., Salnikov, V.S., Drozdov, V.A., Stuken, S.A., Bubnov, A.V., Grigorov, E.I., Kalinkin, A.V., and Zaikovskii, V.I. Investigation of thermal activation of aluminum–manganese total oxidation catalysts. *Kinet. Catal.* 1991, *32*, 439–446.

24. Tsyrulnikov, P.G., Kovalenko, O.N., Gogin, L.L., Starostina, T.G., Noskov, A.S., Kalinkin, A.V., Krukova, G.N., Tsybulya, S.V., Kudrya, E.N., and Bubnov, A.V. Behavior of some deep oxidation catalysts under extreme conditions. I. Comparison of resistance to thermal shock and SO_2 poisoning. *Appl. Catal. A* 1998, *167*, 31–37.

25. Finocchio, E. and Busca, G. Characterization and hydrocarbon oxidation activity of coprecipitated mixed oxides Mn_3O_4/Al_2O_3. *Catal. Today* 2001, *70*, 213–225.

26. Gallardo-Amores, J.M., Armaroli, T., Ramis, G., Finocchio, E., and Busca, G. A study of anatase-supported Mn oxide as catalysts for 2-propanol oxidation. *Appl. Catal. B: Environ.* 1999, *22*, 249–259

27. Tahir, S.F. and Koh, C.A. Catalytic oxidation of ethane over supported metal oxide catalysts. *Chemosphere* 1997, *34*, 1787–1793.

28. Baldi, M., Finocchio, E., Milella, F., and Busca, G. Catalytic combustion of C_3 hydro-carbons and oxygenates over Mn_3O_4. *Appl. Catal B: Environ.* 1998, *16*, 43–51.

29. Baldi, M., Escribano, V.S., Amores, J.M.G., Milella, F., and Busca, G. Characterization of manganese and iron oxides as combustion catalysts for propane and propene. *Appl. Catal B: Environ.* 1998, *17*, L175–L182.

30. Alvarz-Galvan, M.C., de la Pena O'Shea, V.A., Fierro, J.L.G., and Arias, P.L. Alumina-supported manganese and manganese-palladium oxide catalysts for VOCs combustion. *Catal. Commun.* 2003, *4*, 223–228.

31. Gandia, L.M., Vicente, M.A., and Gil, A. Complete oxidation of acetone over manganese oxide catalysts supported on alumina and zirconia-pillared clays. *Appl. Catal. B: Environ.* 2002, *38*, 295–307.

32. Gil, A., Gandia, L.M., and Vincente, M.A. Recent advances in the synthesis and catalytic applications of pillared clays. *Catal. Rev.-Sci. Eng.* 2000, *42*, 145–212.

33. Lopez, E.F., Escribano, V.S., Resini, C., Gallardo-Amores, J.M., and Busca, G. A study of coprecipitated Mn–Zr oxides and their behavior as oxidation catalyts. *Appl. Catal. B: Environ.* 2001, *29*, 251–261.

34. Paulis, M., Gandia, L.M., Gill, A., Sambeth, J., Odriozola, J.A., and Montes, M. Influence of the surface adsorption-desorption processes on the ignition curves of volatile organic compounds (VOCs) complete oxidation over supported catalysts. *Appl. Catal. B: Environ.* 2000, *26*, 37–46.

35. Gandia, L.M., Gil, A., and Korili, S.A. Effects of various alkali-acid additives on the activity of a manganese oxide in the catalytic combustion of ketones. *Appl. Catal. B: Environ.* 2001, *33*, 1–8.

36. Zaki, M.I., Hasan, M.A., Pasupulety, L., Fouad, N.E., and Knozinger, H. CO and CH_4 total oxidation over manganese oxide supported on ZrO_2, TiO_2, TiO_2–Al_2O_3 and SiO_2–Al_2O_3 catalysts. *New J. Chem.* 1999, *23*, 1197–1202.

37. Larsson, P. and Andersson, A. Oxide of copper, ceria promoted copper, manganese and copper manganese on Al_2O_3 for the combustion of CO, ethyl acetate and ethanol. *Appl. Catal. B: Environ.* 2000, *24*, 175–192.

38. Terribile, D., Trovarelli, A., de Leitenburg, C., Primavera, A., and Dolcetti, G. Catalytic combustion of hydrocarbons with Mn and Cu-doped ceria–zirconia solid solutions. *Catal. Today* 1999, *47*, 133–140.

39. Martinez-Arias, A., Cataluña, R., Conesa, J.C., and Soria, J. Effect of copper-ceria interactions on copper reduction in a $Cu/CeO_2/Al_2O_3$ catalyst subjected to thermal treatments in CO. *J. Phys. Chem. B* 1998, *102*, 809–817.

40. Dow, W., Wang, Y., and Huang, T. TPR and XRD studies of yitria-doped ceria/γ-alumina-supported copper oxide catalyst. *Appl. Catal. A* 2000, *190*, 25–34.

41. Hutchings, G.J., Mirzaei, A.A., Joyner, R.W., Siddiqui, M.R.H., and Taylor, S.H. Effect of preparation conditions on the catalytic performance of copper manganese oxide catalysts for CO oxidation. *Appl. Catal. A: Gen.* 1998, *166*, 143–152.

42. Cordi, E.M., O'Neill, P.J., and Falconer, J.L. Transient oxidation of volatile organic compounds on a CuO/Al_2O_3 catalyst. *Appl. Catal. B: Environ.* 1997, *14*, 23–36.

43. Marion, M.C., Garbowski, E., and Primet, M. Physicochemical properties of copper-oxide loaded alumina in methane combustion. *J. Chem. Soc., Faraday Trans.* 1990, *86*, 3027–3032.

44. Park, P.W. and Ledford, J.S. The influence of surface structure on the catalytic activity of alumina supported copper oxide catalysts — Oxidation of carbon monoxide and methane. *Appl. Catal. B: Environ.* 1998, *15*, 221–231.

45. Strohmeier, B.R., Leyden, D.E., Field, R.S., and Hercules, D.M. Surface spectroscopic characterization of Cu/Al_2O_3 catalysts. *J. Catal.* 1985, *94*, 514–530.

46. Tsikoza, L.T., Torasova, D.V., Ketchik, S.V., Maksimov, N.G., and Papovskii, V.V. Physicochemical and catalytic properties of copper–aluminum oxide catalysts. *Kinet. Katal.* 1981, *22*, 1022–1027.

47. Marion, M.C., Garbowski E., and Primet, M. Catalytic properties of copper oxide supported on zinc aluminate in methane combustion. *J. Chem. Soc., Faraday Trans.* 1991, *87*, 1795–1800.

48. Artizzu, P., Garbowski, E., Primet, M., Brulle, Y., and Saint-Just, J. Catalytic combustion of methane on aluminate-supported copper oxide. *Catal. Today* 1999, *47*, 83–93.

49. Guilhaume, N. and Primet, M. Catalytic combustion of methane:copper-oxide supported on high specific area spinels synthesized by a sol–gel process. *J. Chem. Soc., Faraday Trans.* 1994, *90*, 1541–1546.

50. Cheng, W., Shiau, C., Liu, T.H., Tung, H.L., Lu, J., and Hsu, C.C. Promotion of Cu/Cr/Mn catalysts by alkali additives in methanol decomposition. *Appl. Catal. A* 1998, *170*, 215–224.

51. Garbowski, E., Guenin, M., Marion, M.C., and Primet, M. Catalytic properties and surface states of cobalt containing oxidation catalysts. *Appl. Catal. A* 1990, *64*, 209–224.

52. Finocchio, E., Busca, G., Lorenzelli, V., and Escribano, V.S. FTIR studies on the selective oxidation and combustion of light hydrocarbons at metal oxide surfaces. Part 2. Propane and propene oxidation on Co_3O_4. *J. Chem. Soc., Faraday Trans.* 1996, *92*, 1587–1593.

53. Arnone, S., Bagnasco, G., Busca, G., Lisi, L., Russo, G., and Turco, M. Catalytic combustion of methane over transition metal oxides. *Stud. Surf. Sci. Catal.* 1998, *119*, 65–70.

54. Perkas, N., Koltypin, Y., Palchik, O., Gedanken, A., and Chandrasekaran, S. Oxidation of cyclohexane with nonsaturated amorphous catalysts under mild conditions. *Appl. Catal. A* 2001, *209*, 125–130.
55. Younes, M.K., Ghorbel, A., Rives, A., and Hubaut, R. Surface potential study of adsorbed oxygen species on aerogel Cr_2O_3–Al_2O_3 catalyst in toluene oxidation. *J. Chem. Soc., Faraday Trans.* 1998, *94*, 455–458.
56. Finocchio, E., Willey, R.J., Busca, G., and Lorenzelli, V. FTIR studies on the selective oxidation and combustion of light hydrocarbons at metal oxide surfaces. Part 3. Comparison of the oxidation of C_3 organic compounds over Co_3O_4, $MgCr_2O_4$ and CuO. *J. Chem. Soc., Faraday Trans.* 1997, *93*, 175–180.
57. Wang, X. and Xie, Y. Total oxidation of CH_4 on Sn–Cr composite oxide catalysts. *Appl. Catal. B: Environ.* 2001, *35*, 85–94.
58. Ferreira, R.S.G., de Oliveira, P.G.P., and Noronha, F.B. The effect of the nature of vanadium species on benzene total oxidation. *Appl. Catal. B: Environ.* 2001, *29*, 275–283.
59. Finocchio, E., Baldi, M., Busca, G., Pistarino, C., Romezzano, G., Bregani, F., and Toledo, G.P. A study of the abatement of VOC over V_2O_5–WO_3–TiO_2 and alternative SCR catalysts. *Catal. Today* 2000, *59*, 261–268.
60. Taylor, S.H. and O'Leary, S.R. A study of uranium oxide based catalysts for the oxidative destruction of short chain alkanes. Appl. *Catal. B: Environ.* 2000, *25*, 137–149.
61. Harris, R.H., Boyd, V.J., Hutchings, G.J., and Taylor, S.H. Water as a promoter of the complete oxidation of volatile organic compounds over uranium oxide catalysts. *Catal. Lett.* 2002, *78*, 369–372.
62. Ozawa, Y., Tochihara, Y., Nagai, M., and Omi, S. Effect of addition of Nd_2O_3 and La_2O_3 to PdO/Al_2O_3 in catalytic combustion of methane. *Catal. Commun.* 2003, *4*, 87–90.
63. Lacombe, S., Geantet, C., and Mirodatos, C. Oxidative coupling of methane over lanthana catalysts I. Identification and role of specific active sites. *J. Catal.* 1995, *151*, 439–452.
64. Krylov, O.V. Catalytic reactions of partial methane oxidation. *Catal. Today* 1993, *18*, 209–215.
65. Lunsford, J.H. Catalytic conversion of methane to more useful chemicals and fuels. Challenge for the 21st century. *Catal. Today* 2000, *63*, 165–174.
66. Haber, J., Janas, J., Ciak-Czerwenka, J.K., Machej, T., Sadowska, H., and Hellden, S. Total oxidation of nitrogen-containing organic compounds to N_2, CO_2 and H_2O. *Appl. Catal. A: Gen.* 2002, *229*, 23–34.
67. Pradier, C.M., Rodrigues, F., Marcus, P., Landau, M.V., Kaliya, M.L., Gutman, A., and Herskowitz, M. Supported chromia catalysts for oxidation of organic compounds: the state of chromia phase and catalytic performance. *Appl. Catal. B: Environ.* 2000, *27*, 73–85.
68. Luo, M., Yuan, X., and Zheng, X. Catalytst characterization and activity of Ag–Mn, Ag–Co and Ag–Ce composite oxides for oxidation of volatile organic compounds. *Appl. Catal. A* 1998, *175*, 121–129.
69. Scire, S., Minico, S., Crisafulli, C., Satriano, C., and Pistone, A. Catalytic combustion of volatile organic compounds on gold/cerium oxide catalysts. *Appl. Catal. B: Environ.* 2003, *40*, 43–49.

70. Scire, S., Minico, S., Crisafulli, C., and Galvagno, S. Catalytic combustion of volatile organic compounds over group IB metal catalysts on Fe_2O_3. *Catal. Commun.* 2001, *2*, 229–232.

71. Minico, S., Scire, S., Crisafulli, C., and Galvagno, S. Influence of catalyst pretreatments on volatile organic compounds oxidation over gold/iron oxide. *Appl. Catal. B: Environ.* 2001, *34*, 277–285.

72. Mehandjiev, D., Zhecheva, E., Ivanov, G., and Ioncheva, R. Preparation and catalytic activity of nickel-manganese oxide catalysts with an ilmenite-type structure in the reactions of complete oxidation of hydrocarbons. *Appl. Catal. A* 1998, *167*, 277–282.

73. Tejuca, L.G. and Fierro, J.L.G. *Properties and Applications of Perovskite-Type Oxides*. Marcel Dekker: New York, 1993, pp. 1–408.

74. Rao, C.N.R. and Raveau, B. *Transition Metal Oxides*. John Wiley & Sons New York, 1993, pp. 90–134.

75. Pena, M.A. and Fierro, J.L.G. Chemical structures and performance of perovskite oxides. *Chem. Rev.* 2001, *101*, 1981–2017.

76. Islam, M.S., Cherry, M., and Catlow, C.R.A. Oxygen diffusion in $LaMnO_3$ and $LaCoO_3$ perovskite-type oxides: a molecular dynamics study. *J. Solid State Chem.* 1996, *124*, 230–237.

77. Cherry, M., Islam, M.S., and Catlow, C.R.A. Oxygen ion migration in perovskite-type oxides. *J. Solid State Chem.* 1995, *118*, 125–132.

78. Voorhoeve, R.J.H., Remeika, J.P., Freeland, P.E., and Matthias, B.T. Rare earth oxides of manganese and cobalt rival platinum for the treatment of carbon monoxide in auto exhaust. *Science* 1972, *177*, 353–354.

79. Caldararu, M., Ovenston, A., and Walls, J.R. Catalytic and electrical behaviour of $Li_{0.9}Ni_{0.5}Co_{0.5}O_{2-\delta}$ below 400°C. *Appl. Catal. A: Gen.* 1998, *167*, 225–235.

80. Buciuman, F., Patcas, F., Menezo, J., Barbier, J., Hahn, T., and Lintz, H. Catalytic properties of $La_{0.8}A_{0.2}MnO_3$ (A = Sr, Ba, K, Cs) and $LaMn_{0.8}B_{0.2}O_3$ (B = Ni, Zn, Cu) perovskites: 1. Oxidation of hydrogen and propene. *Appl. Catal. B: Environ.* 2002, *35*, 175–183.

81. Oliva, C. and Forni, L. EPR and XRD as probs for activity and durability of $LaMnO_3$ perovskite-like catalysts. *Catal. Commun.* 2000, *1*, 5–8.

82. Oliva, C., Forni, L., and Vishniakov, A.V. Spin glass formation in $La_{0.9}Sr_{0.1}CoO_3$ catalyst for flameless combustion of methane. *Spectrosc. Acta A* 2000, *56*, 301–307.

83. Stojanovic, M., Haverkamp, R.G., Mims, C.A., Moudallal, H., and Jacobson, A.J. Synthesis and characterization of $LaCr_{1-x}Ni_xO_3$ perovskite oxide catalysts. *J. Catal.* 1997, *166*, 315–323.

84. Oliva, C., Forni, L., D'Ambrosio, A., Navarrini, F., Stepanov, A.D., Kagramanov, Z.D., and Mikhailichenko, A.I. Characterization by EPR and other techniques of $La_{1-x}Ce_xCoO_{3+\delta}$ perovskites-like catalysts for methane flameless combustion. *Appl. Catal. A: Gen.* 2001, *205*, 245–252.

85. Oliva, C., Forni, L., Pasqualin, P., D'Ambrosio, A., and Vishniakov, A.V. EPR analysis of $La_{1-x}M_xMnO_{3+y}$ (M = Ce, Eu, Sr) perovskite catalysts for methane oxidation. *Phys. Chem. Chem. Phys.* 1999, *1*, 355–360.

86. Cimino, S., Colonna, S., De Rossi, M., Faticanti, M., Lisi, L., Pettiti, I., and Porta, P. Methane combustion and CO oxidation on zirconia supported La, Mn oxides and $LaMnO_3$ perovskite. *J. Catal.* 2002, *205*, 309–317.

87. Read, M.S.D., Islam, M.S., King, F., and Hancock, F.E. Defect chemistry of $La_2Ni_{(1-x)}M_xO_4$ (M = Mn, Co, Cu): Relavance to catalytic behavior. *J. Phys. Chem. B* 1999, *103*, 1558–1562.

88. Kotomin, E.A. and Evarestov, R.A. Large-scale ab-initio modeling of defects in perovskites: Fe impurity in $SrTiO_3$. *Comput. Mater. Sci.* 2002, *24*, 14–20.

89. Kojima, I., Adachi, H., and Yasumori, I. Electronic structure of the $LaBO_3$ (B = Co, Fe, Al) perovskite oxides related to their catalysis. *Surf. Sci.* 1983, *130*, 50–62.

90. Leanza, R., Rossetti, I., Fabbrini, L., Oliva, C., and Forni, L. Perovskite catalysts for the catalytic flameless combustion of methane: Preparation by flame-hydrolysis and chracterization by TPD-TPR-MS and EPR. *Appl. Catal. B: Environ.* 2000, *28*, 55–64.

91. Marti, P.E. and Baiker, A. Influence of the A-site cation in $AMnO_{3+x}$ and $AFeO_{3+x}$ (A = La, Pr, Nd and Gd) perovskite type oxides on the catalytic activity for methane combustion. *Catal. Lett.* 1994, *26*, 71–84.

92. Baiker, A., Marti, P.E., Keusch, P., Fritsch, E., and Reller, A. Influence of the A-site cation in $ACoO_3$ (A = La, Pr and Gd) perovskite type oxides on catalytic activity for methane combustion. *J. Catal.* 1994, *146*, 268–276.

93. Zhong, Z., Chen, K., Ji, Y., and Yan, Q. Methane combustion over B-site partially substituted perovskite type $LaFeO_3$ prepared by sol–gel method. *Appl. Catal. A* 1997, *156*, 29–41.

94. Ciambelli, P., Cimino, S., Lasorella, G., Lisi, L., De Rossi, S., Faticanti, M., Minelli, G., and Porta, P. CO oxidation and methane combustion on $LaAl_{1-x}Fe_xO_3$ perovskite solid solutions. *Appl. Catal. B: Environ.* 2002, *37*, 231–241.

95. Ciambelli, P., Cimino, S., Lisi, L., Faticanti, M., Minelli, G., Pettiti, I., and Porta, P. La, Ca and Fe oxide perovskites: preparation, characterization and catalytic properties for methane combustion. *Appl. Catal. B: Environ.* 2001, *33*, 193–203.

96. Ponce, S., Pena, M.A., and Fierro, J.L.G. Surface properties and catalytic performance in methane combustion of Sr-substituted lanthanum manganites. *Appl. Catal. B: Environ.* 2000, *24*, 193–205.

97. Saracco, G., Geobaldo, F., and Baldi, G. Methane combustion on Mg-doped $LaMnO_3$ perovskite catalysts. *Appl. Catal. B: Environ.* 1999, *20*, 277–288.

98. Saracco, G., Scibilia, G., Iannibello, A., and Baldi, G. Methane combustion on Mg-doped $LaCrO_3$ perovskite catalysts. *Appl. Catal. B: Environ.* 1996, *8*, 229–244.

99. Marchetti, L. and Forni, L. Catalytic combustion of methane over perovskites. *Appl. Catal. B: Environ.* 1998, *15*, 179–187.

100. Porta, P., De Rossi, S., Faticanti, M., Minelli, G., Pettiti, I., Lisi, L., and Turco, M. Perovskite type oxides, I. Structural magnetic and morphological properties of $LaMn_{1-x}Cu_xO_3$ and $LaCo_{1-x}Cu_xO_3$ solid solutions with large surface area. *J. Solid State Chem.* 1999, *146*, 291–304.

101. Blasin-Aube, V., Belkouch, J., and Monceaux, L. General study of catalytic oxidation of various VOCs over $La_{0.8}Sr_{0.2}MnO_{3+x}$ perovskite catalyst-influence of mixture. *Appl. Catal. B: Environ.* 2003, *43*, 175–176.

102. Busca, G., Finocchio, E., Ramis, G., and Ricchiardi, G. On the role of acidity in catalytic oxidation. *Catal. Today* 1996, *32*, 133–143.

103. Busca, G., Finocchio, E., Lorenzelli, V., Ramis, G., and Baldi, M. IR studies on the activation of C-H hydrocarbon bonds on oxidation catalysts. *Catal. Today* 1999, *49*, 453–465.

104. Busca, G., Daturi, M., Finocchio, E., Lorenzelli, V., Ramis, G., and Willey, R.J. Transition metal mixed oxides as combustion catalysts: preparation, characterization and activity mechanisms. *Catal. Today* 1997, *33*, 239–249.

105. Lee, Y.N., Lago, R.M., Fierro, J.L.G., Cortes, V., Sapina, F., and Martinez, E. Surface properties and catalytic performance for ethane combustion of $La_{1-x}K_xMnO_{3+\delta}$ perovskites. *Appl. Catal. A* 2001, *207*, 17–24.

106. Teraoka, Y., Yoshimatsu, M., Yamazoe, N., and Seiyama, T. Oxygen-sorptive properties and defect structure of perovskite-type oxides. *Chem. Lett.* 1984, *6*, 893–896.

107. Yamazoe, N. and Teraoka, Y. Oxidation catalysts of perovskites — Relationships to bulk structure and composition (valency, defect). *Catal. Today* 1990, *8*, 175–199.

108. Daturi, M., Busca, G., Groppi, G., and Forzatti, P. Preparation and characterization of $SrTi_{1-x-y}Zr_xMn_yO_3$ solid solution powders in relation to their use in combustion catalysis. *Appl. Catal. B: Environ.* 1997, *12*, 325–337.

109. Wang, W., Zhang, H., Lin, G., and Xiong, Z. Study of $Ag/La_{0.6}Sr_{0.4}MnO_3$ catalysts for complete oxidation of methanol and ethanol at low concentrations. *Appl. Catal. B: Environ.* 2000, *24*, 219–232.

110. Ciambelli, P., Cimino, S., De Rossi, S., Lisi, L., Minelli, G., and Russo, G. $AFeO_3$ (A = La, Nd, Sm) and $LaFe_{1-x}Mg_xO_3$ perovskites as methane combustion and CO oxidation catalysts: structural, redox and catalytic properties. *Appl. Catal. B: Environ.* 2001, *29*, 239–250.

111. Ciambelli, P., Cimino, S., De Rossi, S, Faticanti, M., Lisi, L., Minelli, G., Pettiti, I., Russo, G., Turco, M., and Porta, P. $AMnO_3$ (A = La, Nd, Sm) and $Sm_{1-x}Sr_xMnO_3$ perovskites as combustion catalysts: structural, redox and catalytic properties. *Appl. Catal. B: Environ.* 2000, *24*, 243–253.

112. Marti, P.E., Maciejewski, M., and Baiker, A. Methane combustion over $La_{0.8}Sr_{0.2}MnO_{3+x}$ supported on MAl_2O_4 (M = Mg, Ni, and Co) spinels. *Appl. Catal. B: Environ.* 1994, *4*, 225–235.

113. Stojanovic, M., Mims, C.A., Moudallal, H., Yang, Y.L., and Jacobson, A.J. Reaction kinetics of methane oxiation over $LaCr_{1-x}Ni_xO_3$ perovskite catalysts. *J. Catal.* 1997, *166*, 324–332.

114. Ferri, D. and Forni, L. Methane combustion on some perovskite like mixed oxides. *Appl. Catal. B: Environ.* 1998, *16*, 119–126.

115. Forni, L. and Rossetti, I. Catalytic combustion of hydrocarbons over perovskites. *Appl. Catal. B: Environ.* 2002, *38*, 29–37.

116. Kirchnerova, J., Alifanti, M., and Delmon, B. Evidence of phase cooperation in the $LaCoO_3$–CeO_2–Co_3O_4 catalytic system in relation to activity in methane combustion. *Appl. Catal. A* 2002, *231*, 65–80.

117. Belessi, V.C., Ladavos, A.K., and Pomonis, P.J. Methane combustion on La–Sr–Ce–Fe–O mixed oxides: bifunctional synergistic action of $SrFeO_{3-x}$ and CeO_x phases. *Appl. Catal., B: Environ.* 2001, *31*, 183–194.

118. Choudhary, V.R., Uphade, B.S., and Pataskar, S.G. Low temperature complete combustion of methane over Ag-doped $LaFeO_3$ and $LaFe_{0.5}Co_{0.5}O_3$ perovskite oxide catalysts. *Fuel* 1999, *78*, 919–921.

119. Szabo, V., Bassir, M., Neste, A.V., and Kaliaguine, S. Perovskite-type oxides synthesized by reactive grinding: Part II: Catalytic properties of LaCo$_{(1-x)}$Fe$_x$O$_3$ in VOC oxidation. *Appl. Catal. B: Environ.* 2002, *37*, 175–180.

120. Irusta, S., Pina, M.P., Menendez, M., and Santamaria, J. Development and application of perovskite-based catalytic membrane reactors. *Catal. Lett.* 1998, *54*, 69–78.

121. Machida, M., Eguchi, K., and Arai, H. Effect of additives on the surface area of oxide supports for catalytic combustion. *J. Catal.* 1987, *103*, 385–393.

122. Machida, M., Eguchi, K., and Arai, H. High temperature catalytic combustion over cation substituted barium hexaaluminates. *Chem. Lett.* 1987, *16*, 767–770.

123. Machida, M., Kawasaki, H., Eguchi, K., and Arai, H. Surface areas and catalytic activities of Mn-substituted hexaaluminates with various cation compositions. *Chem. Lett.* 1988, *17*, 1461–1464.

124. Machida, M., Eguchi, K., and Arai, H. Effect of additives on the surface area of oxide supports for catalytic combustion. *J. Catal.* 1990, *123*, 477–485.

125. Park, J.G. and Cormack, A.N. Crystal defect structures and phase stability in Ba hexaaluminates. *J. Solid State Chem.* 1996, *121*, 278–290.

126. Machida, M., Eguchi, K., and Arai, H. Preparation and characterization of large surface area barium hexaaluminates. *Chem. Soc. Jpn.* 1998, *61*, 3659–3665.

127. Groppi, G., Belloto, M., Cristiani, C., Forzatti, P., and Villa, P.L. Preparation and characterization of hexaaluminate based materials for catalytic combustion. *Appl. Catal. A* 1993, *104*, 101–108.

128. Ocal, M., Oukaci, R., Marcelin, G., Jang, B.W., and Spivey, J.J. Steady-state isotopic transient kinetic analysis on Pd-supported hexaaluminates used for methane combustion in the presence and absence of NO. *Catal. Today* 2000, *59*, 205–207.

129. Jang, B.W., Nelson, R.M., Spivey, J.J., Ocal, M., Oukaci, R., and Marcelin, G. Catalytic oxidation of methane over hexaaluminates and hexaaluminate-supported Pd catalysts. *Catal. Today* 1999, *47*, 103–113.

130. McCarty, J.G., Gusman, M., Lowe, D.M., Hildenbrand, D.L., and Lau, K.N. Stability of supported metal and supported metal oxide catalysts. *Catal. Today* 1999, *47*, 5–17.

131. Forzatti, P. and Groppi, G. Catalytic combustion for the production of energy. *Catal. Today* 1999, *54*, 165–180.

132. Duart, P.A., Brulle, Y., Gaillard, F., Garbowski, E., Guilhaume, N., and Primet, M. Catalytic combustion of methane over copper and manganese substituted barium hexaaluminates. *Catal. Today* 1999, *54*, 181–190.

133. Duart, P.A., Millet, J.M., Huilhaume, N., Garbowski, E., and Primet, M. Catalytic combustion of methane on substituted barium hexaaluminates. *Catal. Today* 2000, *59*, 163–177.

134. Groppi, G., Lietti, L., Tronconi, E., and Forzatti, P. Catalytic combustion of gasified biomass over Mn-substituted hexaaluminates for gas turbine applications. *Catal. Today* 1998, *45*, 159–165.

135. Groppi, G., Cristiani, C., and Forzatti, P. Preparation, characterization and catalytic activity of pure and substituted La-hexaaluminate systems for high temperature catalytic combustion. *Appl. Catal. B: Environ.* 2001, *35*, 137–148.

136. Lietti, L., Cristiani, C., Groppi, G., and Forzatti, P. Preparation, characterization and reactivity of Me-hexaaluminate (Me = Mn, Co, Fe, Ni, Cr) catalysts in the catalytic combustion of NH_3-containing gasified biomass. *Catal. Today* 2000, *59*, 191–204.

137. Arai, H. and Machida, M. Thermal stabilization of catalyst supports and their application to high-temperature catalytic combustion. *Appl. Catal. A: Gen.* 1996, *138*, 161–176.

138. Machida, M., Sato, A., Kijima, T., Inoue, H., Eguchi, K., and Arai, H. Catalytic properties and surface modification of hexaaluminate microcrystals for combustion catalyst. *Catal. Today* 1995, *26*, 239–245.

139. Machida, M., Eguchi, K., and Arai, H. Catalytic properties of $BaAl_{11}O_{19-\alpha}$ for high temperature catalytic combustion. *J. Catal.* 1989, *120*, 377–386.

140. Berg, M. and Jaas, S. High temperature stable magnesium oxide catalyst for catalytic combustion of methane: a comparison with manganese substituted barium hexaaluminate. *Catal. Today* 1995, *26*, 223–229.

141. Zarur, A.J. and Ying, Y. Reverse microemulsion synthesis of nanostructured complex oxides for catalytic combustion. *Nature* 2000, *403*, 65–67.

142. Liu, Y., Luo, M., Wei, Z., Xin, Q., Ying, P., and Li, C. Catalytic oxidation of chlorobenzene on supported manganese oxide catalysts. *Appl. Catal. B: Environ.* 2001, *29*, 61–67.

143. Padilla, A.M., Corella, J., and Toledo, J.M. Total oxidation of some chlorinated hydrocarbons with commercial chromia based catalysts. *Appl. Catal. B: Environ.* 1999, *22*, 107–121.

144. Kulazynski, M., van Ommen, J.G.V., Trawczynski, J., and Walendziewski, J. Catalytic combustion of trichloroethylene over TiO_2–SiO_2 supported catalysts. *Appl. Catal. B: Environ.* 2002, *36*, 239–247.

145. Gervasini, A., Pirola, C., and Ragaini, V. Destruction of carbon tetrachloride in the presence of hydrogen-supplying compounds with ionisation and catalytic oxidation. *Appl. Catal. B: Environ.* 2002, *38*, 17–28.

146. Krishnamoorthy, S. and Amiridis, M.D. Kinetic and in situ FTIR studies of the catalytic oxidation of 1,2-dichlorobenzene over V_2O_5/Al_2O_3 catalysts. *Catal. Today* 1999, *51*, 203–214.

147. Lai, S.Y., Pan, W., and Ng, C.F. Catalytic hydrolysis of dichlorodifluoromethane (CFC-12) on unpromoted and sulfate promoted TiO_2–ZrO_2 mixed oxide catalysts. *Appl. Catal. B: Environ.* 2000, *24*, 207–217.

148. Doornkamp, C., Clement, M., and Ponec, V. Activity and selectivity patterns in the oxidation of allyl iodide on the period IV metal oxides: the participation of lattice oxygen in selective and total oxidation reactions. *Appl. Catal. A: Gen.* 1999, *188*, 325–336.

149. Sinquin, G., Petit, C., Libs, S., Hindermann, J.P., and Kinnemann, A. Catalytic destruction of chlorinated C_1 volatile organic compounds (CVOCs) reactivity, oxidation and hydrolysis mechanisms. *Appl. Catal. B: Environ.* 2000, *27*, 105–115.

150. Sinquin, G., Petit, C., Libs, S., Hindermann, J.P., and Kiennemann, A. Catalytic destruction of chlorinated C_2 compounds on a $LaMnO_{3+\delta}$ perovskite catalyst. *Appl. Catal. B: Environ.* 2001, *32*, 37–47.

151. Kiebling, D., Schneider, R., Kraak, P., Haftendron, M., and Wendt, G. Perovskite-type-oxides-catalysts for the total oxidation of chlorinated hydrocarbons. *Appl. Catal. B: Environ.* 1998, *19*, 143–151.

152. Otsuka, K. and Jinno, K. Kinetic studies on partial oxidation of methane over samarium oxides. *J. Catal.* 1986, *121*, 237–241.

153. Dissanayake, D., Lunsford, J.H., and Rosynek, M.P. Oxidative coupling of methane over oxide-supported barium catalysts. *J. Catal.* 1993, *143*, 286–298.

154. Buyevskaya, O.V., Rothaemel, H., Zanthoff, H.W., and Baerns, M. Transient studies on reaction steps in the oxidative coupling of methane over catalytic surfaces of MgO and Sm_2O_3. *J. Catal.* 1994, *146*, 346–357.

155. Lacombe, S., Zanthoff, H., and Mirodatos, C. Oxidative coupling of methane over lanthana catalysts — II. A mechanistic study using isotope transient kinetics. *J. Catal.* 1995, *155*, 106–116.

156. Gayko, G., Wolf, D., Kondratenko, E.V., and Baerns, M. Interaction of oxygen with pure and SrO-doped Nd_2O_3 catalysts for the oxidative coupling of methane: study of work function changes. *J. Catal.* 1998, *178*, 441–449.

157. Ling, T., Cheng, Z., and Lee, M. Catalytic behavior and electrical conductivity of $LaNiO_3$ in ethanol oxidation. *Appl. Catal. A* 1996, *136*, 191–203.

158. Ito, T., Wang, J., Lin, C., and Lunsford, J.H. Oxidative dimerization of methane over lithium-promoted magnesium oxide catalyst. *J. Am. Chem. Soc.* 1985, *107*, 5062–5068.

159. Wang, J. and Lunsford, J.H. Characterization of $[Li^+O^-]$ centers in lithium-doped MgO catalysts. *J. Phys. Chem.* 1986, *90*, 5883–5887.

160. Lin, C., Wang, J., and Lunsford, J.H. Oxidative dimerization of methane over sodium promoted calcium oxide. *J. Catal.* 1988, *111*, 302–316.

161. Lehmann, L. and Baerns, M. Kinetics studies of the oxidative coupling of methane over a NaOH/CaO catalyst. *J. Catal.* 1992, *135*, 467–480.

162. Zhang, Z. and Baerns, M. Oxidative coupling of methane over $CaO–CeO_2$ catalysts: effect of oxygen-ion conductivity on C_2 selectivity. *J. Catal.* 1992, *135*, 317–320.

163. Bond, G.C. The role of carbon deposits in metal-catalyzed reactions of hydrocarbons. *Appl. Catal. A* 1997, *149*, 3–25.

164. Somorjai, G.A. *Introduction to Surface Chemistry and Catalysis.* Wiley-Interscience: New York, 1994, pp. 1–220.

165. Davis, S.M., Zaera, F., and Somorjai, G.A. The reactivity and composition of strongly adsorbed carbonaceous deposits on platinum-model of the working hydrocarbon conversion catalyst. *J. Catal.* 1982, *77*, 439–459.

166. Godbey, D., Zaera, F., Yeates, R., and Somorjai, G.A. Hydrogenation of chemisorbed ethylene on clean, hydrogen and ethylidyne covered platinum (111) crystal surfaces. *Surf. Sci.* 1986, *167*, 150–166.

167. Davis, S.M., Zaera, F., and Somorjai, G.A. Surface-structure and temperature dependence of light alkane skeletal rearrangement reactions catalyzed over platinum single-crystal surfaces. *J. Am. Chem. Soc.* 1982, *104*, 7453–7461.

168. Lobo, L.S. and Franco, M.D. Kinetics of catalytic carbon formation on steel surfaces from light hydrocarbons. *Catal. Today* 1990, *7*, 247–256.

169. Krishnankutty, N., Rodriguez, N.M., and Baker, R.T.K. Effect of copper on the decomposition of ethylene over an iron catalyst. *J. Catal.* 1996, *158*, 217–227.

170. Qing, M., Guoxing, X., Shishan, S., Wei, C., Ling, X., and Xiexian, G. Partial oxidation of methane to syngas over nickel-based catalysts modified by alkali metal oxide and rare earth metal oxides. *Appl. Catal. A* 1997, *154*, 17–27.

171. Van Looij, F., Dorrestein, E., Geus, J.W., and Kobussen, S. Partial oxidation of methane to syn gas: the accumulation of carbon on supported nickel catalysts. *Proc. Int. Gas Res. Conf.* 1995, *2*, 2700–2703.

172. Barbier, J., Corro, G., Zhang, Y., Bournonville, J.P., and Frank, J.P. Coke formation on platinum–alumina catalyst of wide varying dispersion. *Appl. Catal. A* 1985, *13*, 245–255.

173. Hettige, C., Mahanama, K.R.R., and Dissanayake, D.P. Cyclohexane oxidation and carbon deposition over metal oxide catalysts. *Chemosphere* 2001, *43*, 1079–1083.

174. Hu, Y.H. and Ruckenstein, E. CH_4 TPR-MS of NiO/MgO solid solution catalysts. *Langmuir* 1997, *13*, 2055–2058.

175. Hu, Y.H. and Ruckenstein, E. The characterization of a highly effective NiO/MgO solid solution in the CO_2 reforming of CH_4. *Catal. Lett.* 1997, *43*, 71–77.

176. Jayathilaka, G.D.L.P., Hettige, C., Mahanama, K.R.R., and Dissanayake, D.P. Carbon formation during volatile organic compound oxidation over a nickel catalyst. *Ind. J. Chem. A* 2001, *40*, 392–395.

177. Battle, P.D., Claridge, J. B., Copplestone, F.A., Carr, S.W., and Tsang, S.C. Partial oxidation of natural gas to synthesis gas over ruthenium perovskite oxides. *Appl. Catal. A* 1994, *118*, 217–227.

178. Zhu, T. and Stephanopoulos, M.F. Catalytic partial oxidation of methane to synthesis gas over Ni-CeO_2. *Appl. Catal. A* 2001, *208*, 403–417.

18 Hydrogenation of Carbon Oxides on Metal Oxides

J.L.G. Fierro
Institute of Catalysis and Petrochemistry, CSIC, Madrid, Spain

CONTENTS

18.1 INTRODUCTION

The catalytic hydrogenation of carbon oxides produces a large variety of products ranging from C_1 containing molecules (CH_4 and CH_3OH) to higher molecular weight paraffins, olefins, and alcohols, a process that requires C–C bond formation through a complex reaction mechanism involving primarily the insertion of methylene units into the hydrocarbon chain [1–4].

In this chapter, recent trends in the study of CO and CO_2 hydrogenation on metal oxides are reviewed. In most cases, and particularly in C–C bond formation via (Equation [18.2]), carbon oxide dissociation occurs. The primarily formed CH_x species between carbon and dissociated hydrogen play an important role in polymerization, and in kinetics leading to high-molecular weight products through secondary reactions. The $CO_x + H_2$ mixtures are converted either to CH_3OH over Cu/ZnO catalysts (Equation [18.1]), and then to gasoline by the methanol-to-gasoline (MTG) process over ZSM-5 zeolites, or to liquid hydrocarbons (HC) through the Fischer–Tropsch (FT) reaction over Group VIII transition metal catalysts (Equation [18.2]). This latter process is a well-established technology that began with the first plant built by Ruhrchemie in 1936, producing 600,000 tons of FT products per year by 1944. From 1955 to the present,

$$CO + 2H_2 \rightarrow CH_3OH \qquad\qquad \Delta H^\circ = -90.7 \text{ kJ/mol} \qquad (18.1)$$

$$nCO + (2n + 1)H_2 \rightarrow C_nH_{2n+2} + nH_2O \quad \Delta H^\circ \text{ depends on } n \qquad (18.2)$$

Sasol has built and operated several FT plants in South Africa. The product usually follows the Anderson–Schulz–Flory distribution and typically consists of linear paraffins and waxes in the range of C_5–C_{40}. It is emphasized that ΔH° values depend on the number of C atoms in the resulting n-paraffin. Thus, for $n = 1, 2, 3,$ and 4, ΔH° values are $-206.1, -347.3, -497.5,$ and -649.9 kJ/mol, respectively. In fact, it can be inferred from these examples that $\Delta H^\circ \approx -148.2 \times n - 54.8$ (in kJ/mol) [5]. Water also plays an important role through the water gas-shift reaction [6]:

$$CO + H_2O \rightarrow CO_2 + H_2 \quad \Delta H^\circ = -41.0 \text{ kJ/mol} \qquad (18.3)$$

and its ability to oxidize the active transition metal that modifies the surface and catalytic chemistry of both hydrogenation and polymerization processes. There is also a very strong influence of metal oxide substrates on the rate and product distribution during CO_x hydrogenation. Thus, the traditional view that CO_x hydrogenation is a transition metal-catalyzed reaction should be abandoned. The oxide–metal interface is often a better catalyst than the transition metal alone. As a result, strong Lewis acid metal oxides such as TiO_x, NbO_x, and TaO_x readily dissociate CO and hence catalyze C–C bond formation.

The focus of this chapter is on the interaction of small molecules such as CO and CO_2, and their transformation in the presence of H_2, with metal ion sites on metal oxide surfaces. There have been many reports of chemisorption, particularly of CO, on metal surfaces; however, it is important to note that while metal oxides are involved in many heterogeneous catalytic processes, there are fewer studies dealing with the interaction of CO_x with metal. Section 18.2 deals with the interaction of

CO (and CO_2) with ZnO in some detail with the aim of establishing the geometric and the electronic requirements of the catalytically relevant surface species. In Section 18.3 and Section 18.4, the nature of the metal oxide promotion of the CO_x will be explored to define electronic structure differences relative to CO binding to pure ZnO, which are related to differences in efficiency in hydrogenation reactions. It is pointed out that the traditional concept of the individual active site involved in the hydrogenation reactions of carbon oxides must be replaced by one that involves a metal–metal oxide interface, since any CO bond responds to an oxide promoter and exhibits very accelerated reaction rates. Section 18.5 describes the performance of metal oxide and metal–metal oxide systems for hydrogenation of CO, CO–CO_2 mixtures, and CO_2.

18.2   ACTIVATION OF H_2, CO, AND CO_2 MOLECULES

When H_2, CO, and CO_2 gas molecules strike the surface of a transition metal or metal oxide they form a precursor state. This precursor state undergoes further electronic restructuring, leading to chemisorption that often results in the dissociation of these molecules, all fragments being adsorbed.

18.2.1 Hydrogen

The dissociation of hydrogen on transition metals and some metal oxides is a simple reaction, as shown by H_2–D_2 exchange at temperatures near 300 K. Although CO_x hydrogenation reactions are usually conducted at much higher temperatures ($\sim$600 K) and under high pressure, some general conclusions can be drawn about H_2 dissociation. Low-coordination number surface sites, steps, and kinks dissociate H_2 upon a single collision.

The adsorption of dihydrogen at low temperature (77 K) on ZnO takes place dissociatively on adjacent O^{2-} and Zn^{2+} sites [7,8]. In the region of low coverages, the infrared (IR) spectra of H_2 adsorbed on ZnO show two IR adsorption bands at 1710 and 3492 cm^{-1}, which correspond, respectively, to the fundamental stretching vibration of hydride (Zn–H) and hydroxyl (O–H) surface species formed upon dissociative chemisorption of the dihydrogen molecule on adjacent $Zn^{2+}O^{2-}$ ion pairs, which act as a Lewis acid–base couple. Upon increasing H_2 coverage, ν(ZnH) decreases to 1690 cm^{-1} while ν(OH) increases to 3512 cm^{-1}. This coverage-dependent frequency shift of the Zn–H and O–H stretching bands very likely arise from a combination of lateral interactions among neighboring adsorbates and through solid-inductive effects. The frequency shifts take place in discrete steps.

Theoretical calculations have shown that the adsorption probability of H_2 depends on its orientation as the molecule approaches the surface. There is a strong preference for an orientation with the H–H molecular axis parallel to a copper surface if adsorption is to occur, since this orientation minimizes the repulsive interaction with the surface [9]. In some cases, dihydrogen complexes have been reported in situations where a reversible equilibrium exists between the dihydrogen form and the dihydride form:

$$M \longleftarrow \begin{matrix} H \\ | \\ H \end{matrix} \quad \rightleftharpoons \quad M \begin{matrix} H \\ \diagup \\ \diagdown \\ H \end{matrix} \qquad (18.4)$$

These complexes are excellent models for studying the detailed mechanism of the oxidative addition of hydrogen to the surface. Generally, the energies involved in the interaction of hydrogen with surface range from -25 to -100 kJ/mol.

The relative ease with which hydrogen chemisorbs on the surface of a metal oxide surface mainly depends on the chemical nature of the oxide and on the O-vacancies. Thus, hydrogen adsorbs dissociatively on a perfect titanium oxide surface [10,11]. The energetically most favorable mode for the adsorption of atomic hydrogen is the adsorption on the outermost O atom, accompanied by the reduction of a Ti atom. In this mode, protons are formally adsorbed while an equivalent amount of Ti(IV) atoms are reduced to Ti(III). Theoretical calculations have demonstrated that H adsorption is less favorable on a defective surface than on a perfect surface. However, the best adsorption mode for the atomic chemisorption on a defective surface is heterolytic adsorption, which involves two different adsorption sites: one $H^+/O^=$ and one H^- on the surface. This adsorption mode is best on irreducible oxides such as MgO; however, it is less favorable than adsorption on the perfect TiO_2 surface [10]. The heat of atomic adsorption in all cases is very weak and dissociation onto the surface is unlikely. The molecular adsorption (physisorption), thus, remains the most stable system.

18.2.2 Carbon Monoxide

The bonding of CO takes place through the σ orbital and back-bonding involving the oxygen π orbital, and it results in a strong chemical bond. CO dissociation does not occur on Cu, Ag, Pd, and Pt metals, however it dissociates in other transition metals and metal oxides. CO adsorption on potassium increases electron donation, which considerably strengthens the CO chemisorption bond. For transition metals that perform CO hydrogenation but do not dissociate the molecule, the formation of HCO, H_2CO, and HCOH species must occur before they undergo further secondary reactions. CO dissociates at certain metal oxide interfaces where the oxide is a strong Lewis acid, such as TiO_x, NbO_x, and TaO_x [4]. The adsorption of CO_2, however, has been harldy studied. Carbon dioxide dissociates to CO and oxygen on group VIII metals and even on copper as one of the elementary steps leading to the water gas-shift reaction (Equation [18.3]) [12,13]. The formation of CO_2 anions is energetically favored by charge transfer from the surface upon adsorption. Other more complex structures, such as carbonate (CO_3^{2-}) and oxalate ($C_2O_4^{2-}$) structures may also be formed.

18.2.2.1   CO chemisorption on a ZnO surface

The interaction of CO with a ZnO surface is quite different from that taking place in zerovalent metals. The adsorption of CO on a transition metal involves CO binding to the metal sites to form organometallic surface species, a typical example being the CO/Ni complex [14]. For the CO/ZnO system, the metal ion is in the divalent oxidation state and surrounded by oxide ligands. Associated with this unusual inorganic complex are unique spectral features, including the CO stretching frequency, obtained by IR spectroscopic studies of powders [8]. The binding of CO to a metal surface results in a decrease in the CO stretching frequency and is associated with the participation of metal d electrons in back-bonding into the $2\pi*$ orbital of the CO molecule. For the CO/Ni$_{(s)}$ complex, the stretching frequency is 2069 cm^{-1} and the gas-phase frequency of CO is 2143 cm^{-1}. However, for CO binding to ZnO powder the CO stretching frequency is found to be 2212 cm^{-1}; that is, far above (69 cm^{-1}) its gas-phase value [8]. The upward shift of 69 cm^{-1} with respect to free CO (2143 cm^{-1}) was interpreted as being mainly due to the vibrational effect induced in the CO molecule by the (axial) electric field created by the Zn^{2+} ion [15]. However, a contribution due to σ-donation from the 5σ molecular orbital of CO to the cation is also likely to exist, as suggested from ultraviolet photoelectron spectroscopy measurements [16]. Note that the asymmetry and broadness of the 2190 cm^{-1} band can be explained in terms of CO adsorption on Zn^{2+} ions belonging to both (1010) and (1120) faces, which have a slightly different configuration of coordinatively unsaturated cations [17].

This increase in the CO stretching frequency was interpreted based on different adsorption models, including CO binding of the carbon end to the surface oxide, forming a pseudo-CO$_2$ surface species, and CO binding of the oxygen end down to the coordinatively unsaturated zinc ion.

Figure 18.1 shows a schematic depiction of how CO binds the carbon end down to the coordinatively unsaturated Zn^{2+} site. It acts as a σ-donor, stabilizing the 5σ molecular orbital and donating electron density to the surface. This bonding results in a net dipole moment of 0.61 D, with a positive charge on the C atom, and withdraws the electron density from the 5σ molecular orbital, which is weakly antibonding with respect to the CO bond. Unlike CO bonding on most transition metals, there is no evidence for π back-bonding for the [CO–ZnO] complex, which is due to the high effective nuclear charge on the Zn^{2+} ion that contracts the orbital and stabilizes the d band to higher binding energies.

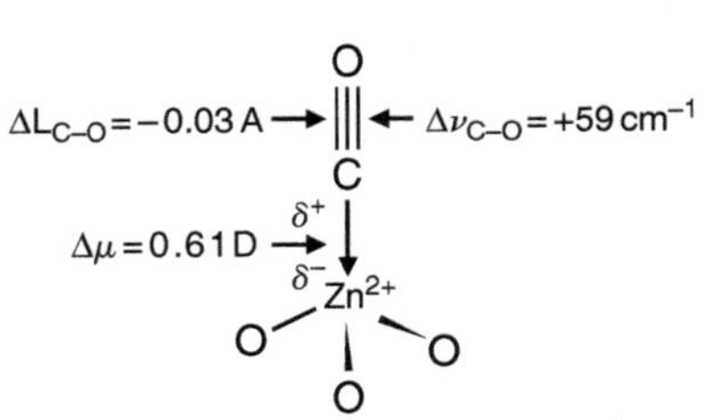

FIGURE 18.1   CO/ZnO active site complex

This unique inorganic complex also has an unusual reactivity. Hydrogenation of CO on transition metals (Ni, Pd, Fe) results in the formation of methane, while hydrogenation of CO on ZnO produces methanol, where no C–O bond scission occurs. There are two promoters involved in this reaction, including M_2O_3 (M = Cr, Al) and copper. M_2O_3 is an intercrystalline promoter that is added in sub-stoichiometric amounts to ensure the formation of a zinc spinel structure, which will inhibit ZnO crystallites from sintering and losing surface area [18]. In these systems, ZnO is the catalytically active phase in the hydrogenation reaction. In commercial methanol synthesis catalysts, copper is added as an intracrystalline promoter, which greatly reduces the activation energy barrier of the reaction, allowing milder operation conditions.

18.2.2.2 CO adsorption on Cu$^+$–ZnO oxides

CO surface bonding on Cu^+ is stronger relative to Zn^{2+} sites due to not only the stronger σ-donor interaction but also the additional presence of π back-bonding. As a consequence of this electronic structure, the concentration of surface bound CO on Cu^+ containing copper–zinc catalysts would increase and hence the CO conversion rate would also increase. The heat of adsorption of CO on the Cu^+ site is almost twice that of the Zn^{2+} site (Table 18.1). For CO bonding to both Cu^+ and Zn^{2+} sites, charge is transferred from the C atom to the surface, thus leading to a decrease in its work function ($\Delta\phi$) and a parallel, shorter, and stronger CO bond, and a net positive charge on the C atom. For these reasons, both Cu^+ and Zn^{2+} sites are suited for the reduction of CO into CH_3OH. The increased strength of the CO bond upon chemisorption is consistent with CO bond retention in the products after the hydrogenation reaction. Additionally, CO chemisorption on the Cu^+ and Zn^{2+} sites leads to a positive charge (ΔQ) on the carbon, which facilitates its nucleophilic attack by hydride species present on the ZnO surface because of the heterolytic dissociation of the hydrogen molecule [7]:

$$H_2 + ZnO \longrightarrow \overset{\displaystyle H \quad H}{\underset{\textstyle O-Zn-O-Zn-O-Zn-O}{\mid \quad \mid}} \tag{18.5}$$

TABLE 18.1

Comparison of Electronic and Physical Properties of CO Binding to Zn^{2+} and Cu^+ Metal Ions

Physical property	ZnO(1010)	CuCl(111)
Adsorption energy (ΔH_a) (kJ/mol)	50.1	96.1
Work function ($\Delta\phi$) (eV)	−1.2	−0.8
CO bond length (ΔL_{C-O}) (A)	−0.03	−0.02
Net charge (ΔQ) (unit charge)	+0.14	+0.10

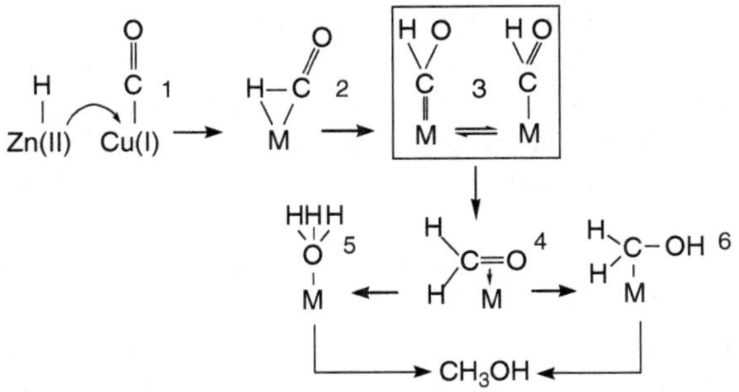

FIGURE 18.2 One of the mechanisms proposed for methanol formation on a Cu/ZnO catalyst

This behavior is indeed opposite to that of Group VIII metals (Ni, Co, or Fe), which split the CO bond and hence produce CH and CC bond containing species upon hydrogenation. Photoelectron spectroscopy allows one to estimate the relative positive charge on the C atom for CO bound to the Cu^+ and Zn^{2+} sites. The C atom is more positive for CO chemisorbed on the Zn^{2+} sites, indicating that the role of Cu^+ as a promoter of the methanol synthesis reaction is not a mere electrostatic activation of the attack by the hydride (see Figure 18.2). The role of Cu^+ would appear to be involved in the stabilization of a further transition state in the reaction. The rate-determining step is the generation of formaldehyde (Figure 18.2, species 4), which is a known uphill reaction [19]. The difference in the bonding of CO to Cu^+ relative to Zn^{2+} sites is the stronger σ donation, together with some π back-bonding. Since both Cu^+ and Zn^{2+} ions have a d^{10} electron configuration, they do not appear to play a significant role in the generation of formyl species (Figure 18.2, species 3). On the other hand, the stabilization of the formyl intermediate (Figure 18.2, species 3) and the η^2-bidentate formaldehyde bound to the Cu^+ site (Figure 18.2, species 4) is achieved by the presence of d-π back-bonding into the unoccupied π^* orbitals of the ligand. Because of the d-π back-bonding, an increase in the carbene character (Figure 18.2, species 3 in the box) would favor the attack by the proton to yield formaldehyde.

18.2.2.3 Methanol formation

The production of methanol at industrial scale takes place in the presence of H_2, CO, and CO_2 mixtures at pressures close to 50 bar in the 550 to 640 K temperature range. The catalyst of choice is copper dispersed on the surface of zinc oxide [20–22]. Recent studies on CO_2 hydrogenation have received much attention in an attempt to rekindle interest in the role of CO_2 in industrial conditions of methanol synthesis. This issue is closely related to the fact that kinetic experiments, using isotope-labeled carbon oxides [22] and spectroscopic experiments [23], have demonstrated that under industrial conditions methanol is produced by the hydrogenation of CO_2, CO merely providing a source of CO_2 and acting as a reducing agent. Nevertheless, the CuO–ZnO system is not so active for CO_2-rich reaction mixtures [24,25]. Attempts have been made to improve catalyst performance for methanol synthesis from H_2/CO_2 feeds. CuO–ZnO catalysts have been variously modified with oxides of different metals, such as chromium [26], zirconium [27–29], vanadium [29], cerium [30], titanium [31,32], gallium [33,34], and palladium [35–38]. Moreover, the methodology employed in catalyst preparation has a strong influence on catalytic performance. It has been observed that the incorporation of palladium to the CuO–ZnO system by impregnation causes a dramatic reduction in the methanol yield and this can be explained in term of the genesis, decomposition, and reconstruction of hydrotalcite-like precursor structures.

A major portion of work is aimed at the elucidation of the reaction mechanism and the roles played by zinc oxide and the oxidation states copper. Metallic copper is the dominant copper species during the reaction, although the presence of Cu^+ species is also important since a small amount of oxygen increases the reaction rate. $^{13}CO_2$ isotope-labeling studies have shown that CO_2 is the primary source of methanol, especially at low conversion since its hydrogenation rate is ~20 times faster than that of CO. In the presence of water, the copper surface becomes restructured, which should influence CO_2 hydrogenation [39]. One of the roles of zinc oxide is to prevent restructuring by initially reacting with water. As the rate of CO_2 hydrogenation decreases due to water adsorption, methanol synthesis from CO hydrogenation becomes more important, particularly at high conversion. Hydrogenation of the formate reaction intermediate that forms from CO_2 is thought to be the rate-determining step in methanol production. This reaction is thought to occur via the following individual steps:

$$COO_{(a)} + H_{(a)} \rightarrow HCOO_{(a)} \qquad (18.6)$$

$$2H_{(a)} + HCOO_{(a)} \rightarrow CH_3O_{(a)} + O_{(a)} \qquad (18.7)$$

$$CH_3O_{(a)} + H_{(a)} \rightarrow CH_3OH_{(g)} \qquad (18.8)$$

Another beneficial effect of zinc oxide is hydrogen and oxygen spillover, which helps hydrogenation and keeps the surface of copper partially oxidized. Upon increasing the reduction temperature, the specific hydrogenation activity for CO_2 also increases while CO hydrogenation follows the opposite trend [40]. This finding is due to the migration of mobile zinc sub-oxide species onto copper and is also believed to stabilize the Cu^+ state [40]. In this sense, it has been argued that the copper–zinc oxide interface is one of the active sites for methanol formation. The zinc oxide patches covering the copper particles are physical barriers that inhibit copper diffusion across the surface and maximize the oxide metal interface area where the reaction takes place.

18.3 CARBON MONOXIDE HYDROGENATION

Transition metals and metal oxide promoted metals have been used for the synthesis of hydrocarbon [41–45] and oxygenated compounds [41,46–48] from syngas

$(CO + H_2)$ mixtures. Regarding these reactions, $\Delta G°$ values at 298 K are in most cases negative and a broad variety of products (paraffins, olefins, alcohols, ethers, and esters) can be obtained under appropriate reaction conditions. From an analysis of the vast existing body of work, it can be concluded that there is a close relationship between the nature and the chemical state of the elements responsible for the activity and the type of surface C containing intermediates and product distributions [21,49]. Thus, ABO_3 perovskite-type oxides with reducible B^{3+} ions under typical reaction conditions are excellent model catalysts for gaining meaningful insight into the surface processes taking place during CO (and CO_2) activation and carbon chain growth.

18.3.1 CO Hydrogenation on Chromia–Zinc Oxides

Both zinc oxide and chromium oxides are described in the former BASF patents as the preferred catalytic components for the synthesis of alcohols from syngas mixtures [50]. Later, Cr_2O_3–ZnO-based catalysts have been explicitly investigated for alcohol synthesis [18,51–54]. In addition, mixed Cr_2O_3–ZnO/HZSM5 systems have been studied for the direct conversion of syngas into aromatics [54,55]. A common feature of both higher alcohol and aromatic synthesis catalysts is the necessity of the formation of surface formate and methoxy species on Cr_2O_3–ZnO catalyst function.

   Through the use of areal rates, Bradford et al. [18] recently reported that specific activities can be described quite well by a simple linear combination of areal rate data for the pure component oxides (Cr_2O_3 and ZnO), indicating that no catalytic synergy is likely to occur between Cr_2O_3 and ZnO in these systems. The specific activity of Cr_2O_3–ZnO catalysts, prepared by co-precipitation with $(NH_4)_2CO_3$, shows a maximum for Cr/Zn ratios of $\sim$0.39 (Figure 18.3). Comparison of rates on an areal basis reveals that the activity of ZnO is effectively identical within the experimental error. Although the limited analysis cannot preclude the formation and presence of active sites for methanol synthesis on the Cr_2O_3–ZnO catalysts that do not exist on the surfaces of either Cr_2O_3 or ZnO, the data do indicate that, if such sites exist, they are unlikely to exert an intrinsic activity for methanol synthesis that exceeds the activity of sites present on the partially reduced ZnO surface. Thus, a rough approach to express global reaction rates, on a specific activity basis as a linear combination of the intrinsic areal rates of the pure component oxides would be:

$$r = r_{ZnO} \cdot S_{ZnO} + r_{Cr_2O_3} \cdot (S_{tot} - S_{ZnO}) \qquad (18.9)$$

where r is the global reaction rate on a specific basis; r_{ZnO} and $r_{Cr_2O_3}$ are the intrinsic areal rates for ZnO and Cr_2O_3, respectively, and S_{ZnO} and S_{tot} are the specific areas of ZnO and the total catalyst area. This simple calculation indicates that the increase in specific activity for methanol synthesis might primarily be due to the increase in accessible ZnO surface area rather than, for example, to the development of new active sites [56].

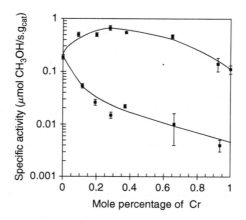

FIGURE 18.3 Specific activity of $(NH_4)CO_3$ precipitated ($\bullet$) and K_2CO_3 precipitated ($\blacksquare$) Cr_2O_3/ZnO catalysts for CH_3OH synthesis at 573 K and 34 bar as a function of mole percentage of Cr. $N_2:H_2:CO = 1:2:1$ molar ratio; gas hourly space velocity $= 5134\ cm^3/h.g_{cat}$. Bars indicate standard deviation (Reproduced with permission from Elsevier)

18.3.2 Strong Metal-Support Interactions

The rate of CO hydrogenation can be increased by almost two orders of magnitude by using titanium dioxide as a support for the hydrogenation function. Additionally, product distribution can also be modified in the presence of certain transition metal oxides that alone are not active in the CO hydrogenation reaction. It has been shown that the reactivity of the oxide–metal interface is much higher than that of the metal alone for this reaction [57]. This is because any CO bond in CO containing molecules such as CO_2 and CH_3OH is very sensitive to the presence of an oxide promoter such as TiO_x, which leads to an enhancement of the reaction rate.

The effect of oxide promoters on CO and CO_2 hydrogenation was experimentally verified by depositing different amounts of MO_x (M = Ti, Nb, Ta, Zr, V, W, and Fe) oxides on several transition metals [58,59]. Rhodium films, on which sub-monolayer amounts of MO_x had been deposited, were used for this purpose (Figure 18.4). Under fixed reaction conditions, a dramatic increase in the hydrogenation rate of carbon dioxide was found for the TiO_x, NbO_x, and TaO_x oxides at coverages ranging from 0.2 to 0.7 monolayers. However, for FeO_x the CO_2 hydrogenation rate was found to decrease monotonously with increasing FeO_x coverage. The complete inhibition of the hydrogenation activity at coverage of one monolayer was explained based on a simple site-blocking mechanism that occurs as the surface area of active rhodium decreases. The maximum in the CO hydrogenation rate at MO_x coverages of about 0.5 monolayers was explained in terms of the enhanced catalytic activity at the metal–metal oxide interface. Surface analysis of deposited MO_x moieties before and after the hydrogenation reaction revealed a correlation between the reducibility and the Lewis acidity of the metal oxide under reaction conditions and the rate enhancement. This improvement in rate was interpreted as a result of the formation of Lewis acid–base complexes of

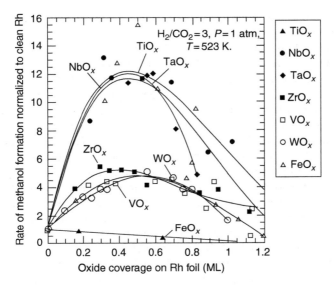

FIGURE 18.4 Effect of the metal oxide coverage on the rate of methane formation from CO_2 and H_2 over Rh foil. $H_2:CO_2 = 3$, $P = 1$ bar, $T = 523$ K (Reproduced with permission from Elsevier.)

adsorbed CO or H_2CO and an exposed metal cation at the Rh/MO_x boundary, which facilitates the cleavage of the C=O bond [59]. It has been shown that successive reduction–oxidation cycles generate an optimized periphery of the metal–metal oxide interface.

18.3.3 Mixed Oxides

The performance in CO hydrogenation of the $LaRhO_3$ mixed oxide was investigated in some detail by Watson and Somorjai [60]. The reaction products were CH_3OH and HC and the relative proportion of CH_3OH/HC was found to be highly temperature dependent. These results were interpreted assuming that two different reaction mechanisms would be involved: CH_3OH formation would take place through a nondissociative adsorption of CO, whereas HC would be formed following a dissociative mechanism. The change in the CH_3OH/HC ratio with temperature is due to the competing H- and CO-insertion reactions and variable proportions of molecularly and dissociatively adsorbed CO and H_2 on the surface. At reaction temperatures below 500 K, methanol is the major reaction product since hydrogenation of adsorbed CO species predominates. However, the selectivity trend changes strikingly at temperatures above 620 K; the C–O bond splits and CH_3OH formation is strongly inhibited, while methane and other $C_2 + HC$ are the major reaction products. In the mid-temperature range, C–O is broken, yielding CH_x fragments that can then recombine to form either CH_4 or other HC through hydrogenation reactions, or can undergo CO insertion to yield oxygenates. Surface analysis of the activated catalysts revealed that the active catalyst contains Rh^+ species and a small proportion of Rh°.

Gysling et al. [61] and Monnier and Apai [62] also investigated the hydrogenation of CO on H_2-reduced $LaRhO_3$ mixed oxide in the temperature range of 493 to 623 K. These authors found that the formation rates of HC decreased monotonously with the number of C atoms, suggesting the participation of a common mechanism for carbon chain growth. However, at temperatures above 548 K the formation rates of $C_2 + OH$ were found to be greater than for CH_3OH, indicating that the mechanism is different in both cases. The $C_2 + OH$ oxygenates are probably formed by CO insertion into adsorbed alkyl fragments, as previously reported by Watson and Somorjai [60]. The similarity in product distribution observed during the hydrogenation of CO on $LaRhO_3$ and a conventional Rh/SiO_2 catalyst at the same reaction temperatures suggests that the same catalytic Rh species would be present in both catalysts. Because the rhodium species present on Rh/SiO_2 systems is metallic Rh° under such reaction conditions [61,62], it may be inferred that the active rhodium present in $LaRhO_3$ must also be Rh°.

Mixed metal oxides of the type $BaBO_3$ (B = Rh, Ru, Ir, Pt) have also been used in CO hydrogenation [62]. $BaRhO_3$ and $BaPtO_3$ were found to selectively yield oxygenates: almost exclusively methanol (maximum selectivities of 62 and 54%, respectively). In contrast, the $BaRuO_3$ system showed the lowest oxygenate selectivity. All these perovskites were pure crystalline phases prior to catalyst testing. The bulk oxidation state of the noble metal was the expected B^{4+}, although partially reduced species instead of, or in addition to, the Ru^{4+} state were also observed. The crystalline and amorphous $BaBO_3$ structures become unstable under the conditions imposed by the hydrogenation reaction; the B^{4+} ion is reduced to M°, and the barium is converted to $BaCO_3$.

Copper-substituted mixed oxides also show good performance in CO hydrogenation [48]. Substitution of Mn by Cu in $LaMnO_3$ samples shifted the product distribution from HC to CH_3OH and small amounts of C_{2+} oxygenates. Substituted oxides of the type $LaTi_{1-x}Cu_xO_3$ showed that CO hydrogenation activity is strongly dependent on Cu substitution, the highest values were attained in the $0.5 \leq x \leq 1.0$ composition range and the lowest for $x = 0$. $LaTiO_3$ produced methanol, CH_4, and CO_2, with selectivities of about 30%, and a substantially lower proportion of dimethylether. The methanol synthesis rate was found to pass through a maximum at substitution $x = 0.6$. All these findings clearly indicate the importance of copper in an oxide matrix for both activity and selectivity toward methanol.

Surface analysis of used catalysts reveals that copper becomes reduced, and that for $x = 1.0$, and to a lesser extent for $x = 0.6$, the reduced Cu-phase coexists with a minor proportion of Cu^{2+} ions (Figure 18.5). This is consistent with the presence of both CuO and La_2CuO_4 phases in the calcined sample; only CuO is reduced in H_2 at 573 K, while La_2CuO_4 remains unreduced up to 673 K:

$$CuO + La_2CuO_4 + H_2 \rightarrow Cu^\circ + La_2CuO_4 + H_2O \qquad (18.10)$$

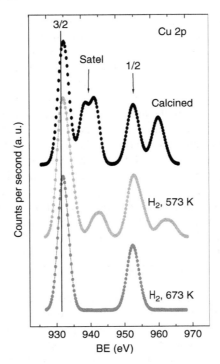

FIGURE 18.5 Cu 2p core-level spectra of $LaTi_{0.4}Cu_{0.6}O_3$ sample subjected to different treatments

The Auger parameter of copper in a representative $LaMn_{0.5}Cu_{0.5}O_3$ catalyst prereduced and used in CO hydrogenation at 573 K appeared at 1849.2 eV [48], in agreement with that at 1848.8 eV observed by Sheffer and King [46] for a K promoted copper catalyst exposed to a CO:H_2 mixture, and by Anewalt et al. [63] for Cu_2O (Cu^+). This indicates that Cu^+ species are present in the used catalyst and are presumably involved in the CH_3OH synthesis reaction. As copper is stabilized in the La_2CuO_4 phase, Cu^+ ions can be developed on its surface under on-stream operation. The fact that Cu^+ ions remain dispersed and embedded in the La_2CuO_4 matrix also explains the high resistance to sintering of the resulting reduced phase, thus explaining the good stability of these catalysts over long periods.

18.4 HYDROGENATION OF CO/CO₂ MIXTURES

Methanol is synthesized from $H_2/(CO+CO_2)$ mixtures over copper–zinc oxide systems appropriately modified with structural promoters (Al_2O_3, Cr_2O_3). Working with these catalysts, Klier [21] has proposed that methanol would be produced from the CO component of the reactant mixture, carbon dioxide being present to maintain the catalyst surface partially oxidized. The site responsible for the reaction was suggested to be a Cu^+ion, located in substitution positions in the ZnO lattice, on which CO adsorbs strongly and then undergoes hydrogenation to methanol. A quite different picture of the mechanism of methanol synthesis emerged from radiolabeling experiments [22] carried out to identify the source of methanol when mixed $CO/CO_2/H_2$ feeds are reacted with copper–zinc oxide (alumina) catalysts. At low conversion levels, the methanol had a specific radioactivity identical to that of the carbon dioxide feed, while at high conversions the specific radioactivity of the methanol was roughly the average of the inlet and the exit carbon dioxide activities. This is direct and irrefutable evidence that it is the CO_2 molecule of the feed that is the immediate precursor to methanol when synthesized on copper–zinc oxide (alumina) catalysts. In addition, ZnO alone produces methanol directly from carbon dioxide and not from carbon monoxide. The role of the CO component of the gas feed is that of a reducing agent, scavenging the oxygen produced in the synthesis of methanol from CO_2, from the surface of the copper [64]. The amount of oxygen existing on the surface of the copper was found to be a function of the CO_2/CO ratio. Therefore, it is agreed that it is the carbon dioxide component of the $CO/CO_2/H_2$ feed that is responsible for the synthesis of methanol, and it is the total surface area of the copper that is the active component of the catalyst.

The temperature-programmed surface reaction (TPSR) technique provided valuable information on the mechanism of the interaction of carbon dioxide and hydrogen on the surface of the copper–zinc oxide (alumina) catalyst [23]. TPRS experiments indicated that the energetics of formate (–OCOH) hydrogenation/hydrogenolysis on the copper component of the catalyst, which is the rate-determining step in methanol synthesis, is unaffected by the intimate mixing of copper and zinc oxide in the catalyst. The copper component of the catalyst is covered with oxygen, and, coexisting with this, formate species also adsorbed on the copper surface. The ZnO component of the catalyst acts neither by providing CO_2 to the copper component nor as a catalyst for the synthesis of methanol by a route parallel to that on the copper; under the reaction conditions, it is hydrided (H–Zn) or contains interstitial hydrogen in amounts exceeding the monolayer.

18.5 Carbon Dioxide Hydrogenation

18.5.1 Active Site and Role of ZnO

There is broad consensus that the copper metal area determines the activity of Cu–ZnO catalysts, in which the role of ZnO is to increase Cu dispersion. However, intrinsic activity, expressed per square meter of copper area, depends on the support of the copper phase. This is clearly seen in the case of the Cu–ZnO catalyst, which — under conditions that approach industrial practice — displays activity in CO_2 hydrogenation of at least one order of magnitude larger than Cu alone. Thus, the enhancement of the activity of copper in the presence of ZnO can be ascribed to the creation of active species on the Cu surface, in addition to the role of dispersing Cu particles.

The study of CO_2 hydrogenation on a physical mixture of Cu/SiO_2 and ZnO/SiO_2 catalyst provides a different picture about the nature of the active site [65]. The methanol synthesis activity of Cu/SiO_2 and the physical mixtures of Cu/SiO_2 + ZnO/SiO_2 as a function of the reduction temperature follow different trends. While the activity of Cu/SiO_2 remains almost unchanged regardless of the reduction temperature, the activity of the physical mixtures increases significantly with increasing reduction temperature in the range of 573 to 723 K. The lattice constant of Cu for Cu/SiO_2 coincides with the value of 3.615 Å of pure copper at any temperature; however, for the physical mixture it increases from 3.62 to 3.64 Å with the reduction temperature. The increase in the lattice constant is indicative of the formation of a CuZn alloy. It is thought that ZnO_x species migrate onto the surface of Cu and dissolve into the Cu particle to form such a CuZn alloy [65]. Elemental analysis of the physical mixture by transmission electron microscopy (TEM) coupled with energy-dispersive x-ray (EDX) provides support for the migration of ZnO_x species onto Cu particles. The coexistence of Zn with Cu was observed on the sample reduced at 723 K in the presence of both Cu/SiO_2 and ZnO/SiO_2. However, migration of ZnO onto the Cu particles does not take place when Cu/SiO_2 and ZnO/SiO_2 are reduced separately at 723 K and then mixed physically. Therefore, the increase in methanol synthesis activity can be attributed to the migration of the ZnO_x species onto the Cu surface.

Valuable information relating to the ZnO_x promotional effect on copper can be gathered from model catalysts [66]. The use of ultrahigh vaccum (UHV) techniques to monitor methanol synthesis from CO_2:H_2 gas mixtures over the Zn-deposited Cu(111) model system have shed some light on the described picture. CO_2 hydrogenation proceeds over the Zn-deposited Cu(111) sample under a pressure of 18 bar. The turnover frequency (TOF) for methanol synthesis at 523 K as a function of the Zn coverage on Cu was measured. The methanol yield (MTY) was of the order of $3 \times 10^{-4}\%$ at a TOF of 2×10^{-2} molecules site^{-1} sec^{-1}, which is lesser than the equilibrium conversion of 6.2%, implying that the reaction occurring is far from equilibrium. The TOF increases linearly with the Zn coverage up to coverages of 0.19 and then decreases at coverages above 0.20. Note that the optimum TOF obtained at a Zn coverage of 0.19 is 13-fold greater than that of the Zn-free Cu(111) surface. Thus, the promotional effect of Zn is also confirmed upon using a model catalyst.

18.5.2   Cu–ZnO-Based Systems

The hydrogenation of CO_2 was recently investigated on Cu–ZnO-based catalysts [37,38]. The reaction mainly produced methanol, although the formation of CO was also observed through the reverse water–gas shift (RWGS) side reaction. C containing products other than methanol and CO were not detected, except for some small fluctuating traces of methane at high temperatures. The MTY, expressed as mole MeOH/(h·kg cat), is shown as a function of reaction temperature in Figure 18.6. The MTY increased with increasing temperature, without any decrease or limiting value, indicating that thermodynamic restrictions or diffusion

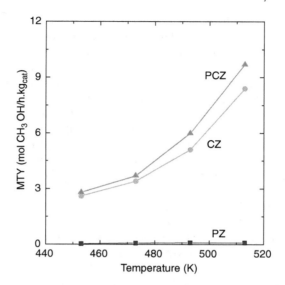

FIGURE 18.6 CH_3OH yield (mol $CH_3OH/h.g_{cat}$) as a function of the reaction temperature for Cu/ZnO (CZ) and Pd-modified Cu/ZnO (PCZ-CP and PCZ-SP) catalysts as a function of CO_2 conversion. $P = 60$ bar, $CO_2:H_2 = 3$, contact time $= 0.0675$ kg·h/m^3, and $T = 453$ to 513 K (Reproduced with permission from Elsevier.)

limitations are absent. The fact that CO production increased faster than the MTY with temperature caused a decrease in methanol selectivity with temperature and, therefore, with CO_2 conversion. This result is depicted in Figure 18.7, where methanol selectivity can be seen to decrease with CO_2 conversion.

The Pd containing catalysts (PCZ) behaved in a very different manner. While the PCZ-CP co-precipitated system is almost inactive, the catalyst prepared by sequential precipitation (PCZ-SP) increased the MTY with respect to the CZ base catalyst. In order to rule out the possibility that the increase in the MTY might be simply due to an additive effect in which palladium would act as an independent catalytic site, a reference catalyst (2/98 wt% PdO/ZnO) was employed. The absence of activity for this catalyst indicates that palladium, in the PCZ-SP catalyst, does not act as an independent and additional catalytic site for methanol synthesis, but gives rise to a synergetic effect. Given that the selectivity toward MeOH for the same CO_2 conversion was higher for the PCZ-SP catalyst rather than for CZ (Figure 18.7), it is clear that the enhancement in MTY was due to an increase in selectivity and not in CO_2 conversion. Therefore, the active sites are more selective for methanol synthesis. This finding is also consistent with the fact that the rate-determining step for methanol synthesis involves the hydrogenolysis of adsorbed formate (H_2COO^*), a situation where palladium would exert its hydrogenating role, whereas the RWGS side reaction involves a nonhydrogenation step as the rate-limiting step, namely the dissociation of adsorbed water to OH* and H*.

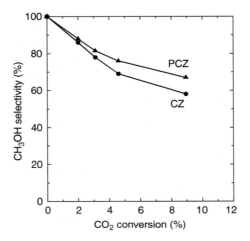

FIGURE 18.7 Methanol selectivity against CO_2 conversion. Catalysts and reaction conditions as in Figure 18.6 (Reproduced with permission from Elsevier.)

18.5.2.1 Surface structures

Palladium incorporation into a conventional CuO–ZnO methanol synthesis catalyst should not lead to a decrease in the copper area. Although there is no complete agreement on the final role of ZnO in methanol synthesis [37,38,67–70], all of them confirm the importance of metallic copper. Copper is required to be highly dispersed and any decrease in copper dispersion would cause a lowering of activity. Thus, the aforementioned decrease in the catalytic activity of the PCZ-CP catalyst can be interpreted in terms of a loss of copper area, as revealed by the N_2O chemisorption measurements. This poor copper dispersion for PCZ-CP is a consequence of two combined effects: the formation of the oxide phases from the decomposition of independent Cu and Zn phases in the precursor (see Reference 38 and references therein), and the high level of residual Na after calcination. Both factors inhibit the interaction between CuO and ZnO and result in an increase in the particle size of both oxide phases. Temperature programmed reduction (TPR) experiments showed a concomitant result in terms of the increased difficulty in CuO reduction (wide reduction peak) because of the large size of CuO crystallites [38]. The loss of interaction between Cu and Zn oxides leads not only to a low Cu dispersion in the reduced state, but also to a decrease in thermal stability during the reaction, as revealed by N_2O chemisorption measurements after reaction. In addition, the TPR technique confirms the low degree of interaction between Pd and Cu–Zn, since the Pd phase is reduced as isolated and large PdO particles. This catalyst configuration involving separate and weak interacting Pd, Cu, and Zn phases, which is obtained by co-precipitation, means that the effect of palladium is not capable of overcoming the loss in copper surface area.

The incorporation of palladium by the sequential precipitation methodology improved the methanol performance of the PCZ-SP catalyst for the direct hydrogenation of CO_2, as compared with the base CZ catalyst. This enhancement is not due to additional active palladium sites, but to a synergetic effect of Pd on the active Cu sites. At the calcined level, the oxide phases are formed by the decomposition of an aurichalcite phase [38]. This structure, in which Cu and Zn are atomically dispersed in a hydroxy-carbonate matrix, gives rise to well interdispersed oxide phases similar to those already found for the CZ precursor. In the reduced state, the copper area for the H_2-reduced PCZ-SP catalyst was almost the same as for the CZ sample. Thus, the enhancement in the methanol yield for PCZ-SP cannot be explained simply by a modification of the copper dispersion associated with the incorporation of Pd. However, the fact that the copper dispersion was maintained in a similar way to the base CZ sample is a key factor in the preparation of an optimized Pd-modified system. Therefore, the good performance obtained for the sequential precipitation methodology is because the aurichalcite precursor structure is preserved upon palladium addition: the copper dispersion was maintained at the same level as in the CZ catalyst and the effect of palladium was manifested.

X-ray photoelectron (XPS) and TPR experiments provided support to clarify the promoting effect of palladium in the PCZ-SP catalyst: an enhancement in CuO reducibility and an increase in the hydrogen consumption were found. The increase in the amount of hydrogen consumed (H_2/M) during CuO reduction for PCZ-SP, with respect to the CZ counterpart [38], indicated that hydrogen is easily dissociated on Pd particles and then spills over the Cu–ZnO phase. Furthermore, an enhancement in CuO reducibility was observed by TPR and supported by XPS. Such a reducibility effect for Pd-containing CuO–ZnO-based catalysts has attracted little attention in the literature but is generally interpreted as a result of direct interaction between Pd and Cu. The good interaction between Pd and Cu means that the H_2 spillover became effective in terms of improvement of CuO reducibility and, consequently, it is expected to improve the performance for methanol synthesis. However, other mechanisms of promotion on the working Pd-modified catalysts cannot be ruled out. A novel mechanism in which Pd, either by an "ensemble" or by a "ligand effect," modifies the surface redox properties of Cu has been advanced [71]. It is thus reasonable to infer that several mechanisms might operate during the reaction.

The hypothesis of an H_2 spillover mechanism, which is responsible for the increase in methanol yield, was also proposed by Inui and Takeguchi [72] and Sahibzada et al. [36], who reported that H_2 spillover influences the transportation rate of hydrogen to the active sites. Moreover, this mechanism keeps the copper surface in a more reduced state, thus counteracting the oxidizing effect of the CO_2 and water byproduct. The higher H_2/M ratio obtained for PCZ-CP catalyst suggested that the spillover effect of palladium is working, but becomes ineffective during the synthesis of methanol owing to the low copper area. The effect of palladium is not capable of counterbalancing this large loss of exposed copper. In fact, palladium cannot enhance the reducibility of CuO owing to the large dimensions of the CuO particles.

To summarize, the structure–activity relationship reveals the importance of the method of palladium incorporation and clearly demonstrates that the precipitation order has a remarkable effect on the properties of the active phases and, therefore, on the catalytic performance for the hydrogenation of CO_2 to methanol.

18.5.2.2 Catalyst deactivation on-stream

The Cu–ZnO-based catalysts deactivate progressively during the CO_2 hydrogenation reaction. The activation rate can be further accelerated when steam is added to the feed stream. Among the possible factors responsible for catalyst deactivation, structural changes in the catalysts appear to play a major role. The x-ray diffraction patterns of the catalysts used in methanol synthesis from CO_2-rich feed streams, in either the presence or the absence of steam, reveal an increase in Cu and ZnO particle sizes [73]. Thus, the water produced during methanol synthesis from a CO_2-rich feed accelerates the crystallization of Cu and ZnO contained in the catalyst, leading to deactivation of the catalyst. A small amount of silica added to the catalyst greatly improves the catalyst stability by suppressing the crystallization of Cu and ZnO. If the methanol synthesis reaction is performed from a CO-rich feed, instead from CO_2, the catalysts both with and without silica deactivate very little because only a small amount of water is produced during the reaction. Accordingly, no remarkable crystallization of Cu and ZnO contained in the catalyst occurs.

18.6 IN SITU STUDIES

As stated in the preceding sections, Cu–ZnO-based methanol synthesis catalysts have been discussed extensively and different views concerning the nature of the active site have emerged. One important recent development has been the application of EXAFS to study such catalysts. Detailed EXAFS studies have addressed the issue of whether Cu metal or a Cu–Zn interacting phase similar to an alloy may be present and to what extent it depends on the reaction conditions [69]. Although EXAFS assays have some limitations for the detection of small amounts of Cu–Zn alloys because amplitude functions and back-scattering amplitudes are quite similar for Cu and Zn, and the introduction of Zn into the Cu lattice will mainly result in small changes in the interatomic distances, Cu/ZnO catalysts heated at 873 K in a reducing atmosphere result in the formation of a Cu–Zn bulk alloy. Apart from this limitation, the temperature is much higher than that required in industrial practice in *in situ* Fourier transform infrared (FTIR) studies [74], and density functional theory (DFT) calculations [75] have provided evidence that Zn or ZnO species may migrate onto the Cu surfaces, especially under very reducing conditions.

The EXAFS spectra reveal large structural and morphological changes in the metallic Cu particles [76]. Strong reversible changes in the Cu–Cu coordination numbers occur along oxidizing–reducing cycles with CO + H_2 mixtures. Since the Cu–Cu distances were the same as in Cu°, the results may be understood by assuming reversible changes in the morphology of the metal particles. The influence on ZnO in the Cu–ZnO system is essential because similar changes did not occur in the Cu/SiO_2 system. A simple and reliable interpretation is that under the most reducing conditions, more oxygen vacancies are formed at the Cu–ZnO interface, and this favors the interaction between Cu and ZnO and a tendency for the Cu particles to spread over the ZnO to form flat structures (with decreased average Cu–Cu coordination numbers) [76]. Under more oxidizing conditions, the O vacancies disappear and the interaction between Cu and ZnO becomes weak. Thus, more spherical equilibrium-like Cu° structures are expected to result. Intuitively, these morphological changes may influence the kinetics of methanol synthesis. Indeed, the impact of both structural and morphological changes brought about by the reaction conditions has been emphasized in transient kinetic experiments [77].

In addition to *operando* EXAFS studies, *in situ* HRTEM allowed getting direct images of the structure and morphology of Cu/ZnO catalysts in different gaseous environments [78]. The HRTEM results not only confirmed the dynamic behavior of the system, but also provided significant additional insight. In a strongly reducing (H_2 and $H_2 + CO$) and in a less reducing environment ($H_2 + H_2O$), the copper is present as metallic Cu particles. In the most reducing atmosphere, the Cu particles appear, as in EXAFS, quite flat and spread over the ZnO support, whereas in the most oxidizing gas the Cu particles adopt a more spherical shape, indicative of decreased interaction with the support.

Transient methanol synthesis experiments have provided an understanding of how the catalytic activity is influenced by the structural transformation [77]. Briefly, the rates of methanol production change when Cu/ZnO/Al_2O_3 and Cu/Al_2O_3 catalysts were exposed to a CO:H_2 = 5:95 (molar) gas mixture and after switching to a less reducing gas CO:CO_2:H_2 = 5:5:90 (molar) mixture [77]. The steady-state activity for both systems is higher in the latter gas since methanol is mainly produced from CO_2. The transient behaviors observed for the two systems are markedly different. For the ZnO containing catalyst, exposure to the very reducing CO:H_2 = 5:95 (molar) mixture gives rise to flat Cu structures, with relatively high surface areas. As a result, a large increase in the methanol synthesis rate occurs after switching to the CO_2 containing gas. However, methanol production slowly decreased since these flat structures are not stable in the more oxidizing gas. For the Cu/Al_2O_3 catalyst, the transient effects are much smoother, since system changes in the interfacial energies are expected to be small. In addition, surface-induced reconstructions caused by chemisorption onto the active Cu surfaces are expected to be minor in view of the low concentration of adsorbed species [39].

18.7 SUMMARY

Some trends in the study of carbon monoxide and carbon dioxide hydrogenations over metal oxides and metal–metal oxide systems were examined in this chapter. These reactions usually take place over multicomponent and often multiphase catalyst systems with bifunctional sites, and produce HC and alcohols. The selectivity in these reactions is basically determined by the nature of the catalytic function, which determines largely the interaction of carbon oxides and hydrogen on the catalyst surface. CO binds to coordinatively unsaturated Zn^{2+} sites and acts as a

σ-donor, stabilizing the 5σ molecular orbital and donating electron density to the surface. This bonding results in a net dipole with positive charge on the C atom, and withdraws electron density from the 5σ molecular orbital, which is weakly antibonding with respect to the CO bond. Unlike CO bonding on most transition metals, there is no evidence for π back-bonding for the [CO–ZnO] complex, due to the high effective nuclear charge on the Zn^{2+} ion, which contracts the orbitals and stabilizes the d band to higher binding energies. CO surface bonding on Cu^+ is stronger, relative to Zn^{2+} sites due to not only the stronger σ-donor interaction but also the additional presence of π back-bonding. As a consequence of this electronic structure, the concentration of surface bound CO on Cu^+ containing copper–zinc catalysts would increase and hence the CO conversion rate would also increase.

The large body of work developed in recent years to elucidate the reaction mechanism and the roles of copper indicated that metallic Cu is the major species during the reaction; however, the presence of Cu^+ is also important because small amounts of oxygen increases the reaction rate. Isotope-labeling investigations showed that CO_2 is the primary source of methanol, particularly at low conversion as its hydrogenation rate is by about 20 times faster than that of CO. Water produced during methanol synthesis from CO_2-rich feed accelerates the crystallization of Cu and ZnO contained in the catalyst to lead to the deactivation of the catalyst. A small amount of silica added to the catalyst greatly improved the catalyst stability by suppressing the crystallization of Cu and ZnO. In the case of methanol synthesis from a CO-rich feed, the catalysts both with and without silica become less deactivated, because only a small amount of water is produced during the reaction, so no remarkable crystallization of Cu and ZnO contained in the catalyst occurs.

Recently, EXAFS, FTIR, and DFT calculations have provided evidence that Zn or ZnO species may migrate onto the Cu surfaces, especially under very reducing conditions. *In situ* HRTEM studies not only confirmed the dynamic behavior of the system, but also provided additional insight on the morphology of copper particles, which appeared spherical, with weak interaction with the ZnO substrate, under oxidizing environment and rather flat under strong reducing conditions.

ACKNOWLEDGMENTS

This contribution was partly supported by the MICYT under grant No. MAT2001-2215-C03-01. I would also like to thank Dr. I. Melian-Cabrera, Dr. M. Lopez-Granados, Dr. P. Terreros, Dr. S. Rojas, Dr. J.M. Campos-Martin, Dr. M. Ojeda, Dr. F.J. Perez Alonso, and Dr. T. Herranz for fruitful discussions.

REFERENCES

1. Somorjai, G.A. *Introduction to Surface Chemistry and Catalysis*. John Wiley Sons, New York, 1994.
2. Dry, M.E. Practical and theoretical aspects of the catalytic Fischer–Tropsch process. *Appl. Catal. A: Gen.* 1996, *138*, 319–344.

3. Iglesia, E. Design, synthesis, and use of cobalt-based Fischer–Tropsch synthesis catalysts. *Appl. Catal. A: Gen.* 1997, *161*, 59–78.

4. Borodko, Y. and Somorjai, G.A. Catalytic hydrogenation of carbon oxides. *Appl. Catal. A: Gen.* 1999, *186*, 355–362.

5. Tabata, K., Teng, Y., Takemoto, T., Suzuki, E., Bañares, M.A., Peña, M.A., and Fierro, J.L.G. Homogeneous activation of methane by oxygen and nitrogen oxides. *Catal. Rev. Sci. Eng.* 2002, *44*, 1–56.

6. Satterfield, C.N. *Heterogeneous Catalysis in Practice*. McGraw Hill, New York, 1980.

7. Kokes, R.J. Characterization of adsorbed intermediates on zinc oxide by infrared spectroscopy. *Acc. Chem. Res.* 1973, *6*, 226–233.

8. Scarano, D., Bertarione, S., Spoto, G., Zecchina, A., and Otero Arean, C. FTIR spectroscopy of hydrogen, carbon monoxide, and methane adsorbed and co-adsorbed on zinc oxide. *Thin Solid Films* 2001, *400*, 50–55.

9. Day, J.Q. and Zhang, J.Z.H. Steric effect in dissociative chemisorption of hydrogen on Cu. *Surf. Sci.* 1994, *319*, 193–198.

10. Leconte, J., Markovits, A., Skalli, M.K., Minot, C., and Belmajdoub, A. Periodic *ab initio* study of the hydrogenated rutile TiO$_2$(110) surface. *Surf. Sci.* 2002, *497*, 194–204.

11. Menetrey, M., Markovits, A., and Minot, C. Reactivity of a reduced metal oxide surface: hydrogen, water and carbon monoxide adsorption on oxygen defective rutile TiO$_2$(110). *Surf. Sci.* 2003, *524*, 49–62.

12. Yoshihara, J., Parker, S.C., Schafer, A., and Campbell, C.T. Methanol synthesis and reverse water gas-shift kinetics over clean polycrystalline copper. *Catal. Lett.* 1995, *31*, 313–324.

13. Yoshihara, J. and Campbell, C.T. Methanol synthesis and reverse water–gas shift kinetics over Cu(110) model catalysts: structural sensitivity. *J. Catal.* 1996, *161*, 776–782.

14. Solomon, E.I., Jones, P.M., and May, J.A. Electronic structures of active sites on metal oxide surfaces: definition of the Cu/ZnO methanol synthesis catalyst by photoelectron spectroscopy. *Chem. Rev.* 1993, *93*, 2623–2644.

15. Zecchina, A. and Otero, C. Diatomic molecular probes for mid-IR studies of zeolites. *Chem. Soc. Rev.* 1996, *25*, 187–207.

16. Anderson, A.B. and Nichols, J.A. Relaxation in ZnO (1010), (0001), and (0001) surface and the adsorption of CO. *J. Am. Chem. Soc.* 1986, *108*, 1385–1388.

17. Duke, C.B. Structure and bonding of tetrahedrally coordinated compound semiconductor cleavage. *J. Vac. Sci. Technol.* 1992, *10*, 2032–2040.

18. Bradford, M.C.J., Konduru, M.V., and Fuentes, D.X. Preparation, characterization and application of Cr$_2$O$_3$/ZnO catalysts for methanol synthesis, *Fuel Process. Technol.* 2003, *83*, 11–25.

19. Collman, J.P., Hegedus, L.S., Norton, J.R., and Finke, R.G. *Principles and Applications of Organotransition Metal Chemistry*. University Science Books, Mill Valley, CA, 1987.

20. Rozovskii, A.Ya. New data on the mechanism of catalytic reactions with the participation of carbon oxides. *Kinet. Catal.* 1980, *21*, 78–87.

21. Klier, K. Methanol synthesis. *Adv. Catal.* 1982, *31*, 243–313.

22. Chinchen, G.C., Denny, P.J., Parker, D.G., and Spencer, M.S. Mechanism of methanol synthesis from CO$_2$/CO/H$_2$ mixtures over copper/zinc oxide/alumina catalysts — use of C-14 labeled reactants. *Appl. Catal.* 1987, *30*, 333–338.

23. Bowker, M., Hadden, R.A., Houghton, H., Hyland, J.N.K., and Waugh, K.C. The mechanism of methanol synthesis on copper–zinc oxide–alumina catalysts. *J. Catal.* 1988, *109*, 263–273.

24. Saito, M., Fujitani, T., Takeuchi, M., and Watanabe, T. Development of copper/zinc oxide-based multicomponent catalysts for methanol synthesis from carbon dioxide and hydrogen. *Appl. Catal. A: Gen.* 1996, *138*, 311–318.

25. Inui, T., Hara, H., Takeguchi, T., and Kim, J.B. Structure and function of Cu-based composite catalysts for highly effective synthesis of methanol by hydrogenation of CO_2 and CO. *Catal. Today* 1997, *36*, 25–32.

26. Tagawa, T., Pleizier, G., and Amenomiya, Y. Methanol synthesis from $CO_2 + H_2$. 1. Characterization of catalysts by TPD. *Appl. Catal.* 1985, *18*, 285–293.

27. Amenomiya, Y. Methanol synthesis from $CO_2 + H_2$. 2. Copper-based binary and ternary catalysts. *Appl. Catal.* 1987, *30*, 57–68.

28. Koeppel, R.A., Baiker, A., Schild, C., and Wokaun, W. Carbon dioxide hydrogenation over Au/ZrO_2 catalysts from amorphous precursors — catalytic reaction mechanism. *J. Chem. Soc., Faraday Trans. I* 1991, *87*, 2821–2828.

29. Kanoun, N., Astier, M.P., and Pajonk, G.M. Catalytic properties of new Cu based catalysts containing Zr and or V for methanol synthesis from a carbon dioxide and hydrogen mixture. *Catal. Lett.* 1992, *15*, 231–236.

30. Walker, A.P., Lambert, R.M., Nix, R.M., and Jennings, J.R. Methanol synthesis over catalysts derived from $CeCu_2$ — transient studies with isotopically labeled reactants. *J. Catal.* 1992, *138*, 694–713.

31. Ghazi, M., Menezo, J.C., and Barrault, J. Characterization of nickel–molybdenum catalysts by x-ray diffraction spectroscopy and Fourier transform infrared spectroscopy. *New J. Chem.* 1991, *15*, 21–26.

32. Arakawa, H., Sayama, K., Okabe, K., and Murakami, A. Promoting effect of TiO_2 addition to CuO–ZnO catalyst on methanol synthesis by catalytic hydrogenation of CO_2. *Stud. Surf. Sci. Catal.* 1993, *77*, 389–392.

33. Fujitani, T., Saito, M., Kanai, Y., Kakumoto, T., Watanabe, T., Nakamura, J., and Uchijima, T. The role of metal oxides in promoting a copper catalyst for methanol synthesis. *Catal. Lett.* 1994, *25*, 271–276.

34. Toyir, J., de la Piscina, P.R., Fierro, J.L.G., and Homs, N. Highly effective conversion of CO_2 to methanol over supported and promoted copper-based catalysts: influence of support and promoter. *Appl. Catal. B: Environ.* 2001, *29*, 207–215.

35. Gotti, A. and Prins, R. Basic metal oxides as co-catalysts in the conversion of synthesis gas to methanol on supported palladium catalysts. *J. Catal.* 1998, *175*, 302–311.

36. Sahibzada, M., Chadwick D., and Metcalfe, I.S. Hydrogenation of carbon dioxide to methanol over palladium-promoted $Cu/ZnO/Al_2O_3$ catalysts. *Catal. Today* 1996, *29*, 367–372.

37. Melián-Cabrera, I., López Granados, M., and Fierro, J.L.G. Reverse topotactic transformation of a Cu–Zn–Al catalysts during wet Pd impregnation: relevance for the performance in methanol synthesis from CO_2/H_2 mixtures. *J. Catal.* 2002, *210*, 273–284.

38. Melián-Cabrera, I., López Granados, M., and Fierro, J.L.G. Pd-modified Cu–Zn catalysts for methanol synthesis from CO_2/H_2 mixtures: catalyst structure and performance. *J. Catal.* 2002, *210*, 285–294.

39. Ovesen, C.V., Clausen, B.S., Schotz, J., Stolze, P., Topsøe, H., and Norskov, J.K. Kinetic implications of dynamical changes in catalyst morphology during methanol synthesis over Cu/ZnO catalysts. *J. Catal.* 1997, *168*, 133–142.

40. Kanai, Y., Watanabe, T., Fujitani, T., Saito, M., Nakamura, J., and Uchijima, T. A surface science investigation of methanol synthesis over a Zn-deposited polycrystalline Cu surface. *J. Catal.* 1996, *160*, 65–75.

41. Fierro, J.L.G. Hydrogenation of carbon oxides over perovskite-type oxides. *Catal. Rev. Sci. Eng.* 1992, *34*, 321–336.

42. Fierro, J.L.G. Catalysis in C_1 chemistry — future and prospect. *Catal. Lett.* 1993, *22*, 67–91.

43. Fox, J.M., III. The different catalytic routes for methane valorization — an assessment of processes for liquid fuels. *Catal. Rev. Sci. Eng.* 1993, *35*, 169–212.

44. Schulz, H. Short history and present trends of Fischer–Tropsch synthesis. *Appl. Catal. A: Gen.* 1999, *186*, 3–12.

45. Van der Laan, G.P. and Beenackers, A.A.C.M. Kinetics and selectivity of the Fischer-Tropsch synthesis: a literature review. *Catal. Rev. Sci. Eng.* 1999, *41*, 255–318.

46. Sheffer, G.R. and King, T.S. Potassium promotional effect of unsupported copper catalysts for methanol synthesis. *J. Catal.* 1989, *115*, 376–387.

47. Brown-Bourzutschky, J.A., Homs, H., and Bell, A.T. Conversion of synthesis gas over $LaMn_{1-x}Cu_xO_3$ perovsites and related copper catalysts. *J. Catal.* 1990, *124*, 52–72.

48. Van Grieken, R., Peña, J.L., Lucas, A., Calleja, G., Rojas, M.L., and Fierro, J.L.G. Selective production of methanol from syngas over $LaTi_{1-x}Cu_xO_3$ mixed oxides. *Catal. Lett.* 1991, *8*, 335–344.

49. Diagne, C., Idriss, H., Pepin, I., Hindermann, J.P., and Kiennemann, A. Temperature-programmed desorption studies on Pd/CeO_2 after methanol and formic acid adsorption and carbon monoxide-hydrogen reaction. *Appl. Catal.* 1989, *50*, 43–53.

50. Mittasch, A. U.S. patent 1,569,775, 1926.

51. Lietti, L., Botta, D., Forzatti, P., Mantica, E., Tronconi, E., and Pasquon, I. Synthesis of alcohols from carbon oxides and hydrogen. 8. A temperature-programmed reaction study of *n*-butanol on a Zn–Cr–O catalyst. *J. Catal.* 1988, *111*, 360–373.

52. Calverley, E.M. and Smith, K.J. Kinetic model for alcohol synthesis over a promoted $Cu/ZnO/Cr_2O_3$ catalyst. *Ind. Eng. Chem. Res.* 1992, *31*, 792–803.

53. Campos-Martin, J.M., Guerrero, A., and Fierro, J.L.G. Structural and surface properties of $CuO–ZnO–Cr_2O_3$ catalysts and their relationship with selectivity to higher alcohol synthesis. *J. Catal.* 1995, *156*, 208–218.

54. Simard, F., Sedran, U.A., Sepulveda, J., Figoli, N., and de Lasa, H. $ZnO–Cr_2O_3 +$ ZSM-5 catalyst with very low Zn/Cr ratio for the transformation of synthesis gas to hydrocarbons. *Appl. Catal. A: Gen.* 1995, *125*, 81–98.

55. Ereña, J., Arandes, J.M., Bilbao, J., Gayubo, A.G., and de Lasa, H. Conversion of syngas to liquid hydrocarbons over a two-component ($Cr_2O_3–ZnO$ and ZSM-5 zeolite) catalyst: kinetic modeling and catalyst deactivation. *Chem. Eng. Sci.* 2000, *55*, 1845–1855.

56. Riva, A., Trifiro, F., Vaccari, A., Mintchev, L., and Busca, G. Structure and reactivity of zinc chromium mixed oxides. 2. Study of the surface reactivity by

temperature-programmed desorption of methanol. *J. Chem. Soc., Faraday Trans. I* 1988, *84*, 1423–1435.

57. Williams, K.J., Boffa, A.B., Salmeron, M., Bell, A.T., and Somorjai, G.A. The kinetics of CO_2 hydrogenation on a Rh foil promoted by titania overlayers. *Catal. Lett.* 1991, *9*, 415–426.

58. Boffa, A.B., Lin, C., Bell, A.T., and Somorjai, G.A. Promotion of CO and CO_2 hydrogenation over Rh by metal oxides — the influence of oxide Lewis acidity and reducibility. *J. Catal.* 1994, *149*, 149–158.

59. Boffa, A.B., Liu, C., Bell, A.T., and Somorjai, G.A. Lewis acidity as an explanation for oxide promotion of metals — implications of its importance and limits for catalytic reactions. *Catal. Lett.* 1994, *27*, 243–249.

60. Watson, P.R. and Somorjai, G.A. The formation of oxygen containing organic molecules by the hydrogenation of carbon monoxide using a lanthanum rhodate catalyst. *J. Catal.* 1982, *74*, 282–295.

61. Gysling, H.J., Monnier, J.R., and Apai, G. Synthesis, characterization and catalytic activity of $LaRhO_3$. *J. Catal.* 1987, *103*, 407–418.

62. Monnier, J.R. and Apai, G. Effect of oxidation states on the syngas activity of transition metal oxide catalysts. *Prepr. Div. Petro. Chem. Am. Chem. Soc.* 1986, *31*, 239.

63. Anewalt, M.R., Brown, D.M., and Karwacki, E.J. An investigation of the chemical states of copper in methanol catalysts and their relevance to activity maintenance. *Prepr. Div. Fuel. Chem. Am. Chem. Soc.* 1984, *29*, 210–214.

64. Chinchen, G.C., Waugh, K.C., and Whan, D.A. The activity and state of the copper surface in methanol synthesis catalysts. *Appl. Catal.* 1986, *25*, 101–107.

65. Fujitani, T. and Nakamura, J. The effect of ZnO in methanol synthesis catalysts on Cu dispersion and the specific activity. *Catal. Lett.* 1998, *56*, 119–124.

66. Fujitani, T. and Nakamura, J. The chemical modification seen in the Cu/ZnO methanol synthesis catalysts. *Appl. Catal. A: Gen.* 2000, *191*, 111–129.

67. Spencer, M.S. Precursors of copper/zinc oxide catalysts. *Catal. Lett.* 2000, *66*, 255–257.

68. Waugh, K.C. Comments on "The effect of ZnO in methanol synthesis catalysts on Cu dispersion and the specific activity" — (by T. Fujitani and J. Nakamura). *Catal. Lett.* 1999, *58*, 163–165.

69. Grunwaldt, J.D., Molenbroek, A.M., Topsøe, N.Y., Topsøe, H., and Clausen, B.S. *In situ* investigations of structural changes in Cu/ZnO catalysts. *J. Catal.* 2000, *194*, 452–460.

70. Choi, Y., Futagami, K., Fujitani, T., and Nakamura, J. The role of ZnO in Cu/ZnO methanol synthesis catalysts — morphology effect or active site model? *Appl. Catal. A: Gen.* 2001, *208*, 163–167.

71. Melian-Cabrera, I., Lopez Granados, M., and Fierro, J.L.G. Effect of Pd on Cu–Zn catalysts for the hydrogenation of CO_2 to methanol: stabilization of Cu metal against CO_2 oxidation. *Catal. Lett.* 2002, *79*, 165–170.

72. Inui, T. and Takeguchi, T. Effective conversion of carbon dioxide and hydrogen to hydrocarbons. *Catal. Today* 1991, *10*, 95–106.

73. Wu, J., Saito, M., Takeuchi, M., and Watanabe, T. The stability of Cu/ZnO-based catalysts in methanol synthesis from a CO_2-rich feed and from a CO-rich feed. *Appl. Catal. A: Gen.* 2001, *218*, 235–240.

74. Topsøe, N.Y. and Topsøe, H. FTIR studies of dynamic surface structural changes in Cu-based methanol synthesis catalysts. *J. Mol. Catal.* A 1999, *141*, 95–105.

75. Greeley, J., Gokhale, A.A., Kreuser, J., Dumesic, J.A., Topsøe, H., Topsøe, N.Y., and Mavrikakis, M. CO vibrational frequencies on methanol synthesis catalysts: a DFT study. *J. Catal.* 2003, *213*, 63–72.

76. Clausen, B.S., Topsøe, H., Hansen, L.B., Stoltze, P., and Nørskov, J.K. Determination of metal-particle sizes from EXAFS. *Catal. Today* 1994, *21*, 49–55.

77. Topsøe, H., Ovesen, C.V., Clausen, B.S., Topsøe, N.Y., Højlund Nielsen, P.E., Törnqvist, E., and Nørskov, J.K. Importance of dynamics in real catalyst systems. *Stud. Surf. Sci. Catal.* 1997, *109*, 121–139.

78. Hansen, P.L., Wagner, J.B., Helveg, S., Rostrup-Nielsen, J.R., Clausen, B.S., and Topsøe, H. Atom-resolved imaging of dynamic shape changes in supported copper nanocrystals. *Science* 2002, *295*, 2053–2055.

19 Photocatalysis: Photocatalysis on Titanium Oxide-Based Catalysts

M. Anpo*, S. Dohshi, M. Kitano, and Y. Hu
Department of Applied Chemistry, Graduate School of
Engineering, Osaka Prefecture University, Sakai, Osaka, Japan

CONTENTS

19.1 INTRODUCTION

Environmental pollution and destruction on a global scale as well as the lack of sufficient clean and natural energy sources have drawn much attention and concern to the vital need for ecologically clean chemical technology, materials, and processes — one of the most urgent challenges that the chemical scientists face. Since the photosensitized decomposition of water into H_2 and O_2 using a electrochemical cell consisted of Pt electrode and TiO_2 semiconductor electrode was first reported by Fujishima and Honda in 1972 [1], photocatalysis using various powdered semiconductors has received much attention for their potential in the conversion of light energy into useful chemical energy. Semiconductor photocatalysis, with its early focus on TiO_2 as an effective photocatalyst, has been applied to a variety of

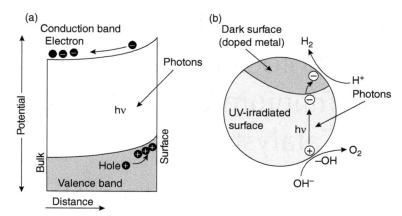

FIGURE 19.1 Photoproduction of electrons and holes in the n-type semiconductor TiO_2 particles (a), and photocatalytic production of H_2 and O_2 on TiO_2 particles suspended in water (b)

reactions, significantly, to address the reduction and elimination of environmental pollutants in water and air. It has been shown to be useful for the decomposition of microorganisms such as bacteria [2] and viruses [3], the deactivation of cancer cells [4,5], the degradation and elimination of offensive odors [6], the photosplitting of water to produce H_2 gas [7–12], the fixation of nitrogen [13–16], and the clean up of oil spills [17–19].

Semiconductors such as TiO_2, ZnO, and Fe_2O_3 can act as sensitizers for light-induced redox processes due to their electronic structure, which is characterized by a filled valence band and an empty conduction band [20]. As shown in Figure 19.1, when the energy of a photon corresponds or exceeds the band gap energy of the semiconductor, an electron is promoted from the valence band into the conduction band leaving a hole. The electrons in the conduction band and holes in the valence band formed in this way can recombine and dissipate the input energy as heat, and become trapped in metastable surface states, or they can react with electron acceptors and electron donors adsorbed on the semiconductor surface or within the surrounding electrical double layer of the charged particles.

In the absence of suitable electron and hole scavengers, the stored electrons and holes are dissipated within a few nanoseconds by their recombination [21]. If a suitable scavenger or surface defect state is available to trap the electron and the hole, recombination is prevented and subsequent redox reactions may occur by the remaining free electrons and the holes. The valence band holes act as powerful oxidants, while the conduction band electrons are good reductants [22]. Most organic photodegradation reactions can be attributed to the high oxidizing power of the holes either directly or indirectly. However, to prevent a buildup of charge one must also provide a reducible species such as oxygen to react with the electrons.

Applications of illuminated semiconductors in the remediation of contaminants have been successful for a wide variety of compounds [23–30]. In many cases, complete mineralization of the organic compounds has been reported. Recently, the discovery of photoinduced super-hydrophilicity and self-cleaning effect of TiO_2 thin films have led to even wider applications for TiO_2 photocatalysts [31].

In this chapter, advanced research work on such photofunctional catalysts and photocatalytic systems are introduced: (i) Design and development of second-generation TiO_2 photocatalysts that can operate under visible light irradiation. (ii) Photocatalytic performance of titanium oxide-based thin film catalysts. (iii) Photocatalytic decomposition of water into H_2 and O_2 using TiO_2 thin film photocatalysts. (iv) Characterizations of the local structures of the active sites of Ti–MCM-41 catalysts and their photocatalytic reactivity for the decomposition of NO into N_2 and O_2.

19.2 DESIGN AND DEVELOPMENT OF SECOND-GENERATION TiO_2 PHOTOCATALYSTS THAT CAN OPERATE UNDER VISIBLE LIGHT IRRADIATION

A highly advanced metal ion-implantation method was applied to modify the electronic properties of bulk TiO_2 photocatalysts by bombarding them with high energy metal ions, and it was discovered that such implantation with various transition metal ions, such as V, Cr, Mn, Fe, and Ni, accelerated by high voltage enabled a large shift in the absorption band of the titanium oxide catalysts toward the visible light region, with differing levels of effectiveness. However, Ar, Mg, or Ti ion-implanted TiO_2 exhibited no shift in the absorption spectra, showing that such a shift is not caused by the high energy implantation process itself but due to some interactions between the transition metal ions and the TiO_2 catalyst. The absorption band of the Cr ion-implanted TiO_2 smoothly shifts toward the visible light region, the extent of the red shift depending on the amount with the absorption maximum and minimum values always remaining constant, can be seen in Figure 19.2(b)–(d). The order of effectiveness in the red shift was found to be as follows: V > Cr > Mn > Fe > Ni ions. Such a shift allows the metal ion-implanted titanium oxide to use solar irradiation more effectively with efficiencies in the range of 20 to 30%.

Furthermore, such red shifts in the absorption band of the metal ion-implanted TiO_2 catalysts were observed for any kind of titanium oxide except the amorphous types, the extent of the shift differing from sample to sample. It was also found that such shifts in the absorption band could be observed only after calcination of the metal ion-implanted TiO_2 samples in O_2 at around 723 to 823 K. Therefore, it is clear that calcination under O_2 atmosphere in combination with metal ion-implantation was found to be instrumental in the red shift of the absorption spectrum toward visible light regions. These results clearly show that such shifts in the absorption band of the TiO_2 catalysts by metal ion-implantation are a general phenomenon and not a special feature of certain kinds of bulk TiO_2 catalysts.

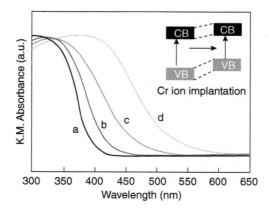

FIGURE 19.2   The UV-Vis absorption spectra of the (a) TiO_2 and (b)–(d) Cr ion-implanted TiO_2 photocatalysts. The amount of implanted Cr ions (μmol/g): (a) 0, (b) 0.22, (c) 0.66, and (d) 1.3

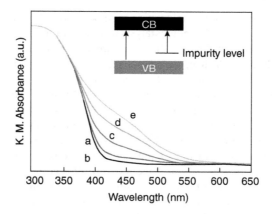

FIGURE 19.3   The UV-Vis absorption spectra of (a) TiO_2 and (b)–(e) Cr ion-doped TiO_2 photocatalysts prepared by an impregnation method. The amount of doped Cr ions (wt%): (a) 0, (b) 0.01, (c) 0.1, (d) 0.5, and (e) 1.0 (0.1 wt% equals to 4.9 μmol/g-TiO_2)

Figure 19.3 shows the absorption spectra of the TiO_2 catalysts impregnated or chemically doped with Cr ions in large amounts as compared with those for the Cr ion-implanted samples. The Cr ion-doped catalysts showed no shift in the absorption edge of TiO_2, however, a new absorption band appears at around 420 nm as a shoulder peak due to the formation of the impurity energy level within the band gap, its intensity increasing with the amounts of Cr ions chemically doped. These results indicate that the method of doping causes the electronic properties of the TiO_2 catalyst to be modified in completely different ways, thus confirming that only metal ion-implanted TiO_2 catalysts show such shifts in the absorption band toward visible light regions, even with much lower amounts of the ions, as compared with those for the chemically doped systems.

With the unimplanted original TiO_2 or with the chemically doped TiO_2 catalysts, the photocatalytic reaction does not proceed under visible light irradiation ($\lambda > 450$ nm). However, visible light irradiation of such advanced metal ion-implanted TiO_2 catalysts enabled them to initiate various significant photocatalytic reactions. Visible light irradiation ($\lambda > 450$ nm) of the Cr ion-implanted TiO_2 catalysts in the presence of NO at 275 K led to the decomposition of NO into N_2, O_2, and N_2O with a good linearity against the light irradiation time. Under the same conditions of visible light irradiation, the unimplanted original pure TiO_2 catalyst did not exhibit any photocatalytic reactivity. Moreover, the action spectrum for this reaction on the metal ion-implanted TiO_2 was in good agreement with the absorption spectrum of the catalyst shown in Figure 19.2, indicating that only metal ion-implanted TiO_2 catalysts were effective for the photocatalytic decomposition reaction of NO under visible light irradiation. Thus, it was found that metal ion-implanted TiO_2 catalysts could absorb visible light up to a wavelength of 400 to 600 nm and could, thus, operate effectively as photocatalysts, hence the name, "second-generation TiO_2 photocatalysts" [2,32–44].

It is important to emphasize that the photocatalytic reactivity of the metal ion-implanted TiO_2 catalysts retained the same photocatalytic efficiency as the unimplanted original TiO_2 catalyst under UV light irradiation ($\lambda < 380$ nm). When metal ions were chemically doped into the TiO_2 catalyst, the photocatalytic efficiency decreased dramatically even under UV irradiation due to the quick recombination of the photoformed electrons and holes through the impurity energy levels formed by the doped metal ions within the band gap of the catalyst (Figure 19.3). These results clearly show that metal ions that are physically implanted do not work as electron and hole recombination centers but work to modify the electronic property of the catalyst.

Various fieldwork experiments were conducted to test the photocatalytic reactivity of the newly developed TiO_2 catalysts under solar beam irradiation. Under outdoor solar light at ordinary temperatures, the Cr and V ion-implanted TiO_2 catalysts showed several times higher photocatalytic reactivity for the decomposition of NO than the unimplanted original TiO_2 catalysts. It was also found that the V ion-implanted TiO_2 catalysts showed several times higher photocatalytic reactivity for the hydrogenation of C_3H_4 with H_2O than the original TiO_2. These results, together with the results shown in Figure 19.2, clearly indicate that by using second-generation titanium oxide photocatalysts developed by applying the metal ion-implantation method, the effective and efficient use of visible and solar light energy by photocatalysis is now possible.

The relationship between the depth profiles of the metal ions implanted within TiO_2 having the same amounts of metal ions, such as V or Cr ions, and their photocatalytic efficiency under visible light irradiation was investigated. It was found that, when the same amount of metal ions were implanted into the deep bulk of the TiO_2 catalyst by applying high acceleration energy, the catalyst exhibited high photocatalytic efficiency under visible light irradiation. On the other hand, when low acceleration energy was applied, the catalyst exhibited low photocatalytic efficiency under the same conditions of visible light irradiation.

Increasing the number (or amounts) of metal ions implanted into the deep bulk of the TiO_2 catalyst was found to cause the photocatalytic efficiency to increase under visible light irradiation, passing through a maximum at around 6×10^{16} ions/cm^2, then decreasing with a further increase in the number of metal ions implanted. The presence of ions at the near surfaces could be observed by electron spectroscopy for chemical analysis (ESCA) measurements only on samples implanted with a large amount of metal ions. These results clearly show that there are optimal conditions in the depth and the amount of metal ions implanted to achieve high photocatalytic reactivity under visible light irradiation.

The electron spin resonance (ESR) spectra of the V ion-implanted TiO_2 catalysts were measured before and after calcination of the samples in O_2 at around 723 to 823 K, respectively. Distinct and characteristic reticular V^{4+} ions were detected only after calcination at around 723 to 823 K. When a shift in the absorption band toward visible light regions by the V ion-implantation was observed, the presence of the reticular V^{4+} ions could be detected by ESR measurements. No such V ions having the same local structure or shifts in the absorption band could be observed with TiO_2 catalysts chemically doped with V ions.

The XANES and Fourier transforms of the EXAFS oscillations for the TiO_2 catalysts chemically doped and also physically implanted with Cr ions were measured in order to clarify the local structures of these Cr ions. Results obtained from the analyses of the x-ray absorption fine structure (XAFS) spectra showed that, for the TiO_2 catalysts chemically doped with Cr ions by an impregnation or sol–gel method, the ions were present as aggregated Cr oxides in octahedral coordination similar to Cr_2O_3 and in tetrahedral coordination similar to CrO_3, respectively. On the other hand, in the catalysts physically implanted with Cr, the ions were found to be present in a highly dispersed and isolated state in octahedral coordination, clearly suggesting that the Cr ions are incorporated in the lattice positions of the TiO_2 catalyst in place of the Ti ions [2,32–44].

These findings clearly show that the success in modifying the electronic state of TiO_2 by metal ion-implantation, enabling the absorption of visible light even longer than 550 nm, is closely associated with the strong and long distance interaction that arises between the TiO_2 and the metal ions implanted, and not by the formation of impurity energy levels within the band gap of the catalysts, which result from the formation of aggregated metal oxide clusters that are often observed in the chemical doping of metal ions and oxides, as shown in Figure 19.3.

These results obtained in the photocatalytic reactions and various spectroscopic measurements of the catalysts clearly indicate that the implanted metal ions are highly dispersed within the deep bulk of the TiO_2 catalysts and work to modify the electronic nature of the catalysts without any changes in the chemical properties of the surface. These modifications were found to be closely associated with an improvement in the reactivity and sensitivity of the TiO_2 photocatalyst, thus, enabling the TiO_2 catalyst to absorb and operate effectively not only under UV but also under visible light irradiation. As a result, under outdoor solar light irradiation at ordinary temperatures, transition metal ion-implanted TiO_2 catalysts showed several times higher photocatalytic efficiency than the unimplanted original TiO_2 catalysts.

19.3 PHOTOCATALYTIC PERFORMANCE OF TITANIUM OXIDE-BASED THIN FILM CATALYSTS

TiO_2 thin films have been widely investigated not only for their high photocatalytic reactivity but also for their unique photoinduced super-hydrophilic properties [31,45–54]. Some of the main preparation methods reported for TiO_2 thin films so far have been the sol–gel method and metal oxide chemical vapor deposition (MOCVD) method. However, recently, we have shown that an ionized cluster beam (ICB) deposition method can be applied as a good technique to produce highly homogeneous and efficient TiO_2 thin film photocatalysts without the need for special treatment or calcination at high temperatures for the crystallization process after deposition [35,36,55,56]. We have also reported that binary oxide thin films of different compositions can be easily prepared by the ICB method using multi-ion sources by the control of each deposition rate [57–59].

In previous literature, we have reported that the photocatalytic reactivity of Ti/Si binary oxides prepared by a coprecipitation method is strongly affected by changes in the composition of the catalyst, that is, by changing the Ti/Si ratios [60]. In addition, Imamura et al. have reported that Ti/Si binary oxides prepared by the sol–gel method exhibit a high catalytic activity for the selective epoxidation of alkenes [61]. On the other hand, to improve the photoinduced super-hydrophilic properties of TiO_2 thin films, a combination or mixing of the Ti oxide moiety with other oxides such as SiO_2 or B_2O_3 has been reported. Significantly, it has been shown that the addition of SiO_2 fine particles or the deposition of the SiO_2 thin layer onto the surface of the TiO_2 thin films is effective in maintaining super-hydrophilic properties even under dark conditions for long periods after light irradiation has been ceased [62].

In this section, Ti/Si and Ti/B binary oxide thin films were prepared by the ICB method using multi-ion sources. Characterization studies were carried out by spectroscopic measurements such as XRD, XAFS and UV-Vis, in order to elucidate the local structure of the Ti oxide species of these thin films. The photocatalytic reactivity was also evaluated for significant reactions such as the decomposition of NO to produce N_2 and O_2 as well as N_2O as a minor product under UV light irradiation. Moreover, the photoinduced super-hydrophilic properties were evaluated by measuring the changes in the contact angle of water droplets under UV irradiation and under dark conditions at 298 K.

Ti/Si and Ti/B binary oxide thin films with various TiO_2 compositions were prepared by the ICB method on a quartz substrate. As a pretreatment, the quartz substrates were ultrasonically cleansed for 15 min in acetone, then dried and annealed at 723 K for 5 h in air. The source materials, Ti metal and SiO (or B_2O_3) powder, were heated at high temperatures, then Ti and SiO (or B_2O_3) vapors were

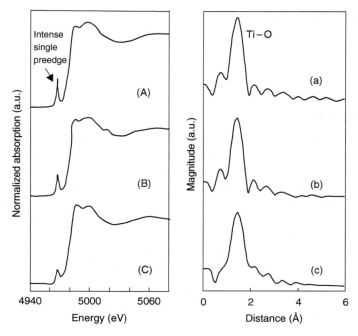

FIGURE 19.4   XAFS (XANES and Fourier transforms of EXAFS) spectra of Ti/Si binary oxide thin films. TiO_2 content (%): (a, A) 6.6, (b, B) 9.5, and (c, C) 50.1

introduced into the high vacuum chamber to produce Ti and SiO (or B_2O_3) clusters. Stoichiometric TiO_2 and SiO_2 (or B_2O_3) nanoclusters were formed on the substrate by the impingement of the ionized and neutral Ti and SiO (or B_2O_3) clusters and oxygen gas (2×10^{-4} Torr), respectively. By controlling each deposition rate, Ti/Si and Ti/B ratios could be accurately determined.

The XAFS spectra (XANES and Fourier transforms of EXAFS) of the Ti/Si and Ti/B binary oxide thin films were measured in order to elucidate the local structure of the Ti oxide species within these thin films. Figure 19.4 shows the XAFS spectra of the Ti/Si binary oxide thin films. The XANES spectra of these thin films with low TiO_2 compositions (Figure 19.4, left) exhibit an intense single preedge peak, which could be attributed to the tetrahedrally coordinated TiO_2 species. Moreover, the Fourier transforms of the EXAFS spectra of the Ti/Si thin films (Figure 19.4, right) exhibit a strong peak at around 1.6 Å which can be attributed to the Ti–O peak.

On the other hand, for the Ti/B binary oxide thin films with low TiO_2 compositions, the XANES spectra (Figure 19.5, left) exhibit three small characteristic preedge peaks which can be attributed to the TiO_2 with an anatase crystalline structure. In the Fourier transforms of the EXAFS spectra for the Ti/B thin films (Figure 19.5, right), not only the Ti–O peak but also the Ti–O–Ti peak can be seen.

The results of the curve fitting analysis of the FT-EXAFS spectra to obtain the coordination number and the bond distance between the Ti and O atoms of the

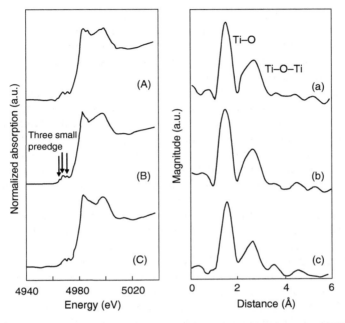

FIGURE 19.5 XAFS (XANES and Fourier transforms of EXAFS) spectra of Ti/B binary oxide thin films. TiO$_2$ content (%): (a, A) 5.0, (b, B) 10.0, and (c, C) 50.0

TABLE 19.1
The Results of the Curve Fitting Analysis of the FT-EXAFS Data for TiO$_2$ Thin Films as well as Ti/Si and Ti/B Binary Oxide Thin Films

Catalyst	Shell	Bond distance (Å)	Coordination number
Pure TiO$_2$	Ti–O	1.91	6.0
Ti/Si (6.6/93.4)	Ti–O	1.81	4.3
Ti/Si (9.5/90.5)	Ti–O	1.82	4.4
Ti/Si (50.1/49.9)	Ti–O	1.85	4.9
Ti/B (5/95)	Ti–O	1.91	5.98
Ti/B (10/90)	Ti–O	1.90	5.97
Ti/B (50/50)	Ti–O	1.90	5.97

Ti oxide species in the Ti/Si and Ti/B binary oxides are summarized in Table 19.1. It is clear that the coordination numbers of the Ti/Si binary oxides with lower TiO$_2$ compositions are close to 4.0, indicating that the major Ti oxide species are located in tetrahedral coordination, whereas the coordination numbers for the Ti/B binary oxides are always close to 6.0 even for the Ti/B with low TiO_2 content, showing that the major Ti oxide moiety are present in octahedral coordination. These results clearly indicate that for Ti/Si binary oxide thin films with low TiO_2 composition, the tetrahedrally coordinated Ti oxide species are highly dispersed in the host SiO_2 matrices, whereas, for the Ti/B binary oxide thin films with low TiO_2 composition, the octahedrally coordinated ultrafine TiO_2 nanoparticles are formed in the host B_2O_3 matrices. Furthermore, the bond distance between the Ti and O atoms of the Ti oxide species in the Ti/Si binary oxide thin films varies from 1.81 to 1.85 Å with an increase in the TiO_2 compositions from 6.6 to 50.1 wt%, reflecting the changes in the structures of the Ti oxides from highly dispersed isolated tetrahedrally coordinated Ti oxide species to TiO_2 anatase nanoparticles dispersed within the SiO_2 matrices. The bond distance between the Ti and O atoms of the Ti oxide species in the Ti/B binary oxide thin films scarcely changes and is almost constant at around 1.90 Å, being in a good agreement with the fact that Ti oxides in Ti/B are always present as TiO_2 nanoparticles within the B_2O_3 matrices.

The UV-Vis absorption spectra using a transmittance method for the Ti/Si and Ti/B binary oxide thin films with lower TiO_2 content are shown in Figure 19.6. For both, Ti/Si and Ti/B binary oxide thin films, when the TiO_2 composition is low, the absorption band of TiO_2 shifts toward shorter wavelength regions. This can be attributed to the presence of the tetrahedrally coordinated Ti oxide species in the SiO_2 matrices and the quantum size effect due to the formation of extremely small octahedrally coordinated TiO_2 nanoparticles in the B_2O_3 matrices, respectively. It is also found that these binary oxide thin films are highly transparent and colorless while having a high transmittance in the visible light region due to their highly homogeneous crystallinity and surface morphology. It should also be noted that the interference fringes, which could be remarkably observed for the pure TiO_2 thin

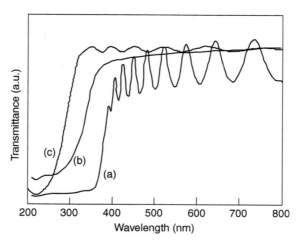

FIGURE 19.6   UV-Vis absorption spectra by transmittance method of Ti/Si and Ti/B binary oxide thin films: (a) TiO_2, (b) Ti/B (5/95), and (c) Ti/Si (6.6/93.4)

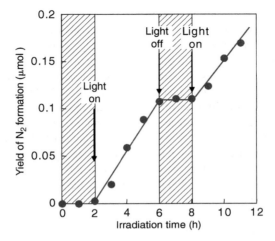

FIGURE 19.7 The reaction time profile of the photocatalytic decomposition of NO on Ti/Si binary oxide thin films (TiO$_2$ content: 6.6%) under UV light irradiation at 275 K

film almost disappear for Ti/Si and Ti/B binary oxide thin films, especially for Ti/B as compared with TiO$_2$ thin films with the same film thickness, by combination or mixing with SiO$_2$ and B$_2$O$_3$. This is a great advantage for the application of these thin films on various substrates, significantly, in that they do not lose their original transparency and color.

UV light irradiation of Ti/Si and Ti/B binary oxide thin films in the presence of NO led to the photocatalytic decomposition of NO into N$_2$ and O$_2$ as well as N$_2$O at 298 K. Figure 19.7 shows the reaction time profile of the photocatalytic decomposition of NO on the Ti/Si (Ti/Si ratio = 6.6/93.4) binary oxide thin films. As shown in this figure, the reaction proceeds with a good linearity against the irradiation time, clearly demonstrating that the decomposition reaction proceeds photocatalytically. We found the effect of the TiO$_2$ composition on the selectivity for the formation of N$_2$ as well as the yields of N$_2$ and N$_2$O in the photocatalytic decomposition of NO on Ti/Si. As mentioned earlier, when the TiO$_2$ composition is low, the tetrahedrally coordinated Ti oxide species are formed as the major Ti oxide species. Upon the excitation of these Ti oxide species by UV light, the charge transfer excited state of these species, (Ti^{3+}–O$^-$)*, are formed. These excited (Ti^{3+}–O$^-$)* species have a high and unique photocatalytic activity for various reactions [63], interacting with the NO molecules to decompose into N$_2$ and O$_2$ with high efficiency and selectivity at 298 K. However, when the TiO$_2$ content is increased, the nanoclusters and even nanoparticles of TiO$_2$ having octahedral coordination are formed, their photocatalytic behavior being the same as that of pure TiO$_2$ semiconducting photocatalysts where UV irradiation induces the formation of electrons in the conduction band and holes in the valence band, respectively, and these species react with NO to lead to the formation of N$_2$O and NO$_2$. Thus, when the TiO$_2$ content was increased, the yield and the selectivity for the formation of N$_2$ decreased. These results are in good agreement with our previous study in which it was found that only highly dispersed tetrahedrally coordinated Ti oxide species decomposes NO into N_2 and O_2, while octahedrally aggregated Ti oxides or nanoparticles decompose NO into N_2O and NO_2 [37,64,65].

On the other hand, for the Ti/B binary oxide thin films of low TiO_2 composition, ultrafine TiO_2 nanoparticles of octahedral coordination are formed in the host B_2O_3. Moreover, in contrast to the characteristics for Ti/Si binary oxide systems, it was found that UV irradiation of the Ti/B binary oxide thin films with low TiO_2 content in the presence of NO leads to the formation of N_2 and O_2 with a high selectivity [57,58]. It was thus observed that in the host B_2O_3, highly dispersed octahedrally coordinated Ti oxide nanoclusters can decompose NO molecules into N_2 and O_2 with high efficiency and selectivity at 298 K [66]. The study on the true nature and mechanisms behind the photocatalytic activity of highly dispersed octahedrally coordinated Ti oxide nanoclusters formed in the host B_2O_3 is now underway.

The contact angle of water droplets for the TiO_2 thin films was found to dramatically decrease under UV light irradiation and finally reach zero degree, that is, highly efficient super-hydrophilicity could be achieved. However, for the Ti/Si binary oxide thin films, the contact angle of the water droplets did not change dramatically, while, as shown in Figure 19.8, for the Ti/B binary oxide thin films, all of the samples except the pure B_2O_3 achieved super-hydrophilicity, reaching a water contact angle of zero degree under UV light irradiation. The figure shows

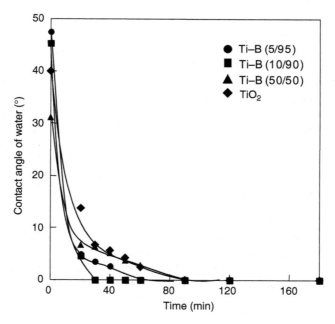

FIGURE 19.8　The changes in the contact angle of water droplets under UV light irradiation on Ti/B binary oxide thin films

the initial decrease rate for the photoinduced hydrophilicity of these binary oxide thin films. The decrease rate in the contact angle for Ti/B was found to be much larger than that of the pure TiO_2 thin films. Especially, for the binary oxide thin films containing more than 90% B_2O_3, the rate was several times larger than that for pure TiO_2. Moreover, as soon as UV light irradiation was discontinued, the contact angle of the water droplets for the TiO_2 thin films immediately increased and returned to its initial value, clearly showing that it is a light-induced phenomenon. However, for the Ti/B binary oxide thin films, only half the value could be recovered.

19.4 PHOTOCATALYTIC DECOMPOSITION OF WATER INTO H_2 AND O_2 USING TiO_2 THIN FILM PHOTOCATALYSTS

In recent years, the production of H_2 and O_2 through the photocatalytic decomposition of water using solar energy has received a great deal of attention in studies for the design and development of artificial photosynthesis systems since it would enable the conversion of solar energy into useful chemical energy. Intensive studies of photocatalysis on semiconducting materials were initiated by the pioneering work of Fujishima and Honda [1], which reports on the first photo-assisted production of H_2 and O_2 from water by utilizing a photoelectrochemical cell consisting of a Pt and TiO_2 electrode under a small electric charge. It was then observed that TiO_2 semiconducting nano-powdered photocatalysts loaded with small amounts of Pt could work to decompose H_2O into H_2 and O_2 as a mixture of gas under UV light irradiation at wavelengths shorter than 380 nm [67–72]. It was also observed that in such powdered photocatalytic systems, the back reaction of the photoformed H_2 and O_2 to form the original H_2O could proceed on these Pt catalysts, resulting in low reaction yields [73]. In addition to the decomposition reaction of H_2O into H_2 and O_2, various other significant applications of TiO_2 photocatalysts for the purification of toxic compounds in polluted water and air [74,75], properties such as their photoinduced super-hydrophilicity [47,57] are also being investigated. TiO_2 photocatalytic materials are attractive for their nontoxic, clean, safe properties as well as for their thermal stability and the abundance of the raw materials. However, until recently, they have worked only under UV light irradiation of wavelengths shorter than 380 nm. However, new types of TiO_2 photocatalysts that are able to absorb and operate effectively and efficiently even under visible or solar light are being developed [38–40,76,77].

Along these lines, we were able to modify the electronic properties of TiO_2 photocatalysts by applying an advanced metal ion-implantation method in which ion-implanted transition metal cations, such as V and Cr, were found to locate at the lattice position to replace the Ti^{4+} of the TiO_2 [41,42,75,78–80]. Although this method is theoretically very interesting and was successful in actual experiments, it was not practical for mass production due to the high cost of using such an advanced, high velocity ion-implantation apparatus to develop these TiO_2 photocatalysts.

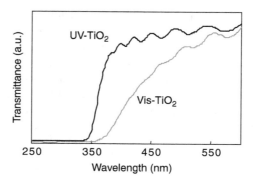

FIGURE 19.9 UV-Vis absorption spectra (in transmittance) of TiO_2 thin films prepared on a quartz substrate by a RF magnetron sputtering deposition method at different substrate temperatures. Substrate temperatures in K: UV-TiO_2, 473 K; Vis-TiO_2, 873 K

In this section, a radio frequency (RF) magnetron sputtering method is introduced as a novel and effective way to develop visible light-responsive TiO_2 thin film photocatalysts [43,75]. This method has great potential for the practical production of new types of thin film photocatalysts since it enables the high speed deposition of uniform thin films with large areas and the precise control of the morphologies and chemical composition of the thin films.

Figure 19.9 shows the UV-Vis absorption spectra of the TiO_2 thin film photocatalysts prepared on quartz substrates with a TiO_2 thickness of 1.2 μm under differing substrate temperatures. As can be seen in Figure 19.9, the thin films prepared at lower substrate temperatures ($T = 473$ K) are colorless and transparent in the visible light regions, enabling the absorption only of UV light shorter than 380 nm (hereafter, referred to as UV-TiO_2 thin films). X-ray diffraction (XRD) analysis of these UV-TiO_2 thin films showed that it is made up of a homogeneous anatase crystalline structure. On the other hand, TiO_2 thin films prepared on substrates at higher temperatures ($T = 873$ K) are observed to be yellow colored, enabling the absorption of visible light, its extent strongly depending on the variations in the substrate temperature (hereafter referred to as Vis-TiO_2 thin films). The thin film photocatalyst prepared at a substrate temperature of 873 K enables the absorption of visible light up to 600 nm, as can be seen in Figure 19.9. XRD analysis of these Vis-TiO_2 thin films showed that they consist of a mixture of the anatase and rutile crystalline structural phases, with different composition ratios for anatase and rutile. Visible light-responsive TiO_2 thin film photocatalysts thus, could be prepared on various substrates by the control of the substrate temperature during the TiO_2 deposition process [42,75].

The photocatalytic decomposition of water using these UV-TiO_2 and Vis-TiO_2 thin film photocatalysts onto which small amounts of Pt were loaded was then successfully carried out. Both the UV-TiO_2 and Vis-TiO_2 thin film photocatalysts could split water into H_2 and O_2 under UV light irradiation. It was found that the Pt loaded Vis-TiO_2 thin film photocatalysts show much higher photocatalytic reactivity than the Pt loaded UV-TiO_2 thin film photocatalysts under UV light irradiation ($\lambda < 380$ nm), although the surface areas of Vis-TiO$_2$ (22.7 m^2/g) was much smaller than UV-TiO$_2$ (50.9 m^2/g).

The effect of the wavelength of the irradiation light (action spectra) on the decomposition of H$_2$O using these types of TiO$_2$ thin film photocatalysts were investigated for the evolution of H$_2$ in water involving methanol as the sacrificial reagent. H$_2$ evolution could be seen when the Pt loaded UV-TiO$_2$ thin films were irradiated under UV light of $\lambda < 380$ nm but not under visible light of $\lambda > 400$ nm, which was quite similar to the reactions observed on the Pt loaded TiO$_2$ thin films prepared by a sol–gel method. On the other hand, Pt loaded Vis-TiO$_2$ exhibited reactivity for the evolution of H$_2$ even under the visible light irradiation of $\lambda > 420$ nm. Moreover, the photocatalytic activity of Pt loaded Vis-TiO$_2$ was also higher than that of Pt loaded UV-TiO$_2$ under UV light irradiation.

These results clearly suggest that not only the band gap value of the Vis-TiO$_2$ is much smaller than that of UV-TiO$_2$ but also the conduction band edge or valence band edge of Vis-TiO$_2$ locates at a position of sufficient potential for the decomposition of H$_2$O into H$_2$ or O$_2$. In fact, Pt loaded Vis-TiO$_2$ could decompose water into H$_2$ and O$_2$ under light irradiation of longer than 390 nm.

In order to clarify the origin or cause for the absorption of visible light up to 600 nm by Vis-TiO$_2$, various spectroscopic investigations using XRD, ESCA, SIMS, TEM, and STM were undertaken. Analyses of these studies showed that Vis-TiO$_2$ exhibits column structures which stand perpendicularly on the substrate while the depth profiles for the concentration of the Ti^{4+} and O^{2-} ions which constitute Vis-TiO$_2$ gradually changes from the top surface to the deep inside bulk, that is, a O^{2-}/Ti^{4+} ratio starting from 2.00 at the top surface to 1.933 in the inside bulk can be observed. For UV-TiO$_2$, a smooth wall-like structure standing on the substrate could be observed with the O^{2-}/Ti^{4+} ratio scarcely changing and keeping a constant of 2.00 from the top surface to the deep inside bulk, showing a remarkable contrast with the Vis-TiO$_2$ thin film photocatalysts. With Vis-TiO$_2$, we believe that such a decline in the values in the Ti^{4+} and O^{2-} composition of the structure causes a profound perturbation in the electronic structure of TiO$_2$, enabling absorption and operation under visible light. The results obtained by SIMS analyses are shown together with a cross-sectional TEM image of the Vis-TiO$_2$ thin films in Figure 19.10.

Recently, many studies have been devoted to the production of H$_2$ and O$_2$ by water splitting using various types of powdered photocatalysts under UV light irradiation [81–83], and in some cases, it has been achieved even under visible light irradiation [84]. However, powdered photocatalytic systems always yield a mixture of H$_2$ and O$_2$ because of the water splitting reaction. The separate evolution of H$_2$ and O$_2$ from water [85,86] is strongly desired to obtain pure H$_2$ gas which can be easily and safely utilized as fuel on a large scale. For this purpose, an H-type cell was constructed consisting of two water phases separated by a proton-exchange membrane and the Vis-TiO$_2$/Ti/Pt photocatalyst. The Vis-TiO$_2$/Ti/Pt photocatalyst was prepared by depositing Vis-TiO$_2$ on one side of a Ti metal foil (1.2 μm) and Pt on the other side of the Ti foil by a RF magnetron sputtering deposition method.

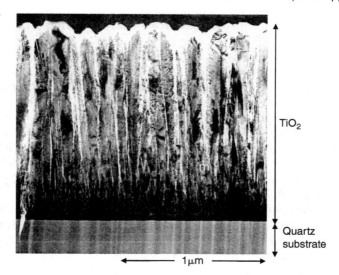

FIGURE 19.10 Cross-sectional TEM image of the Vis-TiO$_2$ thin film photocatalyst prepared on a quartz substrate and the changes in the ratio of Ti^{4+} and O^{2-} composition from the top surface to the inside bulk

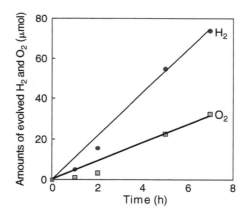

FIGURE 19.11 The photocatalytic decomposition of H$_2$O and the separate evolution of H$_2$ and O$_2$ under light irradiation with a 500 W Xe arc lamp on the Vis-TiO$_2$/Ti/Pt thin film photocatalyst having an H-type cell with two aqueous phases of different pH values, at 298 K

Irradiation of Vis-TiO$_2$ by a 500 W Xe arc lamp led to the stoichiometrical evolution of H$_2$ and O$_2$ from the Pt and Vis-TiO$_2$ sides, respectively (Figure 19.11). The stoichiometrical and separate evolution of H$_2$ and O$_2$ were, thus, successively carried out by the development of a Vis-TiO$_2$/Ti/Pt photocatalyst used in a H-type cell. The yields of the evolved H$_2$ and O$_2$ were found to increase remarkably as the pH difference between the two aqueous phases increased. This suggests that the pH difference improves the charge separation of the photoformed electrons and holes, that is, the electron transfer from the photoirradiated visible-type TiO_2 to the Pt deposited on the other side of the Ti metal foil. We have, in fact, shown that this system can be applied for the separate evolution of H_2 and O_2 under near-visible light irradiation of wavelengths longer than 390 nm and even under sun light beam irradiation.

The Vis-TiO_2 thin film photocatalysts were thus characterized to have a unique columnar structure with a declined O/Ti ratio from the surface to the deep bulk near the substrate, prepared by controlling the substrate temperature and the Ar pressure during the RF magnetron sputtering deposition process. The band gap of Vis-TiO_2 is smaller than that of UV-TiO_2, while the conduction band edge and valence band edge of Vis-TiO_2 have enough potential to decompose water into H_2 or O_2, respectively. We have also succeeded in the stoichiometric and separate evolution of H_2 and O_2 by utilizing the unique H-type photocatalytic cell combined with Vis-TiO_2/Ti/Pt photocatalyst and proton-exchange membrane.

19.5 LOCAL STRUCTURES OF THE ACTIVE SURFACE SITES OF TI-MCM-41 CATALYSTS AND THEIR PHOTOCATALYTIC REACTIVITY

NO_x is an especially harmful atmospheric pollutant and the main cause of acid rain and photochemical smog. The removal of nitrogen oxides (NO_x: NO, N_2O, and NO_2), that is, the direct decomposition of NO_x into N_2 and O_2, has been a great challenge for many researchers [37,87–89]. As a promising way to address such concerns, mesoporous materials incorporating transition metal ions (TMI) such as Ti^{4+} and V^{5+} have attracted much attention not only as active catalysts for the partial oxidation of alkenes but also as effective photocatalysts for the decomposition of NO_x into N_2 and O_2 or the reduction of CO_2 with H_2O to produce CH_4 and CH_3OH [4,90–94]. Such systems exhibit the advantage of having a high dispersion of TMI due to their high internal surface area and nanoscaled pore reaction fields. In fact, these mesoporous materials incorporated with TMI are found to exhibit unique and high photocatalytic activity for various reactions, however, the relationship between the local structure of TMI and their photocatalytic reactivity is yet to be clarified. Here, we have investigated the relationship between the local structure of the Ti^{4+} centers in MCM-41 type materials and their photophysical and photocatalytic characteristics by applying various spectroscopic investigative techniques.

In this chapter, MCM-41 and Ti-MCM-41(x), (content of Ti as wt%, $x = 0.15$, 0.60, 0.85, and 2.00), were synthesized in accordance with previous literature [95], using tetraethyl orthosilicate and tetraisopropyl orthotitanate as the starting materials and cetyltrimethylammonium bromide as the template. After the as-synthesized products were recovered by filtration, washed thoroughly with deionized water, and dried at 373 K for 12 h, calcination of the samples was performed in air at 823 K for 6 h. The titanium content in these materials was determined by atomic absorption analysis.

Prior to photocatalytic reactions and spectroscopic measurements, the catalysts were degassed at 723 K for 2 h, heated in O_2 at the same temperature for 2 h, and finally degassed at 473 K for 2 h. The samples were then characterized by XAFS (XANES and FT-EXAFS, Photon Factory, Tsukuba, Japan) with the Ti K-edge absorption spectra recorded in the transmission or fluorescence mode using photoluminescence (Spex 1943D3) and IR (adsorbed NH_3; Jasco FT/IR 660) spectroscopic investigations. UV irradiation of the catalysts under NO was carried out using a 100 W Hg lamp ($\lambda > 240$ nm) at 295 K. The reaction products were analyzed by gas chromatography.

The XRD patterns showed that the Ti-MCM-41 catalysts have a MCM-41 mesoporous structure. The XAFS experiments indicated that in all of the Ti-MCM-41 catalysts the overwhelming part of the Ti centers exist in tetrahedral coordination. Especially, in Ti-MCM-41 having a Ti content of less than 2.00 wt%, the Ti^{4+} centers are highly dispersed and located in tetrahedral coordination with the 4 oxygen atoms having a Ti–O bond length of 1.81 Å. Nevertheless, as can be seen in Figure 19.12, their diffuse reflectance UV–Vis spectra exhibited different degrees of dependence on the Ti content. In fact, besides a progressive increase in intensity, the increase in the Ti content resulted in a progressive shift of the absorption maximum from 205–208 nm [Ti-MCM-41(0.15) and Ti-MCM-41(0.60)] (Figure 19.12[a,b]) to 215 nm [Ti-MCM-41(2.00)] (Figure 19.12[d]), while a shoulder at ~230 nm became evident in the spectrum of Ti-MCM-41(0.85) (Figure 19.12[c]), contributing to a large extent to the spectral profile of the Ti-MCM-41(2.00) sample. All the absorptions cited here are in the ranges where the ligand-to-metal charge transfer involving an electron transfer from O^{2-} to the Ti^{4+} of the tetrahedrally coordinated Ti oxide species to form a charge transfer excited state, $(Ti^{3+}–O^-)^*$. The bands at 205 to 215 nm can be attributed to the isolated tetrahedrally coordinated Ti oxide species, while the band

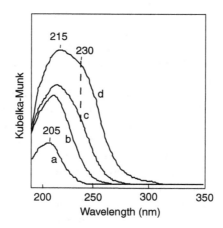

FIGURE 19.12   Diffuse reflectance UV-Vis spectra of Ti-MCM-41 with different Ti content: (a) 0.15, (b) 0.60, (c) 0.85, and (d) 2.00 wt%

at 230–250 nm may be due to the dimeric or oligomeric tetrahedrally coordinated Ti oxide species, the formation of which seems to be favored at higher Ti loadings. Moreover, the tail at $\lambda > 250$ nm observed for Ti-MCM-41 (2.00) can be attributed to a minor moiety of the Ti(IV) centers in penta- and octahedral coordination [96].

In order to monitor the relative amount of the Ti oxide species exposed at the surface walls of the various Ti-MCM-41 materials, the IR spectra of the ammonia irreversibly adsorbed at room temperature were recorded. No bands due to the adsorbed species were observed for pure MCM-41. Conversely, the spectra of the Ti-MCM-41 samples in contact with ammonia exhibited, in the 1800 to 1350 cm^{-1} range, a characteristic band at 1608 cm^{-1} (δ_{asym} NH$_{3ads}$) due to the NH$_3$ molecules irreversibly adsorbed on the Ti(IV) centers of the tetrahedrally coordinated Ti oxide species [97,98]. The increase in the Ti content led to an increase in the intensity of this band, indicating that the amount of tetra-coordinated Ti(IV) sites exposed at the surface walls of the channels of the Ti-MCM-41 materials increased with the Ti content. The bands at 1708, 1660, 1453 cm^{-1} can be assigned to the vibrational bending modes of NH$_4^+$, formed by a protonation of the ammonia molecules by the acidic surface hydroxyl groups. The band at 1553 cm^{-1} is attributed to the Ti–NH$_2$ or Si–NH$_2$ bending mode formed by the irreversible reaction of NH$_3$ with the Si–O–Ti bridges or distorted surface Si–O–Si bridges formed due to the incorporation of Ti into the Si–O–Si networks [98].

Nevertheless, the photoluminescence spectra exhibited different degrees of dependence on the Ti content. As shown in Figure 19.13, Ti-MCM-41 exhibits a typical photoluminescence spectrum at around 450 to 550 nm upon excitation of its charge transfer band at around 220 to 260 nm at 295 K, which coincides well with those previously observed for isolated tetrahedrally coordinated Ti oxide species highly dispersed in silica matrices [44,65,96]. The photoluminescence spectra are attributed to the reverse radiative decay process from the charge transfer excited state to the ground state of the isolated Ti oxides in tetrahedral coordination [44]. As can be seen in Figure 19.13, the intensity of this photoluminescence increases with an increase in the Ti content up to 0.60 wt%, and then sharply decreases with a higher Ti content. Furthermore, it was found that an increase in the Ti content from 0.60 to 0.85 wt% leads to a decrease in the phosphorescence lifetime from 0.1 to 0.025 msec. On the basis of these data, when the Ti-MCM-41 materials are photoexcited by UV light, only the isolated tetrahedrally coordinated Ti oxide species, which are more abundant for Ti content up to 0.60 wt%, can induce an excited state with a lifetime long enough to allow the appearance of a photoluminescence as a radiative decay to the ground state. Moreover, the dimeric and oligomeric tetrahedrally coordinated Ti oxide species, which are most likely the overwhelming species present at higher Ti contents, decay quickly to the ground state through nonradiative, vibrational processes, which are apparently favored due to their clustered structure.

It was found that UV irradiation of Ti-MCM-41 in the presence of NO leads to the formation of N$_2$, O$_2$, and N$_2$O, their yields increasing in proportion to the irradiation time, while under dark conditions, no products could be detected. These

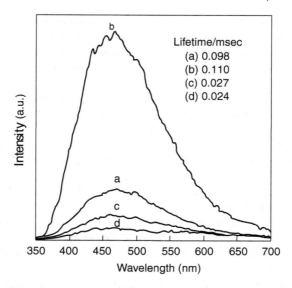

FIGURE 19.13 The photoluminescence spectra of Ti-MCM-41 with different Ti content measured at 295 K: (a) 0.15, (b) 0.60, (c) 0.85, and (d) 2.00 wt%

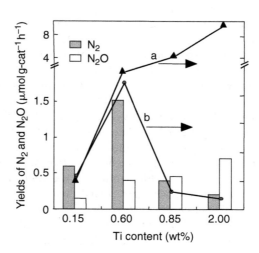

FIGURE 19.14 Relationship of the yields of N_2, N_2O, (a) the intensities of the IR band of $\delta asymNH_3$ adsorbed on Ti(IV) and (b) the photoluminescence spectra of Ti-MCM-41 having differing Ti content

results clearly indicate that the reaction proceeds photocatalytically on Ti-MCM-41 at 295 K. Figure 19.14 shows the effect of the Ti content on the reactivity of the photocatalytic decomposition of NO. As shown in Figure 19.14, the yield and selectivity of N_2 in the decomposition of NO is the highest in the case of

Ti-MCM-41 (0.60 Ti wt%), and an increase in Ti content led to a decrease in the reactivity, showing a good correspondence with the intensities of the photoluminescence spectra due to the isolated tetrahedrally coordinated Ti oxide species. Thus, the amount of isolated tetrahedrally coordinated Ti oxide species, not the total amount of the Ti oxide species exposed at the surface walls of the Ti-MCM-41 channels as monitored by the intensity of the $\delta_{asym}NH_3$ band, appeared to play an important role in this reaction. These results also indicate that only the highly dispersed isolated tetrahedrally coordinated Ti oxide species act as active sites in the photocatalytic decomposition of NO into N_2 and O_2.

It was found that mesoporous Ti-MCM-41 catalysts prepared at ambient temperatures with Ti contents of up to 0.60 wt% involve isolated tetrahedrally coordinated Ti oxide species as the major species, while dimeric or oligomeric Ti oxide species with Ti(IV) in tetrahedral coordination becomes overwhelming within the Ti-MCM-41 with high Ti contents. The photocatalytic reactivity of this Ti-MCM-41 for the decomposition of NO was found to strongly depend on the local structure of the Ti oxide species. Furthermore, the yield and selectivity of N_2 corresponded with the yield of the photoluminescence of the isolated tetrahedrally coordinated Ti oxide species, indicating that only these species are responsible for such photocatalytic reactivity.

19.6 Summary

In this chapter, recent research progress on the photocatalytic reactivity and photoinduced super-hydrophilic property of titanium oxide-based catalysts were summarized centering on the studies in our laboratory. Special attention was focused on the preparation of nano-sized TiO_2 particles, highly dispersed titanium oxide species within zeolite cavities, titanium oxide-based binary catalysts, and second-generation TiO_2 photocatalysts that can operate under visible light irradiation by an advanced metal ion-implantation method. Furthermore, a new and more low-cost RF magnetron sputtering method was presented and detailed characterizations of these visible light-responsive TiO_2 thin film photocatalysts were carried out at the molecular level along with investigations into the various significant photocatalytic reactions that could be initiated.

A combination of the conventional chemical preparation methods and the advanced physical ion-engineering techniques will provide new approaches in the utilization of solar energy as the most abundant and safe energy source, significantly to address environmental pollution on a large global scale as well as realizing the production of H_2 in the photocatalytic splitting of H_2O using unique visible light-responsive titanium oxide photocatalysts and solar irradiation.

References

1. Fujishima, A. and Honda, K. Electrochemical photolysis of water at a semiconductor electrode. *Nature* 1972, *238*, 37.

2. Anpo, M. *Green Chemistry; Challenging Perspectives*. Oxford University Press, Oxford, 2000, p. 1.

3. Anpo, M., Takeuchi, M., Yamashita, H., and Kishiguchi, S. Design and development of new titanium dioxide semiconductor photocatalysts. In *Semiconductor Photochemistry and Photophysics*, Ramamurthy, V. and Schanze, K.S., Eds. Marcel Dekker, Inc., 2003.

4. Anpo, M., Takeuchi, M., Ikeue K., and Dohshi, S. Design and development of titanium oxide photocatalysts operating under visible and UV light irradiation: The applications of metal ion-implantation techniques to semiconducting TiO_2 and Ti/zeolite catalysts. *Curr. Opin. Solid State Mater. Sci.* 2002, *6*, 381.

5. Ireland, J.C., Klostermann, P., Rice, E.W., and Clark, R.M. *Inactivation of Escherichia coli by titanium-dioxide photocatalytic oxidation*. *Appl. Environ. Microbiol.* 1993, *59*, 1668.

6. Sjogren, J.C. and Sierka, R.A. Inactivation of phage MS2 by iron-aided titanium-dioxide photocatalysis. *Appl. Environ. Microbiol.* 1994, *60*, 344.

7. Cai, R.X., Kubota, Y., Shuin, T., Sakai, H., Hashimoto, K., and Fujishima, A. Induction of cytotoxicity by photoexcited TiO_2 particles. *Cancer Res.* 1992, *52*, 2346.

8. Cai, R.X., Hashimoto, K., Kubota, Y., and Fujishima, A. Increment of photocatalytic killing of cancer-cells using TiO_2 with the aid of superoxide-dismutase. *Chem. Lett.* 1992, 427.

9. Suzuki, K. In *Photocatalytic Purification and Treatment of Water and Air*, Ollis, D.F. and Al-Ekabi, H., Eds. Elsevier, Amsterdam, 1993.

10. Karakitsou, K.E. and Verykios, X.E. Effect of altervalent cation doping of TiO_2 on its performance as a photocatalyst for water cleavage. *J. Phys. Chem.* 1993, *97*, 1184.

11. Gratzel, M. Artificial photosynthesis: Water cleavage into hydrogen and oxgen by visible light. *Acc. Chem. Res.* 1981, *14*, 376.

12. Kalyanasundaram, K., Borgarello, E., Duonghong, D., and Gratzel, M. Cleavage of water by visible light irradiation of colloidal CdS solutions. Inhibition of photocorrosion by RuO_2. *Angew. Chem., Int. Ed. Engl.* 1981, *20*, 987.

13. Duonghong, D., Borgarello, E., and Gratzel, M. Dynamics of light-induced water cleavage in colloidal systems. *J. Am. Chem. Soc.* 1981, *103*, 4685.

14. Borgarello, E., Kiwi, J., Pelizzetti, E., Visca, M., and Gratzel, M. Photochemical cleavage of water by photocatalysis. *Nature* 1981, *289*, 158.

15. Wold, A. Photocatalytic properties of TiO_2. *Chem. Mater.* 1993, *5*, 280.

16. Khan, M. and Rao, N.N. Stepwise reduction of coordinated dinitrogen to ammonia via diazinido and hydrazido intermediates on a visible-light irradiated $Pt/CdS/Ag_2S/RuO_2$ particulate system suspended in an aqueous-solution of $K[Ru(EDTA-H)Cl]_2H_2O$. *J. Photochem. Photobiol. A: Chem.* 1991, *56*, 101.

17. Schiavello, M. Some working principles of heterogeneous photocatalysis by semiconductors. *Electrochim. Acta.* 1993, *38*, 11.

18. Khan, M., Chatterjee, D., Krishnaratnam, M., and Bala, M. Photosensitized reduction of N_2 by $Ru(II)(bipy)_3^{2+}$ adsorbed on the surface of $Pt/TiO_2/RuO_2$ semiconductor particulate system containing Ru(III)-EDTA complex and L-ascorbic-acid. *J. Mol. Catal.* 1992, *72*, 13.

19. Khan, M., Chatterjee, D., and Bala, M. Photocatalytic reduction of N_2 to NH_3 sensitized by the [Ru(III)-ethylendiaminetetraacetate-2,2′-bipyridyl] complex in

a Pt-TiO$_2$ semiconductor particulate system. *J. Photochem. Photobiol. A: Chem.* 1992, *67*, 349.

20. Gerischer, H. and Heller, A. Photocatalytic oxidation of organic molecules at TiO$_2$ particles by sunlight in aerated water. *J. Electrochem. Soc.* 1992, *139*, 113.

21. Jackson, N.B., Wang, C.M., Luo, Z., Schwitzgebel, J., Ekerdt, J.G., Brock, J.R., and Heller, A. Attachment of TiO$_2$ powders to hollow glass microbeads-activity of the TiO$_2$ coated beads in the photoassisted oxidation of ethanol to acetaldehyde. *J. Electrochem. Soc.* 1991, *138*, 3660.

22. Nair, M., Luo, Z.H., and Heller, A. Rates of photocatalytic oxidation of crude-oil on salt-water on buoyant, cenosphere-attached titanium-dioxide. *Ind. Eng. Chem. Res.* 1993, *32*, 2318.

23. Boer, K.W. *Survey of Semiconductor Physics*. Van Nostrand Reinhold, New York, 1990, p. 249.

24. Rothenberger, G., Moser, J., Gratzel, M., Serpone, N., and Sharma, D.K. Charge carrier trapping and recombination dynamics in small semiconductor particles. *J. Am. Chem. Soc.* 1985, *107*, 8054.

25. Gratzel, M. *Heterogeneous Photochemical Electron Transfer*. CRC Press, Boca Raton, FL, 1989.

26. Mills, G. and Hoffmann, M.R. Photocatalytic degradation of pentachlorophenol on TiO$_2$ particles-Identification of intermediates and mechanism of reaction. *Environ. Sci. Technol.* 1993, *27*, 1681.

27. Kormann, C., Bahnemann, D.W., and Hoffmann, M.R. Photolysis of chloroform and other organic molecules in aqueous TiO$_2$ suspensions. *Environ. Sci. Technol.* 1991, *25*, 494.

28. Carraway, E.R., Hoffmann, A.J., and Hoffmann, M.R. Photocatalytic oxidation of organic acids on quantum-sized semiconductor colloids. *Environ. Sci. Technol.* 1994, *28*, 786.

29. Chemseddine, A. and Boehm, H.P. A study of the primary step in the photochemical degradation of acetic acid and chloroacetic acids on a TiO$_2$ photocatalyst. *J. Mol. Catal.* 1990, *60*, 295.

30. D'Oliveira, J.C., Minero, C., Pelizzetti, E., and Pichat, P. Photodegradation of dichlorophenols and trichlorophenols in TiO$_2$ aqueous suspention -kinetic effects of the positions of the Cl atoms and identification of the intermediates. *J. Photochem. Photobiol. A: Chem.* 1993, *72*, 261.

31. D'Oliveira, J.C., Al-Sayyed, G., and Pichat, P. Photodegradation of 2-chlorophenol and 3-chlorophenol in TiO$_2$ aqueous suspentions. *Environ. Sci. Technol.* 1990, *24*, 990.

32. Anpo, M., Takeuchi, M., Yamashita, H., Hirao, T., Itoh, N., and Iwamoto, N. The design and development of TiO$_2$ thin film photocatalysts which work under visible light irradiation by the application of ion engineering techniques. In *Proceedings of the 4th International Conference on ECOMATERIAL*, 1999, p. 333.

33. Anpo, M., Takeuchi, M., Kishiguchi, S., and Yamashita, H. Preparation of unique titanium oxide photocatalysts capable of operating under visible light irradiation by an advanced metal ion-implantation method. *Surf. Sci. Jpn* 1999, *20*, 60.

34. Yamashita, H., Harada, M., Misaka, J., Takeuchi, M., Ikeue, K., and Anpo, M. Photocatalytic degradation of propanol diluted in water under visible light irradiation using metal ion-implanted titanium dioxide photocatalysts. *J. Photochem. Photobiol. A: Chem.* 2002, *148*, 257.

35. Takeuchi, M., Yamashita, H., Matsuoka, M., Anpo, M., Hirao, T., Itoh, N., and Iwamoto, N. Photocatalytic decomposition of NO on titanium oxide thin film photocatalysts prepared by an ionized cluster beam technique. *Catal. Lett.* 2000, *66*, 185.

36. Yamashita, H., Harada, M., Misaka, J., Takeuchi, M., Ichihashi, Y., Goto, F., Ishida, M., Sasaki, T., and Anpo, M. Application of ion beam techniques for preparation of metal ion-implanted TiO_2 thin film photocatalyst available under visible light irradiation: Metal ion-implantation and ionized cluster beam method. *J. Synchrotron Rad.* 2001, *8*, 569.

37. Anpo, M. Photocatalysis on titanium oxide catalysts: Approaches in achieving highly efficient reactions and realizing the use of visible light. *Catal, Surv. Jpn.* 1997, *1*, 169.

38. Anpo, M. Utilization of TiO_2 photocatalysts in green chemistry. *Pure Apple. Chem.* 2000, *72*, 1265.

39. Anpo M. Applications of titanium oxide photocatalysts and unique second-generation TiO_2 photocatalysts able to operate under visible light irradiation for the reduction of environmental toxins on a global scale. *Stud. Surf. Sci. Catal.* 2000, *130*, 157.

40. Anpo M. and Takeuchi M. Design and development of second-generation titanium oxide photocatalysts to better our environment-approaches in realizing the use of visibl light. *Int. J. Photoenergy* 2001, *3*, 1

41. Anpo, M., Ichihashi, Y., Takeuchi, M., and Yamashita, H. Design of unique titanium oxide photocatalysts by an advanced metal ion-implantation method and photocatalytic reactions under visible light irradiation. *Res. Chem. Intermed.* 1998, *24*, 143.

42. Takeuchi, M., Yamashita, H., Matsuoka, M., Anpo, M., Hirao, T., Itoh, N., and Iwamoto, N. Photocatalytic decomposition of NO under visible light irradiation on the Cr-ion-implanted TiO_2 thin film photocatalyst. *Catal. Lett.* 2000, *67*, 135.

43. Takeuchi, M., Anpo, M., Hirao, T., Itoh, N., and Iwamoto, N. Preparation of TiO_2 thin film photocatalysts working under visible light irradiation by applying a RF magnetron sputtering deposition method. *Surf. Sci. Jpn.* 2001, *22*, 561.

44. Anpo, M. and Che, M. Applications of photoluminescence techniques to the characterization of solid surfaces in relation to adsorption, catalysis and photocatalysis. *Adv. Catal.* 2000, *44*, 119, and references therein.

45. Hidaka, H., Zhao, J., Pelizzetti, E., and Serpone, N. Photodegradation of surfactants. 8. Comparison of photocatalytic processes between anionic sodium dodecylbenzenesulfonate and cationic benzyldodecyldimetylammonium chloride on the TiO_2 surface. *J. Phys. Chem.* 1992, *96*, 2226.

46. Pelizzetti, E., Minero, C., Piccinini, P., and Vincenti, M. Phototransformations of nitrogen-containing organic compounds over irradiated semiconductor-metal oxides–nitrobenzene and atrazine over TiO_2 and ZnO. *Coord. Chem. Rev.* 1993, *125*, 183.

47. Wang, R., Hashimoto, K., Fujishima, A., Chikuni, M., Kojima, E., Kitamura, A., Shimohigoshi, M., and Watanabe, T. Light-induced amphiphilic surfaces. *Nature* 1997, *388*, 431.

48. Wang, R., Sakai, N., Fujishima, A., Watanabe, T., and Hashimoto, K. Studies of surface wettability conversion on TiO_2 single-crystal surfaces. *J. Phys. Chem. B* 1999, *103*, 2188.

49. Miyauchi, M., Nakajima, A., Fujishima, A., Hashimoto, K., and Watanabe, T. Photoinduced surface reactions on TiO_2 and $SrTiO_3$ films: Photocatalytic oxidation and photoinduced hydrophilicity. *Chem. Mater.* 2000, *12*, 3.
50. Sun, R-D., Nakajima, A., Fujishima, A., Watanabe, T., and Hashimoto, K. Photoinduced surface wettability conversion of ZnO and TiO_2 thin films. *J. Phys. Chem. B* 2001, *105*, 1984.
51. Yu, J.C., Yu, J., Tang, H.Y., and Zhang, L. Effect of surface microstructure on the photoinduced hydrophilicity of porous TiO_2 thin films. *J. Mater. Chem.* 2002, *12*, 81.
52. Miyauchi, M., Nakajima, A., Watanabe, T., and Hashimoto, K. Photocatalysis and photoinduced hydrophilicity of various metal oxide thin films. *Chem. Mater.* 2002, *14*, 2812.
53. Nakamura, M., Sirghi, L., Aoki, T., and Hatanaka, Y. Study on hydrophilic property of hydro-oxygenated amorphous TiO_x:OH thin films. *Surf. Sci.* 2002, *507–510*, 778.
54. Nakamura, M., Kato, S., Aoki, T., Sirghi, L., and Hatanaka, Y. Role of terminal OH groups on the electrical and hydrophilic properties of hydro-oxygenated amorphous TiO_x:OH thin films. *J. Appl. Phys.* 2001, *90*, 3391.
55. Yamashita, H., Harada, M., Tanii, A., Honda, M., Takeuchi, M., Ichihashi, Y., Anpo, M., Iwamoto, N., Itoh, N., and Hirao, T. Preparation of efficient titanium oxide photocatalysts by an ionized cluster beam (ICB) method and their photocatalytic reactivities for the purification of water. *Catal. Today* 2000, *63*, 63.
56. Harada, M., Tanii, A., Yamashita, H., and Anpo, M. Preparation of titanium oxide photocatalysts loaded on activated carbon by an ionized cluster beam method and their photocatalytic reactivities for the degradation of 2-propanol diluted in water. *Z. Phys. Chem.* 1999, *213*, 59.
57. Dohshi, S., Takeuchi, M., and Anpo, M. Photoinduced superhydrophilic properties of Ti–B binary oxide thin films and their photocatalytic reactivity for the decomposition of NO. *J. Nanosci. Nanotech.* 2001, *1*, 337.
58. Anpo, M., Dohshi, S., and Takeuchi, M. Preparation of Ti/B binary oxide thin films by the ionized cluster beam (ICB) method: Their photocatalytic reactivity and photoinduced superhydrophilic properties. *J. Ceram. Process. Res.* 2002, *34*, 258.
59. Takeuchi, M., Matsuoka, M., Yamashita, H., and Anpo, M. Preparation of Ti-Si binary oxide thin film photocatalysts by the application of an ionized cluster beam method. *J. Synchrotron Rad.* 2001, *8*, 643.
60. Anpo, M., Nakaya, H., Kodama, S., Kubokawa, Y., Domen, K., and Onishi, T. Photocatalysis over binary metal-oxides: Enhancement of the photocatalytic activity of TiO_2 in titanium–silicon oxides. *J. Phys. Chem.* 1986, *90*, 1633.
61. Imamura, S., Nakai, T., Kanai, H., and Itoh, T. Effect of tetrahedral Ti in titania-silica mixed oxides on epoxidation activity and lewis acidity. *J. Chem. Soc., Faraday Trans.* 1995, *91*, 1261.
62. Hattori, A., Kawahara, T., Uemoto, T., Suzuki, F., Tada, H., and Ito, S., Ultrathin SiO_x film coating effect on the wettability change of TiO_2 surfaces in the presence and absence of UV light illumination. *J. Colloid Interface Sci.* 2000, *232*, 410.
63. Notari, B. Microporous crystalline titanium silicates. *Adv. Catal.* 1996, *41*, 253.
64. Anpo, M., Aikawa, N., and Kubokawa, Y. Photocatalytic hydrogenation of alkynes and alkenes with water over TiO_2-Pt loading effect on the primary processes. *J. Phys. Chem.* 1984, *88*, 3998.

65. Anpo, M., Aikawa, N., Kubokawa, Y., Che, M., Louis, C., and Giamello, E. Photoluminescence and photocatalytic activity of highly dispersed titanium oxide anchored onto porous vycor glass. *J. Phys. Chem.* 1985, *89*, 5017.

66. Anpo, M., Shima, T., Kodama, S., and Kubokawa, Y. Photocatalytic hydrogenation of CH_3CCH with H_2O on small-particle TiO_2: Size quantization effects and reaction intermediates. *J. Phys. Chem.* 1987, *91*, 4305.

67. Kiwi, J. and Grätzel, M. Optimization of conditions for photochemical water cleavage. Aqueous Pt/TiO_2 (Anatase) dispersions under ultraviolet light. *J. Phys. Chem.* 1984, *88*, 1302.

68. Yamaguchi, K. and Sato, S. Photocatalysis of water over metallized powdered titanium-oxide. *J. Chem. Soc., Faraday Trans.* 1985, *81*, 1237

69. Sayama, K. and Arakawa, H. Effect of carbonate salt addition on the photocatalytic decomposition of liquid water over $Pt-TiO_2$ catalyst. *J. Chem. Soc., Farady Trans.* 1997, *93*, 1647.

70. Moon, S.C., Mametsuka, H., Suzuki, E., and Anpo, M. Stoichiometric decomposition of pure water over Pt-loaded Ti/B binary oxide under UV-irradiation. *Chem. Lett.* 1998, 117.

71. Yoshida, Y., Matsuoka, M., Moon, S.C., Mametsuka, H., Suzuki, E., and Anpo, M. Photocatalytic decomposition of liquid-water on the Pt-loaded TiO_2 catalysts: Effects of the oxidation states of Pt species on the photocatalytic reactivity and the rate of the back reaction. *Res. Chem. Intermed.* 2000, 26, 567.

72. Moon, S.C., Matsumura, Y., Kitano, M., Matsuoka, M., and Anpo, M. Hydrogen production using semiconducting oxide photocatalysts. *Res. Chem. Intermed.* 2003, *29*, 233.

73. Tabata, S., Nishida, H., Masaki, Y., and Tabata, K. Stoichiometric photocatalytic decomposition of pure water in Pt/TiO_2 aqueous suspension system. *Catal. Lett.* 1995, *34*, 245.

74. Hoffmann, M.R., Martin, S.T., Choi, W., and Bahnemann, D.W. Environmental applications of semiconductor photocatalysis. *Chem. Rev.* 1995, *95*, 69.

75. Anpo, M. and Takeuchi, M. The design and development of highly reactive titanium oxide photocatalysts operating under visible light irradiation. *J. Catal.* 2003, *216*, 505.

76. Choi, W., Termin, A., and Hoffmann, M.R. The role of metal-ion dopants in quantum-sized TiO_2-correlation between photoreactivity and charge-carrier recombination dynamics. *J. Phys. Chem.* 1994, *98*, 13669.

77. Asahi, R., Morikawa, T., Ohwaki, T., Aoki, K., and Taga, Y. Visible-light photocatalysis in nitrogen-doped titanium oxides. *Science* 2001, *293*, 269.

78. Yamashita, H., Ichihashi, Y., Takeuchi, M., Kishiguchi, S., and Anpo, M. Characterization of metal ion-implanted titanium oxide photocatalysts operating under visible light irradiation. *J. Synchrotron Rad.* 1999, *6*, 451.

79. Anpo, M. Use of visible light. Second-generation titanium oxide photocatalysts prepared by the application of an advanced metal ion-implantation method. *Pure Apple. Chem.* 2000, *72*, 1787.

80. Yamashita, H., Harada, M., Misaka, J., Takeuchi, M., Ikeue, K., and Anpo, M. Degradation of propanol diluted in water under visible light irradiation using metal ion-implanted titanium dioxide photocatalysts. *J. Photochem. Photobiol. A: Chem.* 2002, *148*, 257.

81. Ishihara, T., Nishiguchi, H., Fukamachi, K., and Takita, Y. Effects of acceptor doping to $KTaO_3$ on photocatalytic decomposition of pure H_2O. *J. Phys. Chem. B* 1999, *103*, 1.

82. Domen, K., Kondo, J.N., Hara, M., and Takata, T. Photo- and mechano-catalytic overall water splitting reactions to form hydrogen and oxygen on heterogeneous catalysts. *Bull. Chem. Soc. Jpn.* 2000, *73*, 1307.

83. Kato, H. and Kudo, A. Water splitting into H_2 and O_2 on alkali tantalate photocatalysts $ATaO_3$ (A = Li, Na, and K). *J. Phys. Chem. B* 2001, *105*, 4285.

84. Zou, Z., Ye, J., Sayama, K., and Arakawa, H. Direct splitting of water under visible light irradiation with an oxide semiconductor photocatalyst. *Nature* 2001, *414*, 625.

85. Fujishima, A., Kohayakawa, K., and Honda, K. Hydrogen production under sunlight with an electrochemical photocell. *J. Electrochem. Soc.* 1975, *122*, 1487.

86. Fukamachi, J., Ohnishi, Y., Miyamoto, C., and Izumi, I. Water decomposition using a p–n type photochemical diode in a dual electrolyte cell. *J. Technol. Educ.* 1996, *5*, 113.

87. Zhang, J., Minagawa, M., Ayusawa, T., Natarajan, S., Yamashita, H., Matsuoka, M., and Anpo, M. In situ investigation of the photocatalytic decomposition of NO on the Ti-HMS under flow and closed reaction systems. *J. Phys. Chem. B* 2000, 11501.

88. Zhang, J., Minagawa, M., Matsuoka, M., Yamashita, H., and Anpo, M. Photocatalytic decomposition of NO on Ti-HMS mesoporous zeolite catalysts. *Catal. Lett.* 2000, *66*, 241.

89. Zhang, J., Hu, Y., Matsuoka, M., Yamashita, H., Minagawa, M., Hidaka, H., and Anpo, M. Relationship between the local structures of titanium oxide photocatalysts and their reactivities in the decomposition of NO. *J. Phys. Chem. B* 2001, *105*, 8395.

90. Anpo, M. In *Studies Surface Science Catalysis*, Vol. 130, *12th International Congress Catalysis, Part A*; Corma, A., Melo, F.V., Mendioroz, S., and Fierro, J.L.G. Eds., Elsevier, Amsterdam, 2000, p. 157.

91. Anpo, M., Higashimoto, S., et al. In *Studies Surface Science Catal.* Gamba, A., Colella, C., and Coluccia, S. Eds., Elsevier, Amsterdam, 2001, Vol. 140, p. 27.

92. Matsuoka, M. and Anpo, M. Local structures, exited states, and photocatalytic reactivities of highly dispersed catalysts constructed within zeolites. *J. Photochem. Photobiol.* 2003, *C3*, 225.

93. Ikeue, K., Yamashita, H., Takewaki, T., and Anpo, M. Photocatalytic reduction of CO_2 with H_2O on Ti-β zeolite photocatalysts: effect of the hydrophobic and hydrophilic properties. *J. Phys. Chem. B* 2001, *105*, 8350.

94. Ikeue, K., Nozaki, S., Ogawa, M., and Anpo, M. Characterization of the self-standing Ti-containing porous silica thin films and their reactivity for the photocatalytic reduction of CO_2 with H_2O. *Catal. Today* 2002, *74*, 241.

95. Zhang, W., Froba, M., Wang, J., Tanev, P.T., Wong, J., and Pinnavaia, T.J., Mesoporous titanosilicate molecular sieves prepared at ambient temperature by electrostatic (SI^-, $S^+X^-I^+$) and neutral ($S° I°$) assembly pathways: a comparison of physical properties and catalytic activity for peroxide oxidations. *J. Am. Chem. Soc.* 1996, *118*, 9164.

96. Marchese, L., Gianotti, E., Dellarocca, V., Maschmeyer, T., Rey, F., Coluccia,S., and Thomas, J.M. Structure–functionality relationships of grafted Ti-MCM41

silicas. Spectroscopic and catalytic studies. *Phys. Chem. Chem. Phys.* 1999, *1*, 585.

97. Marchese, L., Gianotti, E., Maschmeyer, T., Martra, G., Coluccia, S., and Thomas, J.M. Spectroscopic tools for probing the isolated titanium centers in MCM41 mesoporous catalysts. *Il Nuovo Cimento* 1997, *19D*, 1707.

98. Raimondi, M.E., Gianotti, E., Marchese, L., Martra, G., Maschmeyer, T., Seddon, J.M., and Coluccia, S. A spectroscopic study of group IV transition metal incorporated direct templated mesoporous catalysts Part 1: A comparison between materials synthesized using hydrophobic and hydrophilic Ti precursors. *J. Phys. Chem. B* 2000, *104*, 7102.

20 Photocatalytic Activity for Water Decomposition of RuO_2-Dispersed p-Block Metal Oxides with d^{10} Electronic Configuration

Y. Inoue
Department of Chemistry, Nagaoka University of Technology, Nagaoka, Japan

CONTENTS

The p-block metal oxides that involve typical metal ions (In^{3+}, Ga^{3+}, Ge^{4+}, Sn^{4+}, and Sb^{5+}) with d^{10} electronic configuration have been demonstrated to make a stable photocatalyst for the overall splitting of water to produce H_2 and O_2 when combined with RuO_2 as a promoter. The influences of preparation methods, calcination temperature, the amount of RuO_2 dispersed, and the states of RuO_2 particles on the activity of RuO_2-dispersed p-block metal oxides showed that crystallization of p-block metal oxides and high dispersion of RuO_2 particles on them led to high photocatalytic performance: H_2 and O_2 were stably produced as the stoichiometric ratio under UV irradiation. Alkaline metal and alkaline earth metal oxides consisting of distorted MO_6 octahedra and MO_4 tetrahedra (M = metal ion) with dipole moment were photocatalytically active, whereas the metal oxides with distortion-free units exhibited negligible activity. A mechanism has been proposed that internal fields due to the dipole moment, promote the charge separation upon photoexcitation. For the electronic structures, density function theory (DFT) calculation showed that the p-block metal oxides with d^{10} configuration had the valence bands of O2p orbitals and the conduction band of hybridized sp orbitals. The nature of the conduction bands was different from that of conventional transition metal oxides involving metal ions (Ti^{4+}, Zr^{4+}, Nb^{5+}, and Ta^{5+}) with d^0 configuration. The broad hybridized s and p orbitals have large band dispersions which are able to produce photoexcited electrons with large mobility. This is useful for the transfer of photoexcited electrons without recombination to fine RuO_2 particles dispersed on the surface as a promoter. The p-block metal oxide group with d^{10} configuration is concluded to form a new series of the photocatalysts different from the conventional transition metal oxides with d^0 configuration.

20.1 INTRODUCTION

The overall splitting of water using solar energy has drawn considerable attention in view of the current importance of hydrogen as a clean chemical source, and the discovery of new kinds of efficient photocatalysts decomposing water to H_2 and O_2 is among the most desirable issues. Extensive research has been performed, but most of the metal oxide photocatalysts developed in the past three decades have been confined to NiO- or RuO_2-loaded titanates, niobates, zirconates, and tantalates. The representative examples were TiO_2 [1], $SrTiO_3$ [2], $A_2Ti_6O_{13}$ (A = Na, K, Rb) [3,4], $BaTi_4O_9$ [5,6], $A_2La_2Ti_3O_{10}$ (A = K, Rb, Cs) [7,8], $Na_2Ti_3O_7$ [9], $K_2Ti_4O_9$ [10], ZrO_2 [11], $A_4Nb_6O_{17}$ (A = K, Rb) [12], $Sr_2Nb_2O_7$ [13], $ATaO_3$ (A = Na, K) [14,15], MTa_2O_6 (M = Ca, Sr, Ba) [16,17], and $Sr_2Ta_2O_7$ [13]. Note that these transition metal oxides are composed of the octahedrally coordinated d^0 transition metal ions of Ti^{4+}, Zr^{4+}, Nb^{5+}, and Ta^{5+} as a component ion. To our knowledge, for the overall splitting of water to produce hydrogen and oxygen, no stable photocatalysts with electronic configurations other than d^0 configuration have been known. It would be beneficial to demonstrate that metal oxides having other electronic structures are useful for photoassisted water decomposition.

In an attempt to establish a new kind of the photocatalysts, we have paid attention to p-block metal oxides with d^{10} electronic configuration based on the simple consideration that d^0 and d^{10} configurations are similar from quantum chemistry viewpoint. We also found that the later metal oxides form a new group of photocatalysts for the overall splitting of water when combined with RuO_2 as a promoter. In the present review, the photocatalytic properties of RuO_2-loaded p-block metal oxides (indates, gallates, gemanates, stannates, and antoimonates) involving In^{3+}, Ga^{3+}, Ge^{4+}, Sn^{4+}, and Sb^{5+} ions have been stated.

The fundamental steps for water decomposition by metal oxide photocatalysts are (i) photoexcitation of electrons to conduction bands, (ii) the separation of the electrons and holes and transfer to the surface without recombination, and (iii) transfer of electrons and holes to adsorbed species at the reduction and the oxidation sites, respectively. The first and second step are closely related to the electronic and geometric structures of the metal oxides which absorb light energy, whereas the promoters, usually fine particles of RuO$_2$ and NiO dispersed on the metal oxide surfaces, accelerate the third step. Thus, it is evident that the third step is controlled by interactions between the metal oxides and the promoters and by the efficiency of promoters. However, no clear concept to tackle the problems of improvement of the first and second steps have so far been established. For the better design of photocatalysts for water decomposition, it is important to propose a model for these steps. The first step is related to the efficiency of photoexcited charge separation, and the second is associated with the prevention of the recombination of photoexcited charges. Our simple assumption is that the local geometric structure such as a metal–oxygen octahedron and tetrahedron as a fundamental unit and the nature of conduction bands play an important role in the first and second step. In this regard, photocatalysis for water decomposition by p-block metal oxides with d^{10} electronic configuration is discussed based on the roles of local geometric structures and the electronic structures.

20.2 EXPERIMENTAL CONDITIONS

The p-block metal oxides were synthesized by solid-state reactions at high temperatures. Usually a mixture of carbonates and component single metal oxides that were mechanically mixed was used as a starting material and calcined in air for 16 h at temperatures between 1273 and 1723 K. In some cases, coprecipitates were prepared as a mixture. For example, for the synthesis of calcium indate, CaIn$_2$O$_4$, Ca(NO$_3$)$_2$·4H$_2$O and In(NO$_3$)$_3$·3H$_2$O were dissolved in a water–ethanol mixture, and an oxalic acid ethanol solution was added [18]. The coprecipitate was aged at 353 K, dried at 393 K, and calcined at different temperatures from 1273 to 1573 K. The formation of p-block metal oxides was confirmed by the x-ray diffraction patterns reported previously in the literatures.

For RuO$_2$-loading, the p-block metal oxides were impregnated with various ruthenium compounds such as ruthenium chloride, RuCl$_3$, aqueous solution, ruthenium carbonyl complex, Ru$_3$(CO)$_{12}$, in THF, and ruthenium acetyl acetate, Ru(C$_5$H$_7$O$_2$)$_3$ in THF. The Ru compounds loaded metal oxides were dried at 333 K and oxidized at 673 to 773 K in air to convert the ruthenium species to ruthenium oxide.

The photocatalytic water decomposition was performed in a closed gas-circulating apparatus equipped with a quartz reaction cell. About 250 mg photocatalyst was placed in ion exchanged and distilled water in the quartz reaction cell. The reaction system was filled with 1.33 kPa of Ar gas that was circulated with a glass piston pump during the photocatalytic reaction. Powdered photocatalysts were dispersed in the water by stirring of Ar and were irradiated with a 400 W Xe lamp with a wavelength of 280 to 700 nm or a 200 W Hg–Xe lamp with a wavelength of 230 to 436 nm. The products in the gas phase were analyzed by an online gas chromatograph.

20.3 INDATES

It is important to reveal the photocatalytic properties and the electronic and geometric structures that control the characteristic photocatalysis. Alkaline metal indates ($LiInO_2$, $NaInO_2$, and $Na_{0.9}K_{0.1}InO_2$), alkaline earth metal indates ($CaIn_2O_4$, $SrIn_2O_4$, $Sr_{1-x}Ca_xIn_2O_4$ ($x = 0.25, 0.50, 0.75$), $Sr_{1-x}Ba_xIn_2O_4$ ($x = 0.07$)), and lanthanum indate, $LaInO_3$, and gadolinium indate, $NdInO_3$, were investigated.

20.3.1 Alkaline Metal Indates [19]

In the UV diffuse reflectance spectra, $LiInO_2$ showed threshold absorption at around 440 nm. The absorption occurred gradually until 350 nm and steeply in 350 to 300 nm, and reached the largest level at 290 nm. $NaInO_2$ had absorption at around 430 nm and showed a bump in the wavelength range 400 to 350 nm. The maximum absorption occurred at 300 nm. The absorption spectrum of $Na_{0.95}K_{0.05}InO_2$ was analogous to that of $NaInO_2$ except for 10–20 nm shift toward longer wavelength.

The alkaline metal indates, $NaInO_2$, $LiInO_2$, $Na_xLi_{1-x}InO_2$, and $Na_xK_{1-x}InO_2$ ($x = 0.05, 0.1, 0.2$), were combined with 1 wt% RuO_2, and water decomposition was examined under Xe lamp and Hg–Xe lamp irradiation. The photocatalytic activity for water decomposition is shown in Figure 20.1. Under Xe lamp irradiation, the photocatalytic activity of $RuO_2/NaInO_2$ was negligible, but, under Hg–Xe lamp irradiation, $RuO_2/NaInO_2$ produced both hydrogen and oxygen. For $RuO_2/LiInO_2$, neither hydrogen nor oxygen was evolved in the gas phase even for long irradiation. The activity of RuO_2-loaded $Na_{1-x}Li_xInO_2$ ($x = 0.05, 0.1, 0.2$) was lowered dramatically by the addition of Li, whereas the activity of RuO_2-loaded $Na_{1-x}K_xInO_2$ ($x = 0.05, 0.1, 0.2$) decreased gradually with increasing amount of K.

20.3.2 Alkaline Earth Metal Indates [18]

Figure 20.2 shows the UV reflectance spectra of $CaIn_2O_4$ calcined at different temperatures from 1273 to 1573 K. $CaIn_2O_4$ calcined at 1273 K showed threshold absorption at around 450 nm, slightly steep absorption between 420 and 310 nm, and the largest level at 305 nm. Nearly the same absorption spectra were observed for $CaIn_2O_4$ calcined in the temperature range 1373 to 1573 K, although a broad bump appeared at around 400 nm for 1573 K-calcined $CaIn_2O_4$.

Figure 20.3 shows the production of hydrogen and oxygen from water on 1 wt% RuO_2-dispersed $CaIn_2O_4$ under Xe lamp illumination. From the initial stage of

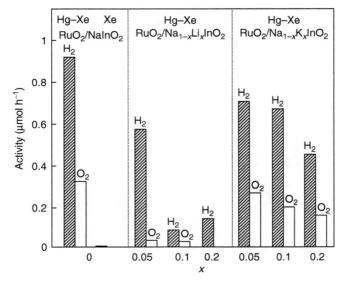

FIGURE 20.1 Photocatalytic activity of RuO$_2$-dispersed Na$_{1-x}$A$_x$InO$_2$ (A = Li, K) under Xe and Hg–Xe lamp irradiation

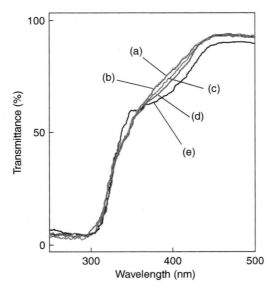

FIGURE 20.2 UV diffuse reflectance spectra of CaIn$_2$O$_4$ calcined at (a) 1273, (b) 1373, (c) 1473, (d) 1523, and (e) 1573 K

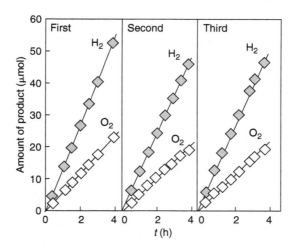

FIGURE 20.3 Production of H_2 and O_2 from water on 1 wt% RuO_2-dispersed $CaIn_2O_4$ under Xe lamp irradiation

irradiation in the first run, both hydrogen and oxygen were produced, and their amounts increased in proportion to irradiation time. Although the photocatalytic activity slightly decreased in the second run, no further drop occurred in the third run, indicating that RuO_2-dispersed $CaIn_2O_4$ photocatalysts were stable. The ratio of H_2 to O_2 was close to the stoichiometric ratio.

The effects of alkaline earth metal ions were examined by replacing Ca of $CaIn_2O_4$ by Sr and Ba ions. Figure 20.4 shows changes in the photocatalytic activity for H_2 and O_2 production of RuO_2-dispersed $Sr_{1-x}Ca_xIn_2O_4$ with x. The activity of RuO_2-dispersed $SrIn_2O_4$ ($x = 0$) was three times lower than that of RuO_2-dispersed $CaIn_2O_4$ ($x = 1$). The activity of RuO_2-dispersed $Sr_xCa_{1-x}In_2O_4$ decreased monotonously with increasing x. The activity of RuO_2-dispersed $Sr_{0.93}Ba_{0.07}In_2O_4$ was nearly the same as that of RuO_2-dispersed $SrIn_2O_4$: the production of hydrogen and oxygen were observed in the repeated run, and the ratio of H_2/O_2 was 2.2. The UV reflectance spectrum of $Sr_{0.93}Ba_{0.07}In_2O_4$ was analogous to that observed for MIn_2O_4 (M = Ca, Sr).

To clarify the photocatalyst conditions for the achievement of high activity, the influences of preparation methods, calcination temperature, and the amount of dispersed RuO_2 on the photocatalytic activities were investigated in detail for $CaIn_2O_4$. Figure 20.5 shows the dependence of photocatalytic activity of 1 wt% RuO_2-dispersed $CaIn_2O_4$ on the calcination temperature of $CaIn_2O_4$. The activity increased considerably with increasing calcination temperatures from 1273 K, reached a maximum at around 1473 K, and decreased sharply above 1500 K. The x-ray diffraction patterns showed that the major diffraction peaks due to (320), (121), and (401) planes appearing at $2\theta = 32.0$, 33.4, and 47.1°, respectively, became narrower with increasing calcination temperatures. The surface area of $CaIn_2O_4$ was 3.3 $m^2 g^{-1}$ for 1273 K, and decreased to 1.8 $m^2 g^{-1}$ for 1373 K, to 0.8 $m^2 g^{-1}$ for 1473 K, and to 0.3 $m^2 g^{-1}$ for 1573 K calcined $CaIn_2O_4$.

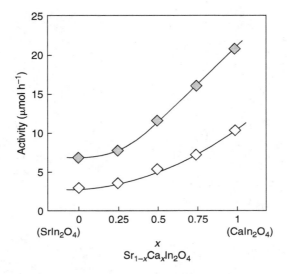

FIGURE 20.4 Monotonous change in photocatalytic activity of Sr$_{1-x}$Ca$_x$In$_2$O$_4$ with x

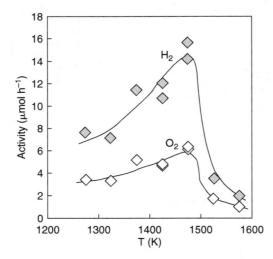

FIGURE 20.5 Changes in photocatalytic activity of 1 wt% RuO$_2$-dispersed CaIn$_2$O$_4$ with calcination temperature of CaIn$_2$O$_4$

Figure 20.6 shows the scanning electron microscopy (SEM) images of CaIn$_2$O$_4$ calcined at various temperatures. The particles of CaIn$_2$O$_4$ calcined at 1273 K were small and had irregular rugged shapes. The particles grew with calcination temperatures from 1323 to 1473 K, loosing the rugged feature. The particle shape changed to spherical at around 1473 K, which was indicative of the agglomeration

Figure 20.6   SEM images of $CaIn_2O_4$ calcined at (a) 1373, (b) 1473, (c) 1523, and (d) 1573

of particles. The particles grew markedly in the temperature range 1523 to 1573 K. The growing of particle size as well as the narrowing of the x-ray diffraction peaks indicated that crystallization proceeded in the lower temperature range 1323 to 1473 K, whereas marked sintering occurred in the higher temperature range 1523 to 1573 K. Thus, a change in the photocatalytic activity with calcination temperature of $CaIn_2O_4$ (Figure 20.5) is explained as follows. The activity rise in the lower calcination temperature is due to elimination, through crystallization, of impurities and structural imperfections that work as recombination sites for photoexcited charges. On the other hand, in the higher calcination temperature region where a dramatic activity decrease occurred, the extraordinary growth of $CaIn_2O_4$ particles was accompanied by a sharp drop of surface area. Thus, it is evident that the activity decrease is related to significant decreases in the surface area of $CaIn_2O_4$.

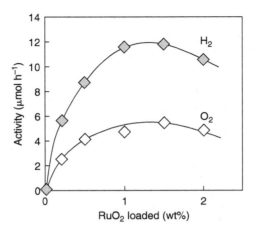

FIGURE 20.7   Photocatalytic activity of RuO$_2$-dispersed CaIn$_2$O$_4$ as a function of RuO$_2$ amount loaded

Figure 20.7 shows the photocatalytic activity as a function of the amount of RuO$_2$ loaded on CaIn$_2$O$_4$. Without RuO$_2$, only a small amount of hydrogen was produced. The loading of RuO$_2$ on 1423 K calcined CaIn$_2$O$_4$ caused a significant increase in photocatalytic activity for both H$_2$ and O$_2$ production. The activity increased markedly in loading range 0.25 to 1.0 wt% and reached a plateau at around 1 to 1.5 wt%. However, further loading to 2 wt% induced a slight activity decrease. Interestingly, for 1573 K calcined CaIn$_2$O$_4$, the photocatalytic activity was larger by a factor of 2.9 for 0.17 wt% than for 1 wt% loading. The RuO$_2$ amount dependence of activity was opposite between low (1473 K) and high (1573 K) temperature-calcined CaIn$_2$O$_4$. As observed in a correlation between RuO$_2$ amount and photocatalytic activity, the facts that activity decreased in a higher RuO$_2$ concentration regime and that the activity increased conversely when the amount of RuO$_2$ was reduced for CaIn$_2$O$_4$ with a small surface area indicate that the excess amount of RuO$_2$ caused a low photocatalytic activity. This is due to the agglomeration and growth of RuO$_2$ particles, which decreases the concentration of photocatalytic active sites. Thus, a high dispersion of small RuO$_2$ particles is essential for high photocatalytic performance of RuO$_2$/CaIn$_2$O$_4$. It turns out that the high photocatalytic activity for water decomposition can be obtained by the combination of well crystallized metal oxides with a highly dispersed RuO$_2$ as a promoter.

20.3.3 Lanthanum and Gadolinium Indates

The UV diffuse reflectance absorption spectrum of lanthanum indate, LaInO$_3$, showed threshold absorption wavelength at 480 nm, main absorption in the wavelength region 450 to 350 nm, and the highest level at 280 nm [20]. RuO$_2$-dispersed LaInO$_3$ had the photocatalytic activity under Hg–Xe lamp irradiation.

The ratio of H_2/O_2 was 2.8. However, little activity was observed under the conditions of Xe lamp irradiation. RuO_2-loaded $NdInO_3$ showed neither hydrogen nor oxygen production even under Hg–Xe lamp irradiation.

20.3.4 Comparison of Photocatalytic Activities among Various Kinds of Indates

It is of interest to compare the photocatalytic activities of RuO_2 loaded different kinds of the alkaline metal, alkaline earth metal, and lanthanum indates investigated under similar reaction conditions. Under Hg–Xe lamp irradiation, the activity was larger in the order $CaIn_2O_4 > SrIn_2O_4 \gg LaInO_3 > NaInO_2 \gg LiInO_2$. $LiInO_2$ was inactive, and the activity of $NaInO_2$ was approximately 31- and 10-fold smaller than that of $CaIn_2O_4$ and $SrIn_2O_4$, respectively. However, under Xe lamp illumination $NaInO_2$ showed a little photocatalytic activity, but MIn_2O_4 ($M = Ca$, Sr) still reserved high photocatalytic activity. RuO_2-dispersed $CaIn_2O_4$ showed the highest activity. $SrIn_2O_4$ and $Sr_{0.93}Ba_{0.07}In_2O_4$ exhibited similar activity. On the other hand, RuO_2-dispersed $LaInO_3$ produced a small amount of hydrogen only. RuO_2-dispersed $LiInO_2$ and $NaInO_2$ showed negligible photocatalytic activity. These results exhibited that alkaline earth metal indates possessed high ability to photocatalytically decompose water and also that there were clear intrinsic differences in the photocatalytic properties among the indates, even though they were commonly composed of an octahedrally coordinated In^{3+} (d^{10}) metal ion.

20.3.5 Role of the Local Structure of InO_6

Figure 20.8 shows the Raman spectra of $CaIn_2O_4$, $SrIn_2O_4$, $Sr_{0.93}Ba_{0.07}In_2O_4$, and $NaInO_2$ [20]. The spectrum of $CaIn_2O_4$ consisted of five major peaks at 200, 262, 328, 406, and 545 cm^{-1} in the wave number region 200 to 600 cm^{-1}. The strongest peak was observed at 545 cm^{-1}, which was assigned to a vibration mode having symmetry of $v_1(A_{1g})$. $SrIn_2O_4$ showed a similar spectrum in which the peak positions were 191, 236, 321, 395, and 542 cm^{-1}: the highest wave number peak at 542 cm^{-1} provided the strongest peak. The Raman spectrum for $Sr_{0.93}Ba_{0.07}In_2O_4$ showed a close similarity to that of $SrIn_2O_4$: the major five peaks were observed at 188, 231, 317, 391, and 540 cm^{-1}. On the other hand, $NaInO_2$ showed two peaks at 330 and 504 cm^{-1}.

$SrIn_2O_4$ has an orthorhombic structure with the lattice parameter of $a = 0.9809$, $b = 1.1449$, and $c = 0.3265$ nm [21], and $CaIn_2O_4$ has an orthorhombic structure with a unit cell of $a = 0.965$, $b = 1.13$, and $c = 0.321$ nm [22]. $SrIn_2O_4$ and $CaIn_2O_4$ have nearly the same crystal structure, but the size of the unit cell is slightly smaller for $CaIn_2O_4$. It is reasonable to consider that a continuous change in the unit cell with x is responsible for the monotonous activity enhancement (Figure 20.8). $Sr_{0.93}Ba_{0.07}In_2O_4$ has a similar orthorhombic crystal structure with a lattice parameter of $a = 0.9858$, $b = 1.152$, and $c = 0.3273$ nm [23]. The Raman peaks appearing at nearly the same positions demonstrate the close similarity in

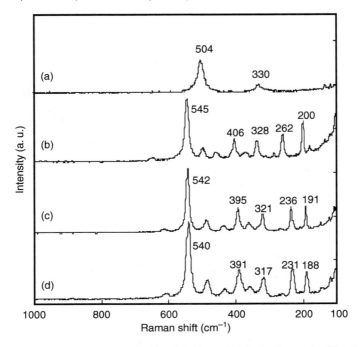

FIGURE 20.8 Raman spectra of (a) NaInO$_2$, (b) CaIn$_2$O$_4$, (c) SrIn$_2$O$_4$, and (d) Sr$_{0.93}$Ba$_{0.07}$In$_2$O$_4$

the crystal structures among them. LaInO$_3$ consists of an orthorhombic structure with a unit cell of $a = 0.5701$, $b = 0.5898$, and $c = 0.8198$ nm [24]. NaInO$_2$ has a hexagonal structure with a unit of $a = 0.3232$, $b = 1.639$ nm, $\alpha = \beta = 90°$, $\gamma = 120°$ nm [25], and is characterized by the macroscopic morphology of a layer structure. LiInO$_2$ has a tetragonal structure with a unit cell of $a = 0.4312$ and $b = 0.9342$ nm [26].

Using the x-ray diffraction data for single crystals in the references, the geometry of InO$_6$ octahedra for LiInO$_2$, NaInO$_2$, Sr$_{0.93}$Ba$_{0.07}$In$_2$O$_4$, and SrIn$_2$O$_4$ was analyzed. The calculation showed that the InO$_6$ octahedra were so heavily distorted that the center of the gravity of six oxygens surrounding an In^{3+} ion deviated from the position of the In^{3+} ion, generating dipole moments inside the octahedral unit. SrIn$_2$O$_4$ has two kinds of the InO$_6$ octahedral units: one octahedron had 2.80 D (D = debye), and the other 1.11 D, as shown in Figure 20.9(a). For Sr$_{0.93}$Ba$_{0.07}$In$_2$O$_4$, the dipole moments were 1.70 and 2.58 D. On the other hand, LiInO$_2$ and NaInO$_2$ had the normal InO$_6$ free from distortion, for which the dipole moment was zero. It is interesting to see how the distortion of InO$_6$ octahedron is correlated with photocatalytic properties.

Figure 20.10 compares photocatalytic activity with dipole moment. The indates with dipole moments are photocatalytically active, whereas the distortion-free

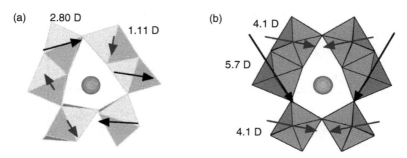

FIGURE 20.9 Pentagonal-prism tunnel structures and dipole moment inside (a) InO_6 of $CaIn_2O_4$ and (b) TiO_6 of $BaTi_4O_9$

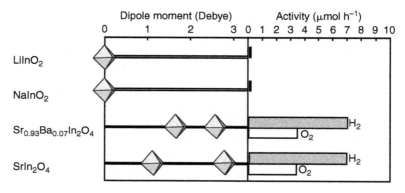

FIGURE 20.10 A correlation between photocatalytic activity and dipole moment of InO_6 octahedra

indates exhibited negligible activity, indicating that there exists a close correlation between the photocatalytic activity and the dipole moment. Unfortunately, no crystallographic data regarding the atom positions were available for $LaInO_3$ and $CaIn_2O_4$. For $CaIn_2O_4$, however, the In–O bond length was reported. $CaIn_2O_4$ has two kinds of InO_6 octahedra: one octahedron has the In–O bond lengths of 0.219 (two), 0.218 (two), 0.213, and 0.206 nm, and the other of 0.222, 0.219, 0.218 (two), and 0.209 (two) nm [14]. The broad distributions of In–O bond lengths and the resemblance of Raman peaks to that of $SrIn_2O_4$ indicate that the InO_6 octahedra in $CaIn_2O_4$ are significantly distorted. Thus, it is apparent that the correlation between photocatalytic activity and dipole moment is true for $CaIn_2O_4$. The local internal fields due to the dipole moment are considered to be useful for electron-hole separation upon photoexcitation.

In the development of transition metal oxide photocatalyts, a target has been placed on the macrostructural effects in an attempt to improve the photocatalytic activity. Layer structure $A_4Nb_6O_{17}$ (A = K, Rb) consisting of macropolyanion sheets of niobates ($N_6O_{17}^{4-}$) has anisotropy along the stacking direction and alternatively arranged two different types of layers. One of the layers intercalated fine Ni particles, as a promoter, and worked to produce H_2 [12]. The other layer produced O_2. A model was proposed in which the layer structures accelerated the charge separation. RuO_2-dispersed $BaTi_4O_9$ and $Na_2Ti_6O_{13}$ had high photocatalytic activities for water decomposition [17–20]. $BaTi_4O_9$ and $A_2Ti_6O_{13}$ (A = Na, K) are composed of tunnel structures, on which fine RuO_2 particles are accommodated, forming a large number of photocatalytic active sites. In addition, $BaTi_4O_9$ [5,6] and $A_2Ti_6O_{13}$ (A = Na, K) [3,4] have distorted TiO_6 octahedra. The out-of-center Ti ions generate dipole moment: $BaTi_4O_9$ has two kinds of TiO_6 octahedra with dipole moment of 5.7 and 4.1 D [27,28] (Figure 20.9[b]). $Na_2Ti_6O_{13}$ has distorted TiO_6 octahedra with dipole moment of 6.7, 5.8, and 5.3 D [29]. In electron pragmatic resonance (EPR) measurements, a signal with $g = 2.018$ and $g = 2.004$ was observed when $BaTi_4O_9$ was irradiated with UV light at 77 K in the presence of gaseous molecules such as Ar, He, O_2, and H_2 [30,31]. A signal with $g = 2.020$, $g = 2.018$, and $g = 2.004$ was also observed for $Na_2Ti_6O_{13}$ under the similar experimental conditions [32]. These EPR signals were assigned to lattice O^-, indicating that the separation of photoexcited charges occurs efficiently. A good correlation was observed between the photocatalytic activity and the O^- radical concentration. These results clearly indicate that the distorted TiO_6 octahedra have high ability to promote the separation of the photoexcited electrons and holes. The dipole moment is likely to form a local field in the interior of TiO_6 octahedra. Thus, a model can be proposed that, the local electric fields due to dipole moment promote the charge separation upon photoexcitation and hence increase the efficiency for photoexcited charge formation.

The same situation is true for the indates. The fact that undistorted normal InO_6 octahedra exhibit no photocatalytic activity explains different photocatalytic activity between MIn_2O_4 (M = Ca, Sr) and $AInO_2$ (A = Li, Na). Note that the correlation between the photocatalytic activity and dipole moment can be observed for both meal oxides with d^{10} and d^0 electronic configurations. For the development of excellent photocatalysts, it is important to take the roles of the distorted metal–oxygen octahedral structures into account.

20.3.6 Electronic Structures

The energy band dispersion diagram and the density of state (DOS) were calculated by the plane wave DFT method, which was performed using the CASTEP program [33]. The core electrons were replaced by the ultra-soft core potentials. For $SrIn_2O_4$, the valence electronic configurations were $4s^24p^65s^2$ for Sr, $4d^{10}5s^25p^1$ for In atom, and $2s^22p^4$ for O atom. The primitive unit cell consisted of $[SrIn_2O_4]_4$, and the number of occupied orbitals was 120. To see the difference and the similarity in the electronic structures between d^{10} and d^0 configurations, the band structures were compared with those of a photocatalytically active transition metal oxide, $BaTi_4O_9$, with d^0 configuration. For $BaTi_4O_9$, the valence electronic configurations were $5s^25p^66s^2$ for Ba, $3s^23p^63d^24s^2$ for Ti, and $2s^22p^4$ for O atom.

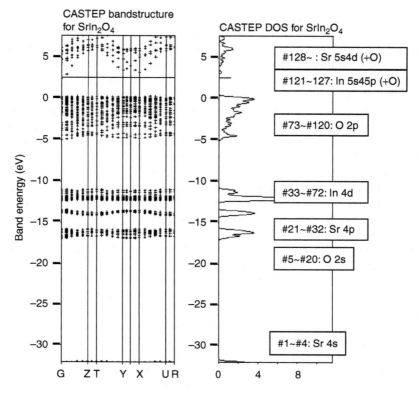

FIGURE 20.11   Energy band diagram and DOS for $SrIn_2O_4$

The unit cell size was $[BaTi_4O_9]_2$, and the number of occupied orbitals was 112. The kinetic energy cutoff was 260 eV for the both systems.

Figure 20.11 shows the energy band dispersion diagram and DOS for $SrIn_2O_4$ [20]. The lowest band consisted of the Sr 4s atomic orbital (AO). The second, third, and fourth bands from the bottom were formed by the O 2s, Sr 4p, and In 4d AOs, respectively. The valence band consisted of 48 orbitals, which was the number that all the O 2p AOs for 16 O atoms were fully occupied (#73 through #120 in this numbering as shown in Figure 20.11).

Figure 20.12 shows the density contour maps of orbitals with the highest and lowest energies of the valence band [20]. The highest energy orbital (#120) showed the p orbital lobes on the O atom, indicating that the orbital (HOMO) was purely composed of the O 2p orbital. The lowest part of the band (#73) was also formed by the O 2p orbital, but with the In 5s5p orbitals mix to some degree. Thus, the valence band was purely composed of the O 2p AOs only, except for a few orbitals in the lowest level. Since the In 4d orbital was located at deep core levels far from the O 2p orbital region, there were no bonding interactions between Sr 4p–O 2p and In 4d–O 2p.

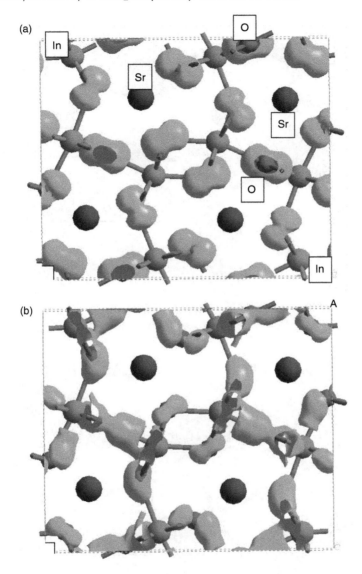

FIGURE 20.12   Density contour maps of orbitals at the top and bottom of the valence band for SrIn$_2$O$_4$. (a) The highest energy (HOMO) level (#120) and (b) the lowest energy level (#70) of the valence band

Figure 20.13 shows the density contour maps of the lowest orbital as well as a little higher level orbital in the conduction band of SrIn$_2$O$_4$ [20]. The bottom of conduction band (LUMO) was composed of the hybridized In 5s and 5p AOs, with a small contribution from the O 2p AO, which is characterized by large

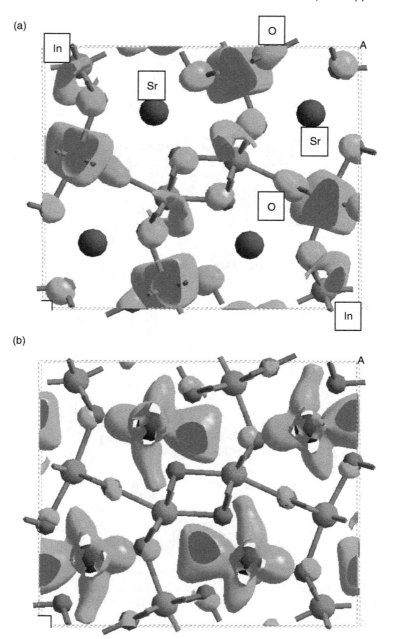

FIGURE 20.13   Density contour maps of orbitals at the lower part of the conduction band for SrIn$_2$O$_4$. (a) The lowest energy (LUMO) level (#121) and (b) a little higher level (#129) of the conduction band

dispersion in the k space (Figure 20.11). The nonspherical shape of the orbital located at a little higher energy (#129 in Figure 20.11) showed significant contribution from the Sr 5s and also 4d AOs to this and higher orbitals. The energy band diagram shows that the electron transfer upon illumination occurs from the O 2p orbital to the hybridized orbitals of In 5s and 5p. The band gap was calculated to be 2.09 eV. The large overlap among the In 5s and 5p AOs leads to large dispersion of the LUMO and hence the excited electrons in the conduction band have large mobility, which are considered to be related to better photocatalytic performance.

Figure 20.14 shows the energy band diagram and the DOS for $NaInO_2$ calculated by the DFT calculation [19]. The energy band diagram and the DOS for $LiInO_2$ were similar to those for $NaInO_2$. The O 2s and In 4d orbitals were located at deep core levels for both $NaInO_2$ and $LiInO_2$. The valence bands are essentially composed of the O 2p orbital in both cases, and the lower part of conduction bands consisted of In 5s+5p hybridized orbitals. The electron transfer upon light absorption occurs from the O 2p to the In 5s+5p orbitals. $NaInO_2$ had the extremely small band gap (0.51 eV), compared with the normal band gap of $LiInO_2$ (2.03 eV). In the UV spectra of $NaInO_2$ and $LiInO_2$, only $NaInO_2$ had absorption in the wavelength region 400 to 360 nm, which is consistent with the small band gap. The small band gap is associated with the large overlap of the In orbitals, indicative of In–In interactions.

Alkaline metal indates, $AInO_2$ (A = Li, Na), and alkaline earth metal indates, MIn_2O_4 (M = Ca, Sr), have similar electronic structures in which the valence bands consist of the O 2p orbitals, and the bottom of the conduction bands were composed of the In 5s and 5p orbitals. However, the photocatalytic activities between them are remarkably different: the latter had a significantly large activity, whereas the former exhibited negligible activity under Xe irradiation. From a viewpoint of electronic structures, there were differences in the degrees of mixing with the In orbitals at conduction bands between the alkaline metal and alkaline earth metal atom orbitals. The energy difference between the top of valence band and Sr 5s AO was 5.5 eV, whereas differences between the top of valence and Li 2s and Na 3s were 7.2 and 10.0, respectively. Thus, there was a considerably large overlap of Sr 5s with In 5s5p AO, compared with the Li 2s and Na 3s orbitals. The smallest mixing for $NaInO_2$ was rather strange, but might be ascribable to a little longer Na–In distance, 0.331 nm, than 0.305 nm for $LiInO_2$ [20].

Figure 20.15 shows the energy band dispersion diagram and DOS for $BaTi_4O_9$ [20]. The band gap was calculated to be 2.84 eV. The density contour maps showed that the valence band was composed purely of the O 2p AOs, which was the same as that for $SrIn_2O_4$. However, the orbitals of the conduction bands exclusively consisted of the Ti3d AOs. The mixing of Ti 3d AO with O 2p AO was extremely small. Thus, the degree of hybridization in the valence and conduction bands was negligible for $BaTi_4O_9$, and the dispersion in k space was quite small for the conduction band of $BaTi_4O_9$. Thus, there was an important difference in the conduction bands between d^{10} and d^0 metal oxides: and the characteristic feature of the former is broad s and p orbitals with large dispersion.

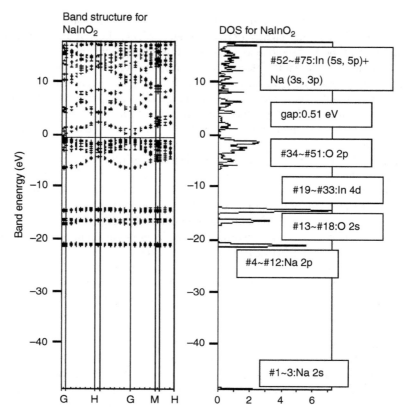

FIGURE 20.14   Energy band diagram and DOS for NaInO$_2$

20.4 ZINC GALLATES, ZNGA$_2$O$_4$ [34]

The UV diffuse reflectance spectra of ZnGa$_2$O$_4$ showed that light absorption started at around 300 nm, increased steeply at around 280 nm and reached a maximum level at 250 nm. The absorption band slightly shifted toward higher wavelength with increasing calcination temperatures.

Figure 20.16 shows the production of hydrogen and oxygen from water on a 1 wt% RuO$_2$-dispersed ZnGa$_2$O$_4$ photocatalyst under Hg–Xe lamp irradiation. The amount of the products increased in proportion to irradiation time. The second and third run showed nearly the same production. The H$_2$/O$_2$ ratio in the third run approached the stoichiometric value of 2.0. About 50-fold turnover number was obtained for 20 h irradiation. These results indicate that RuO$_2$-dispersed ZnGa$_2$O$_4$ makes a stable photocatalyst for water decomposition. In the absence of RuO$_2$, only hydrogen was produced. The photocatalytic activity for both hydrogen and oxygen production increased with increasing amount of RuO$_2$, passed through a maximum at around 1 wt%, and decreased considerably with higher loading.

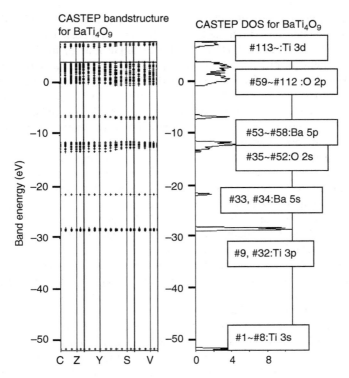

FIGURE 20.15   Energy band diagram and DOS for BaTi₄O₉

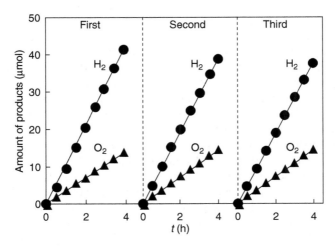

FIGURE 20.16   Production of H₂ and O₂ from water on 1 wt% RuO₂-dispersed ZnGa₂O₄ under Hg–Xe lamp light

$ZnGa_2O_4$ has the structure that Ga^{3+} ions have octahedral coordination, and Zn^{2+} ions are coordinated tetrahedrally [35]. There is a possibility that the photocatalytic activity results from the ZnO_4 tetrahedra in addition to GaO_6 octahedra or without them. In order to clarify this question, the photocatalytic activity of strontium gallate, $SrGa_2O_4$, was examined. For 1 wt% RuO_2-loaded $SrGa_2O_4$, hydrogen and oxygen were produced under UV illumination, and its photocatalytic activity was compatible with that of RuO_2-loaded $ZnGa_2O_4$, indicating that the GaO_6 unit in MGa_2O_4 (M = Zn, Sr) is dominantly responsible for photocatalytic activity of water decomposition.

The dependence of photocatalytic activity on calcination temperature of $ZnGa_2O_4$ showed a maximum at a temperature of 1373 K, between 1273 and 1473 K. In the lower temperature range, the size of $ZnGa_2O_4$ particles observed in the SEM images increased moderately, and the surface area was 2.3 $m^2\,g^{-1}$ for $ZnGa_2O_4$ calcined at 1273 K, 1.7 for 1373 K, and 0.7 for 1473 K. The x-ray diffraction peak became sharp with remarkable enhancement of peak intensity and narrow width. Thus, an increase in the activity is associated with the crystallization of $ZnGa_2O_4$, since crystallization eliminates impurities and structural imperfections that frequently work as a trap site for photoexcited charges. Above 1473 K, when a sharp decrease in the photocatalytic activity occurred, the SEM images showed the extraordinary growth of $ZnGa_2O_4$ particles. Thus, the activity decrease is attributed to the reduction of surface area of $ZnGa_2O_4$. In the correlation between RuO_2 amount and photocatalytic activity, a high concentration of RuO_2 loaded resulted in a considerable decrease in the activity. When the amount of RuO_2 loaded was reduced from 1 to 0.6 wt% for 1473 K calcined $ZnGa_2O_4$ with small surface area, the activity increased five-fold. These results demonstrate that the excess amount of RuO_2 leads to low activity. Thus, the appearance of the maximum is associated with the crystallization and the sintering of $ZnGa_2O_4$. This conclusion is the same as obtained for RuO_2-loaded $CaIn_2O_4$.

Figure 20.17(a) shows photocatalytic activity as a function of oxidation temperature that converts a loaded Ru metal carbonyl complex to ruthenium oxide on a $ZnGa_2O_4$ surface. For $Ru_3(CO)_{12}$ impregnation, the activity was negligible with an oxidation temperature of 373 K. The activity increased with increasing temperature, reached a maximum by oxidation at 573 K and then decreased dramatically by oxidation at the higher temperatures. For $Ru(C_5H_7O_2)_3$ impregnation, nearly the same correlation between the photocatalytic activity and the oxidation temperature was obtained as shown in Figure 20.17(b). Table 20.1 shows the binding energy of the Ru $3d_{5/2}$ level in x-ray photoelectron spectra taken for surface Ru species obtained by different oxidation temperatures. The binding energy was 281.1 eV for Ru species oxidized at 373 K and decreased to 280.7 eV by oxidation at 573 K. In $RuCl_3$ impregnated $Na_2Ti_6O_{13}$, oxidation at 573 to 773 K produced Ru species with the binding energy of 280.6 to 280.7 eV [3]. These values were the same, within experimental accuracy, as that for RuO_2 powder and Ru powder oxidized at 823 K. Thus, the Ru species with 280.7 eV is assigned to Ru^{4+}. Since the highest photocatalytic activity was achieved by oxidation at 573 K, irrespective of employment of the different ruthenium compounds as starting materials, it is

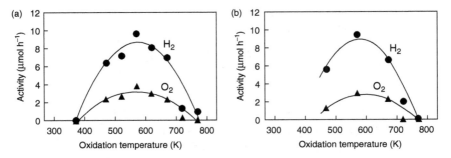

FIGURE 20.17 Changes in photocatalytic activity with oxidation temperature of (a) Ru$_3$(CO)$_{12}$ and (b) Ru(C$_5$H$_7$O$_2$)$_3$ loaded on ZnGa$_2$O$_4$

TABLE 20.1
Binding Energy of Ru 3d$_{5/2}$

Samples	Oxidation temperature (K)	Binding energy of Ru 3d$_{5/2}$ level (eV)	Reference
Ru impregnated	373	281.1	
ZnGa$_2$O$_4$	573	280.7	
	723	280.9	
Ru powder	—	280.1	[2]
	823	280.6	[2]
RuO$_2$ powder	—	280.7	[2]

evident that the active Ru species is RuO$_2$. The binding energy of the Ru species oxidized at 773 K, at which a drastic decrease in the photocatalytic activity occurred was 280.9 eV. The positive shift, although the value is small, suggests that the states of RuO$_2$ varied with the coagulation of RuO$_2$ to produce the assembly of large RuO$_2$ particles.

Figure 20.18 shows the energy band diagram and the DOS for ZnGa$_2$O$_4$ evaluated by the DFT calculation. In the DOS, the first and the second bands with the lowest energies were due to the O 2s and Ga 3d orbitals, respectively. The broad third band was the valence band whose lower part was composed of O 2p, Ga 4s and 4p, Zn 3d, 4s and 4p orbitals. Figure 20.19 shows the density contour maps of orbitals for the lower and upper energy parts of the valence band. The lowest energy orbitals (#29 to #31) were composed of the O 2p and Zn 3d AOs, and contribute to the Zn–O bonding as shown in Figure 20.19(a). In the next orbital #32, the electron density distributes around the Ga and O atoms (Figure 20.19[b]). The higher orbitals from #33 through #45 (not illustrated) contribute to the Zn–O bonding. Thus, the covalent bond character existed between Zn and O atoms,

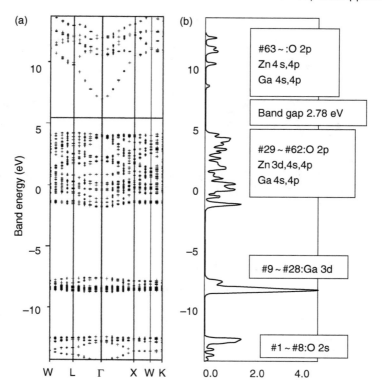

FIGURE 20.18   (a) Energy band diagram and (b) DOS for ZnGa$_2$O$_4$

and also between Ga and O atoms, to a less degree. The top of the valence band (HOMO) is formed by the O 2p AOs only, as shown in Figure 20.19(c).

Figure 20.20 shows the density contour maps of typical orbitals with lower energies in the conduction band. The bottom of conduction band (LUMO) is characterized by its large dispersion in the k-space (Figure 20.16). Figure 20.20(a) shows that this orbital is composed of the Zn 4s4p+Ga 4s4p AOs as well as O 2p AOs. The triangle shaped electron density on Zn atom is a clear evidence for inclusion of Zn 4p AOs. The next orbital (#64) and the one following (#65) have major electron density on Zn and Ga atoms, respectively. The O 2p AOs always mix into the Zn 4s4p or Ga 4s4p AOs with *out of phase*, which is the counterpart of the *in phase* mixing at the lower energy region of the valance band. Figure 20.20(d) shows a higher orbital #69, which involves dominant contribution from the Zn 4p AOs.

ZnGa$_2$O$_4$ had a main light absorption at around 300 nm, which appeared at longer wavelength by 20 nm, compared with that of SrGa$_2$O$_4$. Thus, the band gap for ZnGa$_2$O$_4$ was 4.3 eV, which was smaller than 4.7 eV for SrGa$_2$O$_4$. Sampath et al. [36] showed that the hybridization of Zn 3d with the O 2p orbitals shifts the

(a)

(b)

(c)

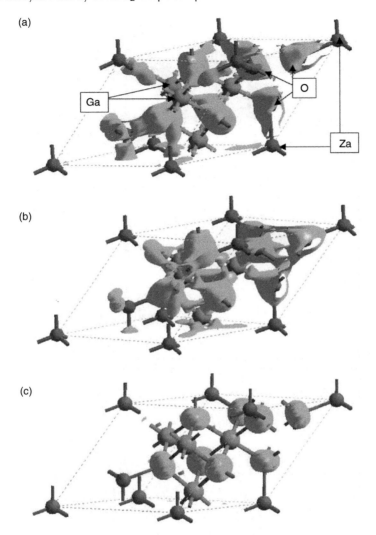

FIGURE 20.19 Density contour maps for orbitals in the valence band. (a) Density map collected for orbitals from #29 to #31. Electron density is located in the bonding region between Zn 3d and O 2p orbitals, (b) density map for orbital #32. Electron density is present between the Ga 4s and 4p and O 2p orbitals, and (c) density map for orbital #62, the HOMO level. Electron density is localized on the O 2p orbital

valence band maximum upwards. For the II–VI semiconductors, Wei and Zunger [37] pointed out that the p–d repulsion repels the valence band maximum upwards without affecting the conduction minimum. Thus, one possibility for the narrower band gap for ZnGa$_2$O$_4$ than for SrGa$_2$O$_4$ is the p–d repulsion. In addition, since the bottom of conduction band is composed of the Zn–Ga hybridized orbitals for

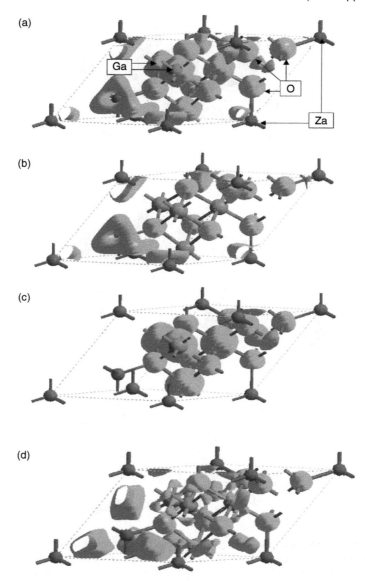

FIGURE 20.20   Density contour maps for orbitals in the conduction band. (a) Density map for #63, the LUMO level. Both 4p(+4s) and Ga 4s(+4p) AOs are included. (b) Density map for orbital #64. Major electron density is on Zn atoms. (c) Density map for orbital #65. Major electron density is on Ga atoms. (d) Density map for orbital #69. Electron is localized on Zn 4p AO and Ga 4p AOs

ZnGa$_2$O$_4$ but of Ga 4s4p AOs only for SrGa$_2$O$_4$, the strong mixing of Zn and Ga orbitals in the conduction band of ZnGa$_2$O$_4$ is responsible for the narrow band gap, as indicated by large dispersion in the DOS.

The energy diagram shows that the electron transfer upon illumination occurs from the O 2p AOs to the hybridized orbital of Ga 4s4p and Zn 4s4p AOs. The band gap was calculated to be 2.78 eV which was in good agreement with 2.79 eV calculated by Sampath et al. [36] based on the tight-binding muffin-tin orbital method. The large overlap among these AOs caused a large dispersion of the LUMO, and hence the excited electrons in the conduction band have large mobility. This is responsible for high photocatalytic performance.

20.5 ZINC GERMANATE, Zn$_2$GeO$_4$ [37]

Zn$_2$GeO$_4$ calcined at a temperature between 1373 and 1573 K had the main threshold wavelength at around 310 nm. Figure 20.21 shows large H$_2$ and O$_2$ productions with a constant rate from water on 1.0 wt% RuO$_2$-dispersed ZnGe$_2$O$_4$ under Hg–Xe lamp irradiation. Through the first to third run, nearly the same production was observed, and a little deterioration of photocatalytic activity was observed. The total amount of H$_2$ produced was larger by a factor of 250 than the number of surface Ge ions.

A change in the photocatalytic activity with the amounts of RuO$_2$ loaded showed that the photocatalytic activity increased with increasing amount of RuO$_2$ in a lower dispersion regime below 1 wt%, reached a maximum at around 1 wt%, followed by a sharp decrease with further RuO$_2$ amount, and attained at a constant level above 2 wt%. In the activity dependence on calcination temperature of

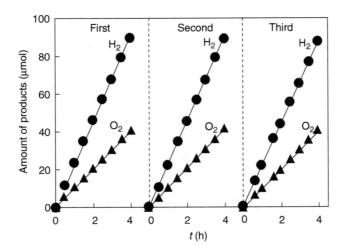

FIGURE 20.21 Production of H$_2$ and O$_2$ from water on 1 wt% RuO$_2$ loaded Zn$_2$GeO$_4$ under Hg–Xe lamp irradiation

Zn_2GeO_4 from 1323 to 1573 K, the activity increased significantly between 1323 and 1473 K. The ratio of H_2/O_2 was 2.8 at calcination temperature at 1323 K, decreased to 2.2 at 1373 K, and to 2.0 at 1473 K. The SEM images and the x-ray diffraction patterns of Zn_2GeO_4 treated in the 1323 to 1473 K temperature regime showed that the size of particles increased considerably, accompanied by diffraction peak narrowing. Evidently, the crystallization of Zn_2GeO_4 occurs markedly over the temperature regime. In a temperature regime above 1473 K, Zn_2GeO_4 particles grew extraordinarily, indicative of a marked decrease in the surface area. The photocatalytic activity decreased at a high RuO_2 concentration, the excess amount of RuO_2 on the small surface area of Zn_2GeO_4 reduces the photocatalytic activity. These findings conclude that the high dispersion of RuO_2 particles on well-crystallized Zn_2GeO_4 is essential for better photocatalytic performance. This conclusion is similar to that obtained for other p-block metal oxides such as MIn_2O_4 (M = Ca, Sr) [18,20] and $ZnGa_2O_4$ [34].

Zn_2GeO_4 has a willemite structure, in which the atoms are arranged in layers at six equally separated levels along the c axis, and each of Ge and Zn ion is in a tetrahedral environment of four oxygen atoms. As shown in Figure 20.22, one GeO_4 tetrahedron and two kinds of ZnO_4 tetrahedra are combined each other through edge oxygen. The rhombohedral cell of Zn_2GeO_4 [38] has $a = 0.8836$ nm and $\alpha = 107.42°$. Two ZnO_4 tetrahedra in Zn_2GeO_4 have the distortion with dipole moment of 1.1 and 1.2 D. Zn_2SiO_4 obtained by replacement of a Ge ion by a Si ion has nearly the same structure of willemite. The rhombohedral cell of Zn_2SiO_4 has $a = 0.8626$ nm and $\alpha = 107.52°$. Zn_2SiO_4 has two kinds of distorted ZnO_4 tetrahedral unit, whose dipole moment is 1.1 and 1.4 D. However, RuO_2-loaded $ZnSiO_4$ was photocatalytically inactive for water decomposition, indicating that only the tetrahedral GeO_4 is responsible for photocatalytic activity

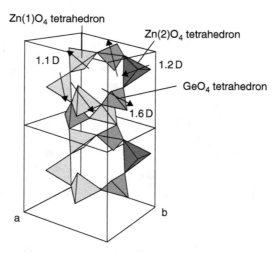

FIGURE 20.22   A schematic representation of local structure of Zn_2GeO_4

for Zn$_2$GeO$_4$. It should be noted that Zn$_2$GeO$_4$ is the first photocatalyst consisting of a tetrahedron unit, MO$_4$ (M = metal ion). The GeO$_4$ tetrahedron of Zn$_2$GeO$_4$ is distorted so heavily that the center of the Ge cation deviates from the gravity of four surrounding oxygen anions and has dipole moment of 1.6 D inside of the tetrahedron. This is in line with the correlation between the photocatalytic activity and the dipole moment that has so far been demonstrated for metal oxides consisting of octahedral unit, MO$_6$ [18]. Figure 20.23 summarizes the active photocatalysts for water decomposition of d^{10} and d^0 metal oxides with pentagonal prism-like tunnel structures. All the metal oxides, irrespective of d^0 and d^{10} electronic configuration and of being formed by octahedral and tetrahedral units, are photocatalytically active when dipole moment is present inside the units.

In band calculation by DFT method, the valence atomic configurations, Zn: 3d^{10}4s^2, Ge: 4s^{2}4p^2, and O: 2s^{2}2p are taken into consideration. The number of occupied bands is 156 for the unit cell construction of (Zn$_2$GeO$_4$)$_6$. Figure 20.24 shows the band dispersion and projected density of states (PDOS). The occupied bands are the O 2s, Ge 4s, and (Zn 3d + O 2p) in the increasing order of energy, and the last one is the valence band. The bottom of conduction band is composed of the Ge 4p orbitals and also small amount of Zn 4s4p orbitals. The band gap is estimated to be 2.0 eV. To see the atom-specific character in each band more clearly, PDOS is further decomposed into AO (more precisely angular momentum) contributions. Figure 20.25 shows AO PDOS for Zn, Ge, and O atoms. As shown in Figure 20.24, the band dispersion is small for the occupied bands, but became very large at the bottom of conduction band. In the electron density contour map

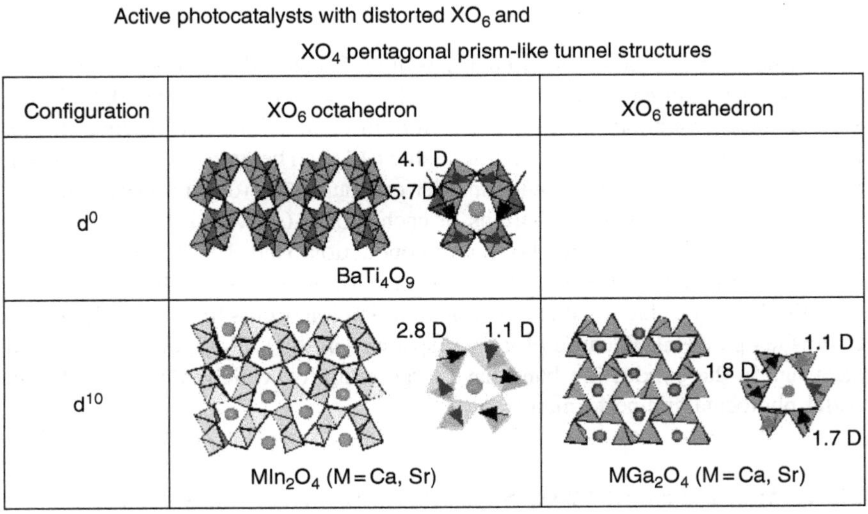

Active photocatalysts with distorted XO$_6$ and
XO$_4$ pentagonal prism-like tunnel structures

Configuration	XO$_6$ octahedron	XO$_6$ tetrahedron
d^0	4.1 D 5.7 D BaTi$_4$O$_9$	
d^{10}	2.8 D 1.1 D MIn$_2$O$_4$ (M = Ca, Sr)	1.1 D 1.8 D 1.7 D MGa$_2$O$_4$ (M = Ca, Sr)

FIGURE 20.23 Photocatalytically active d^{10} and d^0 metal oxides consisting of octahedral and tetrahedral units

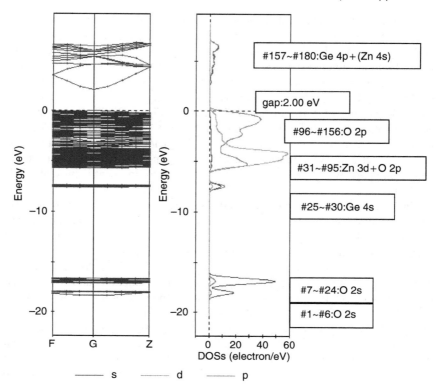

FIGURE 20.24   Band dispersion diagram and projected density of states for Zn_2GeO_4

of the valence band, the lower part consists of the Zn 3d and O 2p orbitals, whereas the upper part is composed of the O 2p orbitals, as shown in Figure 20.26. In the electron density contour map for the bottom of conduction band (Figure 20.27), the electron density with larger contour value (0.1) can be seen only on a Ge atom, whereas the electron density with smaller contour value (0.05) appears on both Ge and Zn atoms. The contributions of each orbital (s, p, and d orbital) of Zn, Ge, and O atom show that the bottom of conduction band (LUMO) is formed by the Ge 4p orbitals, and a little higher level is occupied by Ge 4s and Zn 4s4p orbitals. Thus, the electron transfer upon illumination occurs from the O 2p orbital to the Ge 4p orbital. Because of large dispersion of the LUMO, the photoexcited electrons in the conduction band have large mobility, which is associated with better photocatalytic performance.

20.6   ALKALINE EARTH METAL STANANTES AND ANTIMONATES [39]

In the UV diffuse reflectance spectra of $M_2Sb_2O_7$ (M = Ca, Sr), $CaSb_2O_6$, and $NaSbO_3$, light absorption of $Ca_2Sb_2O_7$ began at around 380 nm and attained

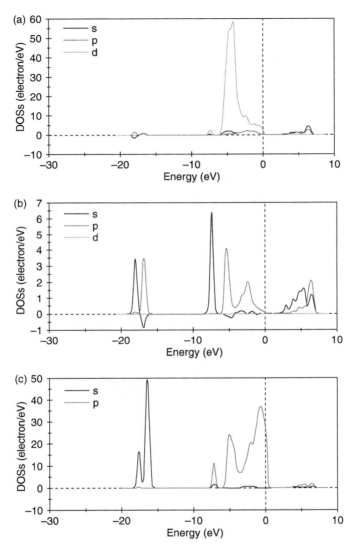

FIGURE 20.25   AO projected DOS for (a) Zn, (b) Ge, and (c) O atom of Zn$_2$GeO$_4$

the maximum level at 280 nm. The absorption characteristics of Sr$_2$Sb$_2$O$_7$ were similar to that of Ca$_2$Sb$_2$O$_7$ except for slightly shorter onset wavelength. For NaSbO$_3$, abrupt absorption occurred at around 280 nm. For CaSb$_2$O$_6$, the absorption occurred gradually at about 400 nm, and the main absorption was similar to that of M$_2$Sb$_2$O$_7$ (M = Ca, Sr).

Figure 20.28(a) shows water decomposition on 1 wt% RuO$_2$-loaded Ca$_2$Sb$_2$O$_7$ under Hg–Xe lamp irradiation. In the initial stage of the first run, the evolution of

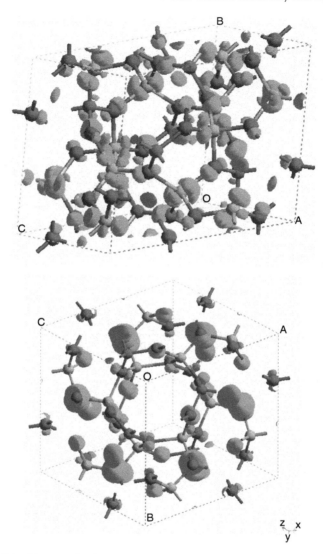

FIGURE 20.26 Electron density contour map for band #90 and the top of valence band. Zn (light gray), Ge (dotted area), and O (black)

H_2 was initially fast, followed by a gradual decrease as the reaction proceeded, whereas oxygen was produced in nearly linear manner with reaction time. The reaction was repeated by evacuating gas phase products, and good reproducibility of both H_2 and O_2 production was observed for the second to the fourth run. In the fourth run, the production of H_2 and O_2 showed a linear manner with reaction time. As shown in Figure 20.28(b), for 1 wt% RuO_2-loaded $Sr_2Sb_2O_7$, H_2 and

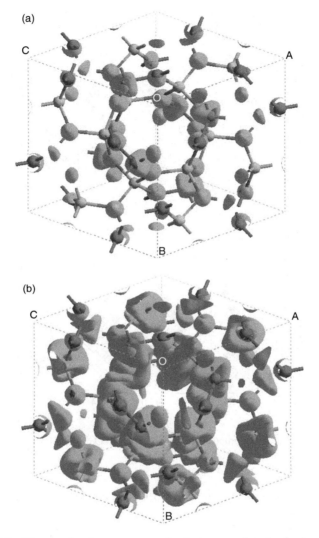

FIGURE 20.27 Electron density contour map for the bottom of conduction band at contour of (a) 0.1 and (b) 0.05

O$_2$ production was in nearly proportion to irradiation time from the initial stage, and stable and reproducible evolution was observed through the first to the fourth run. Little deterioration of the photocatalytic efficiency was observed for both RuO$_2$-loaded Ca$_2$Sb$_2$O$_7$ and Sr$_2$Sb$_2$O$_7$.

Figure 20.29 shows the photocatalytic activity of 1 wt% RuO$_2$-loaded M$_2$Sb$_2$O$_7$ (M = Ca, Sr), CaSb$_2$O$_6$, and NaSbO$_3$. The order of the activity was Sr$_2$Sb$_2$O$_7$ > Ca$_2$Sb$_2$O$_7$ > NaSbO$_3$ > CaSb$_2$O$_6$. Figure 20.30 shows the schematic

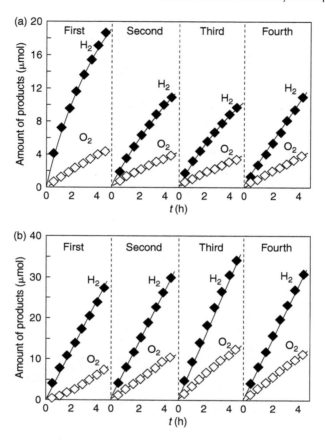

FIGURE 20.28 Water decomposition on 1 wt% RuO_2 loaded (a) $Ca_2Sb_2O_7$ and (b) $Sr_2Sb_2O_7$

representation of crystal structures. $M_2Sb_2O_7$ has two kinds of SbO_6 in a weberite structure. One type $SbO_6(1)$ is surrounded by four SbO_6 octahedra, and the other type $SbO_6(2)$ is connected to six octahedral [39]. $SbO_6(1)$ is a compressed octahedron in which there are two shorter Sb–O bonds and the remaining four bonds with the same bond length. In contrast, $SbO_6(2)$ is an elongated octahedron with two longer Sb–O bonds. $NaSbO_3$ of ilmenite contains alternating layers of edge-sharing SbO_6 octahedra [40], which are distorted in such a way that the Sb ion is in the out-of-center position. $CaSb_2O_6$ consists of infinite sheets of edge-sharing SbO_6 octahedra alternating with layers of Ca ion [41]. The octahedra are distorted by stretch in the one plane and contract in the other direction. Note that photocatalytically active antimonates are composed of deformed SbO_6 octahedra. In RuO_2-loaded M_2SnO_4 (M = Ca, Sr, Ba), the photocatalytic activity was remarkably large for M = Ca and M = Sr, whereas little activity was observed for M = Ba. The crystal structure analysis showed that the SnO_6 octahedra of

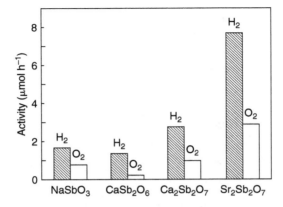

FIGURE 20.29 Comparison of photocatalytic activity among 1 wt% RuO$_2$ loaded Ca$_2$Sb$_2$O$_7$, Sr$_2$Sb$_2$O$_7$, NaSbO$_3$, and CaSb$_2$O$_6$

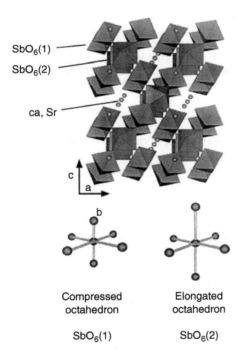

FIGURE 20.30 A schematic representation of weberite structure of M$_2$Sb$_2$O$_7$ (M = Ca, Sr) and two kinds of SbO$_6$ octahedra

Ca$_2$SnO$_4$ and Sr$_2$SnO$_4$ were distorted, whereas Ba$_2$SnO$_4$ had normal SnO$_6$ octahedra without distortion. Based on these results, it is rationally considered that deformed SnO$_6$ octahedra plays an important role in the generation of photocatalysis by the stannates. The important view that the distorted octahedral units are

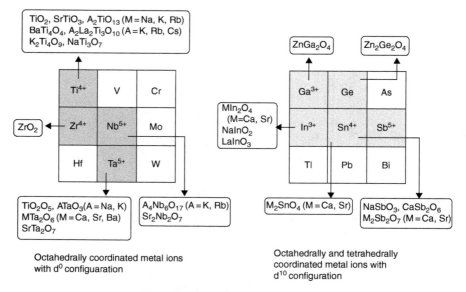

FIGURE 20.31 A list of active photocatalysts. In the two groups, the conventional transition metal oxides (left side) are composed of octahedrally coordinated metal ions with d^0 configuration. The new one (right side) is formed by octahedra and tetrahedra with d^{10} configuration

useful for photocatalysis is also confirmed for both alkaline earth metal antimonates and stannates.

20.7 CONCLUSION

A group of p-block metal oxides with d^{10} configuration have been shown to be photocatalytically active for water decomposition in the presence of RuO_2 as a promoter. Figure 20.31 shows two groups of photocatalysts that are active for water decomposition. One is a conventional photocatalyst consisting of transition metal oxides with d^0 configuration, and the other is a new photocatalyst of p-block metal oxides with d^{10} configuration. The number of active p-block metal oxide photocatalyst discovered in recent years is almost compatible to that of transition metal oxide phtocatalyst established in the past three decades. The advantages of p-block metal oxides is that the conduction bands are composed of largely dispersed broad s and p orbitals. This contrasted the narrow flat d band of transition metal oxides. Thus, the electrons photoexcited to the conduction bands have larger mobility and are effective for photocatalytic reaction. Furthermore, the concept that metal oxides with distorted octahedral make good photocatalysts for water decomposition, which has previously been proposed for various kinds titanates with d^0 configuration, has been established for the p-block metal oxides consisting of not only distorted tetrahedral but also distorted octahedral units. Based on the findings, we can predict that any alkaline metal and alkaline earth metal oxides with d^{10} configuration, will be a good photocatalyst for water decomposition in the case that the metal–oxygen octahedral and tetrahedral units are distorted to an extent to which dipole moment is generated. In this regards, it is more promising for water decomposition reaction to employ composite metal oxides involving two d^{10} metal ions from a viewpoint of changing the nature of conduction bands and the extent of distortion.

ACKNOWLEDGMENTS

The author sincerely thanks Prof. H. Kobayashi, Kurashiki University of Science and The Arts, for DFT calculation. The present work was financially supported by CREST of JST and by Grant-in-Aid for Scientific Research on Priority Area (17029022) from The Ministry of Education, Science, and Culture.

REFERENCES

1. Fujishima, A. and Honda, K. Electrochemical photolysis of water at a semiconductor electrode. *Nature* 1972, *238*, 37–39.
2. Domen, K., Kudo, A., and Onishi, T. Mechanism of photodecomposition of water into H_2 and O_2 over NiO–$SrTiO_3$. *J. Catal.* 1986, *102*, 92–98.
3. Inoue, Y., Kubokawa, T., and Sato, K. Photocatalytic activity of alkali-metal titanates combined with Ru in the decomposition of water. *J. Phys. Chem.* 1991, *95*, 4059–4063.
4. Ogura, S., Kohno, M., Sato, K., and Inoue, Y. Photocatalytic activity for water decomposition of RuO_2-combined $M_2Ti_6O_{13}$ (M = Na, K, Rb, Cs). *Appl. Surf. Sci.* 1997, *121/123*, 521–524.
5. Inoue, Y., Asai, Y., and Sato, K. Photocatalysis with tunnel structures for decomposition of water. 1. $BaTi_4O_9$, a pentagonal prism tunnel structure, and its combination with various promoters. *J. Chem. Soc., Faraday Trans.* 1994, *90*, 797–802.
6. Kohno, M., Kaneko, T., Ogura, S., Sato, K., and Inoue, Y. Dispersion of ruthenium oxide on barium titanates ($Ba_6Ti_{17}O_{40}$, $Ba_4Ti_{13}O_{30}$, $BaTi_4O_9$, and $Ba_2Ti_9O_{20}$) and photocatalytic activity for water decomposition. *J. Chem. Soc., Faraday Trans.* 1998, *94*, 89–94.
7. Takata, T., Furumi, Y., Shinohara, K., Tanaka, A., Hara, M., Kondo, J.N., and Domen, K. Photocatalytic decomposition of water on spontaneously hydrated layered perovskites. *Chem. Mater.* 1997, *9*, 1063–1068.
8. Takata, T., Shinohara, K., Tanaka, A., Hara, M., Kondo, J.N., and Domen, K. A highly active photocatalyst for overall water splitting with a hydrated layered perovskite structure. *J. Photochem. Photobiol. A: Chem.* 1997, *106*, 45–49.
9. Ogura, S., Kohno, M., Sato, K., and Inoue, Y. Preparation of $BaTi_4O_9$ by a sol–gel method and its photocatalytic activity for water decomposition. *J. Mater. Chem.* 1996, *8*, 1921–1924.
10. Ogura, S., Sato, K., and Inoue, Y. Effects of RuO_2 dispersion on photocatalytic activity for water decomposition of $BaTi_4O_9$ with a pentagonal prism tunnel

and $K_2Ti_4O_9$ with a zigzag layer structure. *Phys. Chem. Chem. Phys.* 2000, *2*, 2449–2454.

11. Sayama, K. and Arakawa, H. Photocatalytic decomposition of water and photocatalytic reduction of carbon dioxide over ZrO_2 catalyst. *J. Phys. Chem.* 1993, *97*, 531–533.

12. Kudo, A., Tanaka, A., Domen, K., Maruya, K., Aika, K., and Onishi, T. Photocatalytic decomposition of water over $NiO–K_4Nb_6O_{17}$ catalyst. *J. Catal.* 1988, *111*, 67–76.

13. Kudo, A., Kato, H., and Nakagawa, S. Water splitting into H_2 and O_2 on new $Sr_2M_2O_7$ (M = Nb and Ta) photocatalysts with layered perovskite structures: Factors affecting the photocatalytic activity. *J. Phys. Chem. B* 2000, *104*, 571–575.

14. Kato, H. and Kudo, A. Highly efficient decomposition of pure water into H_2 and O_2 over $NaTaO_3$ photocatalysts. *Catal. Lett.* 1999, *58*, 153–155.

15. Ishihara, T., Nishiguchi, H., Fukamachi, K., and Takita, Y. Effects of acceptor doping to $KTaO_3$ on photocatalytic decomposition of pure H_2O. *J. Phys. Chem. B* 1999, *103*, 1–3.

16. Kato, H. and Kudo, A. New tantalate photocatalysts for water decomposition into H_2 and O_2. *Chem. Phys. Lett.* 1998, *295*, 487–492.

17. Kato, H. and Kudo, A. Photocatalytic decomposition of pure water into H_2 and O_2 over $SrTa_2O_6$ prepared by a flux method. *Chem. Lett.* 1999, *11*, 1207–1208.

18. Sato, J., Saito, N., Nishiyama, H., and Inoue, Y. Photocatalytic activity for water decomposition of indates with octahedrally coordinated d^{10} configuration. I. Influences of preparation conditions on activity. *J. Phys. Chem. B* 2003, *107*, 7965–7969.

19. Sato, J., Kobayashi, H., Saito, N., Nishiyama, H., and Inoue, Y. Photocatalytic activities for water decomposition of RuO_2-loaded $AInO_2$ (A = Li, Na) with d^{10} configuration. *J. Photochem. Photobiol. A. Chem.* 2003, *158*, 139–144.

20. Sato, J., Kobayashi, H., and Inoue, Y. Photocatalytic activity for water decomposition of indates with octahedrally coordinated d^{10} configuration. II. Roles of geometric and electronic structures. *J. Phys. Chem. B* 2003, *107*, 7970–7995.

21. von Schenck, R. and Muller-Buschbaum, H. Alkaline-earth metal oxoindates. 3. Crystal structure investigation on $SrIn_2O_4$. *Z. Anorg. Allg. Chem.* 1973, *398*, 24–30.

22. Criuckshank, F.R., Taylor, D.M., and Glasser, F.P. Crystal chemistry of indates — Calcium, cadmium, strontium, and barium indates *J. Inorg. Nucl. Chem.* 1964, *26*, 937–941.

23. Lalla, A. and Muller-Buschbaum, H. On the atomic distribution in $Ba_2SrIn_2O_6$ with a contribution to the existence of the calcium ferrite-type of oxoindates. *Z. Anorg. Allg. Chem.* 1990, *588*, 117–122.

24. Geller, S. Crystallographic studies of perovskite-life compounds. V. Relative ionic sizes. *Acta Crystallogr.* 1957, *10*, 248–251.

25. Hubbert-Paletta, E., Hoppe, R., and Kreuzburg, G. System $NaInO_2/Na_2SnO_3$. *Z. Anorg. Allgem. Chem.* 1970, *379*, 255–258.

26. Glaum, H., Voigt, S., and Hoppe, R. Two representatives of the α-$LiFeO_2$ type — $LiInO_2$ and α-$LiYbO_2$. *Z. Anorg. Allg. Chem.* 1991, *598*, 129–138.

27. Hofmeister, W. and Tillmanns, E. Refinement of barium tetratitanate, $BaTi_4O_9$, and hexabarium 17-titanate, $Ba_6Ti_{17}O_{40}$. *Acta Crystallogr.* 1984, *C40*, 1510–1512.

28. Templeton, D.H. and Dauben, C.H. Polarized octahedral in barium tetratitanate. *J. Chem. Phys.* 1960, *32*, 1515–1518.

29. Andersson, S. and Wadsley, A.D. Structures of Na$_2$Ti$_6$O$_{13}$ and Rb$_2$Ti$_6$O$_{13}$ and alkañi metal titanates. *Acta Crystallogr.* 1962, *15*, 194–197.

30. Kohno, M., Ogura, S., Sato, K., and Inoue, Y. Effect of tunnel structures of BaTi$_4$O$_9$ and Na$_2$Ti$_6$O$_{13}$ on photocatalytic activity and photoexcited charge separation. *Stud. Surf. Sci. Catal. A* 1996, *101*, 143–152.

31. Kohno, M., Ogura, S., and Inoue, Y. Properties of photocatalysts with tunnel structures: formation of a surface lattice O-radical by the UV irradiation of BaTi$_4$O$_9$ with a pentagonal-prism tunnel structure. *Chem. Phys. Lett.* 1997, *267*, 72–76.

32. Ogura, S., Kohno, M., Sato, K., and Inoue, Y. Photocatalytic activity for water decomposition of RuO$_2$-combined M$_2$Ti$_6$O$_{13}$ (M = Na, K, Rb, Cs). *Appl. Surf. Sci.* 1997, *121*, 521–524.

33. Payne, M.C., Teter, M.P., Allan, D.C., Arias, T.A., and Joannopoulos, J.D. Iterative minimization techniques for abinitio total-energy calculations. Molecular-dynamics and conjugated gradients. *Rev. Mod. Phys.* 1992, *64*, 1045–1097.

34. Ikarashi, K., Sato, J., Kobayashi, H., Saito, N., Nishiyama, H., and Inoue, Y. Photocatalysis for water decomposition by RuO$_2$-dispersed ZnGa$_2$O$_4$ with d^{10} configuration. *J. Phys. Chem. B* 2002, *106*, 9048–9053.

35. Hornstra, J. and Keulen, E. Oxygen parameter of spinel ZnGa$_2$O$_4$. *Philips Res. Rep.* 1972, *27*, 76–80.

36. Sampath, S.K., Kanhere, D.G., and Pandey, R. Electronic structure of spinel oxides: zinc aluminate and zinc gallate. *J. Phys.: Condens. Mat.* 1999, *11*, 3635–3644.

37. Wei, S.H. and Zunger, A. Role of metal d-states in II–VI semiconductors. *Phys. Rev. B* 1988, *37*, 8958–8981.

38. Hang, C., Simonov, M.A., and Belov, N.V. Crystal structures of willemite Zn$_2$SiO$_4$ and its germanium analog Zn$_2$GeO$_4$. *Soviet Phys. Crystallogr.* 1970, *15*, 387–392.

39. Groen, W.A. and Ijdo, D.J.W. Distrontium diantimonate(V) — A Rietveld refiement of neutron powder diffraction data. *Acta Crystallogr.* 1988, *C44*, 782–784.

40. DeBoer, B.G., Young, R.A., and Sakthivel, A. X-ray Rietveld structure refinement of Ca, Sr, and Ba meta-antiminates. *Acta Crystallogr.* 1994, *C50*, 476–482.

41. Wang, B., Chen, S.C., and Greenblatt, M. The crystal structure and ionic conductivity of the ilmenite polymorph of NaSbO$_3$. *J. Solid State Chem.* 1994, *108*, 184–188.

21 Selective Catalytic Reduction (SCR) Processes on Metal Oxides

Gabriele Centi and Siglinda Perathoner

Department of Industrial Chemistry and Engineering of Materials and ELCASS (European Laboratory for Surface Science and Catalysis), University of Messina, Italy

CONTENTS

21.1 INTRODUCTION

The acronym SCR (Selective Catalytic Reduction) is commonly used to indicate the selective reduction of nitrogen oxides (NO, NO_2, or N_2O) in the presence of gaseous oxygen and a reducing agent, either inorganic (ammonia principally) or organic (saturated or unsaturated hydrocarbons, oxygenated hydrocarbons such as methanol, or nitrogen-containing chemicals such as urea). Originally, the SCR concept was used to indicate not only the abatement of NO_x emissions from stationary sources using ammonia as the selective reducing agent (SCR-NH_3) and metal oxide catalysts such as vanadia supported on titania [1], but also the selective reduction of NO_x with hydrocarbons (SCR-HC) or urea (SCR-urea) in the exhaust emissions from lean-burn or gasoline engines [2–4] as well as the selective reduction of N_2O with hydrocarbons in the tail-gas from nitric acid plants [5]. While the SCR process with ammonia for the reduction of NO_x emissions from stationary sources is a well-established commercial process to treat emissions of industrial and utility plants (gas-, oil-, and coal-fired applications), industrial and municipal waste incinerators, chemical plants (HNO_3 tail-gases, FCC regenerators, facilities for the manufacture of explosives), and in glass, steel, and cement industries [6,7], the use of SCR with hydrocarbons or urea (for both stationary and mobile sources) has not yet been commercially established. However, SCR with ammonia will soon be introduced in heavy-duty diesel powered commercial vehicles [8,9]. Note that in this case urea is just a precursor that is hydrolyzed on site (on a first catalyst layer) to generate ammonia and CO_2. On-board storage as well as distribution infrastructure (tank refillers) of urea is simpler and safer using an aqueous solution of urea rather than ammonia (compressed tanks must be used, and ammonia itself is a toxic chemical), compensating for the higher costs related to vaporization and mixing of urea as well as the necessity of an additional catalyst layer for the hydrolysis of the urea. Another case where urea is preferred instead of ammonia, as the selective reducing agent, is when urea is already available, such as in boiler services for hothouses.

Apart from the hydrolysis step, the SCR-urea process is equivalent to that of the stationary sources and the key idea behind the development of SCR-urea for diesel powered cars was the necessity to have a catalyst (i) active in the presence of O_2, (ii) active at very high space-velocities (around $500.000\,h^{-1}$ based on the washcoat of a monolith) and low reaction temperatures (the temperature of the emissions in the typical diesel cycles used in testing are in the range 120 to 200°C for over half the duration of the testing cycle), and (iii) resistant to sulfur and phosphorus deactivation. $V–TiO_2$-based catalysts for $SCR-NH_3$ have these characteristics and for this reason, their application has been developed for mobile sources. However, there are additional aspects (not present in typical $SCR-NH_3$ applications for stationary sources) which make the problem more complex: (i) very fast transients in the concentration of NO (and CO and hydrocarbons) which require a fast parallel dosing of the reducing agent (in addition, current engines do not have a NO_x sensor and commercially available NO_x sensors do not have a sufficiently fast response to closely follow the transients in NO_x concentration), (ii) the necessity of having a postcatalyst layer which can eliminate slipping ammonia (in addition, since CO and HC must be eliminated, current catalytic SCR-urea systems applied to diesel engine emissions are composed typically of five catalytic layers, making the size of the catalytic converter quite large and, therefore, applicable essentially only to heavy-duty trucks and buses), and (iii) the necessity of completely avoiding the possible formation of harmful by-products, even in sub-ppm concentrations (especially at low temperatures during cold starts and when the catalyst is deactivated due to either long mileage or deposition of particulate matter). Due to these reasons, for light-duty diesel engine emissions it is expected that another solution based on the concept of NO_x storage-reduction will probably be applied [10,11]. In this case, NO_x accumulates over the catalyst and is periodically reduced by short transient excursions from lean (excess O_2) to rich (excess reductants, i.e., HC, CO, and H_2) conditions. Although in a broader sense such an NO_x storage-reduction process can also be considered a type of SCR process, the specific stage of NO_x reduction occurs in conditions of oxygen deficiency, similarly to three-way catalysts. This discussion is limited to the case where NO reduction occurs directly in the presence of gaseous O_2.

A specific case of mobile sources where nearly the same conventional system used for stationary sources (SCR-NH_3) can be applied for the reduction of NO in the exhaust gas from marine engines and gas turbines. The first marine installation was in 1989 on a 37,000 tdw deep-sea bulk carrier [12]. Many installations exist today, but this remains a niche application.

Several contributions and review papers have been published on the SCR processes on metal oxides; however, there are still several points that deserve to be commented upon. In this chapter, we will first discuss some general aspects of SCR with special focus on some issues and questions that should be clarified regarding future prospects and new directions of research in this field, which is centred on the use of metal oxide catalysts. Then, a more detailed discussion regarding the type of catalysts and reaction mechanism will be made for the SCR-NH_3 and the SCR-HC processes.

21.2 GENERAL ASPECTS, ISSUES, AND QUESTIONS

21.2.1 Ammonia versus Hydrocarbon as the Selective Reductant

Since the discovery of the possibility of using hydrocarbons for the selective reduction of NO in the presence of gaseous oxygen [13,14] and later of using methane as the selective reductant [15], considerable debate in the literature has regarded the use of hydrocarbons instead of ammonia as the selective reducing agent. This aspect is a key question regarding gaps and needs in the area of SCR processes on metal oxides and it is necessary to try to clarify this problem.

Ammonia is a cheap reactant that can be reasonably, easily stored in a compressed (liquid) form. The main concerns regard safety aspects (ammonia toxicity and risks of explosion) and ammonia slip. To minimize the latter problem it is necessary either to maintain the NH_3/NO ratio in the feed below the stoichiometric value or to use a final catalyst layer to selectively oxidize the ammonia that slips from the reactor [16]. The latter solution is of increasing interest to improve the overall efficiency in the conversion of NO together with minimization of the ammonia slip (benefits in terms of reduced impact on the environment as well as minimization of the possible deposit of ammonium–sulfate on the heat exchanger walls downstream from the SCR reactor).

Hydrocarbons are slightly less expensive than ammonia, and have (usually) a lower intrinsic toxicity. Hydrocarbon slip is a less critical issue, but the possibility that harmful by-products (HCN, acrylonitrile, aldehydes, etc.) may form indicates the necessity to use a final posttreatment catalyst layer. Methane is more difficult to store, but usually is already available in the plants. A clear advantage of hydrocarbons (with respect to ammonia) is that the formation of ammonium sulfate downstream from the SCR reactor (when SO_2 is present in the feed) is avoided, although this problem is not very critical with the new generation of SCR catalysts. On the other hand, CO usually forms when hydrocarbons are used as the reducing agent, and therefore a postcatalyst layer to oxidize CO to CO_2 is required.

It may be concluded that there are no great incentives in terms of cost, safety, and manageability in using hydrocarbons instead of ammonia in the abatement of NO from stationary emissions. If the reaction rates, selectivity in the conversion of NO to N_2 (with respect to either oxidation to NO_2 or partial reduction to N_2O), and catalyst productivity are comparable using either hydrocarbons or ammonia as the reducing agent, then the hydrocarbon is preferable. However, the current best catalysts for SCR-HC are at least one order of magnitude (in terms of reaction rate) less active compared with SCR-NH_3, and are usually less selective. In these conditions, economics do not indicate the suitability of the SCR-HC process, apart from special cases (when the use of ammonia could be unsatisfactory for other reasons).

Between the light alkanes, methane is less suitable than propane, due to (i) the significantly lower activity of the catalysts when methane is used instead of propane (usually the reaction rate is at least one order of magnitude lower), (ii) comparable raw material costs, and (iii) easy propane storage in the liquid form under moderate pressure (which makes the possibility of having service methane already available in the plant not economically relevant). However, most studies on this topic have focused on the reduction of NO using methane as the reducing agent [17]. Also other hydrocarbons such as isobutene have been claimed as interesting for SCR-HC in stationary sources, but all alkenes and particularly isobutene are too expensive to be considered. Oxygenated hydrocarbons are also typically too expensive as raw materials. Methanol could be used, but the possibility that harmful by-products such as formaldehyde may form is a negative drawback.

Clearly, the case of mobile sources is different from stationary sources, because hydrocarbons are already present in the emissions (in this case, ethane and propene are among the main hydrocarbons present, at least in the case of gasoline combustion; the hydrocarbon composition of diesel engine emissions is quite different) or could be added relatively easily by secondary injection. In this case, however, the problem is the low reaction rate (especially at low temperatures) and low selectivity, as well as the insufficient range of temperature application. However, with respect to alternative solutions for mobile sources (including SCR-urea), SCR-HC requires minimal hardware and could use the fuel itself as the reducing agent, avoiding the need for an additional reagent tank. SCR-HC would be the desired technology to attain future legislated NO_x emission limits, if better catalyst design can be developed [18].

A different case is the SCR of N_2O to eliminate this greenhouse chemical from nitric acid plant emissions or waste fluid bed combustors [19]. In this case, ammonia is often not active as a selective reducing agent different from hydrocarbons (propane, e.g., see Reference 20), although recent results show the possibility of using ammonia as a selective reducing agent for N_2O [21]. However, reaction rates and process economics are still in favor of hydrocarbons instead of ammonia for the SCR of N_2O. When the combined abatement of NO and N_2O must be realized [5], hydrocarbons (propane) also appear to be the preferable choice.

In summary, SCR-NH_3 is still the preferable process for the abatement of NO in stationary sources, but when N_2O or combined N_2O/NO abatement is the target of the process, hydrocarbons, and especially propane could be preferable. Still considerable improvement is required for the SCR-CH_4 process, but developments in this field are very fast [22,23]. Better catalyst design based on understanding the relationship between nanostructure and reactivity is expected to improve performances of metal oxide catalysts in the conversion of nitrogen oxides [18].

21.2.2 SCR on Metal Oxide Coupled to NonThermal Plasma

NonThermal Plasma (NTP) has received considerable attention in recent years as an enabling technology for SCR-HC, especially in mobile applications [24]. Dielectric Barrier Discharge (DBD) devices have been commercially available for over 100 years, but recently their design has been considerably improved. They are based on the passage of the gas flow through parallel plates covered with a dielectric barrier (commonly alumina). Upon application of a high-voltage alternating current, a corona is created between the plates. The plasma which is generated causes the fast oxidation (at low temperatures) of NO to NO_2 together with partial hydrocarbon oxidation (using propylene, the formation of CO, CO_2, acetaldehyde, and formaldehyde was observed). Both effects cause a large promotion of the activity of the downstream catalyst [25,26]. For example, a γ-alumina catalyst that is essentially inactive in the SCR of NO with propene at temperatures around 200°C allows a conversion of NO of ~80% (in the presence of NTP). Formation of aldehydes follows the trend of NO concentration suggesting their role in the reaction mechanism. Metal oxides such as alumina, zirconia, or metal-containing zeolites (e.g., Ba/Y) have been used [24–27], but a systematic screening of the catalysts to be used together with NTP was not carried out. Therefore, considerable improvements may still be expected.

Problems observed regard catalyst stability and the tendency to form organic deposits on the catalyst surface. From the application point of view the main technical problem is the energy required to maintain the plasma (typically, 20 to 60 J/L). However, significant improvements are also expected in this regard. A further aspect, which deserves attention is the formation of toxic by-products such as CH_2O and HCN within the NTP plumes. Nevertheless, this route appears promising, especially when a better integrated design between the NTP device and the catalyst becomes available.

NonThermal Plasma promoted catalytic removal of NO_x has also been tested as a new SCR-NH_3 process [28,29]. V_2O_5/TiO_2 [28] and V_2O_5–WO_3/TiO_2 monolithic catalysts [29] have been evaluated. At low reaction temperatures (around 150°C) under conditions in which without plasma treatment the removal of NO_x was minimal, conversions of NO of ~70% could be obtained. Similar effects were observed for the SCR in gas mixtures containing equal amounts of NO and NO_2, indicating that the role of plasma is not only that of promoting the conversion of NO to NO_2. The conversion of NO to NO_2 is a key aspect of SCR-HC [23], while it may or may not be an important factor in SCR-NH_3, depending on the type of catalyst. Copper-based catalysts are significantly promoted using NO_2/NO mixtures [30], while the effect is less important on vanadia on titania type catalysts under nonNTP conditions. Both classes of catalysts have been found to be highly active under NTP conditions and a synergistic combination of NTP and both SCR-NH_3 and -HC was observed to be present under real diesel engine exhaust conditions [31].

Low temperature activity promotion is an issue in mobile (diesel) applications, but may not be a critical issue in several stationary applications, apart from those where the temperature of the emissions to be treated is below 200°C (e.g., when a retrofitting SCR process must be located downstream from secondary exchangers, or in the tail-gas of expanders in a nitric acid plant). In the latter case, a plasmacatalytic process [32] could be interesting. In the other cases, the use of NTP together with the SCR catalyst is not economically valuable. However, the synergetic combination of plasma and catalysts has been shown to significantly promote the conversion of hazardous chemicals such as dioxins [33]. Although this field has not yet been explored, it may be considered a new plasmacatalytic SCR process for the combined elimination of NO_x, CO, and dioxins in the emissions from incinerators.

21.2.3 NH₃-SCR Processes for the Emissions from Combustion and Nitric Acid Plants

There are four main applications of the SCR-NH_3 process for the reduction of NO in the emissions: (a) power plants, (b) gas turbines, (c) waste incineration, and (d) nitric acid plants [34–35]. While often specific distinction is not made between these cases and the same catalysts are assumed applicable in all cases, there are significant differences in terms of composition of the emissions and space-velocities. A specific difference between the first three cases (combustion) and the fourth (nitric acid plants) regards the NO/NO_2 ratio, which is typically close to 20 for combustion processes and close to 1 for the nitric acid plants. Furthermore, no SO_2 is present in the fourth case.

While vanadia on titania-based catalysts can be used for both classes of applications, there are other types of catalysts such as those based on copper [30] that show good performances in the case of mixtures of NO/NO_2 (nitric acid plants), while performances are worse when applied to emissions from catalytic processes.

The presence of NO_2 in the feed (in nitric acid plant emissions, or when the feed is pretreated by NTP) induces a significant change in the reactivity and in the reaction mechanism. Koebel et al. [36] studied the low temperature behavior of the SCR process with feed gases containing both NO and NO_2 and observed the presence of two main reactions:

$$4NH_3 + 2NO + 2NO_2 \rightarrow 4N_2 + 6H_2O \tag{21.1}$$

$$2NH_3 + 2NO_2 \rightarrow NH_4NO_3 + N_2 + H_2O \tag{21.2}$$

The "fast SCR reaction" exhibits a reaction rate at least 10 times higher than that of the well-known standard SCR reaction with pure NO and dominates at temperatures above 200°C. At lower temperatures, the "ammonium nitrate route" becomes increasingly important. At temperatures ranging from 140 to 180°C (typical, however, for SCR-NH_3 in mobile applications), the ammonium nitrate route may be responsible for the whole NO_x conversion observed. Ammonium nitrate decomposition may lead to N_2O. Avila et al. [37] also observed that the NO conversion over a V_2O_5–WO_3/TiO_2 catalyst shows a maximum for a NO_2/NO ratio close to 1 for the presence of an additional reaction pathway involving the formation of ammonium complexes that react with adsorbed NO and NO_2 [37], differently from the mechanism occurring in the absence of NO_2 involving weak adsorption of NO over the catalyst [38].

Although it is often considered that a single reaction mechanism occurs in the selective reduction of NO by ammonia, data show that instead different mechanisms are possible and that depending on the type of catalyst and reaction conditions (feed composition, reaction temperature) one mechanism may prevail over the others [30]. However, not considering this aspect and making extrapolations regarding the reaction mechanism from one catalyst to another or to different reaction conditions, may lead to erroneous conclusions. In addition, it is important to consider all possible opportunities to develop new kinds of catalysts, for example, for the combined removal of NO_x and N_2O from nitric acid plant emissions [39].

21.2.4 The Question of Catalyst Activity in the NH₃-SCR Process

The superior activity of a new type of catalyst with respect to "conventional" SCR catalysts is often claimed in the literature. Long et al. [40], for example, claimed the superiority of pillared clay catalysts for SCR of nitrogen oxides to control power plant emissions by comparing Ce–Fe–TiO_2-PILC (Pillared InterLayer Clay) with a V_2O_5–WO_3/TiO_2 catalyst. There are many more examples of this type of claim in the literature.

While it is usually worthwhile to improve the rate of reaction, this is not a specific issue of the "conventional" V_2O_5–WO_3/TiO_2 type of catalyst for power plant applications. Catalyst activity can easily be improved by increasing the vanadium loading, which, however, is maintained low (typically below 1 wt%) to limit the oxidation of SO_2 to SO_3, a source of problems downstream from the reactor such as plugging (formation of ammonium sulfate) or corrosion. Depending on the SO_2 content in the feed, the catalyst activity can be tuned by changing the V_2O_5 and WO_3 contents in the catalyst to maximize activity in the reduction of NO, while minimizing SO_2 oxidation activity. The ratio between these two reaction rates is the key factor for the choice of the catalyst, not the activity itself in the conversion of NO.

Another important parameter is the selectivity in the conversion of NO with respect to ammonia, always present as a side reaction of ammonia combustion, the minimization of which is a key factor for both process economics and NO efficiency. Related to this aspect, another important parameter is the efficiency of the reduction of NO when the ammonia present is less than the stoichiometric value.

A final relevant parameter is the temperature window in which the conversion of NO is at the maximum value. In fact, due to the presence of the side combustion of ammonia, the conversion of NO typically decreases at temperatures above the maximum. The presence of a sharp or broader maximum is related to the rate of reaction of ammonia with NO (or NO_2) to give N_2 (SCR) with respect to the reaction with O_2 to give either N_2 or NO_x (ammonia oxidation). A broader maximum is helpful in applications where there are fluctuations in the temperature of the emissions.

Resistance to deactivation is also an issue, because commercial SCR catalysts have a life span of over 10 years.

21.3   SCR-NH$_3$

Several excellent reviews have discussed the SCR-NH$_3$ process over metal oxides [6,7,30,34,38,41–43], in terms of both fundamental and technological aspects. Therefore, the discussion here is focused on some additional aspects which deserve attention and are useful for a better understanding of this field. Some background information necessary for the discussion is also included. There are three different classes of commercial catalysts for the SCR-NH$_3$ process (noble metals, metal oxides, and zeolites), but only metal oxide catalysts will be discussed here.

21.3.1   Metal Oxide Catalysts for the SCR-NH$_3$ Process

The largest part of commercial applications use monolith type catalysts based on V_2O_5–WO_3 or V_2O_5–MoO_3 oxides supported on TiO_2 in the anatase crystalline form. The choice of these catalysts does not rely on their superior activity performances, as sometimes thought, but rather on their higher resistance to deactivation by SO_2 and their limited rate of SO_2 oxidation with respect to the rate of NO reduction. This derives from the fact that TiO_2 is only weakly and reversibly sulfated in the presence of SO_2 and O_2, and that the vanadia is well spread over the titania surface. The vanadia content is generally kept low and is reduced below 1 wt% in the presence of high SO_2 concentrations in the feed. WO_3 or MoO_3

(around 10 and 6 wt%, respectively) are typically added to increase the surface acidity, the activity, and the thermal stability of the catalyst, and limit the rate of SO_2 oxidation. MoO_3 also improves catalyst stability in the presence of poisons such as As. The catalysts are further improved by addition of silico-aluminate and fiberglass to increase their mechanical resistance and strength.

The active phase is coated over ceramic (typically cordierite) or sometimes metallic monoliths (thin foils coated with an oxide washcoat), which make it possible to reduce pressure drop, improve resistance to attrition, and lower the plugging rate due to fly ash. The density of the cells as well as geometry of the monolith depends on the type of application (amount of fly ash, SO_2 concentration, and location of the SCR unit in the plant). Proper design of the monolith is very important to maximize the performance and reduce deactivation. Usually, honeycomb monoliths have a wall thickness of 0.5 to 0.6 mm and a channel width of 3 to 4 mm in the case of low dust gas and greater wall thickness (> 1 mm) and larger channels (about 7 mm) for flue gas with high dust content. Increasing cell density (i.e., decreasing the wall thickness and size of the channels) increases the external geometric surface area and therefore monolith productivity.

The thickness of the active phase coating is usually maintained low, because under the typical high space-velocities of common applications (linear velocities in the 4 to 10 m/sec range), the reaction of NO conversion is controlled by interphase gas diffusion, while the control is kinetic for the oxidation of SO_2. In other words, the conversion of SO_2 to SO_3 depends on the catalyst volume and can be reduced by decreasing the thickness of the active phase layer (as well as providing a more uniform thickness at the edges of the square-shaped channel), while the reduction of NO depends mainly on the external geometric surface area and therefore is influenced little by the reduction of the catalyst layer thickness, being the internal effectiveness factor very low [6,43,44]. A simple pseudohomogeneous 1D model of the reactor based on these assumptions provides a good description of the results [6,43]. The effects of inter- and intraphase mass transfer limitations are lumped into an effective pseudo first-order rate constant. However, more complex models that explicitly take into account (i) the surface roughness and turbulence effects at the monolith entrance, for example, using CFD (Computer Fluid Dynamic) models (therefore, determining local fluidodynamic conditions with respect to the use of the same gas–solid mass transfer coefficient along all the monolith walls) and (ii) intraparticle diffusion (using the effective porosity distribution and considering the lack of homogeneity of the washcoat thickness especially in the square-shaped channels) would be preferable for better design of the catalysts (optimal porosity, use of additives such as fibers which increase the surface roughness, optimal catalyst thickness profile, etc.) and to improve performances (reduce ammonia slip and rate of SO_2 oxidation, improve productivity). Therefore, even though SCR-NH_3 over V–(W, Mo)/TiO_2-based monolith catalysts is a well-studied process, there is still room for improvement, especially in the area of understanding the interactions between catalyst and reactor engineering aspects.

Analysis of the dynamics of SCR catalysts is also very important. It has been shown that surface heterogeneity must be considered to describe transient kinetics of NH_3 adsorption–desorption and that, the rate of NO conversion does not depend on the ammonia surface coverage above a critical value [45]. There is probably a reservoir of adsorbed species that may migrate during the catalytic reaction to the active vanadium sites. It was also noted in these studies that ammonia desorption is a much slower process than ammonia adsorption, the rate of which is comparable with that of the surface reaction. In the SO_2 oxidation on the same catalysts, it was also noted in transient experiments [46] that the build up/depletion of sulfates at the catalyst surface is rate controlling in SO_2 oxidation.

Analysis and modeling of the dynamic behavior of the catalyst is useful to closely describe the performance during start up, shut down, and load variation of stationary applications and of critical relevance for SCR-NH_3 of mobile diesel engine emissions. Use of this tool has not been extensively reported in the literature for the design of improved catalysts, although it is a very valuable method. On the contrary, as will be discussed later, use of this tool to derive mechanistic implications is less convincing.

Other metal oxide catalysts studied for the SCR-NH_3 reaction include iron, copper, chromium, and manganese oxides supported on various oxides, introduced into zeolite cavities or added to pillared-type clays. Copper– and copper–nickel catalysts, in particular, show some advantages when NO–NO_2 mixtures are present in the feed and SO_2 is absent [30], such as in the case of nitric acid plant tail emissions. Copper-on-alumina catalysts are also interesting catalysts for the simultaneous removal of NO_x and SO_x from combustion emissions [47]. TiO_2 supported transition metal oxide catalysts (Mn, Cu, and Cr) can effectively reduce NO using NH_3 and excess oxygen with 100% NO conversion and N_2 selectivity at $\leq 120°C$ [48]. However, generally speaking, note that there are no great incentives in using these alternative catalysts with respect to vanadia–titania-based catalysts.

However, note that the mechanism of NO reduction over copper- and manganese-based catalysts is quite different from that over vanadia–titania-based catalysts [30]. Scheme 21.1 reports the proposed mechanism of SCR-NH_3 over Cu–alumina catalysts [30]. The reaction occurs either via formation of an amide (NH_2) species which then reacts with NO to form a nitrosamide intermediate, similarly to that proposed for vanadia–titania catalysts [49] (in these catalysts the NO does not strongly chemisorb or oxidize to NO_2 differently from copper-based catalysts), or via oxidation of NO to NO_2 or nitrite species which then reacts with chemisorbed ammonia. Formation of nitrate (NO_3^-) species is also possible (e.g., as detected by Fourier Transform Infrared (FTIR)), leading to the formation of ammonium nitrate that is responsible for the formation of N_2O. Data also showed that the relative importance of the different pathways of reaction (which are always present) depends on the reaction conditions (feed composition, reaction temperature). Therefore, a single pathway and mechanism of reaction do not exist, but rather multiple pathways coexist and their rates of reaction depend on the experimental conditions. This concept is valid more in general for other type of catalysts. In general, little attention has been given in the literature to considerations that no single reaction mechanism occurs over a specific catalyst and that multiple pathways of reaction are possible. Even though the SCR-NH_3 process

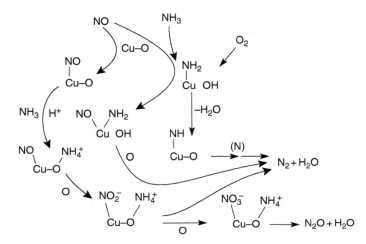

SCHEME 21.1   Proposed mechanism for the NO–NH₃ reaction over Cu-based catalysts in the presence of oxygen (Elaborated from Centi, G. and Perathoner, S. *Appl. Catal. A: Gen.* 1995, *132*, 179–259. With permission.)

may seem to be well established, the need to focus attention on investigation of the reaction mechanism is still required.

21.3.2 Reaction Mechanisms and Microkinetic Approaches to the SCR-NH₃ Reaction

The chemical and mechanistic aspects of the SCR-NH₃ process over metal oxide catalysts have been discussed in details by Busca et al. [38]. Their conclusions are that the mechanism on V(W, Mo)–TiO₂ catalysts is based on the dissociative chemisorption of ammonia on Lewis (vanadium) acid sites and that NO reacts with the amide chemisorbed species to form a nitrosamide key intermediate which then decomposes to N₂ and H₂O. The catalytic cycle is closed by reoxidation of the reduced catalyst by gas-phase oxygen.

This mechanism can be converted to a kinetic scheme to describe the steady state or transient performances of the catalyst. Assuming (i) an Eley–Rideal mechanism (reaction between adsorbed NH₃ and gas-phase NO), (ii) that ammonia and water compete for adsorption onto the active sites, and (iii) that adsorption equilibrium is established for both species, the following Rideal rate expression is eventually obtained [6]:

$$r_{NO} = \frac{k_c K_{NH_3} C_{NH_3} C_{NO}}{(1 + K_{NH_3} C_{NH_3} + K_{H_2O} C_{H_2O})} \tag{21.3}$$

where k_c is the intrinsic chemical rate constant, C_{NO}, C_{NH_3}, and C_{H_2O} are the NO, NH₃, and H₂O gas-phase concentrations, respectively, and K_{NH_3} and K_{H_2O} are the adsorption equilibrium constants for NH₃ and H₂O, respectively.

Steady-state performances of SCR-NH₃ reactors can be correctly described using this microkinetic approach derived from the reaction mechanism outlined previously, as shown in Figure 21.1(a) [50].

On the basis of *in situ* FTIR studies under steady-state conditions, Topsøe et al. [51] have proposed a mechanism by which ammonia is instead adsorbed on a Brønsted acid site associated with a V^{5+}–OH site, followed by activation of the adsorbed ammonia by a nearby V^{5+}=O group (which is reduced to a V^{4+}–OH species). Then NO reacts from the gas-phase with the activated ammonia complex leading to the formation of an intermediate, which then decomposes to nitrogen and water. Regeneration of the active sites (i.e., oxidation of the reduced V^{4+}–OH sites to V^{5+}=O groups) occurs by gas-phase oxygen. Accordingly, the proposed catalytic cycle consists of both acid–base and redox functions.

The resulting kinetic expression from this reaction mechanism is the following:

$$r_{NO} = k_3 P_{NO} \frac{(k_2 K_{NH_3}/k_3)(P_{NH_3}/P_{NO})}{1 + (k_2 K_{NH_3}/k_3)(P_{NH_3}/P_{NO}) + (k_{-2}/k_3)(1/P_{NO}) + K_{NH_3} P_{NH_3}}$$

(21.4)

This equation also provides a good fitting of the experimental data (Figure 21.1[b]) [52,53]. Therefore, both reaction mechanisms lead to a reaction rate that correctly fits the experimental data, even though it is based on different types of surface reactions.

It may be argued that the mechanistic information for the first type of mechanism comes from spectroscopic studies performed under "clean" conditions (i.e., absence of water vapor); while in the second case, the experiments were performed *in situ* on a hydroxylated surface. Therefore, Lewis acid sites (detected under clean conditions) are not present when water vapor is present under practical conditions. This may suggest that the mechanism involving ammonia coordinated over Brønsted acid sites is more correct. On the other hand, it should also be considered that in addition to vanadium, which is recognized as the active element for the reaction, other catalyst components that are inactive or poorly active in the SCR (e.g., the W and Ti surface sites), may play a role in the reaction. Indeed these sites strongly adsorb ammonia and accordingly they participate in the reaction as "reservoirs" of adsorbed NH₃ species, as mentioned earlier. Accordingly, a distinction should be made between the ammonia "adsorption" and "reaction" sites, and care must be taken when discussing, for example, spectroscopic data of the SCR mechanism (indeed in this case the contribution of the bands arising from NH₃ adsorbed on the TiO₂ support are superimposed on those corresponding to the active sites) or in the derivation of kinetic models (the NH₃ adsorption sites may differ from the NH₃ reaction sites).

In conclusion, notwithstanding the many studies on the SCR-NH₃ mechanism, the topic can still be considered not closed and further studies with *in situ* combined methods can perhaps clarify this intriguing question. Modern theoretical methods offer more tools for the research. Miyamoto and coworkers [54], for

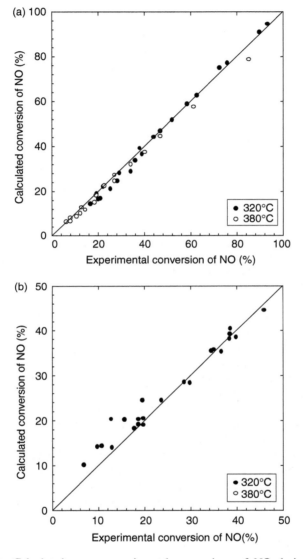

FIGURE 21.1 Calculated versus experimental conversions of NO during SCR-NH$_3$. (a) Based on rate Equation (21.3) elaborated from data reported by Forzatti, P. and Lietti, L. *Heter. Chem. Rev.* 1996, 3, 33–51. With permission. (b) Based on rate Equation (21.4) (Elaborated from the data reported by Dumesic, J.A., Topsøe, H., Chen, Y., and Slabiak, T. *J. Catal.* 1996, *163*, 409–417. Dumesic, J.A.; Topsøe, N.Y.; Slabiak, T., Morsing, P., Clausen, B.S., Tornqvist, E., and Topsøe, H. *Stud. Surf. Sci. Catal.* 1993, *75*, 1325–1337. With permission.)

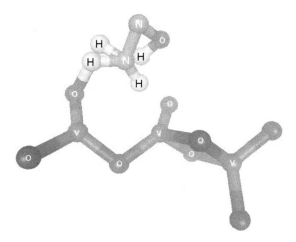

FIGURE 21.2 Scheme of the reaction intermediate formed by the interaction of gaseous NO with the adsorbed ammonia species on the V_2O_5 (010) surface (Elaborated from Yin, X., Han, H., and Miyamoto, A. *Phys. Chem. Chem. Phys.* 2000, *2*, 4243–4248. With permission.)

example, studied the process by periodic first-principle Density Functional Theory (DFT) calculations and showed that over the V_2O_5 (010) surface ammonia is adsorbed preferentially on the Brønsted acid sites, and gaseous NO interacts with the preadsorbed NH_4^+ species to release N_2 and H_2O. A vanadyl (V=O) group lying near to the Brønsted acid site (V–OH) assists in the coordination and activation of the chemisorbed ammonium ion.

Figure 21.2 reports a scheme of the proposed key reaction intermediate [54]. Hydroxylation of NO is the driving force to form the nitrogen–nitrogen bond. The further transformation of the chemisorbed complex requires participation of near lying vanadium atoms. It is interesting to note that this step is analogous to that suggested as the characterizing step in the mechanism of NO reduction by propane over copper-based catalysts [55], pointing out that notwithstanding some differences, there are several similarities between the SCR-NH_3 and -HC mechanisms.

21.4 SCR-HC

The chemistry and reaction mechanisms as well as the prospects for application of the SCR-HC process on metal oxide catalysts have been reviewed by various authors [2,3,23,39,56–59], due to the large and still growing interest in this reaction, although the outlook for application to either emissions of lean-burn or diesel engines (limiting discussion to the direct SCR-HC reaction and not including the so-called NO_x storage-reduction mode [11]) or stationary sources appears much less promising than when the reaction was discovered. Since the discovery of this reaction (around 1990), much debate has been focused on the reaction mechanism, although often clear distinction between general aspects of

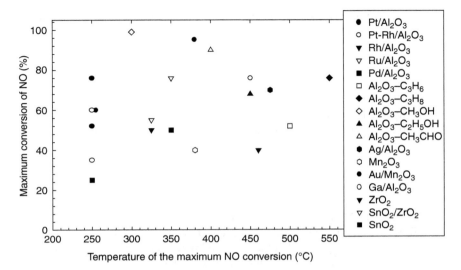

FIGURE 21.3 Maximum conversion temperature of NO and relative temperature of the maximum conversion of NO for a series of metal oxide-based catalysts in the presence of O_2 and using propene (C_3H_6) as the selective reductant or other hydrocarbons, when indicated in the legend

the reaction mechanism and specificity of the catalyst(s) and reaction conditions investigated were not made. It should also be mentioned that there are significant differences in the performances, and probably in the reaction mechanisms between the different classes of catalysts as well as between the effects of different types of hydrocarbons.

In the SCR-HC reaction, similar to the SCR-NH$_3$ reaction, there is a maximum in the conversion of NO as a function of the temperature, although typically the maximum is much sharper using HC instead of NH$_3$ as the selective reducing agent. Figure 21.3 reports some examples of the maximum conversion of NO (and relative temperature of the maximum conversion of NO) for a series of samples in the SCR with propene or other hydrocarbons. The data show that (i) the maximum in NO conversion, in terms of both reaction temperature and maximum conversion of NO, may change in a broad range depending on the catalyst and (ii) for pure alumina, and also for other metal oxide catalysts, the performances may also change considerably depending on the type of hydrocarbon used as the selective reducing agent. It is not very reasonable that the same mechanism operates in all these cases, but little attention has been given toward understanding this question.

Three types of reaction mechanism have been proposed in the literature:

1. NO reacts directly with adsorbed partially oxidized hydrocarbon species.

$$\text{HC} \xrightarrow{\text{O}_2} \text{HC--O} \xrightarrow{\text{NO}} \text{N}_2$$

2. NO is first oxidized to NO_2 which then (adsorbed or in the gas phase) reacts with HC.

$$NO \xrightarrow{O_2} NO_2 \xrightarrow{HC} N_2$$

3. NO decomposes directly on the reduced catalyst surface, while the hydrocarbon maintains the catalyst surface reduced by reacting with the oxygen atoms formed during the decomposition of NO.

$$NO \xrightarrow{cat} N_2 + cat-O$$
$$\uparrow cat + CO_2 \xleftarrow{} HC$$

For example, Misono [23] suggests that the second mechanism is the principal one. Burch et al. [57,58] indicated that the third mechanism dominates, while Sadykov et al. [56] reported that the determining role belongs to the formation of organic nitro compounds, that is, that the first mechanism is the dominating one. All these authors extensively reviewed the literature data and of course reported a series of supporting evidence for their hypotheses. It is not the scope of this chapter to discuss which mechanism is preferable, but clearly these contradictory hypotheses point out that much more research effort is still needed to understand the SCR-HC process.

We may note, however, that often the different suggested mechanisms are based on the identification of surface adspecies by InfraRed (IR) spectroscopy. Since the spectra are usually quite complex with several overlapping bands, the analysis is based on the identification of the nature of the adspecies by feeding single reactants (e.g., the interaction of NO with metal oxide catalysts usually gives rise to the formation of nitrosyl, nitrite, and nitrate species) and the analysis of which species disappear when another reactant (hydrocarbon) is fed to the catalyst on the surface of which the former species (NO_x) are chemisorbed. The chemisorption of HC, however, may give rise to a change in the nature of NO_x adspecies before they eventually react with the hydrocarbon. In other words, there is a change in the nature of the NO_x adspecies before their reaction and therefore the reactive species are not those identified by IR spectroscopy, which instead mainly monitors those species that act as a "reservoir" for the reactive NO_x species generated when the other reactant is chemisorbed. The process is very fast and hard to detect under standard conditions. It is also evident that there are considerable analogies with the question of the effective SCR-NH_3 mechanism under working conditions in which the questions of "reservoir," and reactive and spectator species exist.

The problem of reactive versus spectator species in the SCR-HC mechanism was recently investigated for noble metals (Pt or Rh) supported on TiO_2 [60]. Catalysts based on platinum and rhodium supported on titania show comparable activity, but quite a different nature of chemisorbed NO_x species after interaction with a flow of NO + O_2 in helium. With Pt–TiO_2, the IR spectrum is characterized by the intense bands due to nitrate, while with Rh–TiO_2 the spectrum is

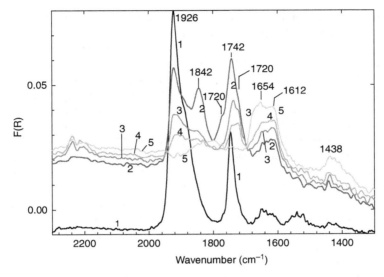

FIGURE 21.4 *In situ* DRIFT tests at 330°C after 30 min in a flow of 0.5% NO + 5% O$_2$ (1) and after addition of propene (0.5% in helium) (2 → 5; after 1 in a flow of propene/He for 1, 2, 3, and 10 min, respectively) on 1% Rh–TiO$_2$ (Elaborated from Arena, G.E. and Centi, G. *Top. Catal.* 2004, 30/31, 147–153. With permission.)

characterized by the intense bands due to nitrosyl species coordinated on Rh. However, in both cases, it was possible to show [60] that the chemisorption of propene first induces a change in coordinated NO$_x$ adspecies and then their conversion. On Rh–TiO$_2$, propene gives rise to the formation of negatively charged nitrosyl and gem-dinitrosyl species, whereas on Pt–TiO$_2$, propene converts nitrate species (probably at the metal-support interface) to weakly bound NO$_x^{\delta+}$ species.

Figure 21.4 shows an example of the *in situ* Diffuse Reflection Infrared Fourier Transform (DRIFT) spectra observed when propene in helium is fed at 330°C to an Rh–TiO$_2$ catalyst with preadsorbed NO$_x$ species. After addition of propene, an immediate decrease in the intensity of the band at 1925 cm^{-1} (Rh–NO$^{\delta+}$) and a simultaneous increase in the intensity of two bands at 1842 and 1755 cm^{-1} (gem-dinitrosyl complex on Rh) was observed. For longer times on stream, the intensity of these bands also decreases, while new bands form at about 1720 and 1654 cm^{-1} which could be attributed to νC=O in coordinated acrylic acid and acrolein, respectively. When acrolein and acrylic acid started to be detected, bands near 2200 cm^{-1} (2235 and 2200 cm^{-1}, attributed to isocyanate coordinated to titanium and rhodium ions, respectively) also were detected.

The formation of bands due to acrolein/acrylic acid and isocyanate could be better evidenced working at 200°C, but using different concentrations in the feed. In this case, the appearance of the bands indicating the formation of acrolein and acrylic acid (νC=O at 1670 and 1720 cm^{-1}, respectively) occurs together with the formation of a series of low intensity bands near 2200 cm^{-1} (2215 and 2175 cm^{-1})

and only for the longer times on stream when bands in the region near to $3000 \, \text{cm}^{-1}$ (νCH) due to coordinated propene also started to be detected.

The initial difference in the nature of the NO_x adspecies on platinum- and rhodium-based catalysts upon $NO_x + O_2$ interaction can be related to the different natures of the noble metal surfaces. In addition, NO_x adspecies tend to be preferentially located on the support (or noble metal/support interface) in Pt/TiO_2, but on the noble metal in Rh/TiO_2. The noble metal influences the interface region and this modification is responsible for the fact that minimal amounts of nitrate (one to two orders of magnitude less) are formed on Rh/TiO_2 with respect to Pt/TiO_2.

Upon interaction with propene this TiO_2 becomes anion deficient causing the formation of weakly coordinated $NO_x^{\delta+}$ species. On Rh/TiO_2, the interaction of propene with the oxidic type overlayer causes its reduction, which is incomplete due to the difficult reduction. Ions or islands of positively charged Rh ions (surrounded by negatively charged oxygen atoms) are created in this process of reduction. These ions are responsible for the negatively charged mono or gem-dinitrosyl species.

For longer times on stream, carbon monoxide (formed during the combustion of propene) tends to inhibit the noble metal surface activity and propene was observed to accumulate over the catalyst surface. In these conditions, partially oxidized species (coordinated acrolein and acrylic acid) and isocyanate started to be detected. DRIFT data indicate that this pathway leading to partially oxidized hydrocarbon species is slower than the oxidation of hydrocarbon leading to the reduction of the noble metal surface.

On Rh–TiO_2, the following reaction mechanism may thus be indicated:

1. Propene reaction with Rh particles generates $Rh^{\delta+}$ ions that chemisorb negatively charged nitrosyl and gem-dinitrosyl species. These chemisorbed species are the precursors for NO dissociation.
2. Oxygen produced by NO dissociation may react with propene activated by the metal surface either to form CO_2 (redox decomposition mechanism) or to form an allyl alcoholate species. When the activity of the surface of the noble metal is inhibited by chemisorption of CO and/or other species the second pathway becomes the prevailing one and detectable by IR. Further, allyl alcoholate quickly transforms to chemisorbed acrolein and then to acrylic acid.
3. N atoms produced in step 2 may react with hydrocarbon fragments or CO to give isocyanate species which further react with adsorbed NO_x to give both N_2 and N_2O. This mechanism is simultaneous to the mechanism described in step 2, because a partial deactivation of the metal surface is also required.

Therefore, these DRIFT *in situ* data indicate that multiple pathways of transformation coexist during catalytic reaction and this explains why discordant opinions about the reaction mechanism are found in the literature. In addition, they show that the chemisorption/conversion of propene induces a considerable change in both the nature and the bond strength with the surface of NO_x adspecies, an effect that should be considered in analyzing reactive versus spectator NO_x adspecies.

This chapter does not intend to indicate that these observations are of general validity for the SCR-HC reaction and catalysts. Indeed, the results show that still further effort is necessary to understand the surface chemistry of this process. There are several factors affecting NO_x removal efficiency with hydrocarbons, such as dispersion, coordination, and local electronic states of the metal cations. The oxide matrix surrounding the active sites is not inert, but plays both an indirect role (through electronic influence on the active centres) and a direct role (role of interface sites, surface acido-base centres — both Brønsted and Lewis sites — and oxygen vacancies). The role of the oxide matrix shows several analogies with the role of the zeolite matrix in metal cation containing zeolites for SCR-HC [61].

Understanding these aspects and the reaction mechanism will probably lead to the design development of a new generation of catalysts for the SCR-HC process.

REFERENCES

1. Nojiri, N., Sakai, Y., and Watanabe, Y. Two catalytic technologies of much influence on progress in chemical process development in Japan. *Catal. Rev. Sci. Eng.* 1995, *37*, 145–178.
2. Amiridis, M.D., Zhang, T., and Farrauto, R.J. Selective catalytic reduction of nitric oxide by hydrocarbons. *Appl. Catal. B: Environ.* 1996, *10*, 203–227.
3. Hamada, H. Selective reduction of NO by hydrocarbons and oxygenated hydrocarbons over metal oxide catalysts. *Catal. Today* 1994, *22*, 21–40.
4. Koebel, M., Elsener, M., and Madia, G. Recent Advances in the Development of SCR-Urea for Automotive Applications. Diesel Emission Control System, Soc. Automotive Eng. Papers, Special publication, SP-1641, 2001, pp. 101–110.
5. van den Brink, R.W., Booneveld, S., Verhaak, M.J.F.M., and de Bruijn, F.A. Selective catalytic reduction of N_2O and NO_x in a single reactor in the nitric acid industry. *Catal. Today* 2002, *75*, 227–232.
6. Forzatti, P., Lietti, L., and Tronconi, E. Nitrogen oxides removal — Industrial. In *Encyclopedia of Catalysis*; Horváth I.T., Chief Ed. John Wiley & Sons: Chichester (UK), 2003.
7. Topsøe, N.-Y., Catalysis for NO_x abatement. Selective catalytic reduction of NO_x by ammonia: fundamental and industrial aspects. *CATTECH* 1997, *1*, 125–134.
8. Saito, S., Shinozaki, R., Suzuki, A., Jyoutaki, H., and Takeda, Y. Development of SCR-urea system for commercial vehicle — basic characteristics and improvement of NO_x conversion at low load operation. Emissions: Advanced Catalyst and Substrates, Measurement and Testing, and Diesel Gaseous Emissions. Soc. Automotive Eng. Papers, Special publication, SP-1801, 2003, pp. 209–214.
9. Koebel, M., Elsener, M., and Kleemann, M. SCR-urea: a promising technique to reduce NO_x emissions from automotive diesel engines. *Catal. Today* 2000, *59*, 335–345.
10. Matsumoto, S. Catalytic reduction of nitrogen oxides in automotive exhaust containing excess oxygen by NO_x storage-reduction catalyst. *CATTECH* 2001, *4*, 102–109.

11. Matsumoto, S., Ikeda, Y., Suzuki, H., Ogai, M., and Miyoshi, N. NO_x storage-reduction catalyst for automotive exhaust with improved tolerance against sulfur poisoning. *Appl. Catal. B: Environ.* 2000, *25*, 115–124.

12. Jensen, A.B. Selective catalytic reduction for maximum NO_x emission. *Scand. Shipping Gaz.* 2000, *37–38*, B82–B85.

13. Iwamoto, M., Yahiro, H., Yu-u, Y., Shundo, S., Mizuno, N. Selective reduction of HO by lower hydrocarbons in the presence of O_2 and SO_2 over copper ion-exchanged zeolites, 1990, *32*, 430–433.

14. Held, W., Konig, A., Richter, T., and Puppe, L. Recent Trends in Automotive Emission Control, Soc. Automotive Eng. Papers, Special publication, SP-810, 1990, pp. 13–20.

15. Li, Y. and Armor, J.N. Catalytic reduction of nitrogen oxides with methane in the presence of excess oxygen. *Appl. Catal. B: Environ.* 1992, *1*, L31–L40.

16. MacKenzie, S. and Morrill, M. Novel catalyst development for the selective conversion of NH_3 to N_2. In *Proceedings of the Air & Waste Management Association's Annual Conference & Exhibition*, 94th, Orlando, FL, United States, June 24–28, 2001, pp. 2097–2104.

17. Fokema, M.D. and Ying, J.Y. The selective catalytic reduction of nitric oxide with methane over nonzeolitic catalysts. *Catal. Rev. Sci. Eng.* 2001, *43*, 1–29.

18. Centi, G., Arena, G.E., and Perathoner, S. Nanostructured catalysts for NO_x storage-reduction and N_2O decomposition. *J. Catal.* 2003, *216*, 443–454.

19. Centi, G., Perathoner, S., and Vazzana, F. Control of non-CO_2 greenhouse gas emissions by catalytic treatments. *CHEMTECH* 1999, *29*, 48–55.

20. Centi, G. and Vazzana, F. Selective catalytic reduction of N_2O in industrial emissions containing O_2, H_2O and SO_2: Behaviour of Fe/ZSM-5 catalysts. *Catal. Today* 1999, *53*, 683–693.

21. Coq, B., Mauvezin, M., Delahay, G., and Kieger, S. Kinetics and mechanism of the N_2O reduction by NH_3 on a Fe-zeolite-beta Catalyst. *J. Catal.* 2000, *195*, 298–303.

22. Shimizu, K., Satsuma, A., and Hattori, T. Metal oxide catalysts for selective reduction of NO_x by hydrocarbons: Toward molecular basis for catalyst design. *Catal. Surv. Jpn.* 2000, *4*, 115–123.

23. Misono, M. Catalytic reduction of nitrogen oxides by bifunctional catalysts. *CATTECH* 1998, *2*, 183–196.

24. Miessner, H., Francke, K.-P., and Rudolph, R. Plasma-enhanced SCR-HC of NO_x in the presence of excess oxygen. *Appl. Catal. B: Environ.* 2002, *36*, 53–62.

25. Chen, Z. and Mathur, V.K. Nonthermal plasma electrocatalytic reduction of nitrogen oxide. *Ind. Eng. Chem. Res.* 2003, *42*, 6682–6687.

26. Kwak, Ja Hun, Szanyi, Janos, and Peden Charles H.F. Nonthermal plasma-assisted catalytic NO_x reduction over Ba-Y,FAU: The effect of catalyst preparation. *J. Catal.* 2003, *220*, 291–298.

27. Yoon, S., Panov, A.G., Tonkyn, R.G., Ebeling, A.C., Barlow, S.E., and Balmer, M.L. An examination of the role of plasma treatment for lean NO_x reduction over sodium zeolite Y and gamma alumina. Part 1. Plasma assisted NO_x reduction over NaY and Al_2O_3. *Catal. Today* 2002, *72*, 243–250.

28. Mok, Y.S., Koh, D.J., Kim, K.T., and Nam, I.-S. Nonthermal plasma-enhanced catalytic removal of nitrogen oxides over V_2O_5/TiO_2 and Cr_2O_3/TiO_2. *Ind. Eng. Chem. Res.* 2003, *42*, 2960–2967.

29. Broer, S. and Hammer, T. Selective catalytic reduction of nitrogen oxides by combining a non-thermal plasma and a V_2O_5-WO_3/TiO_2 catalyst. *Appl. Catal. B: Environ.* 2000, *28*, 101–111.

30. Centi, G. and Perathoner, S. Nature of active species in copper-based catalysts and their chemistry of transformation of nitrogen oxides. *Appl. Catal. A: Gen.* 1995, *132*, 179–259.

31. Miessner, H., Francke, K.-P., Rudolph, R., and Hammer, Th. NO_x removal in excess oxygen by plasma-enhanced selective catalytic reduction. *Catal. Today* 2002, *75*, 325–330.

32. Francke, K.-P., Miessner, H., and Rudolph, R. Plasmacatalytic processes for environmental problems. *Catal. Today* 2000, *59*, 411–416.

33. Hayashi, Y. Technology for decomposition of hazardous gases using both plasma and catalyst effects at one atmospheric pressure. *Oyo Butsuri* 2003, *72*, 448–452.

34. Janssen, F.J. Environmental catalysis — Stationary sources. In *Environmental Catalysis*, Ertl, G., Knözinger, H., and Weitkamp, J. Eds. Wiley-VCH: Weinheimm, 1999, Chap. 2, pp. 119–179.

35. Blanco, J., Avila, P., Marzo, L., Suarez, S., and Knapp, C. Low temperature monolithic SCR catalysts for tail gas treatment in nitric acid plants. *Stud. Surf. Sci. Catal.* 2000, *130*, 1391–1396.

36. Koebel, M., Elsener, M., and Madia, G. Reaction pathways in the selective catalytic reduction process with NO and NO_2 at low temperatures. *Ind. Eng. Chem. Res.* 2001, *40*, 52–59.

37. Avila, P., Barthelemy, C., Bahamonde, A., and Blanco, J.. Catalyst for nitrogen oxide (NO_x) removal in nitric-acid plant gaseous effluents. *Atmospheric Environ., Part A: Gen. Top.* 1993, *27A*, 443–447.

38. Busca, G., Lietti, L., Ramis, G., and Berti, F. Chemical and mechanistic aspects of the selective catalytic reduction of NO_x by ammonia over oxide catalysts. A review. *Appl. Catal., B: Environ.* 1998, *18*, 1–36.

39. Delahay, G., Berthomieu, D., Goursot, A., and Coq, B. Zeolite-based catalysts for the abatement of NO_X and N_2O emissions from man-made activities. *Surfactant Sci. Ser.*, 2003, *108*, 1–24.

40. Long, R.Q., Yang, R.T., and Zammit, K.D. Superior pillared clay catalysts for selective catalytic reduction of nitrogen oxides for power plant emission control. *J. Air Waste Manag. Assoc.* 2000, *50*, 436–442.

41. Heck, R.M. Catalytic abatement of nitrogen oxides — Stationary applications. *Catal. Today* 1999, *53*, 519–523.

42. Nakajima, F., and Hamada, I. The state-of-the-art technology of NO_x control. *Catal. Today* 1996, *29*, 109–115.

43. Forzatti, P. Present status and perspectives in de-NO_x SCR catalysis. *Appl. Catal. A: Gen.* 2001, *222*, 221–236.

44. Tronconi, E., Forzatti, P., Gomez Martin, J.P., and Malloggi, S. Selective catalytic removal of nitrogen oxide (NO_x): A mathematical model for design of catalyst and reactor. *Chem. Eng. Sci.* 1992, *47*, 2401–2406.

45. Lietti, L., Nova, I., Camurri, S., Tronconi, E., and Forzatti, P. Dynamics of the SCR-DeNO$_x$ reaction by the transient-response method. *AIChE J.* 1997, *43*, 2559–2570.

46. Tronconi, E., Cavanna, A., Orsenigo, C., and Forzatti, P. Transient kinetics of SO_2 oxidation over SCR-DeNO$_x$ monolith catalysts. *Ind. Eng. Chem. Res.* 1999, *38*, 2593–2598.

47. Centi, G., Passarini, N., Perathoner, S., Riva, A., and Stella, G. Combined DeSO$_x$/DeNO$_x$ on a copper-oxide on alumina sorbent-catalyst. 3. DeNO$_x$ behavior as a function of the surface coverage with sulfate species. *Ind. Eng. Chem. Res.* 1992, *31*, 1963–1970.

48. Smirniotis, P.G., Pena, D.A., and Uphade, B.S. Low-temperature selective catalytic reduction (SCR) of NO with NH_3 by using Mn, Cr, and Cu oxides supported on hombikat TiO_2. *Ang. Chem. Int. Ed.* 2001, *40*, 2479–2482.

49. Ramis, G., Busca, G., Bregani, F., and Forzatti, P. Fourier transform-infrared study of the adsorption and coadsorption of nitric oxide, nitrogen dioxide and ammonia on vanadia–titania and mechanism of selective catalytic reduction. *Appl. Catal.* 1990, *64*, 259–278.

50. Forzatti, P. and Lietti, L. Recent advances in de-NO$_x$ing catalysis for stationary applications. *Heter. Chem. Rev.*, 1996, *3*, 33–51.

51. Topsøe, N.Y., Topsøe, H., and Dumesic, J.A.. Vanadia/titania catalysts for selective catalytic reduction (SCR) of nitric oxide by ammonia. II. Studies of active sites and formulation of catalytic cycles. *J. Catal.* 1995, *151*, 241–52.

52. Dumesic, J.A., Topsøe, H., Chen, Y., and Slabiak, T. Kinetics of selective catalytic reduction of nitric oxide by ammonia over vanadia/titania. *J. Catal.* 1996, *163*, 409–417.

53. Dumesic, J.A., Topsøe, N.Y., Slabiak, T., Morsing, P., Clausen, B.S., Tornqvist, E., and Topsoe, H. Microkinetic analysis of the selective catalytic reduction (SCR) of nitric oxide over vanadia/titania-based catalysts. *Stud. Surf. Sci. Catal.* 1993, *75*, 1325–1337.

54. Yin, X., Han, H., and Miyamoto, A. Active sites and mechanism of the selective catalytic reduction of NO by NH_3 over V_2O_5: A periodic first-principle study. *Phys. Chem. Chem. Phys.* 2000, *2*, 4243–4248.

55. Centi, G., Galli, A., and Perathoner, S. Reaction pathways of propane and propene conversion in the presence of NO and O_2 on Cu/MFI. *J. Chem. Soc., Faraday Trans.* 1996, *92*, 5129–5140.

56. Sadykov, V.A., Lunin, V.V., Matyshak, V.A., Paukshtis, E.A., Rozovskii, A.Ya., Bulgakov, N.N., and Ross, J.R.H. The reaction mechanism of selective catalytic reduction of nitrogen oxides by hydrocarbons in excess oxygen: intermediates, their reactivity, and routes of transformation. *Kinet. Catal.* 2003, *44*, 379–400.

57. Burch, R., Breen, J.P., and Meunier, F.C. A review of the selective reduction of NO$_x$ with hydrocarbons under lean-burn conditions with non-zeolitic oxide and platinum group metal catalysts. *Appl. Catal. B: Environ.* 2002, *39*, 283–303.

58. Burch, R., Breen, J.P., and Meunier, F.C. Nitrogen oxides removal with supported metals and oxides. In *Encyclopedia of Catalysis*; Horváth I.T., Chief Ed. J. Wiley & Sons: Chichester (UK), 2003.

59. Shimizu, K, Satsuma, A., and Hattori, T. Metal oxide catalysts for selective reduction of NO$_x$ by hydrocarbons: Toward molecular basis for catalyst design. *Catal. Surv. Jpn.* 2001, *4*, 115–123.

60. Arena, G.E., and Centi, G. Influence of hydrocarbon chemisorption on the NO$_x$ adspecies over supported noble-metal catalysts. *Top. Catal.* 2004, 30/31, 147–153.

61. Satsuma, A., Shichi, A., and Hattori, T. The zeolite micropore as a unique reaction field. *CATTECH* 2003, *7*, 42–51.

22 Gas Sensors Based on Semiconducting Metal Oxides

Alexander Gurlo, Nicolae Bârsan, and Udo Weimar
Institute of Physical and Theoretical Chemistry, University of Tübingen, Tübingen, Germany

CONTENTS

22.1 INTRODUCTION

In general, a gas sensor can be defined as a device that informs about the composition of its ambient atmosphere (i.e., responds to the stimulus, Figure 22.1). More specifically, upon interaction with chemical species (adsorption, chemical reaction, and charge transfer), the physicochemical properties of the metal oxide sensitive layer (such as its mass, temperature, and electrical resistance) reversibly change. These changes are translated into an electrical signal such as frequency,

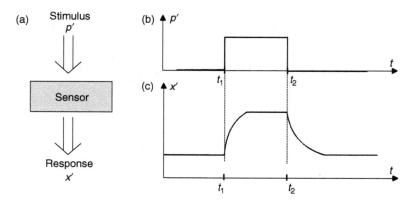

FIGURE 22.1 Sensor principle. (a) Principle functioning of a sensor. (b) Evolution of the stimulus (sensor input) p over time. A stimulus is applied at time t_1 and removed at time t_2. (c) Evolution of the sensor response $x'(t)$ over time. The changes in the stimulus determine correlated changes in the sensor response (sensor output). Changes in the sensor response need some time before an equilibrated sensor response value is achieved reproduced from [30] by permissions of Shaker Verlag

current, voltage, or impedance/conductance, which is then read out and subjected to further data treatment and processing [1–10, 159, 160] (Figure 22.2).

Three main types of the metal oxide-based gas sensors can be distinguished [7]: (i) electrochemical (e.g., potential or resistance changes through charge transfer) (see Reference 8), (ii) chemomechanical or mass-sensitive sensors (see Reference 9, e.g., mass change due to the adsorption, for example, indium–tin oxide coated quartz microbalance) [11], (iii) thermal sensors (e.g., temperature changes through chemical interaction), a classic example is "pellistor," for example, alumina pellet covered with a catalysts (see [7,12]; temperature changes are also recently detected on semiconducting metal oxides [13,14]). Electrochemical sensors are the largest group of chemical sensors, and they can be divided, according to their mode of measurement, into potentiometric (measurement of voltage), amperometric (measurement of current, a typical example for both these categories is yttria-stabilized zirconia-based oxygen sensor), and conductometric (measurement of conductivity, e.g., semiconducting gas sensors) sensors (see reviews [7,8,15–17]). Conductometric gas sensors based on semiconducting metal oxides (e.g., semiconducting gas sensors or MOX gas sensors) are actually the most investigated group of gas sensors. They have attracted the attention of many users and scientists interested in gas sensing under atmospheric conditions [18–32]. (Generally, for metal oxides, an electrical signal [e.g., Hall mobility [23,24], thermovoltage [33,34], and impedance/capacitance [35–37]] can be monitored; however, in most cases the sensor resistance R or conductance G are typically chosen as a sensor response.)

Semiconductor oxide-based gas sensors are classified according to the direction of the conductance change due to the exposure to reducing gases (e.g., CO,

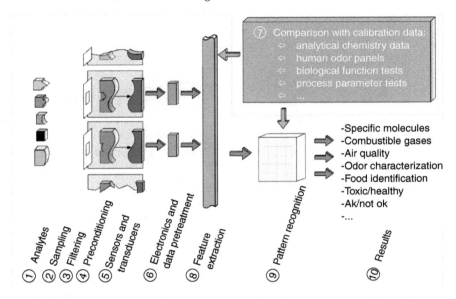

FIGURE 22.2 Sensing principle: the gas detection unit is exposed to a (complex) gas mixture. After sampling, some of the analyte molecules can be selected by a filter, and subsequent to an optional preconditioning of the sample, the remaining molecules come into contact with the sensor(s). Here, some of the molecules will trigger a characteristic change in the sensor properties. The transducer(s) transform these changes into electric signals. The data obtained can then be processed and the characteristic features can be extracted. Depending on the application, a more or less sophisticated pattern recognition may follow in order to gain the information required

EtOH, and H_2O vapors) as "n"-type (conductance increases, e.g., In_2O_3, ZnO, and SnO_2) or "p"-type (conductance decreases, e.g., Cr_2O_3 and CuO) [18]. This classification is related to the (surface) conductivity type of the oxides, which is determined by the nature of the dominant charge carriers at the surface, that is, electrons or holes. In some cases, the changing of the behavior of the oxides, that is, from p to n or the opposite were recorded for single (pure or differently doped) oxides or for mixtures of n- and p-type oxides. The explanation of the switching from one type to the other is a matter of debate; in most cases the electronic origin of the switching, that is, inversion of the type of the carriers due to the formation of an inversion layer, should be considered (for the full discussion see Reference 38).

The idea of using semiconductors as gas sensitive devices leads back to 1953 when Brattain and Bardeen [39] first reported gas sensitive effects on Ge. Later, Heiland [40], Bielanski et al. [41] and Seiyama et al. [42] found gas sensing effects on metal oxides. Taguchi [43] finally brought semiconductor sensors based on metal oxides to an industrial product (Taguchi-type sensors). Today, there are many companies offering semiconducting gas sensors, such as FigaroSensors,

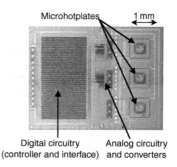

FIGURE 22.3 Examples of a commercially available TGS semiconductor gas sensor (left) and of a advanced research prototype CMOS based gas sensor system–micrograph of the complete chip with microhotplate array and circuitry [49,50] (right), reproduced by permissions of Prof. A. Hierlemann, Institute of Quantum Electronics Laboratory, ETH Zurich

FIS, MICS, CityTech, AppliedSensor, UST, Microsens and Paragon [44–48]. In addition, since they were used in the automotive applications (indoor air quality control), one can say that they became relevant for mass-market applications [23]. Examples of semiconductor gas sensors are shown in Figure 22.3 where commercial available sensors are present together with actual research prototypes.

Although many different metal oxides (among them especially WO_3, see Reference 51 to Reference 53 and references therein, and In_2O_3, see Reference 54 to Reference 56 and references therein) are investigated as sensing materials, SnO_2 sensors are the best-understood prototype of oxide-based gas sensors [20]. Therefore, usually SnO_2 is used as a model system for the basic understanding of metal oxide semiconducting sensors [21,24,57] and in this chapter, we will use the results obtained on SnO_2 as examples to illustrate some concepts.

Highly specific and sensitive metal oxide sensors are not yet available. It is well known that sensor selectivity can be fine-tuned over a wide range by varying the oxide crystal structure and morphology, dopants, contact geometries, operation temperature, or mode of operation, etc. In addition, practical sensor systems may contain a combination of a filter (such as charcoal) in front of the semiconductor sensor to avoid major impact from unwanted gases (e.g., low concentrations of organic volatiles which influence CO detection) (Figure 22.4). The understanding of real sensor signals, as they are measured in practical application, is hence quite difficult. It may even be necessary to separate filter and sensor influences for an unequivocal modeling of sensor responses.

If gas selectivity cannot be achieved by improving the sensor setup itself, it is possible to use several nonselective sensors and predict the concentration by model based, such as multilinear regression (MLR), principle component analysis (PCA), principle component regression (PCR), partial least squares (PLS), and multivariate adaptive regression splines (MARS), or data-based algorithms, such as cluster analysis (CA) and artificial neural networks (ANN) (for details see Reference 10) (Figure 22.5). For common applications of pattern recognition and multi component analysis of gas mixtures, arrays of sensors are usually chosen

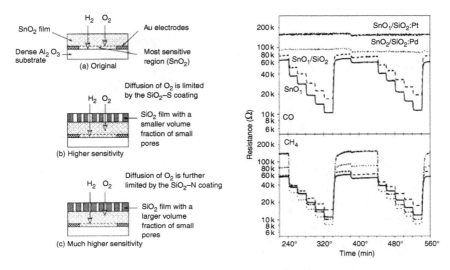

FIGURE 22.4 (a) Schematic diagram of the effect of the SiO_2 coating on the diffusivity of H_2 and O_2 into the interior region of an SnO_2 sensing layer reproduced from [58] by permissions of Wiley-VCH. (b) Transitory resistances of the SnO_2, SnO_2/SiO_2, SnO_2/SiO_2:Pd, and SnO_2/SiO_2:Pt gas sensors when exposed to different concentrations of CO and CH_4: from 20 to 400 and 200 to 4000 ppm, respectively. Two different relative humidity conditions are considered: 30 and 50%, reprinted from [59], copyright (2003), with permission from Elsevier

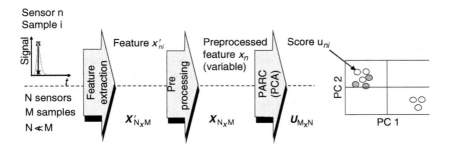

FIGURE 22.5 Schematic diagram of data analysis with PCA (for details see Reference 10, Reference 60, and Reference 61)

which operate at constant temperature. In these cases, a lack of selectivity and therefore overlapping sensitivities of different sensors is of advantage.

Research activity in chemical sensing with metal oxides is currently directed to the design of arrays consisting of different partially selective sensors [62–64] that permit subsequent pattern recognition and multicomponent analysis [60,61,65–67], to the analysis of the dynamic sensor responses due to the temperature modulation [68–75], toward the preseparation and preconcentration of the analytes [59,74–77], toward the miniaturization of the sensors [78–85], especially toward

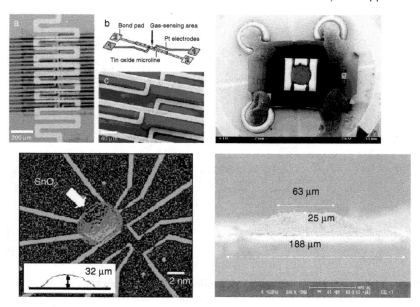

FIGURE 22.6 (Up) left (a) Top view of the gas sensor set up seen through the photoresist mask before deposition of the Pt-thin film wires for the resistance measurements. Each substrate contained one heating element onto which ten connected ceramic microlines were applied (dark horizontal lines). (b) Isometric drawing representing the contact layout for one tin oxide microline. Up to four contacts per microline were available. The inner contacts labeled "1" were connected for the sensor response measurements. (c) SEM of micro-thick film gas-sensors ready for measurement, reproduced from [78] by permissions of Wiley-VCH. (On the right): Single micromachined sensor with Pd-doped SnO$_2$, reprinted from [13], copyright (1999), with permission from Elsevier. (Bottom) (left) SEM image of a miniaturized SnO$_2$ sensitive layer trapped between electrodes and deposited by dip-pen nanolitography. Inset is an AFM section profile from a deposited structure, reprinted with permission from [64], copyright (2003) American Chemical Society. (right) SEM image of a cross-section drop of SnO$_2$ deposited by microdropping, reprinted from [80], copyright (2001), with permission from Elsevier

the development of the deposition techniques (see Figure 22.6) [80,81] compatible with micromachining technologies, MEMS — MicroElectroMechanicalSystems [86,87] and to the full integration of fabricated chemical sensors onto silicon (preferably on CMOS transducers, CMOS: Complementary Metal Oxide Semiconductors) [49,50], see also Figure 22.3. An important part of the improvements in the field of the basic scientific understanding of MOX gas sensors is the development of the suitable theoretical models describing gas interaction with metal oxides [88–94], improvement of the *in situ* spectroscopic techniques for the characterization of the working sensors in the operating conditions [95–103] and the theoretical modeling of the metal oxide surfaces using the quantum chemistry methods [104–125].

22.2 SENSOR PERFORMANCE

The quality of the sensor obviously depends on the magnitude of the sensor response caused by a certain stimulus, the speed with which the equilibrium steady-state sensor response is achieved, the minimum stimulus or stimulus change that can be detected and the cross sensitivity to other stimuli. Measurands that rate the quality of a gas sensor according to its detection properties are sensor signal S, sensitivity m, analytical sensitivity γ, lower detection limit LDL, response/recovery time $t_{response}/t_{recover}$, and selectivity m_{ij} (Table 22.1, Figure 22.7 to Figure 22.10). As mentioned earlier, in the case of MOX gas sensors, the sensor resistance R or the conductance G are typically chosen as the sensor response. The concentration c of a specified gas is usually the stimulus. For most applications, the carrier gas will be synthetic air with certain humidity (usually relative humidity, r.h.).

The *sensor signal S* is used to create a relation between the sensor response R (or G) and the zero response R_0 (or G_0) in the absence of the stimulus (e.g., the sensor resistance in a carrier gas). The sensor response R, on the other hand, is usually related to the resistance in an ambient atmosphere consisting of a carrier gas, plus a defined concentration of the test gas. For gas sensors based on n-type semiconducting metal oxides, one has to deal with two cases corresponding to reducing gases, which decrease the resistance and oxidizing gases, which lead to a resistance increase. In order to be able to compare the performance of gas sensors for oxidizing gases with sensors for reducing gases, the sensor signal is defined differently for the two gas types.

The *(partial) sensitivity m* describes the change in the sensor response (R or G) due to a specified change in the stimulus (gas concentration). The higher the value of the sensor's sensitivity, the more significant is the change in sensor response (R or G) initiated by a small change in the gas concentration. Metal oxide sensors are nonlinear sensors. Consequently, the change in sensor response due to a defined change in gas concentration depends strongly on the gas concentration itself.

The response (R or G) of a sensor for any other property cannot be measured with absolute precision and therefore the therewith-derived value of the stimulus c is to some extent inaccurate. The origin of the inaccuracy might be for example, the noise of the measurement or the limited reproducibility of the measurement due to insufficient stability of the sensors. The *analytical sensitivity γ* can be a powerful tool for measuring the precision with which a stimulus (gas concentration) can be detected. It is very easy to give a practical meaning to the analytical sensitivity of MOX sensors [126]. A gas sensor with an analytical sensitivity of 0.1 ppm^{-1} in a certain concentration range allows the detection of the gas in this concentration range with a precision of 10 ppm. In contrast, for example, the sensitivity of a sensor, the analytical sensitivity of a sensor is independent of the response type of the sensor. Therefore, it is possible to compare quantitatively the sensor performance of sensors with sensor responses that are different in nature and magnitude by means of analytical sensitivity. The LDL is the minimum gas concentration, which can be detected by a given sensor. It relates the sensor response (R or G) to the statistical fluctuations in the zero response (R_0 or G_0).

TABLE 22.1

Measurands Used to Characterize the Sensor Performance

Definition	Definition/MOS
Stimulus	c (gas concentration)
Sensor response	R (resistance) or G (conductance)
Calibration function f	$G = f(c)$ and $R = f(c)$
Sensor signal S	$S_{\text{red}} = \dfrac{R_0}{R} = \dfrac{G}{G_0} \geq 1$ and $S_{\text{ox}} = \dfrac{R}{R_0} = \dfrac{G_0}{G} \geq 1$
(Partial) sensitivity m	$m_{\text{red}}(c) = -\dfrac{\partial R}{\partial c}; m_{\text{red}}(c) = \dfrac{\partial G}{\partial c}$ and $m_{\text{ox}}(c) = \dfrac{\partial R}{\partial c}; m_{\text{ox}}(c) = -\dfrac{\partial G}{\partial c}$
Analytical sensitivity γ	$\gamma = \dfrac{m}{\sigma_R}$, where $\sigma_R = \left(\dfrac{\partial R}{\partial c}\right) \sigma_C$ and $dR = \left(\dfrac{\partial R}{\partial c}\right) dc$, consequently $\gamma = \dfrac{1}{\sigma_C}$, where, σ_R is the standard deviation of the sensor response, that is, the uncertainty with which the sensor response can be measured, and σ_C is the standard deviation of the stimulus, that is, the precision with which the gas concentration can be determined
Lower detection limit (LDL)	$\text{LDL} = c_{\min} = f^{-1}(R_{\min}), R_{\min} = \overline{R}_0 + 3\sigma_0$, whereby $f^{-1}(R)$ is the inverse function of the calibration function $f(c)$ and $c_{\min}$ the minimum detectable gas concentration. The minimum sensor response $R_{\min}$ which is certainly detected is usually chosen as three times the standard deviation of the zero response σ_0; which is considered to be the noise of the measurement
Response time t_{response}	The response time is defined as the time $t_{90\%}$, that is, the time it takes for 90% of the sensor response change after an increase in the stimulus is accomplished
Recovery time or decay time t_{recover}	The recovery time is the time needed for 90% of the sensor response change after stimulus removal is accomplished
Selectivity m_{ij}	$m_{ij}(c_i, c_j) = \dfrac{S_i}{S_j}$ and $m_{ij}(c_i, c_j) = \dfrac{m_i}{m_j}$. The selectivity m_{ij} of a sensor compares sensor signal or the sensitivity to be monitored (S_i/m_i) to the sensor signal/sensitivity of the interfering stimulus (S_j/m_j).
Drift D	$D(c_i, \Delta t) = \dfrac{\partial R}{\partial t}\bigg|_{\Delta t}^{c_i}$. To be able to compare drift values of different sensors, the measurement conditions (c_i, the operation temperature T, the ambient relative humidity (r.h.) and measurement time Δt) must be comparable
Stability ζ	$\zeta(c_i, \Delta t) = \left(\dfrac{(1/n)\sum_{i=1}^{n} R_i}{R_{\max}}\bigg|_{\Delta t}^{c_i}\right) \cdot 100$, where n is the number of measurements, $R_{\max}$ the maximum sensor response value of the determined sensor response values R_i. As in the case of drift, the measurement conditions must be comparable if one wants to compare stability values
Reproducibility Q	$Q_{x'}(c_i) = \left(\dfrac{(1/n)\sum_{k=1}^{n} R_k}{R_{\max}}\bigg|^{c_{i_j}}\right) \cdot 100$, where n is the the number of characterized sensors, R_k the sensor response of the sensor k

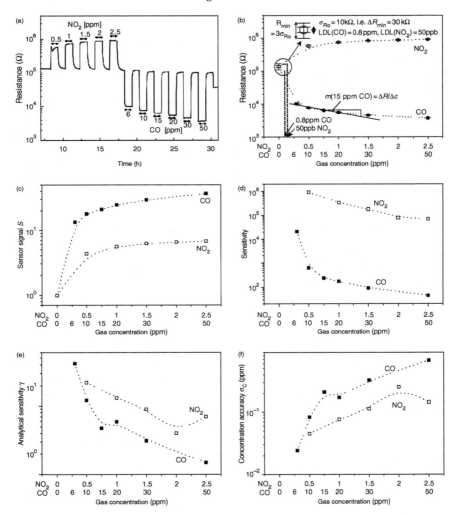

FIGURE 22.7 Characterization of the detection properties of a single MOX sensor. (a) Sensor response (sensor resistance) to an exposure to NO_2 (0.5 to 2.5 ppm) and CO (6 to 50 ppm). Each gas exposure was maintained for one hour before the ambient gas atmosphere was switched back to carrier gas. NO_2 leads to a sensor resistance increase, CO decreases the resistance. (b) Calibration curves for CO and NO_2 as derived from (a). The data points represent the sensor resistance values averaged over the last 10 min of gas exposure. The error bars represent the corresponding standard deviations. The calibration curves are used to estimate the LDL for NO_2 and CO detection. The partial derivative of each calibration curve is used to determine the sensitivity values (see [d]). (c) Sensor signals S for CO and NO_2 and the mean resistance values from (b). (d) Sensitivity m for CO and NO_2 as calculated by determining the partial derivative of the calibration curves (b). (e) Analytical sensitivity γ as determined for NO_2 and CO using the sensitivity values (d) and the standard deviation of the resistance values (b). (f) Accuracy of the CO and NO_2 detection as a function of the gas concentration, reproduced from [30] by permissions of Shaker-Verlag

Figure 22.7 illustrates the definitions given earlier using typical measurement data from MOX sensors. A typical sensor response of a metal oxide sensor to a reducing gas (CO) and an oxidizing gas (NO_2) is shown. Using these measurement data, the sensor signal S, the sensitivity m, the analytical sensitivity γ, and the lower detection limit LDL have been derived.

Two measurands are usually used to measure the speed of a sensor response. The first one is the so-called *response time* $t_{response}$, which refers to the time needed to reach a stable sensor response after a stepwise increase in the stimulus. Hence, it measures the minimum time needed to measure a stimulus. The second often-used measurand is the *recovery time* or decay time $t_{recover}$. It refers to the time the sensor needs for the sensor response to resume zero after removing the stimulus, that is, the time a sensor needs to recover from the effect of the precedent stimulus. In the case of a MOX gas sensor the sensor speed will be determined by the progression of the gas interaction, which occurs at the tin oxide surface. As soon as a steady state at the SnO_2 surface has been achieved, the sensor response reaches its equilibrium value. As a consequence of this, the sensor speed depends strongly on the actual measurement conditions such as the operation temperature T, the ambient relative humidity (r.h.), the gas concentrations c_i and the performance of the gas mixing station. The determination of the response and the recovery time for MOX sensors is illustrated in Figure 22.8.

Sensors are normally sensitive to more than one stimulus and usually show cross-sensitivities. The *selectivity* is a measurand for evaluating the specificity of a sensor by comparing the effects of different gases on a sensor. Apart from the ability to sense a stimulus quickly and with high precision, additional obvious key demands on a sensor are stability, that is, the *reliability* of a sensor over time, and the reproducibility of sensors.

A monotonous change in the sensor response over time in spite of the absence of changes in the ambient atmosphere is called *drift*. One can distinguish between short-term drift, which is typically observed after switching on, and long-term drift, which is observed due to the instability of the sensor.

Apart from monotonous changes, the sensor response might show changes over time which cannot be described by a drift. In order to consider such phenomena, one monitors the evolution of the sensor response under identical conditions. The determination of the standard deviation of the sensor response σ_R of a number of measurements in a specified time period is one way to describe the stability of a sensor. Based on this, a value for the analytical sensitivity can be determined. Another possibility for measuring the stability of a sensor is to calculate the so-called *stability* ζ of a sensor. The stability of a sensor quantitatively describes the variation of the sensor response over time. For an ideal sensor, the stability will be 100, while for all other sensors the stability values will be between 0 and 100.

Typical examples for drift and stability are illustrated in Figure 22.9. In these figures, measurement results obtained on rather unstable sensors are shown and based on these measurement data the drift and the stability values are calculated.

All the definitions discussed earlier were used for classifying the performance of individual sensors. Another important issue for the quality of a sensor type is of

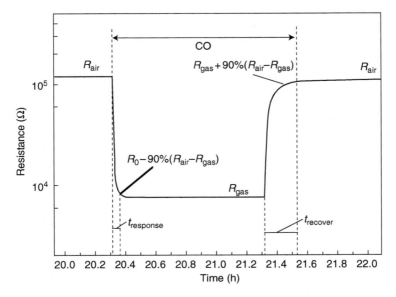

FIGURE 22.8 Speed of sensor response. A stepwise exposure to CO (from t_1 to t_2) causes a resistance decrease. The corresponding response and recovery times, which are in this case mainly related to the time needed to exchange the gas in an voluminous sensor chamber, are highlighted, reproduced from [30] by permissions of Shaker-Verlag

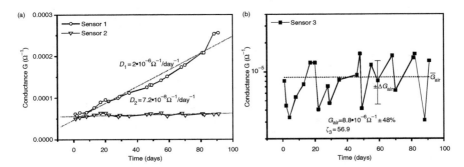

FIGURE 22.9 Drift and stability. (a) Drift of the sensor conductance G. Sensor 1 shows a strong drift, sensor 2 is rather stable. The conductance evolution over time can be described by linear functions. The resulting drift values are given in the figure. (b) Stability of a sensor. Sensor 3 shows no drift but high fluctuations in the conductance in air G_{air} around the average value $\bar{G}_{air}$. The resulting stability value ζ is given in the figure, reproduced from [30] by permissions of Shaker-Verlag

course the similarity of individual sensors of one type. This quality issue is called *reproducibility Q*. It can be treated in a similar way to stability. In order to compare the reproducibility of two sensor batches one can either calculate the mean value and the standard deviation of a sensor property for both batches and compare them

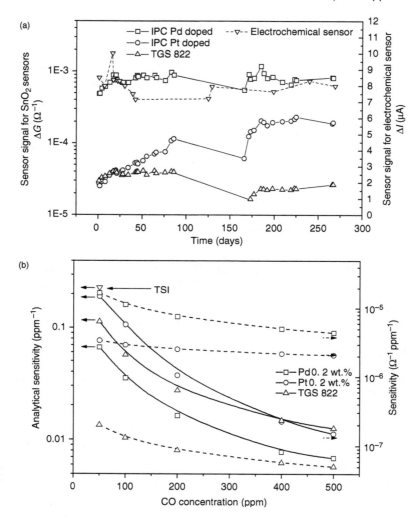

FIGURE 22.10 (a) Sensor signal for SnO$_2$-based sensors and the electrochemical cell for 50 ppm CO. The sensor response has been monitored for 270 days. A total of 35 measurements have been performed in the case of the SnO$_2$-based sensors, 14 measurements with electrochemical cells have been performed during this period of time. The standard deviation for the electrochemical sensor amounts to 9% of the sensor signal taking all the data over time, even the initial start up period. The standard deviations of the sensor signal for the SnO$_2$ sensors are 17, 68, and 24%, for the Pd doped, the Pt doped, and the TGS822 sensor, respectively. (b) The solid lines represent the fitting curve and the error bars represent the standard deviation of a series of 15 measurements within 100 days of characterization. Sensitivity m and analytical sensitivity γ of the aforementioned sensors. The sensitivity decreases with increasing CO concentration in the investigated concentration range. Since the electrochemical cell (TSI) is a linear sensor and the absolute precision of the sensor signal is quite independent of the CO concentration one can assume that this value is representative for the whole concentration range, reprinted with permission from [126], copyright (1999) American Chemical Society.

or calculate the reproducibility Q for each batch. The reproducibility values range from 0 (completely irreproducible sensor) to 100 (perfectly reproducible sensor). In the case of MOXs, the monitored sensor property is often either the sensor resistance or the sensor signal [98].

As an example, Figure 22.10 shows detection properties of SnO_2-based sensors in the CO concentration range between 50 to 500 ppm in comparison with TGS882 sensor of Figaro and commercial electrochemical cell (TSI) [126].

22.3 PRINCIPLE OF OPERATION

The simplest way in which the operation of semiconducting gas sensors described is: the conductivity of metal oxide semiconductor materials changes according to gas concentration changes. This is caused by adsorption/desorption of oxygen and reaction between surface oxygen and gases. These reactions change the electric potential on SnO_2 crystal and results in the decrease of the sensor resistance under the presence of reducing gases like CO (Figure 22.11).

However, in reality, the mechanism of gas sensing on metal oxide-based gas sensors is much more complicate. A sensor element comprises the following parts (Figure 22.12): *sensitive layer* deposited over a *substrate* provided with *electrodes* for the measurement of the electrical characteristics. The device is generally heated by its own *heater*; this one is separated from the sensing layer and the electrodes by an electrically insulating layer. (A useful toll for the identification of different elements of a complex device, such as a thick film gas sensor, is the impedance spectroscopy. In Figure 22.13 the different elements, possible to be present in our gas sensors, are shown together with their equivalent circuits and their dependence of the main parameters that might be influenced by the different ambient atmosphere conditions, for example, band bending V_S and dielectric constant ε, for detail see discussion in the following paragraph.)

The elementary steps of gas sensing will be transduced in electrical signals measured by appropriate electrode structures. The sensing itself can take place at different sites of the structure depending on the morphology. They will play different roles, according to the sensing layer morphology. The overall conduction in a sensor element is determined by the surface reactions, the resulting charge transfer processes with the underlying semiconducting material, and the transport mechanism within the sensing layer:

- *Surface chemistry*: This means the interaction of the reacting gas species at the surface of the metal oxide and the associated charge transfer. This relates to the specific adsorbed oxygen species and the way in which the oxidation of reducing gases (such as CO, hydrocarbons) or adsorption of oxidizing gases (such as NO_2, O_3) will take place. From the modeling point of view, it is described by quasi-chemical equations. The base of the gas detection is the interaction of the gaseous species at the surface of the semiconducting sensitive metal oxide layer. It is important to identify the reaction partners and the input for this is based upon spectroscopic

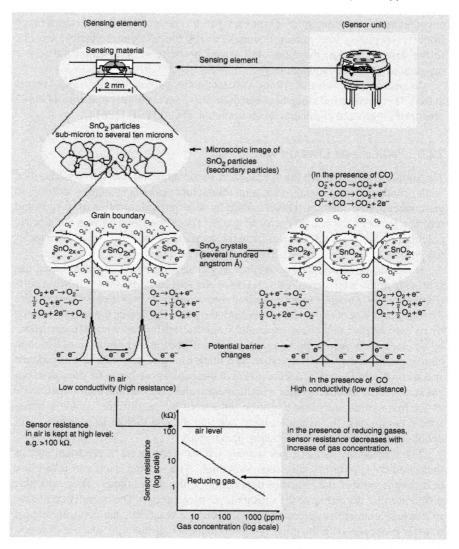

FIGURE 22.11 FIS Sensors: the sensing mechanism of the thick film SnO_2 gas sensors, reproduced from [45] by permissions of FIS INC.

information. Using this input, one can model the interaction using the quasi-chemical formalism (for details see Reference 21).

- As a consequence of this surface interaction, *charge transfer* takes place between the adsorbed species and the semiconducting sensitive material. This charge transfer can take place either with the conduction band or in a localized manner. In the first case, the concentration of the free charge carriers will be influenced. For the understanding of the detection, it is important to deepen our insight in the localized charge transfer case.

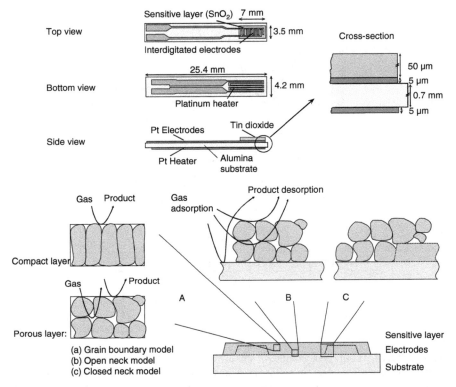

FIGURE 22.12 An example of a thick film SnO$_2$ sensor. (Up) Layout of the planar alumina substrate with Pt electrodes and Pt heater. The SnO$_2$ layer is printed on top of the interdigitated electrodes. The heater on the back keeps the sensor at the operation temperature, reproduced from [22] by permissions of Institute of Physics Publishing. (Bottom) Schematic layout of a typical resistive gas sensor. The sensitive metal oxide layer is deposited over the metal electrodes onto the substrate. In the case of compact layers, the gas cannot penetrate into the sensitive layer and the gas interaction is only taking place at the geometric surface. In the case of porous layers, the gas penetrates into the sensitive layer down to the substrate. The gas interaction can therefore take place at the surface of individual grains, at grain–grain boundaries and at the interface between grains and electrodes and grains and substrates, copyright [21], with kind permission of Springer Science and Business Media

The latter case (in contrast to the first one) will have no direct impact on the conduction. The former determines the appearance of a depletion layer at the surface of the semiconductor material, due to the equilibrium between the trapping of electrons in the surface states (associated with the adsorbed species) and their release due to desorption and the reaction with test gase. From the modeling point of view, it is described by the Poisson and electroneutrality equations. Out of the first two factors, one can calculate the dependence of the electron concentration n_S in the

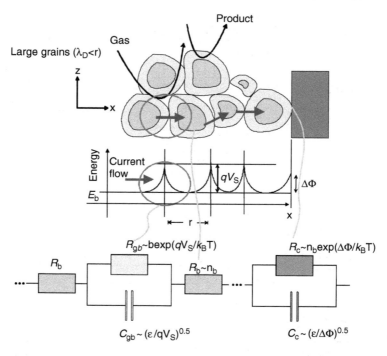

FIGURE 22.13 Equivalent circuit for the different contributions; intergranular contact, bulk, and electrode contact. Intergranular contact: the ionosorption of oxygen at the grain surface results in the creation of a potential barrier and the corresponding depletion layers at the intergranular contacts. An intergranular contact can be represented electronically by a resistor R_{gb} (due to the high resistive depletion layers) and a capacitor C_{gb} (due to the sandwiching of the high resistive depletion layers between two high conductive "plates" of bulk material) in parallel. The electrode contact can also be represented by a (RC)-element. The values of the resistor R_c and the capacitor C_c are independent of the ambient gas atmosphere. The bulk contribution can be represented by a resistor R_b, whose resistance value is hardly influenced by changes in the ambient atmosphere, reproduced from [22] by permissions of Institute of Physics Publishing

    depletion layer near the surface of the semiconductor as a function of the concentration of the test gas (for details see Reference 21).

- The *conduction in the sensitive layer*, which translates the sensing into the measurable electrical signal. This strongly depends on the morphology of the sensitive layer. In turn, the change of the concentration of the free charge carriers is translated into a change of the overall resistance of the sensing layer. The transfer (or calibration) function depends on the morphology of the layer.

    The overall resistance of the sensor will comprise the above-mentioned phenomena combined with the influence of the electrodes. Their influence depends on the morphology of the sensing layer, the geometrical arrangement, and the possible chemical effects.

A complete description of the gas-sensing mechanism for MOX-based sensors should start from elementary steps governing those surface–molecule interactions, which lead to charge transfer (adsorption, reaction, desorption, etc.). The model must show how these steps are linked with the macroscopic parameters describing the sensor response. For this, spectroscopic data, quantum mechanics-based numerical calculations, thermodynamics, kinetics, semiconductor physics are applied. Surface spectroscopy is essential to obtain detailed information about adsorption complexes. Using this input quantum mechanics, calculations determine the associated energetic (electronic) levels of the surface, adsorption complex, precursor state, and free molecule. Statistical thermodynamics and kinetics allow one to link those calculated energy levels with surface coverage of adsorbed species corresponding to real experimental conditions (dependent on temperature, pressure, etc.). Finally, changes in concentrations and mobilities of free charge carriers and the sensor resistance are correlated.

One of the directions in the field of gas sensors is the development of the suitable theoretical models describing gas interaction with metal oxides. In the last years, different phenomenological models were published, among them are conduction and gas surface reaction modeling [88], surface state trapping models for dynamic conductance response [68], models based on Wolkenstein theory of chemisorption [90], numerical simulations of chemisorption isoterms for both non-dissociative and dissociative chemisorption [93], analysis of thickness dependence of the sensitivity and the effects of the structural inhomogeneties of the sensitive film [92], diffusion equation-based studies describing concentration profile inside the porous layer [89], diffusion and reaction model of analytes in sensor [91], and theory explaining drift in metal oxide based sensors on the basis of oxygen diffusion [94].

Also mechanistic models, describing the interaction of different gases with metal oxide sensors, were published, among them are the role of oxygen and water vapor in CO sensing with SnO_2 thick film sensors [127], sensing of hydrocarbons on SnO_2 sensors [128], NO_2 and O_3 interaction with In_2O_3 based sensors [55], O_3 interaction with WO_3 based sensors [52], NO_2 sensing on SnO_2 sensors [129].

22.3.1 Surface Chemistry and Charge Transfer

The donor and acceptor type interaction of a gas and solid are often described by a charge transfer model, which assumes that adsorbed particles induce extrinsic two-dimensional surface states. As mentioned earlier, a distinction is to be made between localized chemisorption and delocalized chemosorption also called ionosorption. Localized chemisorption is due to a charge transfer between an adsorbent (e.g., surface site = surface atom or group of atoms) and an adsorbate. It is important to state that the changes induced by localized chemisorption processes are not easy to read out because they are not influencing the resistance (conductance) of the sensing layer.

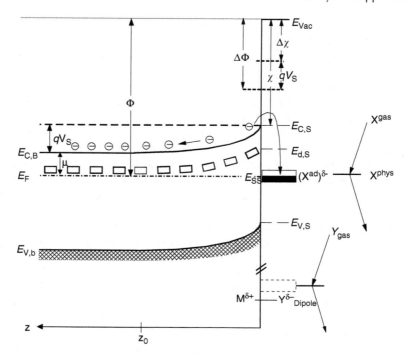

FIGURE 22.14 Ionosorption of an electron acceptor molecule X, such as O_2, on an n-type semiconductor creates surface states E_{SS} that are filled with electrons of the conduction band. This leads to a negative charge of the surface introducing an electric field that prevents a further charge of the surface. A band bending (qV_S) results that leads to a resistance increase. For chemisorption of an electron donor one obtains a band bending toward lower energies and thus a decrease of the sensor resistance. The localized chemisorption of a gaseous species Y determines the appearance of a dipole ($M^{\delta+} - Y^{\delta-}$) through the direct charge transfer from the metal atom

Ionosorption is the so-called "delocalized" chemisorption because the charge transfer involves free charge carriers id. est. the collective properties of the solid [130]. It causes a band bending by changing this surface resistance of the sensing material. The ionosorption can be interpreted as illustrated for an electron acceptor interaction and an n-type semiconductor in Figure 22.14. An electron acceptor molecule X^{gas} gets physisorbed at the surface; hence, it creates an unoccupied surface level. Consequently, an electron transfer from the sensing material will occur and a partial charge $(X^{ad})^{\delta-}$ will be trapped at the surface and a depletion layer will result illustrated by a band bending. This band bending will increase with the concentration of chemisorbed ions at the surface until a steady state is achieved. The width of the depletion layer depends on the surface charge and therefore on the gas concentration of the gas X.

As known, the work function Φ of semiconductors contain three contributions: the energy difference between the Fermi level and conduction band in the bulk

$(E_C - E_F)_b$, band bending qV_S (q denotes elementary charge), and electron affinity χ (due to the definition, $qV_S = E_{C,S} - E_{C,B}$) [131–136]:

$$\Phi = (E_C - E_F)_b + qV_S + \chi \qquad (22.1)$$

All three contributions may change upon gas exposure:

$$\Delta\Phi = \Delta(E_C - E_F)_b + q\Delta V_S + \Delta\chi \qquad (22.2)$$

however, in almost all relevant cases, the change in the bulk can be neglected. Therfore, the work function changes induced by gas exposure follow the change in band bending and electron affinity:

$$\Delta\Phi = q\Delta V_S + \Delta\chi \qquad (22.3)$$

The appearance of a dipole ($M^{\delta+} - Y^{\delta-}$) through the direct charge transfer from/to the metal atom ("localised chemisorption") changes *only electron affinity* χ, the appearance of the depletion/accumulation layer ("ionosorption") changes *only band bending* $q\Delta V_S$. (Separating the contributions from these two effects can be made by performing simultaneous conductance and work function change measurements in more complex cases (such as CO interaction in humid air), where both type of changes are to be expected) [22,134].

Figure 22.15 illustrates typical results, which can be obtained from the work function change measurements.

The atomistic understanding of molecule–MOX interaction requires additional spectroscopic information. Spectroscopic techniques may be applied either under *in situ* real operation conditions of the sensors or under ideal conditions far away from the real practical world (such as the ultra high vacuum conditions and at low temperatures). The latter may lead to extremely detailed results, but extrapolation of the data from ideal to real conditions is not straightforward. Frequently

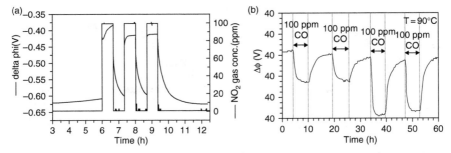

FIGURE 22.15 (a) Work function changes on (a) SnO_2 at 130°C exposed to 100 ppm NO_2 in 0% r.h. air, reprinted from [137], copyright (2001), with permission from Elsevier and (b) on Pd-complex/SnO_2 at 200°C exposed to different CO concentrations in humid air (50% r.h.), reprinted from [138], copyright (2003), with permission from Elsevier

used *in situ* measurements (or "operando spectroscopy" — the shortened version of spectra of a working [*operando* in Latin]) include IR [98,99] and DRIFT [102,103], EPR [139,140], Mössbauer [96,97,101], UV [109], and Raman spectroscopy [141]. Usually, samples with large specific surfaces are required, and sample preparation may differ from the preparation of real sensors. The second type of measurements performed under ideal conditions include spectroscopies such as thermodesorption spectroscopy [57], X-ray photoemission spectroscopy [142,143], and ultraviolet photoemission spectroscopy. Atomic resolution may be obtained with STM [95] or AFM. A short summary of typical surface species (including oxygen, water, and CO) of SnO_2 sensors, which have been identified with both type of techniques are shown in Figure 22.16 to Figure 22.18.

22.3.1.1 Oxygen adsorption

At temperatures between 100 and 500°C, interaction with atmospheric oxygen leads to its ionosorption as molecular (O_2^-) and atomic (O^-, O^{2-}) species. It is proved by TPD, FTIR, and ESR that below 150°C the molecular and above this temperature the atomic species dominate (Figure 22.16). The effect of oxygen (and other oxidizing gases such as NO_2 and O_3) will be the building of a negative charge at the surface and the increase in the band bending ($q \Delta V_S > 0$) and work function ($\Delta \Phi > 0$), consequently (see Figure 22.15[a]). For an n-type semiconductor this will result in the creation of a depletion layer and the decrease of the surface conductance.

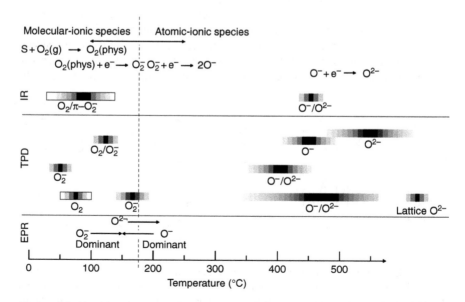

FIGURE 22.16 Literature survey of oxygen species detected at different temperatures at SnO_2 surfaces with IR, TPD, and EPR. Reproduced from [20] by permissions of Springer-Verlag

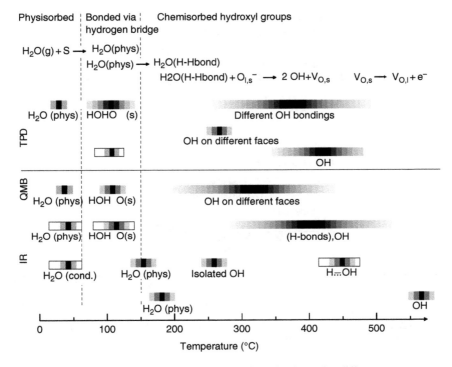

FIGURE 22.17 Literature survey of water-related species formed at different temperatures at SnO$_2$ surfaces. The results have been obtained by means of IR and TPD. Reproduced from [20] by permissions of Springer-Verlag

Reviews on theoretical calculations for metal oxides surfaces are available in the literature [105–125]. The main body of results was obtained for SnO$_2$ (110); the interest is generated by the fact that SnO$_2$ sensors are the best-understood prototype of oxide-based gas sensors and (110) is the thermodynamically most stable rutile surface. For SnO$_2$ (110) a layer of bridging oxygen ions is present on the perfect surface that can easily be removed and the resulting "compact" surface is stable. The reason for the stability is that the Sn ion has two stable oxidation states, Sn^{4+} and Sn^{2+}. It was both proved experimentally and calculated that the removal of bridging oxygen atoms does not provide additional donors, on the opposite; the removal of in-plane oxygen ions does provide donors. These experimental results were obtained under UHV conditions. Cluster as well as first-principle density-functional calculations dealing with oxygen and CO adsorption on SnO$_2$ (110) and M/SnO$_2$ (110) (M = Pd, Pt) as well as the interaction of oxygen with bridging oxygen vacancies were presented recently (see Figure 22.19).

22.3.1.2 The role of water vapor

The role of water vapor should also be taken into consideration because of its proven influence on the sensing mechanism. The interaction with water vapor

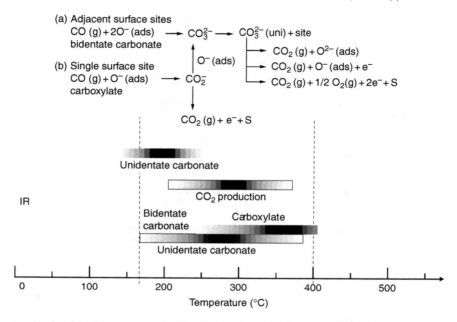

Figure 22.18 Literature survey of CO-related species found by means of IR at different temperatures on a (O_2) preconditioned SnO_2 surface, reproduced from [20] by permissions of Springer-Verlag.

leads to molecular water and hydroxyl groups. Water molecules may be adsorbed by physisorption or hydrogen bonding. Some temperature programmed desorption (TPD) and infrared (IR) studies suggest that at temperatures above 200°C molecular water is no longer present at the surface. Hydroxyl groups appear due to an acid/base reaction with the OH$^-$ sharing its electronic pair with the Lewis acid site (Sn) and leaving the weakly bonded proton, H$^+$, ready for reactions with lattice oxygen (Lewis base) or with adsorbed oxygen. IR studies indicate the presence of hydroxyl groups bound to Sn atoms (Figure 22.17).

For theoretical understanding, clusters containing adsorbed hydroxyl groups and oxygen ions have to be constructed for SnO_2 (110) surfaces to be exposed to ambient atmosphere and then used for theoretical calculations [109,116].

There are three types of mechanisms that were proposed for explaining the experimentally proven increase of surface conductivity in the presence of water vapor. Two, direct mechanisms are proposed by Heiland and Kohl [144] and the third, indirect, is suggested by Morrison [130] and by Henrich and Cox [145].

The first mechanism of Heiland and Kohl attributes the role of electron donor to the "rooted" OH group, the one including lattice oxygen. The equation proposed is:

$$H_2O^{gas} + Sn_{Sn} + O_O \rightleftharpoons \left(Sn_{Sn}^{\delta+} - OH^{\delta-}\right) + (OH)_O^+ + e^- \qquad (22.4)$$

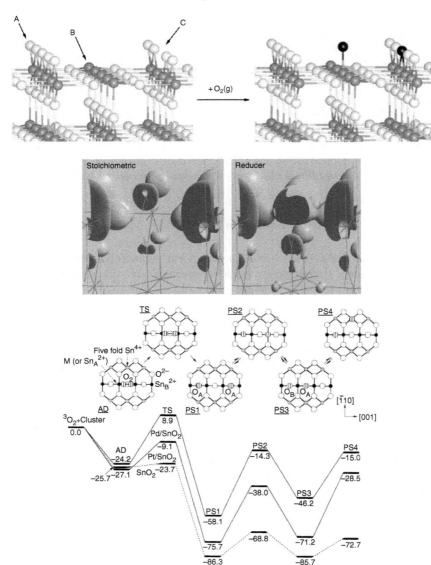

FIGURE 22.19 (Up) Schematic oxygen dissociation process on the reduced SnO_2 (110) surface. The darker atoms in the final state come from the O_2 molecule: one atom fills the O vacancy (C), and the other is adsorbed on the five-coordinated Sn site (B), reproduced from [115] by permissions of the Royal Society of Chemistry. (In the middle) Distribution of lowest unoccupied orbitals of stoichiometric (top) and reduced surface models (bottom) calculated by using USP pseudopential. Significant localization is observed at the site of the fivefold coordinated surface tin atom, reprinted with permission from [111], copyright (2002) by the American Physical Society. (Bottom) Potential energy diagram for adsorption (AD), dissociation (TS), and migration (PS1-PS4) of oxygen on the three-layer reduced M/SnO_2 (M = Pd, Pt) and SnO_2 clusters. Relative energies are given in kcal/mol. The top view of the clusters with the adsorbed oxygen species, reprinted from [114] copyright (2003) with permission from Elsevier

where $(Sn_{Sn}^{\delta+} - OH^{\delta-})$ is denominated as an isolated hydroxyl or OH group (dipole) and $(OH)_O^+$ is the rooted one. In the upper equation, the latter is already ionized.

The reaction implies the homolytic dissociation of water and the reaction of the neutral H atom with the lattice oxygen. The latter is normally in the lattice fixing two electrons consequently being in the 2-state. The built-up rooted OH group, having a lower electron affinity and consequently can be ionized and become a donor (with the injection of an electron in the conduction band).

The second mechanism takes into account the possibility of the reaction between the hydrogen atom and the lattice oxygen and the binding of the resulting hydroxyl group to the Sn atom. The resulting oxygen vacancy will produce, by ionization, the additional electrons. The equation proposed by Heiland and Kohl is:

$$H_2O^{gas} + 2 \cdot Sn_{Sn} + O_O \rightleftharpoons 2 \cdot \left(Sn_{Sn}^{\delta+} - OH^{\delta-} \right) + V_O^{++} + 2 \cdot e^- \qquad (22.5)$$

Morrison, as well as Henrich and Cox consider an indirect effect more probable. This effect could be the interaction either between the hydroxyl group or between the hydrogen atom originating from the water molecule with an acid or basic group, which are also acceptor surface states. Their electronic affinity could change after the interaction. It could also be the influence of the coadsorption of water on the adsorption of another adsorbate, which could be an electron acceptor. Henrich and Cox suggested that the preadsorbed oxygen could be displaced by water adsorption. In any of these mechanisms, the particular state of the surface has a major role, because it is considered that steps and surface defects will increase the dissociative adsorption. The surface dopants could also influence these phenomena; Egashira and Nakashima [146] showed by TPD and isotopic tracer studies combined with TPD that the oxygen adsorbates are rearranged in the presence of adsorbed water. The rearrangement was different in the case of Ag and Pd surface doping.

In choosing between one of the proposed mechanisms, one has to keep in mind that: (i) in all reported experiments, the effect of water vapor was the increase of surface conductance, (ii) the effect is reversible, generally with a time constant in the range of around 1 h. It is not easy to quantify the effect of water adsorption on the charge carrier concentration, n_S (which is normally proportional to the measured conductance). For the first mechanism of water interaction proposed by Heiland and Kohl ("rooted," Equation [22.4]), one could include the effect of water by considering the effect of an increased background of free charge carriers on the adsorption of oxygen. For the second mechanism proposed by Heiland and Kohl ("isolated," Equation [22.5]) one can examine the influence of water adsorption as an electron injection combined with the appearance of new sites for oxygen chemisorption; this is valid if one considers oxygen vacancies as good candidates for oxygen adsorption. In this case, one has to introduce the change in the total concentration of adsorption sites $[S_t]$:

$$[S_t] = [S_{t0}] + k_0 \cdot pH_2O \qquad (22.6)$$

obtained by applying the mass action law to Equation (22.5). $[S_{t0}]$ is the intrinsic concentration of adsorption sites and k_0 is the adsorption constant for water vapor.

In the case of the interaction with surface acceptor states, not related to oxygen adsorption, one can proceed as in the case of the first mechanism proposed by Kohl. In the case of an interaction with oxygen adsorbates, one can consider that the dissociation of oxygen ions is increased and examine the implications.

22.3.1.3 CO sensing mechanism

Reducing gases such as CO react with the surface oxygen ions freeing electrons that can return to the bands. Thus, the effect is the decrease in the band bending ($q\Delta V_S < 0$ and $\Delta\Phi < 0$) and, for an n-type semiconductor, the increase of the surface conductance (see Figure 22.15[b]). One can easily model the dependence of the resistance on the CO concentration by making the following assumptions, supported by the already established knowledge in the field: (i) the reaction of CO takes place with the previously adsorbed oxygen ions (well documented for the temperature and pressure range in which the gas sensors operate); (ii) the adsorption of CO is proportional with the CO concentration in the gas phase (quite reasonable but never really experimentally proved). Based on the a.m. assumptions, one can combine quasi-chemical reactions formalism with semiconductor physics calculations and one obtains power-law dependencies on the form:

$$G \sim p_{CO}^n \qquad (22.7)$$

the value of n depends on the morphology of the sensing layer and on the actual bulk properties of the sensing materials (complete analysis in Reference 21). The relationship described by Equation (22.7) is well supported by experiments (see Figure 22.20 and Table 22.2).

The IR studies identified CO-related species, that is, (i) unidentate and bidentate carbonate between 150 and 400°C, and (ii) carboxylate between 250 and 400°C (Figure 22.18). Simultaneous *in situ* conductance and Mössbauer spectroscopy measurements [96,97] on Pd doped (1% wt.) and undoped SnO$_2$ samples, carried out at different temperatures (50 to 380°C) and at a constant CO concentration of 1% in nitrogen, showed that CO interaction with SnO$_2$ is associated with the process announcing the beginning of Sn(IV) to Sn(II) reduction.

The influence of water vapor on CO detection is well documented (see Reference 22). Generally, the papers are just describing the effects without going too deep with the modeling/explanation of the effects. When they provide models, they are not based on the kind combination of spectroscopic and phenomenological measurement techniques applied in realistic sensor operation conditions, presented here. The explanations span between considering OH groups as weak acceptors, competing with oxygen for the same adsorption sites and more easily accessible to the reaction with CO to a complete decoupling between the CO and water surface reactions. However, the active surface species related to water vapor are considered the hydroxyl groups.

TABLE 22.2

Empirical phenomenological formulae describing the dependence of the responses of SnO_2-based gas sensors (conductance G or resistance R with base conductance G_0 or base resistance R_0 in air) on concentrations of reducing gases. Different authors use different analytical approaches to describe the gas response of different devices. SC, semiconductor. A_s, with various indices, constants used in the calculations or fittings, K_s reaction constants, and p_s partial pressures of gases or vapours, reproduced from [20] by permissions of Springer-Verlag.

Formula	Procedure	Comments
$G = Ac^\beta$	Empirical	Large data base, deviation at high concentration
$G = G_0 + A_1 P_{CO}^{1/2}$	Rate equations and SC physics	Reaction of CO and chemisorbed oxygen
$G = G_0 + A_1 p^\beta$	Empirical	Thin film
$G = G_0 P_{O_2}^{-\beta}\left(1 + K_{CO}P_{H_2O}P_{CO}\right)^\beta$	Empirical	for CO, CH_4, H_2O, CH_4/H_2, constant and modulated temperature
$\frac{G}{G_0} = P_{O_2}^{-\beta}\left\{1 + S_j K_j \Pi_i (A_i p_i)_{ij}^N\right\}^\beta \quad \beta = k_B T/E_0$	Rate equations and SC physics	Physisorbed oxygen, E_0 describes surface disorder, depends on preparation
$G \sim (p_R\, P_{H_2O})^{1/3}$	Rate equations and SC physics	For n-type metal oxides, oxygen species O^-
$\frac{G}{G_0} = x_i \sim (1 + K_{ik}\, p_k)^{\beta_{ik}}$	Empirical	i denotes the sensor, k the gas
$G \approx P_{O_2}^{-m/2.18}\left(A_0 + A_1 p_R^m\right)^{0.92}$	Rate equations and SC physics	Oxygen species (O_m^-) specified by $m = 0.5, 1, 2$
$G = A_0 + A_1 \cdot P_{CO}^{-\beta} + \frac{1}{A_2 p_{CO}^\beta + 1/(G_0 - A_0)}$	Band bending and SC physics	Account for hole contribution in surface conductance
$G = p_R^{-\beta}, \beta = -0.5\left(1 + \frac{1}{(1 - 2\alpha R/b)}\right)$	Rate equations and SC physics	Oxygen species O^-, exponent depends on the actual resistance
$G^2 - G_0^2 \sim P_{CO}$	Rate equations and SC physics	Fully depleted layer at the surface
$\frac{G}{G_0} = \Pi\left(\frac{p_i}{p_{i,0}}\right)^{\beta_i}$	Empirical, Clifford '82	In gas mixtures, normalized to a reference
$G^\beta - G_0^\beta \sim P_{CO}, \beta \geq 2$	Rate equations and SC physics	Procedure for reduced humidity interference
$\frac{G}{G_0} = \left(\frac{p_{CO}}{p_{0,CO}}\right)^{\beta_{CO}} \cdot \left(\frac{p_{H_2O}}{p_{0,H_2O}}\right)^{\beta_{H_2O}}$	Empirical	Interaction of water vapour, $V_O^{\cdot\cdot}$ and dipoles associated with hydroxyl groups
$G = p_R^{2m/(s+1)}$	Rate equations and SC physics	Interacting oxygen species (O_m^{-s}), m and $s = 1, 2$, mobility effects, for completely depleted grains
$G = A_0 \cdot (A_1 \cdot p_R)^\beta$	Empirical	Long-term properties by calibration with reference gas

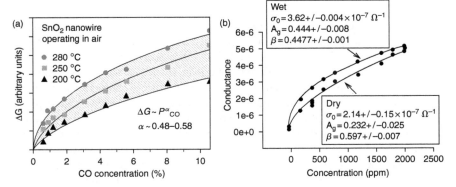

FIGURE 22.20 (a) The change in conductance of individual SnO_2 nanowires as a function of CO concentration at three temperatures. The data was fit to $\Delta G \propto P_{CO}^{\alpha}$. The solid lines represent the best fit with exponent values ranging between 0.48 and 0.58, reproduced from [82] by permissions of Wiley-VCH. (b) Representative fit of the response to $G = A_g P_g^{\beta}$ for the SnO_2 opal in dry and wet air, reproduced from [147] by permissions of Wiley-VCH

As mentioned earlier, by performing simultaneous conductance/resistance and work function measurements one can get information about the changes in surface concentration species that are not carrying a net charge, such as dipoles. This is of special interest when one studies the effect of water vapor because their surface interaction can lead to such species.

The corresponding conductance G may formally be described by equation:

$$G = G_0 \cdot \exp\left\{\frac{(E_F - E_C)_b - qV_S}{kT}\right\} = G_0 \cdot \exp\left\{\frac{\chi - \Phi}{kT}\right\} \qquad (22.8)$$

It is worth mentioning that changes in the electron affinity χ influence the work function Φ but not the electrical conductance G. By combining Equations (22.8) and (22.2),

$$kT \ln(G/G_0) = \chi - \Phi \qquad (22.9)$$

If the sensors are exposed to two different gas atmospheres (initial gas atmosphere I and final gas atmosphere F) and the related resistances (R_I, R_F) and work functions (Φ_I, Φ_F) are monitored, the different contributions to the work function change $\Delta\Phi$ can be determined according to:

$$\Phi_{I,F} = -kT \ln(G_{I,F}/G_0) + \chi_{I,F} \qquad (22.10)$$

hence

$$\Delta\Phi = \Phi_F - \Phi_I = kT \ln(G_I/G_F) + \chi_F - \chi_I = kT \ln(R_F/R_I) + \Delta\chi \qquad (22.11)$$

whereby Φ_I, R_I, and χ_I are the work function, the resistance and the electron affinity in the initial gas atmosphere and Φ_F, R_F, and χ_F the corresponding values in the final gas atmosphere. In all our experiments $\Delta\Phi$ and the electrical resistance were measured and, by using Equation (22.11), $\Delta\chi$ was calculated.

An example of the systematic characterization of CO–water interaction on SnO_2 thick film sensors is given in Reference 22. In order to characterize the influence of water only, the sensors (operated at 270°C) were exposed to an atmosphere with different concentrations of water vapor. An increase in humidity results in a decreased resistance and an increased electron affinity χ. Affinity changes may formally be attributed to changes in the coverage and absolute value of dipoles oriented perpendicular to the surface. The observed increase in electron affinity and the decrease in resistance (band bending) with increasing humidity can be explained by an increased coverage of surface hydroxyl groups as suggested by the mechanisms proposed by Heiland and Kohl and the appearance of donors. It is still not possible to decide between the two models because one cannot differentiate, with phenomenological techniques, between the different ways in which the donors are produced; one can just record the effects, namely the decrease of the resistance.

To further investigate the CO–water interaction, simultaneous work function and resistance measurements have been performed in a background of both humidified and dry synthetic air. Figure 22.21 shows typical results of the influence of water during CO exposure. For zero humidity the electron affinity χ did not (within the precision of the measurement) change. In the presence of water, the electron affinity is changed dramatically upon exposure to CO even at very low concentrations and remains almost constant for higher CO concentrations. The decrease of the electronic affinity can be attributed to a decrease of the concentration of isolated surface hydroxyl groups associated to a reaction with CO. It was proposed that, following a reaction of CO with the oxygen of the isolated hydroxyl group the freed H atom could build up a donor in a similar way as the one in which donors were built by the hydrogen atom resulted from the dissociation of the water molecule. In the case of the first mechanism, this decrease in electron affinity may be explained by the decrease of the concentration of surface dipoles, attributed to $(Sn_{Sn}^+ - OH^-)$, and the observed resistance decrease by the generation of additional "rooted" OH groups:

$$CO^{gas} + \left(Sn_{Sn}^+ - OH^-\right) + O_O \rightarrow CO_2^{gas} + Sn_{Sn} + (OH)_O^+ + e^- \qquad (22.12)$$

whereas in the case of second mechanism (equation), one would have to assume:

$$CO^{gas} + \left(Sn_{Sn}^+ - OH^-\right) + O_O \rightarrow CO_2^{gas} + \left(Sn_{Sn}^+ - OH^-\right) + V_O^{++} + 2 \cdot e^-$$
$$(22.13)$$

For the second case, the electron affinity would not change, which contradicts the experimental observations. This fact strongly suggests that the mechanism of water adsorption is the one described by Equation (22.4). The reaction of

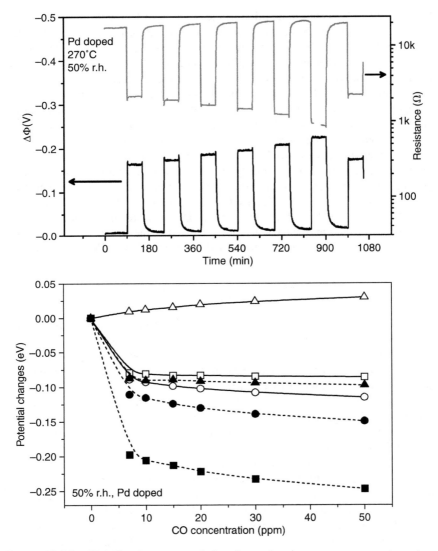

FIGURE 22.21 (Up) Simultaneous work function and resistance measurements on two "identical" Pd doped sensors (0.2% wt. Pd, $T_{cal} = 450°C$ for 8 h, $T_{anneal} = 700°C$) for $T_{op} = 270°C$ and 50% r.h. The sensors were exposed to CO (7 to 50 ppm). The sensors were purged for 90 min with synthetic air between every two successive CO exposures, reproduced from [30] by permissions of Shaker Verlag. (Bottom) Typical potential changes upon CO exposure determined at 270°C using Pd doped sensors (a) in dry air ($\square$ indicates $\Delta\Phi$, $\bigcirc$ indicates $kT \ln(R_F/R_I)$, $\triangle$ indicates $\Delta\chi$) and (b) at 50% r.h. ($\blacksquare$ indicates $\Delta\Phi$, $\bullet$ indicates $kT \ln(R_F/R_I)$, $\blacktriangle$ indicates $\Delta\chi$), reproduced from [22] by permissions of Institute of Physics Publishing

Equation (22.12) occurs in addition to the reaction of CO with ionosorbed oxygen. The proposed mechanism explains the sensitization effect of water on the CO detection by the fact that an increased humidity results in an increased concentration of surface dipoles and, thus, in an increased number of reaction partners (hydroxyl groups) and an increased efficiency in the CO reaction.

The impact of the CO–water reaction seems to decrease strongly with increasing CO concentration; above 10 ppm CO, there are only small changes in the electron affinity. The same behavior is also observed at higher temperatures of up to 400°C, but due to the thermal decomposition of OH groups, the corresponding absolute changes in $\Delta\chi$ as well as the sensitivity to CO decrease. It looks like the difference in CO sensitivity for this type of sensors, doped with Pd, when the temperature or the dopant concentration changes, is mainly related to the changes in the interaction with water vapor. The results of work function changes measurements provide a more solid basis for the findings of DC measurements alone.

For a confirmation of the results, one can evaluate the change in surface dipole coverage ($\Delta\theta$); this one can be determined by means of the following equation:

$$\Delta\chi = \frac{q}{\varepsilon_S \cdot \varepsilon_0} \Delta\theta \cdot N_{(S)\,max}^{ad} \overline{\mu}^{ad} \tag{22.14}$$

With the assumptions that $N_{(s)\,max}^{ad} \approx 10^{19}$ m^{-2} (surface density of tin atoms), $\overline{\mu}^{ad} \approx 1.66$ Debye $= 1.66 \times 3.3 \times 10^{-30}$ cm and $\varepsilon_S = \varepsilon_{SnO_2}/(2)^{0.5} = 8.5$, the changes in dipole coverage are in the range of fractions of monolayers (ML). Bearing in mind that for hydroxyl groups TDS coverage experiments indicate a surface coverage of about 15% of a ML.

Systematic characterization of CO–water interaction on SnO$_2$ thick film sensors by AC impedance spectroscopy confirms the main conclusions of simultaneous conductance and work function changes measurements, and namely (for the full discussion, see Reference 22): (i) for dry background air, the CO dependence of all resistive and capacitive elements of the equivalent circuit resemble those of an inter-granular Schottky contact, that is, the resistance decreases with increasing CO concentration, whereas the capacitance increases; the decrease of the band bending V_S decreases the resistance and increases the capacitance; (ii) For humid background air, the resistance behavior resembles that for dry air, but the capacitive behavior is different. For small CO concentrations, the capacitance is decreased whereas for higher CO concentrations it is increasing like in the case of dry background air. The "strange" capacitive behavior can be explained if one considers that in the value of the capacitance of the intergranular contacts one also has the dielectric constant. For low CO concentrations, the water–CO reaction dominates, that is, surface dipoles (OH groups) are removed from the surface, and electrons are released into the grains. This is in line with the observed resistance decrease. The reason for the observed capacitance decrease is the decrease of ε associated to the decrease of surface dipoles concentration, which is dominant at low CO concentrations. One assumes that in the capacitance change the driving factor is the change in the dielectric constant ε (which decreases the capacitance)

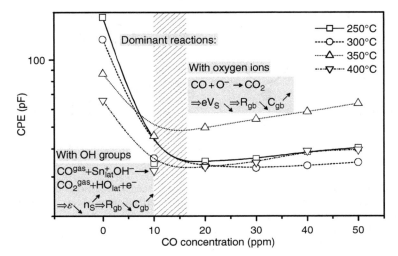

FIGURE 22.22 CO reaction mechanisms as determined by impedance measurements. In the low CO concentration range, the reaction with hydroxyl groups is dominant. The corresponding decrease in the dielectric constant ε results in a decrease in the CPE value. For higher CO concentrations CO reacts predominantly with oxygen. The increase in CPE is related to the decrease in the potential barrier height with increasing CO concentration [22]

and not the change of band bending V_S (which, as already described, increases the capacitance). For higher CO concentrations, the CO–oxygen reaction dominates, that means that the dominant factor is the decrease of the band bending V_S. Accordingly, the resistance is decreased and the capacitance is increased exactly as in the case of dry background, that the curves in dry background gas and in humid background gas become parallel with increasing CO concentration.

The different behavior in dry and humid background air confirms the hypothesis resulting out of the work function measurements of a dominant OH–CO interaction for low CO concentrations (ε-effect) and a dominant CO–O interaction for high CO concentrations (V_S-effect) (see Figure 22.22).

Recently, for the first time, Diffuse Reflectance Infrared Fourier Transform (DRIFT) spectroscopy was applied to characterize tin oxide-based thick film gas sensors under normal working conditions (between 25 and 300°C in dry and humid air) [103]. The obtained results confirm the role of surface hydroxil groups in the CO detection:

$$CO + OH_{ter} + H_2O \rightarrow CO_2^g + H_3O_S^+ + e^- \tag{22.15}$$

The results of DRIFT studies suggest that the role of the donor is played by the hydrated proton and not by the rooted hydroxyl group as it was considered by only taking into account the phenomenological studies. This again indicates that the spectroscopic knowledge is crucial to the understanding of gas sensing. Out of

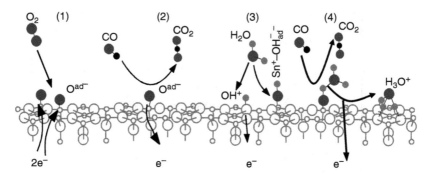

FIGURE 22.23 Possible reaction mechanism for the CO and water adsorption. The reaction between CO and different oxygen species results intermediate products carbonates and carboxylates. The reaction between CO, surface hydroxyl groups in the presence of adsorbed water molecules leads to creation of hydrated protons (H_3O^+) and their higher homologeous ($H_5O_2^+$) (based on Reference 102 and Reference 103)

only electrical studies, it is not possible to distinguish between the surface species, which can play the same "electrical" role. The complete picture of the CO sensing mechanism on SnO_2 thick film sensors is given in Figure 22.23.

22.3.1.4 Effect of doping

In most cases, sensitive layers are not made from pure materials. Additives, so-called dopants (usually Pd, Pt, Au), are employed to enhance the sensing properties. These additives can significantly affect the sensing behavior of a gas sensor. They can result in faster response and recovery times, in an enhancement of sensitivity or selectivity, in a smaller drift or in a better reproducibility. Depending on the application in mind, the doping material and its quantity have to be chosen carefully. In principle, a catalyst either increases the concentration of reactants at the metal oxide surface or lowers the activation energy for the reaction, or both.

To explain the influence of the dopants on the semiconductor, two models are commonly used (Figure 22.24). Both models assume that small clusters of the additive are located on the surface of a much bigger grain of tin dioxide. The distribution of these small dopant particles on the surface is assumed to be more or less homogeneous. (Recently, cluster as well as first-principle density-functional calculations dealing with oxygen and CO adsorption on pure SnO_2 (110) and doped M/SnO_2 (110) [M = Pd, Pt] as well as the interaction of oxygen with bridging oxygen vacancies were published [see Figure 22.19] [111,114,115]).

The first model is called the spillover or catalytic effect. In this case, the catalyst facilitates the activation of certain gas particles, for example, the dissociation of oxygen or hydrogen. Then the activated reactants reach the semiconductor, where the final reaction takes place. Due to spillover, reactions can be accelerated and

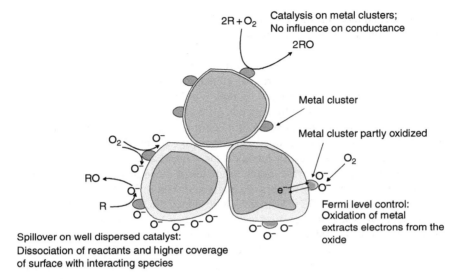

$2R + O_2$ Catalysis on metal clusters;
No influence on conductance

$2RO$

Metal cluster

Metal cluster partly oxidized

O_2 O^-

O^-

O_2

$RO \leftarrow$ O^-

$R \rightarrow$ O^- e^- O^-

O^-

O^- O^- O^-O^- O^- O^-

Fermi level control:
Oxidation of metal
extracts electrons from the
oxide

Spillover on well dispersed catalyst:
Dissociation of reactants and higher coverage
of surface with interacting species

Influence on conductance if reaction
takes place on sensitive oxide surface

FIGURE 22.24   Possible effect of doping. Top: Catalysis on metal clusters. The gas reaction takes place entirely on the cluster. Hence, the conductance remains unaffected. Left: Spillover: Metallic clusters promote the dissociation of certain ambient gases. The reactive particles reach the tin dioxide surface, where the reaction takes place. The conductance is affected due to this. Right: Fermi level control. Metallic clusters are partially oxidised in the gaseous atmosphere. The ambient gases affect the stoichiometry. Changes in the Fermi level position of the clusters change the depletion layer and the band bending inside the tin dioxide, reproduced from [30] by permissions of Shaker Verlag

shorter response time and larger sensitivity can result. If a given catalyst facilitates the activation for only a few gases, a higher selectivity can be obtained. Well-known examples for spillover are the spillover of hydrogen and oxygen from metal catalysts onto the semiconductor support. In the case of the spillover of oxygen and hydrogen due to the presence of Pt, an explanation for the lowering of the activation energy can easily be found. The bonding energy of Pt atoms to hydrogen atoms is not so different from the bonding energy of one hydrogen atom to another. Therefore, comparably little energy is needed to dissociate hydrogen molecules. Similar arguments hold for Pt and O_2. Consequently, the catalyst reduces the energy normally needed for dissociation to a great extent. The subsequent spillover onto SnO_2 is possible after breaking the rather weak bonds between hydrogen or, oxygen and Pt, respectively. Hence, the presence of the catalyst changes the initial dissociation process with rather high activation energy into a process with two lower energy barriers and hence increases the probability of the process.

The second model is the so-called Fermi energy control. In this case, the close electronic contact of the semiconductor with the catalyst dominates the sensor signal. Oxygen species at the surface of the catalyst trap electrons, which are refilled by electrons from the semiconductor. A depletion layer inside the semiconductor and a band bending result. Under the influence of the ambient gas atmosphere, the catalyst particles become oxidized. The stoichiometry of the catalyst oxide (MO_{2-x}) hereby depends on the composition of the ambient gas atmosphere and so does the position of its Fermi level. At equilibrium, the Fermi level of the catalyst and the semiconductor are at the same height. Since the gas reacts via the catalyst with the metal oxide, the chosen catalyst can strongly change the selectivity of pure tin oxide. Some examples of the influence of the doping on the characteristics of SnO_2 sensors are in given in Figure 22.25.

Recently, the *in situ* DRIFT studies on Pd-doped SnO_2 sensors showed that in the presence of water, the reaction between CO and hydroxyl groups determines the creation of the hydrated protons, which additionally increases the sensor signal [103]. The exact role of the Pd in the reaction mechanism is still unclear but, due to the fact that for undoped sensors we record a different effect of humidity, Pd should play a determining role. It is also important to note that hydroxyl groups present at the sensor surface do not seem to react with CO in the absence of the water vapor; this fact indicates that the association between OH groups and H_2O is the condition necessary for reaction with CO.

22.3.1.5 Difference between good catalyst and good sensor

Here, we also have to note that a good catalyst does not always mean a good sensor. The surface of a solid oxide contains complex imperfections, which play the role of local surface donor or acceptor levels. These levels, located usually in the band gap, may intermediate the transfer of electrons between the solid and the adsorbed molecules. The transfer of electrons between the solid catalyst and the reactants remains one of the important factors (e.g., for the breaking of the C–H bond, in which it is involved, is usually the rate-determining step of the oxidation process). However, catalytic transformations are usually only local phenomena, in which active sites are involved. The charge transfer usually takes place only in a localized manner and will have no direct impact on the conduction. In the case of sensors, the charge transfer should take place with the conduction band. Only in these cases, the concentration of the free charge carriers will be influenced (see Figure 22.26).

On the other hand, for the gases involved in the sensing processes, two possibilities should be considered (i) adsorption–desorption of gases (the molecules remain unchanged during the interaction, this kind of interaction is usually observed for oxidizing gases such as NO_2, O_3 on sensors operated at low temperatures) and (ii) catalytic oxidation of test gases (usually observed for the detection of reducing gases). In both cases, for an easy detection the charge transfer should take place with the conduction band and the concentration of free charge carriers will be influenced (see Figure 22.25 and Figure 22.27).

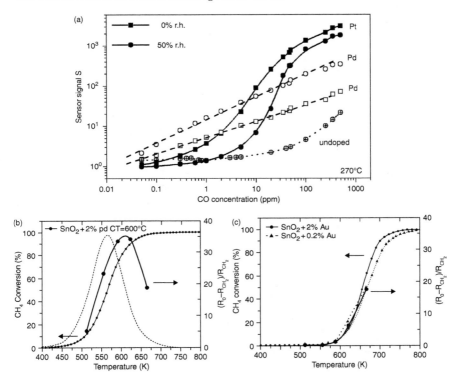

FIGURE 22.25 (a) Sensor signal as a function of the CO concentration for Pd doped sensors (0.2% wt., open symbols), Pt doped sensors (0.2% wt., full symbols) and undoped sensors (cross in center) for 0% r.h. (squares) and 50% r.h. (circles), reproduced from [30] by permissions of Shaker Verlag. Comparison of the sensor response to 1000 ppm CH_4 and the catalytic combustion curves for the different modified materials: (b) SnO_2 with 2%Pd (c) SnO_2 with 0.2 and 2% Au. On graph (b) is also plotted the derivative of the combustion curve as a measure of the variation with the temperature of the percentage of CH_4 molecules reacting with/on the oxide surface. Gold is recognized to present strong segregation phenomena (clustering) shows only small influence on the sensor response, reprinted from [148], copyright (2002), with permission from Elsevier

Kinetic models are applied for time-dependent reactions and, in particular, for catalytic reactions. These reactions are irreversible in the gas phase since the sensors operate in an open flow system (p = constant, $V \neq$ constant). The sensor itself reacts reversibly as an ideal catalyst. Under flow conditions steady-state equilibria influence the conductance and hence the sensor response. Since catalytic oxidation has been proved experimentally, only kinetic models are appropriate to explain, for example, the CO sensing mechanism.

The parameters required to understand *kinetic behavior* on an atomistic level have not yet been published in the literature. These parameters may be determined from the combined results of thermal desorption spectroscopy (TDS), reactive

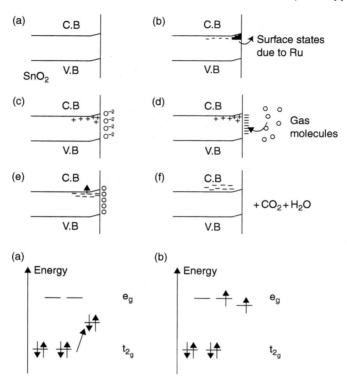

FIGURE 22.26 Difference between catalyst and sensor. (Up) Collective effects: gas sensing mechanism of surface ruthenated tin oxide indicating the role of surface states in enhancing the sensitivity toward butane. The energy released during the decomposition of the products is sufficient for the electrons to jump into the conduction band of tin oxide thus causing an increase in the conductivity of the sensor, reprinted with permission from [100], copyright (2001) American Chemical Society. (Bottom) Local effects: CO oxidation over Co_3O_4. (a) One of the t_{2g} energy levels involved in the CO adsorption becomes destabilized due to the symmetry lowering, resulting from the adsorption. (b) A transition from diamagnetic to paramagnetic Co (III) occurs due to the reduced energy splitting between the two original e_g orbitals and one of the t_{2g} components, reprinted from [149], copyright (2002), with permission from Elsevier

scattering, and IR experiments. The experimental results obtained in this way can be used for semiempirical calculations of surface reactions, and reaction paths can be derived. A survey of the results obtained by applying the above-mentioned methods to oxidic semiconductors is presented in Reference 57. Methane and oxygen reaction schemes on a sputtered SnO_2 film, derived from reactive sputtering experiments, are presented in Figure 22.28. Also, a theoretical analysis of adsorption and dissociation of CH_3OH on the stoichiometric SnO_2 (110) surface [118], a reaction model for methane oxidation (Figure 22.29, compare with Figure 22.28) [113] and NH_3 chemisorption on reduced SnO_2 (110) surfaces [117] were published.

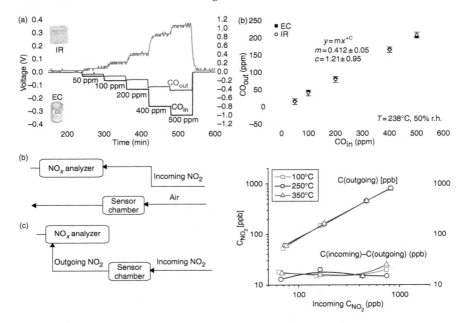

FIGURE 22.27 (Up) CO combustion on SnO_2-based sensors as determined by EC and IR. (a) Raw signals of EC and IR sensors for incoming CO concentrations between 50 and 500 ppm, reproduced from [30] by permissions of Shaker Verlag. (b) Comparison of the amounts of combusted CO and produced CO_2 as determined by EC and IR. Principle of simultaneous consumption and resistivity measurements, reprinted from [162], copyright (2001), with permission from Elsevier. (Bottom) Incoming (supplied by gas mixing system) and outgoing (through the test chamber) concentrations of NO_2/NO are compared. NO_2 combustion as determined by NO_x chemiluminescence analyzer [150]

22.3.2 Transduction: Contribution of Different Sensor Parts in the Sensing Process and Subsequent Transduction

22.3.2.1 Conduction in the sensing layer

The elementary steps of gas sensing will be transduced in electrical signals measured by appropriate electrode structures. The sensing itself can take place at different sites of the structure depending on the morphology. They will play different roles, according to the sensing layer morphology.

A simple distinction can be made between:

- *Compact layers* (Figure 22.30): the interaction with gases takes place only at the geometric surface (such layers are obtained with most of the physical techniques used for thin film deposition).
- *Porous layers* (Figure 22.31): the volume of the layer is also accessible to the gases and in this case the active surface is much higher than

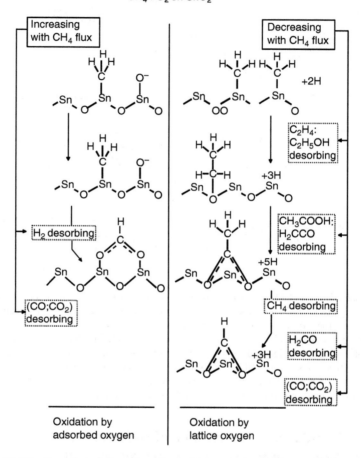

Reactive scattering at 500°C
CH$_4$ + O$_2$ on SnO$_2$

FIGURE 22.28 Reaction scheme of methane (CH$_4$) with oxygen on a sputtered SnO$_2$ film derived from reactive scattering data. Hydrogen atoms, released or consumed in all steps, are not sketched, reprinted from [57], copyright (1989), with permission from Elsevier.

the geometric one (such layers are characteristic to thick film techniques and RGTO [161]).

For compact layers, there are at least two possibilities: completely or partly depleted layers, depending on the ratio between layer thickness and Debye length λ_D. For partly depleted layers (an example here can be the photoresponse of a SnO$_2$ nanoribbon, see Figure 22.30), when surface reactions do not influence the conduction in the entire layer ($z_g < z_0$), the conduction process takes place in the bulk region (of thickness $z_0 - z_g$, much more conductive that the surface

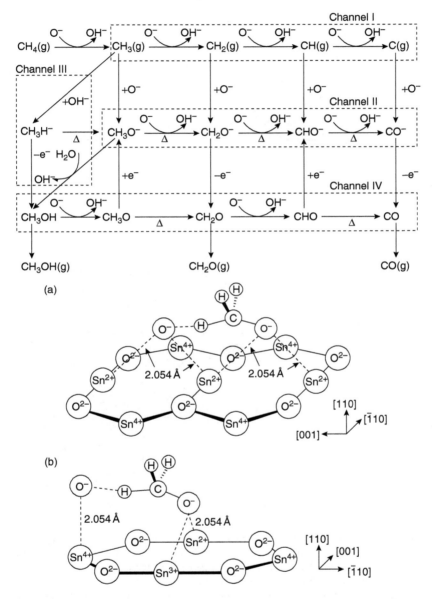

FIGURE 22.29 (Up) Reaction model for methane oxidation on the reduced SnO_2 (110) surface. Schematic structures for the initial reaction steps (a) channel I, (b) channel II, on the reduced (110) surface, reproduced from [113] by permissions of Wiley-VCH.

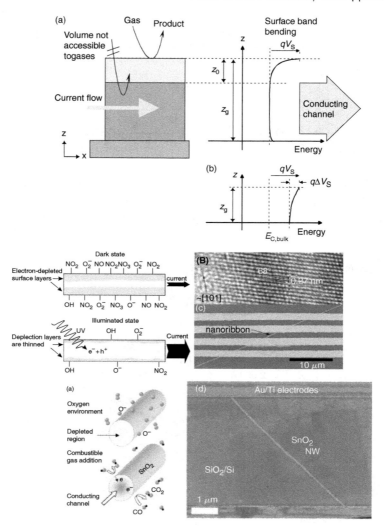

FIGURE 22.30 (Up) Schematic representation of a compact sensing layer with geometry and energy band representations. z_0 thickness of the depleted surface layer, z_g layer thickness, and qV_s band bending. (a) Represents a partly depleted compact layer ("thicker"), (b) represents a completely depleted layer ("thinner"), copyright [21], with kind permission of Springer Science and Business Media. (In the middle): A schematic longitudinal cross-section of a nanoribbon in the dark and in UV light. In the illuminated sate, photogenerated holes recombine with trapped electrons at the surface, desorbing NO_2 and other electron-trapping species: $h^+ + NO_{2(ads)}^{2-} \rightarrow NO_{2(gas)}$. The space charge layers becomes thinner, and the nanoribbon conductivity increases. Ambient NO_2 levels are tracked by monitoring changes in the illuminated state, reproduced from [83] by permissions of Wiley-VCH. (Bottom) (a) and (d) the sensing mechanism of a SnO_2 nanowire involves a completely depleted, hence nonconductive state under an oxidizing ambient and sharply increased conductance due to electron transfer from a surface states back into the nanowire's interior when a reducing gas(CO) is admitted, reproduced from [82] by permissions of Wiley-VCH

depleted layer). Formally two resistances occur in parallel, one influenced by surface reactions and the other not; the conduction is parallel to the surface, and this explains the limited sensitivity. Such a case is generally treated as a conductive layer with a reaction-dependent thickness. For the case of completely depleted layers in the absence of reducing gases (an example here can be the sensing on a SnO_2 nanowire, see Figure 22.30), it is possible that exposure to reducing gases acts as a switch to the partly depleted layer case (due to the injection of additional free charge carriers). It is also possible that exposure to oxidizing gases acts as a switch between partly depleted and completely depleted layer cases.

For porous layers, the situation may be complicated further by the presence of necks between grains. It may be possible to have all three types of contribution presented in a porous layer: surface/bulk (for large enough necks $z > z_0$,), grain boundary (for large grains not sintered together), and flat bands (for small grains and small necks) (Figure 22.31). Of course, what was mentioned for compact layers, that is, the possible switching role of reducing gases, is valid also for porous layers. For small grains and narrow necks, when the mean free path of free charge carriers becomes comparable with the dimension of the grains, a surface influence on mobility should be taken into consideration. This happens because the number of collisions experienced by the free charge carriers in the bulk of the grain becomes comparable with the number of surface collisions; the latter may be influenced by adsorbed species acting as additional scattering centers.

22.3.2.2 Role of contacts

As shown in Figure 22.32 there is a resistance associated with the interface between the semiconducting sensitive layer and the metallic electrode. The importance of this resistance to the overall sensor resistance value depends on the sensing layer and electrode morphology. The possible dependence of the contact resistance on the ambient atmosphere conditions can be described in terms of two contributions: (i) the first is dealing only with the electrical contribution of the semiconducting sensitive layer — electrode interface to the overall sensor resistance, (ii) the second describes the possible chemical influence of, for example, the catalytic activity of the contact material in the region close to the contacts.

Figure 22.32 illustrates the way in which the metal-semiconductor junction, built at electrode-sensitive layer interfaces, influences the overall conduction process. For compact layers they appear as a contact resistance (R_C) in series with the resistance of the metal oxide layer (see Figure 22.13). For partly depleted layers R_C could be dominant, and the reactions taking place at the three-phase boundary, electrode-metal oxide-atmosphere, control the sensing properties.

In porous layers, the influence of R_C may be minimized because it will be connected in series with a large number of resistances, typically thousands, which may have comparable values. Transmission line measurements (TLMs) performed with thick SnO_2 layers exposed to CO and NO_2 did not result in values of R_C clearly distinguishable from the noise, while in the case of thin films the existence of R_C was proven. Again, the relative importance played by different terms may

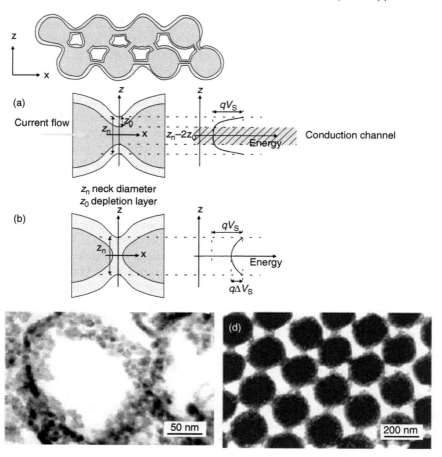

FIGURE 22.31 (Up) Schematic representation of a porous sensing layer with geometry and surface energy band-case with necks between grains. z_n is the neck diameter; z_0 is the thickness of the depletion layer. (a) represents the case of only partly depleted necks whereas (b) represents large grains where the neck contact is completely depleted, reproduced from [21] by permissions of Springer Verlag. (Bottom) An example of the opal-like SnO_2 structures: on the left: A TEM image of the nanocrystallites within the walls; (on the right): a TEM image of the nanocrystallites within the SnO_2 spheres and neck between the spheres, reproduced from [147] by permissions of Wiley-VCH.

be influenced by the presence of reducing gases because one can expect different effects for grain–grain interfaces when compared with electrode–grain interfaces.

As demonstrated in Reference 151 and Reference 152, one can state that the changes of the electrical resistance attributed to the contacts are negligible during the operation of the sensor. The value of the contact resistance is established during the preparation. This value, in contrast to its change, might be important in the overall resistance of the sensor and could even decrease the sensor response

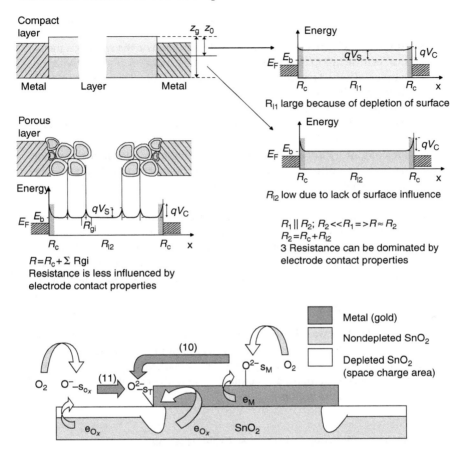

FIGURE 22.32 (Up) Schematic representation of compact and porous sensing layers with geometry and energetic bands, which shows the possible influence of electrode-sensing layers contacts. R_C is the resistance of the electrode-SnO_2 contact, R_{l1} is the resistance of the depleted region of the compact layer, R_{l2} is the resistance of the bulk region of the compact layer, R_1 is the equivalent series resistance of R_{l1} and R_C, R_2 is the equivalent series resistance of R_{l2} and R_C, R_{gi} is the average intergrain resistance in the case of porous layer, E_b is the minimum of the conduction band in the bulk, qV_S is the band bending associated with surface phenomena on the layer, and qV_C also contains the band bending induced at the electrode-SnO_2 contact, copyright [21], with kind permission of Springer Science and Business Media. (Bottom) Proposed mechanism based on space charge area at the three boundary point (gas/metal/oxide) to explain metal effect on tin oxide electrical conduction, reprinted from [153], copyright (2004), with permission from Elsevier.

by being a "dead" series element, especially for the case of compact films where $z_g > z_0$.

The reasons for chemical effects at the electrode sensing layer interface are related to the catalytic nature of the electrode material. The materials used are often platinum and gold. Several effects could be taken into consideration:

- Surface species that can be more easily adsorbed on the electrode metal may diffuse fast to the three-phase boundary, where it can react with the partner adsorbed on the metal oxide sensitive layer. Thus, the increased diffusion will lead to a higher catalytic conversion rate, which will be monitored by the electrical readout.
- Another effect is the increased production ("Catalysis") of reaction partners by the metal electrode material. This can happen by, for example, breaking of hydrocarbons in more active radicals. Hence the reaction partners can diffuse to the three phase boundary (electrode/sensing layer/gas phase) and consequently this region becomes "more active" in gas detection.
- In contrast to the above-mentioned effects, which will enhance the sensor response, it is possible that an increased catalytic reaction on the electrode material with a direct desorption from there, will lead to a gas consumption, which is not monitored by the electrical readout. In consequence, this gas consumption may lead to an overall lowering of the analyte (depending on the given setup) and may thus even lead to a lowering of the sensor signal.

Recently [153], it was demonstrated that in the absence of the noble metal electrode boundary with the reacting gas (oxygen in this case) the change in the resistance of the sensing layer during exposure is markedly diminished. This fact proves that the electrode plays a "chemical" role in the overall performance of a sensor (see Figure 22.32 bottom).

22.4 MULTIPLE FEATURE EXTRACTION APPROACH FOR GAS SENSING

In practical applications, several attempts are usually made to overcome disadvantages of metal oxide-base sensors concern their lack of stability and selectivity, by, for example, using chromatographic columns to separate the components, by operating at different temperatures, by choosing different burning-in procedures, dopants, measuring frequencies, etc. As mentioned earlier (see also Figure 22.5), for common applications of pattern recognition (PARC) and multicomponent analysis (MCA) of gas mixtures, arrays of sensors are usually chosen, which operate at constant temperature. In these cases, a lack of selectivity and therefore overlapping sensitivities of different sensors is of advantage [10,60,61]. Another approach, instead of using several sensors of the same type and reading out one measurement parameter (e.g., the resistance change), is to use several parameters which are extracted from one (or several) sensors.

One of the most popular approach is to periodically change the operation temperature, for example, metal oxide gas sensors or pellistors (see also Figure 22.33). Sears et al. [154] proposed to monitor conductance/time curves for different gases under conditions of thermal cycling. Kunt et al. [155] described an optimization

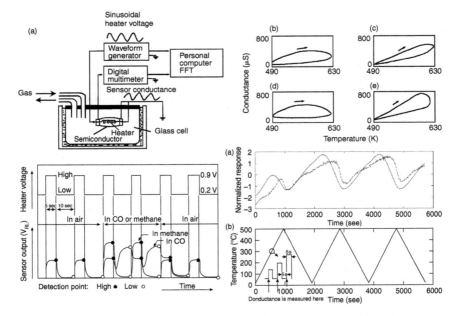

FIGURE 22.33 (Up) (a) Experimental apparatus for detecting the dynamic response of a gas sensor (Figaro TGS 813). Sensor conductance G versus sensor temperature T curve for 1000 ppm aromatic vapours: (b) benzene, (c) chlorobenzene, (d) toluene, and (e) nitrobenzene, reprinted with permission from [70], copyright (1996) American Chemical Society. (Bottom, left) FIS SB-95. When the sensor is operated with high/low periodic operation, sesnor signal changes according to the temperature dependency characteristics. By detecting the sensor signal at sufficient timings (at a high temperature for methane and at low temperature for CO) selective detection of both methane and CO has been achieved, reproduced from [45] by permissions of FIS INC. (Bottom, right) (a) Two-temperature mode of operation is used to improve the speed of response and recovery of the sensor and to minimize the effects of drift. The comparison between the predicted and experimental normalized temperature programming sensing responses to argon with the duration of 5 sec for both puls and delay and (b) the modulation temperature profile, reprinted from [68], copyright (2001), with permission from Elsevier.

of micro-hot plate chemical gas sensor temperature program for the discrimination of ethanol and methanol. Feature extraction by Fast Fourier transform analysis (FFT) of the resistance/conductance of sinusoidal temperature modulated tin oxide sensors was performed by Nakata et al. [24,69,70].

Additional information can be provided by measurements performed at the same conditions, with results so far not commonly used as sensor signals. Examples include monitoring work function changes [135,136], catalytic conversion of CO to CO_2 [98], mobilities [156,157], AC impedance spectroscopy and frequency dependent resistance data [35,36], simultaneous calorimetric and resistive measurements [13,14], and Seebeck effect [33,34,158].

Recently the advantages of using "hybrid sensors" id est. combinations between thick film sensing layers and micromachined hotplates when one tries to develop feature extraction procedures were demonstrated [13,73]. The combination of thick film technology for layer deposition of nanocrystalline SnO_2 and thin film technology for substrate structuring based on conventional Si technology is particularly promising for the construction of cheap, high performance oxide-based sensors. The better functionality, resulting out of the micro-sensor particularities was demonstrated in two cases where they made possible two different multiple feature extraction procedures:

- First, simultaneous monitoring of temperature and resistance changes make it possible to discriminate between CO, CH_4, and C_2H_5OH. This result is extremely important for practical applications, for example, to fulfill the requirements of common safety standards for CO and CH_4 sensors. Qualitative and quantitative analysis by using artificial neural network algorithms was performed with good results. An explanation is given for the observed changes in resistances and temperatures and for their gas specificities.
- Second, the combination between temperature modulation and FFT show that it is possible to identify CO and NO_2 in the ambient atmosphere by using only one sensor operated in the modulated temperature mode. Additional studies are now on the way to extend this approach to identify other gas mixtures with more components and to understand the basic phenomena. Future work will be devoted to the development of appropriate feature extraction procedures for this nonlinear frequency-time problem.

22.5 Conclusions

The scope of this work was to demonstrate that gas sensors based on metal oxides are complex devices and that it is not possible to understand them in the absence of a systematic approach. This one should take into consideration all factors influencing the two basic processes contributing to the final sensor performance, namely:

- Surface chemical reactions and the associated charge transfer. Here it is important to clarify the role of surface dopants and of the interfaces related to the presence of electrodes and substrate.
- Transduction of the associated charge transfer into a measurable parameter. Here one should model the role of morphology of the sensing layer and quantify the importance of measurements parameters either electrical or geometrical.

The key factor is to understand that for describing/modeling a metal oxide based gas sensor one should coherently combine knowledge from two different disciplines, chemistry, and solid state physics and by trying to impose only one kind of viewpoint one will obtain, in the best case, only a part of the picture. In addition, the knowledge to be used in the modeling has to be gathered in conditions that are relevant to the normal sensor operation.

The basics for such an approach are existing due to the research efforts spent in the last two decades in which the focus was shifting from empirical performance improvement toward real basic understanding; this was happening also due to the fact that it was becoming clear that the limits of empirical improvements were reached and there is a need for a qualitative change. It was proved that one can even perform spectroscopy in conditions similar to the real-life conditions for sensors, one has the theoretical modeling apparatus, and one can explain with a reasonable degree of accuracy application relevant phenomena. So, one has now the first steps in the understanding of the gas sensing with metal oxide sensors starting with the mechanistic level and continuing with the transducer function. We should continue and consolidate the already acquired knowledge by emphasizing relevant research targets and concentrating on the needed resources; from the point of view of applications, there is a renewed interest in metal oxide-based gas sensors in new and challenging fields. It seems to us that sensor science has to be elaborated by combining the interests and knowledge of sensor developers, sensor users and scientists aiming for the basic understanding; all these experts should talk to each other for mutual benefit. The users can clearly identify the relevant objectives, the developers can optimize the samples/sensors, and the scientists will be able to build the knowledge needed for the further advancements. The latter will have to be validated by the development of better performing real-world sensors able to be applied for solving relevant practical problems.

REFERENCES

1. Moseley, P. and Tofield, B. *Solid State Gas Sensors*. Adam Hilger: Bristol, 1987, 245 pp.
2. Janata, J. *Principle of Chemical Sensors*. Plenum Press: New York, 1989, 317 pp.
3. Madou, M.J. and Morrison, S.R. *Chemical Sensing with Solid State Devices*. Academic Press: Boston, 1989, 556 pp.
4. Moseley, P.T., Norris, J.O.W., and Williams, D.E. *Techniques and Mechanisms in Gas Sensing*. Adam Hilger: Bristol, 1991, 408 pp.
5. Göpel, W., Hesse, J., and Zemel, J.N., Eds. *Sensors: A Comprehensive Survey*. VCH: Weinheim, 1991, Vol. 1–3.
6. Sberveglieri, G., Ed. *Gas Sensors*. Kluwer: Dordrecht, 1992, 420 pp.
7. Janata, J., Josowicz, M., Vanysek, P., and DeVaney, D.M. Chemical sensors. *Anal. Chem.* 1998, *70*, 179R–208R.
8. Bakker, E. and Telting-Diaz, M. Electrochemical sensors. *Anal. Chem.* 2002, *74*, 2781–2800.
9. Hierlemann, A. and Baltes, H. CMOS-based chemical microsensors. *Analyst* 2002, *128*, 15–28.
10. Pierce, T.C., Schiffman, S.S., Nagle, H.T., and Gardner. J.W., Eds. *Handbook of Machine Olfaction: Electronic Nose Technology*. Wiley-VCH: Weinheim, 2003, 592 pp.

11. Hu, J., Zhu, F., Zhang, J., and Gong, H. A room temperature indium tin oxide/quartz crystal microbalance gas sensor for nitric oxide. *Sensors Actuators B* 2003, *93*, 175–180.

12. Simon, I., Bârsan, N., Bauer, M., and Weimar, U. Micromachined metal oxide gas sensors: opportunities to improve sensor performance. *Sensors Actuators B* 2001, *73*, 1–26.

13. Heilig, A., Bârsan, N., Weimar, U., and Göpel, W. Selectivity enhancement of SnO_2 gas sensors: simultaneous monitoring of resistances and temperatures. *Sensors Actuators B* 1999, *58*, 302–309.

14. Takada, T. A temperature drop on exposure to reducing gases for various metal oxide thin films. *Sensors Actuators B* 2001, *77*, 307.

15. Göpel, W. and Reinhardt, G. Metal oxide sensors: new devices through tailoring interfaces on the atomic scale. In *Sensors Update Sensor Technology — Applications — Markets*, Baltes, H., Göpel, W., and Hesse, J., Eds. VCH: Weinheim, 1996, Vol. 1, pp. 49–120.

16. Lampe, U., Fleisher, M., Reitmeier, N., Meixner, H., McMonagle, J.B., and Marsh, A. New materials for metal oxide sensors. In *Sensors Update Sensor Technology — Applications — Markets*, Baltes, H., Göpel, W., and Hesse, J., Eds. VCH: Weinheim, 1996, Vol. 2, pp. 1–36.

17. Kohl, D. Function and application of gas sensors. *J. Phys. D: Appl. Phys.* 2001, *34*, R125–R149.

18. Williams, D. Semiconducting oxides as gas-sensitive resistors. *Sensors Actuators B* 1999, *57*, 1–16.

19. Ihokura, K. and Watson, J. *Stannic Oxide Gas Sensors, Principles and Applications*. CRC Press: Boca Raton, FL, 1994, 187 pp.

20. Bârsan, N., Schweizer-Berberich, M., and Göpel, W. Fundamentals and practical applications to design nanoscaled SnO_2 gas sensors: a status report. *Fresen. J. Anal. Chem.* 1999, *365*, 287–304.

21. Bârsan, N. and Weimar, U. Conduction model of metal oxide gas sensors. *J. Electroceram.* 2001, *7*, 143–167.

22. Bârsan, N. and Weimar, U. Understanding the fundamental principles of metal oxide based gas sensors, the example of CO sensing with SnO_2 sensors in the presence of humidity. *J. Phys. Condens. Mat.* 2003, *15*, R813–R839.

23. Marek, J., Trah, H.-P., Suzuki, Y., and Yokomori, I., Eds. *Sensors for Automotive Technology*. VCH: Weinheim, 2003, 562 pp.

24. Nakata, S., Ed. *Chemical Analysis Based on Nonlinearity*. Nova Science Publishers: New York, 2003, 156 pp.

25. Weimar, U. *Gas Sensing with Tin Oxide: Elementary Steps and Signal Transduction*. Habilitation thesis, University of Tübingen, 2001, 262 pp.

26. Schweizer-Berberich, M. Gas Sensors Based on Stannic Oxide. Ph.D. thesis, University of Tübingen, Shaker Verlag: Aachen, 1998, 186 pp.

27. Heilig, A. Selektivitätssteigerung von SnO_2-Gassensoren. Ph.D. thesis, University of Tübingen, 1999, 148 pp.

28. Diéguez, A. Structural Analysis for the Improvement of SnO_2 Based Gas Sensors. University of Barcelona, 1999, 297 pp.

29. Sinner-Hettenbach. $SnO_2(110)$ and Nano-SnO_2: Characterization by Surface Analytical Techniques. Ph.D. thesis, University of Tübingen, 2000, 166 pp.

30. Kappler, J. Characterisation of High-Performance SnO_2 Gas Sensors for CO Detection by In-Situ Techniques. Ph.D. thesis, University of Tübingen, Shaker Verlag: Aachen, 2001, 198 pp.
31. Strathmann, S. Sample Conditioning for Multi-Sensor Systems. Ph.D. thesis, University of Tübingen, 2001, 176 pp.
32. Hahn, S. SnO_2 Thick Film Sensors at Ultimate Limits: Performance at Low O_2 and H_2O Concentrations, Size Reduction by CMOS Technology. Ph.D. thesis, University of Tübingen, 2002, 152 pp.
33. Siroky, K. Use of the Seebeck effect for sensing flammable gas and vapours. *Sensors Actuators B* 1993, *17*, 13–17.
34. Vlachos, D.S., Papadopoulos, C.A., and Avaritsiotis, J.N. A technique for suppressing ethanol interference employing Seebeck effect devices with carrier concentration modulation. *Sensors Actuators B* 1997, *44*, 239–242.
35. Weimar, U. and Göpel, W.A.C. Measurements on tin oxide sensors to improve selectivities and sensitivities. *Sensors Actuators B* 1995, *26–27*, 13–18.
36. Vargnese, O.K. and Malhotra, L.K. Studies of ambient dependent electrical behaviour of nanocrystalline SnO_2 thin films using impedance spectroscopy. *J. Appl. Phys.* 2000, *87*, 7457–7465.
37. Scott, R.W.J., Mamak, M., Coombs, N., and Ozin, G.A. Making sense out of sulfated tin dioxide mesostructures. *J. Mater. Chem.* 2003, *13*, 1406–1412.
38. Gurlo, A., Sahm, M., Oprea, A., Bârsan, N., and Weimar, U. A p- to n-transition on α-Fe_2O_3-based thick film sensors studied by conductance and work function change measurements. *Sensors Actuators B*, 2004, *102*, 292–298.
39. Brattain, W. and Bardeen, J. Surface Properties of Germanium. *Bell Telephone Syst. Tech. Publs. Monogr.* 1953, *2086*, 1–41.
40. Heiland, G. Zum Einfluss von Wasserstoff auf die elektrische Leitfähigkeit von ZnO-Kristallen. *Z. Phys.* 1954, *138*, 459–464.
41. Bielanski, A., Deren, J., and Haber, J. Electric conductivity and catalytic activity of semiconducting oxide catalysts. *Nature* 1957, *179*, 668–669.
42. Seiyama, T., Kato, A., Fujiishi, K., and Nagatani, M. A new detector for gaseous components using semiconductive thin films. *Anal. Chem.* 1962, *34*, 1502f.
43. Taguchi, N. Gas Detecting Device. U.S. patent 3,631,436, 1971.
44. http://www.figarosens.com
45. http://www.fisinc.co.jp
46. http://www.appliedsensor.com
47. http://www.microchemical.com
48. http://www.microsens.ch
49. Graf, M., Taschini, S., Käser, P., Hagleitner, C., Hierlemann, A., and Baltes, H. Integrated metal-oxide microsensor array of micro-hotplates with MOS-transistor heater. In Proceedings of the *Eurosensors XVII, European Conference on Solid-State Transducers*, University of Minho, Guimaraes, Portugal, September 21–24, 2003, pp. 1103–1104 (CD-ROM).
50. Barrettino, D., Graf, M., Zimmermann, M., , Hierlemann, A., Baltes, H., Hahn, S., Barsan, N., and Weimar, U. A system architecture of micro-hotplate-based chemical sensors in CMOS technology. In Proceedings of the *Eurosensors XVI, European Conference on Solid-State Transducers*, Prague, Czech Republic, September 15–18, 2002, pp. 419–420 (CD-Rom).

51. Aliwell, S.R., Halsall, J.F., Pratt, K.F.E., Sullivan, J.O., Jones, R.L., Cox, R.A., Utembe, S.R., Hansford, G.M., and Williams, D.E. Ozone sensors based on WO_3: a model for sensor drift and a measurement correction method. *Meas. Sci. Technol.* 2001, 12, 684–690.

52. Williams, D.E., Aliwell, S.R., Pratt, K.F.E., Caruana, D.J., Jones, R.L., Cox, R.A., Hansford, G.M., and Halsall, J.F. Modelling the response of a tungsten oxide semiconductor as a gas sensor for the measurement of ozone. *Meas. Sci. Technol.* 2002, 13, 923–931.

53. Cantalini, C., Wlodarski, W., Li, Y., Passacantando, M., Santucci, S., Comini, E., Faglia, G., and Sberveglieri, G. Investigation on the O_3 sensitivity properties of WO_3 thin films prepared by sol–gel, thermal evaporation and r.f. sputtering techniques. *Sensors Actuators* B 2000, 64, 182–188.

54. Gurlo, A., Ivanovskaya, M., Bârsan, N., Schweizer-Berberich, M., Weimar, U., and Göpel, W. Grain-size control in nanocrystalline In_2O_3 semiconductor gas sensors. *Sensors Actuators* B 1997, 44, 327–333.

55. Gurlo, A., Bârsan, N., Ivanovskaya, M., Weimar, U., and Göpel, W. In_2O_3 and In_2O_3–MoO_3 thin film semiconductor sensors: interaction with NO_2 and O_3. *Sensors Actuators* B 1998, 47, 92–99.

56. Soulantica, K., Erades, L., Sauvan, M., Senocq, F., Maisonnat, A., and Chaudret, B. Synthesis of indium oxide nanoparticles from indium cyclopentadienyl precursor and their application for gas sensing. *Adv. Funct. Mater.* 2003, 13, 553–557.

57. Kohl, D. Surface processes in the detection of reducing gases with SnO_2-based devices. *Sensors Actuators* 1989, 18, 71–118.

58. Shimizu, Y. and Egashira, M. Sensitization of odor semiconductor gas sensors by employing a heterolayer structure. In *Sensors Update Sensor Technology — Applications — Markets*, Baltes, H., Göpel, W., and Hesse, J., Eds. VCH: Weinheim, 1999, Vol. 6, pp. 211–229.

59. Cabot, A., Arbiol, L., Cornet, A., Morante, J.R., Chen, F., and Liu, M. Mesoporous catalytic filters for semiconductor gas sensors. *Thin Solid Films* 2003, 436, 64–69.

60. Gardner, J.W. and Bartlett, P.N. *Electronic Noses. Principles and Applications.* Oxford University Press: New York, 1999, 245 pp.

61. Einax, J.W., Zwanziger, H.W., and Geiß, S. *Chemometrics in Environmental Analysis.* VCH: Weinheim, 1997, 384 pp.

62. Aronova, M.A., Chang, K.S., Takeuchi, I., Jabs, H., Westerheim, D., Gonzalez-Martin, A., Kim, J., and Lewis, B. Combinatorial libraries of semiconductor gas sensors as inorganic electronic noses. *Appl. Phys. Lett.* 2003, 83, 1255–1257.

63. Wöllenstein, J., Plaza, J.A., Cane, C., Min, Y., Böttner, H., and Tuller, H.L. A novel single chip thin film metal oxide array. *Sensors Actuators* B 2003, 93, 350–355.

64. Su, M., Li, S., and Dravid, V.P. Miniaturized chemical multiplexed sensor array. *J. Am. Chem. Soc.* 2003, 125, 9930–9931.

65. Sundic, T., Marco, S., Perera, A., Pardo, A., Hahn, S., Bârsan, N., and Weimar, U. Fuzzy inference system for sensor array calibration: prediction of CO and CH_4 levels in variable humidity conditions. *Chemometrics* 2003, 64, 103–122.

66. Hierlemann, A., Schweizer-Berberich, M., Weimar, U., Kraus, G., Pfau, A., and Göpel, W. Pattern recognition and multicomponent analysis. In *Sensors*

Update Sensor Technology — Applications — Markets, Baltes, H., Göpel, W., and Hesse, J., Eds. VCH: Weinheim, 1996, Vol. 2, pp. 119–180.

67. Schierbaum, K.D., Weimar, U., and Göpel,W. Multicomponent gas analysis: an analytical approach applied to modified SnO_2 sensors. *Sensors Actuators B* 1990, *2*, 71–78.

68. Ding, J., McAvoy, T.J., Cavicchi, R.E., and Semanchik, S. Surface state trapping models for SnO_2-based microhotplate sensors. *Sensors Actuators B* 2001, *77*, 597–613.

69. Nakata, S., Hashimoto, T., and Okunishi, H. Evaluation of the responses of a semiconductor gas sensor to gaseous mixtures under the application of temperature modulation. *Analyst* 2002, *127*, 1642–1648.

70. Nakata, S., Akakabe, S., Nakasuji, M., and Yoshikawa, K. Gas sensing based on a nonlinear response: discrimination between hydrocarbons and quantifications of individual components in a gas mixture. *Anal. Chem.* 1996, *68*, 2067–2072.

71. Maziarz, W., Potempa, P., Sutor, A., and Pisarkiewicz, T. Dynamic response of a semiconductor gas sensor analysed with the help of fuzzy logic. *Thin Solid Films* 2003, *436*, 127–131.

72. Ionescu, R., Llobet, E., Vilanova, X., Brezmes, J., Sueiras, J.E., Calderer, J., and Correig, X. Quantative analysis of NO_2 in the presence of CO using a single tungsten oxide semiconductor sensor and dynamic signal processing. *Analyst* 2002, *127*, 1237–1246.

73. Heilig, A., Barsan, N., Weimar, U., Schweizer-Berberich, M., Gardner, J.W., and Göpel, W. Gas identification by modulating temperatures of SnO_2 based thick film Sensors. *Sensors Actuators B* 1997, *43*, 45–51.

74. Heberle, I., Liebminger, A., Weimar, U., and Göpel, W. Optimised sensor arrays with chromatographic preseparation: characterisation of alcoholic beverages. *Sensors Actuators B* 2000, *68*, 53–57.

75. Frank, M., Ulmer, H., Ruiz, J., Visani, P., and Weimar, U. Complementary analytical measurements based upon gas chromatography/mass spectroscopy, sensor system and human sensory panel: a case study dealing with packaging materials. *Anal. Chim. Acta* 2001, *431*, 11–29.

76. Morris, L., Caruana, D.J., and Williams, D.E. Simple system for part-per-billion-level volatile organic compound analysis in grounwater and urban air. *Meas. Sci. Technol.* 2002, *13*, 603–612.

77. Schweizer-Berberich, M., Strathmann, S., Weimar, U., Sharma, R., Seube, A., Peyre-Lavigne, A., and Göpel, W. Strategies to avoid VOC cross-sensitivity of SnO_2-based CO sensors. *Sensors Actuators B* 1999, *58*, 318–324.

78. Heule, M. and Gauckler, L.J. Gas sensors fabricated from ceramic suspensions by micromolding in capillaries. *Adv. Mater.* 2001, *13*, 1790–1793.

79. Heule, M. and Gauckler, L.J. Miniaturised arrays of tin oxide gas sensors on single microhotplate substrates fabricated by micromolding in capillaries. *Sensors Actuators B* 2003, *93*, 100–106.

80. Cerda, J., Cirera, A., Vila, A., Cornet, A., and Morante, J.R. Deposition on micromachined silicon substrates of gas sensitive layers obtained by a wet chemical route: a CO/CH_4 high performance sensor. *Thin Solid Films* 2001, *391*, 265–269.

81. Rivieri, B., Viricelle, J.-P., and Pijolat, C. Development of tin oxide material by screen-printing technology for micro-machined gas sensors. *Sensors Actuators B* 2003, *93*, 531–537.

82. Kolmakov, A., Zhang, Y., Cheng, G., and Moskovits, M. Detection of CO and O_2 using tin oxide nanowire sensors. *Adv. Mater.* 2003, *15*, 997–1000.

83. Law, M., King, H., Messer, B., Kim, F., and Yang, P. Photochemical sensing of NO_2 with SnO_2 nanoribbon nanosensors at room temperature. *Angew. Chem.* 2002, *114*, 2511–2514.

84. Zhang, D., Li, Ch., Han, S., Tang, T., and Zhou, Ch. Doping dependent NH_3 sensing of indium oxide nanowires. *Appl. Phys. Lett.* 2003, *83*, 1845–1847.

85. Comini, E., Faglia, G., Sberveglieri, G., Pan, Z., and Wang, Z.L. Stable and highly sensitive gas sesnors based on semiconducting oxide nanobelts. *Appl. Phys. Lett.* 2002, *81*, 1869–1871.

86. Müller, G., Fridberger, A., Kreisl, P., Ahlers, S., Schulz, O., and Becker, T. A MEMS toolkit for metal-oxide-based gas sensing system. *Thin Solid Films* 2003, *436*, 34–45.

87. Friedberger, A., Kreisl, P., Rose, E., Müller, G., Kühner, G., Wöllenstein, J., and Böttner, H. Micromechanical fabrication of robust low-power metal oxide gas sensors. *Sensors Actuators B* 2003, *93*, 345–349.

88. Chwieroth, B. and Patton, B.R. Conduction and gas-surface reaction modelling in metal oxide gas sensors. *J. Electroceram.* 2001, *6*, 27–41.

89. Matsunagi, N., Sakai, G., Shimanoe, K., and Yamazoe, N. Diffusion equation-based study of thin film semiconductor gas sensor-response transient. *Sensors Actuators B* 2002, *83*, 216–221.

90. Kissine, V.V., Sysoev, V.V., and Voroshilov, S.A. Conductivity of SnO_2 thin films in the presence of surface adsorbed species. *Sensors Actuators B* 2001, *79*, 163–170.

91. Boeker, P., Wallenafng, O., Horner, G. Mechanistis model of diffusion and reaction in thin sensor layers — The DIRMAS model. *Sensors Actuators B* 2002, *83*, 202–208.

92. Hossein-Babaei, F. and Orvatinia, M. Analysis of thickness dependence of the sensitivity in thin film resistive gas sensors. *Sensors Actuators B* 2003, *89*, 256–261.

93. Rothschild, A. and Komem, Y. Numerical computation of chemisorption isoterms for device modelling of semiconductor gas sensors. *Sensors Actuators B* 2003, *93*, 362–369.

94. Kamp, B., Merkle, R., and Maier, J. Chemical diffusion of oxygen in tin dioxide. *Sensors Actuators B* 2001, *77*, 534–542.

95. Arbiol, J., Corostiza, P., Cirera, A., Cornet, A., and Morante, J.R. *In situ* analysis of the conductance of SnO_2 crytsalline nanoparticles in the presence of oxidizing or reducing atmosphere by scanning tunneling microscopy. *Sensors Actuators B* 2001, *78*, 57–63.

96. Safonova, O., Bezverkhy, I., Fabrichnyi, P., Rumyantseva, M., and Gaskov, A. Mechanism of sensing CO in nitrogen by nanocrystalline SnO_2 and SnO_2 (Pd) studied by Mössbauer spectroscopy and conductance measurements. *J. Mater. Chem.* 2002, *12*, 1174–1178.

97. Maddock, A.G. *Mössbauer Spectroscopy, Principles and Applications.* Horwood: Chichester, 1997, 261 pp.

98. Lenaerts, S., Honore, M., Huyberechts, G., Roggen, J., and Maes, G. In situ infrared and electrical characterization of tin dioxide gas sensors in nitrogen/oxygen mixtures at temperatures up to 720 K. *Sensors Actuators B* 1994, *18–19*, 478–482.

99. Lenaerts, S., Roggen, J., and Maes, G. FTIR characterization of tin dioxide gas sensor materials under working conditions. *Spectrochim. Acta A — Mol. Spectrosc.* 1995, *51*, 883–894.

100. Chaudhary, V.A., Mulla, I.S., Vijayamohanan, K., Hedge, S.G., and Srinivas, D. Hydrocarbon sensing mechanism of surface ruthenated tin oxide: an in situ IR, ESR, and adsorption kinetics study. *J. Phys. Chem.* B 2001, *105*, 2565–2571.

101. Warnken, M., Lazar, K., and Wark, M. Redox behaviour of SnO_2 nanoparticles encapsulated in the pores of zeolites towards reductive gas atmospheres studied by *in situ* diffuse reflectance UV/Vis and Mössbauer spectroscopy. *Phys. Chem. Chem. Phys.* 2001, *3*, 1870–1876.

102. Emiroglu, S., Bârsan, N., Weimar, U., and Hoffmann, V. *In-situ* diffuse reflectance infrared spectroscopy study of CO adsorption on SnO_2. *Thin Solid Films* 2001, *391*, 176–185.

103. Harbeck, S., Szatvanyi, A., Bârsan, N., Weimar, U., and Hoffmann, V. DRIFT studies of thick film un-doped and Pd-doped SnO_2 sensors: temperature changes effect and CO detection mechanism in the presence of water vapour. *Thin Solid Films* 2003, *436*, 76–83.

104. Sinner-Hettenbach, M., Göthelod, M., Weissenrieder, J., von Schenk, H., Weiß, T., Bârsan, N., and Weimar, U. Oxygen-deficient SnO_2 (110): a STM, LEED, and XPS study. *Surf. Sci.* 2001, *477*, 50–58.

105. Rantala, T.S., Lantto, V., and Rantala, T.T. A cluster approach for modelling of surface characteristics of stannic oxide. *Phys. Scripta T* 1994, *T54*, 252–255.

106. Rantala, T.S., Lantto, V., and Rantala, T.T. A cluster approach for the SnO_2 (110) face. *Sensors Actuators* B 1994, *18–19*, 716–719.

107. Rantala, T.S. , Golovanov, V., and Lantto, V. A cluster approach for the adsorption of oxygen and carbon monoxide on SnO_2 and CdS surfaces. *Sensors Actuators B* 1995, *24–25*, 532–536.

108. Goniakowski, J. and Gillan, M.J. The adsorption of H_2O on TiO_2 and SnO_2(110) studied by first-principles calculations. *Surf. Sci.* 1996, *350*, 145–158.

109. Golovanov, V.V., Mäki-Jaskari, M.A., and Rantala, T.T. Semi-empirical and ab initio studies of low-temperature adsorption of oxygen and CO at (110) face of SnO_2. *IEEE Sensors J.* 2002, 2, 416–421.

110. Lantto, V., Rantala, T.T., and Rantala, T.S. Atomic understanding of semiconducting gas sensors. *J. Eur. Ceram. Soc.* 2001, *21*, 1961–1965.

111. Mäki-Jaskari, M.A. and Rantala, T.T. Theoretical study of oxygen-deficient SnO_2 (110) surfaces. *Phys. Rev. B* 2002, *65*, 245428.

112. Yamaguchi, Y., Nagasawa, Y., Tabata, K., and Suzuki, E. The interaction of oxygen with reduced SnO_2 and Ti/SnO_2 (110) surfaces: a density functional theory study. *J. Phys. Chem.* A 2002, *106*, 411–418.

113. Yamaguchi, Y., Nagasawa, Y., Shimomura, S., and Tabata, K. Reaction model for methane oxidation on reduced SnO_2 (110) surfaces. *Int. J. Quantum Chem.* 1999, *74*, 423–433.

114. Yamaguchi, Y., Tabata, K., and Suzuki, E. Density functional theory calculations for the interaction of oxygen with reduced M/SnO_2 (110) (M = Pd, Pt) surfaces. *Surf. Sci.* 2003, *526*, 149–158.

115. Slater, B., Catlow, C.R.A., Williams, D.E., and Stoneham, A.M. Dissociation of O_2 on the reduced SnO_2 (110) surface. *Chem. Commun.* 2000, 1235–1236.

116. Bates, S.P. Full-coverage adsorption of water on SnO_2 (110): the stabilization of the molecular species. *Surf. Sci.* 2002, *512*, 29–36.

117. Abee, M.W. and Cox, D.F. NH_3 chemisorption on stochiometric and oxygen-deficient SnO_2 (110) surfaces. *Surf. Sci.* 2002, *520*, 65–77.

118. Calatayud, M., Andres, J., and Beltran A. A theoretical analysis of adsorption and dissociation of CH_3OH on the stoichiometric SnO_2 (110) surface. *Surf. Sci.* 1999, *430*, 213–222.

119. Oviedo, J. and Gillan, M.J. Energetics and structure of stoichiometric SnO_2 surfaces studied by firts-principles calculations. *Surf. Sci.* 2000, *463*, 93–101.

120. Oviedo, J. and Gillan, M.J. The energetics and structure of oxygen vacancies on the SnO_2 (110) surface. *Surf. Sci.* 2000, *467*, 35–48.

121. Oviedo, J. and Gillan, M.J. First-principles study of the interaction of oxygen with the SnO_2 (110) surface. *Surf. Sci.* 2001, *490*, 221–236.

122. Rodriguez, J.A. Orbital-band interactions and the reactivity of molecules on oxide surfaces: from explanations to predictions. *Theor. Chem. Acc.* 2002, *107*, 117–129.

123. Melle-Franco, M. and Pacchioni, G. CO adsorption on SnO_2 (110): cluster and ab initio calculations. *Surf. Sci.* 2000, *461*, 54–66.

124. Golovanov, V.V. Characterisation of stannic dioxide nanosesnors: experimental and theoretical approaches. *Int. J. Mater. Prod. Technol.* 2003, *18*, 296–312.

125. Morris, L., Williams, D.E., Kaltsoyannis, N., and Tocher, D.A. Surface grafting as route to modifying the gas-sensitive resistor properties of semiconducting oxides: studies of Ru-grafted SnO_2. *Phys. Chem. Chem. Phys.* 2001, *3*, 132–145.

126. Bârsan, N., Stetter, J.R., Findlay, M., and Göpel, W. High performance gas sensing of CO: comparative tests for semiconducting (SnO_2-based) and for amperometric gas sensors. *Anal. Chem.* 1999, *71*, 2512–2517.

127. Hahn, S.H., Bârsan, N., Weimar, U., Ejakov, S.G., Visser, J.H., and Soltis, R.E. CO sensing with SnO_2 thick film sensors: role of oxygen and water vapour. *Thin Solid Films* 2003, *436*, 17–24.

128. Schmid, W., Bârsan, N., and Weimar, U. Sensing of hydrocarbons with tin oxide sensors: possible reaction path as revealed by consumption measurements. *Sensors Actuators* B 2003, *89*, 232–236.

129. Leblanc, E., Perier-Camby, L., Thomas, G., Gibert, R., Primet, M., and Celin, P. NO_x adsorption onto dehydroxylated or hydroxylated tin dioxide surface. Application to SnO_2-based sensors. *Sensors Actuators* B 2000, *62*, 67–72.

130. Morrison, S.R. *The Chemical Physics of Surfaces*, 2nd ed. Plenum Press: New York, 1990, 438 pp.

131. Nowotny, J. and Sloma, M. Work function of oxide ceramic materials. In *Surface and Near-Surface Chemistry of Oxide Materials*, Nowotny, J. and Dufour, L.-C., Eds. Elsevier: Amsterdam, 1988, pp. 281–343.

132. Kappler, J., Weimar, U., and Göpel, W. Potential-controlled gas-sensing devices. In *Advanced Gas Sensing. The Electroadsorptive Effect and Related Techniques*, Doll, T., Ed. Kluwer: Boston, 2003, pp. 55–84.

133. Göpel, W. Chemisorption and charge transfer at semiconductor surfaces: implications for designing gas sensors. *Prog. Surf. Sci.* 1985, *20*, 9–103.

134. Bârsan, N., Heilig, A., Kappler, J., Weimar, U., and Göpel, W. CO–water interaction with Pd-doped SnO_2 gas sensors: simultaneous monitoring of resistances and work functions. In *Proceedings of the Eurosensors XIII, European Conference on Solid-State Transducers*, La Hague, The Netherlands, September 12–15, 1999, pp. 367–369 (CD-Rom).

135. Geistlinger, H., Eisele, I., Flietner, B., and Winter, R. Dipole- and charge transfer contribution to the work function change of semiconducting thin films: experiment and theory. *Sensors Actuators B* 1996, *34*, 499–505.

136. Mizsei, J. and Lantto, V. Simultaneous response of work function and resistivity of some SnO_2-based samples to H_2 and H_2S. *Sensors Actuators B* 1991, *4*, 163–168.

137. Karthigeyan, A., Gupta, R.P., Schargnal, K., Burgmair, M., Zimmer, M., Sharma, S.K., and Eisele, I. NO_2 sensitivity of nanoscaled SnO_2 work function sensors. *Sensors Actuators B* 2001, *78*, 69–72.

138. Kiss, G., Josepovits, V.K., Kovacs, K., Ostrick, B., Fleisher, M., Meixner, H., and Reti F. CO sensitivity of the PtO/SnO_2 and PdO/SnO_2 layer structures: Kelvin probe and XPS analysis. *Thin Solid Films* 2003, *436*, 115–118.

139. Chang, S.C. Oxygen chemisorption on tin oxide: correlation between electrical conductivity and EPR measurements. *J. Vac. Sci. Technol.* 1980, *17*, 366–369.

140. Canevali, C., Chiodini, N., Di Nola, P., Morazzoni, F., Scotti, R., and Bianchi, C.L. Surface reactivity of SnO_2 obtained by sol–gel type condensation: interaction with inert, combustible gases, vapour-phase H_2O and air, as revealed by electron paramagnetic resonance spectroscopy. *J. Mater. Chem.* 1997, *7*, 997–1002.

141. Diéguez, A., Romano-Rodríguez, A., Morante, J.R., Bârsan, N., Weimar, U., and Göpel, W. Nondestructive assessment of the grain size distribution of SnO_2 nanoparticles by low-frequency Raman spectroscopy. *Appl. Phys. Lett.* 1997, *71*, 1957–1959.

142. Szuber, J. and Göpel, W. Photoemission studies of the electronic properties of the space charge layer of SnO_2 (110) surface. *Electron. Technol.* 2000, *33*, 261–281.

143. Cabot, A., Arbiol, J., Morante, J.R., Weimar, U., Barsan, U., and Göpel, W. Analysis of the noble metal catalytic additives introduced by impregnation of as obtained SnO_2 sol–gel nanocrystals for gas sensors. *Sensors Actuators B* 2000, *70*, 87–100.

144. Heiland, G. and Kohl, D. Physical and chemical aspects of oxidic semi-conductor gas sensors. In *Chemical Sensor Technology*, Seiyama, T., Ed. Kodansha/Elsevier: Tokyo/Amsterdam, Vol. 1, pp. 15–38.

145. Henrich, V.A. and Cox, P.A. *The Surface Science of Metal Oxides*. University Press: Cambridge, 1994, 464 pp.

146. Egashira, M. and Nakashima, M. Change of thermal desorption behaviour of adsorbed oxygen with water coadsoption on Ag^+-doped Tin (IV) oxide. *J. Chem. Soc. Chem. Commun.* 1981, 1047–1049.

147. Scott, R.W.J., Yang, S.M., Chabanis, G., Coombs, N., Williams, D.E., and Ozin, G.A. Tin dioxide opals and inverted opals: near-ideal microstructures for has sensors. *Adv. Mater.* 2001, *13*, 1468–1472.

148. Cabot, A., Vila, R., and Morante, J.R. Analysis of the catalytic activity and electrical characteristics of different modified SnO_2 layers for gas sensors. *Sensors Actuators B* 2002, *84*, 12–20.

149. Broqvist, P., Panas, I., and Persson, H. A DFT study on CO oxidation over Co_3O_4. *J. Catal.* 2002, *210*, 198–206.

150. Gurlo, A., Bârsan, N., and Weimar, U. Mechanism of NO_2 sensing on In_2O_3 thick film sensors as revealed by simultaneous consumption and resistivity measurements. In *Proceedings of the Eurosensors XVI, European Conference*

on Solid-State Transducers, Prague, Czech Republic, September 15–18, 2002, pp. 970–973 (CD-Rom).

151. Bauer, M., Bârsan, N., Ingrisch, K., Zeppenfeld, A., Denk, I., Schuman, B., Weimar, U., and Göpel, W. Influence of measuring voltage, geometry and electrodes on the characteristics of thick film SnO_2 gas sensors. In *Proceedings of the 11th European Microelectronics Conference*, Venice, Italy, May 14–16, 1997, pp. 37–44.

152. Schweizer-Berberich, M., Bârsan, N., Weimar, U., Morante, J.R., and Göpel, W. Electrode effects on gas sensing properties of nanocrystalline SnO_2 gas sensors. In *Proceedings of the Eurosensors XI, European on Solid-State Transducers*, Warsaw, Poland, September 21–24, 1997, pp. 1377–1380.

153. Montmeat, P., Lalauze, R., Viricelle, J.-P., Tornier, G., and Pijolat, C. Influence of SnO_2 thick film thickness on the detection properties. In *Proceedings of the Eurosensors XVI, European Conference on Solid-State Transducers*, Prague, Czech Republic, September 15–18, 2002, pp. 1116–1119.

154. Sears, W.M., Colbow, K., and Consadori, F. Algorithms to improve the selectivity of thermally-cycled tin oxide gas sensors. *Sensors Actuators* 1989, *19*, 333–335.

155. Kunt, T.A., McAvoy, T.J., Cavicchi, R.E., and Semancik, S. Optimization of temperature programmed sensing for gas identification using micro-hotplate sensors. *Sensors Actuators* B 1998, *53*, 24–43.

156. Hammond, J.W. and Liu, C.-C. Silicon based microfabricated tin oxide gas sensor incorporating use of Hall effect measurement. *Sensors Actuators B* 2001, *81*, 25–31.

157. Faglia, G., Comini, E., Sberveglieri, G., Rella, R., Siciano, P., and Vasanelli, L. Square and collinear four probe array and Hall measurements on metal oxide thin film gas sensors. *Sensors Actuators B* 1998, *53*, 69–75.

158. Ionescu, R. Combined Seebeck and resistive SnO_2 gas sensor, a new selective device. *Sensors Actuators B* 1998, *48*, 392–394.

159. Doll, T., Ed. *Advanced Gas Sensing. The Electroadsorptive Effect and Related Techniques*. Kluwer: Boston, 2003, 202 pp.

160. Stetter, J.R., Penrose, W.R., and Yao, S. Sensors, chemical sensors, electrochemical sensors, and ECS. *J. Electrochem. Soc.* 2003, *150*, S11–S16.

161. Sberveglieri, G. Classical and novel techniques for the preparation of SnO_2 thin-film gas sensors. *Sensors Actuators B* 1992, *6*, 239–247.

162. Kappler, J., Tomescu, A., Barsan, N. and Weimar, U., Co Consumption of Pd doped SnO_2 based sensors, Thin Solid Films 2001, *391*, 186–191.

23 Fuel Electrodes for Solid Oxide Fuel Cells

S.W. Tao and J.T.S. Irvine
School of Chemistry, University of St Andrews, Scotland, UK

CONTENTS

23.1 INTRODUCTION

Fuel cell technology has been much heralded in recent years as a keystone of the future energy economy. In association with the hydrogen economy, it has been strongly promoted by the governments of most of the world's leading industrialized nations. It is quite likely that the fuel cell's impact on society will be revolutionary. In the long term, fuel cells will be an essential component of any hydrogen or similar clean energy economy, in the short term they promise enhanced conversion efficiencies of more conventional fuels and will deliver large reductions in CO_2 emissions.

Fuel cells can be viewed as devices for electrochemically converting chemical fuels into electricity, essentially batteries with external fuel supplies. Fuel cells

739

offer extremely high chemical to electrical conversion efficiencies due to the absence of the Carnot limitation, and further energy gains can be achieved when the produced heat is utilized in combined heat and power or gas turbine applications. Furthermore, the technology does not produce significant amounts of pollutants such as nitrogen oxides especially when compared with internal combustion engines.

All fuel cells consist of essentially four components: the electrolyte, the air electrode, the fuel electrode, and the interconnect. The solid oxide fuel cell (SOFC) is one of the most exciting systems for future power generation due to its potential fuel flexibility and very high efficiency. In solid oxide fuel cells, which typically operate at temperatures in the range 800 to 950°C, the electrolyte is normally yttria stabilized zirconia (YSZ), which offers good oxide ion transport, while blocking electronic transport. The electrolyte's function is to allow the transport of oxide ions from the electrolyte's interface with the air electrode to its interface with the fuel electrode. The function of the air electrode, which is typically a composite of lanthanum strontium manganese oxide with YSZ, is to facilitate the reduction of oxygen molecules to oxide ions transporting electrons to the electrode/electrolyte interface and allowing gas diffusion to and from this interface. The function of the fuel electrode, which is the main subject of this chapter, is to facilitate the oxidation of the fuel and the transport of electrons from the electrolyte to the fuel electrode interface. In addition, it must allow diffusion of fuel gas to this interface and exhaust gases away from this interface. An appropriate microstructure for the SOFC, with particular attention to the electrode–electrolyte interfaces with a schematic of the electrochemical processes is shown in Figure 23.1. The other element is the interconnect, which has traditionally been lanthanum strontium chromite, but in lower temperature variants attention has been focused on corrosion resistant metallic alloys. The functions of this component are to transfer electrons between individual cells in the stack, while preventing gas crossover between the fuel and oxidant streams.

So far, most SOFC development has been based on YSZ electrolytes due to its high oxygen ion conductivity, good stability under SOFC operating conditions, and high mechanical strength. Although some other electrolyte materials, such as $Ce_{0.9}Gd_{0.1}O_{2-\delta}$ (CGO) [1] and $La_{0.85}Sr_{0.15}Ga_{0.9}Mg_{0.1}O_{3-\delta}$ (LSGM) [2] have been proposed as alternative electrolyte materials, there are still some problems for those materials. For example, gadolinia-doped ceria (CGO) exhibits significant n-type electronic conduction at low oxygen partial pressure (pO_2) above 600°C, which limits its application temperature range. Ceria-based materials may be used for intermediate temperature fuel cells below 600°C with the application of thin film electrolytes; however, the electrode process is rather slow at such a low temperature, which makes it difficult to find a suitable electrode materials. On the other hand, the mechanical strength of CGO is also lower than that of YSZ, which makes it difficult in application as a thin film electrolyte unless strong anode/cathode/interconnector supports are used. LSGM is not very stable under the SOFC operating conditions because of the Ga depletion under reducing atmospheres and the phase segregation at high temperatures [3,4]. In addition, it is not

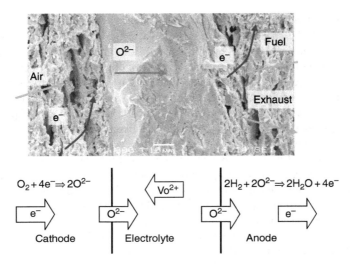

FIGURE 23.1 Schematic of processes occurring within the air electrode, electrolyte, and fuel electrode and at their interfaces, superimposed upon an appropriate microstructure for such an electrode/electrolyte structure (From Atkinson, A., Barnett, S., Gorte, R.J., Irvine, J.T.S., McEvoy, A.J., Mogensen, M., Singhal, C., and Vohs, J.M. *Nat. Mater.* 2004, *3*, 17–27. With permission.)

cost-effective due to the expensive gallium oxide used. Therefore, at this moment, most research is still concentrated on the SOFC system using YSZ as electrolyte. Although all ceramic cathodes $La_{1-x}Sr_xMnO_3$ or $La_{1-x}Sr_xCo_{1-y}Fe_yO_3$ with or without YSZ addition currently function well as SOFC cathodes [5,6], an ideal all ceramic anode material is not available. The materials and catalytic properties for SOFC anode is briefly reviewed in this chapter.

23.2 Choice of Materials for SOFC Anode

The materials selection for an SOFC anode is determined by a number of factors. First, the function required of it as the site for the electrochemical oxidation of the fuel associated with charge transfer to a conducting contact. Second, the environment in which it operates, at high temperature in contact not only with the fuel, including possible impurities and increasing concentrations of oxidation products, but also with other materials, electrolyte, and contact components of the cell, and all this with stability over an adequate commercial lifetime at high efficiency. Third, the processability of the anode, which must be such that an open but well connected framework can be achieved and retained during the fabrication of the fuel cell. While in normal operation the ambient pO_2 is low, it can vary over several orders of magnitude, and to accommodate fault conditions, or even just to provide flexibility of operating parameters, the ability to recover even after brief exposure to air at high temperature would be advantageous. A further aspect of this stability is the maintenance of structural integrity over the whole temperature range to which the component is exposed, from the sintering temperature during fabrication through normal operating conditions and then, repeatedly, cycling down to ambient temperature. Compatibility with other cell component materials implies an absence of solid-state contact reactions, involving interdiffusion of constituent elements of those materials, or formation of reaction product layers, which would interfere with anode functionality. It also requires a match of properties, such as shrinkage during sintering and thermal expansivity to minimize stresses during temperature variations due to operating procedures, start-up, and shutdown.

By definition of its role, it is a requirement that the anode material should be an adequate electronic conductor, and be electrocatalytically sufficiently active to sustain high current density with low overpotential loss. However, the catalytic behavior of anode materials should not extend to the promotion of unwanted side reactions, hydrocarbon pyrolysis followed by deposition of vitreous carbon being an example. An intimate contact between the two solid phases, the electrolyte delivering the oxide ions and the anode on which they are electrically neutralized, is clearly essential, as is access of the fuel and removal of reaction products, these being in the gas phase. In this model, the reaction is therefore sited on a "three-phase boundary" zone. Low loss operation implies that the three-phase boundary is not dimensionally limited to a planar interface of solid materials, but that it be delocalized to provide a "volumetric" reaction region in three dimensions, porous for gas diffusion and permitting both electron and ion transport. One option is to provide a single-phase electrode with mixed conductivity permitting both oxide ion and electron mobility within the anode material. The alternative is to employ a porous composite, as in the nickel-based cermets that have typically been used in SOFCs to date.

23.3 Fuels for High Temperature Fuel Cells

Probably the most important problem to be addressed in the development of the first and succeeding generations of SOFCs is the fuel electrode. While hydrogen would seem the fuel of choice, it is normally derived from hydrocarbon sources. Hydrogen derived from renewable resources would be an appropriate fuel; however, its availability and distribution are very limited. Indeed, there is little environmental benefit in diverting current renewable capacity to hydrogen generation, if the gap created is filled by fossil fuel generated electricity. The only truly clean way to produce hydrogen is via the creation of additional renewable generation capacity.

In the short to medium term, therefore, it is appropriate to look at hydrocarbon fuels, with the environmental benefit arising from the much-improved efficiency of fuel utilization. Natural gas, for example; North Sea gas: CH_4 (95%), C_2H_6 (4%), C_4H_{10} (0.2%), and S (4 ppm), is clearly the most abundant and best-distributed fuel. The development of SOFC anodes capable of operating in natural gas, without suffering from carbon build up due to catalytic cracking, is still far from being achieved.

23.4 CATALYTIC PROCESS AT THE ANODE

23.4.1 Reforming

The reforming Reaction (23.1) is generally associated with a following water gas shift equilibrium Reaction (23.2), additionally methane may be reformed with CO_2 (23.3). One of the most studied reactions is steam methane reforming (23.1).

Steam reforming of methane is highly endothermic and typically occurs at temperatures above 500°C in the presence of suitable catalyst. Two approaches have been developed,

$$CH_4 + H_2O \Leftrightarrow CO + 3H_2 \qquad \Delta H = 206 \text{ kJ mol}^{-1} \qquad (23.1)$$

$$CO + H_2O \Leftrightarrow CO_2 + H_2 \qquad \Delta H = -41 \text{ kJ mol}^{-1} \qquad (23.2)$$

$$CH_4 + CO_2 \Leftrightarrow 2CO + 2H_2O \quad \Delta H = 247 \text{ kJ mol}^{-1} \qquad (23.3)$$

external and internal reforming. In the first case, the reaction occurs in a separate reactor, which consists of heated tubes filled with nickel or noble metal [7]. To obtain high methane conversion and to avoid carbon deposition by methane cracking (23.4) or by the Boudouard Reaction (23.5), steam is introduced in excess. Typical conversion efficiencies for external reforming systems are ~60%. The cracking reaction is even more problematic with the higher hydrocarbons, which also are present in natural gas. An alternative process, partial oxidation (POX) is occasionally used to drive reforming, (23.6), especially when higher hydrocarbons are used as fuel.

$$CH_4 \Leftrightarrow C + 2H_2 \qquad \Delta H = 75 \text{ kJ mol}^{-1} \qquad (23.4)$$

$$2CO \Leftrightarrow C + CO_2 \qquad \Delta H = 86 \text{ kJ mol}^{-1} \qquad (23.5)$$

$$CH_4 + O_2 \Leftrightarrow CO_2 + 2H_2 \quad \Delta H = -320 \text{ kJ mol}^{-1} \qquad (23.6)$$

Internal reforming is a particularly promising concept, where steam reforming of methane takes place directly in the anode compartment. This process not only has no requirement for additional components in the system, but also allows better heat exchange between the endothermic reforming reaction and the exothermic electrochemical reaction within the stack. However, two major problems occur: the risk of carbon deposition and the creation of a thermal gradient that induces stress in the cell. As an example, with the Ni/YSZ cermet, generally used as anode in SOFC, a steam-methane ratio >2 is required to avoid carbon deposition. Several alternatives have been proposed: the use of some prereforming, the distribution of the reforming reaction and the development of new materials.

Such reforming reactions typically occur over Ni or Pt, for $T > 500$°C. These reactions are generally thought to be favored by low pressures, high temperatures, and by large steam partial pressures. The mechanism of the steam methane reforming has been intensively studied; however, the mechanism and the reaction rate depend on the catalyst used (preparation procedure, grain size, surface area) and on the experimental conditions.

In general, the mechanism of steam reforming is described as a succession of elementary steps [8]:

$$CH_4 + ns \rightarrow CH_x - ns + \frac{4-x}{2}H_2 \qquad (23.7)$$

$$H_2O + s \rightarrow O - s + H_2 \qquad (23.8)$$

$$CH_x - ns + O - s \rightarrow CO + \frac{x}{2} + (n+1)s \qquad (23.9)$$

where s is an adsorption site on the catalyst and n the number of sites.

According to Froment [9], hydrocarbon conversion begins with two dissociative adsorption steps. Adsorbed oxygen reacts with methyl species to produce CO or CO_2 by the gas shift reaction. Without insertion of oxygen into the adsorbed radicals, dehydrogenation of methane occurs until carbon is formed. Several parameters influence the mechanism: temperature, steam to methane ratio, reactant, or product partial pressure; however, there are some disagreements in literature. As an example, with the Ni/YSZ cermet, the influence of steam pressure is not fully understood. Achenbach and Riensche [10] found that the rate is not dependent upon steam pressure whereas other studies [11–14] concluded that the rate depends on steam partial pressure. In some cases [11,15], there seems to be a competitive adsorption between methane and steam; the reforming reaction rate has a maximum as a function of steam-methane ratio. Others studies [10,16,17], however show that the rate is not dependent upon steam-methane ratio. The reforming reaction seems to be enhanced by hydrogen partial pressure [11] and with a Ni/Cu alloy, the carbon deposition is enhanced by increase of the total pressure [18]. It seems that methane cracking depends on the dehydrogenation step and on the activities of carbonyl species adsorbed on the catalyst surface [19,20].

To avoid carbon deposition without using excess steam, some authors have proposed to decrease the catalytic activity of nickel. Morimoto and Shimotsu [21] have studied the influence of adding iron to the Ni/YSZ cermet. The iron achieves a decrease in the catalytic activity of nickel and adjusts the thermal expansion coefficient closer to YSZ. Alstrup et al. [22] also added copper to control the catalytic activity of nickel.

Other materials have been studied as catalysts for steam methane reforming. Platinum, rhodium, and ruthenium have high catalytic activity for methane reforming. Suzuki et al. [23] studied the Ru/YSZ cermet. They observed that ruthenium had a great activity and a high resistance to carbon deposition. Some promising results have been obtained with ruthenium inserted into strontium doped lanthanum chromite [24].

A new concept has been proposed, referred to as gradual internal methane reforming [25]. It is based on a local coupling between the steam reforming and the electrochemical oxidation of hydrogen. The steam reforming produces hydrogen,

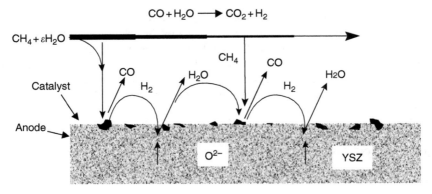

FIGURE 23.2 Schematic diagram illustrating series of processes that lead to the gradual internal methane reforming, where ε is significantly <2 (From Vernoux, P., Guindet, J., and Kleitz, M. *J. Electrochem. Soc.* 1998, *145*, 3487–3492. With permission.)

which is electrochemically oxidized to produce water; in turn, this is used to reform the excess methane. Thus, the reaction is distributed over the entire anode surface. In this concept, as for internal steam reforming, the anode should have catalytic activity for steam reforming and for hydrogen oxidation. Moreover, with gradual internal methane reforming, the catalytic activity of the anode should be carefully controlled and should not be too high to allow the distribution of the reaction, as seen in Figure 23.2.

Besides steam reforming, CO_2 reforming could play an important role at the anode for a direct hydrocarbon SOFC.

$$CH_4 + CO_2 \Leftrightarrow 2CO + 2H_2 \quad \Delta H = 247.05 \text{ kJ mol}^{-1} \tag{23.10}$$

Only few reports were found about CO_2 reforming of methane on under SOFC conditions [26]. The yielded CO and H_2 may be further oxidized by O^{2-} ions.

23.4.2 Direct Oxidation

One of the most exciting possibilities is direct hydrocarbon oxidation. Methane may be partially or completely oxidized according to Reactions (23.11) and (23.12) depending on the methane to oxygen ratio, the temperature and catalytic property of the anode. Direct complete oxidation of methane, Reaction (23.12) has the thermodynamic possibility of 99.2% conversion efficiency.

$$2CH_4 + O_2 \Leftrightarrow 2CO + 4H_2 \qquad \Delta H = -71.86 \text{ kJ mol}^{-1} \tag{23.11}$$

$$CH_4 + 2O_2 \Leftrightarrow CO_2 + 2H_2O \qquad \Delta H = -802 \text{ kJ mol}^{-1} \tag{23.12}$$

$$CH_4 + 4O^{2-} \Leftrightarrow CO_2 + 2H_2O + 8e' \quad \text{Electrochemical} \tag{23.13}$$

If this reaction is to be achieved, it is necessary to avoid or inhibit methane cracking. There is also considerable controversy as to whether the reaction is actually a single step, that is, deep oxidation, or if it involves a series of intermediate reactions with the net result shown in Reaction (23.13). An alternative viewpoint is that direct utilization of methane occurs, whereby the overall process described in Equation (23.12) occurs but in a multistep process. These possibilities are discussed in more detail later in the text.

23.5 SOFC ANODE MATERIALS

23.5.1 Ni-YSZ Anode Materials

Graphite, iron oxide, platinum group, and transition metals have been tested as SOFC anode material [27,28]. Graphite is corroded electrochemically and may be regarded as a fuel for carbon fuel cell rather than the anode. Huge polarization resistance was observed when pure platinum was tested as the SOFC anode [29]. As for the transition metals, iron may be oxidized to oxide under polarization with increasing pO_2 when oxygen is pumped from the cathode side. Cobalt is somewhat more stable and more expensive. Nickel has a significant thermal expansion mismatch to stabilized zirconia, and at high temperatures, the metal aggregates by grain growth, finally obstructing the porosity of the anode and eliminating the three-phase boundaries required for cell operation. An ideal pure metal anode for SOFC has not been and is unlikely to be identified from the point of ionic conduction, which is desired for a good anode material. The introduction of the nickel–zirconia cermet anode by Spacil in the 1960s [30] to some extent solved the thermal mismatch and sintering of nickel as well as the oxygen ionic conduction problems. This 40-year-old composite is inexpensive and exhibit high catalytic activity for hydrogen oxidation [31] and is the current standard anode material for SOFCs. The major disadvantage of the Ni cermet electrode arises from the promotion of competitive catalytic cracking of hydrocarbon reactions [32]. The rapid deposition of carbon at Ni cermets means that direct oxidation of methane is not technically viable in Ni-containing solid oxide fuel cells. In order to utilize natural gas as fuel, the fuel needs to be externally or internally reformed. The poor redox stability of the nickel component may cause cracking of the stack on cycling.

23.5.2 Cu–Zirconia and Cu–Ceria Anode Materials

It was recently reported that the cermet Cu–ceria might be used as a SOFC anode material for direct oxidation of hydrocarbons without carbon deposition, indicating that copper-containing materials might be suitable catalysts for the complete oxidation of hydrocarbon under fuel cell operation [33,34]. In a later report [35], the authors claimed that the slight carbon deposition onto Cu–ceria anode when butane was applied as the fuel might connect the copper particles together, which generates a new conductive path for the anode resulting in increased cell performance (Figure 23.3). The Cu–ceria cermet anode works well when the operating

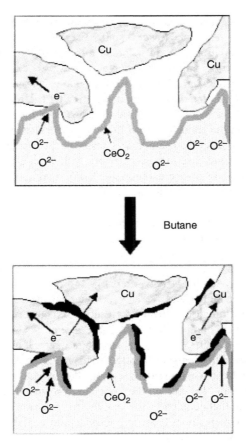

FIGURE 23.3 Schematic diagram of Cu/CeO$_2$ YSZ (or SDC) microstructure showing enhanced current collection following tar deposition after initial exposure to butane under operating conditions (From McIntosh, S., Vohs, J.M., and Gorte, R.J. *J. Electrochem. Soc.* 2003, *150*, A470–A476. With permission.)

temperature is below 800°C. However, the possible sintering of the Cu particles could be a potential problem because of its relatively low melting point (1085°C) during long-term operation at high temperatures. Kiratzis et al. [36] have found that copper was mobile with considerable agglomeration of metallic copper particles when operating the Cu/Y$_{0.2}$Ti$_{0.18}$Zr$_{0.62}$O$_{1.9}$ (YZT) cermet anode above 700°C. In some instances, there was a total segregation of copper from YZT resulting in a copper layer forming at the electrolyte interface with the outer layer of the electrode being essentially YZT. This agglomeration and migration of copper led to a significant degradation in electrochemical performance with large increases in the series resistance and polarization resistance, especially under anodic bias. To avoid the carbon deposition over Ni-YSZ and anode and high temperature sintering of copper particles, some work on Ni-ceria has been done [37]. When a

$Ni/Ce_{0.8}Sm_{0.2}O_{1.9}$ cermet was used as an SOFC anode with 90% H_2/10% H_2O as the fuel, an interface resistance lower than 0.1 Ω cm^2 was obtained below 700°C by using a coprecipitation method to prepare the anode precursor and cofiring it with the electrolyte to improve the contact. However, carbon deposition was observed on the Ni/CGO anode when wet methane was used as the fuel [38,39]. Therefore, the Ni/CGO anode is not suitable for SOFCs using hydrocarbons as fuel. To decrease the activity of nickel for methane cracking, nickel was replaced by a Ni–Cu alloy in the anode cermet [40]. The performance of a fuel cell made with a $Cu_{0.8}Ni_{0.2}$/YSZ cermet anode was tested in dry methane for 500 h and showed a significant increase in power density with time, although small carbon deposits are formed. In a recent report [41], no significant carbon deposition was observed when operating $Ni_{0.52}Cu_{048}/Ce_{0.9}Gd_{0.1}O_{1.95}$ cermet anode in dry methane at 800°C for ∼40 h although the nickel content in the alloy is higher than the $Cu_{0.8}Ni_{0.2}$/YSZ anode. A possible reason is that ceria plays an important role in avoiding carbon deposition.

The influence of adding nickel to $(Ce,Zr)O_2$ or $(Ce,Y,Zr)O_2$ to form a cermet was also studied [42] and cerium oxide is reported as a promising support. In a series of publications, Gorte and coworkers [43,44] reported that noble metals (Pd, Pt, and Rh) had much higher specific rates for water gas shift, steam reforming, and carbon dioxide reforming of methane when supported on cerium oxide than when supported on silica or alumina. Ramirez-Cabrera et al. [45] studied gadolinium-doped ceria, $Ce_{0.9}Gd_{0.1}O_{1.95}$ (CGO) as an anode material at 900°C in 5% CH_4 with steam/methane ratios between 0 and 5.5. This material was resistant to carbon deposition, although the reaction rate was controlled by slow methane adsorption. Marina and Mogensen [46] studied the same catalyst for methane oxidation and it was demonstrated that ceria had a low activity for methane oxidation but a high resistance to carbon deposition. It was proposed to add a catalyst (Ni, Rh, and Ru) to break the C–H bond more easily.

In contrast to the reported inactivity of acceptor-doped ceria, undoped ceria has been reported to be quite an effective oxidation catalyst; this can be attributed to differing electronic functionalities of the doped and undoped materials. One possibility is that the higher concentration of oxygen vacancies makes the doped material more difficult to reduce and hence a less effective catalyst. Care must be taken, however, in comparing activities in oxidizing conditions as frequently encountered in catalysis experiments and under fuel conditions, which are quite reducing. Ceria is a well-known oxidation catalyst and increases the activity of the anode for the electrochemical oxidation of methane. This approach still requires the operating temperature to be maintained below 700°C, in order to suppress carbon deposition reactions, which take place on Ni. Another approach being used in the development of cermet anodes that allow the use of fairly dry hydrocarbon, is to use a relatively inert metal such as Cu for electrical conductivity and a metal oxide to provide catalytic activity and ionic conductivity as described previously.

Although the metal/oxide cermet anode materials exhibit good performance using either hydrogen or methane as the fuel, the sintering of the alloy particles is still a potential problem. On the other hand, for some applications such as portable

SOFCs or SOFC driven engines, a redox stable anode is required to keep the fuel cell system robust. The volume change of the cermet anode during redox cycling will cause the degradation of the stack. Therefore, redox stable anodes are still in high demand. The development of redox stable anode for SOFCs is briefly reviewed in Section 23.5.3.

23.5.3 Alternative Anode Materials

The basic properties of the "standard" Ni/zirconia or Ni/ceria cermet (and to some extent by analogy for Cu cermets) are well established, and for these materials the requirement is to optimize their properties through control of microstructure and minor additives. The situation is different for oxide anodes where the requirement is still at the stage of identifying suitable candidates. Oxides are under investigation as single-phase anodes, single-phase anode current collectors, and components of composites for either of these functions. In addition to electrochemical performance, the candidate materials need to display other characteristic properties, some being essential and some desirable. These properties include electronic conductivity; oxygen diffusivity (ionic conductivity); oxygen surface exchange (reactivity); chemical stability and compatibility; compatible thermal expansion; mechanical strength; and dimensional stability under redox cycling. For application as a current collector, electrochemical activity is not required and therefore oxygen diffusivity and surface exchange properties are not relevant. In this section, we briefly review what is known about these properties for leading candidate materials and identify areas where further work is required. The materials of interest at this stage are ceria (doped and undoped) and transition metal perovskite and fluorite-related structures. From the viewpoint of chemical stability under reducing conditions, oxides containing large amounts of Co and Ni are probably not viable and so the emphasis is on Fe, Mn, Cr, and Ti as transition metal ions that can be used to give electronic conductivity. The relevant properties of some typical candidate oxide anode materials are summarized in Table 23.1.

The target for electronic conductivity for anode materials is often set to be $100 \, S \, cm^{-1}$, but the actual requirement depends on the cell design and particularly the length of the current path to the current collection locations. Thus, this could be relaxed to as low as $1 \, S \, cm^{-1}$ for a well-distributed current collection. Similarly, if the material were used as a porous support and current collector with thickness of 0.5 mm it would also need conductivity greater than approximately $1 \, S \, cm^{-1}$ to maintain losses below $0.1 \, \Omega \, cm^2$. Since these are targets for porous structures, the actual requirement for the intrinsic materials properties, that is, in dense form, would need to be greater by about 1 order of magnitude.

23.5.3.1 CeO$_2$-based mixed conductors

Among the mixed conductors with fluorite structure, rare earth doped ceria exhibits the highest ionic and electronic conductivity; however, its electronic conductivity is still not high enough to match SOFC anode requirements. Up to 70 mol% Nb_2O_5

TABLE 23.1
Comparison of Different Anode Materials for SOFCs

Structure	Typical materials	Stability in reducing atmosphere	Ionic conductivity	Electronic conductivity	Chemical compatibility with YSZ	Thermal compatibility with YSZ	Performance using H_2 as fuel	Performance using CH_4 as fuel	Redox stability
Mixture	Ni-YSZ	✓	✓	✓	✓	✓	✓	×	×
Mixture	Cu-YSZ	?	✓	✓	✓	✓	✓	✓	×
Fluorite	YZT, ScYZT CGO	✓	✓	×	✓	✓	Ok	Ok	✓
Perovskite	$La_{1-x}Sr_xCr_{1-y}TM_yO_3$	✓	?	✓	✓	✓	✓	✓	✓
Pyrochlore	Gd_2TiMoO_7	×	Ok	✓	✓	?	?	?	×
Tungsten bronze	$Sr_{0.6}Ti_{0.2}Nb_{0.8}O_3$	✓	×	✓	✓	?	×	?	✓
Monoclinic S.G. C2/m	Nb_2TiO_7	✓	×	✓	✓	×	?	?	×

Source: From Tao, S.W. and Irvine, J.T.S. *Chem. Rec.* 2004, 4, 83–95. With permission.

may be dissolved into the CeO_2 lattice to form a solid solution. The conductivity of $(Nb_2O_5)_{0.4}(CeO_2)_{0.6}$ is about 2.7 S cm^{-1} at 905°C in a reduced atmosphere. The Nb_2O_5–CeO_2 solid solution might be a potential anode material but the thermal expansion may not match that of YSZ under reduction [47]. The conductivity of $Ce_{0.8}Gd_{0.2}O_{2-\delta}$ is about 8.6×10^{-2} and 0.16 S cm^{-1}, respectively in air and at an pO_2 of 10^{-18} atm at 800°C [48]. $Ce_{0.6}Gd_{0.4}O_{2-\delta}$ has been tested as an SOFC anode and good performance was achieved when H_2 was used as fuel but not for CH_4 [49]; however, it was found that deposition of a thin layer of $(Y_2O_3)_{0.15}(CeO_2)_{0.85}$ between the Ni-YSZ anode and YSZ electrolyte can significantly improve the anode performance when wet methane was used as a fuel [50]. Therefore, doped ceria is a good material for the functional layer between anode and electrolyte. Although doped ceria itself is not good enough as an SOFC anode, it may be mixed with copper or a Cu–Ni alloy to form a cermet, which exhibits good performance using either hydrogen or hydrocarbon as a fuel, as reviewed earlier. The Cu–Ni/Ceria cermet is a good anode disregarding the redox stability.

23.5.3.2 Perovskite anode materials

The perovskite oxide formula can be written as ABO_3 where A is a large cation with a coordination number of 12 and B is a small cation with a coordination number 6 (Figure 23.4) [51]. These materials can accommodate a large content of oxygen vacancies; therefore, some perovskites are good oxygen ionic conductors. The small B-site in the perovskite allows first row transition elements to be introduced in the lattice. These elements exhibit multivalence under different conditions, which may be the source of high electronic conductivity. Good ionic and mixed conductivity is found in several perovskite oxides.

Materials with perovskite structure such as Sr- and Mg-doped $LaGaO_3$ (LSGM), exhibit very high oxygen ionic conductivity [52–54]. Perovskites with transition elements at the B-site also exhibit high electronic or mixed conduction. An early paper has reviewed the p-type perovskite materials for SOFC and other applications [55]. Mixed conducting perovskites such as $La_{1-x}Sr_xMnO_3$ with modest oxide ionic conductivity or $La_{1-x}Sr_xCo_{1-x}Fe_xO_3$ with quite high oxide ionic conductivity have been used as SOFC cathode materials [56]. $La_{1-x}M_xCrO_3$ (M = Ca and Sr), a purely electronic conductor has been widely used as the interconnector for SOFCs [5].

Perovskites have also been widely investigated as potential SOFC anode materials. Among these materials, chromites and titanates are the most promising SOFC anode materials [57,58]. Interesting results have been obtained with lanthanum strontium titanates [59] and especially cerium doped lanthanum strontium titanate [60]; however, it is now thought that the cerium doped anodes are in fact two-phase consisting of a ceria/perovskite assemblage [60]. The structure of $La_{1-x}Sr_xTiO_3$ can be described as perovskite slabs joined by crystallographic shears where the characteristic excess oxygen of these mixed oxides is accommodated [61]. $La_2Sr_4Ti_6O_{19-\delta}$ has been investigated as a potential anode for fuel cells due to the high total conductivity found under reducing conditions. This

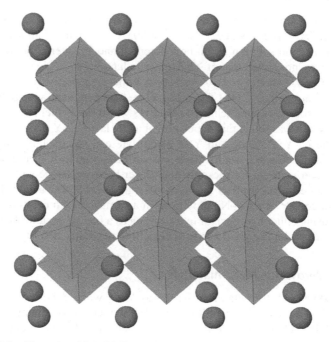

FIGURE 23.4 The perovskite, ABO_3 structure (From Tao, S.W. and Irvine, J.T.S. *Chem. Rec.* 2004, *4*, 83–95. With permission.)

mixed oxide is the $n = 12$ member of the excess oxygen perovskite-related family $La_4Sr_{n-4}Ti_nO_{3n+2}$. Under the most reducing conditions, metallic conductivity, for example 60 S cm^{-1}, is observed and under slightly higher pO$_2$, for example, wet hydrogen, a metal to insulator transition is observed. In addition, initial fuel cell tests were carried out to check the performance of $La_2Sr_4Ti_6O_{19-\delta}$ as an anode for fuel cells. Using $La_2Sr_4Ti_6O_{19-\delta}$ as an anode the polarization resistance (R_p) varies from 2.97 Ω cm^2 at 900°C in wet H$_2$ to 8.93 Ω cm^2 at 900°C operating in wet CH$_4$. A current value of 119 mA cm^2 at 600 mV was found, whereas the maximum power density was 76 mA cm^2 measured in wet H$_2$ at 900°C [61].

It was also reported that Y-doped SrTiO$_3$ exhibits high electrical conduction under the SOFC anodic conditions [62–64]. For example, the optimized composition of $Sr_{0.86}Y_{0.08}TiO_{3-\delta}$ exhibits a conductivity of 82 S cm^{-1} at a pO$_2$ of 10^{-19} atm at 800°C. However, the sample was pre-reduced in pure argon or 7% H$_2$/Ar at 1400°C before conductivity measurements. It is supposed that the conductivity of the materials would be significantly lower if the sample was only reduced below 1000°C in this case less Ti^{4+} would be reduced to Ti^{3+}, which is the source of the high electronic conductivity. The currently used $La_{1-x}Sr_xMnO_3$ or $La_{1-x}Sr_xCo_{1-y}Fe_yO_3$ cathodes will decompose in a reducing atmosphere at 1400°C. The high temperature prereduction process for such titanates makes it difficult to cofire the anode and the cathode. The conductivity of

$Sr_{0.86}Y_{0.08}Ti_{0.9}Sc_{0.1}O_3$ is only about 1 to 2 S cm^{-1} when reduced *in situ* in 5% H$_2$ at 900°C [65]. No phase changes were found for a mixture of Y-doped SrTiO$_3$ (SYT) with YSZ or LSGM on calcining at 1400°C for 10 h indicating good chemical compatibility between the SYT and electrolyte materials. The conductivity of SrTiO$_3$ in a reducing atmosphere can also be improved by replacing titanium with some niobium. For charge compensation, the strontium content at the A-site should decrease. The electrical conductivity of $Sr_{1-x}Ti_{1-x/2}Nb_xO_{3-\delta}$ ($x \leq 0.4$) was investigated [66]. Good electrical conductivity was observed on reduction in low pO$_2$, with a maximum for the sample with $x = 0.25$, $\sigma = 5.6$ S cm^{-1} at 930°C (pO$_2 = 10^{-18}$ atm).

Although SrVO$_3$ shows excellent electronic conductivity of 1000 S cm^{-1} at 800°C and an pO$_2$ of 10^{-20} atm, it is unstable under a more oxidizing atmosphere [67]. The conductivity in a reducing atmosphere may drop rapidly if strontium at A-site is partially or completely replaced by lanthanum, and the stability in oxidizing condition is not improved. A material that has both adequate high temperature conductivity in reducing atmosphere and redox stability has not been found in the valadates. In addition, the reducing process of SrVO$_3$ is rather slow once the material is oxidized.

The perovskite oxide $La_{0.6}Sr_{0.4}Co_{0.2}Fe_{0.8}O_3$ was proposed as an SOFC anode at intermediate temperatures (550 to 700°C). The stability of these materials under the anodic atmosphere is in doubt, however, even at such low temperatures. $SrFeCo_{0.5}O_x$ exhibits both high electronic and ionic conductivities in air and is applicable for ceramic membrane used for gas separation [69]. It was reported that mixed conductors $SrFeCo_{0.5}O_x$, $SrCo_{0.8}Fe_{0.2}O_{3-\delta}$, and $La_{0.6}Sr_{0.4}Fe_{0.8}Co_{0.2}O_{3-\delta}$ might be used as SOFC anode materials. Although there remain some questions about the exact structure of $SrFeCo_{0.5}O_x$, these are generally thought to be related perovskite/brownmillerite intergrowths of the Grenier type [70]. The performance was not ideal when using only these mixed conductors as an anode; however, the performance was improved when these mixed conductors were used as an interlayer between Ni-YSZ anode and YSZ electrolyte. The long-term stability would again be a problem because $SrFeCo_{0.5}O_x$ is unstable under anodic conditions [71].

As stated previously, LaCrO$_3$-based materials have been investigated as interconnect materials for SOFCs [5]; however, they are also potential anode materials for SOFCs due to their relatively good stability in both reducing and oxidizing atmospheres at high temperatures [72]. The reported polarization resistance using these materials is too high for efficient SOFC operation, although significant improvements have been achieved using low level doping of the B-site. As no significant weight loss was observed when LaCrO$_3$ was exposed to a reducing atmosphere (pO$_2 = 10^{-21}$ atm at 1000°C) [73] this indicates that chromium strongly retains its sixfold coordination. Indeed Cr^{III} is well-known to strongly prefer sixfold coordination in its chemistry, thus, it is difficult to introduce the oxygen vacancies that are required for oxygen ion conduction into the LaCrO$_3$ lattice.

When the B-sites are doped by other multivalence transition elements that tolerate reduced oxygen coordination, such as Mn, Fe, Co, Ni, and Cu, oxygen vacancies may be generated at the B-site dopants in a reducing atmosphere at high temperature. Thus, a significant degree of B-site dopant is required to generate a percolation path for oxygen vacancies in order to achieve high oxygen ion conductivity. Quite a lot of attention has been focused on 3% replacement of Cr by V and although methane cracking seems to be avoided [57,74], the polarization resistance is still of the order 10 Ω cm [57,74,75]. The introduction of other transition elements into the B-site of $La_{1-x}Sr_xCr_{1-y}M_yO_3$ (M = Mn, Fe, Co, and Ni) has been shown to improve the catalytic properties for methane reforming [26]. Of the various dopants, nickel seems to be the most successful and the lowest polarization resistances have been reported for 10% Ni-substituted lanthanum chromite [76]; however, other works have found nickel evolution from 10% Ni-doped lanthanum chromites in fuel conditions [77]. Certainly, nickel oxides would not be stable in fuel atmospheres, and although the nickel may be stabilized by the lattice in higher oxidation state, there will always be the suspicion that the activity of nickel-doped perovskites is due to surface evolution of nickel metal and hence questions about long-term stability. In a recent report, a composite anode of 5%Ni with a 50/50 mixture of $La_{0.8}Sr_{0.2}Cr_{0.8}Mn_{0.2}O_3$ and $Ce_{0.9}Gd_{0.1}O_{1.95}$ was successfully used for SOFCs with different fuels [78].

The combination of both titanium and chromium at B-site of a perovskite has also been investigated [58,79]. The highest conductivity in 5% H_2/Ar of 5 S cm^{-1} at 1000°C was observed with composition $La_{0.4}Ca_{0.6}Cr_{0.2}Ti_{0.8}O_{3-\delta}$; however, the catalytic effect of these materials for oxidization of hydrogen at the anode is possibly not ideal because a large anode polarization resistance was observed when $La_{0.7}Sr_{0.3}Cr_{0.8}Ti_{0.2}O_3$ was applied as the SOFC anode [79]. However, data on other relevant properties are rather sparse. $Sr_{0.86}Y_{0.08}TiO_3$, $La_{0.75}Sr_{0.25}Cr_{0.5}Mn_{0.5}O_3$, $La_{0.8}Sr_{0.2}Fe_{0.8}Cr_{0.2}O_3$, and $La_{0.8}Sr_{0.2}Cr_{0.95}Ru_{0.05}O_3$ all have acceptable thermal expansion match to YSZ under reducing conditions, but heavier Sr doping for some of these leads to anomalous expansion coefficients and unacceptably large dimensional changes on redox cycling. Better understanding of the factors that contribute to these dimensional changes would be valuable.

Very little is known about the oxygen diffusivity (ionic conductivity) and surface exchange of oxygen under reducing conditions in any of these materials. Such studies have proved valuable in understanding the behavior of cathodes and now need to be applied to anode materials. So far, only $La_{1-x}Sr_xFe_{0.8}Cr_{0.2}O_3$ has been studied in this way [80].

23.5.3.3 A(B,B')O₃ perovskites

A suitable benchmark material is the perovskite $La_{1-x}Sr_xCrO_3$, which has been thoroughly investigated as an interconnect material for SOFCs [5] and is a potential anode material for SOFCs due to the relatively good stability in both reducing and oxidizing atmospheres at high temperatures [72]. The acceptor doping gives high p-type conductivity in air but, as with all p-type materials, this decreases under reducing conditions. Addition of reducible transition metals such as Ti can introduce significant n-type contribution at low $p(O_2)$s but the dilution of Cr on the B-site generally has a greater effect [58]. Substitution of mid-transition metals such as Mn or Fe does not have such a dilution effect, indicating some complementarity of electronic function and affords an extension of the p-type domain to lower $p(O_2)$s, although conductivities are still below 10 S cm^{-1} under reducing conditions [81]. It is interesting to note that while n-type electronic conductivity would seem more natural form of electronic defect under fuel conditions, p-type conduction, if it can be retained to low partial oxygen pressures offers a significant advantage, in that conductivity will increase under load.

The reported polarization resistance using chromite perovskites is generally too high for efficient SOFC operation, although significant improvements have been achieved using low level doping of the B-site. With 3% replacement of Cr by V methane cracking seems to be avoided, although the polarization resistance is still of the order 10 Ω cm^2 at 900°C [74]. The introduction of other transition elements into the B-site of $La_{1-x}Sr_xCr_{1-y}M_yO_3$ (M = Mn, Fe, Co, and Ni) has been shown to improve the catalytic properties for methane reforming [26]. Of these various dopants, nickel seems to be the most successful and the lowest polarization resistances have been reported for 10% Ni-substituted lanthanum chromite [76]; however, others have found nickel ex-solution from 10% Ni-doped lanthanum chromites in fuel conditions [77]. Certainly, nickel oxides would not be stable in fuel atmospheres, and although nickel may be stabilized by the lattice perovskite in higher oxidation state, there will always be the suspicion that the activity of nickel doped perovskites is due to surface ex-solution of nickel metal, and hence raise questions about long-term stability. A particularly successful approach has been to create perovskites with dual B-site occupancy, such as those based upon lanthanum chromium manganite [81]. This approach has successfully combined the good oxidation catalysis properties of lanthanum manganite with the stability and conductivity of the chromite, without compromising any of these good properties by dilution.

As perovskites with one cation occupying the B-site have not yielded good enough properties for efficient anode operation in SOFCs, we have embarked upon an extensive series of studies looking at the possibility on enhancing performance by using two different B-site ions both with concentration in excess of percolation limit (i.e., >30%). The objective is to obtain complimentary functionality from appropriate cation combinations, hopefully without seriously degrading the good properties induced by the individual ions. Not surprisingly, many of the tested combinations did compromise properties but in some important instances good complimentary functionality was achieved.

Initial efforts focused upon double perovskites with niobium and a first row transition metal or main group ion occupying the B-site. With Nb and Mn occupying the B-site electronic conductivity is fairly low, probably reflecting the rock salt type ordering of the B cations [82]. Using Cu and Nb to a certain extent improves conductivity in air, but in reducing conditions copper metal is exolved and conductivity is impaired, as the resultant perovskite is more resistive and the copper does not form a conducting network [83]. Using Ga with Nb again results in an ordered superstructure that impairs electronic conductivity [84]. The effective difference

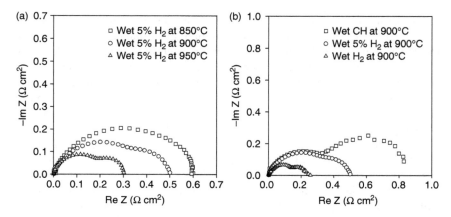

FIGURE 23.5 Electrode impedance of an optimized $La_{0.75}Sr_{0.25}Cr_{0.5}Mn_{0.5}O_3$ anode. (a) in wet 5% H_2, (5% H_2, 3% H_2O, 92% Ar) at 850, 900, and 950°C, (b) at 900°C, in different humidified (3% H_2O) fuel gas compositions at 900°C. The electrode dispersions have been obtained in a three-electrode set up. The electrolyte contribution has been subtracted from the overall impedance (From Tao, S.W. and Irvine, J.T.S. *Nat. Mater.* 2002, 2, 320–323. With permission.)

in coordination demands seem to preclude positive interaction between B-site ions and so we turned our attention to combining first row transition metals.

We recently reported the anode performance of complex perovskites based upon Cr and Mn at the B-sites forming compositions $(La,Sr)Cr_{1-x}M_xO_{3-\delta}$ [81]. Previous works have focused upon doped lanthanum chromite, where doping is used in the solid-state chemical sense of up to 20% dopant on the B-site, usually 5 or 10%. We reported a complex perovskite where two transition metal species occupy the B-site in excess of the percolation limit (e.g., >33%). Such dramatic replacement of an active B-site ion by another element would normally, significantly degrade its functionality; however, if the two elements act in a complementary fashion, then a dramatically improved new material may be achieved. $(La_{0.75}Sr_{0.25})_{1-x}Cr_{0.5}Mn_{0.5}O_3$ (LSCM) exhibits comparable electrochemical performance to Ni/YSZ cermets. Figure 23.5 shows the anode polarization resistance of LSCM in different fuels at 900°C. The electrode polarization resistance approaches 0.2 $\Omega\,cm^2$ at 900°C in 97% H_2/3% H_2O. Very good performance is achieved for methane oxidation without using excess steam. The anode is stable in both fuel and air conditions and shows stable electrode performance in methane. Thus, both redox stability and operation in low steam hydrocarbons have been demonstrated, overcoming two of the major limitations of the current generation of nickel zirconia cermet SOFC anodes.

23.6 ANODE MATERIALS FOR DIRECT HYDROCARBON FUEL CELLS

A subject that is receiving increased attention is the development of new anode materials for direct conversion of hydrocarbon fuels without first reforming those fuels to CO and H_2 [85]. Elimination of the reforming step would decrease system complexity and avoid the necessity of diluting fuels with steam. For higher hydrocarbons, there are also energy losses associated with the need to partially oxidize the fuel. In principle, an SOFC can operate on any combustible fuel that is capable of reacting with oxide ions coming through the electrolyte. In practice, the high operating temperatures in the presence of hydrocarbons can lead to carbon formation, for example, as discussed earlier for Ni based anodes.

Three strategies for direct conversion of hydrocarbon fuels are possible. The first strategy utilizes conventional Ni-cermet anodes but modifies the operating conditions of the fuel cell [50]. For methane at intermediate temperatures, the rate of carbon deposition may be slow enough so that oxygen anions electrochemically driven through the electrolyte and steam generated by oxidation of methane will remove carbon as it is deposited.

A second strategy to avoid reforming involves replacing Ni cermets with composites containing Cu and ceria. Cu is relatively inert towards the carbon-formation reactions that occur on Ni, and stable operation has been observed with fairly large hydrocarbons over Cu-based anodes [33]. Gas-phase pyrolysis reactions can still lead to tar formation on Cu cermets; however, the compounds that form on Cu tend to be poly-aromatics, such as naphthalene and anthracene, rather than graphite. It has been suggested that the poly-aromatic compounds enhance anode performance by providing additional electronic conductivity in the anode [35].

It appears that Cu primarily provides electronic conductivity to the anode and is otherwise catalytically inert. This is confirmed by data showing that Au-ceria-SDC (samaria-doped ceria) composites exhibit a similar performance to that of Cu–ceria-SDC anodes, as Au would not be expected to add catalytic activity [86,87]. It is proposed that the function of ceria is primarily that of an oxidation catalyst, although mixed electronic-ionic conductivity (MEIC) could also enhance anode performance. In general, finely dispersed ceria seems more active than doped ceria ceramics, thus the redox oxygen exchange ability of ceria at fuel conditions might be considered as a key factor.

The third strategy for developing SOFCs that can operate directly on hydrocarbon fuels involves anodes made from electronical conductive ceramics, since these are less likely to promote carbon formation. Reasonable performance can be achieved with direct utilization of methane using an anode made from La-doped strontium titanate [60,61]; however, the addition of ceria to this doped $SrTiO_3$ enhanced the performance significantly [60]. Reasonable performance in methane can also be obtained with anodes made from Gd-doped ceria [88]. Barnett has also reported stable power generation with methane and propane fuels on an anode based on Mn-doped $LaCrO_3$ with YSZ and 5% Ni [78]. The addition of Ni, in amounts small enough to avoid carbon formation, was found to enhance the performance of the $LaCrO_3$ anodes. Similar good performance has been achieved using Ni [26] and Ru [89] substituted lanthanum chromites and most recently the split B-site perovskite $(La,Sr)Mn_{0.5}Cr_{0.5}O_3$ that provides both very good performance and good chemical stability [81].

An important issue for direct utilization of hydrocarbons in all the strategies is the question of mechanism, since this relates directly to the potential of the electrons that are produced in the cell. While it has been demonstrated that there is a balance between the production rate for the total oxidation products and the generation of current [33], the cell potentials that have been reported with hydrocarbon fuels are much lower than would be predicted by direct, electrochemical oxidation of the hydrocarbons in a single step. It seems clear that, at the very least, the direct utilization processes involve an initial activation step, perhaps just involving the breaking of one carbon–hydrogen bond. Indeed, the appearance of tars on Cu/CeO_2 composites utilizing hydrocarbons certainly indicates that some pyrolysis reactions may help initiate the hydrocarbon oxidation process in these systems. The efficient oxidation of produced CO or carbon at ceria may well be the key to the effectiveness of these materials in direct utilization of hydrocarbons, since ceria itself is viewed as being only a moderately good catalyst [33]. Conversely, much of the effort on direct utilization has focused on reforming activity, especially for transition metal doping of lanthanum chromite-based materials. It is worthwhile remembering that one of the main oxidation products is steam, so reforming can still occur even with dry fuel sources. This has led to the concept of gradual internal reforming, which effectively provides a direct utilization mechanism [25]. Preliminary investigations of the activity of the lanthanum strontium chromium manganite perovskite reported by Tao and Irvine [90] indicate poor reforming and good oxidation activity, so this material may act in a manner similar to ceria and so its catalytic properties seem more closely related to a manganite than a chromite. The various processes that can contribute in combination to effectively achieve direct utilization are presented in Figure 23.6.

23.7 MICROSTRUCTURAL COMPROMISE

So far, the discussion has related largely to the compositional nature of candidate anodes; however, the microstructure of the electrode is at least as important as its composition. The optimization of durable efficient nickel cermet anodes in recent

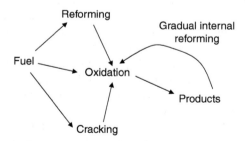

FIGURE 23.6 Diagram showing the contributing processes that can give rise to the direct utilization of hydrocarbon fuels in solid oxide fuel cells (From Atkinson, A., Barnett, S., Gorte, R.J., Irvine, J.T.S., McEvoy, A.J., Mogensen, M., Singhal, C., and Vohs, J.M. *Nat. Mater.* 2004, *3*, 17–27. With permission.)

decades has relied greatly on empirical improvement of materials specifications to control cermet morphology. With modern submicron active ceramic powder, the sintering temperature can be significantly decreased, to 1400°C or lower metal contents can be achieved. Associated with the reduced thermal expansivity of the cermet due to the increased ceramic content, stresses during fabrication, reduction, and operation are minimized, eliminating microfissuring, which contributes to electrode ageing [91]. Nickel oxide of grain size around 1 μm is now used, while the ceramic component may be bidispersed, containing a proportion of coarse powder with grains of 25 μm or larger to form the anode structural skeleton and inhibit the nickel aggregation, mixed with active submicron fine powder to promote sintering. These procedures are applied to the conventional electrolyte-supported configuration where the stabilized zirconia substrate, 150 μm or thicker, also provides the overall structural integrity of the cell. Recent development has extended the function of the material used for the electrochemical anode to become the load- and stress-bearing support for an electrolyte no thicker than 10 μm. This permits lowering the operating temperature to perhaps 650°C, while providing adequate cell performance. Under these conditions materials specifications throughout the system are relaxed, with lower cost metallic structural and interconnect components, and with diminished thermomechanical stresses and reactions between materials, thereby significantly improving durability. The structural cermet, now up to 1 mm thick, provides not only the anode functionality, but can also serve as a reaction volume for fuel processing, such as internal reforming. However, the lower temperatures significantly diminish the thermal activation of the oxidation reactions, implying increased polarization and giving added importance to considerations of electrocatalysis at the anode [92]. This makes a graded anode structure advisable, with a high porosity large grain substrate bearing a finer-structured electrocatalytically active anode layer to contact with the electrolyte.

23.8 Summary and Future Anode Development

Major advances in SOFC anode development have been achieved in recent years and there is good encouragement that new fuel electrode formulations for second-generation commercial fuel cells can be found. Such new anodes will offer improved redox tolerance and better resistance stability in hydrocarbon fuels. Other key strategies have been identified, based on oxidation catalysts such as ceria or lanthanum chromium manganite or reforming catalysts such as nickel or ruthenium-doped lanthanum chromite. Additionally, the function of electronic conductivity in the anode current collector has also been addressed in oxides such as those based upon strontium titanate. Although no single material fulfills all the current collection, electrochemical and catalytic performance indicators required for a supported electrode design to the extent that nickel zirconia cermets do, a number of systems individually meeting these criteria have been attained. Clearly, new composites are possible to fulfill the roles of both the active electrode layer and the current collecting part.

Further research is required to optimize these materials (composition and microstructure); particularly to maximize electronic conductivity without sacrificing essential chemical stability under reducing conditions. Detailed studies are required on the most promising materials to establish and understand dimensional stability on redox cycling and long-term operation and how this can be improved without compromising electrical conductivity. Studies of oxygen diffusion, ionic conductivity, and surface exchange in reducing environments are required on the most promising materials in order to improve understanding of their electro-catalytic properties. Particular care is necessary to allow reproducibility of results between laboratories and especially to ensure that meaningful experiments are performed before their results are presented. Understanding of crystal structure, point defect populations, and electronic structure need to be improved at a fundamental level and related to the key properties mentioned earlier in order to guide the search for new materials. More basic work to probe the mechanistic details of fuel utilization reactions needs to be undertaken. These details are extremely important to rational design of more efficient anodes and thus high-performance SOFCs powered by practical fuels. Practical studies in real fuels need to be expanded, especially addressing issues such as sulfur tolerance and ability to utilize biofuels. The search for totally new oxide materials with even better properties should continue, because even the best ones only meet the requirements. Most importantly, this should include the highest possible electronic conductivity since such a material would also serve as a current collector. Certain materials can match a few requirements but it is very hard to find a material that can match the stringent requirements for SOFC anodes, particularly redox stability and conductivity. Perovskite or perovskite derivatives are promising anode SOFC anode materials.

REFERENCES

1. Steele, B.C.H. Appraisal of $Ce_{1-y}Gd_yO_{2-y/2}$ electrolytes for IT-SOFC operation at 500°C. *Solid State Ionics* 2000, *129*, 95–110.
2. Kuroda, K., Hashimoto, I., Adachi, K., Akikusa, J, Tamou, Y., Komada, N., Ishihara, T., and Takita, Y. Characterization of solid oxide fuel cell using doped lanthanum gallate. *Solid State Ionics* 2000, *132*, 199–208.
3. Yamaji, K., Horita, T., Ishikawa, M., Sakai, N., and Yokokawa, H. Chemical stability of the $La_{0.9}Sr_{0.1}Ga_{0.8}Mg_{0.2}O_{2.85}$ electrolyte in a reducing atmosphere. *Solid State Ionics* 1999, *121*, 217–224.
4. Tao, S.W., Poulsen, F.W., Meng, G.Y., and Sørensen, O.T. High temperature stability study of the oxygen-ion conductor $La_{0.9}Sr_{0.1}Ga_{0.8}Mg_{0.2}O_{3-x}$. *J. Mater. Chem.* 2000, *10*, 1829–1833.
5. Minh, N.Q. Ceramic fuel cells. *J. Am. Ceram. Soc.* 1993, *76*, 563–588.
6. Murray, E.P., Sever, M.J., and Barnett, S.A. Electrochemical performance of $(La,Sr)(Co,Fe)O_3$–$(Ce,Gd)O_3$ composite cathodes. *Solid State Ionics* 2002, *148*, 27–34.
7. Ledjeff, K., Rohrbach, T., and Schaumberg, G. Internal reforming for solid oxide fuel cells. In *Proceedings of the 2nd International Symposium of SOFCs*, Grosz, F., Zegers, P., Singhal, S.C., Yamamoto, O., Eds. The Electrochemical Society: Pennington NJ, 1991, p. 323.

8. Dicks, A.L., Pointon, K.D., and Siddle, A. Intrinsic reaction kinetics of methane steam reforming on a nickel/zirconia anode. *J. Power Sources* 2000, *86*, 523–530.
9. Froment, G.F. Production of synthesis gas by steam- and CO_2-reforming of natural gas. *J. Mol. Catal. A: Chem.* 2000, *163*, 147–156.
10. Achenbach, E. and Riensche, E. Methane steam reforming kinetics for solid oxide fuel cells. *J. Power Sources* 1994, *52*, 283–288.
11. Dicks, A.L. Advances in catalysts for internal reforming in high temperature fuel cells. *J. Power Sources* 1998, *71*, 111–122.
12. Ahmed, K. and Forger, K. Kinetics of internal steam reforming of methane on Ni/YSZ-based anodes for solid oxide fuel cells. *Catal. Today* 2000, *63*, 479–487.
13. Ahmed, K., Seshadri, P., Ramprakashm, Y., Jiang, S.P., and Forger, K. In *Proceeding of SOFC's V*; Stimming., U., Singhal, S.C., Tagawa, H., and Lehnert, W., Eds. The Electrochemical Society Proceedings Series, Pennington NJ, 1997, 97–40, p. 228.
14. Nakagawa, N., Sagara, H., and Kato, K. Catalytic activity of Ni–YSZ–CeO_2 anode for the steam reforming of methane in a direct internal-reforming solid oxide fuel cell. *J. Power Sources* 2001, *92*, 88–94.
15. Yentakakis, I.V., Jiang, Y., Neophytides, S., Bebelis, S., and Vayenas, C.G. In *Proceeding of the 2nd European Solid Oxide Fuel Cell Forum*, Thorstensen, B., Ed. Germany, 1996, p. 131.
16. Onuma, S., Kaimai, A., Kawamura, K., Nigara, Y., Kawasa, T., Mizusaki, J., Iniba, H., and Tagawa, H. Electrochemical oxidation in a CH_4–H_2O system at the interface of a Pt electrode and Y_2O_3-stabilized $ZrO2$ electrolyte. 1. Determination of the predominant reaction process. *J. Electrochem. Soc.* 1998, *145*, 920–925.
17. Ramirez-Cabrera, E., Atkinson, A., and Chadwick, D. In *Proceeding of the 4th European Fuel Cell Forum*, McEvoy, A.J. Ed. Switzerland, 2000, p. 49.
18. Armor, J.N. and Martenak, D.J. Studying carbon formation at elevated pressure. *Appl. Catal. A* 2001, *206*, 231–236.
19. Alstrup, I. and Tavares, M.T. The kinetics of carbon formation from $CH_4 + H_2$ on a silica-supported nickel-catalyst. *J. Catal.* 1992, *135*, 147–155.
20. Rostrup-Nielsen, J.R. New aspects of syngas production and use. *Catal. Today* 2000, *63*, 159–164.
21. Morimoto, K. and Shimotsu, M. In *Proceedings of SOFC's IV*, Dokiya, M., Yamamoto, O., Tagawa, H., and Singhal, S.C., Eds. The Electrochemical Society Proceedings Series, Pennington NJ, 1995, 95–1, p. 269.
22. Alstrup, I., Petersen, U.E., and Rostrup-Nielsen, J.R. Propane hydrogenolysis on sulfur- and copper-modified nickel catalysts. *J. Catal.* 2000, *191*, 401–408.
23. Suzuki, M., Sasaki, H., and Oshi, S. In *Proceeding of the 2nd International Symposium of SOFCs*, Grosz, F., Zegers, P., Singhal, S.C., and Yamamoto, O., Eds. Greece, 1991, p. 323.
24. Sauvet, A.-L. and Fouletier, J. Catalytic properties of new anode materials for solid oxide fuel cells operated under methane at intermediary temperature. *J. Power Sources* 2001, *101*, 259–266.
25. Vernoux, P., Guindet, J., and Kleitz, M. Gradual internal methane reforming in intermediate-temperature solid-oxide fuel cells. *J. Electrochem. Soc.* 1998, *145*, 3487–3492.
26. Sfeir, J., Buffat, P.A., Möckli, P., Xanthopoulos, N., Vasquez, R., Mathieu, H.J., Van, herle J., and Thampi, K.R. Lanthanum chromite based catalysts for oxidation of methane directly on SOFC anode. *J. Catal.* 2001, *202*, 229–244.

27. Baur, E. and Preis, H. Fuel cells with rigid conductors. *Z. Elektrochem.* 1937, *43*, 727–732.

28. Möbius, H.-H. On the history of solid electrolyte fuel cells. *J. Solid State Electrochem.* 1997, *1*, 2–16.

29. Tao, S.W. and Irvine, J.T.S. Investigation of the mixed conducting oxide $Sc_{0.15}Y_{0.05}Zr_{0.62}Ti_{0.18}O_{1.9}$ as a potential SOFC anode material. *J. Electrochem. Soc.* 2004, *151*, A497–A503.

30. Spacil, H.S. U.S. patent 3,558,360; filed October 30, 1964, modified November 2, 1967, granted March 31, 1970.

31. Setoguchi, T., Okamoto, K., Eguchi, K., and Arai, H. Effects of anode material and fuel on anodic reaction of solid oxide fuel cells. *J. Electrochem. Soc.* 1992, *139*, 2875–2880.

32. Steele, B.C.H., Kelly, I., and Middleton, H., Rudkin, R. Oxidation of methane in solid state electrochemical reactions. *Solid State Ionics* 1988, *28–30*, 1547–1552.

33. Park, S.D., Vohs, J.M., and Gorte, R.J. Direct oxidation of hydrocarbons in a solid-oxide fuel cell. *Nature* 2000, *404*, 265–267.

34. Gorte, R.J., Park, S., Vohs, J.M., and Wang, C.H. Anodes for direct oxidation of dry hydrocarbons in a solid-oxide fuel cell. *Adv. Mater.* 2000, *12*, 1465–1469.

35. McIntosh, S., Vohs, J.M., and Gorte, R.J. Role of hydrocarbon deposits in the enhanced performance of direct-oxidation SOFCs. *J. Electrochem. Soc.* 2003, *150*, A470–A476.

36. Kiratzis, N., Holtappels, P., Hatchwell, C.E., Mogensen, M., and Irvine, J.T.S. Preparation and characterization of copper/yttria titania zirconia cermets for use as possible solid oxide fuel cell anodes. *Fuel Cells* 2001, *1*, 211–218.

37. Wang, S., Kato, T., Nagata, S., Honda, T., Kaneko, T., Iwashita, N., and Dokiya, M. NiO/ceria cermet as Anode of reduced-temperature solid oxide fuel cells. *J. Electrochem. Soc.* 2002, *149*, A927–A933.

38. Rosch, B., Tu, H.Y., Stormer, A.O., Muller, A.C., and Stimming, U. Electrochemical behaviour of $Ni-Ce_{0.9}Gd_{0.1}O_{2-\delta}$ SOFC anodes in methane. In *Proceeding of SOFC's VIII*, Singhal, S.C. and Dokiya, M., Eds. Paris, The Electrochemical Society Proceedings Series, Pennington, NJ, 2003, 2003–07, pp. 737–744.

39. Baron, S., Brandon, N., Atkinson, A., and Steele, B.C.H. The impact of wood derived gasification gases on Ni-CGO anodes in IT-SOFCs. In *Proceeding of SOFC's VIII*; Singhal, S.C. and Dokiya, M., Eds. Paris, The Electrochemical Society Proceedings Series, Pennington, NJ, USA, 2003, 2003–07, pp. 762–772.

40. Kim, H., Lu, C., Worrell, W.L., Vohs, J.M., and Gorte, R.J. Cu–Ni cermet anode for direct oxidation of methane in solid-oxide fuel cells. *J. Electrochem. Soc.* 2002, *149*, A247–A250.

41. Sin, A., Tavares, A., Doubitsky, Y., and Zaopo, A. The impact of wood derived gasification gases on Ni-CGO anodes in IT-SOFCs. In *Proceeding of SOFC's VIII*, Singhal, S.C. and Dokiya, M., Eds. Paris, The Electrochemical Society Proceedings Series, Pennington NJ, USA, 2003, 2003–07, pp. 745–751.

42. Dong, W.S., Roh, H.S., Jun, K.W., Park, S.E., and Oh, Y.S. Methane reforming over $Ni/Ce-ZrO_2$ catalysts: effect of nickel content. *Appl. Catal.* A 2002, *226*, 63–72.

43. Bunluesin, T., Gorte, R.J., and Graham, G.W. Studies of the water-gas-shift reaction on ceria-supported Pt, Pd, and Rh: implications for oxygen-storage properties. *Appl. Catal.* B 1998, *15*, 107–114.

44. Sharma, S., Hilaire, S., Vohs, J.M., Gorte, R.J., and Jen, H.W. Evidence for oxidation of ceria by CO_2. *J. Catal.* 2000, *190*, 199–204.

45. Ramirez-Cabrera, E., Atkinson, A., and Chadwick, D. The influence of point defects on the resistance of ceria to carbon deposition in hydrocarbon catalysis. *Solid State Ionics* 2000, *136*, 825–831.

46. Marina, O.A. and Mogensen, M. High-temperature conversion of methane on a composite gadolinia-doped ceria-gold electrode. *Appl. Catal.* A 1999, *189*, 117–126.

47. Naik, I.K. and Tien, T.Y. Electrical conduction in Nb_2O_5-doped cerium dioxide. *J. Electrochem. Soc.* 1979, *126*, 562–566.

48. Yahiro, H., Eguchi, K., and Arai, H. Electrical properties and reducibilities of ceria-rare earth oxide systems and their application to solid oxide fuel cell. *Solid State Ionics* 1989, *36*, 71–75.

49. Marina, O.A., Bagger, C., Primdahl, S., and Mogensen, M. A solid oxide fuel cell with a gadolinia-doped ceria anode: preparation and performance. *Solid State Ionics* 1999, *123*, 199–208.

50. Murray, E.P., Tsai, T., and Barnett, S.A. A direct-methane fuel cell with a ceria-based anode. *Nature* 1999, *400*, 649–651.

51. Tao, S.W. and Irvine, J.T.S. Discovery and characterisation of novel oxide anodes for solid oxide fuel cells. *Chem. Rec.* 2004, *4*, 83–95.

52. Ishihara, T., Matsuda, H., and Takita, Y. Doped $LaGaO_3$ perovskite-type oxide as a new oxide ionic conductor. *J. Am. Chem. Soc.* 1994, *116*, 3801–3803.

53. Ishihara, T., Matsuda, H., and Takita, Y. Effects of rare-earth cations doped for La site on the oxide ionic conductivity of $LaGaO_3$-based perovskite-type oxide. *Solid State Ionics* 1995, *79*, 147–151.

54. Slater, P.R., Irvine, J.T.S., Ishihara, T., and Takita, Y. The structure of the oxide ion conductor $La_{0.9}Sr_{0.1}Ga_{0.8}Mg_{0.2}O_{2.85}$ by powder neutron diffraction. *Solid State Ionics* 1998, *107*, 319–323.

55. Anderson, H.U. Review of p-type doped perovskite materials for SOFC and other applications. *Solid State Ionics* 1992, *52*, 33–41.

56. Tu, H.Y., Takeda, Y., Imanishi, N., and Yamamoto, O. $Ln_{0.4}Sr_{0.6}Co_{0.8}Fe_{0.2}O_{3-\delta}$ (Ln = La, Pr, Nd, Sm, Gd) for the electrode in solid oxide fuel cells. *Solid State Ionics* 1999, *117*, 277–281.

57. Primdahl, S., Hansen, J.R., Grahl-Madsen, L., and Larsen, P.H. Sr-doped $LaCrO_3$ anode for solid oxide fuel cells. *J. Electrochem. Soc.* 2001, *148*, A74–A81.

58. Pudmich, G., Boukamp, B.A., Gonzalez-Cuenca, M., Jungen, W., Zipprich, W., and Tietz, F. Chromite/titanate based perovskites for application as anodes in solid oxide fuel cells. *Solid State Ionics* 2000, *135*, 433–438.

59. Marina, O.A., Canfield, N.L., and Stevenson, J.W. Thermal, electrical, and electrocatalytical properties of lanthanum-doped strontium titanate. *Solid State Ionics* 2002, *149*, 21–28.

60. Marina, O.A. and Pederson, L.R. In *Proceeding of the 5th European Solid Oxide Fuel Cell Forum*, Huijsmans, J., Ed. Switzerland, 2002, p. 481.

61. Canales-Vazquez, J., Tao, S.W., and Irvine, J.T.S. Electrical properties in $La_2Sr_4Ti_6O_{19-\delta}$: a potential anode for high temperature fuel cells. *Solid State Ionics* 2003, *159*, 159–165.

62. Hui, S.Q. and Petric, A. Evaluation of yttrium-doped $SrTiO_3$ as an anode for solid oxide fuel cells. *J. Eur. Ceram. Soc.* 2002, *22*, 1673–1681.

63. Hui, S.Q. and Petric, A. Electrical conductivity of yttrium-doped $SrTiO_3$: influence of transition metal additives. *Mater. Res. Bull.* 2002, *37*, 1215–1231.
64. Hui, S.Q. and Petric, A. Electrical properties of yttrium-doped strontium titanate under reducing conditions. *J. Electrochem. Soc.* 2002, *149*, J1–J10.
65. Tao, S.W. and Irvine, J.T.S. unpublished results.
66. Irvine, J.T.S., Slater, P.R., and Wright, P.A. Synthesis and electrical characterization of the perovskite niobate-titanates $Sr_{1-x/2}Ti_{1-x}Nb_xO_{3-\delta}$. *Ionics* 1996, *2*, 213–216.
67. Hui, S.Q. and Petric, A. Conductivity and stability of $SrVO_3$ and mixed perovskites at low oxygen partial pressures. *Solid State Ionics* 2001, *143*, 275–283.
68. Hartley, A., Sahibzada, M., Weston, M., Metcafe, I.S., and Mantzavinos, D. $La_{0.6}Sr_{0.4}Co_{0.2}Fe_{0.8}O_3$ as the anode and cathode for intermediate temperature solid oxide fuel cells. *Catal. Today* 2000, *55*, 197–204.
69. Balachandran, U., Ma, B., Maiya, P.S., Mieville, R.L., Dusek, J.T., Picciolo, J.J., Guan, J., Dorris, S.E., and Liu, M. Development of mixed-conducting oxides for gas separation. *Solid State Ionics* 1998, *108*, 363–370.
70. Grenier, J.C., Schiffmacher, G., Caro, P., Pouchard, M., and Hagenmuller P. Etude par diffraction X et microscopie electronique du système $CaTiO_3$–$Ca_2Fe_2O_5$. *J. Solid State Chem.* 1977, *20*, 365–379.
71. Ma, B. and Balachandran, U. Phase stability of $SrFeCo_{0.5}O_x$ in reducing environments. *Mater. Res. Bull.* 1998, *33*, 223–236.
72. Yokokawa, H., Sakai, N., Kawada, T., and Dokiya, M. Thermodynamic stability of perovskite oxides and other electrochemical materials. *Solid State Ionics* 1992, *52*, 43–56.
73. Nakamura, T., Petzow, G., and Gauckler, L.J. Stability of the perovskite phase $LaBO_3$ (B = V, Cr, Mn, Fe, Co, Ni) in reducing atmosphere I. Experimental results. *Mater. Res. Bull.* 1979, *14*, 649–659.
74. Vernoux, P., Guillodo, M., Fouletier, J., and Hammou, A. Alternative anode materials for gradual methane reforming in solid oxide fuel cells. *Solid State Ionics* 2000, *135*, 425–431.
75. Matsuzaki, Y. and Yasuda, I. The poisoning effect of sulfur-containing impurity gas on a SOFC anode: Part I. Dependence on temperature, time, and impurity concentration. *Solid State Ionics* 2000, *132*, 261–269.
76. Sfeir, J., Van herle, J., and Vasquez, R. $LaCrO_3$-based anodes for methane oxidation. In *Proceeding of the 5th European Solid Oxide Fuel Cell Forum*, Huijsmans, J., Ed. Switzerland, 2002, pp. 570–577.
77. Sauvet, A.-L. and Irvine, J.T.S. Catalytic and electrocatalytic studies of new type of anode materials for SOFC under methane. In *Proceeding of the 5th European Solid Oxide Fuel Cell Forum*; Huijsmans, J., Ed. Switzerland, 2002, pp. 490–498.
78. Liu, J., Madsen, B.D., Ji, Z.Q., and Barnett, S.A. A fuel-flexible ceramic-based anode for solid oxide fuel cells. *Electrochem. Solid State Lett.* 2002, *5*, A122–A124.
79. Vashook, V., Vasylechko, L., Ullmann, H., and Guth, U. Synthesis, crystal structure, oxygen stoichiometry, and electrical conductivity of $La_{1-a}Ca_aCr_{0.2}Ti_{0.8}O_{3-\delta}$. *Solid State Ionics* 2003, *158*, 317–325.
80. Ramos, T. and Atkinson, A. Mass transport in oxides for membrane reforming of methane. In *Ionic and Mixed Conducting Ceramics IV*, Ramanarayanan, T.A., Worrell, W.L., and Mogensen, M., Eds. Electrochemical Society Inc: Pennington, San Francisco, 2002, pp. 352–367.

81. Tao, S.W. and Irvine, J.T.S. A Redox-stable, efficient anode for solid-oxide fuel cells. *Nat. Mater.* 2003, *2*, 320–323.

82. Tao, S.W. and Irvine, J.T.S. Study on the structural and electrical properties of the perovskite oxide $SrMn_{0.5}Nb_{0.5}O_{3-\delta}$. *J. Mater. Chem.* 2002, *12*, 2356–2360.

83. Tao, S.W. and Irvine, J.T.S. Structure and properties of nonstoichiometric mixed perovskites $A_3B'_{1+x}B''_{2-x}O_{9-\delta}$. *Solid State Ionics* 2002, *154–155*, 659–667.

84. McColm, T.C. and Irvine, J.T.S. Structural and property investigations of strontium galloniobate. *Solid State Ionics* 2002, *152–153*, 615–623.

85. Atkinson, A., Barnett, S., Gorte, R.J., Irvine, J.T.S., McEvoy, A.J., Mogensen, M., Singhal, C., and Vohs, J.M. Advanced anodes for high-temperature fuel cells. *Nat. Mater.* 2004, *3*, 17–27.

86. Lu, C., Worrell, W.L., Vohs, J.M., Gorte, R.J. A comparison of Cu-ceria-SDC and Au-ceria-SDC composites for SOFC anodes, Proceeding of SOFC's VIII, Singhal, S.C., Dokiya, M. Eds. Paris, The Electrochemical Society Proceedings Series, Pennington NJ, 2003, 2003–07, pp. 773–780, ISBN–1–56677–377–6.

87. Lu, C., Worrell, W.L., Gorte, R.J., and Vohs, J.M. SOFCs for direct oxidation of hydrocarbon fuels with samaria-doped ceria electrolyte. *J. Electrochemical. Soc.* 2003, *150*, A354–A358.

88. Sfeir, J. EPFL, Ph.D. thesis 2001.

89. Sauvet, A.-L. and Fouletier, J. Electrochemical properties of a new type of anode material $La_{1-x}Sr_xCr_{1-y}Ru_yO_{3-\delta}$ for SOFC under hydrogen and methane at intermediate temperatures. *Electrochim. Acta.* 2001, *47*, 987–995.

90. Tao, S.W. and Irvine, J.T.S. Synthesis and characterisation of $(La_{0.75}Sr_{0.25})Cr_{0.5}Mn_{0.5}O_{3-\delta}$, a redox-stable, efficient perovskite anode for solid-oxide fuel cells. *J. Electrochem. Soc.* 2004, *151*, A252–A259.

91. Skarmoutsos, D., Teitz, F., and Nikolopoulos, P. Structure–property relationships of Ni/YSZ and $Ni/(YSZ + TiO_2)$ cermets. *Fuel Cells* 2001, *1*, 243.

92. Holtappels, P. Electrocatalysis on Nickel–Cermet Electrodes. Jülich Research Centre report 3414, 1997/ Thesis University of Bonn.

Index

Page numbers in *italics* refer to figures and tables